A COMPUTATIONAL INTRODUCTION
TO NUMBER THEORY AND ALGEBRA

Second Edition

A COMPUTATIONAL INTRODUCTION TO NUMBER THEORY AND ALGEBRA

Second Edition

VICTOR SHOUP

CAMBRIDGE
UNIVERSITY PRESS

CAMBRIDGE UNIVERSITY PRESS
Cambridge, New York, Melbourne, Madrid, Cape Town,
Singapore, São Paulo, Delhi, Mexico City

Cambridge University Press
The Edinburgh Building, Cambridge CB2 8RU, UK

Published in the United States of America by Cambridge University Press, New York

www.cambridge.org
Information on this title: www.cambridge.org/9780521516440

First published 2009

A catalogue record for this publication is available from the British Library

ISBN 978-0-521-51644-0 Hardback

Contents

Preface   *page* x

Preliminaries   xiv

1   Basic properties of the integers   1
    1.1   Divisibility and primality   1
    1.2   Ideals and greatest common divisors   5
    1.3   Some consequences of unique factorization   10

2   Congruences   15
    2.1   Equivalence relations   15
    2.2   Definitions and basic properties of congruences   16
    2.3   Solving linear congruences   19
    2.4   The Chinese remainder theorem   22
    2.5   Residue classes   25
    2.6   Euler's phi function   31
    2.7   Euler's theorem and Fermat's little theorem   32
    2.8   Quadratic residues   35
    2.9   Summations over divisors   45

3   Computing with large integers   50
    3.1   Asymptotic notation   50
    3.2   Machine models and complexity theory   53
    3.3   Basic integer arithmetic   55
    3.4   Computing in $\mathbb{Z}_n$ 64
    3.5   Faster integer arithmetic ($*$)   69
    3.6   Notes   71

4   Euclid's algorithm   74
    4.1   The basic Euclidean algorithm   74
    4.2   The extended Euclidean algorithm   77
    4.3   Computing modular inverses and Chinese remaindering   82

4.4	Speeding up algorithms via modular computation	84
4.5	An effective version of Fermat's two squares theorem	86
4.6	Rational reconstruction and applications	89
4.7	The RSA cryptosystem	99
4.8	Notes	102
5	**The distribution of primes**	**104**
5.1	Chebyshev's theorem on the density of primes	104
5.2	Bertrand's postulate	108
5.3	Mertens' theorem	110
5.4	The sieve of Eratosthenes	115
5.5	The prime number theorem . . . and beyond	116
5.6	Notes	124
6	**Abelian groups**	**126**
6.1	Definitions, basic properties, and examples	126
6.2	Subgroups	132
6.3	Cosets and quotient groups	137
6.4	Group homomorphisms and isomorphisms	142
6.5	Cyclic groups	153
6.6	The structure of finite abelian groups (∗)	163
7	**Rings**	**166**
7.1	Definitions, basic properties, and examples	166
7.2	Polynomial rings	176
7.3	Ideals and quotient rings	185
7.4	Ring homomorphisms and isomorphisms	192
7.5	The structure of $\mathbb{Z}_n^*$	203
8	**Finite and discrete probability distributions**	**207**
8.1	Basic definitions	207
8.2	Conditional probability and independence	213
8.3	Random variables	221
8.4	Expectation and variance	233
8.5	Some useful bounds	241
8.6	Balls and bins	245
8.7	Hash functions	252
8.8	Statistical distance	260
8.9	Measures of randomness and the leftover hash lemma (∗)	266
8.10	Discrete probability distributions	270
8.11	Notes	275

9   **Probabilistic algorithms**   277

9.1   Basic definitions   278

9.2   Generating a random number from a given interval   285

9.3   The generate and test paradigm   287

9.4   Generating a random prime   292

9.5   Generating a random non-increasing sequence   295

9.6   Generating a random factored number   298

9.7   Some complexity theory   302

9.8   Notes   304

10   **Probabilistic primality testing**   306

10.1   Trial division   306

10.2   The Miller–Rabin test   307

10.3   Generating random primes using the Miller–Rabin test   311

10.4   Factoring and computing Euler's phi function   320

10.5   Notes   324

11   **Finding generators and discrete logarithms in $\mathbb{Z}_p^*$**   327

11.1   Finding a generator for $\mathbb{Z}_p^*$ 327

11.2   Computing discrete logarithms in $\mathbb{Z}_p^*$ 329

11.3   The Diffie–Hellman key establishment protocol   334

11.4   Notes   340

12   **Quadratic reciprocity and computing modular square roots**   342

12.1   The Legendre symbol   342

12.2   The Jacobi symbol   346

12.3   Computing the Jacobi symbol   348

12.4   Testing quadratic residuosity   349

12.5   Computing modular square roots   350

12.6   The quadratic residuosity assumption   355

12.7   Notes   357

13   **Modules and vector spaces**   358

13.1   Definitions, basic properties, and examples   358

13.2   Submodules and quotient modules   360

13.3   Module homomorphisms and isomorphisms   363

13.4   Linear independence and bases   367

13.5   Vector spaces and dimension   370

14   **Matrices**   377

14.1   Basic definitions and properties   377

14.2   Matrices and linear maps   381

14.3   The inverse of a matrix   386

14.4 Gaussian elimination 388
14.5 Applications of Gaussian elimination 392
14.6 Notes 398

15 Subexponential-time discrete logarithms and factoring 399
15.1 Smooth numbers 399
15.2 An algorithm for discrete logarithms 400
15.3 An algorithm for factoring integers 407
15.4 Practical improvements 414
15.5 Notes 418

16 More rings 421
16.1 Algebras 421
16.2 The field of fractions of an integral domain 427
16.3 Unique factorization of polynomials 430
16.4 Polynomial congruences 435
16.5 Minimal polynomials 438
16.6 General properties of extension fields 440
16.7 Formal derivatives 444
16.8 Formal power series and Laurent series 446
16.9 Unique factorization domains (∗) 451
16.10 Notes 464

17 Polynomial arithmetic and applications 465
17.1 Basic arithmetic 465
17.2 Computing minimal polynomials in $F[X]/(f)$(I) 468
17.3 Euclid's algorithm 469
17.4 Computing modular inverses and Chinese remaindering 472
17.5 Rational function reconstruction and applications 474
17.6 Faster polynomial arithmetic (∗) 478
17.7 Notes 484

18 Linearly generated sequences and applications 486
18.1 Basic definitions and properties 486
18.2 Computing minimal polynomials: a special case 490
18.3 Computing minimal polynomials: a more general case 492
18.4 Solving sparse linear systems 497
18.5 Computing minimal polynomials in $F[X]/(f)$(II) 500
18.6 The algebra of linear transformations (∗) 501
18.7 Notes 508

19 Finite fields 509
19.1 Preliminaries 509

	19.2	The existence of finite fields	511
	19.3	The subfield structure and uniqueness of finite fields	515
	19.4	Conjugates, norms and traces	516
20		**Algorithms for finite fields**	**522**
	20.1	Tests for and constructing irreducible polynomials	522
	20.2	Computing minimal polynomials in $F[X]/(f)$(III)	525
	20.3	Factoring polynomials: square-free decomposition	526
	20.4	Factoring polynomials: the Cantor–Zassenhaus algorithm	530
	20.5	Factoring polynomials: Berlekamp's algorithm	538
	20.6	Deterministic factorization algorithms ($*$)	544
	20.7	Notes	546
21		**Deterministic primality testing**	**548**
	21.1	The basic idea	548
	21.2	The algorithm and its analysis	549
	21.3	Notes	558
Appendix: Some useful facts			561
Bibliography			566
Index of notation			572
Index			574

Preface

Number theory and algebra play an increasingly significant role in computing and communications, as evidenced by the striking applications of these subjects to such fields as cryptography and coding theory. My goal in writing this book was to provide an introduction to number theory and algebra, with an emphasis on algorithms and applications, that would be accessible to a broad audience. In particular, I wanted to write a book that would be appropriate for typical students in computer science or mathematics who have some amount of general mathematical *experience*, but without presuming too much specific mathematical *knowledge*.

Prerequisites. The mathematical prerequisites are minimal: no particular mathematical concepts beyond what is taught in a typical undergraduate calculus sequence are assumed.

The computer science prerequisites are also quite minimal: it is assumed that the reader is proficient in programming, and has had some exposure to the analysis of algorithms, essentially at the level of an undergraduate course on algorithms and data structures.

Even though it is mathematically quite self contained, the text does presuppose that the reader is comfortable with mathematical formalism and also has some experience in reading and writing mathematical proofs. Readers may have gained such experience in computer science courses such as algorithms, automata or complexity theory, or some type of "discrete mathematics for computer science students" course. They also may have gained such experience in undergraduate mathematics courses, such as abstract or linear algebra. The material in these mathematics courses may overlap with some of the material presented here; however, even if the reader already has had some exposure to this material, it nevertheless may be convenient to have all of the relevant topics easily accessible in one place; moreover, the emphasis and perspective here will no doubt be different from that in a traditional mathematical presentation of these subjects.

Structure of the text. All of the mathematics required beyond basic calculus is developed "from scratch." Moreover, the book generally alternates between "theory" and "applications": one or two chapters on a particular set of purely mathematical concepts are followed by one or two chapters on algorithms and applications; the mathematics provides the theoretical underpinnings for the applications, while the applications both motivate and illustrate the mathematics. Of course, this dichotomy between theory and applications is not perfectly maintained: the chapters that focus mainly on applications include the development of some of the mathematics that is specific to a particular application, and very occasionally, some of the chapters that focus mainly on mathematics include a discussion of related algorithmic ideas as well.

In developing the mathematics needed to discuss certain applications, I have tried to strike a reasonable balance between, on the one hand, presenting the absolute minimum required to understand and rigorously analyze the applications, and on the other hand, presenting a full-blown development of the relevant mathematics. In striking this balance, I wanted to be fairly economical and concise, while at the same time, I wanted to develop enough of the theory so as to present a fairly well-rounded account, giving the reader more of a feeling for the mathematical "big picture."

The mathematical material covered includes the basics of number theory (including unique factorization, congruences, the distribution of primes, and quadratic reciprocity) and of abstract algebra (including groups, rings, fields, and vector spaces). It also includes an introduction to discrete probability theory—this material is needed to properly treat the topics of probabilistic algorithms and cryptographic applications. The treatment of all these topics is more or less standard, except that the text only deals with commutative structures (i.e., abelian groups and commutative rings with unity)—this is all that is really needed for the purposes of this text, and the theory of these structures is much simpler and more transparent than that of more general, non-commutative structures.

The choice of topics covered in this book was motivated primarily by their applicability to computing and communications, especially to the specific areas of cryptography and coding theory. Thus, the book may be useful for reference or self-study by readers who want to learn about cryptography, or it could also be used as a textbook in a graduate or upper-division undergraduate course on (computational) number theory and algebra, perhaps geared towards computer science students.

Since this is an introduction, and not an encyclopedic reference for specialists, some topics simply could not be covered. One such, whose exclusion will undoubtedly be lamented by some, is the theory of lattices, along with algorithms for and applications of lattice basis reduction. Another omission is fast algorithms for

integer and polynomial arithmetic—although some of the basic ideas of this topic
are developed in the exercises, the main body of the text deals only with classical,
quadratic-time algorithms for integer and polynomial arithmetic. However, there
are more advanced texts that cover these topics perfectly well, and they should be
readily accessible to students who have mastered the material in this book.

   Note that while continued fractions are not discussed, the closely related problem of "rational reconstruction" is covered, along with a number of interesting
applications (which could also be solved using continued fractions).

Guidelines for using the text.

- There are a few sections that are marked with a "(∗)," indicating that the
  material covered in that section is a bit technical, and is not needed else-
  where.

- There are many examples in the text, which form an integral part of the
  book, and should not be skipped.

- There are a number of exercises in the text that serve to reinforce, as well
  as to develop important applications and generalizations of, the material
  presented in the text.

- Some exercises are <u>underlined</u>. These develop important (but usually simple) facts, and should be viewed as an integral part of the book. It is highly
  recommended that the reader work these exercises, or at the very least, read
  and understand their statements.

- In solving exercises, the reader is free to use any *previously* stated results
  in the text, including those in previous exercises. However, except where
  otherwise noted, any result in a section marked with a "(∗)," or in §5.5,
  need not and should not be used outside the section in which it appears.

- There is a very brief "Preliminaries" chapter, which fixes a bit of notation
  and recalls a few standard facts. This should be skimmed over by the reader.

- There is an appendix that contains a few useful facts; where such a fact is
  used in the text, there is a reference such as "see §An," which refers to the
  item labeled "An" in the appendix.

The second edition. In preparing this second edition, in addition to correcting
errors in the first edition, I have also made a number of other modifications (hopefully without introducing too many *new* errors). Many passages have been rewrit-
ten to improve the clarity of exposition, and many new exercises and examples
have been added. Especially in the earlier chapters, the presentation is a bit more
leisurely. Some material has been reorganized. Most notably, the chapter on probability now follows the chapters on groups and rings—this allows a number of
examples and concepts in the probability chapter that depend on algebra to be

more fully developed. Also, a number of topics have been moved forward in the text, so as to enliven the material with exciting applications as soon as possible; for example, the RSA cryptosystem is now described right after Euclid's algorithm is presented, and some basic results concerning quadratic residues are introduced right away, in the chapter on congruences. Finally, there are numerous changes in notation and terminology; for example, the notion of a *family* of objects is now used consistently throughout the book (e.g., a pairwise independent family of random variables, a linearly independent family of vectors, a pairwise relatively prime family of integers, etc.).

Feedback. I welcome comments on the book (suggestions for improvement, error reports, etc.) from readers. Please send your comments to

victor@shoup.net.

There is also a web site where further material and information relating to the book (including a list of errata and the latest electronic version of the book) may be found:

www.shoup.net/ntb.

Acknowledgments. I would like to thank a number of people who volunteered their time and energy in reviewing parts of the book at various stages: Joël Alwen, Siddhartha Annapureddy, John Black, Carl Bosley, Joshua Brody, Jan Camenisch, David Cash, Sherman Chow, Ronald Cramer, Marisa Debowsky, Alex Dent, Nelly Fazio, Rosario Gennaro, Mark Giesbrecht, Stuart Haber, Kristiyan Haralambiev, Gene Itkis, Charanjit Jutla, Jonathan Katz, Eike Kiltz, Alfred Menezes, Ilya Mironov, Phong Nguyen, Antonio Nicolosi, Roberto Oliveira, Leonid Reyzin, Louis Salvail, Berry Schoenmakers, Hovav Shacham, Yair Sovran, Panos Toulis, and Daniel Wichs. A very special thanks goes to George Stephanides, who translated the first edition of the book into Greek and reviewed the entire book in preparation for the second edition. I am also grateful to the National Science Foundation for their support provided under grants CCR-0310297 and CNS-0716690. Finally, thanks to David Tranah for all his help and advice, and to David and his colleagues at Cambridge University Press for their progressive attitudes regarding intellectual property and open access.

New York, June 2008 *Victor Shoup*

Preliminaries

We establish here some terminology, notation, and simple facts that will be used throughout the text.

Logarithms and exponentials

We write $\log x$ for the natural logarithm of x, and $\log_b x$ for the logarithm of x to the base b.

We write e^x for the usual exponential function, where $e \approx 2.71828$ is the base of the natural logarithm. We may also write $\exp[x]$ instead of e^x.

Sets and families

We use standard set-theoretic notation: $\emptyset$ denotes the empty set; $x \in A$ means that x is an element, or member, of the set A; for two sets A, B, $A \subseteq B$ means that A is a subset of B (with A possibly equal to B), and $A \subsetneq B$ means that A is a proper subset of B (i.e., $A \subseteq B$ but $A \neq B$). Further, $A \cup B$ denotes the union of A and B, $A \cap B$ the intersection of A and B, and $A \setminus B$ the set of all elements of A that are not in B. If A is a set with a finite number of elements, then we write $|A|$ for its **size**, or **cardinality**. We use standard notation for describing sets; for example, if we define the set $S := \{-2, -1, 0, 1, 2\}$, then $\{x^2 : x \in S\} = \{0, 1, 4\}$ and $\{x \in S : x \text{ is even}\} = \{-2, 0, 2\}$.

We write $S_1 \times \cdots \times S_n$ for the **Cartesian product** of sets $S_1, \ldots, S_n$, which is the set of all n-tuples $(a_1, \ldots, a_n)$, where $a_i \in S_i$ for $i = 1, \ldots, n$. We write $S^{\times n}$ for the Cartesian product of n copies of a set S, and for $x \in S$, we write $x^{\times n}$ for the element of $S^{\times n}$ consisting of n copies of x. (This notation is a bit non-standard, but we reserve the more standard notation S^n for other purposes, so as to avoid ambiguity.)

A **family** is a collection of objects, indexed by some set I, called an **index set**. If for each $i \in I$ we have an associated object x_i, the family of all such objects is denoted by $\{x_i\}_{i \in I}$. Unlike a set, a family may contain duplicates; that is, we may have $x_i = x_j$ for some pair of indices i, j with $i \neq j$. Note that while $\{x_i\}_{i \in I}$ denotes a family, $\{x_i : i \in I\}$ denotes the *set* whose members are the (distinct) x_i's. If the index set I has some natural order, then we may view the family $\{x_i\}_{i \in I}$ as being ordered in the same way; as a special case, a family indexed by a set of integers of the form $\{m, \ldots, n\}$ or $\{m, m+1, \ldots\}$ is a **sequence**, which we may write as $\{x_i\}_{i=m}^{n}$ or $\{x_i\}_{i=m}^{\infty}$. On occasion, if the choice of index set is not important, we may simply define a family by listing or describing its members, without explicitly describing an index set; for example, the phrase "the family of objects a, b, c" may be interpreted as "the family $\{x_i\}_{i=1}^{3}$, where $x_1 := a$, $x_2 := b$, and $x_3 := c$."

Unions and intersections may be generalized to arbitrary families of sets. For a family $\{S_i\}_{i \in I}$ of sets, the union is

$$\bigcup_{i \in I} S_i := \{x : x \in S_i \text{ for some } i \in I\},$$

and for $I \neq \emptyset$, the intersection is

$$\bigcap_{i \in I} S_i := \{x : x \in S_i \text{ for all } i \in I\}.$$

Note that if $I = \emptyset$, the union is by definition $\emptyset$, but the intersection is, in general, not well defined. However, in certain applications, one might define it by a special convention; for example, if all sets under consideration are subsets of some "ambient space," Ω, then the empty intersection is usually taken to be Ω.

Two sets A and B are called **disjoint** if $A \cap B = \emptyset$. A family $\{S_i\}_{i \in I}$ of sets is called **pairwise disjoint** if $S_i \cap S_j = \emptyset$ for all $i, j \in I$ with $i \neq j$. A pairwise disjoint family of non-empty sets whose union is S is called a **partition** of S; equivalently, $\{S_i\}_{i \in I}$ is a partition of a set S if each S_i is a non-empty subset of S, and each element of S belongs to exactly one S_i.

Numbers

We use standard notation for various sets of numbers:

$$\mathbb{Z} := \text{the set of integers} = \{\ldots, -2, -1, 0, 1, 2, \ldots\},$$

$$\mathbb{Q} := \text{the set of rational numbers} = \{a/b : a, b \in \mathbb{Z}, b \neq 0\},$$

$$\mathbb{R} := \text{the set of real numbers},$$

$$\mathbb{C} := \text{the set of complex numbers}.$$

We sometimes use the symbols ∞ and $-\infty$ in simple arithmetic expressions involving real numbers. The interpretation given to such expressions should be obvious: for example, for every $x \in \mathbb{R}$, we have $-\infty < x < \infty$, $x + \infty = \infty$, $x - \infty = -\infty$, $\infty + \infty = \infty$, and $(-\infty) + (-\infty) = -\infty$. Expressions such as $x \cdot (\pm\infty)$ also make sense, provided $x \neq 0$. However, the expressions $\infty - \infty$ and $0 \cdot \infty$ have no sensible interpretation.

We use standard notation for specifying intervals of real numbers: for $a, b \in \mathbb{R}$ with $a \leq b$,

$$[a, b] := \{x \in \mathbb{R} : a \leq x \leq b\}, \qquad (a, b) := \{x \in \mathbb{R} : a < x < b\},$$
$$[a, b) := \{x \in \mathbb{R} : a \leq x < b\}, \qquad (a, b] := \{x \in \mathbb{R} : a < x \leq b\}.$$

As usual, this notation is extended to allow $a = -\infty$ for the intervals $(a, b]$ and (a, b), and $b = \infty$ for the intervals $[a, b)$ and (a, b).

Functions

We write $f : A \to B$ to indicate that f is a function (also called a **map**) from a set A to a set B. If $A' \subseteq A$, then $f(A') := \{f(a) : a \in A'\}$ is the **image** of A' under f, and $f(A)$ is simply referred to as the **image** of f; if $B' \subseteq B$, then $f^{-1}(B') := \{a \in A : f(a) \in B'\}$ is the **pre-image** of B' under f.

A function $f : A \to B$ is called **one-to-one** or **injective** if $f(a) = f(b)$ implies $a = b$. The function f is called **onto** or **surjective** if $f(A) = B$. The function f is called **bijective** if it is both injective and surjective; in this case, f is called a **bijection**, or a **one-to-one correspondence**. If f is bijective, then we may define the **inverse function** $f^{-1} : B \to A$, where for $b \in B$, $f^{-1}(b)$ is defined to be the unique $a \in A$ such that $f(a) = b$; in this case, f^{-1} is also a bijection, and $(f^{-1})^{-1} = f$.

If $A' \subseteq A$, then the **inclusion map** from A' to A is the function $i : A' \to A$ given by $i(a) := a$ for $a \in A'$; when $A' = A$, this is called the **identity map** on A. If $A' \subseteq A$, $f' : A' \to B$, $f : A \to B$, and $f'(a) = f(a)$ for all $a \in A'$, then we say that f' is the **restriction** of f to A', and that f is an **extension** of f' to A.

If $f : A \to B$ and $g : B \to C$ are functions, their **composition** is the function $g \circ f : A \to C$ given by $(g \circ f)(a) := g(f(a))$ for $a \in A$. If $f : A \to B$ is a bijection, then $f^{-1} \circ f$ is the identity map on A, and $f \circ f^{-1}$ is the identity map on B. Conversely, if $f : A \to B$ and $g : B \to A$ are functions such that $g \circ f$ is the identity map on A and $f \circ g$ is the identity map on B, then f and g are bijections, each being the inverse of the other. If $f : A \to B$ and $g : B \to C$ are bijections, then so is $g \circ f$, and $(g \circ f)^{-1} = f^{-1} \circ g^{-1}$.

Function composition is associative; that is, for all functions $f : A \to B$, $g : B \to C$, and $h : C \to D$, we have $(h \circ g) \circ f = h \circ (g \circ f)$. Thus, we can simply write $h \circ g \circ f$ without any ambiguity. More generally, if we have functions $f_i : A_i \to A_{i+1}$ for $i = 1, \ldots, n$, where $n \geq 2$, then we may write their composition as $f_n \circ \cdots \circ f_1$ without any ambiguity. If each f_i is a bijection, then so is $f_n \circ \cdots \circ f_1$, its inverse being $f_1^{-1} \circ \cdots \circ f_n^{-1}$. As a special case of this, if $A_i = A$ and $f_i = f$ for $i = 1, \ldots, n$, then we may write $f_n \circ \cdots \circ f_1$ as f^n. It is understood that $f^1 = f$, and that f^0 is the identity map on A. If f is a bijection, then so is f^n for every non-negative integer n, the inverse function of f^n being $(f^{-1})^n$, which one may simply write as f^{-n}.

If $f : I \to S$ is a function, then we may view f as the family $\{x_i\}_{i \in I}$, where $x_i := f(i)$. Conversely, a family $\{x_i\}_{i \in I}$, where all of the x_i's belong to some set S, may be viewed as the function $f : I \to S$ given by $f(i) := x_i$ for $i \in I$. Really, functions and families are the same thing, the difference being just one of notation and emphasis.

Binary operations

A **binary operation** $\star$ on a set S is a function from $S \times S$ to S, where the value of the function at $(a, b) \in S \times S$ is denoted $a \star b$.

A binary operation $\star$ on S is called **associative** if for all $a, b, c \in S$, we have $(a \star b) \star c = a \star (b \star c)$. In this case, we can simply write $a \star b \star c$ without any ambiguity. More generally, for $a_1, \ldots, a_n \in S$, where $n \geq 2$, we can write $a_1 \star \cdots \star a_n$ without any ambiguity.

A binary operation $\star$ on S is called **commutative** if for all $a, b \in S$, we have $a \star b = b \star a$. If the binary operation $\star$ is both associative and commutative, then not only is the expression $a_1 \star \cdots \star a_n$ unambiguous, but its value remains unchanged even if we re-order the a_i's.

If $\star$ is a binary operation on S, and $S' \subseteq S$, then S' is called **closed under** $\star$ if $a \star b \in S'$ for all $a, b \in S'$.

1

Basic properties of the integers

This chapter discusses some of the basic properties of the integers, including the notions of divisibility and primality, unique factorization into primes, greatest common divisors, and least common multiples.

1.1 Divisibility and primality

A central concept in number theory is *divisibility*.

Consider the integers $\mathbb{Z} = \{\ldots, -2, -1, 0, 1, 2, \ldots\}$. For $a, b \in \mathbb{Z}$, we say that a **divides** b if $az = b$ for some $z \in \mathbb{Z}$. If a divides b, we write $a \mid b$, and we may say that a is a **divisor** of b, or that b is a **multiple** of a, or that b is **divisible by** a. If a does not divide b, then we write $a \nmid b$.

We first state some simple facts about divisibility:

Theorem 1.1. *For all $a, b, c \in \mathbb{Z}$, we have*

   (i) $a \mid a$, $1 \mid a$, and $a \mid 0$;

   (ii) $0 \mid a$ if and only if $a = 0$;

   (iii) $a \mid b$ if and only if $-a \mid b$ if and only if $a \mid -b$;

   (iv) $a \mid b$ and $a \mid c$ implies $a \mid (b + c)$;

   (v) $a \mid b$ and $b \mid c$ implies $a \mid c$.

Proof. These properties can be easily derived from the definition of divisibility, using elementary algebraic properties of the integers. For example, $a \mid a$ because we can write $a \cdot 1 = a$; $1 \mid a$ because we can write $1 \cdot a = a$; $a \mid 0$ because we can write $a \cdot 0 = 0$. We leave it as an easy exercise for the reader to verify the remaining properties. $\square$

We make a simple observation: if $a \mid b$ and $b \neq 0$, then $1 \leq |a| \leq |b|$. Indeed, if $az = b \neq 0$ for some integer z, then $a \neq 0$ and $z \neq 0$; it follows that $|a| \geq 1$, $|z| \geq 1$, and so $|a| \leq |a||z| = |b|$.

1

Theorem 1.2. *For all $a, b \in \mathbb{Z}$, we have $a \mid b$ and $b \mid a$ if and only if $a = \pm b$. In particular, for every $a \in \mathbb{Z}$, we have $a \mid 1$ if and only if $a = \pm 1$.*

Proof. Clearly, if $a = \pm b$, then $a \mid b$ and $b \mid a$. So let us assume that $a \mid b$ and $b \mid a$, and prove that $a = \pm b$. If either of a or b are zero, then the other must be zero as well. So assume that neither is zero. By the above observation, $a \mid b$ implies $|a| \leq |b|$, and $b \mid a$ implies $|b| \leq |a|$; thus, $|a| = |b|$, and so $a = \pm b$. That proves the first statement. The second statement follows from the first by setting $b := 1$, and noting that $1 \mid a$. $\square$

The product of any two non-zero integers is again non-zero. This implies the usual **cancellation law**: if a, b, and c are integers such that $a \neq 0$ and $ab = ac$, then we must have $b = c$; indeed, $ab = ac$ implies $a(b - c) = 0$, and so $a \neq 0$ implies $b - c = 0$, and hence $b = c$.

Primes and composites. Let n be a positive integer. Trivially, 1 and n divide n. If $n > 1$ and no other positive integers besides 1 and n divide n, then we say n is **prime**. If $n > 1$ but n is not prime, then we say that n is **composite**. The number 1 is not considered to be either prime or composite. Evidently, n is composite if and only if $n = ab$ for some integers a, b with $1 < a < n$ and $1 < b < n$. The first few primes are

$$2, 3, 5, 7, 11, 13, 17, \ldots.$$

While it is possible to extend the definition of prime and composite to negative integers, we shall not do so in this text: *whenever we speak of a prime or composite number, we mean a positive integer.*

A basic fact is that every non-zero integer can be expressed as a signed product of primes in an essentially unique way. More precisely:

Theorem 1.3 (Fundamental theorem of arithmetic). *Every non-zero integer n can be expressed as*

$$n = \pm p_1^{e_1} \cdots p_r^{e_r},$$

where $p_1, \ldots, p_r$ are distinct primes and $e_1, \ldots, e_r$ are positive integers. Moreover, this expression is unique, up to a reordering of the primes.

Note that if $n = \pm 1$ in the above theorem, then $r = 0$, and the product of zero terms is interpreted (as usual) as 1.

The theorem intuitively says that the primes act as the "building blocks" out of which all non-zero integers can be formed by multiplication (and negation). The reader may be so familiar with this fact that he may feel it is somehow "self evident," requiring no proof; however, this feeling is simply a delusion, and most of the rest of this section and the next are devoted to developing a proof of this theorem. We shall give a quite leisurely proof, introducing a number of other very important tools and concepts along the way that will be useful later.

To prove Theorem 1.3, we may clearly assume that n is positive, since otherwise, we may multiply n by -1 and reduce to the case where n is positive.

The proof of the existence part of Theorem 1.3 is easy. This amounts to showing that every positive integer n can be expressed as a product (possibly empty) of primes. We may prove this by induction on n. If $n = 1$, the statement is true, as n is the product of zero primes. Now let $n > 1$, and assume that every positive integer smaller than n can be expressed as a product of primes. If n is a prime, then the statement is true, as n is the product of one prime. Assume, then, that n is composite, so that there exist $a, b \in \mathbb{Z}$ with $1 < a < n$, $1 < b < n$, and $n = ab$. By the induction hypothesis, both a and b can be expressed as a product of primes, and so the same holds for n.

The uniqueness part of Theorem 1.3 is the hard part. An essential ingredient in this proof is the following:

Theorem 1.4 (Division with remainder property). *Let $a, b \in \mathbb{Z}$ with $b > 0$. Then there exist unique $q, r \in \mathbb{Z}$ such that $a = bq + r$ and $0 \le r < b$.*

Proof. Consider the set S of non-negative integers of the form $a - bt$ with $t \in \mathbb{Z}$. This set is clearly non-empty; indeed, if $a \ge 0$, set $t := 0$, and if $a < 0$, set $t := a$. Since every non-empty set of non-negative integers contains a minimum, we define r to be the smallest element of S. By definition, r is of the form $r = a - bq$ for some $q \in \mathbb{Z}$, and $r \ge 0$. Also, we must have $r < b$, since otherwise, $r - b$ would be an element of S smaller than r, contradicting the minimality of r; indeed, if $r \ge b$, then we would have $0 \le r - b = a - b(q + 1)$.

That proves the existence of r and q. For uniqueness, suppose that $a = bq + r$ and $a = bq' + r'$, where $0 \le r < b$ and $0 \le r' < b$. Then subtracting these two equations and rearranging terms, we obtain

$$r' - r = b(q - q').$$

Thus, $r' - r$ is a multiple of b; however, $0 \le r < b$ and $0 \le r' < b$ implies $|r' - r| < b$; therefore, the only possibility is $r' - r = 0$. Moreover, $0 = b(q - q')$ and $b \ne 0$ implies $q - q' = 0$. $\square$

Theorem 1.4 can be visualized as follows:

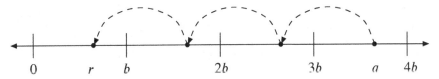

Starting with a, we subtract (or add, if a is negative) the value b until we end up with a number in the interval $[0, b)$.

Floors and ceilings. Let us briefly recall the usual **floor** and **ceiling** functions, denoted $\lfloor \cdot \rfloor$ and $\lceil \cdot \rceil$, respectively. These are functions from $\mathbb{R}$ (the real numbers) to $\mathbb{Z}$. For $x \in \mathbb{R}$, $\lfloor x \rfloor$ is the greatest integer $m \leq x$; equivalently, $\lfloor x \rfloor$ is the unique integer m such that $m \leq x < m + 1$, or put another way, such that $x = m + \varepsilon$ for some $\varepsilon \in [0, 1)$. Also, $\lceil x \rceil$ is the smallest integer $m \geq x$; equivalently, $\lceil x \rceil$ is the unique integer m such that $m - 1 < x \leq m$, or put another way, such that $x = m - \varepsilon$ for some $\varepsilon \in [0, 1)$.

The mod operator. Now let $a, b \in \mathbb{Z}$ with $b > 0$. If q and r are the unique integers from Theorem 1.4 that satisfy $a = bq + r$ and $0 \leq r < b$, we define

$$a \bmod b := r;$$

that is, $a \bmod b$ denotes the remainder in dividing a by b. It is clear that $b \mid a$ if and only if $a \bmod b = 0$. Dividing both sides of the equation $a = bq + r$ by b, we obtain $a/b = q + r/b$. Since $q \in \mathbb{Z}$ and $r/b \in [0, 1)$, we see that $q = \lfloor a/b \rfloor$. Thus,

$$(a \bmod b) = a - b\lfloor a/b \rfloor.$$

One can use this equation to extend the definition of $a \bmod b$ to all integers a and b, with $b \neq 0$; that is, for $b < 0$, we simply define $a \bmod b$ to be $a - b\lfloor a/b \rfloor$.

Theorem 1.4 may be generalized so that when dividing an integer a by a positive integer b, the remainder is placed in an interval other than $[0, b)$. Let x be any real number, and consider the interval $[x, x + b)$. As the reader may easily verify, this interval contains precisely b integers, namely, $\lceil x \rceil, \ldots, \lceil x \rceil + b - 1$. Applying Theorem 1.4 with $a - \lceil x \rceil$ in place of a, we obtain:

Theorem 1.5. *Let $a, b \in \mathbb{Z}$ with $b > 0$, and let $x \in \mathbb{R}$. Then there exist unique $q, r \in \mathbb{Z}$ such that $a = bq + r$ and $r \in [x, x + b)$.*

EXERCISE 1.1. Let $a, b, d \in \mathbb{Z}$ with $d \neq 0$. Show that $a \mid b$ if and only if $da \mid db$.

EXERCISE 1.2. Let n be a composite integer. Show that there exists a prime p dividing n, with $p \leq n^{1/2}$.

EXERCISE 1.3. Let m be a positive integer. Show that for every real number $x \geq 1$, the number of multiples of m in the interval $[1, x]$ is $\lfloor x/m \rfloor$; in particular, for every integer $n \geq 1$, the number of multiples of m among $1, \ldots, n$ is $\lfloor n/m \rfloor$.

EXERCISE 1.4. Let $x \in \mathbb{R}$. Show that $2\lfloor x \rfloor \leq \lfloor 2x \rfloor \leq 2\lfloor x \rfloor + 1$.

EXERCISE 1.5. Let $x \in \mathbb{R}$ and $n \in \mathbb{Z}$ with $n > 0$. Show that $\lfloor \lfloor x \rfloor / n \rfloor = \lfloor x/n \rfloor$; in particular, $\lfloor \lfloor a/b \rfloor / c \rfloor = \lfloor a/bc \rfloor$ for all positive integers a, b, c.

EXERCISE 1.6. Let $a, b \in \mathbb{Z}$ with $b < 0$. Show that $(a \bmod b) \in (b, 0]$.

EXERCISE 1.7. Show that Theorem 1.5 also holds for the interval $(x, x + b]$. Does it hold in general for the intervals $[x, x + b]$ or $(x, x + b)$?

1.2 Ideals and greatest common divisors

To carry on with the proof of Theorem 1.3, we introduce the notion of an **ideal of** $\mathbb{Z}$, which is a non-empty set of integers that is closed under addition, and closed under multiplication by an arbitrary integer. That is, a non-empty set $I \subseteq \mathbb{Z}$ is an ideal if and only if for all $a, b \in I$ and all $z \in \mathbb{Z}$, we have

$$a + b \in I \quad \text{and} \quad az \in I.$$

Besides its utility in proving Theorem 1.3, the notion of an ideal is quite useful in a number of contexts, which will be explored later.

It is easy to see that every ideal I contains 0: since $a \in I$ for some integer a, we have $0 = a \cdot 0 \in I$. Also, note that if an ideal I contains an integer a, it also contains $-a$, since $-a = a \cdot (-1) \in I$. Thus, if an ideal contains a and b, it also contains $a - b$. It is clear that $\{0\}$ and $\mathbb{Z}$ are ideals. Moreover, an ideal I is equal to $\mathbb{Z}$ if and only if $1 \in I$; to see this, note that $1 \in I$ implies that for every $z \in \mathbb{Z}$, we have $z = 1 \cdot z \in I$, and hence $I = \mathbb{Z}$; conversely, if $I = \mathbb{Z}$, then in particular, $1 \in I$.

For $a \in \mathbb{Z}$, define $a\mathbb{Z} := \{az : z \in \mathbb{Z}\}$; that is, $a\mathbb{Z}$ is the set of all multiples of a. If $a = 0$, then clearly $a\mathbb{Z} = \{0\}$; otherwise, $a\mathbb{Z}$ consists of the distinct integers

$$\ldots, -3a, -2a, -a, 0, a, 2a, 3a, \ldots.$$

It is easy to see that $a\mathbb{Z}$ is an ideal: for all $az, az' \in a\mathbb{Z}$ and $z'' \in \mathbb{Z}$, we have $az + az' = a(z + z') \in a\mathbb{Z}$ and $(az)z'' = a(zz'') \in a\mathbb{Z}$. The ideal $a\mathbb{Z}$ is called the **ideal generated by** a, and an ideal of the form $a\mathbb{Z}$ for some $a \in \mathbb{Z}$ is called a **principal ideal**.

Observe that for all $a, b \in \mathbb{Z}$, we have $b \in a\mathbb{Z}$ if and only if $a \mid b$. Also observe that for every ideal I, we have $b \in I$ if and only if $b\mathbb{Z} \subseteq I$. Both of these observations are simple consequences of the definitions, as the reader may verify. Combining these two observations, we see that $b\mathbb{Z} \subseteq a\mathbb{Z}$ if and only if $a \mid b$.

Suppose I_1 and I_2 are ideals. Then it is not hard to see that the set

$$I_1 + I_2 := \{a_1 + a_2 : a_1 \in I_1, a_2 \in I_2\}$$

is also an ideal. Indeed, suppose $a_1 + a_2 \in I_1 + I_2$ and $b_1 + b_2 \in I_1 + I_2$. Then we have $(a_1 + a_2) + (b_1 + b_2) = (a_1 + b_1) + (a_2 + b_2) \in I_1 + I_2$, and for every $z \in \mathbb{Z}$, we have $(a_1 + a_2)z = a_1 z + a_2 z \in I_1 + I_2$.

Example 1.1. Consider the principal ideal $3\mathbb{Z}$. This consists of all multiples of 3; that is, $3\mathbb{Z} = \{\ldots, -9, -6, -3, 0, 3, 6, 9, \ldots\}$. $\square$

Example 1.2. Consider the ideal $3\mathbb{Z} + 5\mathbb{Z}$. This ideal contains $3 \cdot 2 + 5 \cdot (-1) = 1$. Since it contains 1, it contains all integers; that is, $3\mathbb{Z} + 5\mathbb{Z} = \mathbb{Z}$. $\square$

Example 1.3. Consider the ideal $4\mathbb{Z} + 6\mathbb{Z}$. This ideal contains $4 \cdot (-1) + 6 \cdot 1 = 2$, and therefore, it contains all even integers. It does not contain any odd integers, since the sum of two even integers is again even. Thus, $4\mathbb{Z} + 6\mathbb{Z} = 2\mathbb{Z}$. $\square$

In the previous two examples, we defined an ideal that turned out upon closer inspection to be a principal ideal. This was no accident: the following theorem says that all ideals of $\mathbb{Z}$ are principal.

Theorem 1.6. *Let I be an ideal of $\mathbb{Z}$. Then there exists a unique non-negative integer d such that $I = d\mathbb{Z}$.*

Proof. We first prove the existence part of the theorem. If $I = \{0\}$, then $d = 0$ does the job, so let us assume that $I \neq \{0\}$. Since I contains non-zero integers, it must contain positive integers, since if $a \in I$ then so is $-a$. Let d be the smallest positive integer in I. We want to show that $I = d\mathbb{Z}$.

We first show that $I \subseteq d\mathbb{Z}$. To this end, let a be any element in I. It suffices to show that $d \mid a$. Using the division with remainder property, write $a = dq + r$, where $0 \leq r < d$. Then by the closure properties of ideals, one sees that $r = a - dq$ is also an element of I, and by the minimality of the choice of d, we must have $r = 0$. Thus, $d \mid a$.

We have shown that $I \subseteq d\mathbb{Z}$. The fact that $d\mathbb{Z} \subseteq I$ follows from the fact that $d \in I$. Thus, $I = d\mathbb{Z}$.

That proves the existence part of the theorem. For uniqueness, note that if $d\mathbb{Z} = e\mathbb{Z}$ for some non-negative integer e, then $d \mid e$ and $e \mid d$, from which it follows by Theorem 1.2 that $d = \pm e$; since d and e are non-negative, we must have $d = e$. $\square$

Greatest common divisors. For $a, b \in \mathbb{Z}$, we call $d \in \mathbb{Z}$ a **common divisor** of a and b if $d \mid a$ and $d \mid b$; moreover, we call such a d a **greatest common divisor** of a and b if d is non-negative and all other common divisors of a and b divide d.

Theorem 1.7. *For all $a, b \in \mathbb{Z}$, there exists a unique greatest common divisor d of a and b, and moreover, $a\mathbb{Z} + b\mathbb{Z} = d\mathbb{Z}$.*

Proof. We apply the previous theorem to the ideal $I := a\mathbb{Z} + b\mathbb{Z}$. Let $d \in \mathbb{Z}$ with $I = d\mathbb{Z}$, as in that theorem. We wish to show that d is a greatest common divisor of a and b. Note that $a, b, d \in I$ and d is non-negative.

Since $a \in I = d\mathbb{Z}$, we see that $d \mid a$; similarly, $d \mid b$. So we see that d is a common divisor of a and b.

Since $d \in I = a\mathbb{Z} + b\mathbb{Z}$, there exist $s, t \in \mathbb{Z}$ such that $as + bt = d$. Now suppose $a = a'd'$ and $b = b'd'$ for some $a', b', d' \in \mathbb{Z}$. Then the equation $as + bt = d$ implies that $d'(a's + b't) = d$, which says that $d' \mid d$. Thus, any common divisor d' of a and b divides d.

That proves that d is a greatest common divisor of a and b. For uniqueness, note that if e is a greatest common divisor of a and b, then $d \mid e$ and $e \mid d$, and hence $d = \pm e$; since both d and e are non-negative by definition, we have $d = e$. $\square$

For $a, b \in \mathbb{Z}$, we write $\gcd(a, b)$ for the greatest common divisor of a and b. We say that $a, b \in \mathbb{Z}$ are **relatively prime** if $\gcd(a, b) = 1$, which is the same as saying that the only common divisors of a and b are ± 1.

The following is essentially just a restatement of Theorem 1.7, but we state it here for emphasis:

Theorem 1.8. *Let $a, b, r \in \mathbb{Z}$ and let $d := \gcd(a, b)$. Then there exist $s, t \in \mathbb{Z}$ such that $as + bt = r$ if and only if $d \mid r$. In particular, a and b are relatively prime if and only if there exist integers s and t such that $as + bt = 1$.*

Proof. We have

$$as + bt = r \text{ for some } s, t \in \mathbb{Z}$$
$$\Longleftrightarrow r \in a\mathbb{Z} + b\mathbb{Z}$$
$$\Longleftrightarrow r \in d\mathbb{Z} \text{ (by Theorem 1.7)}$$
$$\Longleftrightarrow d \mid r.$$

That proves the first statement. The second statement follows from the first, setting $r := 1$. $\square$

Note that as we have defined it, $\gcd(0, 0) = 0$. Also note that when at least one of a or b are non-zero, $\gcd(a, b)$ may be characterized as the *largest* positive integer that divides both a and b, and as the *smallest* positive integer that can be expressed as $as + bt$ for integers s and t.

Theorem 1.9. *Let $a, b, c \in \mathbb{Z}$ such that $c \mid ab$ and $\gcd(a, c) = 1$. Then $c \mid b$.*

Proof. Suppose that $c \mid ab$ and $\gcd(a, c) = 1$. Then since $\gcd(a, c) = 1$, by Theorem 1.8 we have $as + ct = 1$ for some $s, t \in \mathbb{Z}$. Multiplying this equation by *b*, we obtain

$$abs + cbt = b. \tag{1.1}$$

Since c divides ab by hypothesis, and since c clearly divides cbt, it follows that c divides the left-hand side of (1.1), and hence that c divides b. $\square$

Suppose that p is a prime and a is any integer. As the only divisors of p are ± 1 and $\pm p$, we have

$$p \mid a \implies \gcd(a, p) = p, \text{ and}$$
$$p \nmid a \implies \gcd(a, p) = 1.$$

Combining this observation with the previous theorem, we have:

Theorem 1.10. *Let p be prime, and let $a, b \in \mathbb{Z}$. Then $p \mid ab$ implies that $p \mid a$ or $p \mid b$.*

Proof. Assume that $p \mid ab$. If $p \mid a$, we are done, so assume that $p \nmid a$. By the above observation, $\gcd(a, p) = 1$, and so by Theorem 1.9, we have $p \mid b$. $\square$

An obvious corollary to Theorem 1.10 is that if $a_1, \ldots, a_k$ are integers, and if p is a prime that divides the product $a_1 \cdots a_k$, then $p \mid a_i$ for some $i = 1, \ldots, k$. This is easily proved by induction on k. For $k = 1$, the statement is trivially true. Now let $k > 1$, and assume that statement holds for $k - 1$. Then by Theorem 1.10, either $p \mid a_1$ or $p \mid a_2 \cdots a_k$; if $p \mid a_1$, we are done; otherwise, by induction, p divides one of $a_2, \ldots, a_k$.

Finishing the proof of Theorem 1.3. We are now in a position to prove the uniqueness part of Theorem 1.3, which we can state as follows: if $p_1, \ldots, p_r$ are primes (not necessarily distinct), and $q_1, \ldots, q_s$ are primes (also not necessarily distinct), such that

$$p_1 \cdots p_r = q_1 \cdots q_s, \tag{1.2}$$

then $(p_1, \ldots, p_r)$ is just a reordering of $(q_1, \ldots, q_s)$. We may prove this by induction on r. If $r = 0$, we must have $s = 0$ and we are done. Now suppose $r > 0$, and that the statement holds for $r - 1$. Since $r > 0$, we clearly must have $s > 0$. Also, as p_1 obviously divides the left-hand side of (1.2), it must also divide the right-hand side of (1.2); that is, $p_1 \mid q_1 \cdots q_s$. It follows from (the corollary to) Theorem 1.10 that $p_1 \mid q_j$ for some $j = 1, \ldots, s$, and moreover, since q_j is prime, we must have $p_1 = q_j$. Thus, we may cancel p_1 from the left-hand side of (1.2) and q_j from the right-hand side of (1.2), and the statement now follows from the induction hypothesis. That proves the uniqueness part of Theorem 1.3.

EXERCISE 1.8. Let I be a non-empty set of integers that is closed under addition (i.e., $a + b \in I$ for all $a, b \in I$). Show that I is an ideal if and only if $-a \in I$ for all $a \in I$.

EXERCISE 1.9. Show that for all integers a, b, c, we have:

(a) $\gcd(a, b) = \gcd(b, a)$;

(b) $\gcd(a, b) = |a| \iff a \mid b$;

(c) $\gcd(a, 0) = \gcd(a, a) = |a|$ and $\gcd(a, 1) = 1$;

(d) $\gcd(ca, cb) = |c| \gcd(a, b)$.

EXERCISE 1.10. Show that for all integers a, b with $d := \gcd(a, b) \neq 0$, we have $\gcd(a/d, b/d) = 1$.

EXERCISE 1.11. Let n be an integer. Show that if a, b are relatively prime integers, each of which divides n, then ab divides n.

EXERCISE 1.12. Show that two integers are relatively prime if and only if there is no one prime that divides both of them.

EXERCISE 1.13. Let $a, b_1, \ldots, b_k$ be integers. Show that $\gcd(a, b_1 \cdots b_k) = 1$ if and only if $\gcd(a, b_i) = 1$ for $i = 1, \ldots, k$.

EXERCISE 1.14. Let p be a prime and k an integer, with $0 < k < p$. Show that the binomial coefficient

$$\binom{p}{k} = \frac{p!}{k!(p-k)!},$$

which is an integer (see §A2), is divisible by p.

EXERCISE 1.15. An integer a is called **square-free** if it is not divisible by the square of any integer greater than 1. Show that:

(a) a is square-free if and only if $a = \pm p_1 \cdots p_r$, where the p_i's are distinct primes;

(b) every positive integer n can be expressed uniquely as $n = ab^2$, where a and b are positive integers, and a is square-free.

EXERCISE 1.16. For each positive integer m, let I_m denote $\{0, \ldots, m-1\}$. Let a, b be positive integers, and consider the map

$$\tau : \quad I_b \times I_a \to I_{ab}$$
$$(s, t) \mapsto (as + bt) \bmod ab.$$

Show τ is a bijection if and only if $\gcd(a, b) = 1$.

EXERCISE 1.17. Let a, b, c be positive integers satisfying $\gcd(a, b) = 1$ and $c \geq (a - 1)(b - 1)$. Show that there exist *non-negative* integers s, t such that $c = as + bt$.

EXERCISE 1.18. For each positive integer n, let D_n denote the set of positive divisors of n. Let n_1, n_2 be relatively prime, positive integers. Show that the sets $D_{n_1} \times D_{n_2}$ and $D_{n_1 n_2}$ are in one-to-one correspondence, via the map that sends $(d_1, d_2) \in D_{n_1} \times D_{n_2}$ to $d_1 d_2$.

1.3 Some consequences of unique factorization

The following theorem is a consequence of just the existence part of Theorem 1.3:

Theorem 1.11. *There are infinitely many primes.*

Proof. By way of contradiction, suppose that there were only finitely many primes; call them $p_1, \ldots, p_k$. Then set $M := \prod_{i=1}^{k} p_i$ and $N := M + 1$. Consider a prime p that divides N. There must be at least one such prime p, since $N \geq 2$, and every positive integer can be written as a product of primes. Clearly, p cannot equal any of the p_i's, since if it did, then p would divide M, and hence also divide $N - M = 1$, which is impossible. Therefore, the prime p is not among $p_1, \ldots, p_k$, which contradicts our assumption that these are the only primes. $\square$

For each prime p, we may define the function v_p, mapping non-zero integers to non-negative integers, as follows: for every integer $n \neq 0$, if $n = p^e m$, where $p \nmid m$, then $v_p(n) := e$. We may then write the factorization of n into primes as

$$n = \pm \prod_{p} p^{v_p(n)},$$

where the product is over all primes p; although syntactically this is an infinite product, all but finitely many of its terms are equal to 1, and so this expression makes sense.

Observe that if a and b are non-zero integers, then

$$v_p(a \cdot b) = v_p(a) + v_p(b) \quad \text{for all primes } p, \tag{1.3}$$

and

$$a \mid b \iff v_p(a) \leq v_p(b) \quad \text{for all primes } p. \tag{1.4}$$

From this, it is clear that

$$\gcd(a, b) = \prod_{p} p^{\min(v_p(a), v_p(b))}.$$

Least common multiples. For $a, b \in \mathbb{Z}$, a **common multiple** of a and b is an integer m such that $a \mid m$ and $b \mid m$; moreover, such an m is the **least common multiple** of a and b if m is non-negative and m divides all common multiples of a and b. It is easy to see that the least common multiple exists and is unique, and we denote the least common multiple of a and b by $\mathrm{lcm}(a, b)$. Indeed, for all $a, b \in \mathbb{Z}$, if either a or b are zero, the only common multiple of a and b is 0, and so $\mathrm{lcm}(a, b) = 0$; otherwise, if neither a nor b are zero, we have

$$\mathrm{lcm}(a, b) = \prod_p p^{\max(v_p(a), v_p(b))},$$

or equivalently, $\mathrm{lcm}(a, b)$ may be characterized as the smallest positive integer divisible by both a and b.

It is convenient to extend the domain of definition of v_p to include 0, defining $v_p(0) := \infty$. If we interpret expressions involving "∞" appropriately (see Preliminaries), then for arbitrary $a, b \in \mathbb{Z}$, both (1.3) and (1.4) hold, and in addition,

$$v_p(\gcd(a, b)) = \min(v_p(a), v_p(b)) \text{ and } v_p(\mathrm{lcm}(a, b)) = \max(v_p(a), v_p(b))$$

for all primes p.

Generalizing gcd's and lcm's to many integers. It is easy to generalize the notions of greatest common divisor and least common multiple from two integers to many integers. Let $a_1, \ldots, a_k$ be integers. We call $d \in \mathbb{Z}$ a common divisor of $a_1, \ldots, a_k$ if $d \mid a_i$ for $i = 1, \ldots, k$; moreover, we call such a d the greatest common divisor of $a_1, \ldots, a_k$ if d is non-negative and all other common divisors of $a_1, \ldots, a_k$ divide d. The greatest common divisor of $a_1, \ldots, a_k$ is denoted $\gcd(a_1, \ldots, a_k)$ and is the unique non-negative integer d satisfying

$$v_p(d) = \min(v_p(a_1), \ldots, v_p(a_k)) \text{ for all primes } p.$$

Analogously, we call $m \in \mathbb{Z}$ a common multiple of $a_1, \ldots, a_k$ if $a_i \mid m$ for all $i = 1, \ldots, k$; moreover, such an m is called the least common multiple of $a_1, \ldots, a_k$ if m divides all common multiples of $a_1, \ldots, a_k$. The least common multiple of $a_1, \ldots, a_k$ is denoted $\mathrm{lcm}(a_1, \ldots, a_k)$ and is the unique non-negative integer m satisfying

$$v_p(m) = \max(v_p(a_1), \ldots, v_p(a_k)) \text{ for all primes } p.$$

Finally, we say that the family $\{a_i\}_{i=1}^k$ is **pairwise relatively prime** if for all indices i, j with $i \neq j$, we have $\gcd(a_i, a_j) = 1$. Certainly, if $\{a_i\}_{i=1}^k$ is pairwise relatively prime, and $k > 1$, then $\gcd(a_1, \ldots, a_k) = 1$; however, $\gcd(a_1, \ldots, a_k) = 1$ does not imply that $\{a_i\}_{i=1}^k$ is pairwise relatively prime.

Rational numbers. Consider the rational numbers $\mathbb{Q} = \{a/b : a, b \in \mathbb{Z}, b \neq 0\}$. Given any rational number a/b, if we set $d := \gcd(a, b)$, and define the integers $a_0 := a/d$ and $b_0 := b/d$, then we have $a/b = a_0/b_0$ and $\gcd(a_0, b_0) = 1$. Moreover, if $a_1/b_1 = a_0/b_0$, then we have $a_1 b_0 = a_0 b_1$, and so $b_0 \mid a_0 b_1$; also, since $\gcd(a_0, b_0) = 1$, we see that $b_0 \mid b_1$; writing $b_1 = b_0 c$, we see that $a_1 = a_0 c$. Thus, we can represent every rational number as a fraction in **lowest terms**, which means a fraction of the form a_0/b_0 where a_0 and b_0 are relatively prime; moreover, the values of a_0 and b_0 are uniquely determined up to sign, and every other fraction that represents the same rational number is of the form $a_0 c / b_0 c$, for some non-zero integer c.

EXERCISE 1.19. Let n be an integer. Generalizing Exercise 1.11, show that if $\{a_i\}_{i=1}^{k}$ is a pairwise relatively prime family of integers, where each a_i divides n, then their product $\prod_{i=1}^{k} a_i$ also divides n.

EXERCISE 1.20. Show that for all integers a, b, c, we have:

(a) $\operatorname{lcm}(a, b) = \operatorname{lcm}(b, a)$;

(b) $\operatorname{lcm}(a, b) = |a| \iff b \mid a$;

(c) $\operatorname{lcm}(a, a) = \operatorname{lcm}(a, 1) = |a|$;

(d) $\operatorname{lcm}(ca, cb) = |c| \operatorname{lcm}(a, b)$.

EXERCISE 1.21. Show that for all integers a, b, we have:

(a) $\gcd(a, b) \cdot \operatorname{lcm}(a, b) = |ab|$;

(b) $\gcd(a, b) = 1 \implies \operatorname{lcm}(a, b) = |ab|$.

EXERCISE 1.22. Let $a_1, \ldots, a_k \in \mathbb{Z}$ with $k > 1$. Show that:

$$\gcd(a_1, \ldots, a_k) = \gcd(a_1, \gcd(a_2, \ldots, a_k)) = \gcd(\gcd(a_1, \ldots, a_{k-1}), a_k);$$
$$\operatorname{lcm}(a_1, \ldots, a_k) = \operatorname{lcm}(a_1, \operatorname{lcm}(a_2, \ldots, a_k)) = \operatorname{lcm}(\operatorname{lcm}(a_1, \ldots, a_{k-1}), a_k).$$

EXERCISE 1.23. Let $a_1, \ldots, a_k \in \mathbb{Z}$ with $d := \gcd(a_1, \ldots, a_k)$. Show that $d\mathbb{Z} = a_1\mathbb{Z} + \cdots + a_k\mathbb{Z}$; in particular, there exist integers $z_1, \ldots, z_k$ such that $d = a_1 z_1 + \cdots + a_k z_k$.

EXERCISE 1.24. Show that if $\{a_i\}_{i=1}^{k}$ is a pairwise relatively prime family of integers, then $\operatorname{lcm}(a_1, \ldots, a_k) = |a_1 \cdots a_k|$.

EXERCISE 1.25. Show that every non-zero $x \in \mathbb{Q}$ can be expressed as

$$x = \pm p_1^{e_1} \cdots p_r^{e_r},$$

where the p_i's are distinct primes and the e_i's are non-zero integers, and that this expression in unique up to a reordering of the primes.

EXERCISE 1.26. Let n and k be positive integers, and suppose $x \in \mathbb{Q}$ such that $x^k = n$ for some $x \in \mathbb{Q}$. Show that $x \in \mathbb{Z}$. In other words, $\sqrt[k]{n}$ is either an integer or is irrational.

EXERCISE 1.27. Show that $\gcd(a + b, \text{lcm}(a, b)) = \gcd(a, b)$ for all $a, b \in \mathbb{Z}$.

EXERCISE 1.28. Show that for every positive integer k, there exist k consecutive composite integers. Thus, there are arbitrarily large gaps between primes.

EXERCISE 1.29. Let p be a prime. Show that for all $a, b \in \mathbb{Z}$, we have $v_p(a+b) \geq \min\{v_p(a), v_p(b)\}$, and $v_p(a + b) = v_p(a)$ if $v_p(a) < v_p(b)$.

EXERCISE 1.30. For a given prime p, we may extend the domain of definition of v_p from $\mathbb{Z}$ to $\mathbb{Q}$: for non-zero integers a, b, let us define $v_p(a/b) := v_p(a) - v_p(b)$. Show that:

(a) this definition of $v_p(a/b)$ is unambiguous, in the sense that it does not depend on the particular choice of a and b;

(b) for all $x, y \in \mathbb{Q}$, we have $v_p(xy) = v_p(x) + v_p(y)$;

(c) for all $x, y \in \mathbb{Q}$, we have $v_p(x + y) \geq \min\{v_p(x), v_p(y)\}$, and $v_p(x + y) = v_p(x)$ if $v_p(x) < v_p(y)$;

(d) for all non-zero $x \in \mathbb{Q}$, we have $x = \pm \prod_p p^{v_p(x)}$, where the product is over all primes, and all but a finite number of terms in the product are equal to 1;

(e) for all $x \in \mathbb{Q}$, we have $x \in \mathbb{Z}$ if and only if $v_p(x) \geq 0$ for all primes p.

EXERCISE 1.31. Let n be a positive integer, and let 2^k be the highest power of 2 in the set $S := \{1, \ldots, n\}$. Show that 2^k does not divide any other element in S.

EXERCISE 1.32. Let $n \in \mathbb{Z}$ with $n > 1$. Show that $\sum_{i=1}^{n} 1/i$ is not an integer.

EXERCISE 1.33. Let n be a positive integer, and let C_n denote the number of pairs of integers (a, b) with $a, b \in \{1, \ldots, n\}$ and $\gcd(a, b) = 1$, and let F_n be the number of *distinct* rational numbers a/b, where $0 \leq a < b \leq n$.

(a) Show that $F_n = (C_n + 1)/2$.

(b) Show that $C_n \geq n^2/4$. Hint: first show that $C_n \geq n^2(1 - \sum_{d \geq 2} 1/d^2)$, and then show that $\sum_{d \geq 2} 1/d^2 \leq 3/4$.

EXERCISE 1.34. This exercise develops a characterization of least common multiples in terms of ideals.

(a) Arguing directly from the definition of an ideal, show that if I and J are ideals of $\mathbb{Z}$, then so is $I \cap J$.

(b) Let $a, b \in \mathbb{Z}$, and consider the ideals $I := a\mathbb{Z}$ and $J := b\mathbb{Z}$. By part (a), we know that $I \cap J$ is an ideal. By Theorem 1.6, we know that $I \cap J = m\mathbb{Z}$ for some uniquely determined non-negative integer m. Show that $m = \mathrm{lcm}(a, b)$.

2

Congruences

This chapter introduces the basic properties of congruences modulo n, along with the related notion of residue classes modulo n. Other items discussed include the Chinese remainder theorem, Euler's phi function, Euler's theorem, Fermat's little theorem, quadratic residues, and finally, summations over divisors.

2.1 Equivalence relations

Before discussing congruences, we review the definition and basic properties of equivalence relations.

Let S be a set. A binary relation $\sim$ on S is called an **equivalence relation** if it is

reflexive: $a \sim a$ for all $a \in S$,

symmetric: $a \sim b$ implies $b \sim a$ for all $a, b \in S$, and

transitive: $a \sim b$ and $b \sim c$ implies $a \sim c$ for all $a, b, c \in S$.

If $\sim$ is an equivalence relation on S, then for $a \in S$ one defines its **equivalence class** as the set $\{x \in S : x \sim a\}$.

Theorem 2.1. *Let $\sim$ be an equivalence relation on a set S, and for $a \in S$, let $[a]$ denote its equivalence class. Then for all $a, b \in S$, we have:*

(i) $a \in [a]$;

(ii) $a \in [b]$ implies $[a] = [b]$.

Proof. (i) follows immediately from reflexivity. For (ii), suppose $a \in [b]$, so that $a \sim b$ by definition. We want to show that $[a] = [b]$. To this end, consider any

$x \in S$. We have

$$x \in [a] \implies x \sim a \text{ (by definition)}$$
$$\implies x \sim b \text{ (by transitivity, and since } x \sim a \text{ and } a \sim b)$$
$$\implies x \in [b].$$

Thus, $[a] \subseteq [b]$. By symmetry, we also have $b \sim a$, and reversing the roles of a and b in the above argument, we see that $[b] \subseteq [a]$. $\square$

   This theorem implies that each equivalence class is non-empty, and that each element of S belongs to a unique equivalence class; in other words, the distinct equivalence classes form a *partition* of S (see Preliminaries). A member of an equivalence class is called a **representative** of the class.

EXERCISE 2.1. Consider the relations $=$, $\leq$, and $<$ on the set $\mathbb{R}$. Which of these are equivalence relations? Explain your answers.

EXERCISE 2.2. Let $S := (\mathbb{R} \times \mathbb{R}) \setminus \{(0,0)\}$. For $(x, y), (x', y') \in S$, let us say $(x, y) \sim (x', y')$ if there exists a real number $\lambda > 0$ such that $(x, y) = (\lambda x', \lambda y')$. Show that $\sim$ is an equivalence relation; moreover, show that each equivalence class contains a unique representative that lies on the unit circle (i.e., the set of points (x, y) such that $x^2 + y^2 = 1$).

2.2 Definitions and basic properties of congruences

Let n be a positive integer. For integers a and b, we say that a **is congruent to** b **modulo** n if $n \mid (a - b)$, and we write $a \equiv b \pmod{n}$. If $n \nmid (a - b)$, then we write $a \not\equiv b \pmod{n}$. Equivalently, $a \equiv b \pmod{n}$ if and only if $a = b + ny$ for some $y \in \mathbb{Z}$. The relation $a \equiv b \pmod{n}$ is called a **congruence relation**, or simply, a **congruence**. The number n appearing in such congruences is called the **modulus** of the congruence. This usage of the "mod" notation as part of a congruence is not to be confused with the "mod" operation introduced in §1.1.

   If we view the modulus n as fixed, then the following theorem says that the binary relation "$\cdot \equiv \cdot \pmod{n}$" is an equivalence relation on the set $\mathbb{Z}$.

Theorem 2.2. *Let n be a positive integer. For all $a, b, c \in \mathbb{Z}$, we have:*

   (i) $a \equiv a \pmod{n}$;

   (ii) $a \equiv b \pmod{n}$ implies $b \equiv a \pmod{n}$;

   (iii) $a \equiv b \pmod{n}$ and $b \equiv c \pmod{n}$ implies $a \equiv c \pmod{n}$.

Proof. For (i), observe that n divides $0 = a - a$. For (ii), observe that if n divides

$a - b$, then it also divides $-(a - b) = b - a$. For (iii), observe that if n divides $a - b$ and $b - c$, then it also divides $(a - b) + (b - c) = a - c$. $\Box$

Another key property of congruences is that they are "compatible" with integer addition and multiplication, in the following sense:

Theorem 2.3. *Let $a, a', b, b', n \in \mathbb{Z}$ with $n > 0$. If*

$$a \equiv a' \pmod{n} \text{ and } b \equiv b' \pmod{n},$$

then

$$a + b \equiv a' + b' \pmod{n} \text{ and } a \cdot b \equiv a' \cdot b' \pmod{n}.$$

Proof. Suppose that $a \equiv a' \pmod{n}$ and $b \equiv b' \pmod{n}$. This means that there exist integers x and y such that $a = a' + nx$ and $b = b' + ny$. Therefore,

$$a + b = a' + b' + n(x + y),$$

which proves the first congruence of the theorem, and

$$ab = (a' + nx)(b' + ny) = a'b' + n(a'y + b'x + nxy),$$

which proves the second congruence. $\Box$

Theorems 2.2 and 2.3 allow one to work with congruence relations modulo n much as one would with ordinary equalities: one can add to, subtract from, or multiply both sides of a congruence modulo n by the same integer; also, if b is congruent to a modulo n, one may substitute b for a in any simple arithmetic expression (involving addition, subtraction, and multiplication) appearing in a congruence modulo n.

Now suppose a is an arbitrary, fixed integer, and consider the set of integers z that satisfy the congruence $z \equiv a \pmod{n}$. Since z satisfies this congruence if and only if $z = a + ny$ for some $y \in \mathbb{Z}$, we may apply Theorems 1.4 and 1.5 (with a as given, and $b := n$) to deduce that every interval of n consecutive integers contains exactly one such z. This simple fact is of such fundamental importance that it deserves to be stated as a theorem:

Theorem 2.4. *Let $a, n \in \mathbb{Z}$ with $n > 0$. Then there exists a unique integer z such that $z \equiv a \pmod{n}$ and $0 \leq z < n$, namely, $z := a \bmod n$. More generally, for every $x \in \mathbb{R}$, there exists a unique integer $z \in [x, x + n)$ such that $z \equiv a \pmod{n}$.*

Example 2.1. Let us find the set of solutions z to the congruence

$$3z + 4 \equiv 6 \pmod{7}. \tag{2.1}$$

Suppose that z is a solution to (2.1). Subtracting 4 from both sides of (2.1), we obtain

$$3z \equiv 2 \pmod{7}. \tag{2.2}$$

Next, we would like to divide both sides of this congruence by 3, to get z by itself on the left-hand side. We cannot do this directly, but since $5 \cdot 3 \equiv 1 \pmod{7}$, we can achieve the same effect by multiplying both sides of (2.2) by 5. If we do this, and then replace $5 \cdot 3$ by 1, and $5 \cdot 2$ by 3, we obtain

$$z \equiv 3 \pmod{7}.$$

Thus, if z is a solution to (2.1), we must have $z \equiv 3 \pmod{7}$; conversely, one can verify that if $z \equiv 3 \pmod{7}$, then (2.1) holds. We conclude that the integers z that are solutions to (2.1) are precisely those integers that are congruent to 3 modulo 7, which we can list as follows:

$$\ldots, -18, -11, -4, 3, 10, 17, 24, \ldots \quad \Box$$

In the next section, we shall give a systematic treatment of the problem of solving linear congruences, such as the one appearing in the previous example.

EXERCISE 2.3. Let $a, b, n \in \mathbb{Z}$ with $n > 0$. Show that $a \equiv b \pmod{n}$ if and only if $(a \bmod n) = (b \bmod n)$.

EXERCISE 2.4. Let $a, b, n \in \mathbb{Z}$ with $n > 0$ and $a \equiv b \pmod{n}$. Also, let $c_0, c_1, \ldots, c_k \in \mathbb{Z}$. Show that

$$c_0 + c_1 a + \cdots + c_k a^k \equiv c_0 + c_1 b + \cdots + c_k b^k \pmod{n}.$$

EXERCISE 2.5. Let $a, b, n, n' \in \mathbb{Z}$ with $n > 0$, $n' > 0$, and $n' \mid n$. Show that if $a \equiv b \pmod{n}$, then $a \equiv b \pmod{n'}$.

EXERCISE 2.6. Let $a, b, n, n' \in \mathbb{Z}$ with $n > 0$, $n' > 0$, and $\gcd(n, n') = 1$. Show that if $a \equiv b \pmod{n}$ and $a \equiv b \pmod{n'}$, then $a \equiv b \pmod{nn'}$.

EXERCISE 2.7. Let $a, b, n \in \mathbb{Z}$ with $n > 0$ and $a \equiv b \pmod{n}$. Show that $\gcd(a, n) = \gcd(b, n)$.

EXERCISE 2.8. Let a be a positive integer whose base-10 representation is $a = (a_{k-1} \cdots a_1 a_0)_{10}$. Let b be the sum of the decimal digits of a; that is, let $b := a_0 + a_1 + \cdots + a_{k-1}$. Show that $a \equiv b \pmod{9}$. From this, justify the usual "rules of thumb" for determining divisibility by 9 and 3: a is divisible by 9 (respectively, 3) if and only if the sum of the decimal digits of a is divisible by 9 (respectively, 3).

EXERCISE 2.9. Let e be a positive integer. For $a \in \{0, \ldots, 2^e - 1\}$, let $\tilde{a}$ denote the integer obtained by inverting the bits in the e-bit, binary representation of a (note that $\tilde{a} \in \{0, \ldots, 2^e - 1\}$). Show that $\tilde{a} + 1 \equiv -a \pmod{2^e}$. This justifies the usual rule for computing negatives in 2's complement arithmetic (which is really just arithmetic modulo 2^e).

EXERCISE 2.10. Show that the equation $7y^3 + 2 = z^3$ has no solutions $y, z \in \mathbb{Z}$.

EXERCISE 2.11. Show that there are 14 distinct, possible, yearly (Gregorian) calendars, and show that all 14 calendars actually occur.

2.3 Solving linear congruences

In this section, we consider the general problem of solving linear congruences. More precisely, for a given positive integer n, and arbitrary integers a and b, we wish to determine the set of integers z that satisfy the congruence

$$az \equiv b \pmod{n}. \tag{2.3}$$

Observe that if (2.3) has a solution z, and if $z \equiv z' \pmod{n}$, then z' is also a solution to (2.3). However, (2.3) may or may not have a solution, and if it does, such solutions may or may not be uniquely determined modulo n. The following theorem precisely characterizes the set of solutions of (2.3); basically, it says that (2.3) has a solution if and only if $d := \gcd(a, n)$ divides b, in which case the solution is uniquely determined modulo n/d.

Theorem 2.5. *Let $a, n \in \mathbb{Z}$ with $n > 0$, and let $d := \gcd(a, n)$.*

(i) *For every $b \in \mathbb{Z}$, the congruence $az \equiv b \pmod{n}$ has a solution $z \in \mathbb{Z}$ if and only if $d \mid b$.*

(ii) *For every $z \in \mathbb{Z}$, we have $az \equiv 0 \pmod{n}$ if and only if $z \equiv 0 \pmod{n/d}$.*

(iii) *For all $z, z' \in \mathbb{Z}$, we have $az \equiv az' \pmod{n}$ if and only if $z \equiv z' \pmod{n/d}$.*

Proof. For (i), let $b \in \mathbb{Z}$ be given. Then we have

$$az \equiv b \pmod{n} \text{ for some } z \in \mathbb{Z}$$
$$\Longleftrightarrow az = b + ny \text{ for some } z, y \in \mathbb{Z} \text{ (by definition of congruence)}$$
$$\Longleftrightarrow az - ny = b \text{ for some } z, y \in \mathbb{Z}$$
$$\Longleftrightarrow d \mid b \text{ (by Theorem 1.8).}$$

For (ii), we have

$$n \mid az \Longleftrightarrow n/d \mid (a/d)z \Longleftrightarrow n/d \mid z.$$

All of these implications follow rather trivially from the definition of divisibility, except that for the implication $n/d \mid (a/d)z \implies n/d \mid z$, we use Theorem 1.9 and the fact that $\gcd(a/d, n/d) = 1$.

For (iii), we have

$$az \equiv az' \pmod{n} \iff a(z - z') \equiv 0 \pmod{n}$$
$$\iff z - z' \equiv 0 \pmod{n/d} \text{ (by part (ii))}$$
$$\iff z \equiv z' \pmod{n/d}. \quad \square$$

We can restate Theorem 2.5 in more concrete terms as follows. Let $a, n \in \mathbb{Z}$ with $n > 0$, and let $d := \gcd(a, n)$. Let $I_n := \{0, \ldots, n - 1\}$ and consider the "multiplication by a" map

$$\tau_a : I_n \to I_n$$

$$z \mapsto az \bmod n.$$

The image of τ_a consists of the n/d integers

$$i \cdot d \ (i = 0, \ldots, n/d - 1).$$

Moreover, every element b in the image of τ_a has precisely d pre-images

$$z_0 + j \cdot (n/d) \ (j = 0, \ldots, d - 1),$$

where $z_0 \in \{0, \ldots, n/d - 1\}$. In particular, τ_a is a bijection if and only if a and n are relatively prime.

Example 2.2. The following table illustrates what Theorem 2.5 says for $n = 15$ and $a = 1, 2, 3, 4, 5, 6$.

z	0	1	2	3	4	5	6	7	8	9	10	11	12	13	14
$2z \bmod 15$	0	2	4	6	8	10	12	14	1	3	5	7	9	11	13
$3z \bmod 15$	0	3	6	9	12	0	3	6	9	12	0	3	6	9	12
$4z \bmod 15$	0	4	8	12	1	5	9	13	2	6	10	14	3	7	11
$5z \bmod 15$	0	5	10	0	5	10	0	5	10	0	5	10	0	5	10
$6z \bmod 15$	0	6	12	3	9	0	6	12	3	9	0	6	12	3	9

In the second row, we are looking at the values $2z \bmod 15$, and we see that this row is just a permutation of the first row. So for every b, there exists a unique z such that $2z \equiv b \pmod{15}$. This is implied by the fact that $\gcd(2, 15) = 1$.

In the third row, the only numbers hit are the multiples of 3, which follows from the fact that $\gcd(3, 15) = 3$. Also note that the pattern in this row repeats every five columns; that is, $3z \equiv 3z' \pmod{15}$ if and only if $z \equiv z' \pmod{5}$.

In the fourth row, we again see a permutation of the first row, which follows from the fact that $\gcd(4, 15) = 1$.

In the fifth row, the only numbers hit are the multiples of 5, which follows from the fact that $\gcd(5, 15) = 5$. Also note that the pattern in this row repeats every three columns; that is, $5z \equiv 5z'$ (mod 15) if and only if $z \equiv z'$ (mod 3).

In the sixth row, since $\gcd(6, 15) = 3$, we see a permutation of the third row. The pattern repeats after five columns, although the pattern is a permutation of the pattern in the third row. $\square$

We develop some further consequences of Theorem 2.5.

A cancellation law. Let $a, n \in \mathbb{Z}$ with $n > 0$. Part (iii) of Theorem 2.5 gives us a **cancellation law** for congruences:

if $\gcd(a, n) = 1$ and $az \equiv az'$ (mod n), then $z \equiv z'$ (mod n).

More generally, if $d := \gcd(a, n)$, then we can cancel a from both sides of a congruence modulo n, as long as we replace the modulus by n/d.

Example 2.3. Observe that

$$5 \cdot 2 \equiv 5 \cdot (-4) \ (\text{mod } 6). \tag{2.4}$$

Part (iii) of Theorem 2.5 tells us that since $\gcd(5, 6) = 1$, we may cancel the common factor of 5 from both sides of (2.4), obtaining $2 \equiv -4$ (mod 6), which one can also verify directly.

Next observe that

$$15 \cdot 5 \equiv 15 \cdot 3 \ (\text{mod } 6). \tag{2.5}$$

We cannot simply cancel the common factor of 15 from both sides of (2.5); indeed, $5 \not\equiv 3$ (mod 6). However, $\gcd(15, 6) = 3$, and as part (iii) of Theorem 2.5 guarantees, we do indeed have $5 \equiv 3$ (mod 2). $\square$

Modular inverses. Again, let $a, n \in \mathbb{Z}$ with $n > 0$. We say that $z \in \mathbb{Z}$ is a **multiplicative inverse of** a **modulo** n if $az \equiv 1$ (mod n). Part (i) of Theorem 2.5 says that a has a multiplicative inverse modulo n if and only if $\gcd(a, n) = 1$. Moreover, part (iii) of Theorem 2.5 says that the multiplicative inverse of a, if it exists, is uniquely determined modulo n; that is, if z and z' are multiplicative inverses of a modulo n, then $z \equiv z'$ (mod n). Note that if z is a multiplicative inverse of a modulo n, then a is a multiplicative inverse of z modulo n. Also note that if $a \equiv a'$ (mod n), then z is a multiplicative inverse of a modulo n if and only if z is a multiplicative inverse of a' modulo n.

Now suppose that $a, b, n \in \mathbb{Z}$ with $n > 0$, $a \neq 0$, and $\gcd(a, n) = 1$. Theorem 2.5 says that there exists a unique integer z satisfying

$$az \equiv b \ (\text{mod } n) \ \text{and} \ 0 \leq z < n.$$

Setting $s := b/a \in \mathbb{Q}$, we may generalize the "mod" operation, defining s mod n to be this value z. As the reader may easily verify, this definition of s mod n does not depend on the particular choice of fraction used to represent the rational number s. With this notation, we can simply write a^{-1} mod n to denote the unique multiplicative inverse of a modulo n that lies in the interval $0, \ldots, n-1$.

Example 2.4. Looking back at the table in Example 2.2, we see that

$$2^{-1} \bmod 15 = 8 \quad \text{and} \quad 4^{-1} \bmod 15 = 4,$$

and that neither 3, 5, nor 6 have modular inverses modulo 15. $\square$

Example 2.5. Let $a, b, n \in \mathbb{Z}$ with $n > 0$. We can describe the set of solutions $z \in \mathbb{Z}$ to the congruence $az \equiv b \pmod{n}$ very succinctly in terms of modular inverses.

If $\gcd(a, n) = 1$, then setting $t := a^{-1}$ mod n, and $z_0 := tb$ mod n, we see that z_0 is the unique solution to the congruence $az \equiv b \pmod{n}$ that lies in the interval $\{0, \ldots, n-1\}$.

More generally, if $d := \gcd(a, n)$, then the congruence $az \equiv b \pmod{n}$ has a solution if and only if $d \mid b$. So suppose that $d \mid b$. In this case, if we set $a' := a/d$, $b' := b/d$, and $n' := n/d$, then for each $z \in \mathbb{Z}$, we have $az \equiv b \pmod{n}$ if and only if $a'z \equiv b' \pmod{n'}$. Moreover, $\gcd(a', n') = 1$, and therefore, if we set $t := (a')^{-1}$ mod n' and $z_0 := tb'$ mod n', then the solutions to the congruence $az \equiv b \pmod{n}$ that lie in the interval $\{0, \ldots, n-1\}$ are the d integers $z_0, z_0 + n', \ldots, z_0 + (d-1)n'$. $\square$

EXERCISE 2.12. Let $a_1, \ldots, a_k, b, n$ be integers with $n > 0$. Show that the congruence

$$a_1 z_1 + \cdots + a_k z_k \equiv b \pmod{n}$$

has a solution $z_1, \ldots, z_k \in \mathbb{Z}$ if and only if $d \mid b$, where $d := \gcd(a_1, \ldots, a_k, n)$.

EXERCISE 2.13. Let p be a prime, and let a, b, c, e be integers, such that $e > 0$, $a \not\equiv 0 \pmod{p^{e+1}}$, and $0 \le c < p^e$. Define N to be the number of integers $z \in \{0, \ldots, p^{2e} - 1\}$ such that

$$\left\lfloor \left((az + b) \bmod p^{2e} \right) / p^e \right\rfloor = c.$$

Show that $N = p^e$.

2.4 The Chinese remainder theorem

Next, we consider systems of linear congruences with respect to moduli that are relatively prime in pairs. The result we state here is known as the Chinese remainder theorem, and is extremely useful in a number of contexts.

Theorem 2.6 (Chinese remainder theorem). *Let* $\{n_i\}_{i=1}^{k}$ *be a pairwise relatively prime family of positive integers, and let* $a_1, \ldots, a_k$ *be arbitrary integers. Then there exists a solution* $a \in \mathbb{Z}$ *to the system of congruences*

$$a \equiv a_i \pmod{n_i} \ (i = 1, \ldots, k).$$

Moreover, any $a' \in \mathbb{Z}$ *is a solution to this system of congruences if and only if* $a \equiv a' \pmod{n}$, *where* $n := \prod_{i=1}^{k} n_i$.

Proof. To prove the existence of a solution a to the system of congruences, we first show how to construct integers $e_1, \ldots, e_k$ such that for $i, j = 1, \ldots, k$, we have

$$e_j \equiv \begin{cases} 1 \pmod{n_i} & \text{if } j = i, \\ 0 \pmod{n_i} & \text{if } j \neq i. \end{cases} \tag{2.6}$$

If we do this, then setting

$$a := \sum_{i=1}^{k} a_i e_i,$$

one sees that for $j = 1, \ldots, k$, we have

$$a \equiv \sum_{i=1}^{k} a_i e_i \equiv a_j \pmod{n_j},$$

since all the terms in this sum are zero modulo n_j, except for the term $i = j$, which is congruent to a_j modulo n_j.

To construct $e_1, \ldots, e_k$ satisfying (2.6), let $n := \prod_{i=1}^{k} n_i$ as in the statement of the theorem, and for $i = 1, \ldots, k$, let $n_i^* := n/n_i$; that is, n_i^* is the product of all the moduli n_j with $j \neq i$. From the fact that $\{n_i\}_{i=1}^{k}$ is pairwise relatively prime, it follows that for $i = 1, \ldots, k$, we have $\gcd(n_i, n_i^*) = 1$, and so we may define $t_i := (n_i^*)^{-1} \bmod n_i$ and $e_i := n_i^* t_i$. One sees that $e_i \equiv 1 \pmod{n_i}$, while for $j \neq i$, we have $n_i \mid n_j^*$, and so $e_j \equiv 0 \pmod{n_i}$. Thus, (2.6) is satisfied.

That proves the existence of a solution a to the given system of congruences. If $a \equiv a' \pmod{n}$, then since $n_i \mid n$ for $i = 1, \ldots, k$, we see that $a' \equiv a \equiv a_i \pmod{n_i}$ for $i = 1, \ldots, k$, and so a' also solves the system of congruences.

Finally, if a' is a solution to the given system of congruences, then $a \equiv a_i \equiv a' \pmod{n_i}$ for $i = 1, \ldots, k$. Thus, $n_i \mid (a - a')$ for $i = 1, \ldots, k$. Since $\{n_i\}_{i=1}^{k}$ is pairwise relatively prime, this implies $n \mid (a - a')$, or equivalently, $a \equiv a' \pmod{n}$. $\square$

We can restate Theorem 2.6 in more concrete terms, as follows. For each positive integer m, let I_m denote $\{0, \ldots, m - 1\}$. Suppose $\{n_i\}_{i=1}^{k}$ is a pairwise relatively prime family of positive integers, and set $n := n_1 \cdots n_k$. Then the map

$$\tau : \quad I_n \to I_{n_1} \times \cdots \times I_{n_k}$$

$$a \mapsto (a \bmod n_1, \ldots, a \bmod n_k)$$

is a bijection.

Example 2.6. The following table illustrates what Theorem 2.6 says for $n_1 = 3$ and $n_2 = 5$.

a	0	1	2	3	4	5	6	7	8	9	10	11	12	13	14
$a \bmod 3$	0	1	2	0	1	2	0	1	2	0	1	2	0	1	2
$a \bmod 5$	0	1	2	3	4	0	1	2	3	4	0	1	2	3	4

We see that as a ranges from 0 to 14, the pairs $(a \bmod 3, a \bmod 5)$ range over all pairs (a_1, a_2) with $a_1 \in \{0, 1, 2\}$ and $a_2 \in \{0, \ldots, 4\}$, with every pair being hit exactly once. □

EXERCISE 2.14. Compute the values e_1, e_2, e_3 in the proof of Theorem 2.6 in the case where $k = 3$, $n_1 = 3$, $n_2 = 5$, and $n_3 = 7$. Also, find an integer a such that $a \equiv 1 \pmod 3$, $a \equiv -1 \pmod 5$, and $a \equiv 5 \pmod 7$.

EXERCISE 2.15. If you want to show that you are a real nerd, here is an age-guessing game you might play at a party. You ask a fellow party-goer to divide his age by each of the numbers 3, 4, and 5, and tell you the remainders. Show how to use this information to determine his age.

EXERCISE 2.16. Let $\{n_i\}_{i=1}^k$ be a pairwise relatively prime family of positive integers. Let $a_1, \ldots, a_k$ and $b_1, \ldots, b_k$ be integers, and set $d_i := \gcd(a_i, n_i)$ for $i = 1, \ldots, k$. Show that there exists an integer z such that $a_i z \equiv b_i \pmod{n_i}$ for $i = 1, \ldots, k$ if and only if $d_i \mid b_i$ for $i = 1, \ldots, k$.

EXERCISE 2.17. For each prime p, let $v_p(\cdot)$ be defined as in §1.3. Let $p_1, \ldots, p_r$ be distinct primes, $a_1, \ldots, a_r$ be arbitrary integers, and $e_1, \ldots, e_r$ be arbitrary non-negative integers. Show that there exists an integer a such that $v_{p_i}(a - a_i) = e_i$ for $i = 1, \ldots, r$.

EXERCISE 2.18. Suppose n_1 and n_2 are positive integers, and let $d := \gcd(n_1, n_2)$. Let a_1 and a_2 be arbitrary integers. Show that there exists an integer a such that $a \equiv a_1 \pmod{n_1}$ and $a \equiv a_2 \pmod{n_2}$ if and only if $a_1 \equiv a_2 \pmod d$.

2.5 Residue classes

As we already observed in Theorem 2.2, for any fixed positive integer n, the binary relation "$\cdot \equiv \cdot \pmod{n}$" is an equivalence relation on the set $\mathbb{Z}$. As such, this relation partitions the set $\mathbb{Z}$ into equivalence classes. We denote the equivalence class containing the integer a by $[a]_n$, and when n is clear from context, we simply write $[a]$. By definition, we have

$$z \in [a] \iff z \equiv a \pmod{n} \iff z = a + ny \text{ for some } y \in \mathbb{Z},$$

and hence

$$[a] = a + n\mathbb{Z} := \{a + ny : y \in \mathbb{Z}\}.$$

Historically, these equivalence classes are called **residue classes modulo** n, and we shall adopt this terminology here as well. Note that a given residue class modulo n has many different "names"; for example, the residue class $[n-1]$ is the same as the residue class $[-1]$. Any member of a residue class is called a **representative** of that class.

We define $\mathbb{Z}_n$ to be the set of residue classes modulo n. The following is simply a restatement of Theorem 2.4:

Theorem 2.7. *Let n be a positive integer. Then $\mathbb{Z}_n$ consists of the n distinct residue classes $[0], [1], \ldots, [n-1]$. Moreover, for every $x \in \mathbb{R}$, each residue class modulo n contains a unique representative in the interval $[x, x+n)$.*

When working with residue classes modulo n, one often has in mind a particular set of representatives. Typically, one works with the set of representatives $\{0, 1, \ldots, n-1\}$. However, sometimes it is convenient to work with another set of representatives, such as the representatives in the interval $[-n/2, n/2)$. In this case, if n is odd, we can list the elements of $\mathbb{Z}_n$ as

$$[-(n-1)/2], \ldots, [-1], [0], [1], \ldots, [(n-1)/2],$$

and when n is even, we can list the elements of $\mathbb{Z}_n$ as

$$[-n/2], \ldots, [-1], [0], [1], \ldots, [n/2 - 1].$$

We can "equip" $\mathbb{Z}_n$ with binary operations defining addition and multiplication in a natural way as follows: for $a, b \in \mathbb{Z}$, we define

$$[a] + [b] := [a + b],$$
$$[a] \cdot [b] := [a \cdot b].$$

Of course, one has to check that this definition is unambiguous, in the sense that the sum or product of two residue classes should not depend on which particular representatives of the classes are chosen in the above definitions. More precisely, one must check that if $[a] = [a']$ and $[b] = [b']$, then $[a + b] = [a' + b']$ and $[a \cdot b] = [a' \cdot b']$. However, this property follows immediately from Theorem 2.3.

Observe that for all $a, b, c \in \mathbb{Z}$, we have

$$[a] + [b] = [c] \iff a + b \equiv c \pmod{n},$$

and

$$[a] \cdot [b] = [c] \iff a \cdot b \equiv c \pmod{n},$$

Example 2.7. Consider the residue classes modulo 6. These are as follows:

$$[0] = \{\ldots, -12, -6, 0, 6, 12, \ldots\}$$
$$[1] = \{\ldots, -11, -5, 1, 7, 13, \ldots\}$$
$$[2] = \{\ldots, -10, -4, 2, 8, 14, \ldots\}$$
$$[3] = \{\ldots, -9, -3, 3, 9, 15, \ldots\}$$
$$[4] = \{\ldots, -8, -2, 4, 10, 16, \ldots\}$$
$$[5] = \{\ldots, -7, -1, 5, 11, 17, \ldots\}.$$

Let us write down the addition and multiplication tables for $\mathbb{Z}_6$. The addition table looks like this:

+	[0]	[1]	[2]	[3]	[4]	[5]
[0]	[0]	[1]	[2]	[3]	[4]	[5]
[1]	[1]	[2]	[3]	[4]	[5]	[0]
[2]	[2]	[3]	[4]	[5]	[0]	[1]
[3]	[3]	[4]	[5]	[0]	[1]	[2]
[4]	[4]	[5]	[0]	[1]	[2]	[3]
[5]	[5]	[0]	[1]	[2]	[3]	[4] .

The multiplication table looks like this:

·	[0]	[1]	[2]	[3]	[4]	[5]
[0]	[0]	[0]	[0]	[0]	[0]	[0]
[1]	[0]	[1]	[2]	[3]	[4]	[5]
[2]	[0]	[2]	[4]	[0]	[2]	[4]
[3]	[0]	[3]	[0]	[3]	[0]	[3]
[4]	[0]	[4]	[2]	[0]	[4]	[2]
[5]	[0]	[5]	[4]	[3]	[2]	[1] .

Instead of using representatives in the interval $[0, 6)$, we could just as well use representatives from another interval, such as $[-3, 3)$. Then, instead of naming the residue classes $[0]$, $[1]$, $[2]$, $[3]$, $[4]$, $[5]$, we would name them $[-3]$, $[-2]$, $[-1]$, $[0]$, $[1]$, $[2]$. Observe that $[-3] = [3]$, $[-2] = [4]$, and $[-1] = [5]$. □

These addition and multiplication operations on $\mathbb{Z}_n$ yield a very natural algebraic structure. For example, addition and multiplication are commutative and associative; that is, for all $\alpha, \beta, \gamma \in \mathbb{Z}_n$, we have

$$\alpha + \beta = \beta + \alpha, \quad (\alpha + \beta) + \gamma = \alpha + (\beta + \gamma),$$
$$\alpha\beta = \beta\alpha, \quad (\alpha\beta)\gamma = \alpha(\beta\gamma).$$

Note that we have adopted here the usual convention of writing $\alpha\beta$ in place of $\alpha \cdot \beta$. Furthermore, multiplication distributes over addition; that is, for all $\alpha, \beta, \gamma \in \mathbb{Z}_n$, we have

$$\alpha(\beta + \gamma) = \alpha\beta + \alpha\gamma.$$

All of these properties follow from the definitions, and the corresponding properties for $\mathbb{Z}$; for example, the fact that addition in $\mathbb{Z}_n$ is commutative may be seen as follows: if $\alpha = [a]$ and $\beta = [b]$, then

$$\alpha + \beta = [a] + [b] = [a + b] = [b + a] = [b] + [a] = \beta + \alpha.$$

Because addition and multiplication in $\mathbb{Z}_n$ are associative, for $\alpha_1, \ldots, \alpha_k \in \mathbb{Z}_n$, we may write the sum $\alpha_1 + \cdots + \alpha_k$ and the product $\alpha_1 \cdots \alpha_k$ without any parentheses, and there is no ambiguity; moreover, since both addition and multiplication are commutative, we may rearrange the terms in such sums and products without changing their values.

The residue class $[0]$ acts as an **additive identity**; that is, for all $\alpha \in \mathbb{Z}_n$, we have $\alpha + [0] = \alpha$; indeed, if $\alpha = [a]$, then $a + 0 \equiv a \pmod{n}$. Moreover, $[0]$ is the only element of $\mathbb{Z}_n$ that acts as an additive identity; indeed, if $a + z \equiv a \pmod{n}$ holds for all integers a, then it holds in particular for $a = 0$, which implies $z \equiv 0 \pmod{n}$. The residue class $[0]$ also has the property that $\alpha \cdot [0] = [0]$ for all $\alpha \in \mathbb{Z}_n$.

Every $\alpha \in \mathbb{Z}_n$ has an **additive inverse**, that is, an element $\beta \in \mathbb{Z}_n$ such that $\alpha + \beta = [0]$; indeed, if $\alpha = [a]$, then clearly $\beta := [-a]$ does the job, since $a + (-a) \equiv 0 \pmod{n}$. Moreover, α has a unique additive inverse; indeed, if $a + z \equiv 0 \pmod{n}$, then subtracting a from both sides of this congruence yields $z \equiv -a \pmod{n}$. We naturally denote the additive inverse of α by $-\alpha$. Observe that the additive inverse of $-\alpha$ is α; that is $-(-\alpha) = \alpha$. Also, we have the identities

$$-(\alpha + \beta) = (-\alpha) + (-\beta), \quad (-\alpha)\beta = -(\alpha\beta) = \alpha(-\beta), \quad (-\alpha)(-\beta) = \alpha\beta.$$

For $\alpha, \beta \in \mathbb{Z}_n$, we naturally write $\alpha - \beta$ for $\alpha + (-\beta)$.

The residue class $[1]$ acts as a **multiplicative identity**; that is, for all $\alpha \in \mathbb{Z}_n$, we have $\alpha \cdot [1] = \alpha$; indeed, if $\alpha = [a]$, then $a \cdot 1 \equiv a \pmod{n}$. Moreover, $[1]$ is the only element of $\mathbb{Z}_n$ that acts as a multiplicative identity; indeed, if $a \cdot z \equiv a \pmod{n}$ holds for all integers a, then in particular, it holds for $a = 1$, which implies $z \equiv 1 \pmod{n}$.

For $\alpha \in \mathbb{Z}_n$, we call $\beta \in \mathbb{Z}_n$ a **multiplicative inverse** of α if $\alpha\beta = [1]$. Not all $\alpha \in \mathbb{Z}_n$ have multiplicative inverses. If $\alpha = [a]$ and $\beta = [b]$, then β is a multiplicative inverse of α if and only if $ab \equiv 1 \pmod{n}$. Theorem 2.5 implies that α has a multiplicative inverse if and only if $\gcd(a, n) = 1$, and that if it exists, it is unique. When it exists, we denote the multiplicative inverse of α by α^{-1}.

We define $\mathbb{Z}_n^*$ to be the set of elements of $\mathbb{Z}_n$ that have a multiplicative inverse. By the above discussion, we have

$$\mathbb{Z}_n^* = \{[a] : a = 0, \dots, n - 1, \gcd(a, n) = 1\}.$$

If n is prime, then $\gcd(a, n) = 1$ for $a = 1, \dots, n-1$, and we see that $\mathbb{Z}_n^* = \mathbb{Z}_n \setminus \{[0]\}$. If n is composite, then $\mathbb{Z}_n^* \subsetneq \mathbb{Z}_n \setminus \{[0]\}$; for example, if $d \mid n$ with $1 < d < n$, we see that $[d]$ is not zero, nor does it belong to $\mathbb{Z}_n^*$. Observe that if $\alpha, \beta \in \mathbb{Z}_n^*$, then so are α^{-1} and $\alpha\beta$; indeed,

$$(\alpha^{-1})^{-1} = \alpha \quad \text{and} \quad (\alpha\beta)^{-1} = \alpha^{-1}\beta^{-1}.$$

For $\alpha \in \mathbb{Z}_n$ and $\beta \in \mathbb{Z}_n^*$, we naturally write α/β for $\alpha\beta^{-1}$.

Suppose α, β, γ are elements of $\mathbb{Z}_n$ that satisfy the equation

$$\alpha\beta = \alpha\gamma.$$

If $\alpha \in \mathbb{Z}_n^*$, we may multiply both sides of this equation by α^{-1} to infer that

$$\beta = \gamma.$$

This is the **cancellation law** for $\mathbb{Z}_n$. We stress the requirement that $\alpha \in \mathbb{Z}_n^*$, and not just $\alpha \neq [0]$. Indeed, consider any $\alpha \in \mathbb{Z}_n \setminus \mathbb{Z}_n^*$. Then we have $\alpha = [a]$ with $d := \gcd(a, n) > 1$. Setting $\beta := [n/d]$ and $\gamma := [0]$, we see that

$$\alpha\beta = \alpha\gamma \quad \text{and} \quad \beta \neq \gamma.$$

Example 2.8. We list the elements of $\mathbb{Z}_{15}^*$, and for each $\alpha \in \mathbb{Z}_{15}^*$, we also give α^{-1}:

α	[1]	[2]	[4]	[7]	[8]	[11]	[13]	[14]
α^{-1}	[1]	[8]	[4]	[13]	[2]	[11]	[7]	[14]

. □

For $\alpha_1, \dots, \alpha_k \in \mathbb{Z}_n$, we may naturally write their sum as $\sum_{i=1}^k \alpha_i$. By convention, this sum is $[0]$ when $k = 0$. It is easy to see that $-\sum_{i=1}^k \alpha_i = \sum_{i=1}^k (-\alpha_i)$; that is, the additive inverse of the sum is the sum of the additive inverses. In the special case where all the α_i's have the same value α, we define $k \cdot \alpha := \sum_{i=1}^k \alpha$; thus, $0 \cdot \alpha = [0]$, $1 \cdot \alpha = \alpha$, $2 \cdot \alpha = \alpha + \alpha$, $3 \cdot \alpha = \alpha + \alpha + \alpha$, and so on. The additive inverse of $k \cdot \alpha$ is $k \cdot (-\alpha)$, which we may also write as $(-k) \cdot \alpha$; thus, $(-1) \cdot \alpha = -\alpha$, $(-2) \cdot \alpha = (-\alpha) + (-\alpha) = -(\alpha + \alpha)$, and so on. Therefore, the notation $k \cdot \alpha$, or more simply, $k\alpha$, is defined for all integers k. Note that for all integers k and a, we have $k[a] = [ka] = [k][a]$.

For all $\alpha, \beta \in \mathbb{Z}_n$ and $k, \ell \in \mathbb{Z}$, we have the identities:

$$k(\ell\alpha) = (k\ell)\alpha = \ell(k\alpha), \ (k+\ell)\alpha = k\alpha + \ell\alpha, \ k(\alpha + \beta) = k\alpha + k\beta,$$
$$(k\alpha)\beta = k(\alpha\beta) = \alpha(k\beta).$$

Analogously, for $\alpha_1, \ldots, \alpha_k \in \mathbb{Z}_n$, we may write their product as $\prod_{i=1}^{k} \alpha_i$. By convention, this product is [1] when $k = 0$. It is easy to see that if all of the α_i's belong to $\mathbb{Z}_n^*$, then so does their product, and in particular, $(\prod_{i=1}^{k} \alpha_i)^{-1} = \prod_{i=1}^{k} \alpha_i^{-1}$; that is, the multiplicative inverse of the product is the product of the multiplicative inverses. In the special case where all the α_i's have the same value α, we define $\alpha^k := \prod_{i=1}^{k} \alpha$; thus, $\alpha^0 = [1]$, $\alpha^1 = \alpha$, $\alpha^2 = \alpha\alpha$, $\alpha^3 = \alpha\alpha\alpha$, and so on. If $\alpha \in \mathbb{Z}_n^*$, then the multiplicative inverse of α^k is $(\alpha^{-1})^k$, which we may also write as α^{-k}; for example, $\alpha^{-2} = \alpha^{-1}\alpha^{-1} = (\alpha\alpha)^{-1}$. Therefore, when $\alpha \in \mathbb{Z}_n^*$, the notation α^k is defined for all integers k.

For all $\alpha, \beta \in \mathbb{Z}_n$ and all *non-negative* integers k and ℓ, we have the identities:

$$(\alpha^\ell)^k = \alpha^{k\ell} = (\alpha^k)^\ell, \ \alpha^{k+\ell} = \alpha^k \alpha^\ell, \ (\alpha\beta)^k = \alpha^k \beta^k. \tag{2.7}$$

If $\alpha, \beta \in \mathbb{Z}_n^*$, the identities in (2.7) hold for all $k, \ell \in \mathbb{Z}$.

For all $\alpha_1, \ldots, \alpha_k, \beta_1, \ldots, \beta_\ell \in \mathbb{Z}_n$, the distributive property implies that

$$(\alpha_1 + \cdots + \alpha_k)(\beta_1 + \cdots + \beta_\ell) = \sum_{\substack{1 \le i \le k \\ 1 \le j \le \ell}} \alpha_i \beta_j.$$

One last notational convention. As already mentioned, when the modulus n is clear from context, we usually write $[a]$ instead of $[a]_n$. Although we want to maintain a clear distinction between integers and their residue classes, occasionally even the notation $[a]$ is not only redundant, but distracting; in such situations, we may simply write a instead of $[a]$. For example, for every $\alpha \in \mathbb{Z}_n$, we have the identity $(\alpha + [1]_n)(\alpha - [1]_n) = \alpha^2 - [1]_n$, which we may write more simply as $(\alpha + [1])(\alpha - [1]) = \alpha^2 - [1]$, or even more simply, and hopefully more clearly, as $(\alpha + 1)(\alpha - 1) = \alpha^2 - 1$. Here, the only reasonable interpretation of the symbol "1" is [1], and so there can be no confusion.

In summary, algebraic expressions involving residue classes may be manipulated in much the same way as expressions involving ordinary numbers. Extra complications arise only because when n is composite, some non-zero elements of $\mathbb{Z}_n$ do not have multiplicative inverses, and the usual cancellation law does not apply for such elements.

In general, one has a choice between working with congruences modulo n, or with the algebraic structure $\mathbb{Z}_n$; ultimately, the choice is one of taste and convenience, and it depends on what one prefers to treat as "first class objects": integers and congruence relations, or elements of $\mathbb{Z}_n$.

An alternative, and somewhat more concrete, approach to constructing $\mathbb{Z}_n$ is to directly define it as the set of n "symbols" $[0], [1], \ldots, [n-1]$, with addition and multiplication defined as

$$[a] + [b] := [(a+b) \bmod n], \quad [a] \cdot [b] := [(a \cdot b) \bmod n],$$

for $a, b \in \{0, \ldots, n-1\}$. Such a definition is equivalent to the one we have given here. One should keep this alternative characterization of $\mathbb{Z}_n$ in mind; however, we prefer the characterization in terms of residue classes, as it is mathematically more elegant, and is usually more convenient to work with.

We close this section with a reinterpretation of the Chinese remainder theorem (Theorem 2.6) in terms of residue classes.

Theorem 2.8 (Chinese remainder map). *Let $\{n_i\}_{i=1}^k$ be a pairwise relatively prime family of positive integers, and let $n := \prod_{i=1}^k n_i$. Define the map*

$$\theta : \quad \mathbb{Z}_n \to \mathbb{Z}_{n_1} \times \cdots \times \mathbb{Z}_{n_k}$$

$$[a]_n \mapsto ([a]_{n_1}, \ldots, [a]_{n_k}).$$

(i) *The definition of θ is unambiguous.*

(ii) *θ is bijective.*

(iii) *For all $\alpha, \beta \in \mathbb{Z}_n$, if $\theta(\alpha) = (\alpha_1, \ldots, \alpha_k)$ and $\theta(\beta) = (\beta_1, \ldots, \beta_k)$, then:*

(a) *$\theta(\alpha + \beta) = (\alpha_1 + \beta_1, \ldots, \alpha_k + \beta_k)$;*

(b) *$\theta(-\alpha) = (-\alpha_1, \ldots, -\alpha_k)$;*

(c) *$\theta(\alpha\beta) = (\alpha_1\beta_1, \ldots, \alpha_k\beta_k)$;*

(d) *$\alpha \in \mathbb{Z}_n^*$ if and only if $\alpha_i \in \mathbb{Z}_{n_i}^*$ for $i = 1, \ldots, k$, in which case* $\theta(\alpha^{-1}) = (\alpha_1^{-1}, \ldots, \alpha_k^{-1})$.

Proof. For (i), note that $a \equiv a' \pmod{n}$ implies $a \equiv a' \pmod{n_i}$ for $i = 1, \ldots, k$, and so the definition of θ is unambiguous (it does not depend on the choice of a).

(ii) follows directly from the statement of the Chinese remainder theorem.

For (iii), let $\alpha = [a]_n$ and $\beta = [b]_n$, so that for $i = 1, \ldots, k$, we have $\alpha_i = [a]_{n_i}$ and $\beta_i = [b]_{n_i}$. Then we have

$$\theta(\alpha + \beta) = \theta([a+b]_n) = ([a+b]_{n_1}, \ldots, [a+b]_{n_k}) = (\alpha_1 + \beta_1, \ldots, \alpha_k + \beta_k),$$

$$\theta(-\alpha) = \theta([-a]_n) = ([-a]_{n_1}, \ldots, [-a]_{n_k}) = (-\alpha_1, \ldots, -\alpha_k), \text{ and}$$

$$\theta(\alpha\beta) = \theta([ab]_n) = ([ab]_{n_1}, \ldots, [ab]_{n_k}) = (\alpha_1\beta_1, \ldots, \alpha_k\beta_k).$$

That proves parts (a), (b), and (c). For part (d), we have

$$\alpha \in \mathbb{Z}_n^* \iff \gcd(a, n) = 1$$
$$\iff \gcd(a, n_i) = 1 \ \text{ for } i = 1, \dots, k$$
$$\iff \alpha_i \in \mathbb{Z}_{n_i}^* \ \text{ for } i = 1, \dots, k.$$

Moreover, if $\alpha \in \mathbb{Z}_n^*$ and $\beta = \alpha^{-1}$, then

$$(\alpha_1 \beta_1, \dots, \alpha_k \beta_k) = \theta(\alpha\beta) = \theta([1]_n) = ([1]_{n_1}, \dots, [1]_{n_k}),$$

and so for $i = 1, \dots, k$, we have $\alpha_i \beta_i = [1]_{n_i}$, which is to say $\beta_i = \alpha_i^{-1}$. $\square$

Theorem 2.8 is very powerful conceptually, and is an indispensable tool in many situations. It says that if we want to understand what happens when we add or multiply $\alpha, \beta \in \mathbb{Z}_n$, it suffices to understand what happens when we add or multiply their "components" $\alpha_i, \beta_i \in \mathbb{Z}_{n_i}$. Typically, we choose $n_1, \dots, n_k$ to be primes or prime powers, which usually simplifies the analysis. We shall see many applications of this idea throughout the text.

EXERCISE 2.19. Let $\theta : \mathbb{Z}_n \to \mathbb{Z}_{n_1} \times \cdots \times \mathbb{Z}_{n_k}$ be as in Theorem 2.8, and suppose that $\theta(\alpha) = (\alpha_1, \dots, \alpha_k)$. Show that for every non-negative integer m, we have $\theta(\alpha^m) = (\alpha_1^m, \dots, \alpha_k^m)$. Moreover, if $\alpha \in \mathbb{Z}_n^*$, show that this identity holds for all integers m.

EXERCISE 2.20. Let p be an odd prime. Show that $\sum_{\beta \in \mathbb{Z}_p^*} \beta^{-1} = \sum_{\beta \in \mathbb{Z}_p^*} \beta = 0$.

EXERCISE 2.21. Let p be an odd prime. Show that the numerator of $\sum_{i=1}^{p-1} 1/i$ is divisible by p.

EXERCISE 2.22. Suppose n is square-free (see Exercise 1.15), and let $\alpha, \beta, \gamma \in \mathbb{Z}_n$. Show that $\alpha^2 \beta = \alpha^2 \gamma$ implies $\alpha\beta = \alpha\gamma$.

2.6 Euler's phi function

Euler's phi function (also called **Euler's totient function**) is defined for all positive integers n as

$$\varphi(n) := |\mathbb{Z}_n^*|.$$

Equivalently, $\varphi(n)$ is equal to the number of integers between 0 and $n - 1$ that are relatively prime to n. For example, $\varphi(1) = 1$, $\varphi(2) = 1$, $\varphi(3) = 2$, and $\varphi(4) = 2$.

Using the Chinese remainder theorem, more specifically Theorem 2.8, it is easy to get a nice formula for $\varphi(n)$ in terms of the prime factorization of n, as we establish in the following sequence of theorems.

Theorem 2.9. *Let* $\{n_i\}_{i=1}^k$ *be a pairwise relatively prime family of positive integers, and let* $n := \prod_{i=1}^k n_i$. *Then*

$$\varphi(n) = \prod_{i=1}^k \varphi(n_i).$$

Proof. Consider the map $\theta : \mathbb{Z}_n \to \mathbb{Z}_{n_1} \times \cdots \times \mathbb{Z}_{n_k}$ in Theorem 2.8. By parts (ii) and (iii.d) of that theorem, restricting θ to $\mathbb{Z}_n^*$ yields a one-to-one correspondence between $\mathbb{Z}_n^*$ and $\mathbb{Z}_{n_1}^* \times \cdots \times \mathbb{Z}_{n_k}^*$. The theorem now follows immediately. $\square$

We already know that $\varphi(p) = p - 1$ for every prime p, since the integers $1, \ldots, p - 1$ are not divisible by p, and hence are relatively prime to p. The next theorem generalizes this, giving us a formula for Euler's phi function at prime powers.

Theorem 2.10. *Let* p *be a prime and* e *be a positive integer. Then*

$$\varphi(p^e) = p^{e-1}(p - 1).$$

Proof. The multiples of p among $0, 1, \ldots, p^e - 1$ are

$$0 \cdot p, 1 \cdot p, \ldots, (p^{e-1} - 1) \cdot p,$$

of which there are precisely p^{e-1}. Thus, $\varphi(p^e) = p^e - p^{e-1} = p^{e-1}(p - 1)$. $\square$

If $n = p_1^{e_1} \cdots p_r^{e_r}$ is the factorization of n into primes, then the family of prime powers $\{p_i^{e_i}\}_{i=1}^r$ is pairwise relatively prime, and so Theorem 2.9 implies $\varphi(n) = \varphi(p_1^{e_1}) \cdots \varphi(p_r^{e_r})$. Combining this with Theorem 2.10, we have:

Theorem 2.11. *If* $n = p_1^{e_1} \cdots p_r^{e_r}$ *is the factorization of* n *into primes, then*

$$\varphi(n) = \prod_{i=1}^r p_i^{e_i-1}(p_i - 1) = n \prod_{i=1}^r (1 - 1/p_i).$$

EXERCISE 2.23. Show that $\varphi(nm) = \gcd(n, m) \cdot \varphi(\text{lcm}(n, m))$.

EXERCISE 2.24. Show that if n is divisible by r distinct odd primes, then $2^r \mid \varphi(n)$.

EXERCISE 2.25. Define $\varphi_2(n)$ to be the number of integers $a \in \{0, \ldots, n-1\}$ such that $\gcd(a, n) = \gcd(a + 1, n) = 1$. Show that if $n = p_1^{e_1} \cdots p_r^{e_r}$ is the factorization of n into primes, then $\varphi_2(n) = n \prod_{i=1}^r (1 - 2/p_i)$.

2.7 Euler's theorem and Fermat's little theorem

Let n be a positive integer, and let $\alpha \in \mathbb{Z}_n^*$.

Consider the sequence of powers of α:

$$1 = \alpha^0, \alpha^1, \alpha^2, \ldots.$$

Since each such power is an element of $\mathbb{Z}_n^*$, and since $\mathbb{Z}_n^*$ is a finite set, this sequence of powers must start to repeat at some point; that is, there must be a positive integer k such that $\alpha^k = \alpha^i$ for some $i = 0, \ldots, k - 1$. Let us assume that k is chosen to be the smallest such positive integer. This value k is called the **multiplicative order** of α.

We claim that $\alpha^k = 1$. To see this, suppose by way of contradiction that $\alpha^k = \alpha^i$, for some $i = 1, \ldots, k - 1$; we could then cancel α from both sides of the equation $\alpha^k = \alpha^i$, obtaining $\alpha^{k-1} = \alpha^{i-1}$, which would contradict the minimality of k.

Thus, we can characterize the multiplicative order of α as the smallest positive integer k such that

$$\alpha^k = 1.$$

If $\alpha = [a]$ with $a \in \mathbb{Z}$ (and $\gcd(a, n) = 1$, since $\alpha \in \mathbb{Z}_n^*$), then k is also called the **multiplicative order** of a **modulo** n, and can be characterized as the smallest positive integer k such that

$$a^k \equiv 1 \pmod{n}.$$

From the above discussion, we see that the first k powers of α, that is, $\alpha^0, \alpha^1, \ldots, \alpha^{k-1}$, are distinct. Moreover, other powers of α simply repeat this pattern. The following is an immediate consequence of this observation.

Theorem 2.12. *Let n be a positive integer, and let α be an element of $\mathbb{Z}_n^*$ of multiplicative order k. Then for every $i \in \mathbb{Z}$, we have $\alpha^i = 1$ if and only if k divides i. More generally, for all $i, j \in \mathbb{Z}$, we have $\alpha^i = \alpha^j$ if and only if $i \equiv j \pmod{k}$.*

Example 2.9. *Let $n = 7$. For each value $a = 1, \ldots, 6$, we can compute successive powers of a modulo n to find its multiplicative order modulo n.*

i	1	2	3	4	5	6
$1^i \bmod 7$	1	1	1	1	1	1
$2^i \bmod 7$	2	4	1	2	4	1
$3^i \bmod 7$	3	2	6	4	5	1
$4^i \bmod 7$	4	2	1	4	2	1
$5^i \bmod 7$	5	4	6	2	3	1
$6^i \bmod 7$	6	1	6	1	6	1

So we conclude that modulo 7: 1 has order 1; 6 has order 2; 2 and 4 have order 3; and 3 and 5 have order 6. $\square$

Theorem 2.13 (Euler's theorem). *Let n be a positive integer and $\alpha \in \mathbb{Z}_n^*$. Then $\alpha^{\varphi(n)} = 1$. In particular, the multiplicative order of α divides $\varphi(n)$.*

Proof. Since $\alpha \in \mathbb{Z}_n^*$, for every $\beta \in \mathbb{Z}_n^*$ we have $\alpha\beta \in \mathbb{Z}_n^*$, and so we may define the "multiplication by α" map

$$\tau_\alpha : \ \mathbb{Z}_n^* \to \mathbb{Z}_n^*$$
$$\beta \mapsto \alpha\beta.$$

It is easy to see that τ_α is a bijection:

Injectivity: If $\alpha\beta = \alpha\beta'$, then cancel α to obtain $\beta = \beta'$.

Surjectivity: For every $\gamma \in \mathbb{Z}_n^*$, $\alpha^{-1}\gamma$ is a pre-image of γ under τ_α.

Thus, as β ranges over the set $\mathbb{Z}_n^*$, so does $\alpha\beta$, and we have

$$\prod_{\beta \in \mathbb{Z}_n^*} \beta = \prod_{\beta \in \mathbb{Z}_n^*} (\alpha\beta) = \alpha^{\varphi(n)} \left(\prod_{\beta \in \mathbb{Z}_n^*} \beta \right). \tag{2.8}$$

Canceling the common factor $\prod_{\beta \in \mathbb{Z}_n^*} \beta \in \mathbb{Z}_n^*$ from the left- and right-hand side of (2.8), we obtain

$$1 = \alpha^{\varphi(n)}.$$

That proves the first statement of the theorem. The second follows immediately from Theorem 2.12. $\square$

As a consequence of this, we obtain:

Theorem 2.14 (Fermat's little theorem). *For every prime p, and every $\alpha \in \mathbb{Z}_p$, we have $\alpha^p = \alpha$.*

Proof. If $\alpha = 0$, the statement is obviously true. Otherwise, $\alpha \in \mathbb{Z}_p^*$, and by Theorem 2.13 we have $\alpha^{p-1} = 1$. Multiplying this equation by α yields $\alpha^p = \alpha$. $\square$

In the language of congruences, Fermat's little theorem says that for every prime p and every integer a, we have

$$a^p \equiv a \ (\mathrm{mod}\ p).$$

For a given positive integer n, we say that $a \in \mathbb{Z}$ with $\gcd(a, n) = 1$ is a **primitive root modulo** n if the multiplicative order of a modulo n is equal to $\varphi(n)$. If this is the case, then for $\alpha := [a] \in \mathbb{Z}_n^*$, the powers α^i range over all elements of $\mathbb{Z}_n^*$ as i ranges over the interval $0, \dots, \varphi(n) - 1$. Not all positive integers have primitive roots—we will see in §7.5 that the only positive integers n for which there exists a primitive root modulo n are

$$n = 1, 2, 4, p^e, 2p^e,$$

where p is an odd prime and e is a positive integer.

The following theorem is sometimes useful in determining the multiplicative order of an element in $\mathbb{Z}_n^*$.

Theorem 2.15. *Suppose $\alpha \in \mathbb{Z}_n^*$ has multiplicative order k. Then for every $m \in \mathbb{Z}$, the multiplicative order of α^m is $k / \gcd(m, k)$.*

Proof. Applying Theorem 2.12 to α^m, we see that the multiplicative order of α^m is the smallest positive integer ℓ such that $\alpha^{m\ell} = 1$. But we have

$$\alpha^{m\ell} = 1 \iff m\ell \equiv 0 \pmod{k} \text{ (applying Theorem 2.12 to } \alpha)$$
$$\iff \ell \equiv 0 \pmod{k / \gcd(m, k)} \text{ (by part (ii) of Theorem 2.5). } \square$$

EXERCISE 2.26. Find all elements of $\mathbb{Z}_{19}^*$ of multiplicative order 18.

EXERCISE 2.27. Let $n \in \mathbb{Z}$ with $n > 1$. Show that n is prime if and only if $\alpha^{n-1} = 1$ for every non-zero $\alpha \in \mathbb{Z}_n$.

EXERCISE 2.28. Let $n = pq$, where p and q are distinct primes. Show that if $m := \text{lcm}(p - 1, q - 1)$, then $\alpha^m = 1$ for all $\alpha \in \mathbb{Z}_n^*$.

EXERCISE 2.29. Let p be any prime other than 2 or 5. Show that p divides infinitely many of the numbers 9, 99, 999, etc.

EXERCISE 2.30. Let n be an integer greater than 1. Show that n does not divide $2^n - 1$.

EXERCISE 2.31. Prove the following generalization of Fermat's little theorem: for every positive integer n, and every $\alpha \in \mathbb{Z}_n$, we have $\alpha^n = \alpha^{n-\varphi(n)}$.

EXERCISE 2.32. This exercise develops an alternative proof of Fermat's little theorem.

(a) Using Exercise 1.14, show that for all primes p and integers a, we have $(a + 1)^p \equiv a^p + 1 \pmod{p}$.

(b) Now derive Fermat's little theorem from part (a).

2.8 Quadratic residues

In §2.3, we studied linear congruences. It is natural to study congruences of higher degree as well. In this section, we study a special case of this more general problem, namely, congruences of the form $z^2 \equiv a \pmod{n}$. The theory we develop here nicely illustrates many of the ideas we have discussed earlier, and has a number of interesting applications as well.

We begin with some general, preliminary definitions and general observations about powers in $\mathbb{Z}_n^*$. For each integer m, we define

$$(\mathbb{Z}_n^*)^m := \{\beta^m : \beta \in \mathbb{Z}_n^*\},$$

the set of mth powers in $\mathbb{Z}_n^*$. The set $(\mathbb{Z}_n^*)^m$ is non-empty, as it obviously contains [1].

Theorem 2.16. *Let n be a positive integer, let $\alpha, \beta \in \mathbb{Z}_n^*$, and let m be any integer.*

  (i) *If $\alpha \in (\mathbb{Z}_n^*)^m$, then $\alpha^{-1} \in (\mathbb{Z}_n^*)^m$.*

  (ii) *If $\alpha \in (\mathbb{Z}_n^*)^m$ and $\beta \in (\mathbb{Z}_n^*)^m$, then $\alpha\beta \in (\mathbb{Z}_n^*)^m$.*

  (iii) *If $\alpha \in (\mathbb{Z}_n^*)^m$ and $\beta \notin (\mathbb{Z}_n^*)^m$, then $\alpha\beta \notin (\mathbb{Z}_n^*)^m$.*

Proof. For (i), if $\alpha = \gamma^m$, then $\alpha^{-1} = (\gamma^{-1})^m$.

For (ii), if $\alpha = \gamma^m$ and $\beta = \delta^m$, then $\alpha\beta = (\gamma\delta)^m$.

For (iii), suppose that $\alpha \in (\mathbb{Z}_n^*)^m$, $\beta \notin (\mathbb{Z}_n^*)^m$, and $\alpha\beta \in (\mathbb{Z}_n^*)^m$. Then by (i), $\alpha^{-1} \in (\mathbb{Z}_n^*)^m$, and by (ii), $\beta = \alpha^{-1}(\alpha\beta) \in (\mathbb{Z}_n^*)^m$, a contradiction. $\square$

Theorem 2.17. *Let n be a positive integer. For each $\alpha \in \mathbb{Z}_n^*$, and all $\ell, m \in \mathbb{Z}$ with $\gcd(\ell, m) = 1$, if $\alpha^\ell \in (\mathbb{Z}_n^*)^m$, then $\alpha \in (\mathbb{Z}_n^*)^m$.*

Proof. Suppose $\alpha^\ell = \beta^m \in (\mathbb{Z}_n^*)^m$. Since $\gcd(\ell, m) = 1$, there exist integers s and t such that $\ell s + mt = 1$. We then have

$$\alpha = \alpha^{\ell s + mt} = \alpha^{\ell s} \alpha^{mt} = \beta^{ms} \alpha^{mt} = (\beta^s \alpha^t)^m \in (\mathbb{Z}_n^*)^m. \quad \square$$

We now focus on the squares in $\mathbb{Z}_n^*$, rather than general powers. An integer a is called a **quadratic residue modulo** n if $\gcd(a, n) = 1$ and $a \equiv b^2 \pmod{n}$ for some integer b; in this case, we say that b is a **square root of a modulo** n. In terms of residue classes, a is a quadratic residue modulo n if and only if $[a] \in (\mathbb{Z}_n^*)^2$.

To avoid some annoying technicalities, from now on, we shall consider only the case where n is odd.

2.8.1 Quadratic residues modulo p

We first study quadratic residues modulo an odd prime p, and we begin by determining the square roots of 1 modulo p.

Theorem 2.18. *Let p be an odd prime and $\beta \in \mathbb{Z}_p$. Then $\beta^2 = 1$ if and only if $\beta = \pm 1$.*

Proof. Clearly, if $\beta = \pm 1$, then $\beta^2 = 1$. Conversely, suppose that $\beta^2 = 1$. Write $\beta = [b]$, where $b \in \mathbb{Z}$. Then we have $b^2 \equiv 1 \pmod{p}$, which means that

$$p \mid (b^2 - 1) = (b - 1)(b + 1),$$

and since p is prime, we must have $p \mid (b - 1)$ or $p \mid (b + 1)$. This implies $b \equiv \pm 1 \pmod{p}$, or equivalently, $\beta = \pm 1$. $\square$

This theorem says that modulo p, the only square roots of 1 are 1 and -1, which obviously belong to distinct residue classes (since $p > 2$). From this seemingly trivial fact, a number of quite interesting and useful results may be derived.

Theorem 2.19. *Let p be an odd prime and $\gamma, \beta \in \mathbb{Z}_p^*$. Then $\gamma^2 = \beta^2$ if and only if $\gamma = \pm \beta$.*

Proof. This follows from the previous theorem:

$$\gamma^2 = \beta^2 \iff (\gamma/\beta)^2 = 1 \iff \gamma/\beta = \pm 1 \iff \gamma = \pm \beta. \quad \square$$

This theorem says that if $\alpha = \beta^2$ for some $\beta \in \mathbb{Z}_p^*$, then α has precisely two square roots: β and $-\beta$.

Theorem 2.20. *Let p be an odd prime. Then $|(\mathbb{Z}_p^*)^2| = (p - 1)/2$.*

Proof. By the previous theorem, the "squaring map" $\sigma : \mathbb{Z}_p^* \to \mathbb{Z}_p^*$ that sends β to β^2 is a two-to-one map: every element in the image of σ has precisely two pre-images. As a general principle, if we have a function $f : A \to B$, where A is a finite set and every element in $f(A)$ has exactly d pre-images, then $|f(A)| = |A|/d$. Applying this general principle to our setting, we see that the image of σ is half the size of $\mathbb{Z}_p^*$. $\square$

Thus, for every odd prime p, exactly half the elements of $\mathbb{Z}_p^*$ are squares, and half are non-squares. If we choose our representatives for the residue classes modulo p from the interval $[-p/2, p/2)$, we may list the elements of $\mathbb{Z}_p$ as

$$[-(p - 1)/2], \ldots, [-1], [0], [1], \ldots, [(p - 1)/2].$$

We then see that $\mathbb{Z}_p^*$ consists of the residue classes

$$[\pm 1], \ldots, [\pm (p - 1)/2],$$

and so $(\mathbb{Z}_p^*)^2$ consists of the residue classes

$$[1]^2, \ldots, [(p - 1)/2]^2,$$

which must be distinct, since we know that $|(\mathbb{Z}_p^*)^2| = (p - 1)/2$.

Example 2.10. Let $p = 7$. We can list the elements of $\mathbb{Z}_p^*$ as

$$[\pm 1], [\pm 2], [\pm 3].$$

Squaring these, we see that

$$(\mathbb{Z}_p^*)^2 = \{[1]^2, [2]^2, [3]^2\} = \{[1], [4], [2]\}. \quad \sqcap$$

We next derive an extremely important characterization of quadratic residues.

Theorem 2.21 (Euler's criterion). *Let p be an odd prime and $\alpha \in \mathbb{Z}_p^*$.*

(i) $\alpha^{(p-1)/2} = \pm 1$.

(ii) *If $\alpha \in (\mathbb{Z}_p^*)^2$ then $\alpha^{(p-1)/2} = 1$.*

(iii) *If $\alpha \notin (\mathbb{Z}_p^*)^2$ then $\alpha^{(p-1)/2} = -1$.*

Proof. For (i), let $\gamma = \alpha^{(p-1)/2}$. By Euler's theorem (Theorem 2.13), we have

$$\gamma^2 = \alpha^{p-1} = 1,$$

and hence by Theorem 2.18, we have $\gamma = \pm 1$.

For (ii), suppose that $\alpha = \beta^2$. Then again by Euler's theorem, we have

$$\alpha^{(p-1)/2} = (\beta^2)^{(p-1)/2} = \beta^{p-1} = 1.$$

For (iii), let $\alpha \in \mathbb{Z}_p^* \setminus (\mathbb{Z}_p^*)^2$. We study the product

$$\varepsilon := \prod_{\beta \in \mathbb{Z}_p^*} \beta.$$

We shall show that, on the one hand, $\varepsilon = \alpha^{(p-1)/2}$, while on the other hand, $\varepsilon = -1$.

To show that $\varepsilon = \alpha^{(p-1)/2}$, we group elements of $\mathbb{Z}_p^*$ into pairs of distinct elements whose product is α. More precisely, let $\mathcal{P} := \{S \subseteq \mathbb{Z}_p^* : |S| = 2\}$, and define $C := \{ \{\kappa, \lambda\} \in \mathcal{P} : \kappa\lambda = \alpha\}$. Note that for every $\kappa \in \mathbb{Z}_p^*$, there is a unique $\lambda \in \mathbb{Z}_p^*$ such that $\kappa\lambda = \alpha$, namely, $\lambda := \alpha/\kappa$; moreover, $\kappa \neq \lambda$, since otherwise, we would have $\kappa^2 = \alpha$, contradicting the assumption that $\alpha \notin (\mathbb{Z}_p^*)^2$. Thus, every element of $\mathbb{Z}_p^*$ belongs to exactly one pair in C; in other words, the elements of C form a partition of $\mathbb{Z}_p^*$. It follows that

$$\varepsilon = \prod_{\{\kappa, \lambda\} \in C} (\kappa \cdot \lambda) = \prod_{\{\kappa, \lambda\} \in C} \alpha = \alpha^{(p-1)/2}.$$

To show that $\varepsilon = -1$, we group elements of $\mathbb{Z}_p^*$ into pairs of distinct elements whose product is $[1]$. Define $\mathcal{D} := \{ \{\kappa, \lambda\} \in \mathcal{P} : \kappa\lambda = 1\}$. For every $\kappa \in \mathbb{Z}_p^*$, there exists a unique $\lambda \in \mathbb{Z}_p^*$ such that $\kappa\lambda = 1$, namely, $\lambda := \kappa^{-1}$; moreover, $\kappa = \lambda$ if and only if $\kappa^2 = 1$, and by Theorem 2.18, this happens if and only if $\kappa = \pm 1$. Thus, every element of $\mathbb{Z}_p^*$ except for $[\pm 1]$ belongs to exactly one pair in $\mathcal{D}$; in other words, the elements of $\mathcal{D}$ form a partition of $\mathbb{Z}_p^* \setminus \{[\pm 1]\}$. It follows that

$$\varepsilon = [1] \cdot [-1] \cdot \prod_{\{\kappa, \lambda\} \in \mathcal{D}} (\kappa \cdot \lambda) = [-1] \cdot \prod_{\{\kappa, \lambda\} \in \mathcal{D}} [1] = -1. \quad \square$$

Thus, Euler's criterion says that for every $\alpha \in \mathbb{Z}_p^*$, we have $\alpha^{(p-1)/2} = \pm 1$ and

$$\alpha \in (\mathbb{Z}_p^*)^2 \iff \alpha^{(p-1)/2} = 1.$$

In the course of proving Euler's criterion, we proved the following result, which we state here for completeness:

Theorem 2.22 (Wilson's theorem). *Let p be an odd prime. Then $\prod_{\beta \in \mathbb{Z}_p^*} \beta = -1$.*

In the language of congruences, Wilson's theorem may be stated as follows:

$$(p-1)! \equiv -1 \pmod{p}.$$

We also derive the following simple consequence of Theorem 2.21:

Theorem 2.23. *Let p be an odd prime and $\alpha, \beta \in \mathbb{Z}_p^*$. If $\alpha \notin (\mathbb{Z}_p^*)^2$ and $\beta \notin (\mathbb{Z}_p^*)^2$, then $\alpha\beta \in (\mathbb{Z}_p^*)^2$.*

Proof. Suppose $\alpha \notin (\mathbb{Z}_p^*)^2$ and $\beta \notin (\mathbb{Z}_p^*)^2$. Then by Euler's criterion, we have

$$\alpha^{(p-1)/2} = -1 \quad \text{and} \quad \beta^{(p-1)/2} = -1.$$

Therefore,

$$(\alpha\beta)^{(p-1)/2} = \alpha^{(p-1)/2} \cdot \beta^{(p-1)/2} = [-1] \cdot [-1] = 1,$$

which again by Euler's criterion implies that $\alpha\beta \in (\mathbb{Z}_p^*)^2$. $\square$

This theorem, together with parts (ii) and (iii) of Theorem 2.16, gives us the following simple rules regarding squares in $\mathbb{Z}_p^*$:

square	×	square	=	square,
square	×	non-square	=	non-square,
non-square	×	non-square	=	square.

2.8.2 Quadratic residues modulo p^e

We next study quadratic residues modulo p^e, where p is an odd prime. The key is to establish the analog of Theorem 2.18:

Theorem 2.24. *Let p be an odd prime, e be a positive integer, and $\beta \in \mathbb{Z}_{p^e}$. Then $\beta^2 = 1$ if and only if $\beta = \pm 1$.*

Proof. Clearly, if $\beta = \pm 1$, then $\beta^2 = 1$. Conversely, suppose that $\beta^2 = 1$. Write $\beta = [b]$, where $b \in \mathbb{Z}$. Then we have $b^2 \equiv 1 \pmod{p^e}$, which means that

$$p^e \mid (b^2 - 1) = (b - 1)(b + 1).$$

In particular, $p \mid (b - 1)(b + 1)$, and so $p \mid (b - 1)$ or $p \mid (b + 1)$. Moreover, p cannot divide both $b - 1$ and $b + 1$, as otherwise, it would divide their difference $(b + 1) - (b - 1) = 2$, which is impossible (because p is odd). It follows that $p^e \mid (b - 1)$ or $p^e \mid (b + 1)$, which means $\beta = \pm 1$. $\square$

Theorems 2.19–2.23 generalize immediately from $\mathbb{Z}_p^*$ to $\mathbb{Z}_{p^e}^*$: we really used nothing in the proofs of these theorems other than the fact that ± 1 are the only square roots of 1 modulo p. As such, we state the analogs of these theorems for $\mathbb{Z}_{p^e}^*$ without proof.

Theorem 2.25. *Let p be an odd prime, e be a positive integer, and $\gamma, \beta \in \mathbb{Z}_{p^e}^*$. Then $\gamma^2 = \beta^2$ if and only if $\gamma = \pm\beta$.*

Theorem 2.26. *Let p be an odd prime and e be a positive integer. Then we have $|(\mathbb{Z}_{p^e}^*)^2| = \varphi(p^e)/2$.*

Theorem 2.27. *Let p be an odd prime, e be a positive integer, and $\alpha \in \mathbb{Z}_{p^e}^*$.*

   (i) $\alpha^{\varphi(p^e)/2} = \pm 1$.
   (ii) If $\alpha \in (\mathbb{Z}_{p^e}^*)^2$ then $\alpha^{\varphi(p^e)/2} = 1$.
   (iii) If $\alpha \notin (\mathbb{Z}_{p^e}^*)^2$ then $\alpha^{\varphi(p^e)/2} = -1$.

Theorem 2.28. *Let p be an odd prime and e be a positive integer. Then we have $\prod_{\beta \in \mathbb{Z}_{p^e}^*} \beta = -1$.*

Theorem 2.29. *Let p be an odd prime, e be a positive integer, and $\alpha, \beta \in \mathbb{Z}_{p^e}^*$. If $\alpha \notin (\mathbb{Z}_{p^e}^*)^2$ and $\beta \notin (\mathbb{Z}_{p^e}^*)^2$, then $\alpha\beta \in (\mathbb{Z}_{p^e}^*)^2$.*

It turns out that an integer is a quadratic residue modulo p^e if and only if it is a quadratic residue modulo p.

Theorem 2.30. *Let p be an odd prime, e be a positive integer, and a be any integer. Then a is a quadratic residue modulo p^e if and only if a is a quadratic residue modulo p.*

Proof. Suppose that a is a quadratic residue modulo p^e. Then a is not divisible by p and $a \equiv b^2 \pmod{p^e}$ for some integer b. It follows that $a \equiv b^2 \pmod{p}$, and so a is a quadratic residue modulo p.

Suppose that a is not a quadratic residue modulo p^e. If a is divisible by p, then by definition a is not a quadratic residue modulo p. So suppose a is not divisible by p. By Theorem 2.27, we have

$$a^{p^{e-1}(p-1)/2} \equiv -1 \pmod{p^e}.$$

This congruence holds modulo p as well, and by Fermat's little theorem (applied

$e - 1$ times),

$$a \equiv a^p \equiv a^{p^2} \equiv \cdots \equiv a^{p^{e-1}} \pmod{p},$$

and so

$$-1 \equiv a^{p^{e-1}(p-1)/2} \equiv a^{(p-1)/2} \pmod{p}.$$

Theorem 2.21 therefore implies that a is not a quadratic residue modulo p. □

2.8.3 Quadratic residues modulo n

We now study quadratic residues modulo n, where n is an arbitrary, odd integer, with $n > 1$. Let

$$n = p_1^{e_1} \cdots p_r^{e_r}$$

be the prime factorization of n. Our main tools here are the Chinese remainder map

$$\theta : \mathbb{Z}_n \to \mathbb{Z}_{p_1^{e_1}} \times \cdots \times \mathbb{Z}_{p_r^{e_r}},$$

introduced in Theorem 2.8, together with the results developed so far for quadratic residues modulo odd prime powers.

Let $\alpha \in \mathbb{Z}_n^*$ with $\theta(\alpha) = (\alpha_1, \ldots, \alpha_r)$.

- On the one hand, suppose $\alpha = \beta^2$ for some $\beta \in \mathbb{Z}_n^*$. If $\theta(\beta) = (\beta_1, \ldots, \beta_r)$, we have

$$(\alpha_1, \ldots, \alpha_r) = \theta(\alpha) = \theta(\beta^2) = (\beta_1^2, \ldots, \beta_r^2),$$

  where we have used part (iii.c) of Theorem 2.8. It follows that $\alpha_i = \beta_i^2$ for each i.

- On the other hand, suppose that for each i, $\alpha_i = \beta_i^2$ for some $\beta_i \in \mathbb{Z}_{p_i^{e_i}}^*$. Then setting $\beta := \theta^{-1}(\beta_1, \ldots, \beta_r)$, we have

$$\theta(\beta^2) = (\beta_1^2, \ldots, \beta_r^2) = (\alpha_1, \ldots, \alpha_r) = \theta(\alpha),$$

  where we have again used part (iii.c) of Theorem 2.8, along with the fact that θ is bijective (to define β). Thus, $\theta(\alpha) = \theta(\beta^2)$, and again since θ is bijective, it follows that $\alpha = \beta^2$.

We have shown that

$$\alpha \in (\mathbb{Z}_n^*)^2 \iff \alpha_i \in \left(\mathbb{Z}_{p_i^{e_i}}^*\right)^2 \text{ for } i = 1, \ldots, r.$$

In particular, restricting θ to $(\mathbb{Z}_n^*)^2$ yields a one-to-one correspondence between $(\mathbb{Z}_n^*)^2$ and

$$\left(\mathbb{Z}_{p_1^{e_1}}^*\right)^2 \times \cdots \times \left(\mathbb{Z}_{p_r^{e_r}}^*\right)^2,$$

and therefore, by Theorem 2.26 (and Theorem 2.9), we have

$$|(\mathbb{Z}_n^*)^2| = \prod_{i=1}^{r}(\varphi(p_i^{e_i})/2) = \varphi(n)/2^r.$$

Now suppose that $\alpha = \beta^2$, with $\beta \in \mathbb{Z}_n^*$ and $\theta(\beta) = (\beta_1, \ldots, \beta_r)$. Consider an arbitrary element $\gamma \in \mathbb{Z}_n^*$, with $\theta(\gamma) = (\gamma_1, \ldots, \gamma_r)$. Then we have

$$\gamma^2 = \beta^2 \iff \theta(\gamma^2) = \theta(\beta^2)$$
$$\iff (\gamma_1^2, \ldots, \gamma_r^2) = (\beta_1^2, \ldots, \beta_r^2)$$
$$\iff (\gamma_1, \ldots, \gamma_r) = (\pm\beta_1, \ldots, \pm\beta_r) \quad \text{(by Theorem 2.25).}$$

Therefore, α has precisely 2^r square roots, namely, $\theta^{-1}(\pm\beta_1, \ldots, \pm\beta_r)$.

2.8.4 Square roots of -1 modulo p

Using Euler's criterion, we can easily characterize those primes modulo which -1 is a quadratic residue. This turns out to have a number of nice applications.

Consider an odd prime p. The following theorem says that the question of whether -1 is a quadratic residue modulo p is decided by the residue class of p modulo 4. Since p is odd, either $p \equiv 1 \pmod 4$ or $p \equiv 3 \pmod 4$.

Theorem 2.31. *Let p be an odd prime. Then -1 is a quadratic residue modulo p if and only $p \equiv 1 \pmod 4$.*

Proof. By Euler's criterion, -1 is a quadratic residue modulo p if and only if $(-1)^{(p-1)/2} \equiv 1 \pmod p$. If $p \equiv 1 \pmod 4$, then $(p-1)/2$ is even, and so $(-1)^{(p-1)/2} = 1$. If $p \equiv 3 \pmod 4$, then $(p-1)/2$ is odd, and so $(-1)^{(p-1)/2} = -1$. $\square$

In fact, when $p \equiv 1 \pmod 4$, any non-square in $\mathbb{Z}_p^*$ yields a square root of -1 modulo p, as follows:

Theorem 2.32. *Let p be a prime with $p \equiv 1 \pmod 4$, $\gamma \in \mathbb{Z}_p^* \setminus (\mathbb{Z}_p^*)^2$, and $\beta := \gamma^{(p-1)/4}$. Then $\beta^2 = -1$.*

Proof. This is a simple calculation, based on Euler's criterion:

$$\beta^2 = \gamma^{(p-1)/2} = -1. \quad \square$$

The fact that -1 is a quadratic residue modulo primes $p \equiv 1 \pmod 4$ can be used to prove Fermat's theorem that such primes may be written as the sum of two squares. To do this, we first need the following technical lemma:

Theorem 2.33 (Thue's lemma). *Let* $n, b, r^*, t^* \in \mathbb{Z}$, *with* $0 < r^* \le n < r^*t^*$. *Then there exist* $r, t \in \mathbb{Z}$ *with*

$$r \equiv bt \pmod{n}, \quad |r| < r^*, \quad \text{and} \quad 0 < |t| < t^*.$$

Proof. For $i = 0, \ldots, r^* - 1$ and $j = 0, \ldots, t^* - 1$, we define the number $v_{ij} := i - bj$. Since we have defined r^*t^* numbers, and $r^*t^* > n$, two of these numbers must lie in the same residue class modulo n; that is, for some $(i_1, j_1) \ne (i_2, j_2)$, we have $v_{i_1 j_1} \equiv v_{i_2 j_2} \pmod{n}$. Setting $r := i_1 - i_2$ and $t := j_1 - j_2$, this implies $r \equiv bt \pmod{n}$, $|r| < r^*$, $|t| < t^*$, and that either $r \ne 0$ or $t \ne 0$. It only remains to show that $t \ne 0$. Suppose to the contrary that $t = 0$. This would imply that $r \equiv 0 \pmod{n}$ and $r \ne 0$, which is to say that r is a non-zero multiple of n; however, this is impossible, since $|r| < r^* \le n$. $\square$

Theorem 2.34 (Fermat's two squares theorem). *Let* p *be an odd prime. Then* $p = r^2 + t^2$ *for some* $r, t \in \mathbb{Z}$ *if and only if* $p \equiv 1 \pmod{4}$.

Proof. One direction is easy. Suppose $p \equiv 3 \pmod{4}$. It is easy to see that the square of every integer is congruent to either 0 or 1 modulo 4; therefore, the sum of two squares is congruent to either 0, 1, or 2 modulo 4, and so can not be congruent to p modulo 4 (let alone equal to p).

For the other direction, suppose $p \equiv 1 \pmod{4}$. We know that -1 is a quadratic residue modulo p, so let b be an integer such that $b^2 \equiv -1 \pmod{p}$. Now apply Theorem 2.33 with $n := p$, b as just defined, and $r^* := t^* := \lfloor \sqrt{p} \rfloor + 1$. Evidently, $\lfloor \sqrt{p} \rfloor + 1 > \sqrt{p}$, and hence $r^*t^* > p$. Also, since p is prime, $\sqrt{p}$ is not an integer, and so $\lfloor \sqrt{p} \rfloor < \sqrt{p} < p$; in particular, $r^* = \lfloor \sqrt{p} \rfloor + 1 \le p$. Thus, the hypotheses of that theorem are satisfied, and therefore, there exist integers r and t such that

$$r \equiv bt \pmod{p}, \quad |r| \le \lfloor \sqrt{p} \rfloor < \sqrt{p}, \quad \text{and} \quad 0 < |t| \le \lfloor \sqrt{p} \rfloor < \sqrt{p}.$$

It follows that

$$r^2 \equiv b^2 t^2 \equiv -t^2 \pmod{p}.$$

Thus, $r^2 + t^2$ is a multiple of p and $0 < r^2 + t^2 < 2p$. The only possibility is that $r^2 + t^2 = p$. $\square$

The fact that -1 is a quadratic residue modulo an odd prime p only if $p \equiv 1 \pmod{4}$ can be used so show there are infinitely many such primes.

Theorem 2.35. *There are infinitely many primes* $p \equiv 1 \pmod{4}$.

Proof. Suppose there were only finitely many such primes, $p_1, \ldots, p_k$. Set $M := \prod_{i=1}^{k} p_i$ and $N := 4M^2 + 1$. Let p be any prime dividing N. Evidently, p is not among the p_i's, since if it were, it would divide both N and $4M^2$, and so also $N - 4M^2 = 1$. Also, p is clearly odd, since N is odd. Moreover, $(2M)^2 \equiv -1 \pmod{p}$; therefore, -1 is a quadratic residue modulo p, and so $p \equiv 1 \pmod{4}$, contradicting the assumption that $p_1, \ldots, p_k$ are the only such primes. $\square$

For completeness, we also state the following fact:

Theorem 2.36. *There are infinitely many primes $p \equiv 3 \pmod{4}$.*

Proof. Suppose there were only finitely many such primes, $p_1, \ldots, p_k$. Set $M := \prod_{i=1}^{k} p_i$ and $N := 4M - 1$. Since $N \equiv 3 \pmod{4}$, there must be some prime $p \equiv 3 \pmod{4}$ dividing N (if all primes dividing N were congruent to 1 modulo 4, then so too would be their product N). Evidently, p is not among the p_i's, since if it were, it would divide both N and $4M$, and so also $4M - N = 1$. This contradicts the assumption that $p_1, \ldots, p_k$ are the only primes congruent to 3 modulo 4. $\square$

EXERCISE 2.33. Let $n, m \in \mathbb{Z}$, where $n > 0$, and let $d := \gcd(m, \varphi(n))$. Show that:

(a) if $d = 1$, then $(\mathbb{Z}_n^*)^m = (\mathbb{Z}_n^*)$;

(b) if $\alpha \in (\mathbb{Z}_n^*)^m$, then $\alpha^{\varphi(n)/d} = 1$.

EXERCISE 2.34. Calculate the sets C and $\mathcal{D}$ in the proof of Theorem 2.21 in the case $p = 11$ and $\alpha = -1$.

EXERCISE 2.35. Calculate the square roots of 1 modulo 4, 8, and 16.

EXERCISE 2.36. Let $n \in \mathbb{Z}$ with $n > 1$. Show that n is prime if and only if $(n - 1)! \equiv -1 \pmod{n}$.

EXERCISE 2.37. Let p be a prime with $p \equiv 1 \pmod{4}$, and $b := ((p - 1)/2)!$. Show that $b^2 \equiv -1 \pmod{p}$.

EXERCISE 2.38. Let $n := pq$, where p and q are distinct, odd primes. Show that there exist $\alpha, \beta \in \mathbb{Z}_n^*$ such that $\alpha \notin (\mathbb{Z}_n^*)^2$, $\beta \notin (\mathbb{Z}_n^*)^2$, and $\alpha\beta \notin (\mathbb{Z}_n^*)^2$.

EXERCISE 2.39. Let n be an odd positive integer, and let a be any integer. Show that a is a quadratic residue modulo n if and only if a is a quadratic residue modulo p for each prime $p \mid n$.

EXERCISE 2.40. Show that if p is an odd prime, with $p \equiv 3 \pmod{4}$, then $(\mathbb{Z}_p^*)^4 = (\mathbb{Z}_p^*)^2$. More generally, show that if n is an odd positive integer, where $p \equiv 3 \pmod{4}$ for each prime $p \mid n$, then $(\mathbb{Z}_n^*)^4 = (\mathbb{Z}_n^*)^2$.

EXERCISE 2.41. Let p be an odd prime, and let $e \in \mathbb{Z}$ with $e > 1$. Let a be an integer of the form $a = p^f b$, where $0 \leq f < e$ and $p \nmid b$. Consider the integer solutions z to the congruence $z^2 \equiv a \pmod{p^e}$. Show that a solution exists if and only if f is even and b is a quadratic residue modulo p, in which case there are exactly $2p^f$ distinct solutions modulo p^e.

EXERCISE 2.42. Suppose p is an odd prime, and that $r^2 + t^2 = p$ for some integers r, t. Show that if x, y are integers such that $x^2 + y^2 = p$, then (x, y) must be $(\pm r, \pm t)$ or $(\pm t, \pm r)$.

EXERCISE 2.43. Show that if both u and v are the sum of two squares of integers, then so is their product uv.

EXERCISE 2.44. Suppose $r^2 + t^2 \equiv 0 \pmod{n}$, where n is a positive integer, and suppose p is an odd prime dividing n. Show that:

(a) if p divides neither r nor t, then $p \equiv 1 \pmod 4$;

(b) if p divides one of r or t, then it divides the other, and moreover, p^2 divides n, and $(r/p)^2 + (t/p)^2 \equiv 0 \pmod{n/p^2}$.

EXERCISE 2.45. Let n be a positive integer, and write $n = ab^2$ where a and b are positive integers, and a is square-free (see Exercise 1.15). Show that n is the sum of two squares of integers if and only if no prime $p \equiv 3 \pmod 4$ divides a. Hint: use the previous two exercises.

2.9 Summations over divisors

We close this chapter with a brief treatment of summations over divisors. To this end, we introduce some terminology and notation. By an **arithmetic function**, we simply mean a function from the positive integers into the reals (actually, one usually considers complex-valued functions as well, but we shall not do so here). Let f and g be arithmetic functions. The **Dirichlet product** of f and g, denoted $f \star g$, is the arithmetic function whose value at n is defined by the formula

$$(f \star g)(n) := \sum_{d|n} f(d)g(n/d),$$

the sum being over all positive divisors d of n. Another, more symmetric, way to write this is

$$(f \star g)(n) = \sum_{n=d_1 d_2} f(d_1)g(d_2),$$

the sum being over all pairs (d_1, d_2) of positive integers with $d_1 d_2 = n$.

The Dirichlet product is clearly commutative (i.e., $f \star g = g \star f$), and is associative as well, which one can see by checking that

$$(f \star (g \star h))(n) = \sum_{n=d_1 d_2 d_3} f(d_1)g(d_2)h(d_3) = ((f \star g) \star h)(n),$$

the sum being over all triples (d_1, d_2, d_3) of positive integers with $d_1 d_2 d_3 = n$.

We now introduce three special arithmetic functions: I, 1, and μ. The functions I and 1 are defined as follows:

$$I(n) := \begin{cases} 1 & \text{if } n = 1; \\ 0 & \text{if } n > 1; \end{cases} \qquad 1(n) := 1.$$

The **Möbius function** μ is defined as follows: if $n = p_1^{e_1} \cdots p_r^{e_r}$ is the prime factorization of n, then

$$\mu(n) := \begin{cases} 0 & \text{if } e_i > 1 \text{ for some } i = 1, \ldots, r; \\ (-1)^r & \text{otherwise.} \end{cases}$$

In other words, $\mu(n) = 0$ if n is not square-free (see Exercise 1.15); otherwise, $\mu(n)$ is $(-1)^r$ where r is the number of distinct primes dividing n. Here are some examples:

$$\mu(1) = 1, \ \mu(2) = -1, \ \mu(3) = -1, \ \mu(4) = 0, \ \mu(5) = -1, \ \mu(6) = 1.$$

It is easy to see from the definitions that for every arithmetic function f, we have

$$I \star f = f \quad \text{and} \quad (1 \star f)(n) = \sum_{d \mid n} f(d).$$

Thus, I acts as a multiplicative identity with respect to the Dirichlet product, while "$1 \star$" acts as a "summation over divisors" operator.

An arithmetic function f is called **multiplicative** if $f(1) = 1$ and for all positive integers n, m with $\gcd(n, m) = 1$, we have $f(nm) = f(n)f(m)$.

The reader may easily verify that I, 1, and μ are multiplicative functions. Theorem 2.9 says that Euler's function φ is multiplicative. The reader may also verify the following:

Theorem 2.37. *If f is a multiplicative arithmetic function, and if $n = p_1^{e_1} \cdots p_r^{e_r}$ is the prime factorization of n, then $f(n) = f(p_1^{e_1}) \cdots f(p_r^{e_r})$.*

Proof. Exercise. $\square$

A key property of the Möbius function is the following:

Theorem 2.38. *Let f be a multiplicative arithmetic function. If $n = p_1^{e_1} \cdots p_r^{e_r}$ is the prime factorization of n, then*

$$\sum_{d|n} \mu(d) f(d) = (1 - f(p_1)) \cdots (1 - f(p_r)). \qquad (2.9)$$

Proof. The only non-zero terms appearing in the sum on the left-hand side of (2.9) are those corresponding to divisors d of the form $p_{i_1} \cdots p_{i_\ell}$, where $p_{i_1}, \ldots, p_{i_\ell}$ are distinct; the value contributed to the sum by such a term is $(-1)^\ell f(p_{i_1} \cdots p_{i_\ell}) = (-1)^\ell f(p_{i_1}) \cdots f(p_{i_\ell})$. These are the same as the terms in the expansion of the product on the right-hand side of (2.9). $\square$

If we set $f := 1$ in the previous theorem, then we see that

$$\sum_{d|n} \mu(d) = \begin{cases} 1 & \text{if } n = 1; \\ 0 & \text{if } n > 1. \end{cases}$$

Translating this into the language of Dirichlet products, we have

$$1 \star \mu = I.$$

Thus, with respect to the Dirichlet product, the functions 1 and μ are multiplicative inverses of one another. Based on this, we may easily derive the following:

Theorem 2.39 (Möbius inversion formula). *Let f and F be arithmetic functions. Then $F = 1 \star f$ if and only if $f = \mu \star F$.*

Proof. If $F = 1 \star f$, then

$$\mu \star F = \mu \star (1 \star f) = (\mu \star 1) \star f = I \star f = f,$$

and conversely, if $f = \mu \star F$, then

$$1 \star f = 1 \star (\mu \star F) = (1 \star \mu) \star F = I \star F = F. \quad \square$$

The Möbius inversion formula says this:

$$F(n) = \sum_{d|n} f(d) \text{ for all positive integers } n$$

$$\iff f(n) = \sum_{d|n} \mu(d) F(n/d) \text{ for all positive integers } n.$$

The Möbius inversion formula is a useful tool. As an application, we use it to obtain a simple proof of the following fact:

Theorem 2.40. *For every positive integer n, we have $\sum_{d|n} \varphi(d) = n$.*

Proof. Let us define the arithmetic functions $N(n) := n$ and $M(n) := 1/n$. Our goal is to show that $N = 1 \star \varphi$, and by Möbius inversion, it suffices to show that $\mu \star N = \varphi$. If $n = p_1^{e_1} \cdots p_r^{e_r}$ is the prime factorization of n, we have

$$(\mu \star N)(n) = \sum_{d|n} \mu(d)(n/d) = n \sum_{d|n} \mu(d)/d$$

$$= n \prod_{i=1}^{r}(1 - 1/p_i) \text{ (applying Theorem 2.38 with } f := M)$$

$$= \varphi(n) \text{ (by Theorem 2.11). } \square$$

EXERCISE 2.46. In our definition of a multiplicative function f, we made the requirement that $f(1) = 1$. Show that if we dropped this requirement, the only other function that would satisfy the definition would be the zero function (i.e., the function that is everywhere zero).

EXERCISE 2.47. Let f be a polynomial with integer coefficients, and for each positive integer n, define $\omega_f(n)$ to be the number of integers $x \in \{0, \ldots, n-1\}$ such that $f(x) \equiv 0 \pmod{n}$. Show that ω_f is multiplicative.

EXERCISE 2.48. Show that if f and g are multiplicative, then so is $f \star g$. Hint: use Exercise 1.18.

EXERCISE 2.49. Let $\tau(n)$ be the number of positive divisors of n. Show that:

(a) τ is a multiplicative function;
(b) $\tau(n) = \prod_{i=1}^{r}(e_i + 1)$, where $n = p_1^{e_1} \cdots p_r^{e_r}$ is the prime factorization of n;
(c) $\sum_{d|n} \mu(d)\tau(n/d) = 1$;
(d) $\sum_{d|n} \mu(d)\tau(d) = (-1)^r$, where $n = p_1^{e_1} \cdots p_r^{e_r}$ is the prime factorization of n.

EXERCISE 2.50. Define $\sigma(n) := \sum_{d|n} d$. Show that:

(a) σ is a multiplicative function;
(b) $\sigma(n) = \prod_{i=1}^{r}(p_i^{e_i+1} - 1)/(p_i - 1)$, where $n = p_1^{e_1} \cdots p_r^{e_r}$ is the prime factorization of n;
(c) $\sum_{d|n} \mu(d)\sigma(n/d) = n$;
(d) $\sum_{d|n} \mu(d)\sigma(d) = (-1)^r p_1 \cdots p_r$, where $n = p_1^{e_1} \cdots p_r^{e_r}$ is the prime factorization of n.

EXERCISE 2.51. The **Mangoldt function** $\Lambda(n)$ is defined for all positive integers n as follows: $\Lambda(n) := \log p$, if $n = p^k$ for some prime p and positive integer k, and $\Lambda(n) := 0$, otherwise. Show that $\sum_{d|n} \Lambda(d) = \log n$, and from this, deduce that $\Lambda(n) = -\sum_{d|n} \mu(d) \log d$.

EXERCISE 2.52. Show that if f is multiplicative, and if $n = p_1^{e_1} \cdots p_r^{e_r}$ is the prime factorization of n, then $\sum_{d|n} \mu(d)^2 f(d) = (1 + f(p_1)) \cdots (1 + f(p_r))$.

EXERCISE 2.53. Show that n is square-free if and only if $\sum_{d|n} \mu(d)^2 \varphi(d) = n$.

EXERCISE 2.54. Show that for every arithmetic function f with $f(1) \neq 0$, there is a unique arithmetic function g, called the **Dirichlet inverse** of f, such that $f \star g = I$. Also, show that if $f(1) = 0$, then f has no Dirichlet inverse.

EXERCISE 2.55. Show that if f is a multiplicative function, then so is its Dirichlet inverse (as defined in the previous exercise).

EXERCISE 2.56. This exercise develops an alternative proof of Theorem 2.40 that does not depend on Theorem 2.11. Let n be a positive integer. Define

$$F_n := \{i/n \in \mathbb{Q} : i = 0, \ldots, n - 1\}.$$

Also, for each positive integer d, define

$$G_d := \{a/d \in \mathbb{Q} : a \in \mathbb{Z}, \gcd(a, d) = 1\}.$$

(a) Show that for each $x \in F_n$, there exists a unique positive divisor d of n such that $x \in G_d$.

(b) Show that for each positive divisor d of n, we have

$$F_n \cap G_d = \{a/d : a = 0, \ldots, d - 1, \gcd(a, d) = 1\}.$$

(c) Using (a) and (b), show that $\sum_{d|n} \varphi(d) = n$.

EXERCISE 2.57. Using Möbius inversion, directly derive Theorem 2.11 from Theorem 2.40.

3

Computing with large integers

In this chapter, we review standard asymptotic notation, introduce the formal computational model that we shall use throughout the rest of the text, and discuss basic algorithms for computing with large integers.

3.1 Asymptotic notation

We review some standard notation for relating the rate of growth of functions. This notation will be useful in discussing the running times of algorithms, and in a number of other contexts as well.

Let f and g be real-valued functions. We shall assume that each is defined on the set of non-negative integers, or, alternatively, that each is defined on the set of non-negative reals. Actually, as we are only concerned about the behavior of $f(x)$ and $g(x)$ as $x \to \infty$, we only require that $f(x)$ and $g(x)$ are defined for all sufficiently large x (the phrase "for all sufficiently large x" means "for some x_0 and all $x \geq x_0$"). We further assume that g is **eventually positive**, meaning that $g(x) > 0$ for all sufficiently large x. Then

- $f = O(g)$ means that $|f(x)| \leq cg(x)$ for some positive constant c and all sufficiently large x (read, "f is big-O of g"),

- $f = \Omega(g)$ means that $f(x) \geq cg(x)$ for some positive constant c and all sufficiently large x (read, "f is big-Omega of g"),

- $f = \Theta(g)$ means that $cg(x) \leq f(x) \leq dg(x)$ for some positive constants c and d and all sufficiently large x (read, "f is big-Theta of g"),

- $f = o(g)$ means that $f(x)/g(x) \to 0$ as $x \to \infty$ (read, "f is little-o of g"), and

- $f \sim g$ means that $f(x)/g(x) \to 1$ as $x \to \infty$ (read, "f is asymptotically equal to g").

Example 3.1. Let $f(x) := x^2$ and $g(x) := 2x^2 - 10x + 1$. Then $f = O(g)$ and $f = \Omega(g)$. Indeed, $f = \Theta(g)$. □

Example 3.2. Let $f(x) := x^2$ and $g(x) := x^2 - 10x + 1$. Then $f \sim g$. □

Example 3.3. Let $f(x) := 100x^2$ and $g(x) := x^3$. Then $f = o(g)$. □

Note that by definition, if we write $f = \Omega(g)$, $f = \Theta(g)$, or $f \sim g$, it must be the case that f (in addition to g) is eventually positive; however, if we write $f = O(g)$ or $f = o(g)$, then f need not be eventually positive.

When one writes "$f = O(g)$," one should interpret "$\cdot = O(\cdot)$" as a binary relation between f with g. Analogously for "$f = \Omega(g)$," "$f = \Theta(g)$," and "$f = o(g)$."

One may also write "$O(g)$" in an expression to denote an anonymous function f such that $f = O(g)$. Analogously, $\Omega(g)$, $\Theta(g)$, and $o(g)$ may denote anonymous functions. The expression $O(1)$ denotes a function bounded in absolute value by a constant, while the expression $o(1)$ denotes a function that tends to zero in the limit.

Example 3.4. Let $f(x) := x^3 - 2x^2 + x - 3$. One could write $f(x) = x^3 + O(x^2)$. Here, the anonymous function is $g(x) := -2x^2 + x - 3$, and clearly $g(x) = O(x^2)$. One could also write $f(x) = x^3 - (2 + o(1))x^2$. Here, the anonymous function is $g(x) := -1/x + 3/x^2$. While $g = o(1)$, it is only defined for $x > 0$. This is acceptable, since we will only regard statements such as this asymptotically, as $x \to \infty$. □

As an even further use (abuse?) of the notation, one may use the big-O, big-Omega, and big-Theta notation for functions on an arbitrary domain, in which case the relevant inequalities should hold throughout the entire domain. This usage includes functions of several independent variables, as well as functions defined on sets with no natural ordering.

EXERCISE 3.1. Show that:

    (a) $f = o(g)$ implies $f = O(g)$ and $g \neq O(f)$;

    (b) $f = O(g)$ and $g = O(h)$ implies $f = O(h)$;

    (c) $f = O(g)$ and $g = o(h)$ implies $f = o(h)$;

    (d) $f = o(g)$ and $g = O(h)$ implies $f = o(h)$.

EXERCISE 3.2. Let f and g be eventually positive functions. Show that:

    (a) $f \sim g$ if and only if $f = (1 + o(1))g$;

    (b) $f \sim g$ implies $f = \Theta(g)$;

    (c) $f = \Theta(g)$ if and only if $f = O(g)$ and $f - \Omega(g)$;

(d) $f = \Omega(g)$ if and only if $g = O(f)$.

EXERCISE 3.3. Suppose $f_1 = O(g_1)$ and $f_2 = O(g_2)$. Show that $f_1 + f_2 = O(\max(g_1, g_2))$, $f_1 f_2 = O(g_1 g_2)$, and that for every constant c, $c f_1 = O(g_1)$.

EXERCISE 3.4. Suppose that $f(x) \le c + d g(x)$ for some positive constants c and d, and for all sufficiently large x. Show that if $g = \Omega(1)$, then $f = O(g)$.

EXERCISE 3.5. Suppose f and g are defined on the integers $i \ge k$, and that $g(i) > 0$ for all $i \ge k$. Show that if $f = O(g)$, then there exists a positive constant c such that $|f(i)| \le c g(i)$ for all $i \ge k$.

EXERCISE 3.6. Let f and g be eventually positive functions, and assume that $f(x)/g(x)$ tends to a limit L (possibly $L = \infty$) as $x \to \infty$. Show that:
  (a) if $L = 0$, then $f = o(g)$;
  (b) if $0 < L < \infty$, then $f = \Theta(g)$;
  (c) if $L = \infty$, then $g = o(f)$.

EXERCISE 3.7. Let $f(x) := x^\alpha (\log x)^\beta$ and $g(x) := x^\gamma (\log x)^\delta$, where $\alpha, \beta, \gamma, \delta$ are non-negative constants. Show that if $\alpha < \gamma$, or if $\alpha = \gamma$ and $\beta < \delta$, then $f = o(g)$.

EXERCISE 3.8. Order the following functions in x so that for each adjacent pair f, g in the ordering, we have $f = O(g)$, and indicate if $f = o(g)$, $f \sim g$, or $g = O(f)$:

$$x^3, \; e^x x^2, \; 1/x, \; x^2(x + 100) + 1/x, \; x + \sqrt{x}, \; \log_2 x, \; \log_3 x, \; 2x^2, \; x,$$
$$e^{-x}, \; 2x^2 - 10x + 4, \; e^{x+\sqrt{x}}, \; 2^x, \; 3^x, \; x^{-2}, \; x^2(\log x)^{1000}.$$

EXERCISE 3.9. Show that:
  (a) the relation "$\sim$" is an equivalence relation on the set of eventually positive functions;
  (b) for all eventually positive functions f_1, f_2, g_1, g_2, if $f_1 \sim g_1$ and $f_2 \sim g_2$, then $f_1 \star f_2 \sim g_1 \star g_2$, where "$\star$" denotes addition, multiplication, or division;
  (c) for all eventually positive functions f, g, and every $\alpha > 0$, if $f \sim g$, then $f^\alpha \sim g^\alpha$;
  (d) for all eventually positive functions f, g, and every function h such that $h(x) \to \infty$ as $x \to \infty$, if $f \sim g$, then $f \circ h \sim g \circ h$, where "$\circ$" denotes function composition.

EXERCISE 3.10. Show that all of the claims in the previous exercise also hold when the relation "$\sim$" is replaced with the relation "$\cdot = \Theta(\cdot)$."

EXERCISE 3.11. Let f, g be eventually positive functions. Show that:

(a) $f = \Theta(g)$ if and only if $\log f = \log g + O(1)$;

(b) $f \sim g$ if and only if $\log f = \log g + o(1)$.

EXERCISE 3.12. Suppose that f and g are functions defined on the integers $k, k+1, \ldots$, and that g is eventually positive. For $n \geq k$, define $F(n) := \sum_{i=k}^{n} f(i)$ and $G(n) := \sum_{i=k}^{n} g(i)$. Show that if $f = O(g)$ and G is eventually positive, then $F = O(G)$.

EXERCISE 3.13. Suppose that f and g are piece-wise continuous on $[a, \infty)$ (see §A4), and that g is eventually positive. For $x \geq a$, define $F(x) := \int_{a}^{x} f(t)\, dt$ and $G(x) := \int_{a}^{x} g(t)\, dt$. Show that if $f = O(g)$ and G is eventually positive, then $F = O(G)$.

EXERCISE 3.14. Suppose that f and g are functions defined on the integers $k, k+1, \ldots$, and that both f and g are eventually positive. For $n \geq k$, define $F(n) := \sum_{i=k}^{n} f(i)$ and $G(n) := \sum_{i=k}^{n} g(i)$. Show that if $f \sim g$ and $G(n) \to \infty$ as $n \to \infty$, then $F \sim G$.

EXERCISE 3.15. Suppose that f and g are piece-wise continuous on $[a, \infty)$ (see §A4), and that both f and g are eventually positive. For $x \geq a$, define $F(x) := \int_{a}^{x} f(t)\, dt$ and $G(x) := \int_{a}^{x} g(t)\, dt$. Show that if $f \sim g$ and $G(x) \to \infty$ as $x \to \infty$, then $F \sim G$.

EXERCISE 3.16. Give an example of two non-decreasing functions f and g, each mapping positive integers to positive integers, such that $f \neq O(g)$ and $g \neq O(f)$.

3.2 Machine models and complexity theory

When presenting an algorithm, we shall always use a high-level, and somewhat informal, notation. However, all of our high-level descriptions can be routinely translated into the machine-language of an actual computer. So that our theorems on the running times of algorithms have a precise mathematical meaning, we formally define an "idealized" computer: the **random access machine** or **RAM**.

A RAM consists of an unbounded sequence of **memory cells**

$$m[0], m[1], m[2], \ldots,$$

each of which can store an arbitrary integer, together with a **program**. A program consists of a finite sequence of instructions $I_0, I_1, \ldots$, where each instruction is of one of the following types:

arithmetic This type of instruction is of the form $\gamma \leftarrow \alpha \star \beta$, where $\star$ represents one of the operations addition, subtraction, multiplication, or integer division (i.e., $\lfloor \cdot / \cdot \rfloor$). The values α and β are of the form c, $m[a]$, or $m[m[a]]$, and γ is of the form $m[a]$ or $m[m[a]]$, where c is an integer constant and a is a non-negative integer constant. Execution of this type of instruction causes the value $\alpha \star \beta$ to be evaluated and then stored in γ.

branching This type of instruction is of the form IF $\alpha \diamond \beta$ GOTO i, where i is the index of an instruction, and where $\diamond$ is one of the comparison operations $=, \neq, <, >, \leq, \geq$, and α and β are as above. Execution of this type of instruction causes the "flow of control" to pass conditionally to instruction I_i.

halt The HALT instruction halts the execution of the program.

A RAM works by executing instruction I_0, and continues to execute instructions, following branching instructions as appropriate, until a HALT instruction is reached.

We do not specify input or output instructions, and instead assume that the input and output are to be found in memory cells at some prescribed locations, in some standardized format.

To determine the running time of a program on a given input, we charge 1 unit of time to each instruction executed.

This model of computation closely resembles a typical modern-day computer, except that we have abstracted away many annoying details. However, there are two details of real machines that cannot be ignored; namely, any real machine has a finite number of memory cells, and each cell can store numbers only in some fixed range.

The first limitation must be dealt with by either purchasing sufficient memory or designing more space-efficient algorithms.

The second limitation is especially annoying, as we will want to perform computations with quite large integers — much larger than will fit into any single memory cell of an actual machine. To deal with this limitation, we shall represent such large integers as vectors of digits in some fixed base, so that each digit is bounded in order to fit into a memory cell. This is discussed in more detail in the next section. The only other numbers we actually need to store in memory cells are "small" numbers representing array indices, counters, and the like, which we hope will fit into the memory cells of actual machines. Below, we shall make a more precise, formal restriction on the magnitude of numbers that may be stored in memory cells.

Even with these caveats and restrictions, the running time as we have defined it for a RAM is still only a rough predictor of performance on an actual machine. On a real machine, different instructions may take significantly different amounts of time to execute; for example, a division instruction may take much longer than an addition instruction. Also, on a real machine, the behavior of the cache may significantly affect the time it takes to load or store the operands of an instruction. Finally, the precise running time of an algorithm given by a high-level description will depend on the quality of the translation of this algorithm into "machine code." However, despite all of these problems, it still turns out that measuring the running time on a RAM as we propose here is a good "first order" predictor of performance on real machines in many cases. Also, we shall only state the running time of an algorithm using a big-O estimate, so that implementation-specific constant factors are anyway "swept under the rug."

If we have an algorithm for solving a certain problem, we expect that "larger" instances of the problem will require more time to solve than "smaller" instances, and a general goal in the analysis of any algorithm is to estimate the rate of growth of the running time of the algorithm as a function of the size of its input. For this purpose, we shall simply measure the **size** of an input as the number of memory cells used to represent it. Theoretical computer scientists sometimes equate the notion of "efficient" with "polynomial time" (although not everyone takes theoretical computer scientists very seriously, especially on this point): a **polynomial-time algorithm** is one whose running time on inputs of size n is at most $an^b + c$, for some constants a, b, and c (a "real" theoretical computer scientist will write this as $n^{O(1)}$). Furthermore, we also require that for a polynomial-time algorithm, all numbers stored in memory are at most $a'n^{b'} + c'$ in absolute value, for some constants a', b', and c'. Even for algorithms that are not polynomial time, we shall insist that after executing t instructions, all numbers stored in memory are at most $a'(n + t)^{b'} + c'$ in absolute value, for some constants a', b', and c'.

Note that in defining the notion of polynomial time on a RAM, it is essential that we restrict the magnitude of numbers that may be stored in the machine's memory cells, as we have done above. Without this restriction, a program could perform arithmetic on huge numbers, being charged just one unit of time for each arithmetic operation—not only is this intuitively "wrong," it is possible to come up with programs that solve some problems using a polynomial number of arithmetic operations on huge numbers, and these problems cannot otherwise be solved in polynomial time (see §3.6).

3.3 Basic integer arithmetic

We will need algorithms for performing arithmetic on very large integers. Since such integers will exceed the word-size of actual machines, and to satisfy the formal requirements of our random access model of computation, we shall represent large integers as vectors of digits in some base B, along with a bit indicating the sign. That is, for $a \in \mathbb{Z}$, if we write

$$a = \pm \sum_{i=0}^{k-1} a_i B^i = \pm (a_{k-1} \cdots a_1 a_0)_B,$$

where $0 \leq a_i < B$ for $i = 0, \ldots, k - 1$, then a will be represented in memory as a data structure consisting of the vector of base-B digits $a_0, \ldots, a_{k-1}$, along with a "sign bit" to indicate the sign of a. To ensure a unique representation, if a is non-zero, then the high-order digit a_{k-1} in this representation should be non-zero.

For our purposes, we shall consider B to be a constant, and moreover, a power of 2. The choice of B as a power of 2 is convenient for a number of technical reasons.

A note to the reader: *If you are not interested in the low-level details of algorithms for integer arithmetic, or are willing to take them on faith, you may safely skip ahead to §3.3.5, where the results of this section are summarized.*

We now discuss in detail basic arithmetic algorithms for unsigned (i.e., non-negative) integers — these algorithms work with vectors of base-B digits, and except where explicitly noted, we do not assume that the high-order digits of the input vectors are non-zero, nor do these algorithms ensure that the high-order digit of the output vector is non-zero. These algorithms can be very easily adapted to deal with arbitrary signed integers, and to take proper care that the high-order digit of the vector representing a non-zero number is itself non-zero (the reader is asked to fill in these details in some of the exercises below). All of these algorithms can be implemented directly in a programming language that provides a "built-in" signed integer type that can represent all integers of absolute value less than B^2, and that supports the basic arithmetic operations (addition, subtraction, multiplication, integer division). So, for example, using the C or *Java* programming language's int type on a typical 32-bit computer, we could take $B = 2^{15}$. The resulting software would be reasonably efficient and portable, but certainly not the fastest possible.

Suppose we have the base-B representations of two unsigned integers a and b. We present algorithms to compute the base-B representation of $a + b$, $a - b$, $a \cdot b$, $\lfloor a/b \rfloor$, and $a \bmod b$. To simplify the presentation, for integers x, y with $y \neq 0$, we denote by QuoRem(x, y) the quotient/remainder pair $(\lfloor x/y \rfloor, x \bmod y)$.

3.3.1 Addition

Let $a = (a_{k-1} \cdots a_0)_B$ and $b = (b_{\ell-1} \cdots b_0)_B$ be unsigned integers. Assume that $k \geq \ell \geq 1$ (if $k < \ell$, then we can just swap a and b). The sum $c := a + b$ is of the form $c = (c_k c_{k-1} \cdots c_0)_B$. Using the standard "paper-and-pencil" method (adapted from base-10 to base-B, of course), we can compute the base-B representation of $a + b$ in time $O(k)$, as follows:

> *carry* $\leftarrow 0$
> for $i \leftarrow 0$ to $\ell - 1$ do
>   *tmp* $\leftarrow a_i + b_i + $ *carry*, (*carry*, c_i) $\leftarrow$ QuoRem(*tmp*, B)
> for $i \leftarrow \ell$ to $k - 1$ do
>   *tmp* $\leftarrow a_i + $ *carry*, (*carry*, c_i) $\leftarrow$ QuoRem(*tmp*, B)
> $c_k \leftarrow$ *carry*

Note that in every loop iteration, the value of *carry* is 0 or 1, and the value *tmp* lies between 0 and $2B - 1$.

3.3.2 Subtraction

Let $a = (a_{k-1} \cdots a_0)_B$ and $b = (b_{\ell-1} \cdots b_0)_B$ be unsigned integers. Assume that $k \geq \ell \geq 1$. To compute the difference $c := a - b$, we may use the same algorithm as above, but with the expression "$a_i + b_i$" replaced by "$a_i - b_i$." In every loop iteration, the value of *carry* is 0 or -1, and the value of *tmp* lies between $-B$ and $B - 1$. If $a \geq b$, then $c_k = 0$ (i.e., there is no carry out of the last loop iteration); otherwise, $c_k = -1$ (and $b - a = B^k - (c_{k-1} \cdots c_0)_B$, which can be computed with another execution of the subtraction routine).

3.3.3 Multiplication

Let $a = (a_{k-1} \cdots a_0)_B$ and $b = (b_{\ell-1} \cdots b_0)_B$ be unsigned integers, with $k \geq 1$ and $\ell \geq 1$. The product $c := a \cdot b$ is of the form $(c_{k+\ell-1} \cdots c_0)_B$, and may be computed in time $O(k\ell)$ as follows:

> for $i \leftarrow 0$ to $k + \ell - 1$ do $c_i \leftarrow 0$
> for $i \leftarrow 0$ to $k - 1$ do
>   *carry* $\leftarrow 0$
>   for $j \leftarrow 0$ to $\ell - 1$ do
>   *tmp* $\leftarrow a_i b_j + c_{i+j} + $ *carry*
>   (*carry*, c_{i+j}) $\leftarrow$ QuoRem(*tmp*, B)
> $c_{i+\ell} \leftarrow$ *carry*

Note that at every step in the above algorithm, the value of *carry* lies between 0 and $B - 1$, and the value of *tmp* lies between 0 and $B^2 - 1$.

3.3.4 Division with remainder

Let $a = (a_{k-1} \cdots a_0)_B$ and $b = (b_{\ell-1} \cdots b_0)_B$ be unsigned integers, with $k \geq 1$, $\ell \geq 1$, and $b_{\ell-1} \neq 0$. We want to compute q and r such that $a = bq + r$ and $0 \leq r < b$. Assume that $k \geq \ell$; otherwise, $a < b$, and we can just set $q \leftarrow 0$ and $r \leftarrow a$. The quotient q will have at most $m := k - \ell + 1$ base-B digits. Write $q = (q_{m-1} \cdots q_0)_B$.

At a high level, the strategy we shall use to compute q and r is the following:

$r \leftarrow a$
for $i \leftarrow m - 1$ down to 0 do
$\qquad q_i \leftarrow \lfloor r / B^i b \rfloor$
$\qquad r \leftarrow r - B^i \cdot q_i b$

One easily verifies by induction that at the beginning of each loop iteration, we have $0 \leq r < B^{i+1} b$, and hence each q_i will be between 0 and $B - 1$, as required.

Turning the above strategy into a detailed algorithm takes a bit of work. In particular, we want an easy way to compute $\lfloor r / B^i b \rfloor$. Now, we could in theory just try all possible choices for q_i — this would take time $O(B\ell)$, and viewing B as a constant, this is $O(\ell)$. However, this is not really very desirable from either a practical or theoretical point of view, and we can do much better with just a little effort.

We shall first consider a special case; namely, the case where $\ell = 1$. In this case, the computation of the quotient $\lfloor r / B^i b \rfloor$ is facilitated by the following theorem, which essentially tells us that this quotient is determined by the two high-order digits of r:

Theorem 3.1. *Let x and y be integers such that*

$$0 \leq x = x'2^n + s \ \text{ and } \ 0 < y = y'2^n$$

for some integers n, s, x', y', with $n \geq 0$ and $0 \leq s < 2^n$. Then $\lfloor x/y \rfloor = \lfloor x'/y' \rfloor$.

Proof. We have

$$\frac{x}{y} = \frac{x'}{y'} + \frac{s}{y'2^n} \geq \frac{x'}{y'}.$$

It follows immediately that $\lfloor x/y \rfloor \geq \lfloor x'/y' \rfloor$.

We also have

$$\frac{x}{y} = \frac{x'}{y'} + \frac{s}{y'2^n} < \frac{x'}{y'} + \frac{1}{y'} \leq \left(\left\lfloor \frac{x'}{y'} \right\rfloor + \frac{y'-1}{y'} \right) + \frac{1}{y'} \leq \left\lfloor \frac{x'}{y'} \right\rfloor + 1.$$

Thus, we have $x/y < \lfloor x'/y' \rfloor + 1$, and hence, $\lfloor x/y \rfloor \leq \lfloor x'/y' \rfloor$. $\square$

From this theorem, one sees that the following algorithm correctly computes the quotient and remainder in time $O(k)$ (in the case $\ell = 1$):

> $hi \leftarrow 0$
> for $i \leftarrow k - 1$ down to 0 do
> $tmp \leftarrow hi \cdot B + a_i$
> $(q_i, hi) \leftarrow \text{QuoRem}(tmp, b_0)$
> output the quotient $q = (q_{k-1} \cdots q_0)_B$ and the remainder hi

Note that in every loop iteration, the value of hi lies between 0 and $b_0 \leq B - 1$, and the value of tmp lies between 0 and $B \cdot b_0 + (B - 1) \leq B^2 - 1$.

That takes care of the special case where $\ell = 1$. Now we turn to the general case $\ell \geq 1$. In this case, we cannot so easily get the digits q_i of the quotient, but we can still fairly easily estimate these digits, using the following:

Theorem 3.2. *Let x and y be integers such that*

$$0 \leq x = x'2^n + s \quad and \quad 0 < y = y'2^n + t$$

for some integers n, s, t, x', y' with $n \geq 0$, $0 \leq s < 2^n$, and $0 \leq t < 2^n$. Further, suppose that $2y' \geq x/y$. Then

$$\lfloor x/y \rfloor \leq \lfloor x'/y' \rfloor \leq \lfloor x/y \rfloor + 2.$$

Proof. We have $x/y \leq x/y'2^n$, and so $\lfloor x/y \rfloor \leq \lfloor x/y'2^n \rfloor$, and by the previous theorem, $\lfloor x/y'2^n \rfloor = \lfloor x'/y' \rfloor$. That proves the first inequality.

For the second inequality, first note that from the definitions, we have $x/y \geq x'/(y'+1)$, which implies $x'y - xy' - x \leq 0$. Further, $2y' \geq x/y$ implies $2yy' - x \geq 0$. So we have $2yy' - x \geq 0 \geq x'y - xy' - x$, which implies $x/y \geq x'/y' - 2$, and hence $\lfloor x/y \rfloor \geq \lfloor x'/y' \rfloor - 2$. $\square$

Based on this theorem, we first present an algorithm for division with remainder that works if we assume that b is appropriately "normalized," meaning that $b_{\ell-1} \geq 2^{w-1}$, where $B = 2^w$. This algorithm is shown in Fig. 3.1.

Some remarks are in order.

1. In line 4, we compute q_i, which by Theorem 3.2 is greater than or equal to the true quotient digit, but exceeds this value by at most 2.

2. In line 5, we reduce q_i if it is obviously too big.

3. In lines 6–10, we compute

$$(r_{i+\ell} \cdots r_i)_B \leftarrow (r_{i+\ell} \cdots r_i)_B - q_i b.$$

In each loop iteration, the value of tmp lies between $-(B^2 - B)$ and $B - 1$, and the value *carry* lies between $-(B - 1)$ and 0.

1.   for $i \leftarrow 0$ to $k - 1$ do $r_i \leftarrow a_i$
2. $r_k \leftarrow 0$
3.   for $i \leftarrow k - \ell$ down to 0 do
4. $q_i \leftarrow \lfloor (r_{i+\ell} B + r_{i+\ell-1})/b_{\ell-1} \rfloor$
5.   if $q_i \geq B$ then $q_i \leftarrow B - 1$
6. $carry \leftarrow 0$
7.   for $j \leftarrow 0$ to $\ell - 1$ do
8. $tmp \leftarrow r_{i+j} - q_i b_j + carry$
9. $(carry, r_{i+j}) \leftarrow \text{QuoRem}(tmp, B)$
10. $r_{i+\ell} \leftarrow r_{i+\ell} + carry$
11.   while $r_{i+\ell} < 0$ do
12. $carry \leftarrow 0$
13.   for $j \leftarrow 0$ to $\ell - 1$ do
14. $tmp \leftarrow r_{i+j} + b_i + carry$
15. $(carry, r_{i+j}) \leftarrow \text{QuoRem}(tmp, B)$
16. $r_{i+\ell} \leftarrow r_{i+\ell} + carry$
17. $q_i \leftarrow q_i - 1$
18. output the quotient $q = (q_{k-\ell} \cdots q_0)_B$
          and the remainder $r = (r_{\ell-1} \cdots r_0)_B$

Fig. 3.1. Division with Remainder Algorithm

4. If the estimate q_i is too large, this is manifested by a negative value of $r_{i+\ell}$ at line 10. Lines 11–17 detect and correct this condition: the loop body here executes at most twice; in lines 12–16, we compute

$$(r_{i+\ell} \cdots r_i)_B \leftarrow (r_{i+\ell} \cdots r_i)_B + (b_{\ell-1} \cdots b_0)_B.$$

Just as in the algorithm in §3.3.1, in every iteration of the loop in lines 13–15, the value of *carry* is 0 or 1, and the value *tmp* lies between 0 and $2B - 1$.

It is easily verified that the running time of the above algorithm is $O(\ell \cdot (k - \ell + 1))$.

Finally, consider the general case, where b may not be normalized. We multiply both a and b by an appropriate value $2^{w'}$, with $0 \leq w' < w$, obtaining $a' := a2^{w'}$ and $b' := b2^{w'}$, where b' is normalized; alternatively, we can use a more efficient, special-purpose "left shift" algorithm to achieve the same effect. We then compute q and r' such that $a' = b'q + r'$, using the division algorithm in Fig. 3.1. Observe that $q = \lfloor a'/b' \rfloor = \lfloor a/b \rfloor$, and $r' = r2^{w'}$, where $r = a \bmod b$. To recover r, we simply divide r' by $2^{w'}$, which we can do either using the above "single precision" division algorithm, or by using a special-purpose "right shift" algorithm. All of this normalizing and denormalizing takes time $O(k + \ell)$. Thus, the total running time for division with remainder is still $O(\ell \cdot (k - \ell + 1))$.

EXERCISE 3.17. Work out the details of algorithms for arithmetic on *signed* integers, using the above algorithms for unsigned integers as subroutines. You should give algorithms for addition, subtraction, multiplication, and division with remainder of arbitrary signed integers (for division with remainder, your algorithm should compute $\lfloor a/b \rfloor$ and $a \bmod b$). Make sure your algorithms correctly compute the sign bit of the results, and also strip any leading zero digits from the results.

EXERCISE 3.18. Work out the details of an algorithm that compares two *signed* integers a and b, determining which of $a < b$, $a = b$, or $a > b$ holds.

EXERCISE 3.19. Suppose that we run the division with remainder algorithm in Fig. 3.1 for $\ell > 1$ without normalizing b, but instead, we compute the value q_i in line 4 as follows:

$$q_i \leftarrow \lfloor (r_{i+\ell} B^2 + r_{i+\ell-1} B + r_{i+\ell-2})/(b_{\ell-1} B + b_{\ell-2}) \rfloor.$$

Show that q_i is either equal to the correct quotient digit, or the correct quotient digit plus 1. Note that a limitation of this approach is that the numbers involved in the computation are larger than B^2.

EXERCISE 3.20. Work out the details for an algorithm that shifts a given unsigned integer a to the left by a specified number of bits s (i.e., computes $b := a \cdot 2^s$). The running time of your algorithm should be linear in the number of digits of the output.

EXERCISE 3.21. Work out the details for an algorithm that shifts a given unsigned integer a to the right by a specified number of bits s (i.e., computes $b := \lfloor a/2^s \rfloor$). The running time of your algorithm should be linear in the number of digits of the output. Now modify your algorithm so that it correctly computes $\lfloor a/2^s \rfloor$ for *signed* integers a.

EXERCISE 3.22. This exercise is for *C/Java* programmers. Evaluate the *C/Java* expressions

```
(-17) % 4;   (-17) & 3;
```

and compare these values with $(-17) \bmod 4$. Also evaluate the *C/Java* expressions

```
(-17) / 4;   (-17) >> 2;
```

and compare with $\lfloor -17/4 \rfloor$. Explain your findings.

EXERCISE 3.23. This exercise is also for *C/Java* programmers. Suppose that values of type int are stored using a 32-bit 2's complement representation, and that all basic arithmetic operations are computed correctly modulo 2^{32}, even if an "overflow" happens to occur. Also assume that double precision floating point has 53 bits of precision, and that all basic arithmetic operations give a result with a relative error of at most 2^{-53}. Also assume that conversion from type int to double is exact, and that conversion from double to int truncates the fractional part. Now, suppose we are given int variables a, b, and n, such that $1 < n < 2^{30}$, $0 \le a < n$, and $0 \le b < n$. Show that after the following code sequence is executed, the value of r is equal to $(a \cdot b) \bmod n$:

```
int q;
q  = (int) ((((double) a) * ((double) b)) / ((double) n));
r = a*b - q*n;
if (r >= n)
   r = r - n;
else if (r < 0)
   r = r + n;
```

3.3.5 Summary

We now summarize the results of this section. For an integer a, we define its **bit length**, or simply, its **length**, which we denote by len(a), to be the number of bits in the binary representation of $|a|$; more precisely,

$$\text{len}(a) := \begin{cases} \lfloor \log_2 |a| \rfloor + 1 & \text{if } a \ne 0, \\ 1 & \text{if } a = 0. \end{cases}$$

If len(a) = ℓ, we say that a is an ℓ-**bit integer**. Notice that if a is a positive, ℓ-bit integer, then $\log_2 a < \ell \le \log_2 a + 1$, or equivalently, $2^{\ell-1} \le a < 2^{\ell}$.

Assuming that arbitrarily large integers are represented as described at the beginning of this section, with a sign bit and a vector of base-B digits, where B is a constant power of 2, we may state the following theorem.

Theorem 3.3. *Let a and b be arbitrary integers.*

(i) *We can compute $a \pm b$ in time $O(\text{len}(a) + \text{len}(b))$.*

(ii) *We can compute $a \cdot b$ in time $O(\text{len}(a)\,\text{len}(b))$.*

(iii) *If $b \ne 0$, we can compute the quotient $q := \lfloor a/b \rfloor$ and the remainder $r := a \bmod b$ in time $O(\text{len}(b)\,\text{len}(q))$.*

Note the bound $O(\text{len}(b)\,\text{len}(q))$ in part (iii) of this theorem, which may be significantly less than the bound $O(\text{len}(a)\,\text{len}(b))$. A good way to remember this bound is as follows: the time to compute the quotient and remainder is roughly the same as the time to compute the product bq appearing in the equality $a = bq + r$.

This theorem does not explicitly refer to the base B in the underlying implementation. The choice of B affects the values of the implied big-O constants; while in theory, this is of no significance, it does have a significant impact in practice.

From now on, we shall (for the most part) not worry about the implementation details of long-integer arithmetic, and will just refer directly to this theorem. However, we will occasionally exploit some trivial aspects of our data structure for representing large integers. For example, it is clear that in constant time, we can determine the sign of a given integer a, the bit length of a, and any particular bit of the binary representation of a; moreover, as discussed in Exercises 3.20 and 3.21, multiplications and divisions by powers of 2 can be computed in linear time via "left shifts" and "right shifts." It is also clear that we can convert between the base-2 representation of a given integer and our implementation's internal representation in linear time (other conversions may take longer—see Exercise 3.32).

We wish to stress the point that efficient algorithms on large integers should run in time bounded by a polynomial in the *bit lengths* of the inputs, rather than their *magnitudes*. For example, if the input to an algorithm is an ℓ-bit integer n, and if the algorithm runs in time $O(\ell^2)$, it will easily be able to process 1000-bit inputs in a reasonable amount of time (a fraction of a second) on a typical, modern computer. However, if the algorithm runs in time, say, $O(n^{1/2})$, this means that on 1000-bit inputs, it will take roughly 2^{500} computing steps, which even on the fastest computer available today or in the foreseeable future, will still be running long after our solar system no longer exists.

> **A note on notation: "len" and "log."** In expressing the running times of algorithms in terms of an input a, we generally prefer to write $\text{len}(a)$ rather than $\log a$. One reason is esthetic: writing $\text{len}(a)$ stresses the fact that the running time is a function of the bit length of a. Another reason is technical: for big-O estimates involving functions on an arbitrary domain, the appropriate inequalities should hold throughout the domain, and for this reason, it is very inconvenient to use functions, like log, which vanish or are undefined on some inputs.

EXERCISE 3.24. Let $a, b \in \mathbb{Z}$ with $a \geq b > 0$, and let $q := \lfloor a/b \rfloor$. Show that $\text{len}(a) - \text{len}(b) - 1 \leq \text{len}(q) \leq \text{len}(a) - \text{len}(b) + 1$.

EXERCISE 3.25. Let $n_1, \ldots, n_k$ be positive integers. Show that

$$\sum_{i=1}^{k} \text{len}(n_i) - k \le \text{len}\left(\prod_{i=1}^{k} n_i\right) \le \sum_{i=1}^{k} \text{len}(n_i).$$

EXERCISE 3.26. Show that given integers $n_1, \ldots, n_k$, with each $n_i > 1$, we can compute the product $n := \prod_i n_i$ in time $O(\text{len}(n)^2)$.

EXERCISE 3.27. Show that given integers $a, n_1, \ldots, n_k$, with each $n_i > 1$, where $0 \le a < n := \prod_i n_i$, we can compute $(a \bmod n_1, \ldots, a \bmod n_k)$ in time $O(\text{len}(n)^2)$.

EXERCISE 3.28. Show that given integers $n_1, \ldots, n_k$, with each $n_i > 1$, we can compute $(n/n_1, \ldots, n/n_k)$, where $n := \prod_i n_i$, in time $O(\text{len}(n)^2)$.

EXERCISE 3.29. This exercise develops an algorithm to compute $\lfloor \sqrt{n} \rfloor$ for a given positive integer n. Consider the following algorithm:

$k \leftarrow \lfloor (\text{len}(n) - 1)/2 \rfloor, \ m \leftarrow 2^k$
for $i \leftarrow k - 1$ down to 0 do
      if $(m + 2^i)^2 \le n$ then $m \leftarrow m + 2^i$
output m

(a) Show that this algorithm correctly computes $\lfloor \sqrt{n} \rfloor$.

(b) In a straightforward implementation of this algorithm, each loop iteration takes time $O(\text{len}(n)^2)$, yielding a total running time of $O(\text{len}(n)^3)$. Give a more careful implementation, so that each loop iteration takes time $O(\text{len}(n))$, yielding a total running time is $O(\text{len}(n)^2)$.

EXERCISE 3.30. Modify the algorithm in the previous exercise so that given positive integers n and e, with $n \ge 2^e$, it computes $\lfloor n^{1/e} \rfloor$ in time $O(\text{len}(n)^3/e)$.

EXERCISE 3.31. An integer $n > 1$ is called a **perfect power** if $n = a^b$ for some integers $a > 1$ and $b > 1$. Using the algorithm from the previous exercise, design an efficient algorithm that determines if a given n is a perfect power, and if it is, also computes a and b such that $n = a^b$, where $a > 1$, $b > 1$, and a is as small as possible. Your algorithm should run in time $O(\ell^3 \text{len}(\ell))$, where $\ell := \text{len}(n)$.

EXERCISE 3.32. Show how to convert (in both directions) in time $O(\text{len}(n)^2)$ between the base-10 representation and our implementation's internal representation of an integer n.

3.4 Computing in $\mathbb{Z}_n$

Let n be a positive integer. For every $\alpha \in \mathbb{Z}_n$, there exists a unique integer $a \in \{0, \ldots, n - 1\}$ such that $\alpha = [a]_n$; we call this integer a the **canonical**

representative of α, and denote it by rep(α). For computational purposes, we represent elements of $\mathbb{Z}_n$ by their canonical representatives.

Addition and subtraction in $\mathbb{Z}_n$ can be performed in time $O(\text{len}(n))$: given $\alpha, \beta \in \mathbb{Z}_n$, to compute rep($\alpha + \beta$), we first compute the integer sum rep(α) + rep(β), and then subtract n if the result is greater than or equal to n; similarly, to compute rep($\alpha - \beta$), we compute the integer difference rep(α) $-$ rep(β), adding n if the result is negative. Multiplication in $\mathbb{Z}_n$ can be performed in time $O(\text{len}(n)^2)$: given $\alpha, \beta \in \mathbb{Z}_n$, we compute rep($\alpha \cdot \beta$) as rep($\alpha$) rep($\beta$) mod n, using one integer multiplication and one division with remainder.

> **A note on notation: "rep," "mod," and "$[\cdot]_n$."** In describing algorithms, as well as in other contexts, if α, β are elements of $\mathbb{Z}_n$, we may write, for example, $\gamma \leftarrow \alpha + \beta$ or $\gamma \leftarrow \alpha\beta$, and it is understood that elements of $\mathbb{Z}_n$ are represented by their canonical representatives as discussed above, and arithmetic on canonical representatives is done modulo n. Thus, we have in mind a "strongly typed" language for our pseudo-code that makes a clear distinction between integers in the set $\{0, \ldots, n-1\}$ and elements of $\mathbb{Z}_n$. If $a \in \mathbb{Z}$, we can convert a to an object $\alpha \in \mathbb{Z}_n$ by writing $\alpha \leftarrow [a]_n$, and if $a \in \{0, \ldots, n-1\}$, this type conversion is purely conceptual, involving no actual computation. Conversely, if $\alpha \in \mathbb{Z}_n$, we can convert α to an object $a \in \{0, \ldots, n-1\}$, by writing $a \leftarrow \text{rep}(\alpha)$; again, this type conversion is purely conceptual, and involves no actual computation. It is perhaps also worthwhile to stress the distinction between $a \bmod n$ and $[a]_n$—the former denotes an element of the set $\{0, \ldots, n-1\}$, while the latter denotes an element of $\mathbb{Z}_n$.

Another interesting problem is exponentiation in $\mathbb{Z}_n$: given $\alpha \in \mathbb{Z}_n$ and a non-negative integer e, compute $\alpha^e \in \mathbb{Z}_n$. Perhaps the most obvious way to do this is to iteratively multiply by α a total of e times, requiring time $O(e\ \text{len}(n)^2)$. For small values of e, this is fine; however, a much faster algorithm, the **repeated-squaring algorithm**, computes α^e using just $O(\text{len}(e))$ multiplications in $\mathbb{Z}_n$, thus taking time $O(\text{len}(e)\ \text{len}(n)^2)$.

This method is based on the following observation. Let $e = (b_{\ell-1} \cdots b_0)_2$ be the binary expansion of e (where b_0 is the low-order bit). For $i = 0, \ldots, \ell$, define $e_i := \lfloor e/2^i \rfloor$; the binary expansion of e_i is $e_i = (b_{\ell-1} \cdots b_i)_2$. Also define $\beta_i := \alpha^{e_i}$ for $i = 0, \ldots, \ell$, so $\beta_\ell = 1$ and $\beta_0 = \alpha^e$. Then we have

$$e_i = 2e_{i+1} + b_i \quad \text{and} \quad \beta_i = \beta_{i+1}^2 \cdot \alpha^{b_i} \quad \text{for } i = 0, \ldots, \ell - 1.$$

This observation yields the following algorithm for computing α^e:

The repeated-squaring algorithm. On input α, e, where $\alpha \in \mathbb{Z}_n$ and e is a non-negative integer, do the following, where $e = (b_{\ell-1} \cdots b_0)_2$ is the binary expansion of e:

$$\beta \leftarrow [1]_n$$
for $i \leftarrow \ell - 1$ down to 0 do
$$\beta \leftarrow \beta^2$$
if $b_i = 1$ then $\beta \leftarrow \beta \cdot \alpha$
output β

It is clear that when this algorithm terminates, we have $\beta = \alpha^e$, and that the running-time estimate is as claimed above. Indeed, the algorithm uses ℓ squarings in $\mathbb{Z}_n$, and at most ℓ additional multiplications in $\mathbb{Z}_n$.

Example 3.5. Suppose $e = 37 = (100101)_2$. The above algorithm performs the following operations in this case:

	// computed exponent (in binary)
$\beta \leftarrow [1]$	// 0
$\beta \leftarrow \beta^2, \beta \leftarrow \beta \cdot \alpha$	// 1
$\beta \leftarrow \beta^2$	// 10
$\beta \leftarrow \beta^2$	// 100
$\beta \leftarrow \beta^2, \beta \leftarrow \beta \cdot \alpha$	// 1001
$\beta \leftarrow \beta^2$	// 10010
$\beta \leftarrow \beta^2, \beta \leftarrow \beta \cdot \alpha$	// 100101 . □

The repeated-squaring algorithm has numerous applications. We mention a few here, but we will see many more later on.

Computing multiplicative inverses in $\mathbb{Z}_p$. Suppose we are given a prime p and an element $\alpha \in \mathbb{Z}_p^*$, and we want to compute α^{-1}. By Euler's theorem (Theorem 2.13), we have $\alpha^{p-1} = 1$, and multiplying this equation by α^{-1}, we obtain $\alpha^{p-2} = \alpha^{-1}$. Thus, we can use the repeated-squaring algorithm to compute α^{-1} by raising α to the power $p - 2$. This algorithm runs in time $O(\text{len}(p)^3)$. While this is reasonably efficient, we will develop an even more efficient method in the next chapter, using Euclid's algorithm (which also works with any modulus, not just a prime modulus).

Testing quadratic residuosity. Suppose we are given an odd prime p and an element $\alpha \in \mathbb{Z}_p^*$, and we want to test whether $\alpha \in (\mathbb{Z}_p^*)^2$. By Euler's criterion (Theorem 2.21), we have $\alpha \in (\mathbb{Z}_p^*)^2$ if and only if $\alpha^{(p-1)/2} = 1$. Thus, we can use the repeated-squaring algorithm to test if $\alpha \in (\mathbb{Z}_p^*)^2$ by raising α to the power $(p - 1)/2$. This algorithm runs in time $O(\text{len}(p)^3)$. While this is also reasonably efficient, we will develop an even more efficient method later in the text (in Chapter 12).

Testing for primality. Suppose we are given an integer $n > 1$, and we want to determine whether n is prime or composite. For large n, searching for prime factors of n is hopelessly impractical. A better idea is to use Euler's theorem, combined with the repeated-squaring algorithm: we know that if n is prime, then every non-zero $\alpha \in \mathbb{Z}_n$ satisfies $\alpha^{n-1} = 1$. Conversely, if n is composite, there exists a non-zero $\alpha \in \mathbb{Z}_n$ such that $\alpha^{n-1} \neq 1$ (see Exercise 2.27). This suggests the following "trial and error" strategy for testing if n is prime:

> repeat k times
>> choose $\alpha \in \mathbb{Z}_n \setminus \{[0]\}$
>> compute $\beta \leftarrow \alpha^{n-1}$
>> if $\beta \neq 1$ output "composite" and halt
>
> output "maybe prime"

As stated, this is not a fully specified algorithm: we have to specify the loop-iteration parameter k, and more importantly, we have to specify a procedure for choosing α in each loop iteration. One approach might be to just try $\alpha = [1], [2], [3], \ldots$. Another might be to choose α at random in each loop iteration: this would be an example of a *probabilistic algorithm* (a notion we shall discuss in detail in Chapter 9). In any case, if the algorithm outputs "composite," we may conclude that n is composite (even though the algorithm does not find a non-trivial factor of n). However, if the algorithm completes all k loop iterations and outputs "maybe prime," it is not clear what we should conclude: certainly, we have some reason to suspect that n is prime, but not really a proof; indeed, it may be the case that n is composite, but we were just unlucky in all of our choices for α. Thus, while this rough idea does not quite give us an effective primality test, it is not a bad start, and is the basis for several effective primality tests (a couple of which we shall discuss in detail in Chapters 10 and 21).

EXERCISE 3.33. The repeated-squaring algorithm we have presented here processes the bits of the exponent from left to right (i.e., from high order to low order). Develop an algorithm for exponentiation in $\mathbb{Z}_n$ with similar complexity that processes the bits of the exponent from right to left.

EXERCISE 3.34. Show that given a prime p, $\alpha \in \mathbb{Z}_p$, and an integer $e \geq p$, we can compute α^e in time $O(\text{len}(e)\,\text{len}(p) + \text{len}(p)^3)$.

The following exercises develop some important efficiency improvements to the basic repeated-squaring algorithm.

EXERCISE 3.35. The goal of this exercise is to develop a "2^t-ary" variant of the above repeated-squaring algorithm, in which the exponent is effectively treated as a number in base 2^t, for some parameter t, rather than in base 2. Let $\alpha \in \mathbb{Z}_n$ and let e be a positive integer of length ℓ. Let us write e in base 2^t as $e = (e_k \cdots e_0)_{2^t}$, where $e_k \neq 0$. Consider the following algorithm:

compute a table of values $T[0 \ldots 2^t - 1]$,
      where $T[j] := \alpha^j$ for $j = 0, \ldots, 2^t - 1$
$\beta \leftarrow T[e_k]$
for $i \leftarrow k - 1$ down to 0 do
      $\beta \leftarrow \beta^{2^t} \cdot T[e_i]$

(a) Show that this algorithm correctly computes α^e, and work out the implementation details; in particular, show that it may be implemented in such a
way that it uses at most ℓ squarings and $2^t + \ell/t + O(1)$ additional multiplications in $\mathbb{Z}_n$.

(b) Show that, by appropriately choosing the parameter t, we can bound the
number of multiplications in $\mathbb{Z}_n$ (besides the squarings) by $O(\ell/\operatorname{len}(\ell))$.
Thus, from an asymptotic point of view, the cost of exponentiation is essen-
tially the cost of about ℓ squarings in $\mathbb{Z}_n$.

(c) Improve the algorithm so that it only uses no more than ℓ squarings and
$2^{t-1} + \ell/t + O(1)$ additional multiplications in $\mathbb{Z}_n$. Hint: build a table that
contains only the *odd* powers of α among $\alpha^0, \alpha^1, \ldots, \alpha^{2^t - 1}$.

EXERCISE 3.36. Suppose we are given $\alpha_1, \ldots, \alpha_k \in \mathbb{Z}_n$, along with non-negative
integers $e_1, \ldots, e_k$, where $\operatorname{len}(e_i) \leq \ell$ for $i = 1, \ldots, k$. Show how to compute
$\beta := \alpha_1^{e_1} \cdots \alpha_k^{e_k}$, using at most ℓ squarings and $\ell + 2^k$ additional multiplications
in $\mathbb{Z}_n$. Your algorithm should work in two phases: the first phase uses only the
values $\alpha_1, \ldots, \alpha_k$, and performs at most 2^k multiplications in $\mathbb{Z}_n$; in the second
phase, the algorithm computes β, using the exponents $e_1, \ldots, e_k$, along with the
data computed in the first phase, and performs at most ℓ squarings and ℓ additional
multiplications in $\mathbb{Z}_n$.

EXERCISE 3.37. Suppose that we are to compute α^e, where $\alpha \in \mathbb{Z}_n$, for many
exponents e of length at most ℓ, but with α fixed. Show that for every positive
integer parameter k, we can make a pre-computation (depending on α, ℓ, and k)
that uses at most ℓ squarings and 2^k additional multiplications in $\mathbb{Z}_n$, so that after
the pre-computation, we can compute α^e for every exponent e of length at most ℓ
using at most $\ell/k + O(1)$ squarings and $\ell/k + O(1)$ additional multiplications in
$\mathbb{Z}_n$. Hint: use the algorithm in the previous exercise.

EXERCISE 3.38. Suppose we are given $\alpha \in \mathbb{Z}_n$, along with non-negative integers
$e_1, \ldots, e_r$, where $\operatorname{len}(e_i) \leq \ell$ for $i = 1, \ldots, r$, and $r = O(\operatorname{len}(\ell))$. Using the
previous exercise, show how to compute $(\alpha^{e_1}, \ldots, \alpha^{e_r})$ using $O(\ell)$ multiplications
in $\mathbb{Z}_n$.

EXERCISE 3.39. Suppose we are given $\alpha \in \mathbb{Z}_n$, along with integers $m_1, \ldots, m_r$, with each $m_i > 1$. Let $m := \prod_i m_i$. Also, for $i = 1, \ldots, r$, let $m_i^* := m/m_i$. Show how to compute $(\alpha^{m_1^*}, \ldots, \alpha^{m_r^*})$ using $O(\text{len}(r)\ell)$ multiplications in $\mathbb{Z}_n$, where $\ell := \text{len}(m)$. Hint: divide and conquer. Note that if $r = O(\text{len}(\ell))$, then using the previous exercise, we can solve this problem using just $O(\ell)$ multiplications.

EXERCISE 3.40. Let k be a *constant*, positive integer. Suppose we are given $\alpha_1, \ldots, \alpha_k \in \mathbb{Z}_n$, along with non-negative integers $e_1, \ldots, e_k$, where $\text{len}(e_i) \leq \ell$ for $i = 1, \ldots, k$. Show how to compute the value $\alpha_1^{e_1} \cdots \alpha_k^{e_k}$, using at most ℓ squarings and $O(\ell/\text{len}(\ell))$ additional multiplications in $\mathbb{Z}_n$. Hint: develop a 2^t-ary version of the algorithm in Exercise 3.36.

3.5 Faster integer arithmetic (∗)

The quadratic-time algorithms presented in §3.3 for integer multiplication and division are by no means the fastest possible. The next exercise develops a faster multiplication algorithm.

EXERCISE 3.41. Suppose we have two positive integers a and b, each of length at most ℓ, such that $a = a_1 2^k + a_0$ and $b = b_1 2^k + b_0$, where $0 \leq a_0 < 2^k$ and $0 \leq b_0 < 2^k$. Then

$$ab = a_1 b_1 2^{2k} + (a_0 b_1 + a_1 b_0) 2^k + a_0 b_0.$$

Show how to compute the product ab in time $O(\ell)$, given the products $a_0 b_0$, $a_1 b_1$, and $(a_0 - a_1)(b_0 - b_1)$. From this, design a recursive algorithm that computes ab in time $O(\ell^{\log_2 3})$. (Note that $\log_2 3 \approx 1.58$.)

The algorithm in the previous exercise is also not the best possible. In fact, it is possible to multiply two integers of length at most ℓ *on a RAM* in time $O(\ell)$, but we do not explore this any further for the moment (see §3.6).

The following exercises explore the relationship between integer multiplication and related problems. We assume that we have an algorithm that multiplies two integers of length at most ℓ in time at most $M(\ell)$. It is convenient (and reasonable) to assume that M is a **well-behaved complexity function**. By this, we mean that M maps positive integers to positive real numbers, such that for some constant $\gamma \geq 1$, and all positive integers a and b, we have

$$1 \leq \frac{M(a+b)}{M(a) + M(b)} \leq \gamma.$$

EXERCISE 3.42. Show that if M is a well-behaved complexity function, then it is strictly increasing.

EXERCISE 3.43. Show that if $N(\ell) := M(\ell)/\ell$ is a non-decreasing function, and $M(2\ell)/M(\ell) = O(1)$, then M is a well-behaved complexity function.

EXERCISE 3.44. Let $\alpha > 0$, $\beta \geq 1$, $\gamma \geq 0$, $\delta \geq 0$ be real constants. Show that

$$M(\ell) := \alpha\ell^\beta \operatorname{len}(\ell)^\gamma \operatorname{len}(\operatorname{len}(\ell))^\delta$$

is a well-behaved complexity function.

EXERCISE 3.45. Show that given integers $n > 1$ and $e > 1$, we can compute n^e in time $O(M(\operatorname{len}(n^e)))$.

EXERCISE 3.46. Give an algorithm for Exercise 3.26 whose running time is $O(M(\operatorname{len}(n))\operatorname{len}(k))$. Hint: divide and conquer.

EXERCISE 3.47. In the previous exercise, suppose all the inputs n_i have the same length, and that $M(\ell) = \alpha\ell^\beta$, where α and β are constants with $\alpha > 0$ and $\beta > 1$. Show that your algorithm runs in time $O(M(\operatorname{len}(n)))$.

EXERCISE 3.48. We can represent a "floating point" number $\hat{z}$ as a pair (a, e), where a and e are integers — the value of $\hat{z}$ is the rational number $a2^e$, and we call $\operatorname{len}(a)$ the **precision** of $\hat{z}$. We say that $\hat{z}$ is a k-**bit approximation** of a real number z if $\hat{z}$ has precision k and $\hat{z} = (1 + \varepsilon)z$ for some $|\varepsilon| \leq 2^{-k+1}$. Show that given positive integers b and k, we can compute a k-bit approximation of $1/b$ in time $O(M(k))$. Hint: using Newton iteration, show how to go from a t-bit approximation of $1/b$ to a $(2t - 2)$-bit approximation of $1/b$, making use of just the high-order $O(t)$ bits of b, in time $O(M(t))$. **Newton iteration** is a general method of iteratively approximating a root of an equation $f(x) = 0$ by starting with an initial approximation x_0, and computing subsequent approximations by the formula $x_{i+1} = x_i - f(x_i)/f'(x_i)$, where $f'(x)$ is the derivative of $f(x)$. For this exercise, apply Newton iteration to the function $f(x) = x^{-1} - b$.

EXERCISE 3.49. Using the result of the previous exercise, show that, given positive integers a and b of bit length at most ℓ, we can compute $\lfloor a/b \rfloor$ and $a \bmod b$ in time $O(M(\ell))$. From this we see that, up to a constant factor, division with remainder is no harder than multiplication.

EXERCISE 3.50. Using the result of the previous exercise, give an algorithm for Exercise 3.27 that runs in time $O(M(\operatorname{len}(n))\operatorname{len}(k))$. Hint: divide and conquer.

EXERCISE 3.51. Give an algorithm for Exercise 3.29 whose running time is $O(M(\operatorname{len}(n)))$. Hint: Newton iteration.

EXERCISE 3.52. Suppose we have an algorithm that computes the square of an ℓ-bit integer in time at most $S(\ell)$, where S is a well-behaved complexity function.

Show how to use this algorithm to compute the product of two arbitrary integers of length at most ℓ in time $O(S(\ell))$.

EXERCISE 3.53. Give algorithms for Exercise 3.32 whose running times are $O(M(\ell)\operatorname{len}(\ell))$, where $\ell := \operatorname{len}(n)$. Hint: divide and conquer.

3.6 Notes

Shamir [89] shows how to factor an integer in polynomial time on a RAM, but where the numbers stored in the memory cells may have exponentially many bits. As there is no known polynomial-time factoring algorithm on any realistic machine, Shamir's algorithm demonstrates the importance of restricting the sizes of numbers stored in the memory cells of our RAMs to keep our formal model realistic.

The most practical implementations of algorithms for arithmetic on large integers are written in low-level "assembly language," specific to a particular machine's architecture (e.g., the GNU Multi-Precision library GMP, available at gmplib. org). Besides the general fact that such hand-crafted code is more efficient than that produced by a compiler, there is another, more important reason for using assembly language. A typical 32-bit machine often comes with instructions that allow one to compute the 64-bit product of two 32-bit integers, and similarly, instructions to divide a 64-bit integer by a 32-bit integer (obtaining both the quotient and remainder). However, high-level programming languages do not (as a rule) provide any access to these low-level instructions. Indeed, we suggested in §3.3 using a value for the base B of about half the word-size of the machine, in order to avoid overflow. However, if one codes in assembly language, one can take B to be much closer, or even equal, to the word-size of the machine. Since our basic algorithms for multiplication and division run in time quadratic in the number of base-B digits, the effect of doubling the bit-length of B is to decrease the running time of these algorithms by a factor of *four*. This effect, combined with the improvements one might typically expect from using assembly-language code, can easily lead to a five- to ten-fold decrease in the running time, compared to an implementation in a high-level language. This is, of course, a significant improvement for those interested in serious "number crunching."

The "classical," quadratic-time algorithms presented here for integer multiplication and division are by no means the best possible: there are algorithms that are asymptotically faster. We saw this in the algorithm in Exercise 3.41, which was originally invented by Karatsuba [54] (although Karatsuba is one of two authors on this paper, the paper gives exclusive credit for this particular result to Karatsuba). That algorithm allows us to multiply two integers of length at most ℓ in time

$O(\ell^{\log_2 3})$. The fastest known algorithm for multiplying such integers on a RAM runs in time $O(\ell)$, and is due to Schönhage. It actually works on a very restricted type of RAM called a "pointer machine" (see Exercise 12, Section 4.3.3 of Knuth [56]). See Exercise 17.25 later in this text for a much simpler (but heuristic) $O(\ell)$ multiplication algorithm.

Another model of computation is that of **Boolean circuits**. In this model of computation, one considers families of Boolean circuits (with, say, the usual "and," "or," and "not" gates) that compute a particular function—for every input length, there is a different circuit in the family that computes the function on inputs that are bit strings of that length. One natural notion of complexity for such circuit families is the **size** of the circuit (i.e., the number of gates and wires in the circuit), which is measured as a function of the input length. For many years, the smallest known Boolean circuit that multiplies two integers of length at most ℓ was of size $O(\ell \operatorname{len}(\ell) \operatorname{len}(\operatorname{len}(\ell)))$. This result was due to Schönhage and Strassen [86]. More recently, Fürer showed how to reduce this to $O(\ell \operatorname{len}(\ell) 2^{O(\log^* \ell)})$ [38]. Here, the value of $\log^* n$ is defined as the minimum number of applications of the function $\log_2$ to the number n required to obtain a number that is less than or equal to 1. The function $\log^*$ is an extremely slow growing function, and is a constant for all practical purposes.

It is hard to say which model of computation, the RAM or circuits, is "better." On the one hand, the RAM very naturally models computers as we know them today: one stores small numbers, like array indices, counters, and pointers, in individual words of the machine, and processing such a number typically takes a single "machine cycle." On the other hand, the RAM model, as we formally defined it, invites a certain kind of "cheating," as it allows one to stuff $O(\operatorname{len}(\ell))$-bit integers into memory cells. For example, even with the simple, quadratic-time algorithms for integer arithmetic discussed in §3.3, we can choose the base B to have $\operatorname{len}(\ell)$ bits, in which case these algorithms would run in time $O((\ell/\operatorname{len}(\ell))^2)$. However, just to keep things simple, we have chosen to view B as a constant (from a formal, asymptotic point of view).

In the remainder of this text, unless otherwise specified, we shall always use the classical $O(\ell^2)$ bounds for integer multiplication and division. These have the advantages of being simple and of being reasonably reliable predictors of actual performance for small to moderately sized inputs. For relatively large numbers, experience shows that the classical algorithms are definitely not the best—Karatsuba's multiplication algorithm, and related algorithms for division, are superior on inputs of a thousand bits or so (the exact crossover depends on myriad implementation details). The even "faster" algorithms discussed above are typically not interesting unless the numbers involved are truly huge, of bit length around 10^5–10^6. Thus, the reader should bear in mind that for serious computations involving very large numbers, the faster algorithms are very important, even though this text does not discuss them at great length.

For a good survey of asymptotically fast algorithms for integer arithmetic, see Chapter 9 of Crandall and Pomerance [30], as well as Chapter 4 of Knuth [56].

4

Euclid's algorithm

In this chapter, we discuss Euclid's algorithm for computing greatest common divisors, which, as we will see, has applications far beyond that of just computing greatest common divisors.

4.1 The basic Euclidean algorithm

We consider the following problem: given two non-negative integers a and b, compute their greatest common divisor, $\gcd(a, b)$. We can do this using the well-known **Euclidean algorithm**, also called **Euclid's algorithm**.

The basic idea is the following. Without loss of generality, we may assume that $a \geq b \geq 0$. If $b = 0$, then there is nothing to do, since in this case, $\gcd(a, 0) = a$. Otherwise, $b > 0$, and we can compute the integer quotient $q := \lfloor a/b \rfloor$ and remainder $r := a \bmod b$, where $0 \leq r < b$. From the equation

$$a = bq + r,$$

it is easy to see that if an integer d divides both b and r, then it also divides a; likewise, if an integer d divides a and b, then it also divides r. From this observation, it follows that $\gcd(a, b) = \gcd(b, r)$, and so by performing a division, we reduce the problem of computing $\gcd(a, b)$ to the "smaller" problem of computing $\gcd(b, r)$.

The following theorem develops this idea further:

Theorem 4.1. *Let a, b be integers, with $a \geq b \geq 0$. Using the division with remainder property, define the integers $r_0, r_1, \ldots, r_{\lambda+1}$ and $q_1, \ldots, q_\lambda$, where $\lambda \geq 0$, as follows:*

$$a = r_0,$$
$$b = r_1,$$
$$r_0 = r_1 q_1 + r_2 \qquad (0 < r_2 < r_1),$$
$$\vdots$$
$$r_{i-1} = r_i q_i + r_{i+1} \qquad (0 < r_{i+1} < r_i),$$
$$\vdots$$
$$r_{\lambda-2} = r_{\lambda-1} q_{\lambda-1} + r_\lambda \qquad (0 < r_\lambda < r_{\lambda-1}),$$
$$r_{\lambda-1} = r_\lambda q_\lambda \qquad (r_{\lambda+1} = 0).$$

Note that by definition, $\lambda = 0$ if $b = 0$, and $\lambda > 0$, otherwise. Then we have $r_\lambda = \gcd(a, b)$. Moreover, if $b > 0$, then $\lambda \leq \log b / \log \phi + 1$, where $\phi := (1 + \sqrt{5})/2 \approx 1.62$.

Proof. For the first statement, one sees that for $i = 1, \ldots, \lambda$, we have $r_{i-1} = r_i q_i + r_{i+1}$, from which it follows that the common divisors of r_{i-1} and r_i are the same as the common divisors of r_i and r_{i+1}, and hence $\gcd(r_{i-1}, r_i) = \gcd(r_i, r_{i+1})$. From this, it follows that

$$\gcd(a, b) = \gcd(r_0, r_1) = \cdots = \gcd(r_\lambda, r_{\lambda+1}) = \gcd(r_\lambda, 0) = r_\lambda.$$

To prove the second statement, assume that $b > 0$, and hence $\lambda > 0$. If $\lambda = 1$, the statement is obviously true, so assume $\lambda > 1$. We claim that for $i = 0, \ldots, \lambda - 1$, we have $r_{\lambda-i} \geq \phi^i$. The statement will then follow by setting $i = \lambda - 1$ and taking logarithms.

We now prove the above claim. For $i = 0$ and $i = 1$, we have

$$r_\lambda \geq 1 = \phi^0 \text{ and } r_{\lambda-1} \geq r_\lambda + 1 \geq 2 \geq \phi^1.$$

For $i = 2, \ldots, \lambda - 1$, using induction and applying the fact that $\phi^2 = \phi + 1$, we have

$$r_{\lambda-i} \geq r_{\lambda-(i-1)} + r_{\lambda-(i-2)} \geq \phi^{i-1} + \phi^{i-2} = \phi^{i-2}(1 + \phi) = \phi^i,$$

which proves the claim. $\square$

Example 4.1. Suppose $a = 100$ and $b = 35$. Then the numbers appearing in Theorem 4.1 are easily computed as follows:

i	0	1	2	3	4
r_i	100	35	30	5	0
q_i			2	1	6

So we have $\gcd(a, b) = r_3 = 5$. $\square$

We can easily turn the scheme described in Theorem 4.1 into a simple algorithm:

Euclid's algorithm. On input a, b, where a and b are integers such that $a \geq b \geq 0$, compute $d = \gcd(a, b)$ as follows:

$\quad r \leftarrow a, \ r' \leftarrow b$
$\quad \text{while } r' \neq 0 \text{ do}$
$\quad\quad\quad r'' \leftarrow r \bmod r'$
$\quad\quad\quad (r, r') \leftarrow (r', r'')$
$\quad d \leftarrow r$
$\quad \text{output } d$

We now consider the running time of Euclid's algorithm. Naively, one could estimate this as follows. Suppose a and b are ℓ-bit numbers. The number of divisions performed by the algorithm is the number λ in Theorem 4.1, which is $O(\ell)$. Moreover, each division involves numbers of ℓ bits or fewer in length, and so takes time $O(\ell^2)$. This leads to a bound on the running time of $O(\ell^3)$. However, as the following theorem shows, this cubic running time bound is well off the mark. Intuitively, this is because the cost of performing a division depends on the length of the quotient: the larger the quotient, the more expensive the division, but also, the more progress the algorithm makes towards termination.

Theorem 4.2. *Euclid's algorithm runs in time* $O(\operatorname{len}(a)\operatorname{len}(b))$.

Proof. We may assume that $b > 0$. With notation as in Theorem 4.1, the running time is $O(T)$, where

$$T = \sum_{i=1}^{\lambda} \operatorname{len}(r_i) \operatorname{len}(q_i) \leq \operatorname{len}(b) \sum_{i=1}^{\lambda} \operatorname{len}(q_i)$$

$$\leq \operatorname{len}(b) \sum_{i=1}^{\lambda} (\operatorname{len}(r_{i-1}) - \operatorname{len}(r_i) + 1) \quad \text{(see Exercise 3.24)}$$

$$= \operatorname{len}(b)(\operatorname{len}(r_0) - \operatorname{len}(r_\lambda) + \lambda) \quad \text{(telescoping the sum)}$$

$$\leq \operatorname{len}(b)(\operatorname{len}(a) + \log b / \log \phi + 1) \quad \text{(by Theorem 4.1)}$$

$$= O(\operatorname{len}(a)\operatorname{len}(b)). \quad \square$$

EXERCISE 4.1. With notation as in Theorem 4.1, give a direct and simple proof that for each $i = 1, \ldots, \lambda$, we have $r_{i+1} \leq r_{i-1}/2$. Thus, with every *two* division steps, the bit length of the remainder drops by at least 1. Based on this, give an alternative proof that the number of divisions is $O(\operatorname{len}(b))$.

EXERCISE 4.2. Show how to compute lcm(a, b) in time $O(\text{len}(a) \text{len}(b))$.

EXERCISE 4.3. Let $a, b \in \mathbb{Z}$ with $a \geq b \geq 0$, let $d := \gcd(a, b)$, and assume $d > 0$. Suppose that on input a, b, Euclid's algorithm performs λ division steps, and computes the remainder sequence $\{r_i\}_{i=0}^{\lambda+1}$ and the quotient sequence $\{q_i\}_{i=1}^{\lambda}$ (as in Theorem 4.1). Now suppose we run Euclid's algorithm on input $a/d, b/d$. Show that on these inputs, the number of division steps performed is also λ, the remainder sequence is $\{r_i/d\}_{i=0}^{\lambda+1}$, and the quotient sequence is $\{q_i\}_{i=1}^{\lambda}$.

EXERCISE 4.4. Show that if we run Euclid's algorithm on input a, b, where $a \geq b > 0$, then its running time is $O(\text{len}(a/d) \text{len}(b))$, where $d := \gcd(a, b)$.

EXERCISE 4.5. Let λ be a positive integer. Show that there exist integers a, b with $a > b > 0$ and $\lambda \geq \log b / \log \phi$, such that Euclid's algorithm on input a, b performs at least λ divisions. Thus, the bound in Theorem 4.1 on the number of divisions is essentially tight.

EXERCISE 4.6. This exercise looks at an alternative algorithm for computing $\gcd(a, b)$, called the **binary gcd algorithm**. This algorithm avoids complex operations, such as division and multiplication; instead, it relies only on subtraction, and division and multiplication by powers of 2, which, assuming a binary representation of integers (as we are), can be very efficiently implemented using "right shift" and "left shift" operations. The algorithm takes positive integers a and b as input, and runs as follows:

$r \leftarrow a, \; r' \leftarrow b, \; e \leftarrow 0$
while $2 \mid r$ and $2 \mid r'$ do $r \leftarrow r/2, \; r' \leftarrow r'/2, \; e \leftarrow e + 1$
repeat
        while $2 \mid r$ do $r \leftarrow r/2$
        while $2 \mid r'$ do $r' \leftarrow r'/2$
        if $r' < r$ then $(r, r') \leftarrow (r', r)$
        $r' \leftarrow r' - r$
until $r' = 0$
$d \leftarrow 2^e \cdot r$
output d

Show that this algorithm correctly computes $\gcd(a, b)$, and runs in time $O(\ell^2)$, where $\ell := \max(\text{len}(a), \text{len}(b))$.

4.2 The extended Euclidean algorithm

Let a and b be integers, and let $d := \gcd(a, b)$. We know by Theorem 1.8 that there exist integers s and t such that $as + bt = d$. The **extended Euclidean algorithm** allows us to efficiently compute s and t. The next theorem defines the quantities computed by this algorithm, and states a number of important facts about them; these facts will play a crucial role, both in the analysis of the running time of the algorithm, as well as in applications of the algorithm that we will discuss later.

Theorem 4.3. *Let $a, b, r_0, \ldots, r_{\lambda+1}$ and $q_1, \ldots, q_\lambda$ be as in Theorem 4.1. Define integers $s_0, \ldots, s_{\lambda+1}$ and $t_0, \ldots, t_{\lambda+1}$ as follows:*

$$s_0 := 1, \qquad\qquad\qquad t_0 := 0,$$
$$s_1 := 0, \qquad\qquad\qquad t_1 := 1,$$
$$s_{i+1} := s_{i-1} - s_i q_i, \qquad t_{i+1} := t_{i-1} - t_i q_i \qquad (i = 1, \ldots, \lambda).$$

Then:

(i) *for $i = 0, \ldots, \lambda+1$, we have $a s_i + b t_i = r_i$; in particular, $a s_\lambda + b t_\lambda = \gcd(a, b)$;*

(ii) *for $i = 0, \ldots, \lambda$, we have $s_i t_{i+1} - t_i s_{i+1} = (-1)^i$;*

(iii) *for $i = 0, \ldots, \lambda + 1$, we have $\gcd(s_i, t_i) = 1$;*

(iv) *for $i = 0, \ldots, \lambda$, we have $t_i t_{i+1} \le 0$ and $|t_i| \le |t_{i+1}|$; for $i = 1, \ldots, \lambda$, we have $s_i s_{i+1} \le 0$ and $|s_i| \le |s_{i+1}|$;*

(v) *for $i = 1, \ldots, \lambda + 1$, we have $r_{i-1} |t_i| \le a$ and $r_{i-1} |s_i| \le b$;*

(vi) *if $a > 0$, then for $i = 1, \ldots, \lambda + 1$, we have $|t_i| \le a$ and $|s_i| \le b$; if $a > 1$ and $b > 0$, then $|t_\lambda| \le a/2$ and $|s_\lambda| \le b/2$.*

Proof. (i) is easily proved by induction on i. For $i = 0, 1$, the statement is clear. For $i = 2, \ldots, \lambda + 1$, we have

$$\begin{aligned}
a s_i + b t_i &= a(s_{i-2} - s_{i-1} q_{i-1}) + b(t_{i-2} - t_{i-1} q_{i-1}) \\
&= (a s_{i-2} + b t_{i-2}) - (a s_{i-1} + b t_{i-1}) q_{i-1} \\
&= r_{i-2} - r_{i-1} q_{i-1} \quad \text{(by induction)} \\
&= r_i.
\end{aligned}$$

(ii) is also easily proved by induction on i. For $i = 0$, the statement is clear. For $i = 1, \ldots, \lambda$, we have

$$\begin{aligned}
s_i t_{i+1} - t_i s_{i+1} &= s_i(t_{i-1} - t_i q_i) - t_i(s_{i-1} - s_i q_i) \\
&= -(s_{i-1} t_i - t_{i-1} s_i) \quad \text{(after expanding and simplifying)} \\
&= -(-1)^{i-1} \quad \text{(by induction)} \\
&= (-1)^i.
\end{aligned}$$

(iii) follows directly from (ii).

For (iv), one can easily prove both statements by induction on i. The statement involving the t_i's is clearly true for $i = 0$. For $i = 1, \ldots, \lambda$, we have

$t_{i+1} = t_{i-1} - t_i q_i$; moreover, by the induction hypothesis, t_{i-1} and t_i have opposite signs and $|t_i| \geq |t_{i-1}|$; it follows that $|t_{i+1}| = |t_{i-1}| + |t_i| q_i \geq |t_i|$, and that the sign of t_{i+1} is the opposite of that of t_i. The proof of the statement involving the s_i's is the same, except that we start the induction at $i = 1$.

For (v), one considers the two equations:

$$as_{i-1} + bt_{i-1} = r_{i-1},$$
$$as_i + bt_i = r_i.$$

Subtracting t_{i-1} times the second equation from t_i times the first, and applying (ii), we get $\pm a = t_i r_{i-1} - t_{i-1} r_i$; consequently, using the fact that t_i and t_{i-1} have opposite sign, we obtain

$$a = |t_i r_{i-1} - t_{i-1} r_i| = |t_i| r_{i-1} + |t_{i-1}| r_i \geq |t_i| r_{i-1}.$$

The inequality involving s_i follows similarly, subtracting s_{i-1} times the second equation from s_i times the first.

(vi) follows from (v) and the following observations: if $a > 0$, then $r_{i-1} > 0$ for $i = 1, \ldots, \lambda + 1$; if $a > 1$ and $b > 0$, then $\lambda > 0$ and $r_{\lambda-1} \geq 2$. $\square$

Example 4.2. We continue with Example 4.1. The s_i's and t_i's are easily computed from the q_i's:

i	0	1	2	3	4
r_i	100	35	30	5	0
q_i			2	1	6
s_i	1	0	1	-1	7
t_i	0	1	-2	3	-20

So we have $\gcd(a, b) = 5 = -a + 3b$. $\square$

We can easily turn the scheme described in Theorem 4.3 into a simple algorithm:

The extended Euclidean algorithm. On input a, b, where a and b are integers such that $a \geq b \geq 0$, compute integers d, s, and t, such that $d = \gcd(a, b)$ and $as + bt = d$, as follows:

$r \leftarrow a, \ r' \leftarrow b$
$s \leftarrow 1, \ s' \leftarrow 0$
$t \leftarrow 0, \ t' \leftarrow 1$
while $r' \neq 0$ do
    $q \leftarrow \lfloor r/r' \rfloor, \ r'' \leftarrow r \bmod r'$
    $(r, s, t, r', s', t') \leftarrow (r', s', t', r'', s - s'q, t - t'q)$
$d \leftarrow r$
output d, s, t

Theorem 4.4. *The extended Euclidean algorithm runs in time $O(\text{len}(a)\,\text{len}(b))$.*

Proof. We may assume that $b > 0$. It suffices to analyze the cost of computing the coefficient sequences $\{s_i\}$ and $\{t_i\}$. Consider first the cost of computing all of the t_i's, which is $O(T)$, where $T = \sum_{i=1}^{\lambda} \text{len}(t_i)\,\text{len}(q_i)$. We have $t_1 = 1$ and, by part (vi) of Theorem 4.3, we have $|t_i| \le a$ for $i = 2, \ldots, \lambda$. Arguing as in the proof of Theorem 4.2, we have

$$T \le \text{len}(q_1) + \text{len}(a) \sum_{i=2}^{\lambda} \text{len}(q_i)$$

$$\le \text{len}(a) + \text{len}(a)(\text{len}(r_1) - \text{len}(r_\lambda) + \lambda - 1) = O(\text{len}(a)\,\text{len}(b)).$$

An analogous argument shows that one can also compute all of the s_i's in time $O(\text{len}(a)\,\text{len}(b))$, and in fact, in time $O(\text{len}(b)^2)$. $\square$

For the reader familiar with the basics of the theory of matrices and determinants, it is instructive to view Theorem 4.3 as follows. For $i = 1, \ldots, \lambda$, we have

$$\begin{pmatrix} r_i \\ r_{i+1} \end{pmatrix} = \begin{pmatrix} 0 & 1 \\ 1 & -q_i \end{pmatrix} \begin{pmatrix} r_{i-1} \\ r_i \end{pmatrix}.$$

Recursively expanding the right-hand side of this equation, we have

$$\begin{pmatrix} r_i \\ r_{i+1} \end{pmatrix} = \overbrace{\begin{pmatrix} 0 & 1 \\ 1 & -q_i \end{pmatrix} \cdots \begin{pmatrix} 0 & 1 \\ 1 & -q_1 \end{pmatrix}}^{M_i :=} \begin{pmatrix} a \\ b \end{pmatrix}.$$

This defines the 2×2 matrix M_i for $i = 1, \ldots, \lambda$. If we additionally define M_0 to be the 2×2 identity matrix, then it is easy to see that for $i = 0, \ldots, \lambda$, we have

$$M_i = \begin{pmatrix} s_i & t_i \\ s_{i+1} & t_{i+1} \end{pmatrix}.$$

From these observations, part (i) of Theorem 4.3 is immediate, and part (ii) follows from the fact that M_i is the product of i matrices, each of determinant -1, and the determinant of M_i is evidently $s_i t_{i+1} - t_i s_{i+1}$.

EXERCISE 4.7. In our description of the extended Euclidean algorithm, we made the restriction that the inputs a and b satisfy $a \ge b \ge 0$. Using this restricted algorithm as a subroutine, give an algorithm that works without any restrictions on its input.

EXERCISE 4.8. With notation and assumptions as in Exercise 4.3, suppose that on input a, b, the extended Euclidean algorithm computes the coefficient sequences

$\{s_i\}_{i=0}^{\lambda+1}$ and $\{t_i\}_{i=0}^{\lambda+1}$ (as in Theorem 4.3). Show that the extended Euclidean algorithm on input $a/d, b/d$ computes the same coefficient sequences.

EXERCISE 4.9. Assume notation as in Theorem 4.3. Show that:

(a) for all $i = 2,\ldots,\lambda$, we have $|t_i| < |t_{i+1}|$ and $r_{i-1}|t_i| < a$, and that for all $i = 3,\ldots,\lambda$, we have $|s_i| < |s_{i+1}|$ and $r_{i-1}|s_i| < b$;

(b) $s_i t_i \leq 0$ for $i = 0,\ldots,\lambda+1$;

(c) if $d := \gcd(a,b) > 0$, then $|s_{\lambda+1}| = b/d$ and $|t_{\lambda+1}| = a/d$.

EXERCISE 4.10. One can extend the binary gcd algorithm discussed in Exercise 4.6 so that in addition to computing $d = \gcd(a,b)$, it also computes s and t such that $as + bt = d$. Here is one way to do this (again, we assume that a and b are positive integers):

$r \leftarrow a,\ r' \leftarrow b,\ e \leftarrow 0$
while $2 \mid r$ and $2 \mid r'$ do $r \leftarrow r/2,\ r' \leftarrow r'/2,\ e \leftarrow e+1$
$\tilde{a} \leftarrow r,\ \tilde{b} \leftarrow r',\ s \leftarrow 1,\ t \leftarrow 0,\ s' \leftarrow 0,\ t' \leftarrow 1$
repeat
    while $2 \mid r$ do
        $r \leftarrow r/2$
        if $2 \mid s$ and $2 \mid t$ then $s \leftarrow s/2,\ t \leftarrow t/2$
                      else $s \leftarrow (s+\tilde{b})/2,\ t \leftarrow (t-\tilde{a})/2$
    while $2 \mid r'$ do
        $r' \leftarrow r'/2$
        if $2 \mid s'$ and $2 \mid t'$ then $s' \leftarrow s'/2,\ t' \leftarrow t'/2$
                      else $s' \leftarrow (s'+\tilde{b})/2,\ t' \leftarrow (t'-\tilde{a})/2$
    if $r' < r$ then $(r,s,t,r',s',t') \leftarrow (r',s',t',r,s,t)$
    $r' \leftarrow r'-r,\ s' \leftarrow s'-s,\ t' \leftarrow t'-t$
until $r' = 0$
$d \leftarrow 2^e \cdot r$, output d, s, t

Show that this algorithm is correct and that its running time is $O(\ell^2)$, where $\ell := \max(\text{len}(a),\text{len}(b))$. In particular, you should verify that all of the divisions by 2 performed by the algorithm yield integer results. Moreover, show that the outputs s and t are of length $O(\ell)$.

EXERCISE 4.11. Suppose we modify the extended Euclidean algorithm so that it computes balanced remainders; that is, for $i = 1,\ldots,\lambda$, the values q_i and r_{i+1} are computed so that $r_{i-1} = r_i q_i + r_{i+1}$ and $-|r_i|/2 \leq r_{i+1} < |r_i|/2$. Assume that the s_i's and the t_i's are computed by the same formula as in Theorem 4.3. Give a detailed analysis of the running time of this algorithm, which should include an analysis of the number of division steps, and the sizes of the s_i's and t_i's.

4.3 Computing modular inverses and Chinese remaindering

An important application of the extended Euclidean algorithm is to the problem of computing multiplicative inverses in $\mathbb{Z}_n$.

Theorem 4.5. *Suppose we are given integers n, b, where $0 \le b < n$. Then in time $O(\text{len}(n)^2)$, we can determine if b is relatively prime to n, and if so, compute $b^{-1} \bmod n$.*

Proof. We may assume $n > 1$, since when $n = 1$, we have $b = 0 = b^{-1} \bmod n$. We run the extended Euclidean algorithm on input n, b, obtaining integers d, s, and t, such that $d = \gcd(n, b)$ and $ns + bt = d$. If $d \ne 1$, then b does not have a multiplicative inverse modulo n. Otherwise, if $d = 1$, then t is a multiplicative inverse of b modulo n; however, it may not lie in the range $\{0, \ldots, n-1\}$, as required. By part (vi) of Theorem 4.3, we have $|t| \le n/2 < n$. Thus, if $t \ge 0$, then $b^{-1} \bmod n$ is equal to t; otherwise, $b^{-1} \bmod n$ is equal to $t + n$. Based on Theorem 4.4, it is clear that all the computations can be performed in time $O(\text{len}(n)^2)$. $\square$

Example 4.3. Suppose we are given integers a, b, n, where $0 \le a < n$, and $0 \le b < n$, and we want to compute a solution z to the congruence $az \equiv b \pmod{n}$, or determine that no such solution exists. Based on the discussion in Example 2.5, the following algorithm does the job:

$$d \leftarrow \gcd(a, n)$$
$$\text{if } d \nmid b \text{ then}$$
$$\qquad \text{output ``no solution''}$$
$$\text{else}$$
$$\qquad a' \leftarrow a/d, \; b' \leftarrow b/d, \; n' \leftarrow n/d$$
$$\qquad t \leftarrow (a')^{-1} \bmod n'$$
$$\qquad z \leftarrow tb' \bmod n'$$
$$\qquad \text{output } z$$

Using Euclid's algorithm to compute d, and the extended Euclidean algorithm to compute t (as in Theorem 4.5), the running time of this algorithm is clearly $O(\text{len}(n)^2)$. $\square$

We also observe that the Chinese remainder theorem (Theorem 2.6) can be made computationally effective:

Theorem 4.6 (Effective Chinese remainder theorem). *Suppose we are given integers $n_1, \ldots, n_k$ and $a_1, \ldots, a_k$, where the family $\{n_i\}_{i=1}^{k}$ is pairwise relatively prime, and where $n_i > 1$ and $0 \le a_i < n_i$ for $i = 1, \ldots, k$. Let $n := \prod_{i=1}^{k} n_i$. Then in time $O(\text{len}(n)^2)$, we can compute the unique integer a satisfying $0 \le a < n$ and $a \equiv a_i \pmod{n_i}$ for $i = 1, \ldots, k$.*

Proof. The algorithm is a straightforward implementation of the proof of Theorem 2.6, and runs as follows:

$$n \leftarrow \prod_{i=1}^{k} n_i$$
for $i \leftarrow 1$ to k do
$$n_i^* \leftarrow n/n_i, \; b_i \leftarrow n_i^* \bmod n_i, \; t_i \leftarrow b_i^{-1} \bmod n_i, \; e_i \leftarrow n_i^* t_i$$
$$a \leftarrow \left(\sum_{i=1}^{k} a_i e_i \right) \bmod n$$

We leave it to the reader to verify the running time bound. $\square$

EXERCISE 4.12. In Example 4.3, show that one can easily obtain the quantities d, a', n', and t from the data computed in just a single execution of the extended Euclidean algorithm.

EXERCISE 4.13. In this exercise, you are to make the result of Theorem 2.17 effective. Suppose that we are given a positive integer n, two elements $\alpha, \beta \in \mathbb{Z}_n^*$, and integers ℓ and m, such that $\alpha^\ell = \beta^m$ and $\gcd(\ell, m) = 1$. Show how to compute $\gamma \in \mathbb{Z}_n^*$ such that $\alpha = \gamma^m$ in time $O(\text{len}(\ell) \, \text{len}(m) + (\text{len}(\ell) + \text{len}(m)) \, \text{len}(n)^2)$.

EXERCISE 4.14. In this exercise and the next, you are to analyze an "incremental Chinese remaindering algorithm." Consider the following algorithm, which takes as input integers a_1, n_1, a_2, n_2 satisfying

$$0 \leq a_1 < n_1, \; 0 \leq a_2 < n_2, \; \text{and} \; \gcd(n_1, n_2) = 1.$$

It outputs integers a, n satisfying

$$n = n_1 n_2, \; 0 \leq a < n, \; a \equiv a_1 \pmod{n_1}, \; \text{and} \; a \equiv a_2 \pmod{n_2},$$

and runs as follows:

$$b \leftarrow n_1 \bmod n_2, \; t \leftarrow b^{-1} \bmod n_2, \; h \leftarrow (a_2 - a_1)t \bmod n_2$$
$$a \leftarrow a_1 + n_1 h, \; n \leftarrow n_1 n_2$$
output a, n

Show that the algorithm correctly computes a and n as specified, and runs in time $O(\text{len}(n) \, \text{len}(n_2))$.

EXERCISE 4.15. Using the algorithm in the previous exercise as a subroutine, give a simple $O(\text{len}(n)^2)$ algorithm that takes as input integers $n_1, \ldots, n_k$ and $a_1, \ldots, a_k$, where the family $\{n_i\}_{i=1}^{k}$ is pairwise relatively prime, and where $n_i > 1$ and $0 \leq a_i < n_i$ for $i = 1, \ldots, k$, and outputs integers a and n such that $0 \leq a < n$, $n = \prod_{i=1}^{k} n_i$, and $a \equiv a_i \pmod{n_i}$ for $i = 1, \ldots, k$. The algorithm should be "incremental," in that it processes the pairs (a_i, n_i) one at a time, using time $O(\text{len}(n) \, \text{len}(n_i))$ per pair.

EXERCISE 4.16. Suppose we are given $\alpha_1, \ldots, \alpha_k \in \mathbb{Z}_n^*$. Show how to compute $\alpha_1^{-1}, \ldots, \alpha_k^{-1}$ by computing *one* multiplicative inverse modulo n, and performing fewer than $3k$ multiplications modulo n. This result is useful, as in practice, if n is several hundred bits long, it may take 10–20 times longer to compute multiplicative inverses modulo n than to multiply modulo n.

4.4 Speeding up algorithms via modular computation

An important practical application of the above "computational" version (Theorem 4.6) of the Chinese remainder theorem is a general algorithmic technique that can significantly speed up certain types of computations involving long integers. Instead of trying to describe the technique in some general form, we simply illustrate the technique by means of a specific example: integer matrix multiplication.

Suppose we have two $m \times m$ matrices A and B whose entries are large integers, and we want to compute the product matrix $C := AB$. Suppose that for $r, s = 1, \ldots, m$, the entry of A at row r and column s is a_{rs}, and that for $s, t = 1, \ldots, m$, the entry of B at row s and column t is b_{st}. Then for $r, t = 1, \ldots, m$, the entry of C at row r and column t is c_{rt}, which is given by the usual rule for matrix multiplication:

$$c_{rt} = \sum_{s=1}^{m} a_{rs} b_{st}. \tag{4.1}$$

Suppose further that M is the maximum absolute value of the entries in A and B, so that the entries in C are bounded in absolute value by $M' := M^2 m$. Let $\ell := \text{len}(M)$. To simplify calculations, let us also assume that $m \leq M$ (this is reasonable, as we want to consider large values of M, greater than say 2^{100}, and certainly, we cannot expect to work with $2^{100} \times 2^{100}$ matrices).

By just applying the formula (4.1), we can compute the entries of C using m^3 multiplications of numbers of length at most ℓ, and m^3 additions of numbers of length at most $\text{len}(M')$, where $\text{len}(M') \leq 2\ell + \text{len}(m) = O(\ell)$. This yields a running time of

$$O(m^3 \ell^2). \tag{4.2}$$

Using the Chinese remainder theorem, we can actually do much better than this, as follows.

For every integer $n > 1$, and for all $r, t = 1, \ldots, m$, we have

$$c_{rt} \equiv \sum_{s=1}^{m} a_{rs} b_{st} \pmod{n}. \tag{4.3}$$

Moreover, if we compute integers c'_{rt} such that

$$c'_{rt} \equiv \sum_{s=1}^{m} a_{rs} b_{st} \pmod{n} \tag{4.4}$$

and if we also have

$$-n/2 \leq c'_{rt} < n/2 \quad \text{and} \quad n > 2M', \tag{4.5}$$

then we must have

$$c_{rt} = c'_{rt}. \tag{4.6}$$

To see why (4.6) follows from (4.4) and (4.5), observe that (4.3) and (4.4) imply that $c_{rt} \equiv c'_{rt} \pmod{n}$, which means that n divides $(c_{rt} - c'_{rt})$. Then from the bound $|c_{rt}| \leq M'$ and from (4.5), we obtain

$$|c_{rt} - c'_{rt}| \leq |c_{rt}| + |c'_{rt}| \leq M' + n/2 < n/2 + n/2 = n.$$

So we see that the quantity $(c_{rt} - c'_{rt})$ is a multiple of n, while at the same time this quantity is strictly less than n in absolute value; hence, this quantity must be zero. That proves (4.6).

So from the above discussion, to compute C, it suffices to compute the entries of C modulo n, where we have to make sure that we compute "balanced" remainders in the interval $[-n/2, n/2)$, rather than the more usual "least non-negative" remainders.

To compute C modulo n, we choose a number of small integers $n_1, \ldots, n_k$, such that the family $\{n_i\}_{i=1}^{k}$ is pairwise relatively prime, and the product $n := \prod_{i=1}^{k} n_i$ is just a bit larger than $2M'$. In practice, one would choose the n_i's to be small primes, and a table of such primes could easily be computed in advance, so that all problems up to a given size could be handled. For example, the product of all primes of at most 16 bits is a number that has more than 90,000 bits. Thus, by simply pre-computing and storing a table of small primes, we can handle input matrices with quite large entries (up to about 45,000 bits).

Let us assume that we have pre-computed appropriate small primes $n_1, \ldots, n_k$. Further, we shall assume that addition and multiplication modulo each n_i can be done in *constant* time. This is reasonable from a practical (and theoretical) point of view, since such primes easily "fit" into a machine word, and we can perform modular addition and multiplication using a constant number of built-in machine operations. Finally, we assume that we do not use more n_i's than are necessary, so that $\text{len}(n) = O(\ell)$ and $k = O(\ell)$.

To compute C, we execute the following steps:

1. For each $i = 1, \ldots, k$, do the following:

   (a) compute $\hat{a}_{rs}^{(i)} \leftarrow a_{rs} \bmod n_i$ for $r, s = 1, \ldots, m$,

   (b) compute $\hat{b}_{st}^{(i)} \leftarrow b_{st} \bmod n_i$ for $s, t = 1, \ldots, m$,

   (c) for $r, t = 1, \ldots, m$, compute

$$\hat{c}_{rt}^{(i)} \leftarrow \sum_{s=1}^{m} \hat{a}_{rs}^{(i)} \hat{b}_{st}^{(i)} \bmod n_i.$$

2. For each $r, t = 1, \ldots, m$, apply the Chinese remainder theorem to $\hat{c}_{rt}^{(1)}, \hat{c}_{rt}^{(2)}, \ldots, \hat{c}_{rt}^{(k)}$, obtaining an integer c_{rt}, which should be computed as a balanced remainder modulo n, so that $-n/2 \le c_{rt} < n/2$.

3. Output the matrix C, whose entry in row r and column t is c_{rt}.

Note that in step 2, if our Chinese remainder algorithm happens to be implemented to return an integer a with $0 \le a < n$, we can easily get a balanced remainder by just subtracting n from a if $a \ge n/2$.

The correctness of the above algorithm has already been established. Let us now analyze its running time. The running time of steps 1a and 1b is easily seen to be $O(m^2 \ell^2)$. Under our assumption about the cost of arithmetic modulo small primes, the cost of step 1c is $O(m^3 k)$, and since $k = O(\ell)$, the cost of this step is $O(m^3 \ell)$. Finally, by Theorem 4.6, the cost of step 2 is $O(m^2 \ell^2)$. Thus, the total running time of this algorithm is

$$O(m^2 \ell^2 + m^3 \ell).$$

This is a significant improvement over (4.2); for example, if $\ell \approx m$, then the running time of the original algorithm is $O(m^5)$, while the running time of the modular algorithm is $O(m^4)$.

EXERCISE 4.17. Apply the ideas above to the problem of computing the product of two polynomials whose coefficients are large integers. First, determine the running time of the "obvious" algorithm for multiplying two such polynomials, then design and analyze a "modular" algorithm.

4.5 An effective version of Fermat's two squares theorem

We proved in Theorem 2.34 (in §2.8.4) that every prime $p \equiv 1 \pmod{4}$ can be expressed as a sum of two squares of integers. In this section, we make this theorem computationally effective; that is, we develop an efficient algorithm that takes as input a prime $p \equiv 1 \pmod{4}$, and outputs integers r and t such that $p = r^2 + t^2$.

One essential ingredient in the proof of Theorem 2.34 was Thue's lemma (Theorem 2.33). This lemma asserts the existence of certain numbers, and we proved it using the "pigeonhole principle," which unfortunately does not translate directly into an efficient algorithm to actually find these numbers. However, we can show that these numbers arise as a "natural by-product" of the extended Euclidean algorithm. To make this more precise, let us introduce some notation. For integers a, b, with $a \geq b \geq 0$, let us define

$$\text{EEA}(a, b) := \{(r_i, s_i, t_i)\}_{i=0}^{\lambda+1},$$

where r_i, s_i, and t_i, for $i = 0, \ldots, \lambda + 1$, are defined as in Theorem 4.3.

Theorem 4.7 (Effective Thue's lemma). *Let $n, b, r^*, t^* \in \mathbb{Z}$, with $0 \leq b < n$ and $0 < r^* \leq n < r^* t^*$. Further, let $\text{EEA}(n, b) = \{(r_i, s_i, t_i)\}_{i=0}^{\lambda+1}$, and let j be the smallest index (among $0, \ldots, \lambda + 1$) such that $r_j < r^*$. Then, setting $r := r_j$ and $t := t_j$, we have*

$$r \equiv bt \pmod{n}, \quad 0 \leq r < r^*, \quad \text{and} \quad 0 < |t| < t^*.$$

Proof. Since $r_0 = n \geq r^* > 0 = r_{\lambda+1}$, the value of the index j is well defined; moreover, $j \geq 1$ and $r_{j-1} \geq r^*$. It follows that

$$|t_j| \leq n/r_{j-1} \ \text{(by part (v) of Theorem 4.3)}$$
$$\leq n/r^*$$
$$< t^* \ \text{(since } n < r^* t^*).$$

Since $j \geq 1$, by part (iv) of Theorem 4.3, we have $|t_j| \geq |t_1| > 0$. Finally, since $r_j = ns_j + bt_j$, we have $r_j \equiv bt_j \pmod{n}$. $\square$

What this theorem says is that given n, b, r^*, t^*, to find the desired values r and t, we run the extended Euclidean algorithm on input n, b. This generates a sequence of remainders $r_0 > r_1 > r_2 > \cdots$, where $r_0 = n$ and $r_1 = b$. If r_j is the first remainder in this sequence that falls below r^*, and if s_j and t_j are the corresponding numbers computed by the extended Euclidean algorithm, then $r := r_j$ and $t := t_j$ do the job.

The other essential ingredient in the proof of Theorem 2.34 was Theorem 2.31, which guarantees the existence of a square root of -1 modulo p when p is a prime congruent to 1 modulo 4. We need an effective version of this result as well. Later, in Chapter 12, we will study the general problem of computing square roots modulo primes. Right now, we develop an algorithm for this special case.

Assume we are given a prime $p \equiv 1 \pmod 4$, and we want to compute $\beta \in \mathbb{Z}_p^*$ such that $\beta^2 = -1$. By Theorem 2.32, it suffices to find $\gamma \in \mathbb{Z}_p^* \setminus (\mathbb{Z}_p^*)^2$, since then $\beta := \gamma^{(p-1)/4}$ (which we can efficiently compute via repeated squaring) satisfies

$\beta^2 = -1$. While there is no known efficient, deterministic algorithm to find such a γ, we do know that half the elements of $\mathbb{Z}_p^*$ are squares and half are not (see Theorem 2.20), which suggests the following simple "trial and error" strategy to compute β:

> repeat
> > choose $\gamma \in \mathbb{Z}_p^*$
> > compute $\beta \leftarrow \gamma^{(p-1)/4}$
> until $\beta^2 = -1$
> output β

As an algorithm, this is not fully specified, as we have to specify a procedure for selecting γ in each loop iteration. A reasonable approach is to simply choose γ *at random*: this would be an example of a *probabilistic algorithm*, a notion that we will study in detail in Chapter 9. Let us assume for the moment that this makes sense from a mathematical and algorithmic point of view, so that with each loop iteration, we have a 50% chance of picking a "good" γ, that is, one that is not in $(\mathbb{Z}_p^*)^2$. From this, it follows that with high probability, we should find a "good" γ in just a few loop iterations (the probability that after k loop iterations we still have not found one is $1/2^k$), and that the *expected* number of loop iterations is just 2. The running time of each loop iteration is dominated by the cost of repeated squaring, which is $O(\text{len}(p)^3)$. It follows that the *expected running time* of this algorithm (we will make this notion precise in Chapter 9) is $O(\text{len}(p)^3)$.

Let us now put all the ingredients together to get an algorithm to find r, t such that $p = r^2 + t^2$.

1. Find $\beta \in \mathbb{Z}_p^*$ such that $\beta^2 = -1$, using the above "trial and error" strategy.
2. Set $b \leftarrow \text{rep}(\beta)$ (so that $\beta = [b]$ and $b \in \{0, \dots, p-1\}$).
3. Run the extended Euclidean algorithm on input p, b to obtain $\text{EEA}(p, b)$, and then apply Theorem 4.7 with $n := p$, b, and $r^* := t^* := \lfloor \sqrt{p} \rfloor + 1$, to obtain the values r and t.
4. Output r, t.

When this algorithm terminates, we have $r^2 + t^2 = p$, as required: as we argued in the proof of Theorem 2.34, since $r \equiv bt \pmod{p}$ and $b^2 \equiv -1 \pmod{p}$, it follows that $r^2 + t^2 \equiv 0 \pmod{p}$, and since $0 < r^2 + t^2 < 2p$, we must have $r^2 + t^2 = p$. The (expected) running time of step 1 is $O(\text{len}(p)^3)$. The running time of step 3 is $O(\text{len}(p)^2)$ (note that we can compute $\lfloor \sqrt{p} \rfloor$ in time $O(\text{len}(p)^2)$, using the algorithm in Exercise 3.29). Thus, the total (expected) running time is $O(\text{len}(p)^3)$.

Example 4.4. One can check that $p := 1009$ is prime and $p \equiv 1 \pmod{4}$. Let us express p as a sum of squares using the above algorithm. First, we need to find a square root of -1 modulo p. Let us just try a random number, say 17, and raise this to the power $(p-1)/4 = 252$. One can calculate that $17^{252} \equiv 469 \pmod{1009}$, and $469^2 \equiv -1 \pmod{1009}$. So we were lucky with our first try. Now we run the extended Euclidean algorithm on input $p = 1009$ and $b = 469$, obtaining the following data:

i	r_i	q_i	s_i	t_i
0	1009		1	0
1	469	2	0	1
2	71	6	1	-2
3	43	1	-6	13
4	28	1	7	-15
5	15	1	-13	28
6	13	1	20	-43
7	2	6	-33	71
8	1	2	218	-469
9	0		-469	1009

The first r_j that falls below the threshold $r^* = \lfloor\sqrt{1009}\rfloor + 1 = 32$ is at $j = 4$, and so we set $r := 28$ and $t := -15$. One verifies that $r^2 + t^2 = 28^2 + 15^2 = 1009 = p$. □

It is natural to ask whether one can solve this problem without resorting to randomization. The answer is "yes" (see §4.8), but the only known deterministic algorithms for this problem are quite impractical (albeit polynomial time). This example illustrates the utility of randomization as an algorithm design technique, one that has proved to be invaluable in solving numerous algorithmic problems in number theory; indeed, in §3.4 we already mentioned its use in connection with primality testing, and we will explore many other applications as well (after putting the notion of a probabilistic algorithm on firm mathematical ground in Chapter 9).

4.6 Rational reconstruction and applications

In the previous section, we saw how to apply the extended Euclidean algorithm to obtain an effective version of Thue's lemma. This lemma asserts that for given integers n and b, there exists a pair of integers (r, t) satisfying $r \equiv bt \pmod{n}$, and contained in a prescribed rectangle, provided the area of the rectangle is large enough, relative to n. In this section, we first prove a corresponding uniqueness theorem, under the assumption that the area of the rectangle is not too large; of course, if $r \equiv bt \pmod{n}$, then for any non-zero integer q, we also have $rq \equiv b(tq) \pmod{n}$, and so we can only hope to guarantee that the *ratio* r/t is unique. After proving this uniqueness theorem, we show how to make this theorem computationally effective, and then develop several very neat applications.

The basic uniqueness statement is as follows:

Theorem 4.8. *Let $n, b, r^*, t^* \in \mathbb{Z}$ with $r^* \geq 0$, $t^* > 0$, and $n > 2r^*t^*$. Further, suppose that $r, t, r', t' \in \mathbb{Z}$ satisfy*

$$r \equiv bt \pmod{n}, \quad |r| \leq r^*, \quad 0 < |t| \leq t^*, \tag{4.7}$$

$$r' \equiv bt' \pmod{n}, \quad |r'| \leq r^*, \quad 0 < |t'| \leq t^*. \tag{4.8}$$

Then $r/t = r'/t'$.

Proof. Consider the two congruences

$$r \equiv bt \pmod{n},$$

$$r' \equiv bt' \pmod{n}.$$

Subtracting t times the second from t' times the first, we obtain

$$rt' - r't \equiv 0 \pmod{n}.$$

However, we also have

$$|rt' - r't| \leq |r||t'| + |r'||t| \leq 2r^*t^* < n.$$

Thus, $rt' - r't$ is a multiple of n, but less than n in absolute value; the only possibility is that $rt' - r't = 0$, which means $r/t = r'/t'$. $\square$

Now suppose that we are given $n, b, r^*, t^* \in \mathbb{Z}$ as in the above theorem; moreover, suppose that there exist $r, t \in \mathbb{Z}$ satisfying (4.7), but that these values are *not* given to us. Note that under the hypothesis of Theorem 4.8, Thue's lemma cannot be used to ensure the existence of such r and t, but in our eventual applications, we will have other reasons that will guarantee this. We would like to find $r', t' \in \mathbb{Z}$ satisfying (4.8), and if we do this, then by the theorem, we know that $r/t = r'/t'$. We call this the **rational reconstruction problem**. We can solve this problem efficiently using the extended Euclidean algorithm; indeed, just as in the case of our effective version of Thue's lemma, the desired values of r' and t' appear as "natural by-products" of that algorithm. To state the result precisely, let us recall the notation we introduced in the last section: for integers a, b, with $a \geq b \geq 0$, we defined

$$\text{EEA}(a, b) := \{(r_i, s_i, t_i)\}_{i=0}^{\lambda+1},$$

where r_i, s_i, and t_i, for $i = 0, \ldots, \lambda + 1$, are defined as in Theorem 4.3.

Theorem 4.9 (Rational reconstruction). *Let $n, b, r^*, t^* \in \mathbb{Z}$ with $0 \leq b < n$, $0 \leq r^* < n$, and $t^* > 0$. Further, let $\text{EEA}(n, b) = \{(r_i, s_i, t_i)\}_{i=0}^{\lambda+1}$, and let j be the smallest index (among $0, \ldots, \lambda + 1$) such that $r_j \leq r^*$, and set*

$$r' := r_j, \quad s' := s_j, \quad \text{and} \quad t' := t_j.$$

Finally, suppose that there exist $r, s, t \in \mathbb{Z}$ such that

$$r = ns + bt, \quad |r| \le r^*, \quad \text{and} \quad 0 < |t| \le t^*.$$

Then we have:

(i) $0 < |t'| \le t^*$;

(ii) *if $n > 2r^*t^*$, then for some non-zero integer q,*

$$r = r'q, \quad s = s'q, \quad \text{and} \quad t = t'q.$$

Proof. Since $r_0 = n > r^* \ge 0 = r_{\lambda+1}$, the value of j is well defined, and moreover, $j \ge 1$, and we have the inequalities

$$0 \le r_j \le r^* < r_{j-1}, \quad 0 < |t_j|, \quad |r| \le r^*, \quad \text{and} \quad 0 < |t| \le t^*, \tag{4.9}$$

along with the identities

$$r_{j-1} = ns_{j-1} + bt_{j-1}, \tag{4.10}$$

$$r_j = ns_j + bt_j, \tag{4.11}$$

$$r = ns + bt. \tag{4.12}$$

We now turn to part (i) of the theorem. Our goal is to prove that

$$|t_j| \le t^*. \tag{4.13}$$

This is the hardest part of the proof. To this end, let

$$\varepsilon := s_j t_{j-1} - s_{j-1} t_j, \quad \mu := (t_{j-1}s - s_{j-1}t)/\varepsilon, \quad v := (s_j t - t_j s)/\varepsilon.$$

Since $\varepsilon = \pm 1$, the numbers μ and v are integers; moreover, one may easily verify that they satisfy the equations

$$s_j \mu + s_{j-1} v = s, \tag{4.14}$$

$$t_j \mu + t_{j-1} v = t. \tag{4.15}$$

We now use these identities to prove (4.13). We consider three cases:

(i) Suppose $v = 0$. In this case, (4.15) implies $t_j \mid t$, and since $t \ne 0$, this implies $|t_j| \le |t| \le t^*$.

(ii) Suppose $\mu v < 0$. In this case, since t_j and t_{j-1} have opposite sign, (4.15) implies $|t| = |t_j \mu| + |t_{j-1} v| \ge |t_j|$, and so again, we have $|t_j| \le |t| \le t^*$.

(iii) The only remaining possibility is that $v \ne 0$ and $\mu v \ge 0$. We argue that this is impossible. Adding n times (4.14) to b times (4.15), and using the identities (4.10), (4.11), and (4.12), we obtain

$$r_j \mu + r_{j-1} v = r.$$

If $v \neq 0$ and μ and v had the same sign, we would have $|r| = |r_j\mu| + |r_{j-1}v| \geq r_{j-1}$, and hence $r_{j-1} \leq |r| \leq r^*$; however, this contradicts the fact that $r_{j-1} > r^*$.

That proves the inequality (4.13). We now turn to the proof of part (ii) of the theorem, which relies critically on this inequality. Assume that

$$n > 2r^*t^*. \tag{4.16}$$

From (4.11) and (4.12), we have

$$r_j \equiv bt_j \pmod{n} \quad \text{and} \quad r \equiv bt \pmod{n}.$$

Combining this with the inequalities (4.9), (4.13), and (4.16), we see that the hypotheses of Theorem 4.8 are satisfied, and so we may conclude that

$$rt_j - r_jt = 0. \tag{4.17}$$

Subtracting t_j times (4.12) from t times (4.11), and using the identity (4.17), we obtain $n(st_j - s_jt) = 0$, and hence

$$st_j - s_jt = 0. \tag{4.18}$$

From (4.18), we see that $t_j \mid s_jt$, and since $\gcd(s_j, t_j) = 1$, we must have $t_j \mid t$. So $t = t_jq$ for some q, and we must have $q \neq 0$ since $t \neq 0$. Substituting t_jq for t in equations (4.17) and (4.18) yields $r = r_jq$ and $s = s_jq$. That proves part (ii) of the theorem. $\square$

In our applications in this text, we shall only directly use part (ii) of this theorem; however, part (i) has applications as well (see Exercise 4.18).

4.6.1 Application: recovering fractions from their decimal expansions

It should be a familiar fact to the reader that every real number has a decimal expansion, and that this decimal expansion is unique, provided one rules out those expansions that end in an infinite run of 9's (e.g., $1/10 = 0.1000 \cdots = 0.0999 \cdots$).

Now suppose that Alice and Bob play a game. Alice thinks of a rational number $z := s/t$, where s and t are integers with $0 \leq s < t$, and tells Bob some of the high-order digits in the decimal expansion of z. Bob's goal in the game is to determine z. Can he do this?

The answer is "yes," provided Bob knows an upper bound M on t, and provided Alice gives Bob enough digits. Of course, Bob probably remembers from grade school that the decimal expansion of z is ultimately periodic, and that given enough digits of z so that the periodic part is included, he can recover z; however, this technique is quite useless in practice, as the length of the period can be huge—

$\Theta(M)$ in the worst case (see Exercises 4.21–4.23 below). The method we discuss here requires only $O(\text{len}(M))$ digits.

Suppose Alice gives Bob the high-order k digits of z, for some $k \geq 1$. That is, if

$$z = 0.z_1 z_2 z_3 \cdots \qquad (4.19)$$

is the decimal expansion of z, then Alice gives Bob $z_1, \ldots, z_k$. Now, if 10^k is much smaller than M^2, the number z is not even uniquely determined by these digits, since there are $\Omega(M^2)$ distinct rational numbers of the form s/t, with $0 \leq s < t \leq M$ (see Exercise 1.33). However, if $10^k > 2M^2$, then not only is z uniquely determined by $z_1, \ldots, z_k$, but using Theorem 4.9, Bob can efficiently compute it.

We shall presently describe efficient algorithms for both Alice and Bob, but before doing so, we make a few general observations about the decimal expansion of z. Let e be an arbitrary non-negative integer, and suppose that the decimal expansion of z is as in (4.19). Observe that

$$10^e z = z_1 \cdots z_e . z_{e+1} z_{e+2} \cdots .$$

It follows that

$$\lfloor 10^e z \rfloor = z_1 \cdots z_e . 0 . \qquad (4.20)$$

Since $z = s/t$, if we set $r := 10^e s \bmod t$, then $10^e s = \lfloor 10^e z \rfloor t + r$, and dividing this by t, we have $10^e z = \lfloor 10^e z \rfloor + r/t$, where $r/t \in [0, 1)$. Therefore,

$$\frac{10^e s \bmod t}{t} = 0 . z_{e+1} z_{e+2} z_{e+3} \cdots . \qquad (4.21)$$

Next, consider Alice. Based on the above discussion, Alice may use the following simple, iterative algorithm to compute $z_1, \ldots, z_k$, for arbitrary $k \geq 1$, after she chooses s and t:

$x_1 \leftarrow s$
for $i \leftarrow 1$ to k do
    $y_i \leftarrow 10x_i$
    $z_i \leftarrow \lfloor y_i/t \rfloor$
    $x_{i+1} \leftarrow y_i \bmod t$
output $z_1, \ldots, z_k$

Correctness follows easily from the observation that for each $i = 1, 2, \ldots$, we have $x_i = 10^{i-1}s \bmod t$; indeed, applying (4.21) with $e = i - 1$, we have $x_i/t = 0 . z_i z_{i+1} z_{i+2} \cdots$, and consequently, by (4.20) with $e = 1$ and x_i/t in the role of z, we have $\lfloor 10 x_i/t \rfloor = z_i$. The total time for Alice's computation is $O(k \, \text{len}(M))$, since each loop iteration takes time $O(\text{len}(M))$.

Finally, consider Bob. Given the high-order digits $z_1, \ldots, z_k$ of $z = s/t$, along with the upper bound M on t, he can compute z as follows:

1. Compute $n \leftarrow 10^k$ and $b \leftarrow \sum_{i=1}^{k} z_i 10^{k-i}$.

2. Run the extended Euclidean algorithm on input n, b to obtain $EEA(n, b)$, and then apply Theorem 4.9 with n, b, and $r^* := t^* := M$, to obtain the values r', s', t'.

3. Output the rational number $-s'/t'$.

Let us analyze this algorithm, assuming that $10^k > 2M^2$.

For correctness, we must show that $z = -s'/t'$. To prove this, observe that by (4.20) with $e = k$, we have $b = \lfloor nz \rfloor = \lfloor ns/t \rfloor$. Moreover, if we set $r := ns \bmod t$, then we have

$$r = ns - bt, \ 0 \le r < t \le r^*, \ 0 < t \le t^*, \text{ and } n > 2r^*t^*.$$

It follows that the integers s', t' from Theorem 4.9 satisfy $s = s'q$ and $-t = t'q$ for some non-zero integer q. Thus, $s/t = -s'/t'$, as required. As a bonus, since the extended Euclidean algorithm guarantees that $\gcd(s', t') = 1$, not only do we obtain z, but we obtain z expressed as a fraction in lowest terms.

We leave it to the reader to verify that Bob's computation may be performed in time $O(k^2)$.

We conclude that both Alice and Bob can successfully play this game with k chosen so that $k = O(\text{len}(M))$, in which case, their algorithms run in time $O(\text{len}(M)^2)$.

Example 4.5. Alice chooses integers s, t, with $0 \le s < t \le 1000$, and tells Bob the high-order seven digits in the decimal expansion of $z := s/t$, from which Bob should be able to compute z. Suppose $s = 511$ and $t = 710$. Then $s/t = 0.7197183098591549 \cdots$. Bob receives the digits $7, 1, 9, 7, 1, 8, 3$, and computes $n = 10^7$ and $b = 7197183$. Running the extended Euclidean algorithm on input n, b, Bob obtains the data in Fig. 4.1. The first r_j that meets the threshold $r^* = 1000$ is at $j = 10$, and Bob reads off $s' = 511$ and $t' = -710$, from which he obtains $z = -s'/t' = 511/710$.

Another interesting phenomenon to observe in Fig. 4.1 is that the fractions $-s_i/t_i$ are very good approximations to the fraction $b/n = 7197183/10000000$; indeed, if we compute the error terms $b/n + s_i/t_i$ for $i = 1, \ldots, 5$, we get (approximately)

$$0.72, \quad -0.28, \quad 0.053, \quad -0.03, \quad 0.0054.$$

Thus, we can approximate the "complicated" fraction $7197183/10000000$ by the "very simple" fraction $5/7$, introducing an absolute error of less than 0.006. Exercise 4.18 explores this "data compression" capability of Euclid's algorithm in more generality. $\square$

i	r_i	q_i	s_i	t_i
0	10000000		1	0
1	7197183	1	0	1
2	2802817	2	1	-1
3	1591549	1	-2	3
4	1211268	1	3	-4
5	380281	3	-5	7
6	70425	5	18	-25
7	28156	2	-95	132
8	14113	1	208	-289
9	14043	1	-303	421
10	70	200	511	-710
11	43	1	-102503	142421
12	27	1	103014	-143131
13	16	1	-205517	285552
14	11	1	308531	-428683
15	5	2	-514048	714235
16	1	5	1336627	-1857153
17		0	-7197183	10000000

Fig. 4.1. Bob's data from the extended Euclidean algorithm

4.6.2 Application: Chinese remaindering with errors

One interpretation of the Chinese remainder theorem is that if we "encode" an integer a, with $0 \le a < n$, as the sequence $(a_1, \ldots, a_k)$, where $a_i = a \bmod n_i$ for $i = 1, \ldots, k$, then we can efficiently recover a from this encoding. Here, of course, $n = n_1 \cdots n_k$, and the family $\{n_i\}_{i=1}^{k}$ is pairwise relatively prime.

Suppose that Alice encodes a as $(a_1, \ldots, a_k)$, and sends this encoding to Bob over some communication network; however, because the network is not perfect, during the transmission of the encoding, some (but hopefully not too many) of the values $a_1, \ldots, a_k$ may be corrupted. The question is, can Bob still efficiently recover the original a from its corrupted encoding?

To make the problem more precise, suppose that the original, correct encoding of a is $(a_1, \ldots, a_k)$, and the corrupted encoding is $(b_1, \ldots, b_k)$. Let us define $G \subseteq \{1, \ldots, k\}$ to be the set of "good" positions i with $a_i = b_i$, and $B \subseteq \{1, \ldots, k\}$ to be the set of "bad" positions i with $a_i \ne b_i$. We shall assume that $|B| \le \ell$, where ℓ is some specified parameter.

Of course, if Bob hopes to recover a, we need to build some redundancy into the system; that is, we must require that $0 \le a \le M$ for some bound M that is somewhat smaller than n. Now, if Bob knew the location of bad positions, and if the product of the n_i's at the good positions exceeds M, then Bob could simply discard the errors, and reconstruct a by applying the Chinese remainder theorem to the a_i's and n_i's at the good positions. However, in general, Bob will not know a priori the locations of the bad positions, and so this approach will not work.

Despite these apparent difficulties, Theorem 4.9 may be used to solve the problem quite easily, as follows. Let P be an upper bound on the product of any ℓ of the integers $n_1, \ldots, n_k$ (e.g., we could take P to be the product of the ℓ largest numbers among $n_1, \ldots, n_k$). Further, let us assume that $n > 2MP^2$.

Now, suppose Bob obtains the corrupted encoding $(b_1, \ldots, b_k)$. Here is what Bob does to recover a:

1. Apply the Chinese remainder theorem, obtaining the integer b satisfying $0 \le b < n$ and $b \equiv b_i \pmod{n_i}$ for $i = 1, \ldots, k$.

2. Run the extended Euclidean algorithm on input n, b to obtain EEA(n, b), and then apply Theorem 4.9 with n, b, $r^* := MP$ and $t^* := P$, to obtain values r', s', t'.

3. If $t' \mid r'$, output the integer r'/t'; otherwise, output "error."

We claim that the above procedure outputs a, under our assumption that the set B of bad positions is of size at most ℓ. To see this, let $t := \prod_{i \in B} n_i$. By construction, we have $1 \le t \le P$. Also, let $r := at$, and note that $0 \le r \le r^*$ and $0 < t \le t^*$. We claim that

$$r \equiv bt \pmod{n}. \tag{4.22}$$

To show that (4.22) holds, it suffices to show that

$$at \equiv bt \pmod{n_i} \tag{4.23}$$

for all $i = 1, \ldots, k$. To show this, for each index i we consider two cases:

Case 1: $i \in G$. In this case, we have $a_i = b_i$, and therefore,

$$at \equiv a_i t \equiv b_i t \equiv bt \pmod{n_i}.$$

Case 2: $i \in B$. In this case, we have $n_i \mid t$, and therefore,

$$at \equiv 0 \equiv bt \pmod{n_i}.$$

Thus, (4.23) holds for all $i = 1, \ldots, k$, and so it follows that (4.22) holds. Therefore, the values r', t' obtained from Theorem 4.9 satisfy

$$\frac{r'}{t'} = \frac{r}{t} = \frac{at}{t} = a.$$

One easily checks that both the procedures to encode and decode a value a run in time $O(\text{len}(n)^2)$.

The above scheme is an example of an **error correcting code**, and is actually the integer analog of a **Reed–Solomon code**.

Example 4.6. Suppose we want to encode a 1024-bit message as a sequence of 16-bit blocks, so that the above scheme can correct up to 3 corrupted blocks. Without any error correction, we would need just $1024/16 = 64$ blocks. However, to correct this many errors, we need a few extra blocks; in fact, 7 will do.

Of course, a 1024-bit message can naturally be viewed as an integer a in the set $\{0, \ldots, 2^{1024} - 1\}$, and the ith 16-bit block in the encoding can be viewed as an integer a_i in the set $\{0, \ldots, 2^{16} - 1\}$. Setting $k := 71$, we select k primes, $n_1, \ldots, n_k$, each 16-bits in length. In fact, let us choose $n_1, \ldots, n_k$ to be the *largest* k primes under 2^{16}. If we do this, then the smallest prime among the n_i's turns out to be 64717, which is greater than $2^{15.98}$. We may set $M := 2^{1024}$, and since we want to correct up to 3 errors, we may set $P := 2^{3\cdot16}$. Then with $n := \prod_i n_i$, we have

$$n > 2^{71\cdot15.98} = 2^{1134.58} > 2^{1121} = 2^{1+1024+6\cdot16} = 2MP^2.$$

Thus, with these parameter settings, the above scheme will correct up to 3 corrupted blocks. This comes at a cost of increasing the length of the message from 1024 bits to $71 \cdot 16 = 1136$ bits, an increase of about 11%. $\square$

4.6.3 Applications to symbolic algebra

Rational reconstruction also has a number of applications in symbolic algebra. We briefly sketch one such application here. Suppose that we want to find the solution v to the equation $vA = w$, where we are given as input a non-singular square integer matrix A and an integer vector w. The solution vector v will, in general, have rational entries. We stress that we want to compute the *exact* solution v, and not some floating point approximation to it. Now, we could solve for v directly using Gaussian elimination; however, the intermediate quantities computed by that algorithm would be rational numbers whose numerators and denominators might get quite large, leading to a rather lengthy computation (however, it is possible to show that the overall running time is still polynomial in the input length).

Another approach is to compute a solution vector modulo n, where n is a power of a prime that does not divide the determinant of A. Provided n is large enough, one can then recover the solution vector v using rational reconstruction. With this approach, all of the computations can be carried out using arithmetic on integers not too much larger than n, leading to a more efficient algorithm. More of the details of this procedure are developed later, in Exercise 14.18.

EXERCISE 4.18. Let $n, b \in \mathbb{Z}$ with $0 \le b < n$, and let $EEA(n, b) = \{(r_i, s_i, t_i)\}_{i=0}^{\lambda+1}$. This exercise develops some key properties of the fractions $-s_i/t_i$ as approximations to b/n. For $i = 1, \ldots, \lambda + 1$, let $\varepsilon_i := b/n + s_i/t_i$.

  (a) Show that $\varepsilon_i = r_i/t_i n$ for $i = 1, \ldots, \lambda + 1$.

  (b) Show that successive ε_i's strictly decrease in absolute value, and alternate in sign.

  (c) Show that $|\varepsilon_i| < 1/t_i^2$ for $i = 1, \ldots, \lambda$, and $\varepsilon_{\lambda+1} = 0$.

  (d) Show that for all $s, t \in \mathbb{Z}$ with $t \ne 0$, if $|b/n - s/t| < 1/2t^2$, then $s/t = -s_i/t_i$ for some $i = 1, \ldots, \lambda + 1$. Hint: use part (ii) of Theorem 4.9.

  (e) Consider a fixed index $i \in \{2, \ldots, \lambda + 1\}$. Show that for all $s, t \in \mathbb{Z}$, if $0 < |t| \le |t_i|$ and $|b/n - s/t| \le |\varepsilon_i|$, then $s/t = -s_i/t_i$. In this sense, $-s_i/t_i$ is the unique, best approximation to b/n among all fractions of denominator at most $|t_i|$. Hint: use part (i) of Theorem 4.9.

EXERCISE 4.19. Using the decimal approximation $\pi \approx 3.141592654$, apply Euclid's algorithm to calculate a rational number of denominator less than 1000 that is within 10^{-6} of π. Illustrate the computation with a table as in Fig. 4.1.

EXERCISE 4.20. Show that given integers s, t, k, with $0 \le s < t$, and $k > 0$, we can compute the kth digit in the decimal expansion of s/t in time $O(\text{len}(k) \text{len}(t)^2)$.

  For the following exercises, we need a definition. Let $\Psi = \{z_i\}_{i=1}^{\infty}$ be a sequence of elements drawn from some arbitrary set. For integers $k \ge 0$ and $\ell \ge 1$, we say that Ψ is (k, ℓ)-**periodic** if $z_i = z_{i+\ell}$ for all $i > k$; in addition, we say that Ψ is **ultimately periodic** if it is (k, ℓ)-periodic for some (k, ℓ).

EXERCISE 4.21. Show that if a sequence Ψ is ultimately periodic, then it is (k^*, ℓ^*)-periodic for some uniquely determined pair (k^*, ℓ^*) for which the following holds: for every pair (k, ℓ) such that Ψ is (k, ℓ)-periodic, we have $k^* \le k$ and $\ell^* \mid \ell$.

  The value ℓ^* in the above exercise is called the **period** of Ψ, and k^* is called the **pre-period** of Ψ. If its pre-period is zero, then Ψ is called **purely periodic**.

EXERCISE 4.22. Let z be a real number whose decimal expansion is an ultimately periodic sequence. Show that z is rational.

EXERCISE 4.23. Let $z = s/t \in \mathbb{Q}$, where s and t are relatively prime integers with $0 \le s < t$. Show that:

  (a) there exist integers k, k' such that $0 \le k < k'$ and $s10^k \equiv s10^{k'} \pmod{t}$;

  (b) for all integers k, k' with $0 \le k < k'$, the decimal expansion of z is $(k, k' - k)$-periodic if and only if $s10^k \equiv s10^{k'} \pmod{t}$;

(c) if $\gcd(10, t) = 1$, then the decimal expansion of z is purely periodic with period equal to the multiplicative order of 10 modulo t;

(d) more generally, if k is the smallest non-negative integer such that 10 and $t' := t / \gcd(10^k, t)$ are relatively prime, then the decimal expansion of z is ultimately periodic with pre-period k and period equal to the multiplicative order of 10 modulo t'.

A famous conjecture of Artin postulates that for every integer d, not equal to -1 or to the square of an integer, there are infinitely many primes t such that d has multiplicative order $t - 1$ modulo t. If Artin's conjecture is true, then by part (c) of the previous exercise, there are infinitely many primes t such that the decimal expansion of s/t, for every s with $0 < s < t$, is a purely periodic sequence of period $t - 1$. In light of these observations, the "grade school" method of computing a fraction from its decimal expansion using the period is hopelessly impractical.

4.7 The RSA cryptosystem

One of the more exciting uses of number theory in recent decades is its application to cryptography. In this section, we give a brief overview of the RSA cryptosystem, named after its inventors Rivest, Shamir, and Adleman. At this point in the text, we already have the concepts and tools at our disposal necessary to understand the basic operation of this system, even though a full understanding of the system will require other ideas that will be developed later in the text.

Suppose that Alice wants to send a secret message to Bob over an insecure network. An adversary may be able to eavesdrop on the network, and so sending the message "in the clear" is not an option. Using older, more traditional cryptographic techniques would require that Alice and Bob share a secret key between them; however, this creates the problem of securely generating such a shared secret. The RSA cryptosystem is an example of a **public key cryptosystem**. To use the system, Bob simply places a "public key" in the equivalent of an electronic telephone book, while keeping a corresponding "private key" secret. To send a secret message to Bob, Alice obtains Bob's public key from the telephone book, and uses this to encrypt her message. Upon receipt of the encrypted message, Bob uses his private key to decrypt it, obtaining the original message.

Here is how the RSA cryptosystem works. To generate a public key/private key pair, Bob generates two very large, random primes p and q, with $p \neq q$. To be secure, p and q should be quite large; in practice, they are chosen to be around 512 bits in length. Efficient algorithms for generating such primes exist, and we shall discuss them in detail later in the text (that there are sufficiently many primes of a given bit length will be discussed in Chapter 5; algorithms for generating them will be discussed at a high level in §9.4, and in greater detail in Chapter 10). Next, Bob computes $n := pq$. Bob also selects an integer $e > 1$ such that $\gcd(e, \varphi(n)) = 1$, where φ is Euler's phi function. Here, $\varphi(n) = (p-1)(q-1)$. Finally, Bob computes $d := e^{-1} \bmod \varphi(n)$, using the extended Euclidean algorithm. The public key is the pair (n, e), and the private key is the pair (n, d). The integer e is called the "encryption exponent" and d is called the "decryption exponent." In practice, the integers n and d are about 1024 bits in length, while e is usually significantly shorter.

After Bob publishes his public key (n, e), Alice may send a secret message to Bob as follows. Suppose that a message is encoded in some canonical way as a number between 0 and $n - 1$ — we can always interpret a bit string of length less than $\text{len}(n)$ as such a number. Thus, we may assume that a message is an element α of $\mathbb{Z}_n$. To encrypt the message α, Alice simply computes $\beta := \alpha^e$ using repeated squaring. The encrypted message is β. When Bob receives β, he computes $\gamma := \beta^d$, and interprets γ as a message.

The most basic requirement of any encryption scheme is that decryption should "undo" encryption. In this case, this means that for all $\alpha \in \mathbb{Z}_n$, we should have

$$(\alpha^e)^d = \alpha. \tag{4.24}$$

If $\alpha \in \mathbb{Z}_n^*$, then this is clearly the case, since we have $ed = 1 + \varphi(n)k$ for some positive integer k, and hence by Euler's theorem (Theorem 2.13), we have

$$(\alpha^e)^d = \alpha^{ed} = \alpha^{1+\varphi(n)k} = \alpha \cdot \alpha^{\varphi(n)k} = \alpha.$$

To argue that (4.24) holds in general, let α be an arbitrary element of $\mathbb{Z}_n$, and suppose $\alpha = [a]_n$. If $a \equiv 0 \pmod{p}$, then trivially $a^{ed} \equiv 0 \pmod{p}$; otherwise,

$$a^{ed} \equiv a^{1+\varphi(n)k} \equiv a \cdot a^{\varphi(n)k} \equiv a \pmod{p},$$

where the last congruence follows from the fact that $\varphi(n)k$ is a multiple of $p - 1$, which is a multiple of the multiplicative order of a modulo p (again by Euler's theorem). Thus, we have shown that $a^{ed} \equiv a \pmod{p}$. The same argument shows that $a^{ed} \equiv a \pmod{q}$, and these two congruences together imply that $a^{ed} \equiv a \pmod{n}$. Thus, we have shown that equation (4.24) holds for all $\alpha \in \mathbb{Z}_n$.

Of course, the interesting question about the RSA cryptosystem is whether or not it really is secure. Now, if an adversary, given only the public key (n, e), were able to factor n, then he could easily compute the decryption exponent d himself using the same algorithm used by Bob. It is widely believed that factoring n is computationally infeasible, for sufficiently large n, and so this line of attack is ineffective, barring a breakthrough in factorization algorithms. Indeed, while trying to factor n by brute-force search is clearly infeasible, there are much faster algorithms, but even these are not fast enough to pose a serious threat to the security of the RSA cryptosystem. We shall discuss some of these faster algorithms in some detail later in the text (in Chapter 15).

Can one break the RSA cryptosystem without factoring n? For example, it is natural to ask whether one can compute the decryption exponent d without having to go to the trouble of factoring n. It turns out that the answer to this question is "no": if one could compute the decryption exponent d, then $ed - 1$ would be a multiple of $\varphi(n)$, and as we shall see later in §10.4, given any multiple of $\varphi(n)$, we can easily factor n. Thus, computing the decryption exponent is equivalent to factoring n, and so this line of attack is also ineffective. But there still could be other lines of attack. For example, even if we assume that factoring large numbers is infeasible, this is not enough to guarantee that for a given encrypted message β, the adversary is unable to compute β^d (although nobody actually knows how to do this without first factoring n).

The reader should be warned that the proper notion of security for an encryption scheme is quite subtle, and a detailed discussion of this is well beyond the scope of this text. Indeed, the simple version of RSA presented here suffers from a number of security problems (because of this, actual implementations of public-key encryption schemes based on RSA are somewhat more complicated). We mention one such problem here (others are examined in some of the exercises below). Suppose an eavesdropping adversary knows that Alice will send one of a few, known, candidate messages. For example, an adversary may know that Alice's message is either "let's meet today" or "let's meet tomorrow." In this case, the adversary can encrypt for himself each of the candidate messages, intercept Alice's actual encrypted message, and then by simply comparing encryptions, the adversary can determine which particular message Alice encrypted. This type of attack works simply because the encryption algorithm is deterministic, and in fact, any deterministic encryption algorithm will be vulnerable to this type of attack. To avoid this type of attack, one must use a *probabilistic* encryption algorithm. In the case of the RSA cryptosystem, this is often achieved by padding the message with some random bits before encrypting it (but even this must be done carefully).

EXERCISE 4.24. This exercise develops a method to speed up RSA decryption. Suppose that we are given two distinct ℓ-bit primes, p and q, an element $\beta \in \mathbb{Z}_n$, where $n := pq$, and an integer d, where $1 < d < \varphi(n)$. Using the algorithm from Exercise 3.35, we can compute β^d at a cost of essentially 2ℓ squarings in $\mathbb{Z}_n$. Show how this can be improved, making use of the factorization of n, so that the total cost is essentially that of ℓ squarings in $\mathbb{Z}_p$ and ℓ squarings in $\mathbb{Z}_q$, leading to a roughly four-fold speed-up in the running time.

EXERCISE 4.25. Alice submits a bid to an auction, and so that other bidders cannot see her bid, she encrypts it under the public key of the auction service. Suppose that the auction service provides a public key for an RSA encryption scheme, with a modulus n. Assume that bids are encoded simply as integers between 0 and $n - 1$ prior to encryption. Also, assume that Alice submits a bid that is a "round number," which in this case means that her bid is a number that is divisible by 10. Show how an eavesdropper can submit an encryption of a bid that exceeds Alice's bid by 10%, without even knowing what Alice's bid is. In particular, your attack should work even if the space of possible bids is very large.

EXERCISE 4.26. To speed up RSA encryption, one may choose a very small encryption exponent. This exercise develops a "small encryption exponent attack" on RSA. Suppose Bob, Bill, and Betty have RSA public keys with moduli n_1, n_2, and n_3, and all three use encryption exponent 3. Assume that $\{n_i\}_{i=1}^3$ is pairwise relatively prime. Suppose that Alice sends an encryption of the same message to Bob, Bill, and Betty — that is, Alice encodes her message as an integer a, with $0 \le a < \min\{n_1, n_2, n_3\}$, and computes the three encrypted messages $\beta_i := [a^3]_{n_i}$, for $i = 1, \ldots, 3$. Show how to recover Alice's message from these three encrypted messages.

EXERCISE 4.27. To speed up RSA decryption, one might choose a small decryption exponent, and then derive the encryption exponent from this. This exercise develops a "small decryption exponent attack" on RSA. Suppose $n = pq$, where p and q are distinct primes with $\text{len}(p) = \text{len}(q)$. Let d and e be integers such that $1 < d < \varphi(n)$, $1 < e < \varphi(n)$, and $de \equiv 1 \pmod{\varphi(n)}$. Further, assume that $d < n^{1/4}/3$. Show how to efficiently compute d, given n and e. Hint: since $ed \equiv 1 \pmod{\varphi(n)}$, it follows that $ed = 1 + \varphi(n)k$ for an integer k with $0 < k < d$; let $r := nk - ed$, and show that $|r| < n^{3/4}$; next, show how to recover d (along with r and k) using Theorem 4.9.

4.8 Notes

The Euclidean algorithm as we have presented it here is not the fastest known algorithm for computing greatest common divisors. The asymptotically fastest known algorithm for computing the greatest common divisor of two numbers of bit length at most ℓ runs in time $O(\ell \, \text{len}(\ell))$ on a RAM, which is due to Schönhage [85]. The same algorithm leads to Boolean circuits of size $O(\ell \, \text{len}(\ell)^2 \, \text{len}(\text{len}(\ell)))$, which using Fürer's result [38], can be reduced to $O(\ell \, \text{len}(\ell)^2 \, 2^{O(\log^* n)})$. The same complexity results also hold for the extended Euclidean algorithm, as well as for Chinese remaindering, Thue's lemma, and rational reconstruction.

Experience suggests that such fast algorithms for greatest common divisors are not of much practical value, unless the integers involved are *very* large — at least several tens of thousands of bits in length. The extra "log" factor and the rather large multiplicative constants seem to slow things down too much.

The binary gcd algorithm (Exercise 4.6) is due to Stein [100]. The extended binary gcd algorithm (Exercise 4.10) was first described by Knuth [56], who attributes it to M. Penk. Our formulation of both of these algorithms closely follows that of Menezes, van Oorschot, and Vanstone [66]. Experience suggests that the binary gcd algorithm is faster in practice than Euclid's algorithm.

Schoof [87] presents (among other things) a deterministic, polynomial-time algorithm that computes a square root of -1 modulo p for any given prime $p \equiv 1 \pmod 4$. If we use this algorithm in §4.5, we get a deterministic, polynomial-time algorithm to compute integers r and t such that $p = r^2 + t^2$.

Our Theorem 4.9 is a generalization of one stated in Wang, Guy, and Davenport [103]. One can generalize Theorem 4.9 using the theory of **continued fractions**. With this, one can generalize Exercise 4.18 to deal with rational approximations to irrational numbers. More on this can be found, for example, in the book by Hardy and Wright [46].

The application of Euclid's algorithm to computing a rational number from the first digits of its decimal expansion was observed by Blum, Blum, and Shub [17], where they considered the possibility of using such sequences of digits as a pseudo-random number generator — the conclusion, of course, is that this is not such a good idea.

The RSA cryptosystem was invented by Rivest, Shamir, and Adleman [82]. There is a vast literature on cryptography. One starting point is the book by Menezes, van Oorschot, and Vanstone [66]. The attack in Exercise 4.27 is due to Wiener [110]; this attack was recently strengthened by Boneh and Durfee [19].

5

The distribution of primes

This chapter concerns itself with the question: how many primes are there? In Chapter 1, we proved that there are infinitely many primes; however, we are interested in a more quantitative answer to this question; that is, we want to know how "dense" the prime numbers are.

This chapter has a bit more of an "analytical" flavor than other chapters in this text. However, we shall not make use of any mathematics beyond that of elementary calculus.

5.1 Chebyshev's theorem on the density of primes

The natural way of measuring the density of primes is to count the number of primes up to a bound x, where x is a real number. To this end, we introduce the function $\pi(x)$, whose value at each real number $x \geq 0$ is defined to be the number of primes up to (and including) x. For example, $\pi(1) = 0$, $\pi(2) = 1$, and $\pi(7.5) = 4$. The function $\pi(x)$ is an example of a "step function," that is, a function that changes values only at a discrete set of points. It might seem more natural to define $\pi(x)$ only on the integers, but it is the tradition to define it over the real numbers (and there are some technical benefits in doing so).

Let us first take a look at some values of $\pi(x)$. Table 5.1 shows values of $\pi(x)$ for $x = 10^{3i}$ and $i = 1, \ldots, 6$. The third column of this table shows the value of $x/\pi(x)$ (to five decimal places). One can see that the differences between successive rows of this third column are roughly the same — about 6.9 — which suggests that the function $x/\pi(x)$ grows logarithmically in x. Indeed, as $\log(10^3) \approx 6.9$, it would not be unreasonable to guess that $x/\pi(x) \approx \log x$, or equivalently, $\pi(x) \approx x/\log x$ (as discussed in the Preliminaries, $\log x$ denotes the natural logarithm of x).

The following theorem is a first — and important — step towards making the above guesswork more rigorous (the statements of this and many other results in this chapter make use of the asymptotic notation introduced in §3.1):

Table 5.1. *Some values of* $\pi(x)$

x	$\pi(x)$	$x/\pi(x)$
10^3	168	5.95238
10^6	78498	12.73918
10^9	50847534	19.66664
10^{12}	37607912018	26.59015
10^{15}	29844570422669	33.50693
10^{18}	24739954287740860	40.42045

Theorem 5.1 (Chebyshev's theorem). *We have*

$$\pi(x) = \Theta(x/\log x).$$

It is not too difficult to prove this theorem, which we now proceed to do in several steps. We begin with some elementary bounds on binomial coefficients (see §A2):

Lemma 5.2. *If m is a positive integer, then*

$$\binom{2m}{m} \geq 2^{2m}/2m \quad \text{and} \quad \binom{2m+1}{m} < 2^{2m}.$$

Proof. As $\binom{2m}{m}$ is the largest binomial coefficient in the binomial expansion of $(1+1)^{2m}$, we have

$$2^{2m} = \sum_{i=0}^{2m}\binom{2m}{i} = 1 + \sum_{i=1}^{2m-1}\binom{2m}{i} + 1 \leq 2 + (2m-1)\binom{2m}{m} \leq 2m\binom{2m}{m}.$$

The proves the first inequality. For the second, observe that the binomial coefficient $\binom{2m+1}{m}$ occurs twice in the binomial expansion of $(1+1)^{2m+1}$, and is therefore less than $2^{2m+1}/2 = 2^{2m}$. $\square$

Next, recalling that $v_p(n)$ denotes the power to which a prime p divides an integer n, we continue with the following observation:

Lemma 5.3. *Let n be a positive integer. For every prime p, we have*

$$v_p(n!) = \sum_{k\geq 1}\lfloor n/p^k\rfloor.$$

Proof. For all positive integers j, k, define $d_{jk} := 1$ if $p^k \mid j$, and $d_{jk} := 0$, otherwise. Observe that $v_p(j) = \sum_{k\geq 1} d_{jk}$ (this sum is actually finite, since $d_{jk} = 0$

for all sufficiently large k). So we have

$$v_p(n!) = \sum_{j=1}^{n} v_p(j) = \sum_{j=1}^{n} \sum_{k \geq 1} d_{jk} = \sum_{k \geq 1} \sum_{j=1}^{n} d_{jk}.$$

Finally, note that $\sum_{j=1}^{n} d_{jk}$ is equal to the number of multiples of p^k among the integers $1, \ldots, n$, which by Exercise 1.3 is equal to $\lfloor n/p^k \rfloor$. $\square$

The following theorem gives a lower bound on $\pi(x)$.

Theorem 5.4. $\pi(n) \geq \frac{1}{2}(\log 2)n / \log n$ *for every integer* $n \geq 2$.

Proof. Let m be a positive integer, and consider the binomial coefficient

$$N := \binom{2m}{m} = \frac{(2m)!}{(m!)^2}.$$

It is clear that N is divisible only by primes p up to $2m$. Applying Lemma 5.3 to the identity $N = (2m)!/(m!)^2$, we have

$$v_p(N) = \sum_{k \geq 1} (\lfloor 2m/p^k \rfloor - 2\lfloor m/p^k \rfloor).$$

Each term in this sum is either 0 or 1 (see Exercise 1.4), and for $k > \log(2m)/\log p$, each term is zero. Thus, $v_p(N) \leq \log(2m)/\log p$. So we have

$$\pi(2m)\log(2m) = \sum_{p \leq 2m} \frac{\log(2m)}{\log p} \log p$$

$$\geq \sum_{p \leq 2m} v_p(N) \log p = \log N,$$

where the summations are over the primes p up to $2m$. By Lemma 5.2, we have $N \geq 2^{2m}/2m \geq 2^m$, and hence

$$\pi(2m)\log(2m) \geq m \log 2 = \frac{1}{2}(\log 2)(2m).$$

That proves the theorem for even n. Now consider odd $n \geq 3$, so $n = 2m - 1$ for some $m \geq 2$. It is easily verified that the function $x/\log x$ is increasing for $x \geq 3$; therefore,

$$\pi(2m - 1) = \pi(2m)$$

$$\geq \frac{1}{2}(\log 2)(2m)/\log(2m)$$

$$\geq \frac{1}{2}(\log 2)(2m - 1)/\log(2m - 1).$$

That proves the theorem for odd n. $\square$

As a consequence of the above theorem, we have $\pi(x) = \Omega(x/\log x)$ for real numbers x. Indeed, setting $c := \frac{1}{2}(\log 2)$, for every real number $x \geq 2$, we have

$$\pi(x) = \pi(\lfloor x \rfloor) \geq c\lfloor x \rfloor / \log\lfloor x \rfloor \geq c(x-1)/\log x;$$

from this, it is clear that $\pi(x) = \Omega(x/\log x)$.

To obtain a corresponding upper bound for $\pi(x)$, we introduce an auxiliary function, called **Chebyshev's theta function**:

$$\vartheta(x) := \sum_{p \leq x} \log p,$$

where the sum is over all primes p up to x.

Chebyshev's theta function is an example of a summation over primes, and in this chapter, we will be considering a number of functions that are defined in terms of sums or products over primes (and indeed, such summations already cropped up in the proof of Theorem 5.4). To avoid excessive tedium, we adopt the usual convention used by number theorists: if not explicitly stated, summations and products over the variable p are always understood to be over primes. For example, we may write $\pi(x) = \sum_{p \leq x} 1$.

Theorem 5.5. *We have*

$$\vartheta(x) = \Theta(\pi(x) \log x).$$

Proof. On the one hand, we have

$$\vartheta(x) = \sum_{p \leq x} \log p \leq \log x \sum_{p \leq x} 1 = \pi(x) \log x.$$

On the other hand, we have

$$\vartheta(x) = \sum_{p \leq x} \log p \geq \sum_{x^{1/2} < p \leq x} \log p \geq \frac{1}{2}\log x \sum_{x^{1/2} < p \leq x} 1$$

$$= \frac{1}{2}\log x \left(\pi(x) - \pi(x^{1/2})\right) = \frac{1}{2}\left(1 - \pi(x^{1/2})/\pi(x)\right)\pi(x)\log x.$$

It will therefore suffice to show that $\pi(x^{1/2})/\pi(x) = o(1)$. Clearly, $\pi(x^{1/2}) \leq x^{1/2}$. Moreover, by the previous theorem, $\pi(x) = \Omega(x/\log x)$. Therefore,

$$\pi(x^{1/2})/\pi(x) = O(\log x/x^{1/2}) = o(1),$$

and the theorem follows. $\square$

Theorem 5.6. $\vartheta(x) < 2(\log 2)x$ *for every real number* $x \geq 1$.

Proof. It suffices to prove that $\vartheta(n) < 2(\log 2)n$ for every positive integer n, since then $\vartheta(x) = \vartheta(\lfloor x \rfloor) < 2(\log 2)\lfloor x \rfloor \leq 2(\log 2)x$. We prove this by induction on n.

For $n = 1$ and $n = 2$, this is clear, so assume $n > 2$. If n is even, then using the induction hypothesis for $n - 1$, we have

$$\vartheta(n) = \vartheta(n - 1) < 2(\log 2)(n - 1) < 2(\log 2)n.$$

Now consider the case where n is odd. Write $n = 2m + 1$, where m is a positive integer, and consider the binomial coefficient

$$M := \binom{2m + 1}{m} = \frac{(2m + 1) \cdots (m + 2)}{m!}.$$

Observe that M is divisible by all primes p with $m + 1 < p \le 2m + 1$. Moreover, be Lemma 5.2, we have $M < 2^{2m}$. It follows that

$$\vartheta(2m + 1) - \vartheta(m + 1) = \sum_{m+1 < p \le 2m+1} \log p \le \log M < 2(\log 2)m.$$

Using this, and the induction hypothesis for $m + 1$, we obtain

$$\vartheta(n) = \vartheta(2m + 1) - \vartheta(m + 1) + \vartheta(m + 1)$$
$$< 2(\log 2)m + 2(\log 2)(m + 1) = 2(\log 2)n. \quad \square$$

Another way of stating the above theorem is:

$$\prod_{p \le x} p < 4^x.$$

Theorem 5.1 follows immediately from Theorems 5.4, 5.5 and 5.6. Note that we have also proved:

Theorem 5.7. *We have*

$$\vartheta(x) = \Theta(x).$$

EXERCISE 5.1. For each positive integer n, let p_n denote the nth prime. Show that $p_n = \Theta(n \log n)$.

EXERCISE 5.2. For each positive integer n, let $\omega(n)$ denote the number of distinct primes dividing n. Show that $\omega(n) = O(\log n / \log \log n)$.

EXERCISE 5.3. Show that $\sum_{p \le x} 1 / \log p = \Theta(x/(\log x)^2)$.

5.2 Bertrand's postulate

Suppose we want to know how many primes there are of a given bit length, or more generally, how many primes there are between m and $2m$ for a given positive integer m. Neither the statement, nor our proof, of Chebyshev's theorem imply that there are *any* primes between m and $2m$, let alone a useful density estimate of such primes.

Bertrand's postulate is the assertion that for every positive integer m, there exists a prime between m and $2m$. We shall in fact prove a stronger result: there is at least one prime between m and $2m$, and moreover, the number of such primes is $\Omega(m/\log m)$.

Theorem 5.8 (Bertrand's postulate). *For every positive integer m, we have*

$$\pi(2m) - \pi(m) > \frac{m}{3\log(2m)}.$$

The proof uses Theorem 5.6, along with a more careful re-working of the proof of Theorem 5.4. The theorem is clearly true for $m \le 2$, so we may assume that $m \ge 3$. As in the proof of the Theorem 5.4, define $N := \binom{2m}{m}$, and recall that N is divisible only by primes less than $2m$, and that we have the identity

$$v_p(N) = \sum_{k \ge 1} (\lfloor 2m/p^k \rfloor - 2\lfloor m/p^k \rfloor), \tag{5.1}$$

where each term in the sum is either 0 or 1. We can characterize the values $v_p(N)$ a bit more precisely, as follows:

Lemma 5.9. *Let $m \ge 3$ and $N := \binom{2m}{m}$. For all primes p, we have:*

$$p^{v_p(N)} \le 2m; \tag{5.2}$$

$$\text{if } p > \sqrt{2m}, \text{ then } v_p(N) \le 1; \tag{5.3}$$

$$\text{if } 2m/3 < p \le m, \text{ then } v_p(N) = 0; \tag{5.4}$$

$$\text{if } m < p < 2m, \text{ then } v_p(N) = 1. \tag{5.5}$$

Proof. For (5.2), all terms with $k > \log(2m)/\log p$ in (5.1) vanish, and hence $v_p(N) \le \log(2m)/\log p$, from which it follows that $p^{v_p(N)} \le 2m$.

(5.3) follows immediately from (5.2).

For (5.4), if $2m/3 < p \le m$, then $2m/p < 3$, and we must also have $p \ge 3$, since $p = 2$ implies $m < 3$. We have $p^2 > p(2m/3) = 2m(p/3) \ge 2m$, and hence all terms with $k > 1$ in (5.1) vanish. The term with $k = 1$ also vanishes, since $1 \le m/p < 3/2$, from which it follows that $2 \le 2m/p < 3$, and hence $\lfloor m/p \rfloor = 1$ and $\lfloor 2m/p \rfloor = 2$.

For (5.5), if $m < p < 2m$, it follows that $1 < 2m/p < 2$, so $\lfloor 2m/p \rfloor = 1$. Also, $m/p < 1$, so $\lfloor m/p \rfloor = 0$. It follows that the term with $k = 1$ in (5.1) is 1, and it is clear that $2m/p^k < 1$ for all $k > 1$, and so all the other terms vanish. $\square$

We now have the necessary technical ingredients to prove Theorem 5.8. Define

$$P_m := \prod_{m < p < 2m} p,$$

and define Q_m so that

$$N = Q_m P_m.$$

By (5.4) and (5.5), we see that

$$Q_m = \prod_{p \leq 2m/3} p^{v_p(N)}.$$

Moreover, by (5.3), $v_p(N) > 1$ for at most those $p \leq \sqrt{2m}$, so there are at most $\sqrt{2m}$ such primes, and by (5.2), the contribution of each such prime to the above product is at most $2m$. Combining this with Theorem 5.6, we obtain

$$Q_m < (2m)^{\sqrt{2m}} \cdot 4^{2m/3}.$$

We now apply Lemma 5.2, obtaining

$$P_m = N Q_m^{-1} \geq 2^{2m} (2m)^{-1} Q_m^{-1} > 4^{m/3} (2m)^{-(1+\sqrt{2m})}.$$

It follows that

$$\pi(2m) - \pi(m) \geq \log P_m / \log(2m) > \frac{m \log 4}{3 \log(2m)} - (1 + \sqrt{2m})$$

$$= \frac{m}{3 \log(2m)} + \frac{m(\log 4 - 1)}{3 \log(2m)} - (1 + \sqrt{2m}).$$

Clearly, for all sufficiently large m, we have

$$\frac{m(\log 4 - 1)}{3 \log(2m)} > 1 + \sqrt{2m}. \tag{5.6}$$

That proves Theorem 5.8 for all sufficiently large m. Moreover, a simple calculation shows that (5.6) holds for all $m \geq 13{,}000$, and one can verify by brute force (with the aid of a computer) that the theorem holds for $m < 13{,}000$.

5.3 Mertens' theorem

Our next goal is to prove the following theorem, which turns out to have a number of applications.

Theorem 5.10. *We have*

$$\sum_{p \leq x} \frac{1}{p} = \log \log x + O(1).$$

The proof of this theorem, while not difficult, is a bit technical, and we proceed in several steps.

Theorem 5.11. *We have*

$$\sum_{p \leq x} \frac{\log p}{p} = \log x + O(1).$$

Proof. Let $n := \lfloor x \rfloor$. The idea of the proof is to estimate $\log(n!)$ in two different ways. By Lemma 5.3, we have

$$\log(n!) = \sum_{p \leq n} \sum_{k \geq 1} \lfloor n/p^k \rfloor \log p = \sum_{p \leq n} \lfloor n/p \rfloor \log p + \sum_{k \geq 2} \sum_{p \leq n} \lfloor n/p^k \rfloor \log p.$$

We next show that the last sum is $O(n)$. We have

$$\sum_{p \leq n} \log p \sum_{k \geq 2} \lfloor n/p^k \rfloor \leq n \sum_{p \leq n} \log p \sum_{k \geq 2} p^{-k}$$

$$= n \sum_{p \leq n} \frac{\log p}{p^2} \cdot \frac{1}{1 - 1/p} = n \sum_{p \leq n} \frac{\log p}{p(p-1)}$$

$$\leq n \sum_{k \geq 2} \frac{\log k}{k(k-1)} = O(n).$$

Thus, we have shown that

$$\log(n!) = \sum_{p \leq n} \lfloor n/p \rfloor \log p + O(n).$$

Since $\lfloor n/p \rfloor = n/p + O(1)$, applying Theorem 5.6 (and Exercise 3.12), we obtain

$$\log(n!) = \sum_{p \leq n} (n/p) \log p + O\left(\sum_{p \leq n} \log p\right) + O(n) = n \sum_{p \leq n} \frac{\log p}{p} + O(n). \quad (5.7)$$

We can also estimate $\log(n!)$ by estimating a sum by an integral (see §A5):

$$\log(n!) = \sum_{k=1}^{n} \log k = \int_{1}^{n} \log t \, dt + O(\log n) = n \log n - n + O(\log n). \quad (5.8)$$

Combining (5.7) and (5.8), and noting that $\log x - \log n = o(1)$ (see Exercise 3.11), we obtain

$$\sum_{p \leq x} \frac{\log p}{p} = \log n + O(1) = \log x + O(1),$$

which proves the theorem. □

We shall also need the following theorem, which is a very useful tool in its own right; it is essentially a discrete variant of "integration by parts."

Theorem 5.12 (Abel's identity). *Let* $\{c_i\}_{i=k}^{\infty}$ *be a sequence of real numbers, and for each real number t, define*

$$C(t) := \sum_{k \le i \le t} c_i.$$

Further, suppose that $f(t)$ is a function with a continuous derivative $f'(t)$ on the interval $[k, x]$, where x is a real number, with $x \ge k$. Then

$$\sum_{k \le i \le x} c_i f(i) = C(x)f(x) - \int_k^x C(t)f'(t)\, dt.$$

Note that since $C(t)$ is a step function, the integrand $C(t)f'(t)$ is piece-wise continuous on $[k, x]$, and hence the integral is well defined (see §A4).

Proof. Let $n := \lfloor x \rfloor$. We have

$$\sum_{i=k}^{n} c_i f(i) = C(k)f(k) + \sum_{i=k+1}^{n} [C(i) - C(i-1)]f(i)$$

$$= \sum_{i=k}^{n-1} C(i)[f(i) - f(i+1)] + C(n)f(n)$$

$$= \sum_{i=k}^{n-1} C(i)[f(i) - f(i+1)] + C(n)[f(n) - f(x)] + C(x)f(x).$$

Observe that for $i = k, \ldots, n-1$, we have $C(t) = C(i)$ for all $t \in [i, i+1)$, and so

$$C(i)[f(i) - f(i+1)] = -C(i)\int_i^{i+1} f'(t)\, dt = -\int_i^{i+1} C(t)f'(t)\, dt;$$

likewise,

$$C(n)[f(n) - f(x)] = -\int_n^x C(t)f'(t)\, dt,$$

from which the theorem directly follows. $\square$

Proof of Theorem 5.10. For $i \ge 2$, set

$$c_i := \begin{cases} (\log i)/i & \text{if } i \text{ is prime,} \\ 0 & \text{otherwise.} \end{cases}$$

By Theorem 5.11, we have

$$C(t) := \sum_{2 \le i \le t} c_i = \sum_{p \le t} \frac{\log p}{p} = \log t + R(t),$$

where $R(t) = O(1)$. Applying Theorem 5.12 with $f(t) := 1/\log t$ (and using Exercise 3.13), we obtain

$$\sum_{p \leq x} \frac{1}{p} = \sum_{2 \leq i \leq x} c_i f(i) = \frac{C(x)}{\log x} + \int_2^x \frac{C(t)}{t(\log t)^2} dt$$

$$= 1 + \frac{R(x)}{\log x} + \int_2^x \frac{dt}{t \log t} + \int_2^x \frac{R(t)}{t(\log t)^2} dt$$

$$= 1 + O(1/\log x) + (\log \log x - \log \log 2) + O(1)$$

$$= \log \log x + O(1). \quad \Box$$

Using Theorem 5.10, we can easily show the following:

Theorem 5.13 (Mertens' theorem). *We have*

$$\prod_{p \leq x}(1 - 1/p) = \Theta(1/\log x).$$

Proof. Using parts (i) and (iii) of §A1, for any fixed prime p, we have

$$-\frac{1}{p^2} \leq \frac{1}{p} + \log(1 - 1/p) \leq 0. \tag{5.9}$$

Moreover, since

$$\sum_{p \leq x} \frac{1}{p^2} \leq \sum_{i \geq 2} \frac{1}{i^2} < \infty,$$

summing the inequality (5.9) over all primes $p \leq x$ yields

$$-C \leq \sum_{p \leq x} \frac{1}{p} + \log g(x) \leq 0,$$

where C is a positive constant, and $g(x) := \prod_{p \leq x}(1 - 1/p)$. From this, and from Theorem 5.10, we obtain $\log g(x) = -\log \log x + O(1)$, which implies that $g(x) = \Theta(1/\log x)$ (see Exercise 3.11). That proves the theorem. $\quad \Box$

EXERCISE 5.4. For each positive integer k, let P_k denote the product of the first k primes. Show that $\varphi(P_k) = \Theta(P_k/\log \log P_k)$ (here, φ is Euler's phi function).

EXERCISE 5.5. The previous exercise showed that $\varphi(n)$ could be as small as (about) $n/\log \log n$ for infinitely many n. Show that this is the "worst case," in the sense that $\varphi(n) = \Omega(n/\log \log n)$.

EXERCISE 5.6. Show that for every positive integer constant k,

$$\int_2^x \frac{dt}{(\log t)^k} = \frac{x}{(\log x)^k} + O\left(\frac{x}{(\log x)^{k+1}}\right).$$

This fact may be useful in some of the following exercises.

EXERCISE 5.7. Use Chebyshev's theorem and Abel's identity to prove a stronger version of Theorem 5.5: $\vartheta(x) = \pi(x) \log x + O(x/\log x)$.

EXERCISE 5.8. Use Chebyshev's theorem and Abel's identity to show that

$$\sum_{p \leq x} \frac{1}{\log p} = \frac{\pi(x)}{\log x} + O(x/(\log x)^3).$$

EXERCISE 5.9. Show that

$$\prod_{2 < p \leq x} (1 - 2/p) = \Theta(1/(\log x)^2).$$

EXERCISE 5.10. Show that if $\pi(x) \sim cx/\log x$ for some constant c, then we must have $c = 1$.

EXERCISE 5.11. Strengthen Theorem 5.10: show that for some constant A, we have $\sum_{p \leq x} 1/p = \log \log x + A + o(1)$. You do not need to estimate A, but in fact $A \approx 0.261497212847643$.

EXERCISE 5.12. Use the result from the previous exercise to strengthen Mertens' theorem: show that for some constant B_1, we have $\prod_{p \leq x}(1 - 1/p) \sim B_1/(\log x)$. You do not need to estimate B_1, but in fact $B_1 \approx 0.561459483566885$.

EXERCISE 5.13. Strengthen the result of Exercise 5.9: show that for some constant B_2, we have

$$\prod_{2 < p \leq x} (1 - 2/p) \sim B_2/(\log x)^2.$$

You do not need to estimate B_2, but in fact $B_2 \approx 0.832429065662$.

EXERCISE 5.14. Use Abel's identity to derive **Euler's summation formula**: if $f(t)$ has a continuous derivative $f'(t)$ on the interval $[a, b]$, where a and b are integers, then

$$\sum_{i=a}^{b} f(i) - \int_a^b f(t) \, dt = f(a) + \int_a^b (t - \lfloor t \rfloor) f'(t) \, dt.$$

EXERCISE 5.15. Use Euler's summation formula (previous exercise) to show that

$$\log(n!) = n \log n - n + \tfrac{1}{2} \log n + O(1),$$

and from this, conclude that $n! = \Theta((n/e)^n \sqrt{n})$. This is a weak form of **Stirling's approximation**; a sharper form states that $n! \sim (n/e)^n \sqrt{2\pi n}$.

EXERCISE 5.16. Use Stirling's approximation (previous exercise) to show that

$$\binom{2m}{m} = \Theta(2^{2m}/\sqrt{m}).$$

5.4 The sieve of Eratosthenes

As an application of Theorem 5.10, consider the **sieve of Eratosthenes**. This is an algorithm that generates all the primes up to a given bound n. It uses an array $A[2 \ldots n]$, and runs as follows.

> for $k \leftarrow 2$ to n do $A[k] \leftarrow 1$
> for $k \leftarrow 2$ to $\lfloor \sqrt{n} \rfloor$ do
>   if $A[k] = 1$ then
> $i \leftarrow 2k$
>   while $i \leq n$ do
> $A[i] \leftarrow 0, \ i \leftarrow i + k$

When the algorithm finishes, we have $A[k] = 1$ if and only if k is prime, for $k = 2, \ldots, n$. This can easily be proven using the fact (see Exercise 1.2) that a composite number k between 2 and n must be divisible by a prime that is at most $\sqrt{n}$, and by proving by induction on k that at the beginning of each iteration of the main loop, $A[i] = 0$ if and only if i is divisible by a prime less than k, for $i = k, \ldots, n$. We leave the details of this to the reader.

We are more interested in the running time of the algorithm. To analyze the running time, we assume that all arithmetic operations take constant time; this is reasonable, since all the numbers computed are used as array indices and thus should fit in single machine words. Therefore, we can assume that built-in arithmetic instructions are used for operating on such numbers.

Every time we execute the inner loop of the algorithm, we perform $O(n/k)$ steps to clear the entries of A indexed by multiples of k. Pessimistically, then, we could bound the total running time by $O(n T(n))$, where

$$T(n) := \sum_{k \leq \sqrt{n}} 1/k.$$

Estimating the sum by an integral (see §A5), we have

$$T(n) = \sum_{k=1}^{\lfloor \sqrt{n} \rfloor} 1/k = \int_1^{\lfloor \sqrt{n} \rfloor} \frac{dy}{y} + O(1) \sim \frac{1}{2} \log n.$$

This implies a $O(n \, \text{len}(n))$ bound on the running time of the algorithm. However, this rather crude analysis ignores the fact that the inner loop is executed only for *prime* values of k; taking this fact into account, we see that the running time is $O(n\,T_1(n))$, where

$$T_1(n) := \sum_{p \le \sqrt{n}} 1/p.$$

By Theorem 5.10, $T_1(n) = \log \log n + O(1)$, which implies a $O(n \operatorname{len}(\operatorname{len}(n)))$ bound on the running time of the algorithm. This is a substantial improvement over the above, rather crude analysis.

EXERCISE 5.17. Give a detailed proof of the correctness of the above algorithm.

EXERCISE 5.18. One drawback of the above algorithm is its use of space: it requires an array of size n. Show how to modify the algorithm, without substantially increasing its running time, so that one can enumerate all the primes up to n, using an auxiliary array of size just $O(\sqrt{n})$.

EXERCISE 5.19. Design and analyze an algorithm that on input n outputs the table of values $\tau(k)$ for $k = 1, \ldots, n$, where $\tau(k)$ is the number of positive divisors of k. Your algorithm should run in time $O(n \operatorname{len}(n))$.

5.5 The prime number theorem ... and beyond

In this section, we survey a number of theorems and conjectures related to the distribution of primes. This is a vast area of mathematical research, with a number of very deep results. We shall be stating a number of theorems from the literature in this section without proof; while our intent is to keep the text as self contained as possible, and to avoid degenerating into "mathematical tourism," it nevertheless is a good idea to occasionally have a somewhat broader perspective. In the subsequent chapters, we shall not make any critical use of the theorems in this section.

5.5.1 The prime number theorem

The main theorem in the theory of the density of primes is the following.

Theorem 5.14 (Prime number theorem). *We have*

$$\pi(x) \sim x/\log x.$$

Proof. Literature—see §5.6. $\square$

As we saw in Exercise 5.10, if $\pi(x)/(x/\log x)$ tends to a limit as $x \to \infty$, then the limit must be 1, so in fact the hard part of proving the prime number theorem is to show that $\pi(x)/(x/\log x)$ does indeed tend to some limit.

EXERCISE 5.20. Using the prime number theorem, show that $\vartheta(x) \sim x$.

EXERCISE 5.21. Using the prime number theorem, show that $p_n \sim n \log n$, where p_n denotes the nth prime.

EXERCISE 5.22. Using the prime number theorem, show that Bertrand's postulate can be strengthened (asymptotically) as follows: for every $\varepsilon > 0$, there exist positive constants c and x_0, such that for all $x \geq x_0$, we have

$$\pi((1 + \varepsilon)x) - \pi(x) \geq c\frac{x}{\log x}.$$

5.5.2 The error term in the prime number theorem

The prime number theorem says that

$$|\pi(x) - x/\log x| \leq \delta(x),$$

where $\delta(x) = o(x/\log x)$. A natural question is: how small is the "error term" $\delta(x)$? It can be shown that

$$\pi(x) = x/\log x + O(x/(\log x)^2). \tag{5.10}$$

This bound on the error term is not very impressive, but unfortunately, cannot be improved upon. The problem is that $x/\log x$ is not really the best "simple" function that approximates $\pi(x)$. It turns out that a better approximation to $\pi(x)$ is the **logarithmic integral**, defined for all real numbers $x \geq 2$ as

$$\mathrm{li}(x) := \int_2^x \frac{dt}{\log t}.$$

It is not hard to show (see Exercise 5.6) that

$$\mathrm{li}(x) = x/\log x + O(x/(\log x)^2). \tag{5.11}$$

Thus, $\mathrm{li}(x) \sim x/\log x \sim \pi(x)$. However, the error term in the approximation of $\pi(x)$ by $\mathrm{li}(x)$ is much better. This is illustrated numerically in Table 5.2; for example, at $x = 10^{18}$, $\mathrm{li}(x)$ approximates $\pi(x)$ with a relative error just under 10^{-9}, while $x/\log x$ approximates $\pi(x)$ with a relative error of about 0.025.

The sharpest proven result on the error in approximating $\pi(x)$ by $\mathrm{li}(x)$ is the following:

Theorem 5.15. *Let* $\kappa(x) := (\log x)^{3/5}(\log \log x)^{-1/5}$. *Then for some* $c > 0$, *we have*

$$\pi(x) = \mathrm{li}(x) + O(xe^{-c\kappa(x)}).$$

Proof. Literature — see §5.6. $\square$

Table 5.2. *Values of* $\pi(x)$, li(x), *and* $x/\log x$

x	$\pi(x)$	li(x)	$x/\log x$
10^3	168	176.6	144.8
10^6	78498	78626.5	72382.4
10^9	50847534	50849233.9	48254942.4
10^{12}	37607912018	37607950279.8	36191206825.3
10^{15}	29844570422669	29844571475286.5	28952965460216.8
10^{18}	2473995428740860	24739954309690414.0	24127471216847323.8

Note that the error term $xe^{-c_K(x)}$ is $o(x/(\log x)^k)$ for every fixed $k \geq 0$. Also note that (5.10) follows directly from (5.11) and Theorem 5.15.

Although the above estimate on the error term in the approximation of $\pi(x)$ by li(x) is pretty good, it is conjectured that the actual error term is much smaller:

Conjecture 5.16. *For all* $x \geq 2.01$, *we have*

$$|\pi(x) - \mathrm{li}(x)| < x^{1/2} \log x.$$

Conjecture 5.16 is equivalent to the famous **Riemann hypothesis**, which is a conjecture about the location of the zeros of a certain function, called **Riemann's zeta function**. We give a *very* brief, high-level account of this conjecture, and its connection to the theory of the distribution of primes.

For all real numbers $s > 1$, the zeta function is defined as

$$\zeta(s) := \sum_{n=1}^{\infty} \frac{1}{n^s}. \qquad (5.12)$$

Note that because $s > 1$, the infinite series defining $\zeta(s)$ converges. A simple, but important, connection between the zeta function and the theory of prime numbers is the following:

Theorem 5.17 (Euler's identity). *For every real number* $s > 1$, *we have*

$$\zeta(s) = \prod_{p} (1 - p^{-s})^{-1}, \qquad (5.13)$$

where the product is over all primes p.

Proof. The rigorous interpretation of the infinite product on the right-hand side of (5.13) is as a limit of finite products. Thus, if p_i denotes the ith prime, for $i = 1, 2, \ldots$, then we are really proving that

$$\zeta(s) = \lim_{r \to \infty} \prod_{i=1}^{r} (1 - p_i^{-s})^{-1}.$$

Now, from the identity

$$(1 - p_i^{-s})^{-1} = \sum_{e=0}^{\infty} p_i^{-es},$$

we have

$$\prod_{i=1}^{r} (1 - p_i^{-s})^{-1} = \left(1 + p_1^{-s} + p_1^{-2s} + \cdots\right) \cdots \left(1 + p_r^{-s} + p_r^{-2s} + \cdots\right)$$

$$= \sum_{n=1}^{\infty} \frac{h_r(n)}{n^s},$$

where

$$h_r(n) := \begin{cases} 1 & \text{if } n \text{ is divisible only by the primes } p_1, \ldots, p_r; \\ 0 & \text{otherwise.} \end{cases}$$

Here, we have made use of the fact (see §A7) that we can multiply term-wise infinite series with non-negative terms.

Now, for every $\varepsilon > 0$, there exists n_0 such that $\sum_{n=n_0}^{\infty} n^{-s} < \varepsilon$ (because the series defining $\zeta(s)$ converges). Moreover, there exists an r_0 such that $h_r(n) = 1$ for all $n < n_0$ and $r \geq r_0$. Therefore, for all $r \geq r_0$, we have

$$\left| \sum_{n=1}^{\infty} \frac{h_r(n)}{n^s} - \zeta(s) \right| \leq \sum_{n=n_0}^{\infty} n^{-s} < \varepsilon.$$

It follows that

$$\lim_{r \to \infty} \sum_{n=1}^{\infty} \frac{h_r(n)}{n^s} = \zeta(s),$$

which proves the theorem. □

While Theorem 5.17 is nice, things become much more interesting if one extends the domain of definition of the zeta function to the complex plane. For the reader who is familiar with just a little complex analysis, it is easy to see that the infinite series defining the zeta function in (5.12) converges absolutely for all complex numbers s whose real part is greater than 1, and that (5.13) holds as well for such s. However, it is possible to extend the domain of definition of $\zeta(s)$ even further— in fact, one can extend the definition of $\zeta(s)$ in a "nice way" (in the language of complex analysis, *analytically continue*) to the entire complex plane (except the point $s = 1$, where there is a simple pole). Exactly how this is done is beyond the scope of this text, but assuming this extended definition of $\zeta(s)$, we can now state the Riemann hypothesis:

Conjecture 5.18 (Riemann hypothesis). *Suppose s is a complex number with $s = x + yi$, where $x, y \in \mathbb{R}$, such that $\zeta(s) = 0$ and $0 < x < 1$. Then $x = 1/2$.*

A lot is known about the zeros of the zeta function in the "critical strip," which consists of those points s whose real part is greater than 0 and less than 1: it is known that there are infinitely many such zeros, and there are even good estimates about their density. It turns out that one can apply standard tools in complex analysis, like contour integration, to the zeta function (and functions derived from it) to answer various questions about the distribution of primes. Indeed, such techniques may be used to prove the prime number theorem. However, if one assumes the Riemann hypothesis, then these techniques yield much sharper results, such as the bound in Conjecture 5.16.

EXERCISE 5.23. For any arithmetic function a (mapping positive integers to reals), we can form the **Dirichlet series**

$$F_a(s) := \sum_{n=1}^{\infty} \frac{a(n)}{n^s}.$$

For simplicity we assume that s takes only real values, even though such series are usually studied for complex values of s.

(a) Show that if the Dirichlet series $F_a(s)$ converges absolutely for some real s, then it converges absolutely for all real $s' \geq s$.

(b) From part (a), conclude that for any given arithmetic function a, there is an **interval of absolute convergence** of the form (s_0, ∞), where we allow $s_0 = -\infty$ and $s_0 = \infty$, such that $F_a(s)$ converges absolutely for $s > s_0$, and does not converge absolutely for $s < s_0$.

(c) Let a and b be arithmetic functions such that $F_a(s)$ has an interval of absolute convergence (s_0, ∞) and $F_b(s)$ has an interval of absolute convergence (s_0', ∞), and assume that $s_0 < \infty$ and $s_0' < \infty$. Let $c := a \star b$ be the Dirichlet product of a and b, as defined in §2.9. Show that for all $s \in (\max(s_0, s_0'), \infty)$, the series $F_c(s)$ converges absolutely and, moreover, that $F_a(s)F_b(s) = F_c(s)$.

5.5.3 Explicit estimates

Sometimes, it is useful to have explicit estimates for $\pi(x)$, as well as related functions, like $\vartheta(x)$ and the nth prime function p_n. The following theorem presents a number of bounds that have been proved without relying on any unproved conjectures.

Theorem 5.19. *We have:*

(i) $\dfrac{x}{\log x}\left(1 + \dfrac{1}{2\log x}\right) < \pi(x) < \dfrac{x}{\log x}\left(1 + \dfrac{3}{2\log x}\right), \quad \text{for } x \geq 59;$

(ii) $n(\log n + \log\log n - 3/2) < p_n < n(\log n + \log\log n - 1/2), \quad \text{for } n \geq 20;$

(iii) $x\left(1 - \dfrac{1}{2\log x}\right) < \vartheta(x) < x\left(1 + \dfrac{1}{2\log x}\right), \quad \text{for } x \geq 563;$

(iv) $\log\log x + A - \dfrac{1}{2(\log x)^2} < \sum\limits_{p \leq x} 1/p < \log\log x + A + \dfrac{1}{2(\log x)^2},$

for $x \geq 286$, where $A \approx 0.261497212847643;$

(v) $\dfrac{B_1}{\log x}\left(1 - \dfrac{1}{2(\log x)^2}\right) < \prod\limits_{p \leq x}\left(1 - \dfrac{1}{p}\right) < \dfrac{B_1}{\log x}\left(1 + \dfrac{1}{2(\log x)^2}\right),$

for $x \geq 285$, where $B_1 \approx 0.561459483566885.$

Proof. Literature—see §5.6. □

5.5.4 Primes in arithmetic progressions

In Theorems 2.35 and 2.36, we proved that there are infinitely many primes $p \equiv 1 \pmod 4$ and infinitely many primes $p \equiv 3 \pmod 4$. These results are actually special cases of a much more general result.

Let d be a positive integer, and let a be any integer. An **arithmetic progression** with first term a and common difference d consists of all integers of the form

$$a + dm, \quad m = 0, 1, 2, \ldots.$$

The question is: under what conditions does such an arithmetic progression contain infinitely many primes? An equivalent formulation is: under what conditions are there infinitely many primes $p \equiv a \pmod d$? If a and d have a common factor $c > 1$, then every term in the progression is divisible by c, and so there can be at most one prime in the progression. So a necessary condition for the existence of infinitely many primes $p \equiv a \pmod d$ is that $\gcd(a, d) = 1$. A famous theorem due to Dirichlet states that this is a sufficient condition as well.

Theorem 5.20 (Dirichlet's theorem). *Let $a, d \in \mathbb{Z}$ with $d > 0$ and $\gcd(a, d) = 1$. Then there are infinitely many primes $p \equiv a \pmod d$.*

Proof. Literature—see §5.6. □

We can also ask about the density of primes in arithmetic progressions. One might expect that for a fixed value of d, the primes are distributed in roughly equal measure among the $\varphi(d)$ different residue classes $[a]_d$ with $\gcd(a, d) = 1$ (here, φ is Euler's phi function). This is in fact the case. To formulate such assertions, we define $\pi(x; d, a)$ to be the number of primes p up to x with $p \equiv a \pmod{d}$.

Theorem 5.21. *Let* $a, d \in \mathbb{Z}$ *with* $d > 0$ *and* $\gcd(a, d) = 1$. *Then*

$$\pi(x; d, a) \sim \frac{x}{\varphi(d) \log x}.$$

Proof. Literature—see §5.6. $\square$

The above theorem is only applicable in the case where d and a are fixed as $x \to \infty$. For example, it says that roughly half the primes up to x are congruent to 1 modulo 4, and roughly half the primes up to x are congruent to 3 modulo 4. However, suppose $d \to \infty$, and we want to estimate, say, the number of primes $p \equiv 1 \pmod{d}$ up to d^3. Theorem 5.21 does not help us here. The following conjecture does, however:

Conjecture 5.22. *Let* $x \in \mathbb{R}$, $a, d \in \mathbb{Z}$ *with* $x \geq 2$, $d \geq 2$, *and* $\gcd(a, d) = 1$. *Then*

$$\left| \pi(x; d, a) - \frac{\mathrm{li}(x)}{\varphi(d)} \right| \leq x^{1/2} (\log x + 2 \log d).$$

The above conjecture is in fact a consequence of a generalization of the Riemann hypothesis — see §5.6. This conjecture implies that for every constant $\alpha < 1/2$, if $2 \leq d \leq x^\alpha$, then $\pi(x; d, a)$ is closely approximated by $\mathrm{li}(x)/\varphi(d)$ (see Exercise 5.24). It can also be used to get an upper bound on the least prime $p \equiv a \pmod{d}$ (see Exercise 5.25). The following theorem is the best rigorously proven upper bound on the smallest prime in an arithmetic progression:

Theorem 5.23. *There exists a constant* c *such that for all* $a, d \in \mathbb{Z}$ *with* $d \geq 2$ *and* $\gcd(a, d) = 1$, *the least prime* $p \equiv a \pmod{d}$ *is at most* $cd^{11/2}$.

Proof. Literature—see §5.6. $\square$

EXERCISE 5.24. Assuming Conjecture 5.22, show that for all α, ε satisfying $0 < \alpha < 1/2$ and $0 < \varepsilon < 1$, there exists an x_0, such that for all $x > x_0$, for all $d \in \mathbb{Z}$ with $2 \leq d \leq x^\alpha$, and for all $a \in \mathbb{Z}$ relatively prime to d, the number of primes $p \leq x$ such that $p \equiv a \pmod{d}$ is at least $(1 - \varepsilon) \mathrm{li}(x)/\varphi(d)$ and at most $(1 + \varepsilon) \mathrm{li}(x)/\varphi(d)$.

EXERCISE 5.25. Assuming Conjecture 5.22, show that there exists a constant c such that for all $a, d \in \mathbb{Z}$ with $d \geq 2$ and $\gcd(a, d) = 1$, the least prime $p \equiv a \pmod{d}$ is at most $c\varphi(d)^2 (\log d)^4$.

5.5.5 Sophie Germain primes

A **Sophie Germain prime** is a prime p such that $2p + 1$ is also prime. Such primes are actually useful in a number of practical applications, and so we discuss them briefly here.

It is an open problem to prove (or disprove) that there are infinitely many Sophie Germain primes. However, numerical evidence, and heuristic arguments, strongly suggest not only that there are infinitely many such primes, but also a fairly precise estimate on the density of such primes.

Let $\pi^*(x)$ denote the number of Sophie Germain primes up to x.

Conjecture 5.24. *We have*

$$\pi^*(x) \sim C \frac{x}{(\log x)^2},$$

where C is the constant

$$C := 2 \prod_{p > 2} \frac{p(p-2)}{(p-1)^2} \approx 1.32032,$$

and the product is over all primes $p > 2$.

The above conjecture is a special case of the following, more general conjecture.

Conjecture 5.25 (Dickson's conjecture). *Let $(a_1, b_1), \ldots, (a_k, b_k)$ be distinct pairs of integers, where each a_i is positive. Let $P(x)$ be the number of positive integers m up to x such that $a_i m + b_i$ are simultaneously prime for $i = 1, \ldots, k$. For each prime p, let $\omega(p)$ be the number of integers $m \in \{0, \ldots, p-1\}$ that satisfy*

$$\prod_{i=1}^{k} (a_i m + b_i) \equiv 0 \pmod{p}.$$

If $\omega(p) < p$ for each prime p, then

$$P(x) \sim D \frac{x}{(\log x)^k},$$

where

$$D := \prod_p \frac{1 - \omega(p)/p}{(1 - 1/p)^k},$$

the product being over all primes p.

In Exercise 5.26 below, you are asked to verify that the quantity D appearing in Conjecture 5.25 satisfies $0 < D < \infty$. Conjecture 5.24 is implied by Conjecture 5.25 with $k := 2$, $(a_1, b_1) := (1, 0)$, and $(a_2, b_2) := (2, 1)$; in this case,

$\omega(2) = 1$ and $\omega(p) = 2$ for all $p > 2$. The above conjecture also includes (a strong version of) the famous **twin primes conjecture** as a special case: the number of primes p up to x such that $p + 2$ is also prime is $\sim Cx/(\log x)^2$, where C is the same constant as in Conjecture 5.24.

A heuristic argument in favor of Conjecture 5.25 runs as follows. In some sense, the chance that a large positive integer m is prime is about $1/\log m$. Since $\log(a_i m + b_i) \sim \log m$, the chance that $a_1 m + b_1, \ldots, a_k m + b_k$ are all prime should be about $1/(\log m)^k$. But this ignores the fact that $a_1 m + b_1, \ldots, a_k m + b_k$ are not quite random integers. For each prime p, we must apply a "correction factor" r_p/s_p, where r_p is the chance that for random m, none of $a_1 m + b_1, \ldots, a_k m + b_k$ is divisible by p, and s_p is the chance that for k truly random, large integers, none of them is divisible by p. One sees that $r_p = 1 - \omega(p)/p$ and $s_p = (1 - 1/p)^k$. This implies (using §A5 and Exercise 5.6) that $P(x)$ should be about

$$D \sum_{m \leq x} 1/(\log m)^k \sim D \int_2^x dt/(\log t)^k \sim Dx/(\log x)^k.$$

Although Conjecture 5.25 is well supported by numerical evidence, there seems little hope of it being proved any time soon, even under the Riemann hypothesis or any of its generalizations.

EXERCISE 5.26. Show that the quantity D appearing in Conjecture 5.25 satisfies $0 < D < \infty$. Hint: first show that $\omega(p) = k$ for all sufficiently large p.

EXERCISE 5.27. Derive Theorem 5.21 from Conjecture 5.25.

EXERCISE 5.28. Show that the constant C appearing in Conjecture 5.24 satisfies

$$2C = B_2/B_1^2,$$

where B_1 and B_2 are the constants from Exercises 5.12 and 5.13.

5.6 Notes

The prime number theorem was conjectured by Gauss in 1791. It was proven independently in 1896 by Hadamard and de la Vallée Poussin. A proof of the prime number theorem may be found, for example, in the book by Hardy and Wright [46].

Theorem 5.19, as well as the estimates for the constants A, B_1, and B_2 mentioned in that theorem and Exercises 5.11, 5.12, and 5.13, are from Rosser and Schoenfeld [83].

Theorem 5.15 is from Walfisz [102].

Theorem 5.17, which made the first connection between the theory of prime numbers and the zeta function, was discovered in the 18th century by Euler. The Riemann hypothesis was made by Riemann in 1859, and to this day, remains one of the most vexing conjectures in mathematics. Riemann in fact showed that his conjecture about the zeros of the zeta function is equivalent to the conjecture that for each fixed $\varepsilon > 0$, $\pi(x) = \text{li}(x) + O(x^{1/2+\varepsilon})$. This was strengthened by von Koch in 1901, who showed that the Riemann hypothesis is true if and only if $\pi(x) = \text{li}(x) + O(x^{1/2} \log x)$. See Chapter 1 of the book by Crandall and Pomerance [30] for more on the connection between the Riemann hypothesis and the theory of prime numbers; in particular, see Exercise 1.36 in that book for an outline of a proof that Conjecture 5.16 follows from the Riemann hypothesis.

A warning: some authors (and software packages) define the logarithmic integral using the interval of integration $(0, x)$, rather than $(2, x)$, which increases its value by a constant $c \approx 1.0452$.

Theorem 5.20 was proved by Dirichlet in 1837, while Theorem 5.21 was proved by de la Vallée Poussin in 1896. A result of Oesterlé [73] implies that Conjecture 5.22 for $d \geq 3$ is a consequence of an assumption about the location of the zeros of certain generalizations of Riemann's zeta function; the case $d = 2$ follows from the bound in Conjecture 5.16 under the ordinary Riemann hypothesis. Theorem 5.23 is from Heath-Brown [47]. The bound in Exercise 5.25 can be improved to $c\varphi(d)^2 (\log d)^2$ (see Theorem 8.5.8 of [11]).

Conjecture 5.25 originates from Dickson [33]. In fact, Dickson only conjectured that the quantity $P(x)$ defined in Conjecture 5.25 tends to infinity. The conjectured formula for the rate of growth of $P(x)$ is a special case of a more general conjecture stated by Bateman and Horn [12], which generalizes various, more specific conjectures stated by Hardy and Littlewood [45].

For the reader who is interested in learning more on the topics discussed in this chapter, we recommend the books by Apostol [8] and Hardy and Wright [46]; indeed, many of the proofs presented in this chapter are minor variations on proofs from these two books. Our proof of Bertrand's postulate is based on the presentation in Section 9.2 of Redmond [80]. See also Bach and Shallit [11] (especially Chapter 8), as well as Crandall and Pomerance [30] (especially Chapter 1), for a more detailed overview of these topics.

The data in Tables 5.1 and 5.2 was obtained using the computer program *Maple*.

6

Abelian groups

This chapter introduces the notion of an abelian group. This is an abstraction that models many different algebraic structures, and yet despite the level of generality, a number of very useful results can be easily obtained.

6.1 Definitions, basic properties, and examples

Definition 6.1. *An **abelian group** is a set G together with a binary operation $\star$ on G such that:*

(i) for all $a, b, c \in G$, $a \star (b \star c) = (a \star b) \star c$ (i.e., $\star$ is associative);

*(ii) there exists $e \in G$ (called the **identity element**) such that for all $a \in G$, $a \star e = a = e \star a$;*

*(iii) for all $a \in G$ there exists $a' \in G$ (called the **inverse of** a) such that $a \star a' = e = a' \star a$;*

(iv) for all $a, b \in G$, $a \star b = b \star a$ (i.e., $\star$ is commutative).

While there is a more general notion of a **group**, which may be defined simply by dropping property (iv) in Definition 6.1, we shall not need this notion in this text. The restriction to abelian groups helps to simplify the discussion significantly. Because we will only be dealing with abelian groups, we may occasionally simply say "group" instead of "abelian group."

Before looking at examples, let us state some very basic properties of abelian groups that follow directly from the definition:

Theorem 6.2. *Let G be an abelian group with binary operation $\star$. Then we have:*

(i) G contains only one identity element;

(ii) every element of G has only one inverse.

Proof. Suppose e, e' are both identities. Then we have

$$e = e \star e' = e',$$

where we have used part (ii) of Definition 6.1, once with e' as the identity, and once with e as the identity. That proves part (i) of the theorem.

To prove part (ii) of the theorem, let $a \in G$, and suppose that a has two inverses, a' and a''. Then using parts (i)–(iii) of Definition 6.1, we have

$$a' = a' \star e \text{ (by part (ii))}$$
$$= a' \star (a \star a'') \text{ (by part (iii) with inverse } a'' \text{ of } a)$$
$$= (a' \star a) \star a'' \text{ (by part (i))}$$
$$= e \star a'' \text{ (by part (iii) with inverse } a' \text{ of } a)$$
$$= a'' \text{ (by part (ii))}. \ \square$$

These uniqueness properties justify use of the definite article in Definition 6.1 in conjunction with the terms "identity element" and "inverse." Note that we never used part (iv) of the definition in the proof of the above theorem.

Abelian groups are lurking everywhere, as the following examples illustrate.

Example 6.1. The set of integers $\mathbb{Z}$ under addition forms an abelian group, with 0 being the identity, and $-a$ being the inverse of $a \in \mathbb{Z}$. $\square$

Example 6.2. For each integer n, the set $n\mathbb{Z} = \{nz : z \in \mathbb{Z}\}$ under addition forms an abelian group, again, with 0 being the identity, and $n(-z)$ being the inverse of nz. $\square$

Example 6.3. The set of non-negative integers under addition does not form an abelian group, since additive inverses do not exist for any positive integers. $\square$

Example 6.4. The set of integers under multiplication does not form an abelian group, since inverses do not exist for any integers other than ± 1. $\square$

Example 6.5. The set of integers $\{\pm 1\}$ under multiplication forms an abelian group, with 1 being the identity, and -1 its own inverse. $\square$

Example 6.6. The set of rational numbers $\mathbb{Q} = \{a/b : a, b \in \mathbb{Z}, b \neq 0\}$ under addition forms an abelian group, with 0 being the identity, and $(-a)/b$ being the inverse of a/b. $\square$

Example 6.7. The set of non-zero rational numbers $\mathbb{Q}^*$ under multiplication forms an abelian group, with 1 being the identity, and b/a being the inverse of a/b. $\square$

Example 6.8. The set $\mathbb{Z}_n$ under addition forms an abelian group, where $[0]_n$ is the identity, and where $[-a]_n$ is the inverse of $[a]_n$. $\square$

Example 6.9. The set $\mathbb{Z}_n^*$ of residue classes $[a]_n$ with $\gcd(a, n) = 1$ under multiplication forms an abelian group, where $[1]_n$ is the identity, and if b is a multiplicative inverse of a modulo n, then $[b]_n$ is the inverse of $[a]_n$. □

Example 6.10. For every positive integer n, the set of n-bit strings under the "exclusive or" operation forms an abelian group, where the "all zero" bit string is the identity, and every bit string is its own inverse. □

Example 6.11. The set $\mathcal{F}^*$ of all arithmetic functions f, such that $f(1) \neq 0$, and with the Dirichlet product as the binary operation (see §2.9) forms an abelian group. The special function I is the identity, and inverses are guaranteed by Exercise 2.54. □

Example 6.12. The set of all finite bit strings under concatenation does not form an abelian group. Although concatenation is associative and the empty string acts as an identity element, inverses do not exist (except for the empty string), nor is concatenation commutative. □

Example 6.13. The set of 2×2 integer matrices with determinant ± 1, together with the binary operation of matrix multiplication, is an example of a *non-abelian* group; that is, it satisfies properties (i)–(iii) of Definition 6.1, but not property (iv). □

Example 6.14. The set of all permutations on a given set of size $n \geq 3$, together with the binary operation of function composition, is another example of a non-abelian group (for $n = 1, 2$, it is an abelian group). □

Consider an abelian group G with binary operation $\star$. Since the group operation is associative, for all $a_1, \ldots, a_k \in G$, we may write $a_1 \star \cdots \star a_k$ without parentheses, and there can be no ambiguity as to the value of such an expression: any explicit parenthesization of this expression yields the same value. Furthermore, since the group operation is commutative, reordering the a_i's does not change this value.

Note that in specifying a group, one must specify both the underlying set G as well as the binary operation; however, in practice, the binary operation is often implicit from context, and by abuse of notation, one often refers to G itself as the group. For example, when talking about the abelian groups $\mathbb{Z}$ and $\mathbb{Z}_n$, it is understood that the group operation is addition, while when talking about the abelian group $\mathbb{Z}_n^*$, it is understood that the group operation is multiplication.

Typically, instead of using a special symbol like "$\star$" for the group operation, one uses the usual addition ("$+$") or multiplication ("$\cdot$") operations.

Additive notation. If an abelian group G is written additively, using "$+$" as the group operation, then the identity element is denoted by 0_G (or just 0 if G is clear from context), and is also called the **zero element**. The inverse of an element $a \in G$ is denoted by $-a$. For $a, b \in G$, $a - b$ denotes $a + (-b)$.

Multiplicative notation. If an abelian group G is written multiplicatively, using "·" as the group operation, then the identity element is denoted by 1_G (or just 1 if G is clear from context). The inverse of an element $a \in G$ is denoted by a^{-1}. As usual, one may write ab in place of $a \cdot b$. Also, one may write a/b for ab^{-1}.

For any particular, concrete abelian group, the most natural choice of notation is clear (e.g., addition for $\mathbb{Z}$ and $\mathbb{Z}_n$, multiplication for $\mathbb{Z}_n^*$); however, for a "generic" group, the choice is largely a matter of taste. By convention, **whenever we consider a "generic" abelian group, we shall use** *additive* **notation for the group operation**, unless otherwise specified.

The next theorem states a few simple but useful properties of abelian groups (stated using our default, additive notation).

Theorem 6.3. *Let G be an abelian group. Then for all $a, b, c \in G$, we have:*

   (i) *if $a + b = a + c$, then $b = c$;*

   (ii) *the equation $a + x = b$ has a unique solution $x \in G$;*

   (iii) *$-(a + b) = (-a) + (-b)$;*

   (iv) *$-(-a) = a$.*

Proof. These statements all follow easily from Definition 6.1 and Theorem 6.2. For (i), just add $-a$ to both sides of the equation $a + b = a + c$. For (ii), the solution is $x = b - a$. For (iii), we have

$$(a + b) + ((-a) + (-b)) = (a + (-a)) + (b + (-b)) = 0_G + 0_G = 0_G,$$

which shows that $(-a) + (-b)$ is indeed the inverse of $a + b$. For (iv), we have $(-a) + a = 0_G$, which means that a is the inverse of $-a$. $\square$

Part (i) of the above theorem is the **cancellation law** for abelian groups.

If $a_1, \ldots, a_k$ are elements of an abelian group G, we naturally write $\sum_{i=1}^{k} a_i$ for their sum $a_1 + \cdots + a_k$. By convention, the sum is 0_G when $k = 0$. Part (iii) of Theorem 6.3 obviously generalizes, so that $-\sum_{i=1}^{k} a_i = \sum_{i=1}^{k} (-a_i)$. In the special case where all the a_i's have the same value a, we define $k \cdot a := \sum_{i=1}^{k} a$, whose inverse is $k \cdot (-a)$, which we may write as $(-k) \cdot a$. Thus, the notation $k \cdot a$, or more simply, ka, is defined for all integers k. Observe that by definition, $1a = a$ and $(-1)a = -a$.

Theorem 6.4. *Let G be an abelian group. Then for all $a, b \in G$ and $k, \ell \in \mathbb{Z}$, we have:*

   (i) *$k(\ell a) = (k\ell)a = \ell(ka)$;*

(ii) $(k + \ell)a = ka + \ell a$;

(iii) $k(a + b) = ka + kb$.

Proof. The proof of this is easy, but tedious. We leave the details as an exercise to the reader. □

Multiplicative notation: It is perhaps helpful to translate the above discussion from additive to multiplicative notation. If a group G is written using multiplicative notation, then Theorem 6.3 says that (i) $ab = ac$ implies $b = c$, (ii) $ax = b$ has a unique solution, (iii) $(ab)^{-1} = a^{-1}b^{-1}$, and (iv) $(a^{-1})^{-1} = a$. If $a_1, \ldots, a_k \in G$, we write their product $a_1 \cdots a_k$ as $\prod_{i=1}^{k} a_i$, which is 1_G when $k = 0$. We have $(\prod_{i=1}^{k} a_i)^{-1} = \prod_{i=1}^{k} a_i^{-1}$. We also define $a^k := \prod_{i=1}^{k} a$, and we have $(a^k)^{-1} = (a^{-1})^k$, which we may write as a^{-k}. Theorem 6.4 says that (i) $(a^\ell)^k = a^{k\ell} = (a^k)^\ell$, (ii) $a^{k+\ell} = a^k a^\ell$, and (iii) $(ab)^k = a^k b^k$.

An abelian group G may be **trivial**, meaning that it consists of just the zero element 0_G, with $0_G + 0_G = 0_G$. An abelian group G may be infinite or finite: if the group is finite, we define its **order** to be the number of elements in the underlying set G; otherwise, we say that the group has **infinite order**.

Example 6.15. The order of the additive group $\mathbb{Z}_n$ is n. If $n = 1$, then $\mathbb{Z}_n$ is the trivial group. □

Example 6.16. The order of the multiplicative group $\mathbb{Z}_n^*$ is $\varphi(n)$, where φ is Euler's phi function, defined in §2.6. □

Example 6.17. The additive group $\mathbb{Z}$ has infinite order. □

We close this section with two simple constructions for combining groups to build new groups.

Example 6.18. If $G_1, \ldots, G_k$ are abelian groups, we can form the **direct product** $H := G_1 \times \cdots \times G_k$, which consists of all k-tuples $(a_1, \ldots, a_k)$ with $a_1 \in G_1$, $\ldots, a_k \in G_k$. We can view H in a natural way as an abelian group if we define the group operation component-wise:

$$(a_1, \ldots, a_k) + (b_1, \ldots, b_k) := (a_1 + b_1, \ldots, a_k + b_k).$$

Of course, the groups $G_1, \ldots, G_k$ may be different, and the group operation applied in the ith component corresponds to the group operation associated with G_i. We leave it to the reader to verify that H is in fact an abelian group, where $0_H = (0_{G_1}, \ldots, 0_{G_k})$ and $-(a_1, \ldots, a_k) = (-a_1, \ldots, -a_k)$. As a special case, if $G = G_1 = \cdots = G_k$, then the k-wise direct product of G is denoted $G^{\times k}$. □

Example 6.19. Let G be an abelian group. An element $(a_1, \ldots, a_k)$ of $G^{\times k}$ may be identified with the function $f : \{1, \ldots, k\} \to G$ given by $f(i) = a_i$ for $i = 1, \ldots, k$. We can generalize this, replacing $\{1, \ldots, k\}$ by an arbitrary set I. We define $\mathrm{Map}(I, G)$ to be the set of all functions $f : I \to G$, which we naturally view as a group by defining the group operation point-wise: for $f, g \in \mathrm{Map}(I, G)$, we define

$$(f + g)(i) := f(i) + g(i) \text{ for all } i \in I.$$

Again, we leave it to the reader to verify that $\mathrm{Map}(I, G)$ is an abelian group, where the identity element is the function that maps each $i \in I$ to 0_G, and for $f \in \mathrm{Map}(I, G)$, we have $(-f)(i) = -(f(i))$ for all $i \in I$. $\square$

EXERCISE 6.1. For a finite abelian group, one can completely specify the group by writing down the group operation table. For instance, Example 2.7 presented an addition table for $\mathbb{Z}_6$.

  (a) Write down group operation tables for the following finite abelian groups: $\mathbb{Z}_5$, $\mathbb{Z}_5^*$, and $\mathbb{Z}_3 \times \mathbb{Z}_4^*$.

  (b) Show that the group operation table for every finite abelian group is a **Latin square**; that is, each element of the group appears exactly once in each row and column.

  (c) Below is an addition table for an abelian group that consists of the elements $\{a, b, c, d\}$; however, some entries are missing. Fill in the missing entries.

+	a	b	c	d
a	a			
b	b	a		
c			a	
d				

EXERCISE 6.2. Let $G := \{x \in \mathbb{R} : x > 1\}$, and define $a \star b := ab - a - b + 2$ for all $a, b \in \mathbb{R}$. Show that:

  (a) G is closed under $\star$;

  (b) the set G under the operation $\star$ forms an abelian group.

EXERCISE 6.3. Let G be an abelian group, and let g be an arbitrary, fixed element of G. Assume that the group operation of G is written additively. We define a new binary operation $\odot$ on G, as follows: for $a, b \in G$, let $a \odot b := a + b + g$. Show that the set G under $\odot$ forms an abelian group.

EXERCISE 6.4. Let G be a finite abelian group of even order. Show that there exists $a \in G$ with $a \neq 0_G$ and $2a = 0_G$.

EXERCISE 6.5. Let $\star$ be a binary operation on a non-empty, *finite* set G. Assume that $\star$ is associative, commutative, and satisfies the cancellation law: $a \star b = a \star c$ implies $b = c$. Show that G under $\star$ forms an abelian group.

EXERCISE 6.6. Show that the result of the previous exercise need not hold if G is infinite.

6.2 Subgroups

We next introduce the notion of a subgroup.

Definition 6.5. *Let G be an abelian group, and let H be a non-empty subset of G such that*

(i) *$a + b \in H$ for all $a, b \in H$, and*

(ii) *$-a \in H$ for all $a \in H$.*

*Then H is called a **subgroup of** G.*

In words: H is a subgroup of G if it is closed under the group operation and taking inverses.

Multiplicative notation: if the abelian group G in the above definition is written using multiplicative notation, then H is a subgroup if $ab \in H$ and $a^{-1} \in H$ for all $a, b \in H$.

Theorem 6.6. *If G is an abelian group, and H is a subgroup of G, then H contains 0_G; moreover, the binary operation of G, when restricted to H, yields a binary operation that makes H into an abelian group whose identity is 0_G.*

Proof. First, to see that $0_G \in H$, just pick any $a \in H$, and using both properties of the definition of a subgroup, we see that $0_G = a + (-a) \in H$.

Next, note that by property (i) of Definition 6.5, H is closed under addition, which means that the restriction of the binary operation "+" on G to H induces a well-defined binary operation on H. So now it suffices to show that H, together with this operation, satisfies the defining properties of an abelian group. Associativity and commutativity follow directly from the corresponding properties for G. Since 0_G acts as the identity on G, it does so on H as well. Finally, property (ii) of Definition 6.5 guarantees that every element $a \in H$ has an inverse in H, namely, $-a$. $\square$

Clearly, for an abelian group G, the subsets G and $\{0_G\}$ are subgroups, though not very interesting ones. Other, more interesting subgroups may sometimes be found by using the following two theorems.

Theorem 6.7. *Let G be an abelian group, and let m be an integer. Then*

$$mG := \{ma : a \in G\}$$

is a subgroup of G.

Proof. The set mG is non-empty, since $0_G = m0_G \in mG$. For $ma, mb \in mG$, we have $ma + mb = m(a + b) \in mG$, and $-(ma) = m(-a) \in mG$. □

Theorem 6.8. *Let G be an abelian group, and let m be an integer. Then*

$$G\{m\} := \{a \in G : ma = 0_G\}$$

is a subgroup of G.

Proof. The set $G\{m\}$ is non-empty, since $m0_G = 0_G$, and so $G\{m\}$ contains 0_G. If $ma = 0_G$ and $mb = 0_G$, then $m(a + b) = ma + mb = 0_G + 0_G = 0_G$ and $m(-a) = -(ma) = -0_G = 0_G$. □

Multiplicative notation: if the abelian group G in the above two theorems is written using multiplicative notation, then we write the subgroup of the first theorem as $G^m := \{a^m : a \in G\}$. The subgroup in the second theorem is denoted in the same way: $G\{m\} := \{a \in G : a^m = 1_G\}$.

Example 6.20. We already proved that $(\mathbb{Z}_n^*)^m$ is a subgroup of $\mathbb{Z}_n^*$ in Theorem 2.16. Also, the proof of Theorem 2.17 clearly works for an arbitrary abelian group G: for each $a \in G$, and all $\ell, m \in \mathbb{Z}$ with $\gcd(\ell, m) = 1$, if $\ell a \in mG$, then $a \in mG$. □

Example 6.21. Let p be an odd prime. Then by Theorem 2.20, $(\mathbb{Z}_p^*)^2$ is a subgroup of $\mathbb{Z}_p^*$ of order $(p - 1)/2$, and as we saw in Theorem 2.18, $\mathbb{Z}_p^*\{2\} = \{[\pm 1]\}$. □

Example 6.22. For every integer m, the set $m\mathbb{Z}$ is the subgroup of the additive group $\mathbb{Z}$ consisting of all multiples of m. This is the same as the *ideal of $\mathbb{Z}$ generated by m*, which we already studied in some detail in §1.2. Two such subgroups $m\mathbb{Z}$ and $m'\mathbb{Z}$ are equal if and only if $m = \pm m'$. The subgroup $\mathbb{Z}\{m\}$ is equal to $\mathbb{Z}$ if $m = 0$, and is equal to $\{0\}$ otherwise. □

Example 6.23. Let n be a positive integer, let $m \in \mathbb{Z}$, and consider the subgroup $m\mathbb{Z}_n$ of the additive group $\mathbb{Z}_n$. Now, for every residue class $[z] \in \mathbb{Z}_n$, we have $m[z] = [mz]$. Therefore, $[b] \in m\mathbb{Z}_n$ if and only if there exists $z \in \mathbb{Z}$ such that $mz \equiv b \pmod{n}$. By part (i) of Theorem 2.5, such a z exists if and only if $d \mid b$, where $d := \gcd(m, n)$. Thus, $m\mathbb{Z}_n$ consists precisely of the n/d distinct residue classes

$$[i \cdot d] \quad (i = 0, \ldots, n/d - 1),$$

and in particular, $m\mathbb{Z}_n = d\mathbb{Z}_n$.

Now consider the subgroup $\mathbb{Z}_n\{m\}$ of $\mathbb{Z}_n$. The residue class $[z]$ is in $\mathbb{Z}_n\{m\}$ if and only if $mz \equiv 0 \pmod{n}$. By part (ii) of Theorem 2.5, this happens if and only if $z \equiv 0 \pmod{n/d}$, where $d := \gcd(m, n)$ as above. Thus, $\mathbb{Z}_n\{m\}$ consists precisely of the d residue classes

$$[i \cdot n/d] \quad (i = 0, \dots, d-1),$$

and in particular, $\mathbb{Z}_n\{m\} = \mathbb{Z}_n\{d\} = (n/d)\mathbb{Z}_n$. $\square$

Example 6.24. For $n = 15$, consider again the table in Example 2.2. For $m = 1$, $2, 3, 4, 5, 6$, the elements appearing in the mth row of that table form the subgroup $m\mathbb{Z}_n$ of $\mathbb{Z}_n$, and also the subgroup $\mathbb{Z}_n\{n/d\}$, where $d := \gcd(m, n)$. $\square$

Because the abelian groups $\mathbb{Z}$ and $\mathbb{Z}_n$ are of such importance, it is a good idea to completely characterize all subgroups of these abelian groups. As the following two theorems show, the subgroups in Examples 6.22 and 6.23 are the *only* ones.

Theorem 6.9. *If G is a subgroup of $\mathbb{Z}$, then there exists a unique non-negative integer m such that $G = m\mathbb{Z}$. Moreover, for two non-negative integers m_1 and m_2, we have $m_1\mathbb{Z} \subseteq m_2\mathbb{Z}$ if and only if $m_2 \mid m_1$.*

Proof. Actually, we have already proven this. One only needs to observe that a subset G of $\mathbb{Z}$ is a subgroup if and only if it is an ideal of $\mathbb{Z}$, as defined in §1.2 (see Exercise 1.8). The first statement of the theorem then follows from Theorem 1.6. The second statement follows easily from the definitions, as was observed in §1.2. $\square$

Theorem 6.10. *If G is a subgroup of $\mathbb{Z}_n$, then there exists a unique positive integer d dividing n such that $G = d\mathbb{Z}_n$. Also, for all positive divisors d_1, d_2 of n, we have $d_1\mathbb{Z}_n \subseteq d_2\mathbb{Z}_n$ if and only if $d_2 \mid d_1$.*

Proof. Note that the second statement implies the uniqueness part of the first statement, so it suffices to prove just the existence part of the first statement and the second statement.

Let G be an arbitrary subgroup of $\mathbb{Z}_n$, and let $H := \{z \in \mathbb{Z} : [z] \in G\}$. We claim that H is a subgroup of $\mathbb{Z}$. To see this, observe that if $a, b \in H$, then $[a]$ and $[b]$ belong to G, and hence so do $[a + b] = [a] + [b]$ and $[-a] = -[a]$, and thus $a + b$ and $-a$ belong to H. That proves the claim, and Theorem 6.9 implies that $H = d\mathbb{Z}$ for some non-negative integer d. It follows that

$$G = \{[y] : y \in H\} = \{[dz] : z \in \mathbb{Z}\} = d\mathbb{Z}_n.$$

Evidently, $n \in H = d\mathbb{Z}$, and hence $d \mid n$. That proves the existence part of the first statement of the theorem.

To prove the second statement of the theorem, observe that if d_1 and d_2 are arbitrary integers, then

$$d_1 \mathbb{Z}_n \subseteq d_2 \mathbb{Z}_n \iff d_2 z \equiv d_1 \pmod{n} \text{ for some } z \in \mathbb{Z}$$
$$\iff \gcd(d_2, n) \mid d_1 \text{ (by part (i) of Theorem 2.5).}$$

In particular, if d_2 is a positive divisor of n, then $\gcd(d_2, n) = d_2$, which proves the second statement. $\square$

Of course, not all abelian groups have such a simple subgroup structure.

Example 6.25. Consider the group $G = \mathbb{Z}_2 \times \mathbb{Z}_2$. For every non-zero $\alpha \in G$, $\alpha + \alpha = 0_G$. From this, it is clear that the set $H = \{0_G, \alpha\}$ is a subgroup of G. However, for every integer m, $mG = G$ if m is odd, and $mG = \{0_G\}$ if m is even. Thus, the subgroup H is not of the form mG for any m. $\square$

Example 6.26. Consider the group $\mathbb{Z}_{15}^*$. We can enumerate its elements as

$$[\pm 1], [\pm 2], [\pm 4], [\pm 7].$$

Therefore, the elements of $(\mathbb{Z}_{15}^*)^2$ are

$$[1]^2 = [1], \ [2]^2 = [4], \ [4]^2 = [16] = [1], \ [7]^2 = [49] = [4];$$

thus, $(\mathbb{Z}_{15}^*)^2$ has order 2, consisting as it does of the two distinct elements $[1]$ and $[4]$.

Going further, one sees that $(\mathbb{Z}_{15}^*)^4 = \{[1]\}$. Thus, $\alpha^4 = [1]$ for all $\alpha \in \mathbb{Z}_{15}^*$.

By direct calculation, one can determine that $(\mathbb{Z}_{15}^*)^3 = \mathbb{Z}_{15}^*$; that is, cubing simply permutes $\mathbb{Z}_{15}^*$.

For any given integer m, write $m = 4q + r$, where $0 \le r < 4$. Then for every $\alpha \in \mathbb{Z}_{15}^*$, we have $\alpha^m = \alpha^{4q+r} = \alpha^{4q} \alpha^r = \alpha^r$. Thus, $(\mathbb{Z}_{15}^*)^m$ is either $\mathbb{Z}_{15}^*$, $(\mathbb{Z}_{15}^*)^2$, or $\{[1]\}$.

However, there are certainly other subgroups of $\mathbb{Z}_{15}^*$—for example, the subgroup $\{[\pm 1]\}$. $\square$

Example 6.27. Consider the group $\mathbb{Z}_5^* = \{[\pm 1], [\pm 2]\}$. The elements of $(\mathbb{Z}_5^*)^2$ are

$$[1]^2 = [1], \ [2]^2 = [4] = [-1];$$

thus, $(\mathbb{Z}_5^*)^2 = \{[\pm 1]\}$ and has order 2.

There are in fact no other subgroups of $\mathbb{Z}_5^*$ besides $\mathbb{Z}_5^*$, $\{[\pm 1]\}$, and $\{[1]\}$. Indeed, if H is a subgroup containing $[2]$, then we must have $H = \mathbb{Z}_5^*$: $[2] \in H$ implies $[2]^2 = [4] = [-1] \in H$, which implies $[-2] \in H$ as well. The same holds if H is a subgroup containing $[-2]$. $\square$

Example 6.28. Consider again the abelian group $\mathcal{F}^*$ of arithmetic functions f, such that $f(1) \neq 0$, and with the Dirichlet product as the binary operation, as discussed in Example 6.11. Exercises 2.48 and 2.55 imply that the subset of all multiplicative functions is a subgroup. $\square$

We close this section with two theorems that provide useful ways to build new subgroups out of old ones.

Theorem 6.11. *If H_1 and H_2 are subgroups of an abelian group G, then so is*

$$H_1 + H_2 := \{a_1 + a_2 : a_1 \in H_1, a_2 \in H_2\}.$$

Proof. It is evident that $H_1 + H_2$ is non-empty, as it contains $0_G + 0_G = 0_G$. Consider two elements in $H_1 + H_2$, which we can write as $a_1 + a_2$ and $b_1 + b_2$, where $a_1, b_1 \in H_1$ and $a_2, b_2 \in H_2$. Then by the closure properties of subgroups, $a_1+b_1 \in H_1$ and $a_2+b_2 \in H_2$, and hence $(a_1+a_2)+(b_1+b_2) = (a_1+b_1)+(a_2+b_2) \in H_1 + H_2$. Similarly, $-(a_1 + a_2) = (-a_1) + (-a_2) \in H_1 + H_2$. $\square$

Multiplicative notation: if the abelian group G in the above theorem is written multiplicatively, then the subgroup defined in the theorem is written $H_1 H_2 := \{a_1 a_2 : a_1 \in H_1, a_2 \in H_2\}$.

Theorem 6.12. *If H_1 and H_2 are subgroups of an abelian group G, then so is $H_1 \cap H_2$.*

Proof. It is evident that $H_1 \cap H_2$ is non-empty, as both H_1 and H_2 contain 0_G, and hence so does their intersection. If $a \in H_1 \cap H_2$ and $b \in H_1 \cap H_2$, then since $a, b \in H_1$, we have $a + b \in H_1$, and since $a, b \in H_2$, we have $a + b \in H_2$; therefore, $a + b \in H_1 \cap H_2$. Similarly, $-a \in H_1$ and $-a \in H_2$, and therefore, $-a \in H_1 \cap H_2$. $\square$

Let G be an abelian group and H_1, H_2, H_3 subgroups of G. The reader may verify that $H_1 + H_2 = H_2 + H_1$ and $(H_1 + H_2) + H_3 = H_1 + (H_2 + H_3)$. It follows that if $H_1, \ldots, H_k$ are subgroups of G, then we can write $H_1 + \cdots + H_k$ without any parentheses, and there can be no ambiguity; moreover, the order of the H_i's does not matter. The same holds with "+" replaced by "$\cap$."

A warning: If H is a subgroup of an abelian group G, then in general, we have $H + H \neq 2H$. For example, $\mathbb{Z} + \mathbb{Z} = \mathbb{Z}$, while $2\mathbb{Z} \neq \mathbb{Z}$.

EXERCISE 6.7. Let G be an abelian group.

  (a) Suppose that H is a non-empty subset of G. Show that H is a subgroup of G if and only if $a - b \in H$ for all $a, b \in H$.

(b) Suppose that H is a non-empty, *finite* subset of G such that $a + b \in H$ for all $a, b \in H$. Show that H is a subgroup of G.

EXERCISE 6.8. Let G be an abelian group.

(a) Show that if H is a subgroup of G, $h \in H$, and $g \in G \setminus H$, then $h + g \in G \setminus H$.

(b) Suppose that H is a non-empty subset of G such that for all $h, g \in G$: (i) $h \in H$ implies $-h \in H$, and (ii) $h \in H$ and $g \in G \setminus H$ implies $h + g \in G \setminus H$. Show that H is a subgroup of G.

EXERCISE 6.9. Show that if H is a subgroup of an abelian group G, then a set $K \subseteq H$ is a subgroup of G if and only if K is a subgroup of H.

EXERCISE 6.10. Let G be an abelian group with subgroups H_1 and H_2. Show that every subgroup H of G that contains $H_1 \cup H_2$ must contain all of $H_1 + H_2$, and that $H_1 \subseteq H_2$ if and only if $H_1 + H_2 = H_2$.

EXERCISE 6.11. Let H_1 be a subgroup of an abelian group G_1 and H_2 a subgroup of an abelian group G_2. Show that $H_1 \times H_2$ is a subgroup of $G_1 \times G_2$.

EXERCISE 6.12. Show that if G_1 and G_2 are abelian groups, and m is an integer, then $m(G_1 \times G_2) = mG_1 \times mG_2$.

EXERCISE 6.13. Let G_1 and G_2 be abelian groups, and let H be a subgroup of $G_1 \times G_2$. Define

$$H_1 := \{a_1 \in G_1 : (a_1, a_2) \in H \text{ for some } a_2 \in G_2\}.$$

Show that H_1 is a subgroup of G_1.

EXERCISE 6.14. Let I be a set and G be an abelian group, and consider the group $\text{Map}(I, G)$ of functions $f : I \to G$. Let $\text{Map}^{\#}(I, G)$ be the set of functions $f \in \text{Map}(I, G)$ such that $f(i) \neq 0_G$ for at most finitely many $i \in I$. Show that $\text{Map}^{\#}(I, G)$ is a subgroup of $\text{Map}(I, G)$.

6.3 Cosets and quotient groups

We now generalize the notion of a congruence relation.

Let G be an abelian group, and let H be a subgroup of G. For $a, b \in G$, we write $a \equiv b \pmod{H}$ if $a - b \in H$. In other words, $a \equiv b \pmod{H}$ if and only if $a = b + h$ for some $h \in H$.

Analogous to Theorem 2.2, if we view the subgroup H as fixed, then the following theorem says that the binary relation "$\cdot \equiv \cdot \pmod{H}$" is an equivalence relation on the set G:

Theorem 6.13. *Let G be an abelian group and H a subgroup of G. For all $a, b, c \in G$, we have:*

(i) $a \equiv a \pmod{H}$;

(ii) $a \equiv b \pmod{H}$ *implies* $b \equiv a \pmod{H}$;

(iii) $a \equiv b \pmod{H}$ *and* $b \equiv c \pmod{H}$ *implies* $a \equiv c \pmod{H}$.

Proof. For (i), observe that H contains $0_G = a - a$. For (ii), observe that if H contains $a - b$, then it also contains $-(a - b) = b - a$. For (iii), observe that if H contains $a - b$ and $b - c$, then it also contains $(a - b) + (b - c) = a - c$. $\square$

Since the binary relation "$\cdot \equiv \cdot \pmod{H}$" is an equivalence relation, it partitions G into equivalence classes (see Theorem 2.1). For $a \in G$, we denote the equivalence class containing a by $[a]_H$. By definition, we have

$$x \in [a]_H \iff x \equiv a \pmod{H} \iff x = a + h \text{ for some } h \in H,$$

and hence

$$[a]_H = a + H := \{a + h : h \in H\}.$$

It is also clear that $[0_G]_H = H$.

Historically, these equivalence classes are called **cosets of H in G**, and we shall adopt this terminology here as well. Any member of a coset is called a **representative** of the coset.

Multiplicative notation: if G is written multiplicatively, then $a \equiv b \pmod{H}$ means $ab^{-1} \in H$, and $[a]_H = aH := \{ah : h \in H\}$.

Example 6.29. Let $G := \mathbb{Z}$ and $H := n\mathbb{Z}$ for some positive integer n. Then $a \equiv b \pmod{H}$ if and only if $a \equiv b \pmod{n}$. The coset $[a]_H$ is exactly the same thing as the residue class $[a]_n \in \mathbb{Z}_n$. $\square$

Example 6.30. Let $G := \mathbb{Z}_6$, which consists of the residue classes $[0], [1], [2], [3]$, $[4], [5]$. Let H be the subgroup $3G = \{[0], [3]\}$ of G. The coset of H containing the residue class $[1]$ is $[1] + H = \{[1], [4]\}$, and the coset of H containing the residue class $[2]$ is $[2] + H = \{[2], [5]\}$. The cosets $\{[0], [3]\}$, $\{[1], [4]\}$, and $\{[2], [5]\}$ are the only cosets of H in G, and they clearly partition the set $\mathbb{Z}_6$. Note that each coset of H in G contains two elements, each of which is itself a coset of $6\mathbb{Z}$ in $\mathbb{Z}$ (i.e., a residue classes modulo 6). $\square$

In the previous example, we saw that each coset contained the same number of elements. As the next theorem shows, this was no accident.

Theorem 6.14. *Let G be an abelian group and H a subgroup of G. For all $a, b \in G$, the function*

$$f: \quad G \to G$$
$$x \mapsto b - a + x$$

is a bijection, which, when restricted to the coset $[a]_H$, yields a bijection from $[a]_H$ to the coset $[b]_H$. In particular, every two cosets of H in G have the same cardinality.

Proof. First, we claim that f is a bijection. Indeed, if $f(x) = f(x')$, then $b - a + x = b - a + x'$, and subtracting b and adding a to both sides of this equation yields $x = x'$. That proves that f is injective. To prove that f is surjective, observe that for any given $x' \in G$, we have $f(a - b + x') = x'$.

Second, we claim that for all $x \in G$, we have $x \in [a]_H$ if and only if $f(x) \in [b]_H$. On the one hand, suppose that $x \in [a]_H$, which means that $x = a + h$ for some $h \in H$. Subtracting a and adding b to both sides of this equation yields $b - a + x = b + h$, which means $f(x) \in [b]_H$. Conversely, suppose that $f(x) \in [b]_H$, which means that $b - a + x = b + h$ for some $h \in H$. Subtracting b and adding a to both sides of this equation yields $x = a + h$, which means that $x \in [a]_H$.

The theorem is now immediate from these two claims. $\square$

An incredibly useful consequence of the above theorem is:

Theorem 6.15 (Lagrange's theorem). *If G is a finite abelian group, and H is a subgroup of G, then the order of H divides the order of G.*

Proof. This is an immediate consequence of the previous theorem, and the fact that the cosets of H in G partition G. $\square$

Analogous to Theorem 2.3, we have:

Theorem 6.16. *Suppose G is an abelian group and H is a subgroup of G. For all $a, a', b, b' \in G$, if $a \equiv a' \pmod{H}$ and $b \equiv b' \pmod{H}$, then we have $a + b \equiv a' + b' \pmod{H}$.*

Proof. Now, $a \equiv a' \pmod{H}$ and $b \equiv b' \pmod{H}$ means that $a = a' + x$ and $b = b' + y$ for some $x, y \in H$. Therefore, $a + b = (a' + x) + (b' + y) = (a' + b') + (x + y)$, and since $x + y \in H$, this means that $a + b \equiv a' + b' \pmod{H}$. $\square$

Let G be an abelian group and H a subgroup. Let G/H denote the set of all cosets of H in G. Theorem 6.16 allows us to define a binary operation on G/H in the following natural way: for $a, b \in G$, define

$$[a]_H + [b]_H := [a + b]_H.$$

That this definition is unambiguous follows immediately from Theorem 6.16: if $[a]_H = [a']_H$ and $[b]_H = [b']_H$, then $[a + b]_H = [a' + b']_H$.

We can easily verify that this operation makes G/H into an abelian group. We need to check that the four properties of Definition 6.1 are satisfied:

(i) Associativity:

$$[a]_H + ([b]_H + [c]_H) = [a]_H + [b + c]_H = [a + (b + c)]_H$$
$$= [(a + b) + c]_H = [a + b]_H + [c]_H$$
$$= ([a]_H + [b]_H) + [c]_H.$$

Here, we have used the definition of addition of cosets, and the corresponding associativity property for G.

(ii) Identity element: the coset $[0_G]_H = H$ acts as the identity element, since

$$[a]_H + [0_G]_H = [a + 0_G]_H = [a]_H = [0_G + a]_H = [0_G]_H + [a]_H.$$

(iii) Inverses: the inverse of the coset $[a]_H$ is $[-a]_H$, since

$$[a]_H + [-a]_H = [a + (-a)]_H = [0_G]_H = [(-a) + a]_H = [-a]_H + [a]_H.$$

(iv) Commutativity:

$$[a]_H + [b]_H = [a + b]_H = [b + a]_H = [b]_H + [a]_H.$$

The group G/H is called the **quotient group of G modulo H**. The order of the group G/H is sometimes denoted $[G : H]$ and is called the **index of H in G**. Note that if $H = G$, then the quotient group G/H is the trivial group, and so $[G : H] = 1$.

Multiplicative notation: if G is written multiplicatively, then the definition of the group operation of G/H is expressed $[a]_H \cdot [b]_H := [a \cdot b]_H$; the identity element of G/H is $[1_G]_H = H$, and the inverse of $[a]_H$ is $[a^{-1}]_H$.

Theorem 6.17. *Suppose G is a finite abelian group and H is a subgroup of G. Then $[G : H] = |G|/|H|$. Moreover, if K is a subgroup of H, then*

$$[G : K] = [G : H][H : K].$$

Proof. The fact that $[G : H] = |G|/|H|$ follows directly from Theorem 6.14. The fact that $[G : K] = [G : H][H : K]$ follows from a simple calculation:

$$[G : H] = \frac{|G|}{|H|} = \frac{|G|/|K|}{|H|/|K|} = \frac{[G : K]}{[H : K]}. \quad \Box$$

Example 6.31. For each $n \geq 1$, the group $\mathbb{Z}_n$ is precisely the quotient group $\mathbb{Z}/n\mathbb{Z}$. $\Box$

Example 6.32. Continuing with Example 6.30, let $G := \mathbb{Z}_6$ and $H := 3G = \{[0], [3]\}$. The quotient group G/H has order 3, and consists of the cosets

$$\alpha := \{[0], [3]\}, \quad \beta := \{[1], [4]\}, \quad \gamma := \{[2], [5]\}.$$

If we write out an addition table for G, grouping together elements in cosets of H in G, then we also get an addition table for the quotient group G/H:

+	[0]	[3]	[1]	[4]	[2]	[5]
[0]	[0]	[3]	[1]	[4]	[2]	[5]
[3]	[3]	[0]	[4]	[1]	[5]	[2]
[1]	[1]	[4]	[2]	[5]	[3]	[0]
[4]	[4]	[1]	[5]	[2]	[0]	[3]
[2]	[2]	[5]	[3]	[0]	[4]	[1]
[5]	[5]	[2]	[0]	[3]	[1]	[4]

This table illustrates quite graphically the point of Theorem 6.16: for every two cosets, if we take any element from the first and add it to any element of the second, we always end up in the same coset.

We can also write down just the addition table for G/H:

+	α	β	γ
α	α	β	γ
β	β	γ	α
γ	γ	α	β

Note that by replacing α with $[0]_3$, β with $[1]_3$, and γ with $[2]_3$, the addition table for G/H becomes the addition table for $\mathbb{Z}_3$. In this sense, we can view G/H as essentially just a "renaming" of $\mathbb{Z}_3$. □

Example 6.33. Let us return to Example 6.26. The multiplicative group $\mathbb{Z}_{15}^*$, as we saw, is of order 8. The subgroup $(\mathbb{Z}_{15}^*)^2$ of $\mathbb{Z}_{15}^*$ has order 2. Therefore, the quotient group $\mathbb{Z}_{15}^*/(\mathbb{Z}_{15}^*)^2$ has order 4. Indeed, the cosets are

$$\alpha_{00} := (\mathbb{Z}_{15}^*)^2 = \{[1], [4]\}, \qquad \alpha_{01} := [-1](\mathbb{Z}_{15}^*)^2 = \{[-1], [-4]\},$$
$$\alpha_{10} := [2](\mathbb{Z}_{15}^*)^2 = \{[2], [-7]\}, \quad \alpha_{11} := [-2](\mathbb{Z}_{15}^*)^2 = \{[-2], [7]\}.$$

We can write down the multiplication table for the quotient group:

·	α_{00}	α_{01}	α_{10}	α_{11}
α_{00}	α_{00}	α_{01}	α_{10}	α_{11}
α_{01}	α_{01}	α_{00}	α_{11}	α_{10}
α_{10}	α_{10}	α_{11}	α_{00}	α_{01}
α_{11}	α_{11}	α_{10}	α_{01}	α_{00}

Note that this group is essentially just a "renaming" of the additive group $\mathbb{Z}_2 \times \mathbb{Z}_2$. □

Example 6.34. As we saw in Example 6.27, $(\mathbb{Z}_5^*)^2 = \{[\pm 1]\}$. Therefore, the quotient group $\mathbb{Z}_5^*/(\mathbb{Z}_5^*)^2$ has order 2. The cosets of $(\mathbb{Z}_5^*)^2$ in $\mathbb{Z}_5^*$ are $\alpha_0 := \{[\pm 1]\}$ and $\alpha_1 := \{[\pm 2]\}$, and the multiplication table looks like this:

$\cdot$	α_0	α_1
α_0	α_0	α_1
α_1	α_1	α_0

We see that the quotient group is essentially just a "renaming" of $\mathbb{Z}_2$. □

EXERCISE 6.15. Write down the cosets of $(\mathbb{Z}_{35}^*)^2$ in $\mathbb{Z}_{35}^*$, along with the multiplication table for the quotient group $\mathbb{Z}_{35}^*/(\mathbb{Z}_{35}^*)^2$.

EXERCISE 6.16. Let n be an odd, positive integer whose factorization into primes is $n = p_1^{e_1} \cdots p_r^{e_r}$. Show that $[\mathbb{Z}_n^* : (\mathbb{Z}_n^*)^2] = 2^r$.

EXERCISE 6.17. Let n be a positive integer, and let m be any integer. Show that $[\mathbb{Z}_n : m\mathbb{Z}_n] = n/\gcd(m, n)$.

EXERCISE 6.18. Let G be an abelian group and H a subgroup with $[G : H] = 2$. Show that if $a, b \in G \setminus H$, then $a + b \in H$.

EXERCISE 6.19. Let H be a subgroup of an abelian group G, and let $a, b \in G$ with $a \equiv b \pmod{H}$. Show that $ka \equiv kb \pmod{H}$ for all $k \in \mathbb{Z}$.

EXERCISE 6.20. Let G be an abelian group, and let $\sim$ be an equivalence relation on G. Further, suppose that for all $a, a', b \in G$, if $a \sim a'$, then $a + b \sim a' + b$. Let $H := \{a \in G : a \sim 0_G\}$. Show that H is a subgroup of G, and that for all $a, b \in G$, we have $a \sim b$ if and only if $a \equiv b \pmod{H}$.

EXERCISE 6.21. Let H be a subgroup of an abelian group G, and let $a, b \in G$. Show that $[a + b]_H = \{x + y : x \in [a]_H, y \in [b]_H\}$.

6.4 Group homomorphisms and isomorphisms

In this section, we study maps that relate the structure of one group to another. Such maps are often very useful, as they may allow us to transfer hard-won knowledge about one group to another, perhaps more mysterious, group.

Definition 6.18. *A **group homomorphism** is a function ρ from an abelian group G to an abelian group G' such that $\rho(a + b) = \rho(a) + \rho(b)$ for all $a, b \in G$.*

Note that in the equality $\rho(a + b) = \rho(a) + \rho(b)$ in the above definition, the addition on the left-hand side is taking place in the group G while the addition on the right-hand side is taking place in the group G'.

Two sets play a critical role in the study of a group homomorphism $\rho : G \to G'$. The first set is the **image** of ρ, that is, the set $\rho(G) = \{\rho(a) : a \in G\}$. The second set is the **kernel** of ρ, defined as the set of all elements of G that are mapped to $0_{G'}$ by ρ, that is, the set $\rho^{-1}(\{0_{G'}\}) = \{a \in G : \rho(a) = 0_{G'}\}$. We introduce the following notation for these sets: Im ρ denotes the image of ρ, and Ker ρ denotes the kernel of ρ.

Example 6.35. If H is a subgroup of an abelian group G, then the inclusion map $i : H \to G$ is obviously a group homomorphism. $\square$

Example 6.36. Suppose H is a subgroup of an abelian group G. We define the map

$$\rho : \quad G \to G/H$$
$$a \mapsto [a]_H.$$

It is not hard to see that this is a group homomorphism. Indeed, this follows almost immediately from the way we defined addition in the quotient group G/H:

$$\rho(a + b) = [a + b]_H = [a]_H + [b]_H = \rho(a) + \rho(b).$$

It is clear that ρ is surjective. It is also not hard to see that Ker $\rho = H$; indeed, H is the identity element in G/H, and $[a]_H = H$ if and only if $a \in H$. The map ρ is called the **natural map** from G to G/H. $\square$

Example 6.37. For a given positive integer n, the natural map from $\mathbb{Z}$ to $\mathbb{Z}_n$ sends $a \in \mathbb{Z}$ to the residue class $[a]_n$. This map is a surjective group homomorphism with kernel $n\mathbb{Z}$. $\square$

Example 6.38. Suppose G is an abelian group and m is an integer. The map

$$\rho : \quad G \to G$$
$$a \mapsto ma$$

is a group homomorphism, since

$$\rho(a + b) = m(a + b) = ma + mb = \rho(a) + \rho(b).$$

The image of this homomorphism is the subgroup mG and the kernel is the subgroup $G\{m\}$. We call this map the m-**multiplication map on** G. If G is written multiplicatively, then this map, which sends $a \in G$ to $a^m \in G$, is called the m-**power map on** G, and its image is G^m. $\square$

Example 6.39. Let p be an odd prime. Consider the 2-power, or squaring, map on $\mathbb{Z}_p^*$. Then as we saw in Example 6.21, the image $(\mathbb{Z}_p^*)^2$ of this map is a subgroup of $\mathbb{Z}_p^*$ of order $(p-1)/2$, and its kernel is $\mathbb{Z}_p^*\{2\} = \{[\pm 1]\}$. $\square$

Example 6.40. Consider the m-multiplication map on $\mathbb{Z}$. As we saw in Example 6.22, its image $m\mathbb{Z}$ is equal to $\mathbb{Z}$ if and only if $m = \pm 1$, while its kernel $\mathbb{Z}\{m\}$ is equal to $\mathbb{Z}$ if $m = 0$, and is equal to $\{0\}$ otherwise. $\square$

Example 6.41. Consider the m-multiplication map on $\mathbb{Z}_n$. As we saw in Example 6.23, if $d := \gcd(m, n)$, the image $m\mathbb{Z}_n$ of this map is a subgroup of $\mathbb{Z}_n$ of order n/d, while its kernel $\mathbb{Z}_n\{m\}$ is a subgroup of order d. $\square$

Example 6.42. Suppose G is an abelian group and a is an element of G. It is easy to see that the map

$$\rho : \quad \mathbb{Z} \to G$$

$$z \mapsto za$$

is a group homomorphism, since

$$\rho(z + z') = (z + z')a = za + z'a = \rho(z) + \rho(z'). \quad \square$$

Example 6.43. As a special case of the previous example, let n be a positive integer and let α be an element of $\mathbb{Z}_n^*$. Let $\rho : \mathbb{Z} \to \mathbb{Z}_n^*$ be the group homomorphism that sends $z \in \mathbb{Z}$ to $\alpha^z \in \mathbb{Z}_n^*$. That ρ is a group homomorphism means that $\alpha^{z+z'} = \alpha^z \alpha^{z'}$ for all $z, z' \in \mathbb{Z}$ (note that the group operation is addition in $\mathbb{Z}$ and multiplication in $\mathbb{Z}_n^*$). If the multiplicative order of α is equal to k, then as discussed in §2.7, the image of ρ consists of the k distinct group elements $\alpha^0, \alpha^1, \ldots, \alpha^{k-1}$. The kernel of ρ consists of those integers z such that $\alpha^z = 1$. Again by the discussion in §2.7, the kernel of ρ is equal to the subgroup $k\mathbb{Z}$. $\square$

Example 6.44. Generalizing Example 6.42, the reader may verify that if $a_1, \ldots, a_k$ are fixed elements of an abelian group G, then the map

$$\rho : \qquad \mathbb{Z}^{\times k} \to G$$

$$(z_1, \ldots, z_k) \mapsto z_1 a_1 + \cdots + z_k a_k$$

is a group homomorphism. $\square$

Example 6.45. Suppose that $H_1, \ldots, H_k$ are subgroups of an abelian group G. The reader may easily verify that the map

$$\rho : \quad H_1 \times \cdots \times H_k \to G$$

$$(a_1, \ldots, a_k) \mapsto a_1 + \cdots + a_k$$

is a group homomorphism whose image is the subgroup $H_1 + \cdots + H_k$. $\square$

The following theorem summarizes some of the most important properties of group homomorphisms.

Theorem 6.19. *Let ρ be a group homomorphism from G to G'. Then:*

(i) $\rho(0_G) = 0_{G'}$;

(ii) $\rho(-a) = -\rho(a)$ *for all* $a \in G$;

(iii) $\rho(na) = n\rho(a)$ *for all* $n \in \mathbb{Z}$ *and* $a \in G$;

(iv) *if H is a subgroup of G, then $\rho(H)$ is a subgroup of G'; in particular (setting $H := G$), $\mathrm{Im}\, \rho$ is a subgroup of G';*

(v) *if H' is a subgroup of G', then $\rho^{-1}(H')$ is a subgroup of G; in particular (setting $H' := \{0_{G'}\}$), $\mathrm{Ker}\, \rho$ is a subgroup of G;*

(vi) *for all $a, b \in G$, $\rho(a) = \rho(b)$ if and only if $a \equiv b \pmod{\mathrm{Ker}\, \rho}$;*

(vii) *ρ is injective if and only if $\mathrm{Ker}\, \rho = \{0_G\}$.*

Proof. These are all straightforward calculations.

(i) We have
$$0_{G'} + \rho(0_G) = \rho(0_G) = \rho(0_G + 0_G) = \rho(0_G) + \rho(0_G).$$
Now cancel $\rho(0_G)$ from both sides.

(ii) We have
$$0_{G'} = \rho(0_G) = \rho(a + (-a)) = \rho(a) + \rho(-a),$$
and hence $\rho(-a)$ is the inverse of $\rho(a)$.

(iii) For $n = 0$, this follows from part (i). For $n > 0$, this follows from the definitions by induction on n. For $n < 0$, this follows from the positive case and part (ii).

(iv) For all $a, b \in H$, we have $a + b \in H$ and $-a \in H$; hence, $\rho(H)$ contains $\rho(a + b) = \rho(a) + \rho(b)$ and $\rho(-a) = -\rho(a)$.

(v) $\rho^{-1}(H')$ is non-empty, since $\rho(0_G) = 0'_G \in H'$. If $\rho(a) \in H'$ and $\rho(b) \in H'$, then $\rho(a + b) = \rho(a) + \rho(b) \in H'$, and $\rho(-a) = -\rho(a) \in H'$.

(vi) We have
$$\rho(a) = \rho(b) \iff \rho(a) - \rho(b) = 0_{G'} \iff \rho(a - b) = 0_{G'}$$
$$\iff a - b \in \mathrm{Ker}\, \rho \iff a \equiv b \pmod{\mathrm{Ker}\, \rho}.$$

(vii) If ρ is injective, then in particular, $\rho^{-1}(\{0_{G'}\})$ cannot contain any other element besides 0_G. If ρ is not injective, then there exist two distinct elements $a, b \in G$ with $\rho(a) = \rho(b)$, and by part (vi), $\mathrm{Ker}\, \rho$ contains the element $a - b$, which is non-zero. $\square$

Part (vii) of the above theorem is particularly useful: to check that a group homomorphism is injective, it suffices to determine if $\mathrm{Ker}\, \rho = \{0_G\}$. Thus, the injectivity and surjectivity of a given group homomorphism $\rho : G \to G'$ may be characterized in terms of its kernel and image:

- ρ is injective if and only if its kernel is trivial (i.e. $\mathrm{Ker}\, \rho = \{0_G\}$);

- ρ is surjective if and only if $\mathrm{Im}\, \rho = G'$.

We next present two very easy theorems that allow us to compose group homomorphisms in simple ways.

Theorem 6.20. *If $\rho : G \to G'$ and $\rho' : G' \to G''$ are group homomorphisms, then so is their composition $\rho' \circ \rho : G \to G''$.*

Proof. For all $a, b \in G$, we have

$$\rho'(\rho(a + b)) = \rho'(\rho(a) + \rho(b)) = \rho'(\rho(a)) + \rho'(\rho(b)). \quad \square$$

Theorem 6.21. *Let $\rho_i : G \to G'_i$, for $i = 1, \ldots, k$, be group homomorphisms. Then the map*

$$\rho : \quad G \to G'_1 \times \cdots \times G'_k$$
$$a \mapsto (\rho_1(a), \ldots, \rho_k(a))$$

is a group homomorphism.

Proof. For all $a, b \in G$, we have

$$\rho(a + b) = (\rho_1(a + b), \ldots, \rho_k(a + b)) = (\rho_1(a) + \rho_1(b), \ldots, \rho_k(a) + \rho_k(b))$$
$$= \rho(a) + \rho(b). \quad \square$$

Consider a group homomorphism $\rho : G \to G'$. If ρ is bijective, then ρ is called a **group isomorphism** of G with G'. If such a group isomorphism ρ exists, we say that G **is isomorphic to** G', and write $G \cong G'$. Moreover, if $G = G'$, then ρ is called a **group automorphism** on G.

Theorem 6.22. *If ρ is a group isomorphism of G with G', then the inverse function ρ^{-1} is a group isomorphism of G' with G.*

Proof. For all $a', b' \in G'$, we have

$$\rho(\rho^{-1}(a') + \rho^{-1}(b')) = \rho(\rho^{-1}(a')) + \rho(\rho^{-1}(b')) = a' + b',$$

and hence $\rho^{-1}(a') + \rho^{-1}(b') = \rho^{-1}(a' + b'). \quad \square$

Because of this theorem, if G is isomorphic to G', we may simply say that "G and G' are isomorphic."

We stress that a group isomorphism $\rho : G \to G'$ is essentially just a "renaming" of the group elements. This can be visualized as follows. Imagine the addition table for G written out with rows and columns labeled by elements of G, with the entry in row a and column b being $a + b$. Now suppose we use the function ρ to consistently rename all the elements of G appearing in this table: the label on row a is replaced by $\rho(a)$, the label on column b by $\rho(b)$, and the entry in row a and column b by $\rho(a + b)$. Because ρ is bijective, every element of G' appears exactly once as a label on a row and as a label on a column; moreover, because $\rho(a + b) = \rho(a) + \rho(b)$, what we end up with is an addition table for G'. It follows that all structural properties of the group are preserved, even though the two groups might look quite different syntactically.

Example 6.46. As was shown in Example 6.32, the quotient group G/H discussed in that example is isomorphic to $\mathbb{Z}_3$. As was shown in Example 6.33, the quotient group $\mathbb{Z}_{15}^*/(\mathbb{Z}_{15}^*)^2$ is isomorphic to $\mathbb{Z}_2 \times \mathbb{Z}_2$. As was shown in Example 6.34, the quotient group $\mathbb{Z}_5^*/(\mathbb{Z}_5^*)^2$ is isomorphic to $\mathbb{Z}_2$. □

Example 6.47. If $\gcd(m, n) = 1$, then the m-multiplication map on $\mathbb{Z}_n$ is a group automorphism. □

The next theorem tells us that corresponding to any group homomorphism, there is a natural group isomomorphism. As group isomorphisms are much nicer than group homomorphisms, this is often very useful.

Theorem 6.23 (First isomorphism theorem). *Let $\rho : G \to G'$ be a group homomorphism with kernel K and image H'. Then we have a group isomorphism*

$$G/K \cong H'.$$

Specifically, the map

$$\bar{\rho} : \; G/K \to G'$$
$$[a]_K \mapsto \rho(a)$$

is an injective group homomorphism whose image is H'.

Proof. Using part (vi) of Theorem 6.19, we see that for all $a, b \in G$, we have

$$[a]_K = [b]_K \iff a \equiv b \pmod{K} \iff \rho(a) = \rho(b).$$

This immediately implies that the definition of $\bar{\rho}$ is unambiguous ($[a]_K = [b]_K$ implies $\rho(a) = \rho(b)$), and that $\bar{\rho}$ is injective ($\rho(a) = \rho(b)$ implies $[a]_K = [b]_K$). It is clear that $\bar{\rho}$ maps onto H', since every element of H' is of the form $\rho(a)$ for some $a \in G$, and the map $\bar{\rho}$ sends $[a]_K$ to $\rho(a)$. Finally, to see that $\bar{\rho}$ is a group homomorphism, note that

$$\bar{\rho}([a]_K + [b]_K) = \bar{\rho}([a + b]_K) = \rho(a + b) = \rho(a) + \rho(b) = \bar{\rho}([a]_K) + \bar{\rho}([b]_K). \quad □$$

We can generalize the previous theorem, as follows:

Theorem 6.24. *Let $\rho : G \to G'$ be a group homomorphism. Then for every subgroup H of G with $H \subseteq \text{Ker}\,\rho$, we may define a group homomorphism*

$$\bar{\rho} : \quad G/H \to G'$$

$$[a]_H \mapsto \rho(a).$$

Moreover, $\text{Im}\,\bar{\rho} = \text{Im}\,\rho$, and $\bar{\rho}$ is injective if and only if $H = \text{Ker}\,\rho$.

Proof. Using the assumption that $H \subseteq \text{Ker}\,\rho$, we see that $\bar{\rho}$ is unambiguously defined, since for all $a, b \in G$, we have

$$[a]_H = [b]_H \implies a \equiv b \,(\text{mod } H) \implies a \equiv b \,(\text{mod Ker}\,\rho) \implies \rho(a) = \rho(b).$$

That $\bar{\rho}$ is a group homomorphism, with $\text{Im}\,\bar{\rho} = \text{Im}\,\rho$, follows as in the proof of Theorem 6.23. If $H = \text{Ker}\,\rho$, then by Theorem 6.23, $\bar{\rho}$ is injective, and if $H \subsetneq \text{Ker}\,\rho$, then $\bar{\rho}$ is not injective, since if we choose $a \in \text{Ker}\,\rho \setminus H$, we see that $\bar{\rho}([a]_H) = 0_{G'}$, and hence $\text{Ker}\,\bar{\rho}$ is non-trivial. $\square$

The next theorem gives us another important construction of a group isomorphism.

Theorem 6.25 (Internal direct product). *Let G be an abelian group with subgroups H_1, H_2, where $H_1 \cap H_2 = \{0_G\}$. Then we have a group isomorphism*

$$H_1 \times H_2 \cong H_1 + H_2$$

given by the map

$$\rho : \quad H_1 \times H_2 \to H_1 + H_2$$

$$(a_1, a_2) \mapsto a_1 + a_2.$$

Proof. We already saw that ρ is a surjective group homomorphism in Example 6.45. To see that ρ is injective, it suffices to show that $\text{Ker}\,\rho$ is trivial; that is, it suffices to show that for all $a_1 \in H_1$ and $a_2 \in H_2$, if $a_1 + a_2 = 0_G$, then $a_1 = a_2 = 0_G$. But $a_1 + a_2 = 0_G$ implies $a_1 = -a_2 \in H_2$, and hence $a_1 \in H_1 \cap H_2 = \{0_G\}$, and so $a_1 = 0_G$. Similarly, one shows that $a_2 = 0_G$, and that finishes the proof. $\square$

If H_1, H_2 are as in the above theorem, then $H_1 + H_2$ is sometimes called the **internal direct product** of H_1 and H_2.

Example 6.48. We can use the general theory developed so far to get a quick-and-dirty proof of the Chinese remainder theorem (Theorem 2.6). Let $\{n_i\}_{i=1}^k$ be a pairwise relatively prime family of positive integers, and let $n := \prod_{i=1}^k n_i$. Consider the map

$$\rho : \quad \mathbb{Z} \to \mathbb{Z}_{n_1} \times \cdots \times \mathbb{Z}_{n_k}$$

$$a \mapsto ([a]_{n_1}, \ldots, [a]_{n_k}).$$

It is easy to see that this map is a group homomorphism; indeed, it is the map constructed in Theorem 6.21 applied with the natural maps $\rho_i : \mathbb{Z} \to \mathbb{Z}_{n_i}$, for $i = 1, \ldots, k$. Evidently, $a \in \operatorname{Ker} \rho$ if and only if $n_i \mid a$ for $i = 1, \ldots, k$, and since $\{n_i\}_{i=1}^{k}$ is pairwise relatively prime, it follows that $a \in \operatorname{Ker} \rho$ if and only if $n \mid a$; that is, $\operatorname{Ker} \rho = n\mathbb{Z}$. Theorem 6.23 then gives us an injective group homomorphism

$$\bar{\rho}: \quad \mathbb{Z}_n \to \mathbb{Z}_{n_1} \times \cdots \times \mathbb{Z}_{n_k}$$
$$[a]_n \mapsto ([a]_{n_1}, \ldots, [a]_{n_k}).$$

But since the sets $\mathbb{Z}_n$ and $\mathbb{Z}_{n_1} \times \cdots \times \mathbb{Z}_{n_k}$ have the same size, injectivity implies surjectivity. From this, Theorem 2.6 is immediate.

The map $\bar{\rho}$ is a group isomorphism

$$\mathbb{Z}_n \cong \mathbb{Z}_{n_1} \times \cdots \times \mathbb{Z}_{n_k}.$$

In fact, the map $\bar{\rho}$ is the same as the map θ in Theorem 2.8, and so we also immediately obtain parts (i), (ii), (iii.a), and (iii.b) of that theorem.

Observe that parts (iii.c) and (iii.d) of Theorem 2.8 imply that restricting the map θ to $\mathbb{Z}_n^*$ yields an isomorphism of *multiplicative* groups

$$\mathbb{Z}_n^* \cong \mathbb{Z}_{n_1}^* \times \cdots \times \mathbb{Z}_{n_k}^*.$$

This fact does *not* follow from the general theory developed so far; however, in the next chapter, we will see how this fact fits into the broader algebraic picture.

One advantage of our original proof of Theorem 2.6 is that it gives us an explicit formula for the inverse map θ^{-1}, which is useful in computations. $\square$

Example 6.49. Let n_1, n_2 be positive integers with $n_1 \mid n_2$. Consider the natural map $\rho : \mathbb{Z} \to \mathbb{Z}_{n_1}$. This is a surjective group homomorphism with $\operatorname{Ker} \rho = n_1\mathbb{Z}$. Since $H := n_2\mathbb{Z} \subseteq n_1\mathbb{Z}$, we may apply Theorem 6.24 with the subgroup H, obtaining the surjective group homomorphism

$$\bar{\rho}: \quad \mathbb{Z}_{n_2} \to \mathbb{Z}_{n_1}$$
$$[a]_{n_2} \mapsto [a]_{n_1}. \quad \square$$

Example 6.50. Let us revisit Example 6.23. Let n be a positive integer, and let m be any integer. Let $\rho_1 : \mathbb{Z} \to \mathbb{Z}_n$ be the natural map, and let $\rho_2 : \mathbb{Z}_n \to \mathbb{Z}_n$ be the m-multiplication map. The composed map $\rho := \rho_2 \circ \rho_1$ from $\mathbb{Z}$ to $\mathbb{Z}_n$ is also a group homomorphism. For each $z \in \mathbb{Z}$, we have $\rho(z) = m[z]_n = [mz]_n$. The kernel of ρ consists of those integers z such that $mz \equiv 0 \pmod{n}$, and so part (ii) of Theorem 2.5 implies that $\operatorname{Ker} \rho = (n/d)\mathbb{Z}$, where $d := \gcd(m, n)$. The image of ρ is $m\mathbb{Z}_n$. Theorem 6.23 therefore implies that the map

$$\bar{\rho}: \quad \mathbb{Z}_{n/d} \to m\mathbb{Z}_n$$
$$[z]_{n/d} \mapsto m[z]_n$$

is a group isomorphism. $\square$

Example 6.51. Consider the group $\mathbb{Z}_p^*$ where p is an odd prime, and let $\rho : \mathbb{Z}_p^* \to \mathbb{Z}_p^*$ be the squaring map. By definition, $\mathrm{Im}\,\rho = (\mathbb{Z}_p^*)^2$, and we proved in Theorem 2.18 that $\mathrm{Ker}\,\rho = \{[\pm 1]\}$. Theorem 2.19 says that for all $\gamma, \beta \in \mathbb{Z}_p^*$, $\gamma^2 = \beta^2$ if and only if $\gamma = \pm\beta$. This fact can also be seen to be a special case of part (vi) of Theorem 6.19. Theorem 6.23 says that $\mathbb{Z}_p^*/\mathrm{Ker}\,\rho \cong \mathrm{Im}\,\rho$, and since $|\mathbb{Z}_p^*/\mathrm{Ker}\,\rho| = |\mathbb{Z}_p^*|/|\mathrm{Ker}\,\rho| = (p-1)/2$, we see that Theorem 2.20, which says that $|(\mathbb{Z}_p^*)^2| = (p-1)/2$, follows from this.

Let $H := (\mathbb{Z}_p^*)^2$, and consider the quotient group $\mathbb{Z}_p^*/H$. Since $|H| = (p-1)/2$, we know that $|\mathbb{Z}_p^*/H| = |\mathbb{Z}_p^*|/|H| = 2$, and hence $\mathbb{Z}_p^*/H$ consists of the two cosets H and $\overline{H} := \mathbb{Z}_p^* \setminus H$.

Let α be an arbitrary, fixed element of $\overline{H}$, and consider the map

$$\tau : \quad \mathbb{Z} \to \mathbb{Z}_p^*/H$$
$$z \mapsto [\alpha^z]_H.$$

It is easy to see that τ is a group homomorphism; indeed, it is the composition of the homomorphism discussed in Example 6.43 and the natural map from $\mathbb{Z}_p^*$ to $\mathbb{Z}_p^*/H$. Moreover, it is easy to see (for example, as a special case of Theorem 2.17) that

$$\alpha^z \in H \iff z \text{ is even.}$$

From this, it follows that $\mathrm{Ker}\,\tau = 2\mathbb{Z}$; also, since $\mathbb{Z}_p^*/H$ consists of just the two cosets H and $\overline{H}$, it follows that τ is surjective. Therefore, Theorem 6.23 says that the map

$$\overline{\tau} : \quad \mathbb{Z}_2 \to \mathbb{Z}_p^*/H$$
$$[z]_2 \mapsto [\alpha^z]_H$$

is a group isomorphism, under which $[0]_2$ corresponds to H, and $[1]_2$ corresponds to $\overline{H}$.

This isomorphism gives another way to derive Theorem 2.23, which says that in $\mathbb{Z}_p^*$, the product of two non-squares is a square; indeed, the statement "non-zero plus non-zero equals zero in $\mathbb{Z}_2$" translates via the isomorphism $\overline{\tau}$ to the statement "non-square times non-square equals square in $\mathbb{Z}_p^*$." $\square$

Example 6.52. Let $\mathbb{Q}^*$ be the multiplicative group of non-zero rational numbers. Let H_1 be the subgroup $\{\pm 1\}$, and let H_2 be the subgroup of positive rationals. It is easy to see that $\mathbb{Q}^* = H_1 \cdot H_2$ and that $H_1 \cap H_2 = \{1\}$. Thus, $\mathbb{Q}^*$ is the internal direct product of H_1 and H_2, and Theorem 6.25 gives us a group isomorphism $\mathbb{Q}^* \cong H_1 \times H_2$. $\square$

Let G and G' be abelian groups. Recall from Example 6.19 that $\mathrm{Map}(G, G')$ is the group of all functions $\sigma : G \to G'$, where the group operation is defined point-wise using the group operation of G':

$$(\sigma + \tau)(a) = \sigma(a) + \tau(a) \text{ and } (-\sigma)(a) = -\sigma(a)$$

for all $\sigma, \tau \in \mathrm{Map}(G, G')$ and all $a \in G$. The following theorem isolates an important subgroup of this group.

Theorem 6.26. *Let G and G' be abelian groups, and consider the group of functions $\mathrm{Map}(G, G')$. Then*

$$\mathrm{Hom}(G, G') := \{\sigma \in \mathrm{Map}(G, G') : \sigma \text{ is a group homomorphism}\}$$

is a subgroup of $\mathrm{Map}(G, G')$.

Proof. First, observe that $\mathrm{Hom}(G, G')$ is non-empty, as it contains the map that sends everything in G to $0_{G'}$ (this is the identity element of $\mathrm{Map}(G, G')$).

Next, we have to show that if σ and τ are homomorphisms from G to G', then so are $\sigma + \tau$ and $-\sigma$. But $\sigma + \tau = \rho_2 \circ \rho_1$, where $\rho_1 : G \to G' \times G'$ is the map constructed in Theorem 6.21, applied with σ and τ, and $\rho_2 : G' \times G' \to G'$ is as in Example 6.45. Also, $-\sigma = \rho_{-1} \circ \sigma$, where ρ_{-1} is the (-1)-multiplication map. $\square$

EXERCISE 6.22. Verify that the "is isomorphic to" relation on abelian groups is an equivalence relation; that is, for all abelian groups G_1, G_2, G_3, we have:

(a) $G_1 \cong G_1$;

(b) $G_1 \cong G_2$ implies $G_2 \cong G_1$;

(c) $G_1 \cong G_2$ and $G_2 \cong G_3$ implies $G_1 \cong G_3$.

EXERCISE 6.23. Let $\rho_i : G_i \to G'_i$, for $i = 1, \ldots, k$, be group homomorphisms. Show that the map

$$\rho : \quad G_1 \times \cdots \times G_k \to G'_1 \times \cdots \times G'_k$$
$$(a_1, \ldots, a_k) \mapsto (\rho_1(a_1), \ldots, \rho_k(a_k))$$

is a group homomorphism. Also show that if each ρ_i is an isomorphism, then so is ρ.

EXERCISE 6.24. Let $\rho : G \to G'$ be a group homomorphism. Let H, K be subgroups of G and let m be a positive integer. Show that $\rho(H + K) = \rho(H) + \rho(K)$ and $\rho(mH) = m\rho(H)$.

EXERCISE 6.25. Let $\rho : G \to G'$ be a group homomorphism. Let H be a subgroup of G, and let $\tau : H \to G'$ be the restriction of ρ to H. Show that τ is a group homomorphism and that $\mathrm{Ker}\,\tau = \mathrm{Ker}\,\rho \cap H$.

EXERCISE 6.26. Suppose $G_1, \ldots, G_k$ are abelian groups. Show that for each $i = 1, \ldots, k$, the projection map $\pi_i : G_1 \times \cdots \times G_k \to G_i$ that sends $(a_1, \ldots, a_k)$ to a_i is a surjective group homomorphism.

EXERCISE 6.27. Show that if $G = G_1 \times G_2$ for abelian groups G_1 and G_2, and H_1 is a subgroup of G_1 and H_2 is a subgroup of G_2, then we have a group isomorphism $G/(H_1 \times H_2) \cong G_1/H_1 \times G_2/H_2$.

EXERCISE 6.28. Let G be an abelian group with subgroups H and K.

    (a) Show that we have a group isomorphism $(H + K)/K \cong H/(H \cap K)$.

    (b) Show that if H and K are finite, then $|H + K| = |H||K|/|H \cap K|$.

EXERCISE 6.29. Let G be an abelian group with subgroups H, K, and A, where $K \subseteq H$. Show that $(H \cap A)/(K \cap A)$ is isomorphic to a subgroup of H/K.

EXERCISE 6.30. Let $\rho : G \to G'$ be a group homomorphism with kernel K. Let H be a subgroup of G. Show that we have a group isomorphism $G/(H + K) \cong \rho(G)/\rho(H)$.

EXERCISE 6.31. Let $\rho : G \to G'$ be a surjective group homomorphism. Let S be the set of all subgroups of G that contain $\mathrm{Ker}\,\rho$, and let S' be the set of all subgroups of G'. Show that the sets S and S' are in one-to-one correspondence, via the map that sends $H \in S$ to $\rho(H) \in S'$. Also show that this correspondence preserves inclusions; that is, for all $H_1, H_2 \in S$, we have $H_1 \subseteq H_2 \iff \rho(H_1) \subseteq \rho(H_2)$.

EXERCISE 6.32. Use the previous exercise, together with Theorem 6.9, to get a short proof of Theorem 6.10.

EXERCISE 6.33. Show that the homomorphism of Example 6.44 arises by direct application of Example 6.42, combined with Theorems 6.20 and 6.21.

EXERCISE 6.34. Suppose that G, G_1, and G_2 are abelian groups, and that $\rho : G_1 \times G_2 \to G$ is a group isomorphism. Let $H_1 := \rho(G_1 \times \{0_{G_2}\})$ and $H_2 := \rho(\{0_{G_1}\} \times G_2)$. Show that G is the internal direct product of H_1 and H_2.

EXERCISE 6.35. Let $\mathbb{Z}^+$ denote the set of positive integers, and let $\mathbb{Q}^*$ be the multiplicative group of non-zero rational numbers. Consider the abelian groups $\mathrm{Map}^{\#}(\mathbb{Z}^+, \mathbb{Z})$ and $\mathrm{Map}^{\#}(\mathbb{Z}^+, \mathbb{Z}_2)$, as defined in Exercise 6.14. Show that we have group isomorphisms

    (a) $\mathbb{Q}^* \cong \mathbb{Z}_2 \times \mathrm{Map}^{\#}(\mathbb{Z}^+, \mathbb{Z})$, and

    (b) $\mathbb{Q}^*/(\mathbb{Q}^*)^2 \cong \mathrm{Map}^{\#}(\mathbb{Z}^+, \mathbb{Z}_2)$.

EXERCISE 6.36. Let n be an odd, positive integer whose factorization into primes is $n = p_1^{e_1} \cdots p_r^{e_r}$. Show that:

(a) we have a group isomorphism $\mathbb{Z}_n^*/(\mathbb{Z}_n^*)^2 \cong \mathbb{Z}_2^{\times r}$;

(b) if $p_i \equiv 3 \pmod 4$ for each $i = 1, \ldots, r$, then the squaring map on $(\mathbb{Z}_n^*)^2$ is a group automorphism.

EXERCISE 6.37. Which of the following pairs of groups are isomorphic? Why or why not? (a) $\mathbb{Z}_2 \times \mathbb{Z}_2$ and $\mathbb{Z}_4$, (b) $\mathbb{Z}_{12}^*$ and $\mathbb{Z}_8^*$, (c) $\mathbb{Z}_5^*$ and $\mathbb{Z}_4$, (d) $\mathbb{Z}_2 \times \mathbb{Z}$ and $\mathbb{Z}$, (e) $\mathbb{Q}$ and $\mathbb{Z}$, (f) $\mathbb{Z} \times \mathbb{Z}$ and $\mathbb{Z}$.

6.5 Cyclic groups

Let G be an abelian group. For $a \in G$, define $\langle a \rangle := \{za : z \in \mathbb{Z}\}$. It is easy to see that $\langle a \rangle$ is a subgroup of G; indeed, it is the image of the group homomorphism discussed in Example 6.42. Moreover, $\langle a \rangle$ is the smallest subgroup of G containing a; that is, $\langle a \rangle$ contains a, and every subgroup of G that contains a must contain everything in $\langle a \rangle$. Indeed, if a subgroup contains a, it must contain $a + a = 2a$, $a + a + a = 3a$, and so on; it must also contain $0_G = 0a$, $-a = (-1)a$, $(-a) + (-a) = (-2)a$, and so on. The subgroup $\langle a \rangle$ is called **the subgroup (of G) generated by** a. Also, one defines the **order** of a to be the order of the subgroup $\langle a \rangle$.

More generally, for $a_1, \ldots, a_k \in G$, we define

$$\langle a_1, \ldots, a_k \rangle := \{z_1 a_1 + \cdots + z_k a_k : z_1, \ldots, z_k \in \mathbb{Z}\}.$$

It is easy to see that $\langle a_1, \ldots, a_k \rangle$ is a subgroup of G; indeed, it is the image of the group homomorphism discussed in Example 6.44. Moreover, this subgroup is the smallest subgroup of G that contains $a_1, \ldots, a_k$; that is, $\langle a_1, \ldots, a_k \rangle$ contains the elements $a_1, \ldots, a_k$, and every subgroup of G that contains these elements must contain everything in $\langle a_1, \ldots, a_k \rangle$. The subgroup $\langle a_1, \ldots, a_k \rangle$ is called the **subgroup (of G) generated by** $a_1, \ldots, a_k$.

An abelian group G is called **cyclic** if $G = \langle a \rangle$ for some $a \in G$, in which case, a is called a **generator for** G. An abelian group G is called **finitely generated** if $G = \langle a_1, \ldots, a_k \rangle$ for some $a_1, \ldots, a_k \in G$.

Multiplicative notation: if G is written multiplicatively, then $\langle a \rangle := \{a^z : z \in \mathbb{Z}\}$, and $\langle a_1, \ldots, a_k \rangle := \{a_1^{z_1} \cdots a_k^{z_k} : z_1, \ldots, z_k \in \mathbb{Z}\}$; also, for emphasis and clarity, we use the term **multiplicative order of** a.

Example 6.53. Consider the additive group $\mathbb{Z}$. This is a cyclic group, with 1 being a generator:

$$\langle 1 \rangle = \{z \cdot 1 : z \in \mathbb{Z}\} = \{z : z \in \mathbb{Z}\} = \mathbb{Z}.$$

For every $m \in \mathbb{Z}$, we have

$$\langle m \rangle = \{zm : z \in \mathbb{Z}\} = \{mz : z \in \mathbb{Z}\} = m\mathbb{Z}.$$

It follows that the only elements of $\mathbb{Z}$ that generate $\mathbb{Z}$ are 1 and -1: every other element generates a subgroup that is strictly contained in $\mathbb{Z}$. $\square$

Example 6.54. For $n > 0$, consider the additive group $\mathbb{Z}_n$. This is a cyclic group, with [1] being a generator:

$$\langle [1] \rangle = \{z[1] : z \in \mathbb{Z}\} = \{[z] : z \in \mathbb{Z}\} = \mathbb{Z}_n.$$

For every $m \in \mathbb{Z}$, we have

$$\langle [m] \rangle = \{z[m] : z \in \mathbb{Z}\} = \{[zm] : z \in \mathbb{Z}\} = \{m[z] : z \in \mathbb{Z}\} = m\mathbb{Z}_n.$$

By Example 6.23, the subgroup $m\mathbb{Z}_n$ has order $n/\gcd(m, n)$. Thus, $[m]$ has order $n/\gcd(m, n)$; in particular, $[m]$ generates $\mathbb{Z}_n$ if and only if m is relatively prime to n, and hence, the number of generators of $\mathbb{Z}_n$ is $\varphi(n)$. $\square$

Implicit in Examples 6.53 and 6.54 is the following general fact:

Theorem 6.27. *Let G be a cyclic group generated by a. Then for every $m \in \mathbb{Z}$, we have*

$$\langle ma \rangle = mG.$$

Proof. We have

$$\langle ma \rangle = \{z(ma) : z \in \mathbb{Z}\} = \{m(za) : z \in \mathbb{Z}\} = m\langle a \rangle = mG. \quad \square$$

The following two examples present some groups that are *not* cyclic.

Example 6.55. Consider the additive group $G := \mathbb{Z} \times \mathbb{Z}$. Set

$$\alpha_1 := (1, 0) \in G \quad \text{and} \quad \alpha_2 := (0, 1) \in G.$$

It is not hard to see that $G = \langle \alpha_1, \alpha_2 \rangle$, since for all $z_1, z_2 \in \mathbb{Z}$, we have

$$z_1\alpha_1 + z_2\alpha_2 = (z_1, 0) + (0, z_2) = (z_1, z_2).$$

However, G is not cyclic. To see this, let $\beta = (b_1, b_2)$ be an arbitrary element of G. We claim that one of α_1 or α_2 does not belong to $\langle \beta \rangle$. Suppose to the contrary that both α_1 and α_2 belong to $\langle \beta \rangle$. This would imply that there exist integers z and z' such that

$$zb_1 = 1, \qquad\qquad zb_2 = 0,$$
$$z'b_1 = 0, \qquad\qquad z'b_2 = 1.$$

Multiplying the upper left equality by the lower right, and the upper right by the lower left, we obtain

$$1 = zz'b_1b_2 = 0,$$

which is impossible. □

Example 6.56. Consider the additive group $G := \mathbb{Z}_{n_1} \times \mathbb{Z}_{n_2}$. Set

$$\alpha_1 := ([1]_{n_1}, [0]_{n_2}) \in G \text{ and } \alpha_2 := ([0]_{n_1}, [1]_{n_2}) \in G.$$

It is not hard to see that $G = \langle \alpha_1, \alpha_2 \rangle$, since for all $z_1, z_2 \in \mathbb{Z}$, we have

$$z_1\alpha_1 + z_2\alpha_2 = ([z_1]_{n_1}, [0]_{n_2}) + ([0]_{n_1}, [z_2]_{n_2}) = ([z_1]_{n_1}, [z_2]_{n_2}).$$

However, G may or may not be cyclic: it depends on $d := \gcd(n_1, n_2)$.

If $d = 1$, then G is cyclic, with $\alpha := ([1]_{n_1}, [1]_{n_2})$ being a generator. One can see this easily using the Chinese remainder theorem: for all $z_1, z_2 \in \mathbb{Z}$, there exists $z \in \mathbb{Z}$ such that

$$z \equiv z_1 \pmod{n_1} \text{ and } z \equiv z_2 \pmod{n_2},$$

which implies

$$z\alpha = ([z]_{n_1}, [z]_{n_2}) = ([z_1]_{n_1}, [z_2]_{n_2}).$$

If $d > 1$, then G is not cyclic. To see this, let $\beta = ([b_1]_{n_1}, [b_2]_{n_2})$ be an arbitrary element of G. We claim that one of α_1 or α_2 does not belong to $\langle \beta \rangle$. Suppose to the contrary that both α_1 and α_2 belong to $\langle \beta \rangle$. This would imply that there exist integers z and z' such that

$$zb_1 \equiv 1 \pmod{n_1}, \qquad zb_2 \equiv 0 \pmod{n_2},$$
$$z'b_1 \equiv 0 \pmod{n_1}, \qquad z'b_2 \equiv 1 \pmod{n_2}.$$

All of these congruences hold modulo d as well, and multiplying the upper left congruence by the lower right, and the upper right by the lower left, we obtain

$$1 \equiv zz'b_1b_2 \equiv 0 \pmod{d},$$

which is impossible. □

It should be clear that since a group isomorphism preserves all structural properties of groups, it preserves the property of being cyclic. We state this, along with related facts, as a theorem.

Theorem 6.28. *Let $\rho : G \to G'$ be a group isomorphism.*

(i) For all $a \in G$, we have $\rho(\langle a \rangle) = \langle \rho(a) \rangle$.

(ii) *For all $a \in G$, a and $\rho(a)$ have the same order.*

(iii) *G is cyclic if and only if G' is cyclic.*

Proof. For all $a \in G$, we have

$$\rho(\langle a \rangle) = \{\rho(za) : z \in \mathbb{Z}\} = \{z\rho(a) : z \in \mathbb{Z}\} = \langle \rho(a) \rangle.$$

That proves (i).

(ii) follows from (i) and the fact that ρ is injective.

(iii) follows from (i), as follows. If G is cyclic, then $G = \langle a \rangle$, and since ρ is surjective, we have $G' = \rho(G) = \langle \rho(a) \rangle$. The converse follows by applying the same argument to the inverse isomorphism $\rho^{-1} : G' \to G$. $\square$

Example 6.57. Consider again the additive group $G := \mathbb{Z}_{n_1} \times \mathbb{Z}_{n_2}$, discussed in Example 6.56. If $\gcd(n_1, n_2) = 1$, then one can also see that G is cyclic as follows: by the discussion in Example 6.48, we know that G is isomorphic to $\mathbb{Z}_{n_1 n_2}$, and since $\mathbb{Z}_{n_1 n_2}$ is cyclic, so is G. $\square$

Example 6.58. Consider again the subgroup $m\mathbb{Z}_n$ of $\mathbb{Z}_n$, discussed in Example 6.54. One can also see that this is cyclic of order n/d, where $d := \gcd(m, n)$, as follows: in Example 6.50, we constructed an isomorphism between $\mathbb{Z}_{n/d}$ and $m\mathbb{Z}_n$, and this implies $m\mathbb{Z}_n$ is cyclic of order n/d. $\square$

Classification of cyclic groups. Examples 6.53 and 6.54 are extremely important examples of cyclic groups. Indeed, as we shall now demonstrate, every cyclic group is isomorphic either to $\mathbb{Z}$ or to $\mathbb{Z}_n$ for some $n > 0$.

Suppose that G is a cyclic group with generator a. Consider the map $\rho : \mathbb{Z} \to G$ that sends $z \in \mathbb{Z}$ to $za \in G$. As discussed in Example 6.42, this map is a group homomorphism, and since a is a generator for G, it must be surjective. There are two cases to consider.

Case 1: $\operatorname{Ker} \rho = \{0\}$. In this case, ρ is an isomorphism of $\mathbb{Z}$ with G.

Case 2: $\operatorname{Ker} \rho \neq \{0\}$. In this case, since $\operatorname{Ker} \rho$ is a subgroup of $\mathbb{Z}$ different from $\{0\}$, by Theorem 6.9, it must be of the form $n\mathbb{Z}$ for some $n > 0$. Hence, by Theorem 6.23, the map $\bar{\rho} : \mathbb{Z}_n \to G$ that sends $[z]_n$ to za is an isomorphism of $\mathbb{Z}_n$ with G.

Based on this isomorphism, we immediately obtain:

Theorem 6.29. *Let G be an abelian group and let $a \in G$. If there exists a positive integer m such that $ma = 0_G$, then the least such positive integer n is the order of a; in this case, we have:*

- *for every integer z, $za = 0_G$ if and only if n divides z, and more generally, for all integers z_1, z_2, we have $z_1 a = z_2 a$ if and only if $z_1 \equiv z_2 \pmod{n}$;*

- the subgroup $\langle a \rangle$ consists of the n distinct elements

$$0 \cdot a, 1 \cdot a, \ldots, (n-1) \cdot a.$$

Otherwise, a has infinite order, and every element of $\langle a \rangle$ can be expressed as za for some unique integer z.

In the case where the group is finite, we can say more:

Theorem 6.30. *Let G be a finite abelian group and let $a \in G$. Then $|G|a = 0_G$ and the order of a divides $|G|$.*

Proof. Since $\langle a \rangle$ is a subgroup of G, by Lagrange's theorem (Theorem 6.15), the order of a divides $|G|$. It then follows by Theorem 6.29 that $|G|a = 0_G$. $\square$

Example 6.59. Let $a, n \in \mathbb{Z}$ with $n > 0$ and $\gcd(a, n) = 1$, and let $\alpha := [a] \in \mathbb{Z}_n^*$. Theorem 6.29 implies that the definition given in this section of the multiplicative order of α is consistent with that given in §2.7. Moreover, Euler's theorem (Theorem 2.13) can be seen as just a special case of Theorem 6.30. Also, note that α is a generator for $\mathbb{Z}_n^*$ if and only if a is a primitive root modulo p. $\square$

Example 6.60. As we saw in Example 6.26, all elements of $\mathbb{Z}_{15}^*$ have multiplicative order dividing 4, and since $\mathbb{Z}_{15}^*$ has order 8, we conclude that $\mathbb{Z}_{15}^*$ is not cyclic. $\square$

Example 6.61. The group $\mathbb{Z}_5^*$ is cyclic, with $[2]$ being a generator:

$$[2]^2 = [4] = [-1], \quad [2]^3 = [-2], \quad [2]^4 = [1]. \quad \square$$

Example 6.62. Based on the calculations in Example 2.9, we may conclude that $\mathbb{Z}_7^*$ is cyclic, with both $[3]$ and $[5]$ being generators. $\square$

Example 6.63. Consider again the additive group $G := \mathbb{Z}_{n_1} \times \mathbb{Z}_{n_2}$, discussed in Example 6.56. If $d := \gcd(n_1, n_2) > 1$, then one can also see that G is not cyclic as follows: for every $\beta \in G$, we have $(n_1 n_2 / d)\beta = 0_G$, and hence by Theorem 6.29, the order of β divides $n_1 n_2 / d$. $\square$

The following two theorems completely characterize the subgroup structure of cyclic groups. Actually, we have already proven most of the results in these two theorems, but nevertheless, they deserve special emphasis.

Theorem 6.31. *Let G be a cyclic group of infinite order.*

(i) G is isomorphic to $\mathbb{Z}$.

(ii) There is a one-to-one correspondence between the non-negative integers and the subgroups of G, where each such integer m corresponds to the cyclic group $m G$.

(iii) *For every two non-negative integers m, m', we have $mG \subseteq m'G$ if and only if $m' \mid m$.*

Proof. That $G \cong \mathbb{Z}$ was established in our classification of cyclic groups, and so it suffices to prove the other statements of the theorem for $G = \mathbb{Z}$. As we saw in Example 6.53, for every integer m, the subgroup $m\mathbb{Z}$ is cyclic, as it is generated by m. This fact, together with Theorem 6.9, establishes all the other statements. □

Theorem 6.32. *Let G be a cyclic group of finite order n.*

  (i) *G is isomorphic to $\mathbb{Z}_n$.*

 (ii) *There is a one-to-one correspondence between the positive divisors of n and the subgroups of G, where each such divisor d corresponds to the subgroup dG; moreover, dG is a cyclic group of order n/d.*

(iii) *For each positive divisor d of n, we have $dG = G\{n/d\}$; that is, the kernel of the (n/d)-multiplication map is equal to the image of the d-multiplication map; in particular, $G\{n/d\}$ has order n/d.*

(iv) *For every two positive divisors d, d' of n, we have $dG \subseteq d'G$ if and only if $d' \mid d$.*

 (v) *For every positive divisor d of n, the number of elements of order d in G is $\varphi(d)$.*

(vi) *For every integer m, we have $mG = dG$ and $G\{m\} = G\{d\}$, where $d := \gcd(m, n)$.*

Proof. That $G \cong \mathbb{Z}_n$ was established in our classification of cyclic groups, and so it suffices to prove the other statements of the theorem for $G = \mathbb{Z}_n$.

The one-to-one correspondence in part (ii) was established in Theorem 6.10. By the discussion in Example 6.54, it is clear that $d\mathbb{Z}_n$ is generated by $[d]$ and has order n/d.

Part (iii) was established in Example 6.23.

Part (iv) was established in Theorem 6.10.

For part (v), the elements of order d in $\mathbb{Z}_n$ are all contained in $\mathbb{Z}_n\{d\}$, and so the number of such elements is equal to the number of generators of $\mathbb{Z}_n\{d\}$. The group $\mathbb{Z}_n\{d\}$ is cyclic of order d, and so is isomorphic to $\mathbb{Z}_d$, and as we saw in Example 6.54, this group has $\varphi(d)$ generators.

Part (vi) was established in Example 6.23. □

Since cyclic groups are in some sense the simplest kind of abelian group, it is nice to establish some sufficient conditions under which a group must be cyclic. The following three theorems provide such conditions.

Theorem 6.33. *If G is an abelian group of prime order, then G is cyclic.*

Proof. Let $|G| = p$, which, by hypothesis, is prime. Let $a \in G$ with $a \neq 0_G$, and let k be the order of a. As the order of an element divides the order of the group, we have $k \mid p$, and so $k = 1$ or $k = p$. Since $a \neq 0_G$, we must have $k \neq 1$, and so $k = p$, which implies that a generates G. $\square$

Theorem 6.34. *If G_1 and G_2 are finite cyclic groups of relatively prime order, then $G_1 \times G_2$ is also cyclic. In particular, if G_1 is generated by a_1 and G_2 is generated by a_2, then $G_1 \times G_2$ is generated by (a_1, a_2).*

Proof. We give a direct proof, based on Theorem 6.29. Let $n_1 := |G_1|$ and $n_2 := |G_2|$, where $\gcd(n_1, n_2) = 1$. Also, let $a_1 \in G_1$ have order n_1 and $a_2 \in G_2$ have order n_2. We want to show that (a_1, a_2) has order $n_1 n_2$. Applying Theorem 6.29 to (a_1, a_2), we see that the order of (a_1, a_2) is the smallest positive integer k such that $k(a_1, a_2) = (0_{G_1}, 0_{G_2})$. Now, for every integer k, we have $k(a_1, a_2) = (ka_1, ka_2)$, and

$$(ka_1, ka_2) = (0_{G_1}, 0_{G_2}) \iff n_1 \mid k \text{ and } n_2 \mid k$$
$$\text{(applying Theorem 6.29 to } a_1 \text{ and } a_2)$$
$$\iff n_1 n_2 \mid k \text{ (since } \gcd(n_1, n_2) = 1). \quad \square$$

Theorem 6.35. *Let G be a cyclic group. Then for every subgroup H of G, both H and G/H are cyclic.*

Proof. The fact that H is cyclic follows from part (ii) of Theorem 6.31 in the case where G is infinite, and part (ii) of Theorem 6.32 in the case where G is finite. If G is generated by a, then it is easy to see that G/H is generated by $[a]_H$. $\square$

The next three theorems are often useful in calculating the order of a group element. The first generalizes Theorem 2.15.

Theorem 6.36. *Let G be an abelian group, let $a \in G$ be of finite order n, and let m be an arbitrary integer. Then the order of ma is $n/\gcd(m, n)$.*

Proof. Let $H := \langle a \rangle$, and $d := \gcd(m, n)$. By Theorem 6.27, we have $\langle ma \rangle = mH$, and by Theorem 6.32, we have $mH = dH$, which has order n/d.

That proves the theorem. Alternatively, we can give a direct proof, based on Theorem 6.29. Applying Theorem 6.29 to ma, we see that the order of ma is the smallest positive integer k such that $k(ma) = 0_G$. Now, for every integer k, we have $k(ma) = (km)a$, and

$$(km)a = 0_G \iff km \equiv 0 \pmod{n} \text{ (applying Theorem 6.29 to } a)$$
$$\iff k \equiv 0 \pmod{n/\gcd(m, n)} \text{ (by part (ii) of Theorem 2.5). } \square$$

Theorem 6.37. *Suppose that a is an element of an abelian group, and for some prime p and integer $e \geq 1$, we have $p^e a = 0_G$ and $p^{e-1}a \neq 0_G$. Then a has order p^e.*

Proof. If m is the order of a, then since $p^e a = 0_G$, we have $m \mid p^e$. So $m = p^f$ for some $f = 0, \ldots, e$. If $f < e$, then $p^{e-1}a = 0_G$, contradicting the assumption that $p^{e-1}a \neq 0_G$. $\square$

Theorem 6.38. *Suppose G is an abelian group with $a_1, a_2 \in G$ such that a_1 is of finite order n_1, a_2 is of finite order n_2, and $\gcd(n_1, n_2) = 1$. Then the order of $a_1 + a_2$ is $n_1 n_2$.*

Proof. Let $H_1 := \langle a_1 \rangle$ and $H_2 := \langle a_2 \rangle$ so that $|H_1| = n_1$ and $|H_2| = n_2$.

First, we claim that $H_1 \cap H_2 = \{0_G\}$. To see this, observe that $H_1 \cap H_2$ is a subgroup of H_1, and so $|H_1 \cap H_2|$ divides n_1; similarly, $|H_1 \cap H_2|$ divides n_2. Since $\gcd(n_1, n_2) = 1$, we must have $|H_1 \cap H_2| = 1$, and that proves the claim.

Using the claim, we can apply Theorem 6.25, obtaining a group isomorphism between $H_1 + H_2$ and $H_1 \times H_2$. Under this isomorphism, the group element $a_1 + a_2 \in H_1 + H_2$ corresponds to $(a_1, a_2) \in H_1 \times H_2$, which by Theorem 6.34 (again using the fact that $\gcd(n_1, n_2) = 1$) has order $n_1 n_2$. $\square$

For an abelian group G, we say that an integer k **kills** G if $kG = \{0_G\}$. Consider the set $\mathcal{K}_G$ of integers that kill G. Evidently, $\mathcal{K}_G$ is a subgroup of $\mathbb{Z}$, and hence of the form $m\mathbb{Z}$ for a uniquely determined non-negative integer m. This integer m is called the **exponent** of G. If $m \neq 0$, then we see that m is the least positive integer that kills G.

The following two theorems state some simple properties of the exponent of a group.

Theorem 6.39. *Let G be an abelian group of exponent m.*

   (i) *For every integer k, k kills G if and only if $m \mid k$.*

   (ii) *If G has finite order, then m divides $|G|$.*

   (iii) *If $m \neq 0$, then for every $a \in G$, the order of a is finite and divides m.*

   (iv) *If G is cyclic, then the exponent of G is 0 if G is infinite, and is $|G|$ if G is finite.*

Proof. Exercise. $\square$

Theorem 6.40. *If G_1 and G_2 are abelian groups of exponents m_1 and m_2, then the exponent of $G_1 \times G_2$ is $\mathrm{lcm}(m_1, m_2)$.*

Proof. Exercise. $\square$

Example 6.64. The additive group $\mathbb{Z}$ has exponent 0. $\square$

Example 6.65. The additive group $\mathbb{Z}_n$ has exponent n. $\square$

Example 6.66. The additive group $\mathbb{Z}_{n_1} \times \mathbb{Z}_{n_2}$ has exponent $\mathrm{lcm}(n_1, n_2)$. $\square$

Example 6.67. The multiplicative group $\mathbb{Z}_{15}^*$ has exponent 4 (see Example 6.26). $\square$

The next two theorems develop some crucial properties about the structure of finite abelian groups.

Theorem 6.41. *If an abelian group G has non-zero exponent m, then G contains an element of order m. In particular, a finite abelian group is cyclic if and only if its order equals its exponent.*

Proof. The second statement follows immediately from the first. For the first statement, let $m = \prod_{i=1}^{r} p_i^{e_i}$ be the prime factorization of m.

First, we claim that for each $i = 1, \ldots, r$, there exists $a_i \in G$ such that $(m/p_i)a_i \neq 0_G$. Suppose the claim were false: then for some i, $(m/p_i)a = 0_G$ for all $a \in G$; however, this contradicts the minimality property in the definition of the exponent m. That proves the claim.

Let $a_1, \ldots, a_r$ be as in the above claim. Then by Theorem 6.37, $(m/p_i^{e_i})a_i$ has order $p_i^{e_i}$ for each $i = 1, \ldots, r$. Finally, by Theorem 6.38, the group element

$$(m/p_1^{e_1})a_1 + \cdots + (m/p_r^{e_r})a_r$$

has order m. $\square$

Theorem 6.42. *Let G be a finite abelian group of order n. If p is a prime dividing n, then G contains an element of order p.*

Proof. We can prove this by induction on n.

If $n = 1$, then the theorem is vacuously true.

Now assume $n > 1$ and that the theorem holds for all groups of order strictly less than n. Let a be any non-zero element of G, and let m be the order of a. Since a is non-zero, we must have $m > 1$. If $p \mid m$, then $(m/p)a$ is an element of order p, and we are done. So assume that $p \nmid m$ and consider the quotient group G/H, where H is the subgroup of G generated by a. Since H has order m, G/H has order n/m, which is strictly less than n, and since $p \nmid m$, we must have $p \mid (n/m)$. So we can apply the induction hypothesis to the group G/H and the prime p, which says that there is an element $b \in G$ such that the coset $[b]_H \in G/H$ has order p. If ℓ is the order of b, then $\ell b = 0_G$, and so $\ell b \equiv 0_G \pmod{H}$, which implies that the order of $[b]_H$ divides ℓ. Thus, $p \mid \ell$, and so $(\ell/p)b$ is an element of G of order p. $\square$

As a corollary, we have:

Theorem 6.43. *Let G be a finite abelian group. Then the primes dividing the exponent of G are the same as the primes dividing its order.*

Proof. Since the exponent divides the order, every prime dividing the exponent must divide the order. Conversely, if a prime p divides the order, then since there is an element of order p in the group, the exponent must be divisible by p. $\square$

EXERCISE 6.38. Find $\alpha_1, \alpha_2 \in \mathbb{Z}_{15}^*$ such that $\mathbb{Z}_{15}^* = \langle \alpha_1, \alpha_2 \rangle$.

EXERCISE 6.39. Show that $\mathbb{Q}^*$ is not finitely generated.

EXERCISE 6.40. Let G be an abelian group, $a \in G$, and $m \in \mathbb{Z}$, such that $m > 0$ and $ma = 0_G$. Let $m = p_1^{e_1} \cdots p_r^{e_r}$ be the prime factorization of m. For $i = 1, \ldots, r$, let f_i be the largest non-negative integer such that $f_i \leq e_i$ and $m/p_i^{f_i} \cdot a = 0_G$. Show that the order of a is equal to $p_1^{e_1-f_1} \cdots p_r^{e_r-f_r}$.

EXERCISE 6.41. Let G be an abelian group of order n, and let m be an integer. Show that $mG = G$ if and only if $\gcd(m, n) = 1$.

EXERCISE 6.42. Let H be a subgroup of an abelian group G. Show that:

(a) if H and G/H are both finitely generated, then so is G;

(b) if G is finite, $\gcd(|H|, |G/H|) = 1$, and H and G/H are both cyclic, then G is cyclic.

EXERCISE 6.43. Let G be an abelian group of exponent $m_1 m_2$, where m_1 and m_2 are relatively prime. Show that G is the internal direct product of $m_1 G$ and $m_2 G$.

EXERCISE 6.44. Show how Theorem 2.40 easily follows from Theorem 6.32.

EXERCISE 6.45. As additive groups, $\mathbb{Z}$ is clearly a subgroup of $\mathbb{Q}$. Consider the quotient group $G := \mathbb{Q}/\mathbb{Z}$, and show that:

(a) all elements of G have finite order;

(b) G has exponent 0;

(c) for all positive integers m, we have $mG = G$ and $G\{m\} \cong \mathbb{Z}_m$;

(d) all finite subgroups of G are cyclic.

EXERCISE 6.46. Suppose that G is an abelian group that satisfies the following properties:

(i) for all $m \in \mathbb{Z}$, $G\{m\}$ is either equal to G or is of finite order;

(ii) for some $m \in \mathbb{Z}$, $\{0_G\} \subsetneq G\{m\} \subsetneq G$.

Show that $G\{m\}$ is finite for all non-zero $m \in \mathbb{Z}$.

6.6 The structure of finite abelian groups (∗)

We next state a theorem that classifies all finite abelian groups up to isomorphism.

Theorem 6.44 (Fundamental theorem of finite abelian groups). *A finite abelian group (with more than one element) is isomorphic to a direct product of cyclic groups*

$$\mathbb{Z}_{p_1^{e_1}} \times \cdots \times \mathbb{Z}_{p_r^{e_r}},$$

where the p_i's are primes (not necessarily distinct) and the e_i's are positive integers. This direct product of cyclic groups is unique up to the order of the factors.

An alternative statement of this theorem is the following:

Theorem 6.45. *A finite abelian group (with more than one element) is isomorphic to a direct product of cyclic groups*

$$\mathbb{Z}_{m_1} \times \cdots \times \mathbb{Z}_{m_t},$$

where each $m_i > 1$, and where for $i = 1, \ldots, t - 1$, we have $m_i \mid m_{i+1}$. Moreover, the integers $m_1, \ldots, m_t$ are uniquely determined, and m_t is the exponent of the group.

The statements of these theorems are much more important than their proofs, which are a bit technical. Even if the reader does not study the proofs, he is urged to understand what the theorems actually say.

In an exercise below, you are asked to show that these two theorems are equivalent. We now prove Theorem 6.45, which we break into two lemmas, the first of which proves the existence part of the theorem, and the second of which proves the uniqueness part.

Lemma 6.46. *A finite abelian group (with more than one element) is isomorphic to a direct product of cyclic groups*

$$\mathbb{Z}_{m_1} \times \cdots \times \mathbb{Z}_{m_t},$$

where each $m_i > 1$, and where for $i = 1, \ldots, t - 1$, we have $m_i \mid m_{i+1}$; moreover, m_t is the exponent of the group.

Proof. Let G be a finite abelian group with more than one element, and let m be the exponent of G. By Theorem 6.41, there exists an element $a \in G$ of order m. Let $A = \langle a \rangle$. Then $A \cong \mathbb{Z}_m$. Now, if $A = G$, the lemma is proved. So assume that $A \subsetneq G$.

We will show that there exists a subgroup B of G such that $G = A + B$ and $A \cap B = \{0_G\}$. From this, Theorem 6.25 gives us an isomorphism of G with

$A \times B$. Moreover, the exponent of B is clearly a divisor of m, and so the lemma will follow by induction (on the order of the group).

So it suffices to show the existence of a subgroup B as above. We prove this by contradiction. Suppose that there is no such subgroup, and among all subgroups B such that $A \cap B = \{0_G\}$, assume that B is maximal, meaning that there is no subgroup B' of G such that $B \subsetneq B'$ and $A \cap B' = \{0_G\}$. By assumption $C := A + B \subsetneq G$.

Let d be any element of G that lies outside of C. Consider the quotient group G/C, and let r be the order of $[d]_C \in G/C$. Note that $r > 1$ and $r \mid m$. We shall define a group element d' with slightly nicer properties than d, as follows. Since $rd \in C$, we have $rd = sa + b$ for some $s \in \mathbb{Z}$ and $b \in B$. We claim that $r \mid s$. To see this, note that $0_G = md = (m/r)rd = (m/r)sa + (m/r)b$, and since $A \cap B = \{0_G\}$, we have $(m/r)sa = 0_G$, which can only happen if $r \mid s$. That proves the claim. This allows us to define $d' := d - (s/r)a$. Since $d \equiv d' \pmod C$, we see not only that $[d']_C \in G/C$ has order r, but also that $rd' \in B$.

We next show that $A \cap (B + \langle d' \rangle) = \{0_G\}$, which will yield the contradiction we seek, and thus prove the lemma. Because $A \cap B = \{0_G\}$, it will suffice to show that $A \cap (B + \langle d' \rangle) \subseteq B$. Now, suppose we have a group element $b' + xd' \in A$, with $b' \in B$ and $x \in \mathbb{Z}$. Then in particular, $xd' \in C$, and so $r \mid x$, since $[d']_C \in G/C$ has order r. Further, since $rd' \in B$, we have $xd' \in B$, whence $b' + xd' \in B$. $\square$

Lemma 6.47. *Suppose that* $G := \mathbb{Z}_{m_1} \times \cdots \times \mathbb{Z}_{m_t}$ *and* $H := \mathbb{Z}_{n_1} \times \cdots \times \mathbb{Z}_{n_t}$ *are isomorphic, where the* m_i's *and* n_i's *are positive integers (possibly 1) such that* $m_i \mid m_{i+1}$ *and* $n_i \mid n_{i+1}$ *for* $i = 1, \ldots, t-1$. *Then* $m_i = n_i$ *for* $i = 1, \ldots, t$.

Proof. Clearly, $\prod_i m_i = |G| = |H| = \prod_i n_i$. We prove the lemma by induction on the order of the group. If the group order is 1, then clearly all the m_i's and n_i's must be 1, and we are done. Otherwise, let p be a prime dividing the group order. Now, suppose that p divides $m_r, \ldots, m_t$ but not $m_1, \ldots, m_{r-1}$, and that p divides $n_s, \ldots, n_t$ but not $n_1, \ldots, n_{s-1}$, where $r \le t$ and $s \le t$. Evidently, the groups pG and pH are isomorphic. Moreover,

$$ pG \cong \mathbb{Z}_{m_1} \times \cdots \times \mathbb{Z}_{m_{r-1}} \times \mathbb{Z}_{m_r/p} \times \cdots \times \mathbb{Z}_{m_t/p}, $$

and

$$ pH \cong \mathbb{Z}_{n_1} \times \cdots \times \mathbb{Z}_{n_{s-1}} \times \mathbb{Z}_{n_s/p} \times \cdots \times \mathbb{Z}_{n_t/p}. $$

Thus, we see that $|pG| = |G|/p^{t-r+1}$ and $|pH| = |H|/p^{t-s+1}$, from which it follows that $r = s$, and the lemma then follows by induction. $\square$

EXERCISE 6.47. Show that Theorems 6.44 and 6.45 are equivalent; that is, show that each one implies the other. To do this, give a natural one-to-one correspondence between sequences of prime powers (as in Theorem 6.44) and sequences of integers $m_1, \ldots, m_t$ (as in Theorem 6.45).

EXERCISE 6.48. Using the fundamental theorem of finite abelian groups (either form), give short and simple proofs of Theorems 6.41 and 6.42.

EXERCISE 6.49. In our proof of Euler's criterion (Theorem 2.21), we really only used the fact that $\mathbb{Z}_p^*$ has a unique element of multiplicative order 2. This exercise develops a proof of a generalization of Euler's criterion, based on the fundamental theorem of finite abelian groups. Suppose G is an abelian group of even order n that contains a unique element of order 2.

(a) Show that $G \cong \mathbb{Z}_{2^e} \times \mathbb{Z}_{m_1} \times \cdots \times \mathbb{Z}_{m_k}$, where $e > 0$ and the m_i's are odd integers.

(b) Using part (a), show that $2G = G\{n/2\}$.

EXERCISE 6.50. Let G be a non-trivial, finite abelian group. Let s be the smallest positive integer such that $G = \langle a_1, \ldots, a_s \rangle$ for some $a_1, \ldots, a_s \in G$. Show that s is equal to the value of t in Theorem 6.45. In particular, G is cyclic if and only if $t = 1$.

EXERCISE 6.51. Suppose $G \cong \mathbb{Z}_{m_1} \times \cdots \times \mathbb{Z}_{m_t}$. Let p be a prime, and let s be the number of m_i's divisible by p. Show that $G\{p\} \cong \mathbb{Z}_p^{\times s}$.

EXERCISE 6.52. Suppose $G \cong \mathbb{Z}_{m_1} \times \cdots \times \mathbb{Z}_{m_t}$ with $m_i \mid m_{i+1}$ for $i = 1, \ldots, t-1$, and that H is a subgroup of G. Show that $H \cong \mathbb{Z}_{n_1} \times \cdots \times \mathbb{Z}_{n_t}$, where $n_i \mid n_{i+1}$ for $i = 1, \ldots, t-1$ and $n_i \mid m_i$ for $i = 1, \ldots, t$.

EXERCISE 6.53. Suppose that G is an abelian group such that for all $m > 0$, we have $mG = G$ and $|G\{m\}| = m^2$ (note that G is not finite). Show that $G\{m\} \cong \mathbb{Z}_m \times \mathbb{Z}_m$ for all $m > 0$. Hint: use induction on the number of prime factors of m.

7

Rings

This chapter introduces the notion of a ring, more specifically, a commutative ring with unity. While there is a lot of terminology associated with rings, the basic ideas are fairly simple. Intuitively speaking, a ring is an algebraic structure with addition and multiplication operations that behave as one would expect.

7.1 Definitions, basic properties, and examples

Definition 7.1. *A **commutative ring with unity** is a set R together with addition and multiplication operations on R, such that:*

(i) *the set R under addition forms an abelian group, and we denote the additive identity by 0_R;*

(ii) *multiplication is associative; that is, for all $a, b, c \in R$, we have $a(bc) = (ab)c$;*

(iii) *multiplication distributes over addition; that is, for all $a, b, c \in R$, we have $a(b + c) = ab + ac$ and $(b + c)a = ba + ca$;*

(iv) *there exists a multiplicative identity; that is, there exists an element $1_R \in R$, such that $1_R \cdot a = a = a \cdot 1_R$ for all $a \in R$;*

(v) *multiplication is commutative; that is, for all $a, b \in R$, we have $ab = ba$.*

There are other, more general (and less convenient) types of rings — one can drop properties (iv) and (v), and still have what is called a **ring**. We shall not, however, be working with such general rings in this text. Therefore, to simplify terminology, **from now on, by a "ring," we shall always mean a commutative ring with unity**.

Let R be a ring. Notice that because of the distributive law, for any fixed $a \in R$, the map from R to R that sends $b \in R$ to $ab \in R$ is a group homomorphism with respect to the underlying additive group of R. We call this the a-**multiplication map**.

We first state some simple facts:

Theorem 7.2. *Let R be a ring. Then:*

(i) *the multiplicative identity 1_R is unique;*

(ii) *$0_R \cdot a = 0_R$ for all $a \in R$;*

(iii) *$(-a)b = -(ab) = a(-b)$ for all $a, b \in R$;*

(iv) *$(-a)(-b) = ab$ for all $a, b \in R$;*

(v) *$(ka)b = k(ab) = a(kb)$ for all $k \in \mathbb{Z}$ and $a, b \in R$.*

Proof. Part (i) may be proved using the same argument as was used to prove part (i) of Theorem 6.2. Parts (ii), (iii), and (v) follow directly from parts (i), (ii), and (iii) of Theorem 6.19, using appropriate multiplication maps, discussed above. Part (iv) follows from part (iii), along with part (iv) of Theorem 6.3: $(-a)(-b) = -(a(-b)) = -(-(ab)) = ab$. $\square$

Example 7.1. The set $\mathbb{Z}$ under the usual rules of multiplication and addition forms a ring. $\square$

Example 7.2. For $n \geq 1$, the set $\mathbb{Z}_n$ under the rules of multiplication and addition defined in §2.5 forms a ring. $\square$

Example 7.3. The set $\mathbb{Q}$ of rational numbers under the usual rules of multiplication and addition forms a ring. $\square$

Example 7.4. The set $\mathbb{R}$ of real numbers under the usual rules of multiplication and addition forms a ring. $\square$

Example 7.5. The set $\mathbb{C}$ of complex numbers under the usual rules of multiplication and addition forms a ring. Every $\alpha \in \mathbb{C}$ can be written (uniquely) as $\alpha = a + bi$, where $a, b \in \mathbb{R}$ and $i = \sqrt{-1}$. If $\alpha' = a' + b'i$ is another complex number, with $a', b' \in \mathbb{R}$, then

$$\alpha + \alpha' = (a + a') + (b + b')i \text{ and } \alpha\alpha' = (aa' - bb') + (ab' + a'b)i.$$

The fact that $\mathbb{C}$ is a ring can be verified by direct calculation; however, we shall see later that this follows easily from more general considerations.

Recall the **complex conjugation** operation, which sends α to $\bar{\alpha} := a - bi$. One can verify by direct calculation that complex conjugation is both additive and multiplicative; that is, $\overline{\alpha + \alpha'} = \bar{\alpha} + \bar{\alpha}'$ and $\overline{\alpha \cdot \alpha'} = \bar{\alpha} \cdot \bar{\alpha}'$.

The **norm** of α is $N(\alpha) := \alpha\bar{\alpha} = a^2 + b^2$. So we see that $N(\alpha)$ is a non-negative real number, and is zero if and only if $\alpha = 0$. Moreover, from the multiplicativity of complex conjugation, it is easy to see that the norm is multiplicative as well: $N(\alpha\alpha') = \alpha\alpha'\overline{\alpha\alpha'} = \alpha\alpha'\bar{\alpha}\bar{\alpha}' = \alpha\bar{\alpha}\alpha'\bar{\alpha}' = N(\alpha)N(\alpha')$. $\square$

Example 7.6. Consider the set $\mathcal{F}$ of all arithmetic functions, that is, functions mapping positive integers to reals. Let us define addition of arithmetic functions point-wise (i.e., $(f + g)(n) = f(n) + g(n)$ for all positive integers n) and multiplication using the Dirichlet product, introduced in §2.9. The reader should verify that with addition and multiplication so defined, $\mathcal{F}$ forms a ring, where the all-zero function is the additive identity, and the special function I defined in §2.9 is the multiplicative identity. $\square$

Example 7.7. Generalizing Example 6.18, if $R_1, \ldots, R_k$ are rings, then we can form the **direct product** $S := R_1 \times \cdots \times R_k$, which consists of all k-tuples $(a_1, \ldots, a_k)$ with $a_1 \in R_1, \ldots, a_k \in R_k$. We can view S in a natural way as a ring, with addition and multiplication defined component-wise. The additive identity is $(0_{R_1}, \ldots, 0_{R_k})$ and the multiplicative identity is $(1_{R_1}, \ldots, 1_{R_k})$. When $R = R_1 = \cdots = R_k$, the k-wise direct product of R is denoted $R^{\times k}$. $\square$

Example 7.8. Generalizing Example 6.19, if I is an arbitrary set and R is a ring, then $\mathrm{Map}(I, R)$, which is the set of all functions $f : I \to R$, may be naturally viewed as a ring, with addition and multiplication defined point-wise: for $f, g \in \mathrm{Map}(I, R)$, we define

$$(f + g)(i) := f(i) + g(i) \ \text{ and } \ (f \cdot g)(i) := f(i) \cdot g(i) \ \text{ for all } i \in I.$$

We leave it to the reader to verify that $\mathrm{Map}(I, R)$ is indeed a ring, where the additive identity is the all-zero function, and the multiplicative identity is the all-one function. $\square$

A ring R may be **trivial**, meaning that it consists of the single element 0_R, with $0_R + 0_R = 0_R$ and $0_R \cdot 0_R = 0_R$. Certainly, if R is trivial, then $1_R = 0_R$. Conversely, if $1_R = 0_R$, then for all $a \in R$, we have $a = 1_R \cdot a = 0_R \cdot a = 0_R$, and hence R is trivial. Trivial rings are not very interesting, but they naturally arise in certain constructions.

For $a_1, \ldots, a_k \in R$, the product $a_1 \cdots a_k$ needs no parentheses, because multiplication is associative; moreover, we can reorder the a_i's without changing the value of the product, since multiplication is commutative. We can also write this product as $\prod_{i=1}^{k} a_i$. By convention, such a product is defined to be 1_R when $k = 0$. When $a = a_1 = \cdots = a_k$, we can write this product as a^k. The reader may verify the usual power laws: for all $a, b \in R$, and all non-negative integers k and ℓ, we have

$$(a^\ell)^k = a^{k\ell} = (a^k)^\ell, \ a^{k+\ell} = a^k a^\ell, \ (ab)^k = a^k b^k. \tag{7.1}$$

For all $a_1, \ldots, a_k, b_1, \ldots, b_\ell \in R$, the distributive law implies

$$(a_1 + \cdots + a_k)(b_1 + \cdots + b_\ell) = \sum_{\substack{1 \le i \le k \\ 1 \le j \le \ell}} a_i b_j.$$

A ring R is in particular an abelian group with respect to addition. We shall call a subgroup of the additive group of R an **additive subgroup** of R. The **characteristic** of R is defined as the exponent of this group (see §6.5). Note that for all $m \in \mathbb{Z}$ and $a \in R$, we have

$$ma = m(1_R \cdot a) = (m \cdot 1_R)a,$$

so that if $m \cdot 1_R = 0_R$, then $ma = 0_R$ for all $a \in R$. Thus, if the additive order of 1_R is infinite, the characteristic of R is zero, and otherwise, the characteristic of R is equal to the additive order of 1_R.

Example 7.9. The ring $\mathbb{Z}$ has characteristic zero, $\mathbb{Z}_n$ has characteristic n, and $\mathbb{Z}_{n_1} \times \mathbb{Z}_{n_2}$ has characteristic $\mathrm{lcm}(n_1, n_2)$. $\square$

When there is no possibility for confusion, one may write "0" instead of "0_R" and "1" instead of "1_R." Also, one may also write, for example, 2_R to denote $2 \cdot 1_R$, 3_R to denote $3 \cdot 1_R$, and so on; moreover, where the context is clear, one may use an implicit "type cast," so that $m \in \mathbb{Z}$ really means $m \cdot 1_R$.

EXERCISE 7.1. Show that the familiar **binomial theorem** (see §A2) holds in an arbitrary ring R; that is, for all $a, b \in R$ and every positive integer n, we have

$$(a + b)^n = \sum_{k=0}^{n} \binom{n}{k} a^{n-k} b^k.$$

EXERCISE 7.2. Let R be a ring. For additive subgroups A and B of R, we define their **ring-theoretic product** AB as the set of all elements of R that can be expressed as

$$a_1 b_1 + \cdots + a_k b_k$$

for some $a_1, \ldots, a_k \in A$ and $b_1, \ldots, b_k \in B$; by definition, this set includes the "empty sum" 0_R. Show that for all additive subgroups A, B, and C of R:

(a) AB is also an additive subgroup of R;

(b) $AB = BA$;

(c) $A(BC) = (AB)C$;

(d) $A(B + C) = AB + AC$.

7.1.1 Divisibility, units, and fields

For elements a, b in a ring R, we say that a **divides** b if $ar = b$ for some $r \in R$. If a divides b, we write $a \mid b$, and we may say that a is a **divisor** of b, or that b is a **multiple** of a, or that b is **divisible by** a. If a does not divide b, then we write $a \nmid b$. Note that Theorem 1.1 holds for an arbitrary ring.

We call $a \in R$ a **unit** if $a \mid 1_R$, that is, if $ar = 1_R$ for some $r \in R$. Using the same argument as was used to prove part (ii) of Theorem 6.2, it is easy to see that r is uniquely determined; it is called the **multiplicative inverse** of a, and we denote it by a^{-1}. Also, for $b \in R$, we may write b/a to denote ba^{-1}. Evidently, if a is a unit, then $a \mid b$ for every $b \in R$.

We denote the set of units by R^*. It is easy to see that $1_R \in R^*$. Moreover, R^* is closed under multiplication; indeed, if a and b are elements of R^*, then $(ab)^{-1} = a^{-1}b^{-1}$. It follows that with respect to the multiplication operation of the ring, R^* is an abelian group, called the **multiplicative group of units** of R. If $a \in R^*$ and k is a positive integer, then $a^k \in R^*$; indeed, the multiplicative inverse of a^k is $(a^{-1})^k$, which we may also write as a^{-k} (which is consistent with our notation for abelian groups). For all $a, b \in R^*$, the identities (7.1) hold for *all* integers k and ℓ.

If R is non-trivial and every non-zero element of R has a multiplicative inverse, then R is called a **field**.

Example 7.10. The only units in the ring $\mathbb{Z}$ are ± 1. Hence, $\mathbb{Z}$ is not a field. $\square$

Example 7.11. Let n be a positive integer. The units in $\mathbb{Z}_n$ are the residue classes $[a]_n$ with $\gcd(a, n) = 1$. In particular, if n is prime, all non-zero residue classes are units, and if n is composite, some non-zero residue classes are not units. Hence, $\mathbb{Z}_n$ is a field if and only if n is prime. The notation $\mathbb{Z}_n^*$ introduced in this section for the group of units of the ring $\mathbb{Z}_n$ is consistent with the notation introduced in §2.5. $\square$

Example 7.12. Every non-zero element of $\mathbb{Q}$ is a unit. Hence, $\mathbb{Q}$ is a field. $\square$

Example 7.13. Every non-zero element of $\mathbb{R}$ is a unit. Hence, $\mathbb{R}$ is a field. $\square$

Example 7.14. For non-zero $\alpha = a + bi \in \mathbb{C}$, with $a, b \in \mathbb{R}$, we have $c := N(\alpha) = a^2 + b^2 > 0$. It follows that the complex number $\bar{\alpha}c^{-1} = (ac^{-1}) + (-bc^{-1})i$ is the multiplicative inverse of α, since $\alpha \cdot \bar{\alpha}c^{-1} = (\alpha\bar{\alpha})c^{-1} = 1$. Hence, every non-zero element of $\mathbb{C}$ is a unit, and so $\mathbb{C}$ is a field. $\square$

Example 7.15. For rings $R_1, \ldots, R_k$, it is easy to see that the multiplicative group of units of the direct product $R_1 \times \cdots \times R_k$ is equal to $R_1^* \times \cdots \times R_k^*$. Indeed, by definition, $(a_1, \ldots, a_k)$ has a multiplicative inverse if and only if each individual a_i does. $\square$

Example 7.16. If I is a set and R is a ring, then the units in $\mathrm{Map}(I, R)$ are those functions $f : I \to R$ such that $f(i) \in R^*$ for all $i \in I$. $\square$

Example 7.17. Consider the ring $\mathcal{F}$ of arithmetic functions defined in Example 7.6. By the result of Exercise 2.54, $\mathcal{F}^* = \{f \in \mathcal{F} : f(1) \neq 0\}$. $\square$

7.1.2 Zero divisors and integral domains

Let R be a ring. If a and b are non-zero elements of R such that $ab = 0$, then a and b are both called **zero divisors**. If R is non-trivial and has no zero divisors, then it is called an **integral domain**. Note that if a is a unit in R, it cannot be a zero divisor (if $ab = 0$, then multiplying both sides of this equation by a^{-1} yields $b = 0$). In particular, it follows that every field is an integral domain.

Example 7.18. $\mathbb{Z}$ is an integral domain. $\square$

Example 7.19. For $n > 1$, $\mathbb{Z}_n$ is an integral domain if and only if n is prime. In particular, if n is composite, so $n = ab$ with $1 < a < n$ and $1 < b < n$, then $[a]_n$ and $[b]_n$ are zero divisors: $[a]_n [b]_n = [0]_n$, but $[a]_n \neq [0]_n$ and $[b]_n \neq [0]_n$. $\square$

Example 7.20. $\mathbb{Q}$, $\mathbb{R}$, and $\mathbb{C}$ are fields, and hence are also integral domains. $\square$

Example 7.21. For two non-trivial rings R_1, R_2, an element $(a_1, a_2) \in R_1 \times R_2$ is a zero divisor if and only if a_1 is a zero divisor, a_2 is a zero divisor, or exactly one of a_1 or a_2 is zero. In particular, $R_1 \times R_2$ is not an integral domain. $\square$

The next two theorems establish certain results that are analogous to familiar facts about integer divisibility. These results hold in a general ring, provided one avoids zero divisors. The first is a **cancellation law**:

Theorem 7.3. *If R is a ring, and $a, b, c \in R$ such that $a \neq 0$ and a is not a zero divisor, then $ab = ac$ implies $b = c$.*

Proof. $ab = bc$ implies $a(b - c) = 0$. The fact that $a \neq 0$ and a is not a zero divisor implies that we must have $b - c = 0$, and so $b = c$. $\square$

Theorem 7.4. *Let R be a ring.*

    (i) Suppose $a, b \in R$, and that either a or b is not a zero divisor. Then $a \mid b$ and $b \mid a$ if and only if $ar = b$ for some $r \in R^*$.

    (ii) Suppose $a, b \in R$, $a \mid b$, $a \neq 0$, and a is not a zero divisor. Then there exists a unique $r \in R$ such that $ar = b$, which we denote by b/a.

Proof. For the first statement, if $ar = b$ for some $r \in R^*$, then we also have $br^{-1} = a$; thus, $a \mid b$ and $b \mid a$. For the converse, suppose that $a \mid b$ and $b \mid a$. We may assume that b is not a zero divisor (otherwise, exchange the roles of a and b). We may also assume that b is non-zero (otherwise, $b \mid a$ implies $a = 0$, and so the conclusion holds with any r). Now, $a \mid b$ implies $ar = b$ for some $r \in R$, and $b \mid a$ implies $br' = a$ for some $r' \in R$, and hence $b = ar = br'r$. Canceling b from both sides of the equation $b = br'r$, we obtain $1 = r'r$, and so r is a unit.

For the second statement, $a \mid b$ means $ar = b$ for some $r \in R$. Moreover, this value of r is unique: if $ar = b = ar'$, then we may cancel a, obtaining $r = r'$. $\square$

Of course, in the previous two theorems, if the ring is an integral domain, then there are no zero divisors, and so the hypotheses may be simplified in this case, dropping the explicit requirement that certain elements are not zero divisors. In particular, if a, b, and c are elements of an integral domain, such that $ab = ac$ and $a \neq 0$, then we can cancel a, obtaining $b = c$.

The next two theorems state some facts which pertain specifically to integral domains.

Theorem 7.5. *The characteristic of an integral domain is either zero or a prime.*

Proof. By way of contradiction, suppose that D is an integral domain with characteristic m that is neither zero nor prime. Since, by definition, D is not a trivial ring, we cannot have $m = 1$, and so m must be composite. Say $m = st$, where $1 < s < m$ and $1 < t < m$. Since m is the additive order of 1_D, it follows that $(s \cdot 1_D) \neq 0_D$ and $(t \cdot 1_D) \neq 0_D$; moreover, since D is an integral domain, it follows that $(s \cdot 1_D)(t \cdot 1_D) \neq 0_D$. So we have

$$0_D = m \cdot 1_D = (st) \cdot 1_D = (s \cdot 1_D)(t \cdot 1_D) \neq 0_D,$$

a contradiction. $\square$

Theorem 7.6. *Every finite integral domain is a field.*

Proof. Let D be a finite integral domain, and let a be any non-zero element of D. Consider the a-multiplication map that sends $b \in D$ to ab, which is a group homomorphism on the additive group of D. Since a is not a zero-divisor, it follows that the kernel of the a-multiplication map is $\{0_D\}$, hence the map is injective, and by finiteness, it must be surjective as well. In particular, there must be an element $b \in D$ such that $ab = 1_D$. $\square$

Theorem 7.7. *Every finite field F must be of cardinality p^w, where p is prime, w is a positive integer, and p is the characteristic of F.*

Proof. By Theorem 7.5, the characteristic of F is either zero or a prime, and since F is finite, it must be prime. Let p denote the characteristic. By definition, p is the exponent of the additive group of F, and by Theorem 6.43, the primes dividing the exponent are the same as the primes dividing the order, and hence F must have cardinality p^w for some positive integer w. $\square$

Of course, for every prime p, $\mathbb{Z}_p$ is a finite field of cardinality p. As we shall see later (in Chapter 19), for every prime p and positive integer w, there exists a field of cardinality p^w. Later in this chapter, we shall see some specific examples of finite fields of cardinality p^2 (Examples 7.40, 7.59, and 7.60).

EXERCISE 7.3. Let R be a ring, and let $a, b \in R$ such that $ab \neq 0$. Show that ab is a zero divisor if and only if a is a zero divisor or b is a zero divisor.

EXERCISE 7.4. Suppose that R is a non-trivial ring in which the cancellation law holds in general: for all $a, b, c \in R$, if $a \neq 0$ and $ab = ac$, then $b = c$. Show that R is an integral domain.

EXERCISE 7.5. Let R be a ring of characteristic $m > 0$, and let n be an integer. Show that:

    (a) if $\gcd(n, m) = 1$, then $n \cdot 1_R$ is a unit;

    (b) if $1 < \gcd(n, m) < m$, then $n \cdot 1_R$ is a zero divisor;

    (c) otherwise, $n \cdot 1_R = 0$.

EXERCISE 7.6. Let D be an integral domain, $m \in \mathbb{Z}$, and $a \in D$. Show that $ma = 0$ if and only if m is a multiple of the characteristic of D or $a = 0$.

EXERCISE 7.7. Show that for all $n \geq 1$, and for all $a, b \in \mathbb{Z}_n$, if $a \mid b$ and $b \mid a$, then $ar = b$ for some $r \in \mathbb{Z}_n^*$. Hint: this result *does not* follow from part (i) of Theorem 7.4, as we allow a and b to be zero divisors here; first consider the case where n is a prime power.

EXERCISE 7.8. Show that the ring $\mathcal{F}$ of arithmetic functions defined in Example 7.6 is an integral domain.

EXERCISE 7.9. This exercise depends on results in §6.6. Using the fundamental theorem of finite abelian groups, show that the additive group of a finite field of characteristic p and cardinality p^w is isomorphic to $\mathbb{Z}_p^{\times w}$.

7.1.3 Subrings

Definition 7.8. *A subset S of a ring R is called a **subring** if*

    (i) S is an additive subgroup of R,

    (ii) S is closed under multiplication, and

    (iii) $1_R \in S$.

It is clear that the operations of addition and multiplication on a ring R make a subring S of R into a ring, where 0_R is the additive identity of S and 1_R is the multiplicative identity of S. One may also call R an **extension ring** of S.

Some texts do not require that 1_R belongs to a subring S, and instead require only that S contains a multiplicative identity, which may be different than that of R. This is perfectly reasonable, but for simplicity, we restrict ourselves to the case where $1_R \in S$.

Expanding the above definition, we see that a subset S of R is a subring if and only if $1_R \in S$ and for all $a, b \in S$, we have

$$a + b \in S, \quad -a \in S, \quad \text{and} \quad ab \in S.$$

In fact, to verify that S is a subring, it suffices to show that $-1_R \in S$ and that S is closed under addition and multiplication; indeed, if $-1_R \in S$ and S is closed under multiplication, then S is closed under negation, and further, $1_R = -(-1_R) \in S$.

Example 7.22. $\mathbb{Z}$ is a subring of $\mathbb{Q}$. $\square$

Example 7.23. $\mathbb{Q}$ is a subring of $\mathbb{R}$. $\square$

Example 7.24. $\mathbb{R}$ is a subring of $\mathbb{C}$. Note that for all $\alpha := a + bi \in \mathbb{C}$, with $a, b \in \mathbb{R}$, we have $\bar{\alpha} = \alpha \iff a + bi = a - bi \iff b = 0$. That is, $\bar{\alpha} = \alpha \iff \alpha \in \mathbb{R}$. $\square$

Example 7.25. The set $\mathbb{Z}[i]$ of complex numbers of the form $a + bi$, with $a, b \in \mathbb{Z}$, is a subring of $\mathbb{C}$. It is called the ring of **Gaussian integers**. Since $\mathbb{C}$ is a field, it contains no zero divisors, and hence $\mathbb{Z}[i]$ contains no zero divisors either. Hence, $\mathbb{Z}[i]$ is an integral domain.

Let us determine the units of $\mathbb{Z}[i]$. Suppose $\alpha \in \mathbb{Z}[i]$ is a unit, so that there exists $\alpha' \in \mathbb{Z}[i]$ such that $\alpha\alpha' = 1$. Taking norms, we obtain

$$1 = N(1) = N(\alpha\alpha') = N(\alpha)N(\alpha').$$

Since the norm of any Gaussian integer is itself a non-negative integer, and since $N(\alpha)N(\alpha') = 1$, we must have $N(\alpha) = 1$. Now, if $\alpha = a + bi$, with $a, b \in \mathbb{Z}$, then $1 = N(\alpha) = a^2 + b^2$, which implies that $\alpha = \pm 1$ or $\alpha = \pm i$. Conversely, it is easy to see that ± 1 and $\pm i$ are indeed units, and so these are the only units in $\mathbb{Z}[i]$. $\square$

Example 7.26. Let m be a positive integer, and let $\mathbb{Q}^{(m)}$ be the set of rational numbers which can be written as a/b, where a and b are integers, and b is relatively prime to m. Then $\mathbb{Q}^{(m)}$ is a subring of $\mathbb{Q}$, since for all $a, b, c, d \in \mathbb{Z}$ with $\gcd(b, m) = 1$ and $\gcd(d, m) = 1$, we have

$$\frac{a}{b} + \frac{c}{d} = \frac{ad + bc}{bd} \quad \text{and} \quad \frac{a}{b} \cdot \frac{c}{d} = \frac{ac}{bd},$$

and since $\gcd(bd, m) = 1$, it follows that the sum and product of any two elements of $\mathbb{Q}^{(m)}$ are again in $\mathbb{Q}^{(m)}$. Clearly, $\mathbb{Q}^{(m)}$ contains -1, and so it follows that $\mathbb{Q}^{(m)}$ is a subring of $\mathbb{Q}$. The units of $\mathbb{Q}^{(m)}$ are precisely those rational numbers of the form a/b, where $\gcd(a, m) = \gcd(b, m) = 1$. $\square$

Example 7.27. Suppose R is a non-trivial ring. Then the set $\{0_R\}$ is not a subring of R: although it satisfies the first two requirements of the definition of a subring, it does not satisfy the third. $\square$

Generalizing the argument in Example 7.25, it is clear that every subring of an integral domain is itself an integral domain. However, it is not the case that a subring of a field is always a field: the subring $\mathbb{Z}$ of $\mathbb{Q}$ is a counter-example. If F' is a subring of a field F, and F' is itself a field, then we say that F' is a **subfield** of F, and that F is an **extension field** of F'. For example, $\mathbb{Q}$ is a subfield of $\mathbb{R}$, which in turn is a subfield of $\mathbb{C}$.

EXERCISE 7.10. Show that if S is a subring of a ring R, then a set $T \subseteq S$ is a subring of R if and only if T is a subring of S.

EXERCISE 7.11. Show that if S and T are subrings of R, then so is $S \cap T$.

EXERCISE 7.12. Let S_1 be a subring of R_1, and S_2 a subring of R_2. Show that $S_1 \times S_2$ is a subring of $R_1 \times R_2$.

EXERCISE 7.13. Suppose that S and T are subrings of a ring R. Show that their ring-theoretic product ST (see Exercise 7.2) is a subring of R that contains $S \cup T$, and is the smallest such subring.

EXERCISE 7.14. Show that the set $\mathbb{Q}[i]$ of complex numbers of the form $a + bi$, with $a, b \in \mathbb{Q}$, is a subfield of $\mathbb{C}$.

EXERCISE 7.15. Consider the ring $\mathrm{Map}(\mathbb{R}, \mathbb{R})$ of functions $f : \mathbb{R} \to \mathbb{R}$, with addition and multiplication defined point-wise.

   (a) Show that $\mathrm{Map}(\mathbb{R}, \mathbb{R})$ is not an integral domain, and that $\mathrm{Map}(\mathbb{R}, \mathbb{R})^*$ consists of those functions that never vanish.

   (b) Let $a, b \in \mathrm{Map}(\mathbb{R}, \mathbb{R})$. Show that if $a \mid b$ and $b \mid a$, then $ar = b$ for some $r \in \mathrm{Map}(\mathbb{R}, \mathbb{R})^*$.

   (c) Let C be the subset of $\mathrm{Map}(\mathbb{R}, \mathbb{R})$ of continuous functions. Show that C is a subring of $\mathrm{Map}(\mathbb{R}, \mathbb{R})$, and that all functions in C^* are either everywhere positive or everywhere negative.

   (d) Find elements $a, b \in C$, such that in the ring C, we have $a \mid b$ and $b \mid a$, yet there is no $r \in C^*$ such that $ar = b$.

7.2 Polynomial rings

If R is a ring, then we can form the **ring of polynomials** $R[X]$, consisting of all polynomials $g = a_0 + a_1 X + \cdots + a_k X^k$ in the **indeterminate**, or "formal" variable, X, with coefficients a_i in R, and with addition and multiplication defined in the usual way.

Example 7.28. Let us define a few polynomials over the ring $\mathbb{Z}$:

$$a := 3 + X^2,\ b := 1 + 2X - X^3,\ c := 5,\ d := 1 + X,\ e := X,\ f := 4X^3.$$

We have

$$a+b = 4+2X+X^2-X^3,\ a \cdot b = 3+6X+X^2-X^3-X^5,\ cd+ef = 5+5X+4X^4.\ \square$$

As illustrated in the previous example, elements of R are also considered to be polynomials. Such polynomials are called **constant polynomials**. The set R of constant polynomials forms a subring of $R[X]$. In particular, 0_R is the additive identity in $R[X]$ and 1_R is the multiplicative identity in $R[X]$. Note that if R is the trivial ring, then so is $R[X]$; also, if R is a subring of E, then $R[X]$ is a subring of $E[X]$.

So as to keep the distinction between ring elements and indeterminates clear, we shall use the symbol "X" only to denote the latter. Also, for a polynomial $g \in R[X]$, we shall in general write this simply as "g," and not as "$g(X)$." Of course, the choice of the symbol "X" is arbitrary; occasionally, we may use another symbol, such as "Y," as an alternative.

7.2.1 Formalities

For completeness, we present a more formal definition of the ring $R[X]$. The reader should bear in mind that this formalism is rather tedious, and may be more distracting than it is enlightening. Formally, a polynomial $g \in R[X]$ is an infinite sequence $\{a_i\}_{i=0}^{\infty}$, where each $a_i \in R$, but only finitely many of the a_i's are non-zero (intuitively, a_i represents the coefficient of X^i). For each non-negative integer j, it will be convenient to define the function $\varepsilon_j : R \to R[X]$ that maps $c \in R$ to the sequence $\{c_i\}_{i=0}^{\infty} \in R[X]$, where $c_j := c$ and $c_i := 0_R$ for $i \neq j$ (intuitively, $\varepsilon_j(c)$ represents the polynomial cX^j).

For

$$g = \{a_i\}_{i=0}^{\infty} \in R[X] \text{ and } h = \{b_i\}_{i=0}^{\infty} \in R[X],$$

we define

$$g + h := \{s_i\}_{i=0}^{\infty} \text{ and } gh := \{p_i\}_{i=0}^{\infty},$$

where for $i = 0, 1, 2, \ldots,$

$$s_i := a_i + b_i \tag{7.2}$$

and

$$p_i := \sum_{i=j+k} a_j b_k, \tag{7.3}$$

the sum being over all pairs (j, k) of non-negative integers such that $i = j + k$ (which is a finite sum). We leave it to the reader to verify that $g + h$ and gh are polynomials (i.e., only finitely many of the s_i's and p_i's are non-zero). The reader may also verify that all the requirements of Definition 7.1 are satisfied: the additive identity is the all-zero sequence $\varepsilon_0(0_R)$, and the multiplicative identity is $\varepsilon_0(1_R)$.

One can easily verify that for all $c, d \in R$, we have

$$\varepsilon_0(c + d) = \varepsilon_0(c) + \varepsilon_0(d) \text{ and } \varepsilon_0(cd) = \varepsilon_0(c)\varepsilon_0(d).$$

We shall identify $c \in R$ with $\varepsilon_0(c) \in R[X]$, viewing the ring element c as simply "shorthand" for the polynomial $\varepsilon_0(c)$ in contexts where a polynomial is expected. Note that while c and $\varepsilon_0(c)$ are not the same mathematical object, there will be no confusion in treating them as such. Thus, from a narrow, legalistic point of view, R is not a subring of $R[X]$, but we shall not let such annoying details prevent us from continuing to speak of it as such. Indeed, by appropriately renaming elements, we can make R a subring of $R[X]$ in the literal sense of the term.

We also define $X := \varepsilon_1(1_R)$. One can verify that $X^i = \varepsilon_i(1_R)$ for all $i \geq 0$. More generally, for any polynomial $g = \{a_i\}_{i=0}^{\infty}$, if $a_i = 0_R$ for all i exceeding some value k, then we have $g = \sum_{i=0}^{k} \varepsilon_0(a_i)X^i$. Writing a_i in place of $\varepsilon_0(a_i)$, we have $g = \sum_{i=0}^{k} a_i X^i$, and so we can return to the standard practice of writing polynomials as we did in Example 7.28, without any loss of precision.

7.2.2 Basic properties of polynomial rings

Let R be a ring. For non-zero $g \in R[X]$, if $g = \sum_{i=0}^{k} a_i X^i$ with $a_k \neq 0$, then we call k the **degree** of g, denoted $\deg(g)$, we call a_k the **leading coefficient** of g, denoted $\mathrm{lc}(g)$, and we call a_0 the **constant term** of g. If $\mathrm{lc}(g) = 1$, then g is called **monic**.

Suppose $g = \sum_{i=0}^{k} a_i X^i$ and $h = \sum_{i=0}^{\ell} b_i X^i$ are polynomials such that $a_k \neq 0$ and $b_\ell \neq 0$, so that $\deg(g) = k$ and $\mathrm{lc}(g) = a_k$, and $\deg(h) = \ell$ and $\mathrm{lc}(h) = b_\ell$. When we multiply these two polynomials, we get

$$gh = a_0 b_0 + (a_0 b_1 + a_1 b_0)X + \cdots + a_k b_\ell X^{k+\ell}.$$

In particular, $\deg(gh) \leq \deg(g) + \deg(h)$. If either of a_k or b_ℓ are not zero divisors, then $a_k b_\ell$ is not zero, and hence $\deg(gh) = \deg(g) + \deg(h)$. However, if both a_k

and b_ℓ are zero divisors, then we may have $a_k b_\ell = 0$, in which case, the product gh may be zero, or perhaps $gh \neq 0$ but $\deg(gh) < \deg(g) + \deg(h)$.

For the zero polynomial, we establish the following conventions: its leading coefficient and constant term are defined to be 0_R, and its degree is defined to be $-\infty$. With these conventions, we may succinctly state that

> *for all $g, h \in R[X]$, we have $\deg(gh) \leq \deg(g) + \deg(h)$, with equality guaranteed to hold unless the leading coefficients of both g and h are zero divisors.*

In particular, if the leading coefficient of a polynomial is not a zero divisor, then the polynomial is not a zero divisor. In the case where the ring of coefficients is an integral domain, we can be more precise:

Theorem 7.9. *Let D be an integral domain. Then:*

(i) *for all $g, h \in D[X]$, we have $\deg(gh) = \deg(g) + \deg(h)$;*

(ii) *$D[X]$ is an integral domain;*

(iii) *$(D[X])^* = D^*$.*

Proof. Exercise. $\square$

An extremely important property of polynomials is a division with remainder property, analogous to that for the integers:

Theorem 7.10 (Division with remainder property). *Let R be a ring. For all $g, h \in R[X]$ with $h \neq 0$ and $\mathrm{lc}(h) \in R^*$, there exist unique $q, r \in R[X]$ such that $g = hq + r$ and $\deg(r) < \deg(h)$.*

Proof. Consider the set $S := \{g - ht : t \in R[X]\}$. Let $r = g - hq$ be an element of S of minimum degree. We must have $\deg(r) < \deg(h)$, since otherwise, we could subtract an appropriate multiple of h from r so as to eliminate the leading coefficient of r, obtaining

$$r' := r - h \cdot (\mathrm{lc}(r) \, \mathrm{lc}(h)^{-1} X^{\deg(r) - \deg(h)}) \in S,$$

where $\deg(r') < \deg(r)$, contradicting the minimality of $\deg(r)$.

That proves the existence of r and q. For uniqueness, suppose that $g = hq + r$ and $g = hq' + r'$, where $\deg(r) < \deg(h)$ and $\deg(r') < \deg(h)$. This implies $r' - r = h \cdot (q - q')$. However, if $q \neq q'$, then

$$\deg(h) > \deg(r' - r) = \deg(h \cdot (q - q')) = \deg(h) + \deg(q - q') \geq \deg(h),$$

which is impossible. Therefore, we must have $q = q'$, and hence $r = r'$. $\square$

If $g = hq + r$ as in the above theorem, we define $g \bmod h := r$. Clearly, $h \mid g$ if and only if $g \bmod h = 0$. Moreover, note that if $\deg(g) < \deg(h)$, then $q = 0$ and $r = g$; otherwise, if $\deg(g) \geq \deg(h)$, then $q \neq 0$ and $\deg(g) = \deg(h) + \deg(q)$.

7.2.3 Polynomial evaluation

A polynomial $g = \sum_{i=0}^{k} a_i X^i \in R[X]$ naturally defines a polynomial function on R that sends $x \in R$ to $\sum_{i=0}^{k} a_i x^i \in R$, and we denote the value of this function as $g(x)$ (note that "X" denotes an indeterminate, while "x" denotes an element of R). It is important to regard polynomials over R as formal expressions, and *not* to identify them with their corresponding functions. In particular, two polynomials are equal if and only if their coefficients are equal, while two functions are equal if and only if their values agree at all points in R. This distinction is important, since there are rings R over which two different polynomials define the same function. One can of course define the ring of polynomial functions on R, but in general, that ring has a different structure from the ring of polynomials over R.

Example 7.29. In the ring $\mathbb{Z}_p$, for prime p, by Fermat's little theorem (Theorem 2.14), we have $x^p = x$ for all $x \in \mathbb{Z}_p$. However, the polynomials X^p and X are not the same polynomials (in particular, the former has degree p, while the latter has degree 1). $\square$

More generally, suppose R is a subring of a ring E. Then every polynomial $g = \sum_{i=0}^{k} a_i X^i \in R[X]$ defines a polynomial function from E to E that sends $\alpha \in E$ to $\sum_{i=0}^{k} a_i \alpha^i \in E$, and, again, the value of this function is denoted $g(\alpha)$. We say that α is a **root** of g if $g(\alpha) = 0$.

An obvious, yet important, fact is the following:

Theorem 7.11. *Let R be a subring of a ring E. For all $g, h \in R[X]$ and $\alpha \in E$, if $s := g + h \in R[X]$ and $p := gh \in R[X]$, then we have*

$$s(\alpha) = g(\alpha) + h(\alpha) \quad \text{and} \quad p(\alpha) = g(\alpha)h(\alpha).$$

Also, if $c \in R$ is a constant polynomial, then $c(\alpha) = c$ for all $\alpha \in E$.

Proof. The statement about evaluating a constant polynomial is clear from the definitions. The proof of the statements about evaluating the sum or product of polynomials is really just symbol pushing. Indeed, suppose $g = \sum_i a_i X^i$ and $h = \sum_i b_i X^i$. Then $s = \sum_i (a_i + b_i) X^i$, and so

$$s(\alpha) = \sum_i (a_i + b_i) \alpha^i = \sum_i a_i \alpha^i + \sum_i b_i \alpha^i = g(\alpha) + h(\alpha).$$

Also, we have

$$p = \left(\sum_i a_i X^i\right)\left(\sum_j b_j X^j\right) = \sum_{i,j} a_i b_j X^{i+j},$$

and employing the result for evaluating sums of polynomials, we have

$$p(\alpha) = \sum_{i,j} a_i b_j \alpha^{i+j} = \left(\sum_i a_i \alpha^i\right)\left(\sum_j b_j \alpha^j\right) = g(\alpha)h(\alpha). \quad \square$$

Example 7.30. Consider the polynomial $g := 2X^3 - 2X^2 + X - 1 \in \mathbb{Z}[X]$. We can write $g = (2X^2 + 1)(X - 1)$. For any element α of $\mathbb{Z}$, or an extension ring of $\mathbb{Z}$, we have $g(\alpha) = (2\alpha^2 + 1)(\alpha - 1)$. From this, it is clear that in $\mathbb{Z}$, g has a root only at 1; moreover, it has no other roots in $\mathbb{R}$, but in $\mathbb{C}$, it also has roots $\pm i/\sqrt{2}$. $\square$

Example 7.31. If $E = R[X]$, then evaluating a polynomial $g \in R[X]$ at a point $\alpha \in E$ amounts to polynomial composition. For example, if $g := X^2 + X$ and $\alpha := X + 1$, then

$$g(\alpha) = g(X + 1) = (X + 1)^2 + (X + 1) = X^2 + 3X + 2. \quad \square$$

The reader is perhaps familiar with the fact that over the real or the complex numbers, every polynomial of degree k has at most k distinct roots, and the fact that every set of k points can be interpolated by a unique polynomial of degree less than k. As we will now see, these results extend to much more general, though not completely arbitrary, coefficient rings.

Theorem 7.12. *Let R be a ring, $g \in R[X]$, and $x \in R$. Then there exists a unique polynomial $q \in R[X]$ such that $g = (X - x)q + g(x)$. In particular, x is a root of g if and only if $(X - x)$ divides g.*

Proof. If R is the trivial ring, there is nothing to prove, so assume that R is non-trivial. Using the division with remainder property for polynomials, there exist unique $q, r \in R[X]$ such that $g = (X - x)q + r$, with $q, r \in R[X]$ and $\deg(r) < 1$, which means that $r \in R$. Evaluating at x, we see that $g(x) = (x - x)q(x) + r = r$. That proves the first statement. The second follows immediately from the first. $\square$

Note that the above theorem says that $X - x$ divides $g - g(x)$, and the polynomial q in the theorem may be expressed (using the notation introduced in part (ii) of Theorem 7.4) as

$$q = \frac{g - g(x)}{X - x}.$$

Theorem 7.13. *Let D be an integral domain, and let $x_1, \ldots, x_k$ be distinct elements of D. Then for every polynomial $g \in D[X]$, the elements $x_1, \ldots, x_k$ are roots of g if and only if the polynomial $\prod_{i=1}^{k}(X - x_i)$ divides g.*

Proof. One direction is trivial: if $\prod_{i=1}^{k}(X - x_i)$ divides g, then it is clear that each x_i is a root of g. We prove the converse by induction on k. The base case $k = 1$ is just Theorem 7.12. So assume $k > 1$, and that the statement holds for $k - 1$. Let $g \in D[X]$ and let $x_1, \ldots, x_k$ be distinct roots of g. Since x_k is a root of g, then by Theorem 7.12, there exists $q \in D[X]$ such that $g = (X - x_k)q$. Moreover, for each $i = 1, \ldots, k - 1$, we have

$$0 = g(x_i) = (x_i - x_k)q(x_i),$$

and since $x_i - x_k \neq 0$ and D is an integral domain, we must have $q(x_i) = 0$. Thus, q has roots $x_1, \ldots, x_{k-1}$, and by induction $\prod_{i=1}^{k-1}(X - x_i)$ divides q, from which it then follows that $\prod_{i=1}^{k}(X - x_i)$ divides g. $\square$

Note that in this theorem, we can slightly weaken the hypothesis: we do not need to assume that the coefficient ring is an integral domain; rather, all we really need is that for all $i \neq j$, the difference $x_i - x_j$ is not a zero divisor.

As an immediate consequence of this theorem, we obtain:

Theorem 7.14. *Let D be an integral domain, and suppose that $g \in D[X]$, with $\deg(g) = k \geq 0$. Then g has at most k distinct roots.*

Proof. If g had $k + 1$ distinct roots $x_1, \ldots, x_{k+1}$, then by the previous theorem, the polynomial $\prod_{i=1}^{k+1}(X - x_i)$, which has degree $k + 1$, would divide g, which has degree k — an impossibility. $\square$

Theorem 7.15 (Lagrange interpolation). *Let F be a field, let $x_1, \ldots, x_k$ be distinct elements of F, and let $y_1, \ldots, y_k$ be arbitrary elements of F. Then there exists a unique polynomial $g \in F[X]$ with $\deg(g) < k$ such that $g(x_i) = y_i$ for $i = 1, \ldots, k$, namely*

$$g := \sum_{i=1}^{k} y_i \frac{\prod_{j \neq i}(X - x_j)}{\prod_{j \neq i}(x_i - x_j)}.$$

Proof. For the existence part of the theorem, one just has to verify that $g(x_i) = y_i$ for the given g, which clearly has degree less than k. This is easy to see: for $i = 1, \ldots, k$, evaluating the ith term in the sum defining g at x_i yields y_i, while evaluating any other term at x_i yields 0. The uniqueness part of the theorem follows almost immediately from Theorem 7.14: if g and h are polynomials of degree less than k such that $g(x_i) = y_i = h(x_i)$ for $i = 1, \ldots, k$, then $g - h$ is a polynomial of degree less than k with k distinct roots, which, by the previous theorem, is impossible. $\square$

Again, we can slightly weaken the hypothesis of this theorem: we do not need to assume that the coefficient ring is a field; rather, all we really need is that for all $i \neq j$, the difference $x_i - x_j$ is a unit.

EXERCISE 7.16. Let D be an infinite integral domain, and let $g, h \in D[X]$. Show that if $g(x) = h(x)$ for all $x \in D$, then $g = h$. Thus, for an infinite integral domain D, there is a one-to-one correspondence between polynomials over D and polynomial functions on D.

EXERCISE 7.17. Let F be a field.

(a) Show that for all $b \in F$, we have $b^2 = 1$ if and only if $b = \pm 1$.

(b) Show that for all $a, b \in F$, we have $a^2 = b^2$ if and only if $a = \pm b$.

(c) Show that the familiar **quadratic formula** holds for F, assuming F has characteristic other than 2, so that $2_F \neq 0_F$. That is, for all $a, b, c \in F$ with $a \neq 0$, the polynomial $g := aX^2 + bX + c \in F[X]$ has a root in F if and only if there exists $e \in F$ such that $e^2 = d$, where d is the **discriminant** of g, defined as $d := b^2 - 4ac$, and in this case the roots of g are $(-b \pm e)/2a$.

EXERCISE 7.18. Let R be a ring, let $g \in R[X]$, with $\deg(g) = k \geq 0$, and let x be an element of R. Show that:

(a) there exist an integer m, with $0 \leq m \leq k$, and a polynomial $q \in R[X]$, such that

$$g = (X - x)^m q \text{ and } q(x) \neq 0,$$

and moreover, the values of m and q are uniquely determined;

(b) if we evaluate g at $X + x$, we have

$$g(X + x) = \sum_{i=0}^{k} b_i X^i,$$

where $b_0 = \cdots = b_{m-1} = 0$ and $b_m = q(x) \neq 0$.

Let $m_x(g)$ denote the value m in the previous exercise; for completeness, one can define $m_x(g) := \infty$ if g is the zero polynomial. If $m_x(g) > 0$, then x is called a root of g of **multiplicity** $m_x(g)$; if $m_x(g) = 1$, then x is called a **simple root** of g, and if $m_x(g) > 1$, then x is called a **multiple root** of g.

The following exercise refines Theorem 7.14, taking into account multiplicities.

EXERCISE 7.19. Let D be an integral domain, and suppose that $g \in D[X]$, with $\deg(g) = k \geq 0$. Show that

$$\sum_{x \in D} m_x(g) \leq k.$$

EXERCISE 7.20. Let D be an integral domain, let $g, h \in D[X]$, and let $x \in D$. Show that $m_x(gh) = m_x(g) + m_x(h)$.

7.2.4 Multi-variate polynomials

One can naturally generalize the notion of a polynomial in a single variable to that of a polynomial in several variables.

Consider the ring $R[X]$ of polynomials over a ring R. If Y is another indeterminate, we can form the ring $R[X][Y]$ of polynomials in Y whose coefficients are themselves polynomials in X over the ring R. One may write $R[X, Y]$ instead of $R[X][Y]$. An element of $R[X, Y]$ is called a **bivariate polynomial**.

Consider a typical element $g \in R[X, Y]$, which may be written

$$g = \sum_{j=0}^{\ell} \left(\sum_{i=0}^{k} a_{ij} X^i \right) Y^j. \tag{7.4}$$

Rearranging terms, this may also be written as

$$g = \sum_{\substack{0 \le i \le k \\ 0 \le j \le \ell}} a_{ij} X^i Y^j, \tag{7.5}$$

or as

$$g = \sum_{i=0}^{k} \left(\sum_{j=0}^{\ell} a_{ij} Y^j \right) X^j. \tag{7.6}$$

If g is written as in (7.5), the terms $X^i Y^j$ are called **monomials**. The **total degree** of such a monomial $X^i Y^j$ is defined to be $i + j$, and if g is non-zero, then the **total degree** of g, denoted $\mathrm{Deg}(g)$, is defined to be the maximum total degree among all monomials $X^i Y^j$ appearing in (7.5) with a non-zero coefficient a_{ij}. We define the total degree of the zero polynomial to be $-\infty$.

When g is written as in (7.6), one sees that we can naturally view g as an element of $R[Y][X]$, that is, as a polynomial in X whose coefficients are polynomials in Y. From a strict, syntactic point of view, the rings $R[Y][X]$ and $R[X][Y]$ are not the same, but there is no harm done in blurring this distinction when convenient. We denote by $\deg_X(g)$ the degree of g, viewed as a polynomial in X, and by $\deg_Y(g)$ the degree of g, viewed as a polynomial in Y.

Example 7.32. Let us illustrate, with a particular example, the three different forms — as in (7.4), (7.5), and (7.6) — of expressing a bivariate polynomial. In the ring $\mathbb{Z}[X, Y]$ we have

$$g = (5X^2 - 3X + 4)Y + (2X^2 + 1)$$
$$= 5X^2 Y + 2X^2 - 3XY + 4Y + 1$$
$$= (5Y + 2)X^2 + (-3Y)X + (4Y + 1).$$

We have $\text{Deg}(g) = 3$, $\deg_X(g) = 2$, and $\deg_Y(g) = 1$. □

More generally, we can form the ring $R[X_1, \ldots, X_n]$ of **multi-variate polynomials** over R in the variables $X_1, \ldots, X_n$. Formally, we can define this ring recursively as $R[X_1, \ldots, X_{n-1}][X_n]$, that is, the ring of polynomials in the variable X_n, with coefficients in $R[X_1, \ldots, X_{n-1}]$. A **monomial** is a term of the form $X_1^{e_1} \cdots X_n^{e_n}$, and the **total degree** of such a monomial is $e_1 + \cdots + e_n$. Every non-zero multi-variate polynomial g can be expressed uniquely (up to a re-ordering of terms) as $a_1 \mu_1 + \cdots + a_k \mu_k$, where each a_i is a non-zero element of R, and each μ_i is a monomial; we define the **total degree** of g, denoted $\text{Deg}(g)$, to be the maximum of the total degrees of the μ_i's. As usual, the zero polynomial is defined to have total degree $-\infty$.

Just as for bivariate polynomials, the order of the indeterminates is not important, and for every $i = 1, \ldots, n$, one can naturally view any $g \in R[X_1, \ldots, X_n]$ as a polynomial in X_i over the ring $R[X_1, \ldots, X_{i-1}, X_{i+1}, \ldots, X_n]$, and define $\deg_{X_i}(g)$ to be the degree of g when viewed in this way.

Just as polynomials in a single variable define polynomial functions, so do polynomials in several variables. If R is a subring of E, $g \in R[X_1, \ldots, X_n]$, and $\alpha_1, \ldots, \alpha_n \in E$, we define $g(\alpha_1, \ldots, \alpha_n)$ to be the element of E obtained by evaluating the expression obtained by substituting α_i for X_i in g. Theorem 7.11 carries over directly to the multi-variate case.

EXERCISE 7.21. Let R be a ring, and consider the ring of multi-variate polynomials $R[X_1, \ldots, X_n]$. For $m \geq 0$, define H_m to be the subset of polynomials that can be expressed as $a_1 \mu_1 + \cdots + a_k \mu_k$, where each a_i belongs to R and each μ_i is a monomial of total degree m (by definition, H_m includes the zero polynomial, and $H_0 = R$). Polynomials that belong to H_m for some m are called **homogeneous polynomials**. Show that:

(a) if $g, h \in H_m$, then $g + h \in H_m$;

(b) if $g \in H_\ell$ and $h \in H_m$, then $gh \in H_{\ell+m}$;

(c) every non-zero polynomial g can be expressed uniquely as $g_0 + \cdots + g_d$, where $g_i \in H_i$ for $i = 0, \ldots, d$, $g_d \neq 0$, and $d = \text{Deg}(g)$;

(d) for all polynomials g, h, we have $\text{Deg}(gh) \leq \text{Deg}(g) + \text{Deg}(h)$, and if R is an integral domain, then $\text{Deg}(gh) = \text{Deg}(g) + \text{Deg}(h)$.

EXERCISE 7.22. Suppose that D is an integral domain, and g, h are non-zero, multi-variate polynomials over D such that gh is homogeneous. Show that g and h are also homogeneous.

EXERCISE 7.23. Let R be a ring, and let $x_1, \ldots, x_n$ be elements of R. Show that every polynomial $g \in R[X_1, \ldots, X_n]$ can be expressed as

$$g = (X_1 - x_1)q_1 + \cdots + (X_n - x_n)q_n + g(x_1, \ldots, x_n),$$

where $q_1, \ldots, q_n \in R[X_1, \ldots, X_n]$.

EXERCISE 7.24. This exercise generalizes Theorem 7.14. Let D be an integral domain, and let $g \in D[X_1, \ldots, X_n]$, with $\text{Deg}(g) = k \geq 0$. Let S be a finite, non-empty subset of D. Show that the number of elements $(x_1, \ldots, x_n) \in S^{\times n}$ such that $g(x_1, \ldots, x_n) = 0$ is at most $k|S|^{n-1}$.

7.3 Ideals and quotient rings

Definition 7.16. *Let R be a ring. An **ideal of** R is an additive subgroup I of R such that $ar \in I$ for all $a \in I$ and $r \in R$ (i.e., I is closed under multiplication by elements of R).*

Expanding the above definition, we see that a non-empty subset I of R is an ideal of R if and only if for all $a, b \in I$ and $r \in R$, we have

$$a + b \in I, \quad -a \in I, \quad \text{and} \quad ar \in I.$$

Since R is commutative, the condition $ar \in I$ is equivalent to $ra \in I$. The condition $-a \in I$ is redundant, as it is implied by the condition $ar \in I$ with $r := -1_R$. In the case when R is the ring $\mathbb{Z}$, this definition of an ideal is consistent with that given in §1.2.

Clearly, $\{0_R\}$ and R are ideals of R. From the fact that an ideal I is closed under multiplication by elements of R, it is easy to see that $I = R$ if and only if $1_R \in I$.

Example 7.33. For each $m \in \mathbb{Z}$, the set $m\mathbb{Z}$ is not only an additive subgroup of the ring $\mathbb{Z}$, it is also an ideal of this ring. □

Example 7.34. For each $m \in \mathbb{Z}$, the set $m\mathbb{Z}_n$ is not only an additive subgroup of the ring $\mathbb{Z}_n$, it is also an ideal of this ring. □

Example 7.35. In the previous two examples, we saw that for some rings, the notion of an additive subgroup coincides with that of an ideal. Of course, that is the exception, not the rule. Consider the ring of polynomials $R[X]$. Suppose g is a non-zero polynomial in $R[X]$. The additive subgroup generated by g contains only polynomials whose degrees are at most that of g. However, this subgroup is not an ideal, since every ideal containing g must also contain $g \cdot X^i$ for all $i \geq 0$, and must therefore contain polynomials of arbitrarily high degree. $\square$

Example 7.36. Let R be a ring and $x \in R$. Consider the set

$$I := \{g \in R[X] : g(x) = 0\}.$$

It is not hard to see that I is an ideal of $R[X]$. Indeed, for all $g, h \in I$ and $q \in R[X]$, we have

$$(g + h)(x) = g(x) + h(x) = 0 + 0 = 0 \quad \text{and} \quad (gq)(x) = g(x)q(x) = 0 \cdot q(x) = 0.$$

Moreover, by Theorem 7.12, we have $I = \{(X - x)q : q \in R[X]\}$. $\square$

We next develop some general constructions of ideals.

Theorem 7.17. *Let R be a ring and let $a \in R$. Then $aR := \{ar : r \in R\}$ is an ideal of R.*

Proof. This is an easy calculation. For all $ar, ar' \in aR$ and $r'' \in R$, we have $ar + ar' = a(r + r') \in aR$ and $(ar)r'' = a(rr'') \in aR$. $\square$

The ideal aR in the previous theorem is called the **ideal of R generated by** a. An ideal of this form is called a **principal ideal**. Since R is commutative, one could also write this ideal as $Ra := \{ra : r \in R\}$. This ideal is the smallest ideal of R containing a; that is, aR contains a, and every ideal of R that contains a must contain everything in aR.

Corresponding to Theorems 6.11 and 6.12, we have:

Theorem 7.18. *If I_1 and I_2 are ideals of a ring R, then so are $I_1 + I_2$ and $I_1 \cap I_2$.*

Proof. We already know that $I_1 + I_2$ and $I_1 \cap I_2$ are additive subgroups of R, so it suffices to show that they are closed under multiplication by elements of R. The reader may easily verify that this is the case. $\square$

Let $a_1, \ldots, a_k$ be elements of a ring R. The ideal $a_1 R + \cdots + a_k R$ is called the **ideal of R generated by** $a_1, \ldots, a_k$. When the ring R is clear from context, one often writes $(a_1, \ldots, a_k)$ to denote this ideal. This ideal is that smallest ideal of R containing $a_1, \ldots, a_k$.

Example 7.37. Let n be a positive integer, and let x be any integer. Define $I := \{g \in \mathbb{Z}[X] : g(x) \equiv 0 \pmod{n}\}$. We claim that I is the ideal $(X - x, n)$ of $\mathbb{Z}[X]$. To see this, consider any fixed $g \in \mathbb{Z}[X]$. Using Theorem 7.12, we have $g = (X - x)q + g(x)$ for some $q \in \mathbb{Z}[X]$. Using the division with remainder property for integers, we have $g(x) = nq' + r$ for some $r \in \{0, \ldots, n - 1\}$ and $q' \in \mathbb{Z}$. Thus, $g(x) \equiv r \pmod{n}$, and if $g(x) \equiv 0 \pmod{n}$, then we must have

$r = 0$, and hence $g = (X - x)q + nq' \in (X - x, n)$. Conversely, if $g \in (X - x, n)$, we can write $g = (X - x)q + nq'$ for some $q, q' \in \mathbb{Z}[X]$, and from this, it is clear that $g(x) = nq'(x) \equiv 0 \pmod{n}$. □

Let I be an ideal of a ring R. Since I is an additive subgroup of R, we may adopt the congruence notation in §6.3, writing $a \equiv b \pmod{I}$ to mean $a - b \in I$, and we can form the additive quotient group R/I of cosets. Recall that for $a \in R$, the coset of I containing a is denoted $[a]_I$, and that $[a]_I = a + I = \{a + x : x \in I\}$. Also recall that addition in R/I was defined in terms of addition of coset representatives; that is, for $a, b \in I$, we defined

$$[a]_I + [b]_I := [a + b]_I.$$

Theorem 6.16 ensured that this definition was unambiguous.

Our goal now is to make R/I into a ring by similarly defining multiplication in R/I in terms of multiplication of coset representatives. To do this, we need the following multiplicative analog of Theorem 6.16, which exploits in an essential way the fact that an ideal is closed under multiplication by elements of R; in fact, this is one of the main motivations for defining the notion of an ideal as we did.

Theorem 7.19. *Suppose I is an ideal of a ring R. For all $a, a', b, b' \in R$, if $a \equiv a' \pmod{I}$ and $b \equiv b' \pmod{I}$, then $ab \equiv a'b' \pmod{I}$.*

Proof. If $a = a' + x$ for some $x \in I$ and $b = b' + y$ for some $y \in I$, then $ab = a'b' + a'y + b'x + xy$. Since I is closed under multiplication by elements of R, we see that $a'y, b'x, xy \in I$, and since I is closed under addition, $a'y + b'x + xy \in I$. Hence, $ab - a'b' \in I$. □

Using this theorem we can now unambiguously define multiplication on R/I as follows: for $a, b \in R$,

$$[a]_I \cdot [b]_I := [ab]_I.$$

Once that is done, it is straightforward to verify that all the properties that make R a ring are inherited by R/I — we leave the details of this to the reader. The multiplicative identity of R/I is the coset $[1_R]_I$.

The ring R/I is called the **quotient ring** or **residue class ring of R modulo I**. Elements of R/I may be called **residue classes**.

Note that if $I = dR$, then $a \equiv b \pmod{I}$ if and only if $d \mid (a - b)$, and as a matter of notation, one may simply write this congruence as $a \equiv b \pmod{d}$. We may also write $[a]_d$ instead of $[a]_I$.

Finally, note that if $I = R$, then R/I is the trivial ring.

Example 7.38. For each $n \geq 1$, the ring $\mathbb{Z}_n$ is precisely the quotient ring $\mathbb{Z}/n\mathbb{Z}$. □

Example 7.39. Let f be a polynomial over a ring R with $\deg(f) = \ell \geq 0$ and $\mathrm{lc}(f) \in R^*$, and consider the quotient ring $E := R[X]/fR[X]$. By the division with remainder property for polynomials (Theorem 7.10), for every $g \in R[X]$, there exists a unique polynomial $h \in R[X]$ such that $g \equiv h \pmod{f}$ and $\deg(h) < \ell$. From this, it follows that every element of E can be written uniquely as $[h]_f$, where $h \in R[X]$ is a polynomial of degree less than ℓ. Note that in this situation, we will generally prefer the more compact notation $R[X]/(f)$, instead of $R[X]/fR[X]$. $\square$

Example 7.40. Consider the polynomial $f := X^2 + X + 1 \in \mathbb{Z}_2[X]$ and the quotient ring $E := \mathbb{Z}_2[X]/(f)$. Let us name the elements of E as follows:

$$00 := [0]_f, \quad 01 := [1]_f, \quad 10 := [X]_f, \quad 11 := [X + 1]_f.$$

With this naming convention, addition of two elements in E corresponds to just computing the bit-wise exclusive-or of their names. More precisely, the addition table for E is the following:

+	00	01	10	11
00	00	01	10	11
01	01	00	11	10
10	10	11	00	01
11	11	10	01	00

Note that 00 acts as the additive identity for E, and that as an additive group, E is isomorphic to the additive group $\mathbb{Z}_2 \times \mathbb{Z}_2$.

As for multiplication in E, one has to compute the product of two polynomials, and then reduce modulo f. For example, to compute $10 \cdot 11$, using the identity $X^2 \equiv X + 1 \pmod{f}$, one sees that

$$X \cdot (X + 1) \equiv X^2 + X \equiv (X + 1) + X \equiv 1 \pmod{f};$$

thus, $10 \cdot 11 = 01$. The reader may verify the following multiplication table for E:

$\cdot$	00	01	10	11
00	00	00	00	00
01	00	01	10	11
10	00	10	11	01
11	00	11	01	10

Observe that 01 acts as the multiplicative identity for E. Notice that every non-zero element of E has a multiplicative inverse, and so E is in fact a field. Observe that E^* is cyclic: the reader may verify that both 10 and 11 have multiplicative order 3.

This is the first example we have seen of a finite field whose cardinality is not prime. $\square$

EXERCISE 7.25. Show that if F is a field, then the only ideals of F are $\{0_F\}$ and F.

EXERCISE 7.26. Let a, b be elements of a ring R. Show that

$$a \mid b \iff b \in aR \iff bR \subseteq aR.$$

EXERCISE 7.27. Let R be a ring. Show that if I is a non-empty subset of $R[X]$ that is closed under addition, multiplication by elements of R, and multiplication by X, then I is an ideal of $R[X]$.

EXERCISE 7.28. Let I be an ideal of R, and S a subring of R. Show that $I \cap S$ is an ideal of S.

EXERCISE 7.29. Let I be an ideal of R, and S a subring of R. Show that $I + S$ is a subring of R, and that I is an ideal of $I + S$.

EXERCISE 7.30. Let I_1 be an ideal of R_1, and I_2 an ideal of R_2. Show that $I_1 \times I_2$ is an ideal of $R_1 \times R_2$.

EXERCISE 7.31. Write down the multiplication table for $\mathbb{Z}_2[X]/(X^2 + X)$. Is this a field?

EXERCISE 7.32. Let I be an ideal of a ring R, and let x and y be elements of R with $x \equiv y \pmod{I}$. Let $g \in R[X]$. Show that $g(x) \equiv g(y) \pmod{I}$.

EXERCISE 7.33. Let R be a ring, and fix $x_1, \ldots, x_n \in R$. Let

$$I := \{g \in R[X_1, \ldots, X_n] : g(x_1, \ldots, x_n) = 0\}.$$

Show that I is an ideal of $R[X_1, \ldots, X_n]$, and that $I = (X_1 - x_1, \ldots, X_n - x_n)$.

EXERCISE 7.34. Let p be a prime, and consider the ring $\mathbb{Q}^{(p)}$ (see Example 7.26). Show that every non-zero ideal of $\mathbb{Q}^{(p)}$ is of the form (p^i), for some uniquely determined integer $i \geq 0$.

EXERCISE 7.35. Let p be a prime. Show that in the ring $\mathbb{Z}[X]$, the ideal (X, p) is not a principal ideal.

EXERCISE 7.36. Let F be a field. Show that in the ring $F[X, Y]$, the ideal (X, Y) is not a principal ideal.

EXERCISE 7.37. Let R be a ring, and let $\{I_i\}_{i=0}^{\infty}$ be a sequence of ideals of R such that $I_i \subseteq I_{i+1}$ for all $i = 0, 1, 2, \ldots$. Show that the union $\bigcup_{i=0}^{\infty} I_i$ is also an ideal of R.

EXERCISE 7.38. Let R be a ring. An ideal I of R is called **prime** if $I \subsetneq R$ and if for all $a, b \in R$, $ab \in I$ implies $a \in I$ or $b \in I$. An ideal I of R is called **maximal** if $I \subsetneq R$ and there are no ideals J of R such that $I \subsetneq J \subsetneq R$. Show that:

(a) an ideal I of R is prime if and only if R/I is an integral domain;

(b) an ideal I of R is maximal if and only if R/I is a field;

(c) all maximal ideals of R are also prime ideals.

EXERCISE 7.39. This exercise explores some examples of prime and maximal ideals. Show that:

(a) in the ring $\mathbb{Z}$, the ideal $\{0\}$ is prime but not maximal, and that the maximal ideals are precisely those of the form $p\mathbb{Z}$, where p is prime;

(b) in an integral domain D, the ideal $\{0\}$ is prime, and this ideal is maximal if and only if D is a field;

(c) if p is a prime, then in the ring $\mathbb{Z}[X]$, the ideal (X, p) is maximal, while the ideals (X) and (p) are prime, but not maximal;

(d) if F is a field, then in the ring $F[X, Y]$, the ideal (X, Y) is maximal, while the ideals (X) and (Y) are prime, but not maximal.

EXERCISE 7.40. It is a fact that every non-trivial ring R contain at least one maximal ideal. Showing this in general requires some fancy set-theoretic notions. This exercise develops a simple proof in the case where R is countable (see §A3).

(a) Show that if R is non-trivial but finite, then it contains a maximal ideal.

(b) Assume that R is countably infinite, and let $a_1, a_2, a_3, \ldots$ be an enumeration of the elements of R. Define a sequence of ideals $I_0, I_1, I_2, \ldots$, as follows. Set $I_0 := \{0_R\}$, and for each $i \geq 0$, define

$$I_{i+1} := \begin{cases} I_i + a_i R & \text{if } I_i + a_i R \subsetneq R; \\ I_i & \text{otherwise.} \end{cases}$$

Finally, set $I := \bigcup_{i=0}^{\infty} I_i$, which by Exercise 7.37 is an ideal of R. Show that I is a maximal ideal of R. Hint: first, show that $I \subsetneq R$ by assuming that $1_R \in I$ and deriving a contradiction; then, show that I is maximal by assuming that for some $i = 1, 2, \ldots$, we have $I \subsetneq I + a_i R \subsetneq R$, and deriving a contradiction.

EXERCISE 7.41. Let R be a ring, and let I and J be ideals of R. With the ring-theoretic product as defined in Exercise 7.2, show that:

(a) IJ is an ideal;

(b) if I and J are principal ideals, with $I = aR$ and $J = bR$, then $IJ = abR$, and so is also a principal ideal;

(c) $IJ \subseteq I \cap J$;

(d) if $I + J = R$, then $IJ = I \cap J$.

EXERCISE 7.42. Let R be a subring of E, and I an ideal of R. Show that the ring-theoretic product IE is an ideal of E that contains I, and is the smallest such ideal.

EXERCISE 7.43. Let M be a maximal ideal of a ring R, and let $a, b \in R$. Show that if $ab \in M^2$ and $b \notin M$, then $a \in M^2$. Here, $M^2 := MM$, the ring-theoretic product.

EXERCISE 7.44. Let F be a field, let $f \in F[X, Y]$, and let $E := F[X, Y]/(f)$. Define $V(f) := \{(x, y) \in F \times F : f(x, y) = 0\}$.

(a) Every element α of E naturally defines a function from $V(f)$ to F, as follows: if $\alpha = [g]_f$, with $g \in F[X, Y]$, then for $P = (x, y) \in V(f)$, we define $\alpha(P) := g(x, y)$. Show that this definition is unambiguous, that is, $g \equiv h \pmod{f}$ implies $g(x, y) = h(x, y)$.

(b) For $P = (x, y) \in V(f)$, define $M_P := \{\alpha \in E : \alpha(P) = 0\}$. Show that M_P is a maximal ideal of E, and that $M_P = \mu E + \nu E$, where $\mu := [X - x]_f$ and $\nu := [Y - y]_f$.

EXERCISE 7.45. Continuing with the previous exercise, now assume that the characteristic of F is not 2, and that $f = Y^2 - \phi$, where $\phi \in F[X]$ is a non-zero polynomial with no multiple roots in F (see definitions after Exercise 7.18).

(a) Show that if $P = (x, y) \in V(f)$, then so is $\bar{P} := (x, -y)$, and that $P = \bar{P} \iff y = 0 \iff \phi(x) = 0$.

(b) Let $P = (x, y) \in V(f)$ and $\mu := [X - x]_f \in E$. Show that $\mu E = M_P M_{\bar{P}}$ (the ring-theoretic product). Hint: use Exercise 7.43, and treat the cases $P = \bar{P}$ and $P \neq \bar{P}$ separately.

EXERCISE 7.46. Let R be a ring, and I an ideal of R. Define $\text{Rad}(I)$ to be the set of all $a \in R$ such that $a^n \in I$ for some positive integer n.

(a) Show that $\text{Rad}(I)$ is an ideal of R containing I. Hint: show that if $a^n \in I$ and $b^m \in I$, then $(a + b)^{n+m} \in I$.

(b) Show that if $R = \mathbb{Z}$ and $I = (d)$, where $d = p_1^{e_1} \cdots p_r^{e_r}$ is the prime factorization of d, then $\text{Rad}(I) = (p_1 \cdots p_r)$.

7.4 Ring homomorphisms and isomorphisms

Definition 7.20. *A function ρ from a ring R to a ring R' is called a* **ring homomorphism** *if*

    (i) *ρ is a group homomorphism with respect to the underlying additive groups of R and R',*

    (ii) *$\rho(ab) = \rho(a)\rho(b)$ for all $a, b \in R$, and*

    (iii) *$\rho(1_R) = 1_{R'}$.*

Expanding the definition, the requirements that ρ must satisfy in order to be a ring homomorphism are that for all $a, b \in R$, we have $\rho(a + b) = \rho(a) + \rho(b)$ and $\rho(ab) = \rho(a)\rho(b)$, and that $\rho(1_R) = 1_{R'}$.

Note that some texts do not require that a ring homomorphism satisfies part (iii) of our definition (which is not redundant — see Examples 7.49 and 7.50 below). Since a ring homomorphism is also an additive group homomorphism, we use the same notation and terminology for image and kernel.

Example 7.41. If S is a subring of a ring R, then the inclusion map $i : S \to R$ is obviously a ring homomorphism. □

Example 7.42. Suppose I is an ideal of a ring R. Analogous to Example 6.36, we may define the **natural map** from the ring R to the quotient ring R/I as follows:

$$\rho : \quad R \to R/I$$
$$a \mapsto [a]_I.$$

Not only is this a surjective homomorphism of additive groups, with kernel I, it is a *ring* homomorphism. Indeed, we have

$$\rho(ab) = [ab]_I = [a]_I \cdot [b]_I = \rho(a) \cdot \rho(b),$$

and $\rho(1_R) = [1_R]_I$, which is the multiplicative identity in R/I. □

Example 7.43. For a given positive integer n, the natural map from $\mathbb{Z}$ to $\mathbb{Z}_n$ sends $a \in \mathbb{Z}$ to the residue class $[a]_n$. This is a surjective ring homomorphism, whose kernel is $n\mathbb{Z}$. □

Example 7.44. Let R be a subring of a ring E, and fix $\alpha \in E$. The **polynomial evaluation map**

$$\rho : \quad R[X] \to E$$
$$g \mapsto g(\alpha)$$

is a ring homomorphism (see Theorem 7.11). The image of ρ consists of all polynomial expressions in α with coefficients in R, and is denoted $R[\alpha]$. As the reader may verify, $R[\alpha]$ is a subring of E containing α and all of R, and is the smallest such subring of E. $\square$

Example 7.45. We can generalize the previous example to multi-variate polynomials. If R is a subring of a ring E and $\alpha_1, \ldots, \alpha_n \in E$, then the map

$$\rho: \quad R[X_1, \ldots, X_n] \to E$$
$$g \mapsto g(\alpha_1, \ldots, \alpha_n)$$

is a ring homomorphism. Its image consists of all polynomial expressions in $\alpha_1, \ldots, \alpha_n$ with coefficients in R, and is denoted $R[\alpha_1, \ldots, \alpha_n]$. Moreover, this image is a subring of E containing $\alpha_1, \ldots, \alpha_n$ and all of R, and is the smallest such subring of E. Note that $R[\alpha_1, \ldots, \alpha_n] = R[\alpha_1, \ldots, \alpha_{n-1}][\alpha_n]$. $\square$

Example 7.46. Let $\rho : R \to R'$ be a ring homomorphism. We can extend the domain of definition of ρ from R to $R[X]$ by defining $\rho(\sum_i a_i X^i) := \sum_i \rho(a_i) X^i$. This yields a ring homomorphism from $R[X]$ into $R'[X]$. To verify this, suppose $g = \sum_i a_i X^i$ and $h = \sum_i b_i X^i$ are polynomials in $R[X]$. Let $s := g + h \in R[X]$ and $p := gh \in R[X]$, and write $s = \sum_i s_i X^i$ and $p = \sum_i p_i X^i$, so that

$$s_i = a_i + b_i \quad \text{and} \quad p_i = \sum_{i=j+k} a_j b_k.$$

Then we have

$$\rho(s_i) = \rho(a_i + b_i) = \rho(a_i) + \rho(b_i),$$

which is the coefficient of X^i in $\rho(g) + \rho(h)$, and

$$\rho(p_i) = \rho\left(\sum_{i=j+k} a_j b_k \right) = \sum_{i=j+k} \rho(a_j b_k) = \sum_{i=j+k} \rho(a_j)\rho(b_k),$$

which is the coefficient of X^i in $\rho(g)\rho(h)$.

Sometimes a more compact notation is convenient: we may prefer to write $\bar{a}$ for the image of $a \in R$ under ρ, and if we do this, then for $g = \sum_i a_i X^i \in R[X]$, we write $\bar{g}$ for the image $\sum_i \bar{a}_i X^i$ of g under the extension of ρ to $R[X]$. $\square$

Example 7.47. Consider the natural map that sends $a \in \mathbb{Z}$ to $\bar{a} := [a]_n \in \mathbb{Z}_n$ (see Example 7.43). As in the previous example, we may extend this to a ring homomorphism from $\mathbb{Z}[X]$ to $\mathbb{Z}_n[X]$ that sends $g = \sum_i a_i X^i \in \mathbb{Z}[X]$ to $\bar{g} = \sum_i \bar{a}_i X^i \in \mathbb{Z}_n[X]$. This homomorphism is clearly surjective. Let us determine its kernel. Observe that if $g = \sum_i a_i X^i$, then $\bar{g} = 0$ if and only if $n \mid a_i$ for each i; therefore, the kernel is the ideal $n\mathbb{Z}[X]$ of $\mathbb{Z}[X]$. $\square$

Example 7.48. Let R be a ring of prime characteristic p. For all $a, b \in R$, we have (see Exercise 7.1)

$$(a + b)^p = \sum_{k=0}^{p} \binom{p}{k} a^{p-k} b^k.$$

However, by Exercise 1.14, all of the binomial coefficients are multiples of p, except for $k = 0$ and $k = p$, and hence in the ring R, all of these terms vanish, leaving us with

$$(a + b)^p = a^p + b^p.$$

This result is often jokingly referred to as the "freshman's dream," for somewhat obvious reasons.

Of course, as always, we have

$$(ab)^p = a^p b^p \quad \text{and} \quad 1_R^p = 1_R,$$

and so it follows that the map that sends $a \in R$ to $a^p \in R$ is a ring homomorphism from R into R. □

Example 7.49. Suppose R is a non-trivial ring, and let $\rho : R \to R$ map everything in R to 0_R. Then ρ satisfies parts (i) and (ii) of Definition 7.20, but not part (iii). □

Example 7.50. In special situations, part (iii) of Definition 7.20 may be redundant. One such situation arises when $\rho : R \to R'$ is surjective. In this case, we know that $1_{R'} = \rho(a)$ for some $a \in R$, and by part (ii) of the definition, we have

$$\rho(1_R) = \rho(1_R) \cdot 1_{R'} = \rho(1_R)\rho(a) = \rho(1_R \cdot a) = \rho(a) = 1_{R'}. \quad \square$$

For a ring homomorphism $\rho : R \to R'$, all of the results of Theorem 6.19 apply. In particular, $\rho(0_R) = 0_{R'}$, $\rho(a) = \rho(b)$ if and only if $a \equiv b \pmod{\operatorname{Ker} \rho}$, and ρ is injective if and only if $\operatorname{Ker} \rho = \{0_R\}$. However, we may strengthen Theorem 6.19 as follows:

Theorem 7.21. *Let $\rho : R \to R'$ be a ring homomorphism.*

(i) *If S is a subring of R, then $\rho(S)$ is a subring of R'; in particular (setting $S := R$), $\operatorname{Im} \rho$ is a subring of R'.*

(ii) *If S' is a subring of R', then $\rho^{-1}(S')$ is a subring of R.*

(ii) *If I is an ideal of R, then $\rho(I)$ is an ideal of $\operatorname{Im} \rho$.*

(iv) *If I' is an ideal of $\operatorname{Im} \rho$, then $\rho^{-1}(I')$ is an ideal of R; in particular (setting $I' := \{0_{R'}\}$), $\operatorname{Ker} \rho$ is an ideal of R.*

Proof. In each part, we already know that the relevant object is an additive subgroup, and so it suffices to show that the appropriate additional properties are satisfied.

(i) For all $a, b \in S$, we have $ab \in S$, and hence $\rho(S)$ contains $\rho(ab) = \rho(a)\rho(b)$. Also, $1_R \in S$, and hence $\rho(S)$ contains $\rho(1_R) = 1_{R'}$.

(ii) If $\rho(a) \in S'$ and $\rho(b) \in S'$, then $\rho(ab) = \rho(a)\rho(b) \in S'$. Moreover, $\rho(1_R) = 1_{R'} \in S'$.

(iii) For all $a \in I$ and $r \in R$, we have $ar \in I$, and hence $\rho(I)$ contains $\rho(ar) = \rho(a)\rho(r)$.

(iv) For all $a \in \rho^{-1}(I')$ and $r \in R$, we have $\rho(ar) = \rho(a)\rho(r)$, and since $\rho(a)$ belongs to the ideal I', so does $\rho(a)\rho(r)$, and hence $\rho^{-1}(I')$ contains ar. $\square$

Theorems 6.20 and 6.21 have natural ring analogs—one only has to show that the corresponding group homomorphisms satisfy the additional requirements of a ring homomorphism, which we leave to the reader to verify:

Theorem 7.22. *If $\rho : R \to R'$ and $\rho' : R' \to R''$ are ring homomorphisms, then so is their composition $\rho' \circ \rho : R \to R''$.*

Theorem 7.23. *Let $\rho_i : R \to R_i'$, for $i = 1, \ldots, k$, be ring homomorphisms. Then the map*

$$\rho : \quad R \to R_1' \times \cdots \times R_k'$$
$$a \mapsto (\rho_1(a), \ldots, \rho_k(a))$$

is a ring homomorphism.

If a ring homomorphism $\rho : R \to R'$ is a bijection, then it is called a **ring isomorphism** of R with R'. If such a ring isomorphism ρ exists, we say that R **is isomorphic to** R', and write $R \cong R'$. Moreover, if $R = R'$, then ρ is called a **ring automorphism** on R.

Analogous to Theorem 6.22, we have:

Theorem 7.24. *If ρ is a ring isomorphism of R with R', then the inverse function ρ^{-1} is a ring isomorphism of R' with R.*

Proof. Exercise. $\square$

Because of this theorem, if R is isomorphic to R', we may simply say that "R and R' are isomorphic." We stress that a ring isomorphism is essentially just a "renaming" of elements; in particular, we have:

Theorem 7.25. *Let $\rho : R \to R'$ be a ring isomorphism.*

(i) *For all $a \in R$, a is a zero divisor if and only if $\rho(a)$ is a zero divisor.*

(ii) *For all $a \in R$, a is a unit if and only if $\rho(a)$ is a unit.*

(iii) *The restriction of R to R^* is a group isomorphism of R^* with $(R')^*$.*

Proof. Exercise. □

An injective ring homomorphism $\rho : R \to E$ is called an **embedding** of R in E. In this case, Im ρ is a subring of E and $R \cong$ Im ρ. If the embedding is a natural one that is clear from context, we may simply identify elements of R with their images in E under the embedding; that is, for $a \in R$, we may simply write "a," and it is understood that this really means "$\rho(a)$" if the context demands an element of E. As a slight abuse of terminology, we shall say that R is a subring of E. Indeed, by appropriately renaming elements, we can always make R a subring of E in the literal sense of the term.

This practice of identifying elements of a ring with their images in another ring under a natural embedding is very common. We have already seen an example of this, namely, when we formally defined the ring of polynomials $R[X]$ over R in §7.2.1, we defined the map $\varepsilon_0 : R \to R[X]$ that sends $c \in R$ to the polynomial whose constant term is c, with all other coefficients zero. This map ε_0 is an embedding, and it was via this embedding that we identified elements of R with elements of $R[X]$, and so viewed R as a subring of $R[X]$. We shall see more examples of this later (in particular, Example 7.55 below).

Theorems 6.23 and 6.24 also have natural ring analogs—again, one only has to show that the corresponding group homomorphisms are also ring homomorphisms:

Theorem 7.26 (First isomorphism theorem). *Let $\rho : R \to R'$ be a ring homomorphism with kernel K and image S'. Then we have a ring isomorphism*

$$R/K \cong S'.$$

Specifically, the map

$$\bar{\rho} : \quad R/K \to R'$$
$$[a]_K \mapsto \rho(a)$$

is an injective ring homomorphism whose image is S'.

Theorem 7.27. *Let $\rho : R \to R'$ be a ring homomorphism. Then for every ideal I of R with $I \subseteq$ Ker ρ, we may define a ring homomorphism*

$$\bar{\rho} : \quad R/I \to R'$$
$$[a]_I \mapsto \rho(a).$$

Moreover, Im $\bar{\rho} =$ Im ρ, and $\bar{\rho}$ is injective if and only if $I =$ Ker ρ.

Example 7.51. Returning again to the Chinese remainder theorem and the discussion in Example 6.48, if $\{n_i\}_{i=1}^{k}$ is a pairwise relatively prime family of positive integers, and $n := \prod_{i=1}^{k} n_i$, then the map

$$\rho : \quad \mathbb{Z} \to \mathbb{Z}_{n_1} \times \cdots \times \mathbb{Z}_{n_k}$$
$$a \mapsto ([a]_{n_1}, \ldots, [a]_{n_k})$$

is not just a surjective group homomorphism with kernel $n\mathbb{Z}$, it is also a *ring* homomorphism. Applying Theorem 7.26, we get a *ring* isomorphism

$$\bar{\rho} : \quad \mathbb{Z}_n \to \mathbb{Z}_{n_1} \times \cdots \times \mathbb{Z}_{n_k}$$
$$[a]_n \mapsto ([a]_{n_1}, \ldots, [a]_{n_k}),$$

which is the same function as the function θ in Theorem 2.8. By part (iii) of Theorem 7.25, the restriction of θ to $\mathbb{Z}_n^*$ is a group isomorphism of $\mathbb{Z}_n^*$ with the multiplicative group of units of $\mathbb{Z}_{n_1} \times \cdots \times \mathbb{Z}_{n_k}$, which (according to Example 7.15) is $\mathbb{Z}_{n_1}^* \times \cdots \times \mathbb{Z}_{n_k}^*$. Thus, part (iii) of Theorem 2.8 is an immediate consequence of the above observations. □

Example 7.52. Extending Example 6.49, if n_1 and n_2 are positive integers with $n_1 \mid n_2$, then the map

$$\bar{\rho} : \quad \mathbb{Z}_{n_2} \to \mathbb{Z}_{n_1}$$
$$[a]_{n_2} \mapsto [a]_{n_1}$$

is a surjective *ring* homomorphism. □

Example 7.53. For a ring R, consider the map $\rho : \mathbb{Z} \to R$ that sends $m \in \mathbb{Z}$ to $m \cdot 1_R$ in R. It is easily verified that ρ is a ring homomorphism. Since Ker ρ is an ideal of $\mathbb{Z}$, it is either $\{0\}$ or of the form $n\mathbb{Z}$ for some $n > 0$. In the first case, if Ker $\rho = \{0\}$, then Im $\rho \cong \mathbb{Z}$, and so the ring $\mathbb{Z}$ is embedded in R, and R has characteristic zero. In the second case, if Ker $\rho = n\mathbb{Z}$ for some $n > 0$, then by Theorem 7.26, Im $\rho \cong \mathbb{Z}_n$, and so the ring $\mathbb{Z}_n$ is embedded in R, and R has characteristic n.

Note that Im ρ is the smallest subring of R: any subring of R must contain 1_R and be closed under addition and subtraction, and so must contain Im ρ. □

Example 7.54. We can generalize Example 7.44 by evaluating polynomials at several points. This is most fruitful when the underlying coefficient ring is a field, and the evaluation points belong to the same field. So let F be a field, and let $x_1, \ldots, x_k$ be distinct elements of F. Define the map

$$\rho : \quad F[X] \to F^{\times k}$$
$$g \mapsto (g(x_1), \ldots, g(x_k)).$$

This is a ring homomorphism (as seen by applying Theorem 7.23 to the polynomial evaluation maps at the points $x_1, \ldots, x_k$). By Theorem 7.13, Ker $\rho = (f)$, where

$f := \prod_{i=1}^{k}(X - x_i)$. By Theorem 7.15, ρ is surjective. Therefore, by Theorem 7.26, we get a ring isomorphism

$$\bar{\rho}: \quad F[X]/(f) \to F^{\times k}$$
$$[g]_f \mapsto (g(x_1), \dots, g(x_k)). \quad \square$$

Example 7.55. As in Example 7.39, let f be a polynomial over a ring R with $\deg(f) = \ell$ and $\mathrm{lc}(f) \in R^*$, but now assume that $\ell > 0$. Consider the natural map ρ from $R[X]$ to the quotient ring $E := R[X]/(f)$ that sends $g \in R[X]$ to $[g]_f$. Let τ be the restriction of ρ to the subring R of $R[X]$. Evidently, τ is a ring homomorphism from R into E. Moreover, since distinct polynomials of degree less than ℓ belong to distinct residue classes modulo f, we see that τ is injective. Thus, τ is an embedding of R into E. As τ is a very natural embedding, we can identify elements of R with their images in E under τ, and regard R as a subring of E. Taking this point of view, we see that if $g = \sum_i a_i X^i$, then

$$[g]_f = \left[\sum_i a_i X^i\right]_f = \sum_i [a_i]_f ([X]_f)^i = \sum_i a_i \xi^i = g(\xi),$$

where $\xi := [X]_f \in E$. Therefore, the natural map ρ may be viewed as the polynomial evaluation map (see Example 7.44) that sends $g \in R[X]$ to $g(\xi) \in E$.

Note that we have $E = R[\xi]$; moreover, every element of E can be expressed uniquely as $g(\xi)$ for some $g \in R[X]$ of degree less than ℓ, and more generally, for arbitrary $g, h \in R[X]$, we have $g(\xi) = h(\xi)$ if and only if $g \equiv h \pmod{f}$. Finally, note that $f(\xi) = [f]_f = [0]_f$; that is, ξ is a root of f. $\square$

Example 7.56. As a special case of Example 7.55, let $f := X^2 + 1 \in \mathbb{R}[X]$, and consider the quotient ring $\mathbb{R}[X]/(f)$. If we set $i := [X]_f \in \mathbb{R}[X]/(f)$, then every element of $\mathbb{R}[X]/(f)$ can be expressed uniquely as $a + bi$, where $a, b \in \mathbb{R}$. Moreover, we have $i^2 = -1$, and more generally, for all $a, b, a', b' \in \mathbb{R}$, we have

$$(a + bi) + (a' + b'i) = (a + a') + (b + b')i$$

and

$$(a + bi) \cdot (a' + b'i) = (aa' - bb') + (ab' + a'b)i.$$

Thus, the rules for arithmetic in $\mathbb{R}[X]/(f)$ are precisely the familiar rules of complex arithmetic, and so $\mathbb{C}$ and $\mathbb{R}[X]/(f)$ are essentially the same, as rings. Indeed, the "algebraically correct" way of defining the field of complex numbers $\mathbb{C}$ is simply to define it to be the quotient ring $\mathbb{R}[X]/(f)$ in the first place. This will be our point of view from now on. $\square$

Example 7.57. Consider the polynomial evaluation map

$$\rho: \quad \mathbb{R}[X] \to \mathbb{C} = R[X]/(X^2 + 1)$$
$$g \mapsto g(-i).$$

For every $g \in \mathbb{R}[X]$, we may write $g = (X^2 + 1)q + a + bX$, where $q \in \mathbb{R}[X]$ and $a, b \in \mathbb{R}$. Since $(-i)^2 + 1 = i^2 + 1 = 0$, we have

$$g(-i) = ((-i)^2 + 1)q(-i) + a - bi = a - bi.$$

Clearly, then, ρ is surjective and the kernel of ρ is the ideal of $\mathbb{R}[X]$ generated by the polynomial $X^2 + 1$. By Theorem 7.26, we therefore get a ring automorphism $\bar{\rho}$ on $\mathbb{C}$ that sends $a + bi \in \mathbb{C}$ to $a - bi$. In fact, $\bar{\rho}$ is none other than the complex conjugation map. Indeed, this is the "algebraically correct" way of defining complex conjugation in the first place. $\square$

Example 7.58. We defined the ring $\mathbb{Z}[i]$ of Gaussian integers in Example 7.25 as a subring of $\mathbb{C}$. Let us verify that the notation $\mathbb{Z}[i]$ introduced in Example 7.25 is consistent with that introduced in Example 7.44. Consider the polynomial evaluation map $\rho: \mathbb{Z}[X] \to \mathbb{C}$ that sends $g \in \mathbb{Z}[X]$ to $g(i) \in \mathbb{C}$. For every $g \in \mathbb{Z}[X]$, we may write $g = (X^2 + 1)q + a + bX$, where $q \in \mathbb{Z}[X]$ and $a, b \in \mathbb{Z}$. Since $i^2 + 1 = 0$, we have $g(i) = (i^2 + 1)q(i) + a + bi = a + bi$. Clearly, then, the image of ρ is the set $\{a + bi : a, b \in \mathbb{Z}\}$, and the kernel of ρ is the ideal of $\mathbb{Z}[X]$ generated by the polynomial $X^2 + 1$. This shows that $\mathbb{Z}[i]$ in Example 7.25 is the same as $\mathbb{Z}[i]$ in Example 7.44, and moreover, Theorem 7.26 implies that $\mathbb{Z}[i]$ is isomorphic to $\mathbb{Z}[X]/(X^2 + 1)$.

Therefore, we can directly construct the Gaussian integers as the quotient ring $\mathbb{Z}[X]/(X^2 + 1)$. Likewise the field $\mathbb{Q}[i]$ (see Exercise 7.14) can be constructed directly as $\mathbb{Q}[X]/(X^2 + 1)$. $\square$

Example 7.59. Let p be a prime, and consider the quotient ring $E := \mathbb{Z}_p[X]/(f)$, where $f := X^2 + 1$. If we set $i := [X]_f \in E$, then $E = \mathbb{Z}_p[i] = \{a + bi : a, b \in \mathbb{Z}_p\}$. In particular, E is a ring of cardinality p^2. Moreover, we have $i^2 = -1$, and the rules for addition and multiplication in E look exactly the same as they do in $\mathbb{C}$: for all $a, b, a', b' \in \mathbb{Z}_p$, we have

$$(a + bi) + (a' + b'i) = (a + a') + (b + b')i$$

and

$$(a + bi) \cdot (a' + b'i) = (aa' - bb') + (ab' + a'b)i.$$

The ring E may or may not be a field. We now determine for which primes p we get a field.

If $p = 2$, then $0 = 1 + i^2 = (1 + i)^2$ (see Example 7.48), and so in this case, $1 + i$ is a zero divisor and E is not a field.

Now suppose p is odd. There are two subcases to consider: $p \equiv 1 \pmod 4$ and $p \equiv 3 \pmod 4$.

Suppose $p \equiv 1 \pmod 4$. By Theorem 2.31, there exists $c \in \mathbb{Z}_p$ such that $c^2 = -1$, and therefore $f = X^2 + 1 = X^2 - c^2 = (X - c)(X + c)$, and by Example 7.45, we have a ring isomorphism $E \cong \mathbb{Z}_p \times \mathbb{Z}_p$ (which maps $a + bi \in E$ to $(a + bc, a - bc) \in \mathbb{Z}_p \times \mathbb{Z}_p$); in particular, E is not a field. Indeed, $c + i$ is a zero divisor, since $(c + i)(c - i) = c^2 - i^2 = c^2 + 1 = 0$.

Suppose $p \equiv 3 \pmod 4$. By Theorem 2.31, there is no $c \in \mathbb{Z}_p$ such that $c^2 = -1$. It follows that for all $a, b \in \mathbb{Z}_p$, not both zero, we must have $a^2 + b^2 \neq 0$; indeed, suppose that $a^2 + b^2 = 0$, and that, say, $b \neq 0$; then we would have $(a/b)^2 = -1$, contradicting the assumption that -1 has no square root in $\mathbb{Z}_p$. Therefore, $a^2 + b^2$ has a multiplicative inverse in $\mathbb{Z}_p$, from which it follows that the formula for multiplicative inverses in $\mathbb{C}$ applies equally well in E; that is,

$$(a + bi)^{-1} = \frac{a - bi}{a^2 + b^2}.$$

Therefore, in this case, E is a field. $\square$

In Example 7.40, we saw a finite field of cardinality 4. The previous example provides us with an explicit construction of a finite field of cardinality p^2, for every prime p congruent to 3 modulo 4. As the next example shows, there exist finite fields of cardinality p^2 for all primes p.

Example 7.60. Let p an odd prime, and let $d \in \mathbb{Z}_p^*$. Let $f := X^2 - d \in \mathbb{Z}_p[X]$, and consider the ring $E := \mathbb{Z}_p[X]/(f) = \mathbb{Z}_p[\xi]$, where $\xi := [X]_f \in E$. We have $E = \{a + b\xi : a, b \in \mathbb{Z}_p\}$ and $|E| = p^2$. Note that $\xi^2 = d$, and the general rules for arithmetic in E look like this: for all $a, b, a', b' \in \mathbb{Z}_p$, we have

$$(a + b\xi) + (a' + b'\xi) = (a + a') + (b + b')\xi$$

and

$$(a + b\xi) \cdot (a' + b'\xi) = (aa' + bb'd) + (ab' + a'b)\xi.$$

Suppose that $d \in (\mathbb{Z}_p^*)^2$, so that $d = c^2$ for some $c \in \mathbb{Z}_p^*$. Then $f = (X - c)(X + c)$, and like in previous example, we have a ring isomorphism $E \cong \mathbb{Z}_p \times \mathbb{Z}_p$ (which maps $a + b\xi \in E$ to $(a + bc, a - bc) \in \mathbb{Z}_p \times \mathbb{Z}_p$); in particular, E is not a field.

Suppose that $d \notin (\mathbb{Z}_p^*)^2$. This implies that for all $a, b \in \mathbb{Z}_p$, not both zero, we have $a^2 - b^2 d \neq 0$. Using this, we get the following formula for multiplicative inverses in E:

$$(a + b\xi)^{-1} = \frac{a - b\xi}{a^2 - b^2 d}.$$

Therefore, E is a field in this case.

By Theorem 2.20, we know that $|(\mathbb{Z}_p^*)^2| = (p-1)/2$, and hence there exists $d \in \mathbb{Z}_p^* \setminus (\mathbb{Z}_p^*)^2$ for all odd primes p. Thus, we have a general (though not explicit) construction for finite fields of cardinality p^2 for all odd primes p. $\square$

EXERCISE 7.47. Show that if $\rho : F \to R$ is a ring homomorphism from a field F into a ring R, then either R is trivial or ρ is injective. Hint: use Exercise 7.25.

EXERCISE 7.48. Verify that the "is isomorphic to" relation on rings is an equivalence relation; that is, for all rings R_1, R_2, R_3, we have:

   (a) $R_1 \cong R_1$;

   (b) $R_1 \cong R_2$ implies $R_2 \cong R_1$;

   (c) $R_1 \cong R_2$ and $R_2 \cong R_3$ implies $R_1 \cong R_3$.

EXERCISE 7.49. Let $\rho_i : R_i \to R_i'$, for $i = 1, \ldots, k$, be ring homomorphisms. Show that the map

$$\rho : \quad R_1 \times \cdots \times R_k \to R_1' \times \cdots \times R_k'$$
$$(a_1, \ldots, a_k) \mapsto (\rho_1(a_1), \ldots, \rho_k(a_k))$$

is a ring homomorphism.

EXERCISE 7.50. Let $\rho : R \to R'$ be a ring homomorphism, and let $a \in R$. Show that $\rho(aR) = \rho(a)\rho(R)$.

EXERCISE 7.51. Let $\rho : R \to R'$ be a ring homomorphism. Let S be a subring of R, and let $\tau : S \to R'$ be the restriction of ρ to S. Show that τ is a ring homomorphism and that $\operatorname{Ker} \tau = \operatorname{Ker} \rho \cap S$.

EXERCISE 7.52. Suppose $R_1, \ldots, R_k$ are rings. Show that for each $i = 1, \ldots, k$, the projection map $\pi_i : R_1 \times \cdots \times R_k \to R_i$ that sends $(a_1, \ldots, a_k)$ to a_i is a surjective ring homomorphism.

EXERCISE 7.53. Show that if $R = R_1 \times R_2$ for rings R_1 and R_2, and I_1 is an ideal of R_1 and I_2 is an ideal of R_2, then we have a ring isomorphism $R/(I_1 \times I_2) \cong R_1/I_1 \times R_2/I_2$.

EXERCISE 7.54. Let I be an ideal of R, and S a subring of R. As we saw in Exercises 7.28, and 7.29, $I \cap S$ is an ideal of S, and I is an ideal of the subring $I + S$. Show that we have a ring isomorphism $(I + S)/I \cong S/(I \cap S)$.

EXERCISE 7.55. Let $\rho : R \to R'$ be a ring homomorphism with kernel K. Let I be an ideal of R. Show that we have a ring isomorphism $R/(I + K) \cong \rho(R)/\rho(I)$.

EXERCISE 7.56. Let n be a positive integer, and consider the natural map that sends $a \in \mathbb{Z}$ to $\bar{a} := [a]_n \in \mathbb{Z}_n$, which we may extend coefficient-wise to a ring homomorphism from $\mathbb{Z}[X]$ to $\mathbb{Z}_n[X]$, as in Example 7.47. Show that for every $f \in \mathbb{Z}[X]$, we have a ring isomorphism $\mathbb{Z}[X]/(f, n) \cong \mathbb{Z}_n[X]/(\bar{f})$.

EXERCISE 7.57. Let n be a positive integer. Show that we have ring isomorphisms $\mathbb{Z}[X]/(n) \cong \mathbb{Z}_n[X]$, $\mathbb{Z}[X]/(X) \cong \mathbb{Z}$, and $\mathbb{Z}[X]/(X, n) \cong \mathbb{Z}_n$.

EXERCISE 7.58. Let $n = pq$, where p and q are distinct primes. Show that we have a ring isomorphism $\mathbb{Z}_n[X] \cong \mathbb{Z}_p[X] \times \mathbb{Z}_q[X]$.

EXERCISE 7.59. Let p be a prime with $p \equiv 1 \pmod{4}$. Show that we have a ring isomorphism $\mathbb{Z}[X]/(X^2 + 1, p) \cong \mathbb{Z}_p \times \mathbb{Z}_p$.

EXERCISE 7.60. Let $\rho : R \to R'$ be a surjective ring homomorphism. Let S be the set of all ideals of R that contain $\mathrm{Ker}\,\rho$, and let S' be the set of all ideals of R'. Show that the sets S and S' are in one-to-one correspondence, via the map that sends $I \in S$ to $\rho(I) \in S'$. Moreover, show that under this correspondence, prime ideals in S correspond to prime ideals in S', and maximal ideals in S correspond to maximal ideals in S'. (See Exercise 7.38.)

EXERCISE 7.61. Let n be a positive integer whose factorization into primes is $n = p_1^{e_1} \cdots p_r^{e_r}$. What are the prime ideals of $\mathbb{Z}_n$? (See Exercise 7.38.)

EXERCISE 7.62. Let $\rho : R \to S$ be a ring homomorphism. Show that $\rho(R^*) \subseteq S^*$, and that the restriction of ρ to R^* yields a group homomorphism $\rho^* : R^* \to S^*$.

EXERCISE 7.63. Let R be a ring, and let $x_1, \ldots, x_n$ be elements of R. Show that the rings R and $R[X_1, \ldots, X_n]/(X_1 - x_1, \ldots, X_n - x_n)$ are isomorphic.

EXERCISE 7.64. This exercise and the next generalize the Chinese remainder theorem to arbitrary rings. Suppose I and J are two ideals of a ring R such that $I + J = R$. Show that the map $\rho : R \to R/I \times R/J$ that sends $a \in R$ to $([a]_I, [a]_J)$ is a surjective ring homomorphism with kernel IJ (see Exercise 7.41). Conclude that $R/(IJ)$ is isomorphic to $R/I \times R/J$.

EXERCISE 7.65. Generalize the previous exercise, showing that $R/(I_1 \cdots I_k)$ is isomorphic to $R/I_1 \times \cdots \times R/I_k$, where R is a ring, and $I_1, \ldots, I_k$ are ideals of R, provided $I_i + I_j = R$ for all i, j such that $i \neq j$.

EXERCISE 7.66. Let $\mathbb{Q}^{(m)}$ be the subring of $\mathbb{Q}$ defined in Example 7.26. Let us define the map $\rho : \mathbb{Q}^{(m)} \to \mathbb{Z}_m$ as follows. For $a/b \in \mathbb{Q}$ with b relatively prime to m, $\rho(a/b) := [a]_m([b]_m)^{-1}$. Show that ρ is unambiguously defined, and is a surjective ring homomorphism. Also, describe the kernel of ρ.

EXERCISE 7.67. Let R be a ring, $a \in R^*$, and $b \in R$. Define the map $\rho : R[X] \rightarrow R[X]$ that sends $g \in R[X]$ to $g(aX + b)$. Show that ρ is a ring automorphism.

EXERCISE 7.68. Consider the subring $\mathbb{Z}[1/2]$ of $\mathbb{Q}$. Show that $\mathbb{Z}[1/2] = \{a/2^i : a, i \in \mathbb{Z}, i \geq 0\}$, that $(\mathbb{Z}[1/2])^* = \{2^i : i \in \mathbb{Z}\}$, and that every non-zero ideal of $\mathbb{Z}[1/2]$ is of the form (m), for some uniquely determined, *odd* integer m.

7.5 The structure of $\mathbb{Z}_n^*$

We are now in a position to precisely characterize the structure of the group $\mathbb{Z}_n^*$, for an arbitrary integer $n > 1$. This characterization will prove to be very useful in a number of applications.

Suppose $n = p_1^{e_1} \cdots p_r^{e_r}$ is the factorization of n into primes. By the Chinese remainder theorem (see Theorem 2.8 and Example 7.51), we have the ring isomorphism

$$\theta : \quad \mathbb{Z}_n \rightarrow \mathbb{Z}_{p_1^{e_1}} \times \cdots \times \mathbb{Z}_{p_r^{e_r}}$$

$$[a]_n \mapsto ([a]_{p_1^{e_1}}, \ldots, [a]_{p_r^{e_r}}),$$

and restricting θ to $\mathbb{Z}_n^*$ yields a group isomorphism

$$\mathbb{Z}_n^* \cong \mathbb{Z}_{p_1^{e_1}}^* \times \cdots \times \mathbb{Z}_{p_r^{e_r}}^*.$$

Thus, to determine the structure of the group $\mathbb{Z}_n^*$ for general n, it suffices to determine the structure for $n = p^e$, where p is prime. By Theorem 2.10, we already know the order of the group $\mathbb{Z}_{p^e}^*$, namely, $\varphi(p^e) = p^{e-1}(p-1)$, where φ is Euler's phi function.

The main result of this section is the following:

Theorem 7.28. *If p is an odd prime, then for every positive integer e, the group $\mathbb{Z}_{p^e}^*$ is cyclic. The group $\mathbb{Z}_{2^e}^*$ is cyclic for $e = 1$ or 2, but not for $e \geq 3$. For $e \geq 3$, $\mathbb{Z}_{2^e}^*$ is isomorphic to the additive group $\mathbb{Z}_2 \times \mathbb{Z}_{2^{e-2}}$.*

In the case where $e = 1$, this theorem is a special case of the following, more general, theorem:

Theorem 7.29. *Let D be an integral domain and G a subgroup of D^* of finite order. Then G is cyclic.*

Proof. Suppose G is not cyclic. If m is the exponent of G, then by Theorem 6.41, we know that $m < |G|$. Moreover, by definition, $a^m = 1$ for all $a \in G$; that is, every element of G is a root of the polynomial $X^m - 1 \in D[X]$. But by Theorem 7.14, a polynomial of degree m over an integral domain has at most m distinct roots, and this contradicts the fact that $m < |G|$. $\square$

This theorem immediately implies that $\mathbb{Z}_p^*$ is cyclic for every prime p, since $\mathbb{Z}_p$ is a field; however, we cannot directly use this theorem to prove that $\mathbb{Z}_{p^e}^*$ is cyclic for $e > 1$ (and p odd), because $\mathbb{Z}_{p^e}$ is not a field. To deal with the case $e > 1$, we need a few simple facts.

Lemma 7.30. *Let p be a prime. For every positive integer e, if $a \equiv b \pmod{p^e}$, then $a^p \equiv b^p \pmod{p^{e+1}}$.*

Proof. Suppose $a \equiv b \pmod{p^e}$, so that $a = b + cp^e$ for some $c \in \mathbb{Z}$. Then $a^p = b^p + pb^{p-1}cp^e + dp^{2e}$ for some $d \in \mathbb{Z}$, and it follows that $a^p \equiv b^p \pmod{p^{e+1}}$. $\square$

Lemma 7.31. *Let p be a prime, and let e be a positive integer such that $p^e > 2$. If $a \equiv 1 + p^e \pmod{p^{e+1}}$, then $a^p \equiv 1 + p^{e+1} \pmod{p^{e+2}}$.*

Proof. Suppose $a \equiv 1 + p^e \pmod{p^{e+1}}$. By Lemma 7.30, $a^p \equiv (1 + p^e)^p \pmod{p^{e+2}}$. Expanding $(1 + p^e)^p$, we have

$$(1 + p^e)^p = 1 + p \cdot p^e + \sum_{k=2}^{p-1} \binom{p}{k} p^{ek} + p^{ep}.$$

By Exercise 1.14, all of the terms in the sum on k are divisible by p^{1+2e}, and $1 + 2e \geq e + 2$ for all $e \geq 1$. For the term p^{ep}, the assumption that $p^e > 2$ means that either $p \geq 3$ or $e \geq 2$, which implies $ep \geq e + 2$. $\square$

Now consider Theorem 7.28 in the case where p is odd. As we already know that $\mathbb{Z}_p^*$ is cyclic, assume $e > 1$. Let $x \in \mathbb{Z}$ be chosen so that $[x]_p$ generates $\mathbb{Z}_p^*$. Suppose the multiplicative order of $[x]_{p^e} \in \mathbb{Z}_{p^e}^*$ is m. We have $x^m \equiv 1 \pmod{p^e}$; hence, $x^m \equiv 1 \pmod{p}$, and so it must be the case that $p - 1$ divides m; thus, $[x^{m/(p-1)}]_{p^e}$ has multiplicative order exactly $p - 1$. By Theorem 6.38, if we find an integer y such that $[y]_{p^e}$ has multiplicative order p^{e-1}, then $[x^{m/(p-1)}y]_{p^e}$ has multiplicative order $(p - 1)p^{e-1}$, and we are done. We claim that $y := 1 + p$ does the job. Any integer between 0 and $p^e - 1$ can be expressed as an e-digit number in base p; for example, $y = (0 \cdots 0\,1\,1)_p$. If we compute successive pth powers of y modulo p^e, then by Lemma 7.31 we have

$$
\begin{aligned}
y \bmod p^e &= (0 & \cdots & & 0\,1\,1)_p, \\
y^p \bmod p^e &= (* & \cdots & & *\,1\,0\,1)_p, \\
y^{p^2} \bmod p^e &= (* & \cdots & & *\,1\,0\,0\,1)_p, \\
&\ \ \vdots \\
y^{p^{e-2}} \bmod p^e &= (1\,0 & \cdots & & 0\,1)_p, \\
y^{p^{e-1}} \bmod p^e &= (0 & \cdots & & 0\,1)_p.
\end{aligned}
$$

Here, "$*$" indicates an arbitrary digit. From this table of values, it is clear (see

Theorem 6.37) that $[y]_{p^e}$ has multiplicative order p^{e-1}. That proves Theorem 7.28 for odd p.

We now prove Theorem 7.28 in the case $p = 2$. For $e = 1$ and $e = 2$, the theorem is easily verified. Suppose $e \geq 3$. Consider the subgroup $G \subseteq \mathbb{Z}_{2^e}^*$ generated by $[5]_{2^e}$. Expressing integers between 0 and $2^e - 1$ as e-digit binary numbers, and applying Lemma 7.31, we have

$$
\begin{aligned}
5 \bmod 2^e &= (0 \quad \cdots \quad 0\,1\,0\,1)_2, \\
5^2 \bmod 2^e &= (* \quad \cdots \quad *\,1\,0\,0\,1)_2,
\end{aligned}
$$

$$\vdots$$

$$
\begin{aligned}
5^{2^{e-3}} \bmod 2^e &= (1\,0 \quad \cdots \quad 0\,1)_2, \\
5^{2^{e-2}} \bmod 2^e &= (0 \quad \cdots \quad 0\,1)_2.
\end{aligned}
$$

So it is clear (see Theorem 6.37) that $[5]_{2^e}$ has multiplicative order 2^{e-2}. We claim that $[-1]_{2^e} \notin G$. If it were, then since it has multiplicative order 2, and since every cyclic group of even order has precisely one element of order 2 (see Theorem 6.32), it must be equal to $[5^{2^{e-3}}]_{2^e}$; however, it is clear from the above calculation that $5^{2^{e-3}} \not\equiv -1 \pmod{2^e}$. Let $H \subseteq \mathbb{Z}_{2^e}^*$ be the subgroup generated by $[-1]_{2^e}$. Then from the above, $G \cap H = \{[1]_{2^e}\}$, and hence by Theorem 6.25, $G \times H$ is isomorphic to the subgroup $G \cdot H$ of $\mathbb{Z}_{2^e}^*$. But since the orders of $G \times H$ and $\mathbb{Z}_{2^e}^*$ are equal, we must have $G \cdot H = \mathbb{Z}_{2^e}^*$. That proves the theorem.

Example 7.61. Let p be an odd prime, and let d be a positive integer dividing $p - 1$. Since $\mathbb{Z}_p^*$ is a cyclic group of order $p - 1$, Theorem 6.32, implies that $(\mathbb{Z}_p^*)^d$ is the unique subgroup of $\mathbb{Z}_p^*$ of order $(p - 1)/d$, and moreover, $(\mathbb{Z}_p^*)^d = \mathbb{Z}_p^*\{(p - 1)/d\}$; that is, for all $\alpha \in \mathbb{Z}_p^*$, we have

$$\alpha = \beta^d \text{ for some } \beta \in \mathbb{Z}_p^* \iff \alpha^{(p-1)/d} = 1.$$

Setting $d = 2$, we arrive again at Euler's criterion (Theorem 2.21), but by a very different, and perhaps more elegant, route than that taken in our original proof of that theorem. $\square$

EXERCISE 7.69. Show that if n is a positive integer, the group $\mathbb{Z}_n^*$ is cyclic if and only if

$$n = 1, 2, 4, p^e, \text{ or } 2p^e,$$

where p is an odd prime and e is a positive integer.

EXERCISE 7.70. Let $n = pq$, where p and q are distinct primes such that $p = 2p'+1$ and $q = 2q' + 1$, where p' and q' are themselves prime. Show that the subgroup $(\mathbb{Z}_n^*)^2$ of squares is a cyclic group of order $p'q'$.

EXERCISE 7.71. Let $n = pq$, where p and q are distinct primes such that $p \nmid (q-1)$ and $q \nmid (p-1)$.

(a) Show that the map that sends $[a]_n \in \mathbb{Z}_n^*$ to $[a^n]_{n^2} \in (\mathbb{Z}_{n^2}^*)^n$ is a group isomorphism (in particular, you need to show that this map is unambiguously defined).

(b) Consider the element $\alpha := [1+n]_{n^2} \in \mathbb{Z}_{n^2}^*$; show that for every non-negative integer k, $\alpha^k = [1 + kn]_{n^2}$; deduce that α has multiplicative order n, and also that the identity $\alpha^k = [1 + kn]_{n^2}$ holds for all integers k.

(c) Show that the map that sends $([k]_n, [a]_n) \in \mathbb{Z}_n \times \mathbb{Z}_n^*$ to $[(1 + kn)a^n]_{n^2} \in \mathbb{Z}_{n^2}^*$ is a group isomorphism.

EXERCISE 7.72. This exercise develops an alternative proof of Theorem 7.29 that relies on less group theory. Let n be the order of the group G. Using Theorem 7.14, show that for all $d \mid n$, there are at most d elements in the group whose multiplicative order divides d. From this, deduce that for all $d \mid n$, the number of elements of multiplicative order d is either 0 or $\varphi(d)$. Now use Theorem 2.40 to deduce that for all $d \mid n$ (and in particular, for $d = n$), the number of elements of multiplicative order d is equal to $\varphi(d)$.

8

Finite and discrete probability distributions

To understand the algorithmic aspects of number theory and algebra, and applications such as cryptography, a firm grasp of the basics of probability theory is required. This chapter introduces concepts from probability theory, starting with the basic notions of probability distributions on finite sample spaces, and then continuing with conditional probability and independence, random variables, and expectation. Applications such as "balls and bins," "hash functions," and the "left-over hash lemma" are also discussed. The chapter closes by extending the basic theory to probability distributions on countably infinite sample spaces.

8.1 Basic definitions

Let Ω be a finite, non-empty set. A **probability distribution on** Ω is a function $P : \Omega \to [0, 1]$ that satisfies the following property:

$$\sum_{\omega \in \Omega} P(\omega) = 1. \tag{8.1}$$

The set Ω is called the **sample space of** P.

Intuitively, the elements of Ω represent the possible outcomes of a random experiment, where the probability of outcome $\omega \in \Omega$ is $P(\omega)$. For now, we shall only consider probability distributions on finite sample spaces. Later in this chapter, in §8.10, we generalize this to allow probability distributions on *countably infinite* sample spaces.

Example 8.1. If we think of rolling a fair die, then setting $\Omega := \{1, 2, 3, 4, 5, 6\}$, and $P(\omega) := 1/6$ for all $\omega \in \Omega$, gives a probability distribution that naturally describes the possible outcomes of the experiment. □

Example 8.2. More generally, if Ω is any non-empty, finite set, and $P(\omega) := 1/|\Omega|$ for all $\omega \in \Omega$, then P is called the **uniform distribution on** Ω. □

Example 8.3. A coin toss is an example of a **Bernoulli trial**, which in general is an experiment with only two possible outcomes: *success*, which occurs with probability p; and *failure*, which occurs with probability $q := 1 - p$. Of course, *success* and *failure* are arbitrary names, which can be changed as convenient. In the case of a coin, we might associate *success* with the outcome that the coin comes up *heads*. For a fair coin, we have $p = q = 1/2$; for a biased coin, we have $p \neq 1/2$. □

An **event** is a subset A of Ω, and the **probability of** A is defined to be

$$P[A] := \sum_{\omega \in A} P(\omega). \qquad (8.2)$$

While an event is simply a subset of the sample space, when discussing the probability of an event (or other properties to be introduced later), the discussion always takes place relative to a particular probability distribution, which may be implicit from context.

For events A and B, their union $A \cup B$ logically represents the event that *either* the event A *or* the event B occurs (or both), while their intersection $A \cap B$ logically represents the event that *both* A *and* B occur. For an event A, we define its complement $\overline{A} := \Omega \setminus A$, which logically represents the event that A does *not* occur.

In working with events, one makes frequent use of the usual rules of Boolean logic. **De Morgan's law** says that for all events A and B,

$$\overline{A \cup B} = \overline{A} \cap \overline{B} \text{ and } \overline{A \cap B} = \overline{A} \cup \overline{B}.$$

We also have the **Boolean distributive law**: for all events A, B, and C,

$$A \cap (B \cup C) = (A \cap B) \cup (A \cap C) \text{ and } A \cup (B \cap C) = (A \cup B) \cap (A \cup C).$$

Example 8.4. Continuing with Example 8.1, the event that the die has an odd value is $A := \{1, 3, 5\}$, and we have $P[A] = 1/2$. The event that the die has a value greater than 2 is $B := \{3, 4, 5, 6\}$, and $P[B] = 2/3$. The event that the die has a value that is at most 2 is $\overline{B} = \{1, 2\}$, and $P[\overline{B}] = 1/3$. The event that the value of the die is odd *or* exceeds 2 is $A \cup B = \{1, 3, 4, 5, 6\}$, and $P[A \cup B] = 5/6$. The event that the value of the die is odd *and* exceeds 2 is $A \cap B = \{3, 5\}$, and $P[A \cap B] = 1/3$. □

Example 8.5. If P is the uniform distribution on a set Ω, and A is a subset of Ω, then $P[A] = |A|/|\Omega|$. □

We next derive some elementary facts about probabilities of certain events, and relations among them. It is clear from the definitions that

$$P[\emptyset] = 0 \text{ and } P[\Omega] = 1,$$

and that for every event A, we have

$$P[\overline{A}] = 1 - P[A].$$

Now consider events A and B, and their union $A \cup B$. We have

$$P[A \cup B] \leq P[A] + P[B]; \tag{8.3}$$

moreover,

$$P[A \cup B] = P[A] + P[B] \text{ if } A \text{ and } B \text{ are disjoint}, \tag{8.4}$$

that is, if $A \cap B = \emptyset$. The exact formula for arbitrary events A and B is:

$$P[A \cup B] = P[A] + P[B] - P[A \cap B]. \tag{8.5}$$

(8.3), (8.4), and (8.5) all follow from the observation that in the expression

$$P[A] + P[B] = \sum_{\omega \in A} P(\omega) + \sum_{\omega \in B} P(\omega),$$

the value $P(\omega)$ is counted once for each $\omega \in A \cup B$, except for those $\omega \in A \cap B$, for which $P(\omega)$ is counted twice.

Example 8.6. Alice rolls two dice, and asks Bob to guess a value that appears on either of the two dice (without looking). Let us model this situation by considering the uniform distribution on $\Omega := \{1, \dots, 6\} \times \{1, \dots, 6\}$, where for each pair $(s, t) \in \Omega$, s represents the value of the first die, and t the value of the second.

For $k = 1, \dots, 6$, let A_k be the event that the first die is k, and B_k the event that the second die is k. Let $C_k = A_k \cup B_k$ be the event that k appears on either of the two dice. No matter what value k Bob chooses, the probability that this choice is correct is

$$P[C_k] = P[A_k \cup B_k] = P[A_k] + P[B_k] - P[A_k \cap B_k]$$
$$= 1/6 + 1/6 - 1/36 = 11/36,$$

which is slightly less than the estimate $P[A_k] + P[B_k]$ obtained from (8.3). $\square$

If $\{A_i\}_{i \in I}$ is a family of events, indexed by some set I, we can naturally form the union $\bigcup_{i \in I} A_i$ and intersection $\bigcap_{i \in I} A_i$. If $I = \emptyset$, then by definition, the union is $\emptyset$, and by special convention, the intersection is the entire sample space Ω. Logically, the union represents the event that *some* A_i occurs, and the intersection represents the event that *all* the A_i's occur. De Morgan's law generalizes as follows:

$$\overline{\bigcup_{i \in I} A_i} = \bigcap_{i \in I} \overline{A_i} \text{ and } \overline{\bigcap_{i \in I} A_i} = \bigcup_{i \in I} \overline{A_i},$$

and if B is an event, then the Boolean distributive law generalizes as follows:

$$B \cap \left(\bigcup_{i \in I} A_i\right) = \bigcup_{i \in I} (B \cap A_i) \text{ and } B \cup \left(\bigcap_{i \in I} A_i\right) = \bigcap_{i \in I} (B \cup A_i).$$

We now generalize (8.3), (8.4), and (8.5) from pairs of events to families of events. Let $\{A_i\}_{i \in I}$ be a finite family of events (i.e., the index set I is finite). Using (8.3), it follows by induction on $|I|$ that

$$P\left[\bigcup_{i \in I} A_i\right] \leq \sum_{i \in I} P[A_i], \tag{8.6}$$

which is known as **Boole's inequality** (and sometimes called the **union bound**). Analogously, using (8.4), it follows by induction on $|I|$ that

$$P\left[\bigcup_{i \in I} A_i\right] = \sum_{i \in I} P[A_i] \text{ if } \{A_i\}_{i \in I} \text{ is pairwise disjoint}, \tag{8.7}$$

that is, if $A_i \cap A_j = \emptyset$ for all $i, j \in I$ with $i \neq j$. We shall refer to (8.7) as **Boole's equality**. Both (8.6) and (8.7) are invaluable tools in calculating or estimating the probability of an event A by breaking A up into a family $\{A_i\}_{i \in I}$ of smaller, and hopefully simpler, events, whose union is A. We shall make frequent use of them.

The generalization of (8.5) is messier. Consider first the case of three events, A, B, and C. We have

$$P[A \cup B \cup C] = P[A] + P[B] + P[C] - P[A \cap B] - P[A \cap C] - P[B \cap C]$$
$$+ P[A \cap B \cap C].$$

Thus, starting with the sum of the probabilities of the individual events, we have to subtract a "correction term" that consists of the sum of probabilities of all intersections of pairs of events; however, this is an "over-correction," and we have to correct the correction by adding back in the probability of the intersection of all three events. The general statement is as follows:

Theorem 8.1 (Inclusion/exclusion principle). *Let* $\{A_i\}_{i \in I}$ *be a finite family of events. Then*

$$P\left[\bigcup_{i \in I} A_i\right] = \sum_{\emptyset \subsetneq J \subseteq I} (-1)^{|J|-1} P\left[\bigcap_{j \in J} A_j\right],$$

the sum being over all non-empty subsets J of I.

Proof. For $\omega \in \Omega$ and $B \subseteq \Omega$, define $\delta_\omega[B] := 1$ if $\omega \in B$, and $\delta_\omega[B] := 0$ if $\omega \notin B$. As a function of ω, $\delta_\omega[B]$ is simply the characteristic function of B. One may easily verify that for all $\omega \in \Omega$, $B \subseteq \Omega$, and $C \subseteq \Omega$, we have $\delta_\omega[\bar{B}] = 1 - \delta_\omega[B]$ and $\delta_\omega[B \cap C] = \delta_\omega[B]\delta_\omega[C]$. It is also easily seen that for every $B \subseteq \Omega$, we have $\sum_{\omega \in \Omega} P(\omega)\delta_\omega[B] = P[B]$.

Let $\mathcal{A} := \bigcup_{i \in I} \mathcal{A}_i$, and for $J \subseteq I$, let $\mathcal{A}_J := \bigcap_{j \in J} \mathcal{A}_j$. For every $\omega \in \Omega$,

$$1 - \delta_\omega[\mathcal{A}] = \delta_\omega[\bar{\mathcal{A}}] = \delta_\omega\left[\bigcap_{i \in I} \bar{\mathcal{A}}_i\right] = \prod_{i \in I} \delta_\omega[\bar{\mathcal{A}}_i] = \prod_{i \in I} (1 - \delta_\omega[\mathcal{A}_i])$$

$$= \sum_{J \subseteq I} (-1)^{|J|} \prod_{j \in J} \delta_\omega[\mathcal{A}_j] = \sum_{J \subseteq I} (-1)^{|J|} \delta_\omega[\mathcal{A}_J],$$

and so

$$\delta_\omega[\mathcal{A}] = \sum_{\emptyset \subsetneq J \subseteq I} (-1)^{|J|-1} \delta_\omega[\mathcal{A}_J]. \tag{8.8}$$

Multiplying (8.8) by $P(\omega)$, and summing over all $\omega \in \Omega$, we have

$$P[\mathcal{A}] = \sum_{\omega \in \Omega} P(\omega) \delta_\omega[\mathcal{A}] = \sum_{\omega \in \Omega} P(\omega) \sum_{\emptyset \subsetneq J \subseteq I} (-1)^{|J|-1} \delta_\omega[\mathcal{A}_J]$$

$$= \sum_{\emptyset \subsetneq J \subseteq I} (-1)^{|J|-1} \sum_{\omega \in \Omega} P(\omega) \delta_\omega[\mathcal{A}_J] = \sum_{\emptyset \subsetneq J \subseteq I} (-1)^{|J|-1} P[\mathcal{A}_J]. \quad \square$$

One can also state the inclusion/exclusion principle in a slightly different way, splitting the sum into terms with $|J| = 1$, $|J| = 2$, etc., as follows:

$$P\left[\bigcup_{i \in I} \mathcal{A}_i\right] = \sum_{i \in I} P[\mathcal{A}_i] + \sum_{k=2}^{|I|} (-1)^{k-1} \sum_{\substack{J \subseteq I \\ |J|=k}} P\left[\bigcap_{j \in J} \mathcal{A}_j\right],$$

where the last sum in this formula is taken over all subsets J of I of size k.

We next consider a useful way to "glue together" probability distributions. Suppose one conducts two physically separate and unrelated random experiments, with each experiment modeled separately as a probability distribution. What we would like is a way to combine these distributions, obtaining a single probability distribution that models the two experiments as one grand experiment. This can be accomplished in general, as follows.

Let $P_1 : \Omega_1 \to [0, 1]$ and $P_2 : \Omega_2 \to [0, 1]$ be probability distributions. Their **product distribution** $P := P_1 P_2$ is defined as follows:

$$P : \quad \Omega_1 \times \Omega_2 \to [0, 1]$$
$$(\omega_1, \omega_2) \mapsto P_1(\omega_1) P_2(\omega_2).$$

It is easily verified that P is a probability distribution on the sample space $\Omega_1 \times \Omega_2$:

$$\sum_{\omega_1, \omega_2} P(\omega_1, \omega_2) = \sum_{\omega_1, \omega_2} P_1(\omega_1) P_2(\omega_2) = \left(\sum_{\omega_1} P_1(\omega_1)\right)\left(\sum_{\omega_2} P_2(\omega_2)\right) = 1 \cdot 1 = 1.$$

More generally, if $P_i : \Omega_i \to [0, 1]$, for $i = 1, \ldots, n$, are probability distributions, then their product distribution is $P := P_1 \cdots P_n$, where

$$P : \quad \Omega_1 \times \cdots \times \Omega_n \to [0,1]$$
$$(\omega_1, \ldots, \omega_n) \mapsto P_1(\omega_1) \cdots P_n(\omega_n).$$

If $P_1 = P_2 = \cdots = P_n$, then we may write $P = P_1^n$. It is clear from the definitions that if each P_i is the uniform distribution on Ω_i, then P is the uniform distribution on $\Omega_1 \times \cdots \times \Omega_n$.

Example 8.7. We can view the probability distribution P in Example 8.6 as P_1^2, where P_1 is the uniform distribution on $\{1, \ldots, 6\}$. □

Example 8.8. Suppose we have a coin that comes up *heads* with some probability p, and *tails* with probability $q := 1 - p$. We toss the coin n times, and record the outcomes. We can model this as the product distribution $P = P_1^n$, where P_1 is the distribution of a Bernoulli trial (see Example 8.3) with success probability p, and where we identify *success* with *heads*, and *failure* with *tails*. The sample space Ω of P is the set of all 2^n tuples $\omega = (\omega_1, \ldots, \omega_n)$, where each ω_i is either *heads* or *tails*. If the tuple ω has k *heads* and $n - k$ *tails*, then $P(\omega) = p^k q^{n-k}$, regardless of the positions of the *heads* and *tails* in the tuple.

For each $k = 0, \ldots, n$, let $\mathcal{A}_k$ be the event that our coin comes up *heads* exactly k times. As a set, $\mathcal{A}_k$ consists of all those tuples in the sample space with exactly k *heads*, and so

$$|\mathcal{A}_k| = \binom{n}{k},$$

from which it follows that

$$P[\mathcal{A}_k] = \binom{n}{k} p^k q^{n-k}.$$

If our coin is a fair coin, so that $p = q = 1/2$, then P is the uniform distribution on Ω, and for each $k = 0, \ldots, n$, we have

$$P[\mathcal{A}_k] = \binom{n}{k} 2^{-n}. \quad □$$

Suppose $P : \Omega \to [0,1]$ is a probability distribution. The **support** of P is defined to be the set $\{\omega \in \Omega : P(\omega) \neq 0\}$. Now consider another probability distribution $P' : \Omega' \to [0,1]$. Of course, these two distributions are equal if and only if $\Omega = \Omega'$ and $P(\omega) = P'(\omega)$ for all $\omega \in \Omega$. However, it is natural and convenient to have a more relaxed notion of equality. We shall say that P and P' are **essentially equal** if the restriction of P to its support is equal to the restriction of P' to its support. For example, if P is the probability distribution on $\{1, 2, 3, 4\}$ that assigns probability

1/3 to 1, 2, and 3, and probability 0 to 4, we may say that P is essentially the uniform distribution on $\{1, 2, 3\}$.

EXERCISE 8.1. Show that $P[\mathcal{A} \cap \mathcal{B}] P[\mathcal{A} \cup \mathcal{B}] \leq P[\mathcal{A}] P[\mathcal{B}]$ for all events $\mathcal{A}, \mathcal{B}$.

EXERCISE 8.2. Suppose $\mathcal{A}, \mathcal{B}, \mathcal{C}$ are events such that $\mathcal{A} \cap \bar{C} = \mathcal{B} \cap \bar{C}$. Show that $|P[\mathcal{A}] - P[\mathcal{B}]| \leq P[\mathcal{C}]$.

EXERCISE 8.3. Let m be a positive integer, and let $\alpha(m)$ be the probability that a number chosen at random from $\{1, \ldots, m\}$ is divisible by either 4, 5, or 6. Write down an exact formula for $\alpha(m)$, and also show that $\alpha(m) = 14/30 + O(1/m)$.

EXERCISE 8.4. This exercise asks you to generalize Boole's inequality (8.6), proving **Bonferroni's inequalities**. Let $\{\mathcal{A}_i\}_{i \in I}$ be a finite family of events, where $n := |I|$. For $m = 0, \ldots, n$, define

$$\alpha_m := \sum_{k=1}^{m} (-1)^{k-1} \sum_{\substack{J \subseteq I \\ |J| = k}} P\left[\bigcap_{j \in J} \mathcal{A}_j\right].$$

Also, define

$$\alpha := P\left[\bigcup_{i \in I} \mathcal{A}_i\right].$$

Show that $\alpha \leq \alpha_m$ if m is odd, and $\alpha \geq \alpha_m$ if m is even. Hint: use induction on n.

8.2 Conditional probability and independence

Let P be a probability distribution on a sample space Ω.

For a given event $\mathcal{B} \subseteq \Omega$ with $P[\mathcal{B}] \neq 0$, and for $\omega \in \Omega$, let us define

$$P(\omega \mid \mathcal{B}) := \begin{cases} P(\omega)/P[\mathcal{B}] & \text{if } \omega \in \mathcal{B}, \\ 0 & \text{otherwise.} \end{cases}$$

Viewing $\mathcal{B}$ as fixed, the function $P(\cdot \mid \mathcal{B})$ is a new probability distribution on the sample space Ω, called the **conditional distribution (derived from P) given** $\mathcal{B}$.

Intuitively, $P(\cdot \mid \mathcal{B})$ has the following interpretation. Suppose a random experiment produces an outcome according to the distribution P. Further, suppose we learn that the event $\mathcal{B}$ has occurred, but nothing else about the outcome. Then the distribution $P(\cdot \mid \mathcal{B})$ assigns new probabilities to all possible outcomes, reflecting the partial knowledge that the event $\mathcal{B}$ has occurred.

For a given event $A \subseteq \Omega$, its probability with respect to the conditional distribution given B is

$$P[A \mid B] = \sum_{\omega \in A} P(\omega \mid B) = \frac{P[A \cap B]}{P[B]}.$$

The value $P[A \mid B]$ is called the **conditional probability of A given B**. Again, the intuition is that this is the probability that the event A occurs, given the partial knowledge that the event B has occurred.

For events A and B, if $P[A \cap B] = P[A]P[B]$, then A and B are called **independent** events. If $P[B] \neq 0$, one easily sees that A and B are independent if and only if $P[A \mid B] = P[A]$; intuitively, independence means that the partial knowledge that event B has occurred does not affect the likelihood that A occurs.

Example 8.9. Suppose P is the uniform distribution on Ω, and that $B \subseteq \Omega$ with $P[B] \neq 0$. Then the conditional distribution given B is essentially the uniform distribution on B. □

Example 8.10. Consider again Example 8.4, where A is the event that the value on the die is odd, and B is the event that the value of the die exceeds 2. Then as we calculated, $P[A] = 1/2$, $P[B] = 2/3$, and $P[A \cap B] = 1/3$; thus, $P[A \cap B] = P[A]P[B]$, and we conclude that A and B are independent. Indeed, $P[A \mid B] = (1/3)/(2/3) = 1/2 = P[A]$; intuitively, given the partial knowledge that the value on the die exceeds 2, we know it is equally likely to be either 3, 4, 5, or 6, and so the conditional probability that it is odd is $1/2$.

However, consider the event C that the value on the die exceeds 3. We have $P[C] = 1/2$ and $P[A \cap C] = 1/6 \neq 1/4$, from which we conclude that A and C are *not* independent. Indeed, $P[A \mid C] = (1/6)/(1/2) = 1/3 \neq P[A]$; intuitively, given the partial knowledge that the value on the die exceeds 3, we know it is equally likely to be either 4, 5, or 6, and so the conditional probability that it is odd is just $1/3$, and not $1/2$. □

Example 8.11. In Example 8.6, suppose that Alice tells Bob the sum of the two dice before Bob makes his guess. The following table is useful for visualizing the situation:

6	7	8	9	10	11	12
5	6	7	8	9	10	11
4	5	6	7	8	9	10
3	4	5	6	7	8	9
2	3	4	5	6	7	8
1	2	3	4	5	6	7
	1	2	3	4	5	6

For example, suppose Alice tells Bob the sum is 4. Then what is Bob's best strategy in this case? Let $\mathcal{D}_\ell$ be the event that the sum is ℓ, for $\ell = 2, \ldots, 12$, and consider the conditional distribution given $\mathcal{D}_4$. This conditional distribution is essentially the uniform distribution on the set $\{(1, 3), (2, 2), (3, 1)\}$. The numbers 1 and 3 both appear in two pairs, while the number 2 appears in just one pair. Therefore,

$$P[C_1 \mid \mathcal{D}_4] = P[C_3 \mid \mathcal{D}_4] = 2/3,$$

while

$$P[C_2 \mid \mathcal{D}_4] = 1/3$$

and

$$P[C_4 \mid \mathcal{D}_4] = P[C_5 \mid \mathcal{D}_4] = P[C_6 \mid \mathcal{D}_4] = 0.$$

Thus, if the sum is 4, Bob's best strategy is to guess either 1 or 3, which will be correct with probability $2/3$.

Similarly, if the sum is 5, then we consider the conditional distribution given $\mathcal{D}_5$, which is essentially the uniform distribution on $\{(1, 4), (2, 3), (3, 2), (4, 1)\}$. In this case, Bob should choose one of the numbers $k = 1, \ldots, 4$, each of which will be correct with probability $P[C_k \mid \mathcal{D}_5] = 1/2$. □

Suppose $\{B_i\}_{i \in I}$ is a finite, pairwise disjoint family of events, whose union is Ω. Now consider an arbitrary event A. Since $\{A \cap B_i\}_{i \in I}$ is a pairwise disjoint family of events whose union is A, Boole's equality (8.7) implies

$$P[A] = \sum_{i \in I} P[A \cap B_i]. \tag{8.9}$$

Furthermore, if each B_i occurs with non-zero probability (so that, in particular, $\{B_i\}_{i \in I}$ is a partition of Ω), then we have

$$P[A] = \sum_{i \in I} P[A \mid B_i] P[B_i]. \tag{8.10}$$

If, in addition, $P[A] \neq 0$, then for each $j \in I$, we have

$$P[B_j \mid A] = \frac{P[A \cap B_j]}{P[A]} = \frac{P[A \mid B_j] P[B_j]}{\sum_{i \in I} P[A \mid B_i] P[B_i]}. \tag{8.11}$$

Equations (8.9) and (8.10) are sometimes called the **law of total probability**, while equation (8.11) is known as **Bayes' theorem**. Equation (8.10) (resp., (8.11)) is useful for computing or estimating $P[A]$ (resp., $P[B_j \mid A]$) by conditioning on the events B_i.

Example 8.12. Let us continue with Example 8.11, and compute Bob's overall probability of winning, assuming he follows an optimal strategy. If the sum is 2 or 12, clearly there is only one sensible choice for Bob to make, and it will certainly be correct. If the sum is any other number ℓ, and there are N_ℓ pairs in the sample space that sum to that number, then there will always be a value that appears in exactly 2 of these N_ℓ pairs, and Bob should choose such a value (see the diagram in Example 8.11). Indeed, this is achieved by the simple rule of choosing the value 1 if $\ell \leq 7$, and the value 6 if $\ell > 7$. This is an optimal strategy for Bob, and if C is the event that Bob wins following this strategy, then by total probability (8.10), we have

$$P[C] = \sum_{\ell=2}^{12} P[C \mid D_\ell] P[D_\ell].$$

Moreover,

$$P[C \mid D_2] P[D_2] = 1 \cdot \frac{1}{36} = \frac{1}{36}, \quad P[C \mid D_{12}] P[D_{12}] = 1 \cdot \frac{1}{36} = \frac{1}{36},$$

and for $\ell = 3, \ldots, 11$, we have

$$P[C \mid D_\ell] P[D_\ell] = \frac{2}{N_\ell} \cdot \frac{N_\ell}{36} = \frac{1}{18}.$$

Therefore,

$$P[C] = \frac{1}{36} + \frac{1}{36} + \frac{9}{18} = \frac{10}{18}. \quad \square$$

Example 8.13. Suppose that the rate of incidence of disease X in the overall population is 1%. Also suppose that there is a test for disease X; however, the test is not perfect: it has a 5% false positive rate (i.e., 5% of healthy patients test positive for the disease), and a 2% false negative rate (i.e., 2% of sick patients test negative for the disease). A doctor gives the test to a patient and it comes out positive. How should the doctor advise his patient? In particular, what is the probability that the patient actually has disease X, given a positive test result?

Amazingly, many trained doctors will say the probability is 95%, since the test has a false positive rate of 5%. However, this conclusion is completely wrong.

Let A be the event that the test is positive and let B be the event that the patient has disease X. The relevant quantity that we need to estimate is $P[B \mid A]$; that is, the probability that the patient has disease X, given a positive test result. We use Bayes' theorem to do this:

$$P[B \mid A] = \frac{P[A \mid B] P[B]}{P[A \mid B] P[B] + P[A \mid \bar{B}] P[\bar{B}]} = \frac{0.98 \cdot 0.01}{0.98 \cdot 0.01 + 0.05 \cdot 0.99} \approx 0.17.$$

Thus, the chances that the patient has disease X given a positive test result are just 17%. The correct intuition here is that it is much more likely to get a false positive than it is to actually have the disease.

Of course, the real world is a bit more complicated than this example suggests:

the doctor may be giving the patient the test because other risk factors or symptoms may suggest that the patient is more likely to have the disease than a random member of the population, in which case the above analysis does not apply. □

Example 8.14. This example is based on the TV game show "Let's make a deal," which was popular in the 1970's. In this game, a contestant chooses one of three doors. Behind two doors is a "zonk," that is, something amusing but of little or no value, such as a goat, and behind one of the doors is a "grand prize," such as a car or vacation package. We may assume that the door behind which the grand prize is placed is chosen at random from among the three doors, with equal probability. After the contestant chooses a door, the host of the show, Monty Hall, always reveals a zonk behind one of the two doors not chosen by the contestant. The contestant is then given a choice: either stay with his initial choice of door, or switch to the other unopened door. After the contestant finalizes his decision on which door to choose, that door is opened and he wins whatever is behind it. The question is, which strategy is better for the contestant: to stay or to switch?

Let us evaluate the two strategies. If the contestant always stays with his initial selection, then it is clear that his probability of success is exactly $1/3$.

Now consider the strategy of always switching. Let B be the event that the contestant's initial choice was correct, and let A be the event that the contestant wins the grand prize. On the one hand, if the contestant's initial choice was correct, then switching will certainly lead to failure (in this case, Monty has two doors to choose from, but his choice does not affect the outcome). Thus, $P[A \mid B] = 0$. On the other hand, suppose that the contestant's initial choice was incorrect, so that one of the zonks is behind the initially chosen door. Since Monty reveals the other zonk, switching will lead with certainty to success. Thus, $P[A \mid \bar{B}] = 1$. Furthermore, it is clear that $P[B] = 1/3$. So using total probability (8.10), we compute

$$P[A] = P[A \mid B] P[B] + P[A \mid \bar{B}] P[\bar{B}] = 0 \cdot (1/3) + 1 \cdot (2/3) = 2/3.$$

Thus, the "stay" strategy has a success probability of $1/3$, while the "switch" strategy has a success probability of $2/3$. So it is better to switch than to stay.

Of course, real life is a bit more complicated. Monty did not always reveal a zonk and offer a choice to switch. Indeed, if Monty *only* revealed a zonk when the contestant had chosen the correct door, then switching would certainly be the wrong strategy. However, if Monty's choice itself was a random decision made independently of the contestant's initial choice, then switching is again the preferred strategy. □

We next generalize the notion of independence from pairs of events to families of events. Let $\{A_i\}_{i \in I}$ be a finite family of events. For a given positive integer k, we say that the family $\{A_i\}_{i \in I}$ is k-**wise independent** if the following holds:

$$P\left[\bigcap_{j \in J} A_j\right] = \prod_{j \in J} P[A_j] \text{ for all } J \subseteq I \text{ with } |J| \leq k.$$

The family $\{A_i\}_{i \in I}$ is called **pairwise independent** if it is 2-wise independent. Equivalently, pairwise independence means that for all $i, j \in I$ with $i \neq j$, we have $P[A_i \cap A_j] = P[A_i] P[A_j]$, or put yet another way, that for all $i, j \in I$ with $i \neq j$, the events A_i and A_j are independent.

The family $\{A_i\}_{i \in I}$ is called **mutually independent** if it is k-wise independent for all positive integers k. Equivalently, mutual independence means that

$$P\left[\bigcap_{j \in J} A_j\right] = \prod_{j \in J} P[A_j] \text{ for all } J \subseteq I.$$

If $n := |I| > 0$, mutual independence is equivalent to n-wise independence; moreover, if $0 < k \leq n$, then $\{A_i\}_{i \in I}$ is k-wise independent if and only if $\{A_j\}_{j \in J}$ is mutually independent for every $J \subseteq I$ with $|J| = k$.

In defining independence, the choice of the index set I plays no real role, and we can rename elements of I as convenient.

Example 8.15. Suppose we toss a fair coin three times, which we formally model using the uniform distribution on the set of all 8 possible outcomes of the three coin tosses: (*heads, heads, heads*), (*heads, heads, tails*), etc., as in Example 8.8. For $i = 1, 2, 3$, let A_i be the event that the ith toss comes up *heads*. Then $\{A_i\}_{i=1}^{3}$ is a mutually independent family of events, where each individual A_i occurs with probability $1/2$.

Now let B_{12} be the event that the first and second tosses agree (i.e., both *heads* or both *tails*), let B_{13} be the event that the first and third tosses agree, and let B_{23} be the event that the second and third tosses agree. Then the family of events B_{12}, B_{13}, B_{23} is pairwise independent, but not mutually independent. Indeed, the probability that any given individual event occurs is $1/2$, and the probability that any given pair of events occurs is $1/4$; however, the probability that all three events occur is also $1/4$, since if any two events occur, then so does the third. □

We close this section with some simple facts about independence of events and their complements.

Theorem 8.2. *If A and B are independent events, then so are A and $\bar{B}$.*

Proof. We have

$$P[A] = P[A \cap B] + P[A \cap \bar{B}] \text{ (by total probability (8.9))}$$
$$= P[A] P[B] + P[A \cap \bar{B}] \text{ (since A and B are independent).}$$

Therefore,

$$P[\mathcal{A} \cap \bar{\mathcal{B}}] = P[\mathcal{A}] - P[\mathcal{A}] P[\mathcal{B}] = P[\mathcal{A}](1 - P[\mathcal{B}]) = P[\mathcal{A}] P[\bar{\mathcal{B}}]. \quad \square$$

This theorem implies that

$$\mathcal{A} \text{ and } \mathcal{B} \text{ are independent} \iff \mathcal{A} \text{ and } \bar{\mathcal{B}} \text{ are independent}$$
$$\iff \bar{\mathcal{A}} \text{ and } \mathcal{B} \quad " \quad "$$
$$\iff \bar{\mathcal{A}} \text{ and } \bar{\mathcal{B}} \quad " \quad " \quad .$$

The following theorem generalizes this result to families of events. It says that if a family of events is k-wise independent, then the family obtained by complementing any number of members of the given family is also k-wise independent.

Theorem 8.3. *Let* $\{\mathcal{A}_i\}_{i \in I}$ *be a finite, k-wise independent family of events. Let J be a subset of I, and for each $i \in I$, define $\mathcal{A}_i' := \mathcal{A}_i$ if $i \in J$, and $\mathcal{A}_i' := \bar{\mathcal{A}}_i$ if $i \notin J$. Then $\{\mathcal{A}_i'\}_{i \in I}$ is also k-wise independent.*

Proof. It suffices to prove the theorem for the case where $J = I \setminus \{d\}$, for an arbitrary $d \in I$: this allows us to complement any single member of the family that we wish, without affecting independence; by repeating the procedure, we can complement any number of them.

To this end, it will suffice to show the following: if $J \subseteq I$, $|J| < k$, $d \in I \setminus J$, and $\mathcal{A}_J := \bigcap_{j \in J} \mathcal{A}_j$, we have

$$P[\bar{\mathcal{A}}_d \cap \mathcal{A}_J] = (1 - P[\mathcal{A}_d]) \prod_{j \in J} P[\mathcal{A}_j]. \tag{8.12}$$

Using total probability (8.9), along with the independence hypothesis (twice), we have

$$\prod_{j \in J} P[\mathcal{A}_j] = P[\mathcal{A}_J] = P[\mathcal{A}_d \cap \mathcal{A}_J] + P[\bar{\mathcal{A}}_d \cap \mathcal{A}_J]$$

$$= P[\mathcal{A}_d] \cdot \prod_{j \in J} P[\mathcal{A}_j] + P[\bar{\mathcal{A}}_d \cap \mathcal{A}_J],$$

from which (8.12) follows immediately. $\quad \square$

EXERCISE 8.5. For events $\mathcal{A}_1, \ldots, \mathcal{A}_n$, define $\alpha_1 := P[\mathcal{A}_1]$, and for $i = 2, \ldots, n$, define $\alpha_i := P[\mathcal{A}_i \mid \mathcal{A}_1 \cap \cdots \cap \mathcal{A}_{i-1}]$ (assume that $P[\mathcal{A}_1 \cap \cdots \cap \mathcal{A}_{n-1}] \neq 0$). Show that $P[\mathcal{A}_1 \cap \cdots \cap \mathcal{A}_n] = \alpha_1 \cdots \alpha_n$.

EXERCISE 8.6. Let $\mathcal{B}$ be an event, and let $\{\mathcal{B}_i\}_{i \in I}$ be a finite, pairwise disjoint family of events whose union is $\mathcal{B}$. Generalizing the law of total probability (equations (8.9) and (8.10)), show that for every event A, we have $P[A \cap B] = \sum_{i \in I} P[A \cap B_i]$, and if $P[B] \neq 0$ and $I^* := \{i \in I : P[B_i] \neq 0\}$, then

$$P[A \mid B] P[B] = \sum_{i \in I^*} P[A \mid B_i] P[B_i].$$

Also show that if $P[A \mid B_i] \leq \alpha$ for each $i \in I^*$, then $P[A \mid B] \leq \alpha$.

EXERCISE 8.7. Let B be an event with $P[B] \neq 0$, and let $\{C_i\}_{i \in I}$ be a finite, pairwise disjoint family of events whose union contains B. Again, generalizing the law of total probability, show that for every event A, if $I^* := \{i \in I : P[B \cap C_i] \neq 0\}$, then we have

$$P[A \mid B] = \sum_{i \in I^*} P[A \mid B \cap C_i] P[C_i \mid B].$$

EXERCISE 8.8. Three fair coins are tossed. Let A be the event that at least two coins are *heads*. Let B be the event that the number of *heads* is odd. Let C be the event that the third coin is *heads*. Are A and B independent? A and C? B and C?

EXERCISE 8.9. Consider again the situation in Example 8.11, but now suppose that Alice only tells Bob the value of the sum of the two dice modulo 6. Describe an optimal strategy for Bob, and calculate his overall probability of winning.

EXERCISE 8.10. Consider again the situation in Example 8.13, but now suppose that the patient is visiting the doctor because he has symptom Y. Furthermore, it is known that everyone who has disease X exhibits symptom Y, while 10% of the population overall exhibits symptom Y. Assuming that the accuracy of the test is not affected by the presence of symptom Y, how should the doctor advise his patient should the test come out positive?

EXERCISE 8.11. This exercise develops an alternative proof, based on probability theory, of Theorem 2.11. Let n be a positive integer and consider an experiment in which a number a is chosen uniformly at random from $\{0, \ldots, n-1\}$. If $n = p_1^{e_1} \cdots p_r^{e_r}$ is the prime factorization of n, let A_i be the event that a is divisible by p_i, for $i = 1, \ldots, r$.

(a) Show that $\varphi(n)/n = P[\bar{A}_1 \cap \cdots \cap \bar{A}_r]$, where φ is Euler's phi function.

(b) Show that if $J \subseteq \{1, \ldots, r\}$, then

$$P\left[\bigcap_{j \in J} A_j\right] = 1 \bigg/ \prod_{j \in J} p_j.$$

Conclude that $\{A_i\}_{i=1}^r$ is mutually independent, and that $P[A_i] = 1/p_i$ for each $i = 1, \ldots, r$.

(c) Using part (b), deduce that

$$P[\bar{A}_1 \cap \cdots \cap \bar{A}_r] = \prod_{i=1}^{r}(1 - 1/p_i).$$

(d) Combine parts (a) and (c) to derive the result of Theorem 2.11 that

$$\varphi(n) = n \prod_{i=1}^{r}(1 - 1/p_i).$$

8.3 Random variables

It is sometimes convenient to associate a real number, or other mathematical object, with each outcome of a random experiment. The notion of a random variable formalizes this idea.

Let P be a probability distribution on a sample space Ω. A **random variable** X is a function $X : \Omega \to S$, where S is some set, and we say that X **takes values in** S. We do not require that the values taken by X are real numbers, but if this is the case, we say that X is **real valued**. For $s \in S$, "$X = s$" denotes the event $\{\omega \in \Omega : X(\omega) = s\}$. It is immediate from this definition that

$$P[X = s] = \sum_{\omega \in X^{-1}(\{s\})} P(\omega).$$

More generally, for any predicate ϕ on S, we may write "$\phi(X)$" as shorthand for the event $\{\omega \in \Omega : \phi(X(\omega))\}$. When we speak of the **image** of X, we simply mean its image in the usual function-theoretic sense, that is, the set $X(\Omega) = \{X(\omega) : \omega \in \Omega\}$. While a random variable is simply a function on the sample space, any discussion of its properties always takes place relative to a particular probability distribution, which may be implicit from context.

One can easily combine random variables to define new random variables. Suppose $X_1, \ldots, X_n$ are random variables, where $X_i : \Omega \to S_i$ for $i = 1, \ldots, n$. Then $(X_1, \ldots, X_n)$ denotes the random variable that maps $\omega \in \Omega$ to $(X_1(\omega), \ldots, X_n(\omega)) \in S_1 \times \cdots \times S_n$. If $f : S_1 \times \cdots \times S_n \to T$ is a function, then $f(X_1, \ldots, X_n)$ denotes the random variable that maps $\omega \in \Omega$ to $f(X_1(\omega), \ldots, X_n(\omega))$. If f is applied using a special notation, the same notation may be applied to denote the resulting random variable; for example, if X and Y are random variables taking values in a set S, and $\star$ is a binary operation on S, then $X \star Y$ denotes the random variable that maps $\omega \in \Omega$ to $X(\omega) \star Y(\omega) \in S$.

Let X be a random variable whose image is S. The variable X determines a probability distribution $P_X : S \to [0, 1]$ on the set S, where $P_X(s) := P[X = s]$ for each $s \in S$. We call P_X the **distribution of** X. If P_X is the uniform distribution on S, then we say that X is **uniformly distributed over** S.

Suppose X and Y are random variables that take values in a set S. If $P[X = s] = P[Y = s]$ for all $s \in S$, then the distributions of X and Y are essentially equal even if their images are not identical.

Example 8.16. Again suppose we roll two dice, and model this experiment as the uniform distribution on $\Omega := \{1, \dots, 6\} \times \{1, \dots, 6\}$. We can define the random variable X that takes the value of the first die, and the random variable Y that takes the value of the second; formally, X and Y are functions on Ω, where

$$X(s, t) := s \text{ and } Y(s, t) := t \text{ for } (s, t) \in \Omega.$$

For each value $s \in \{1, \dots, 6\}$, the event $X = s$ is $\{(s, 1), \dots, (s, 6)\}$, and so $P[X = s] = 6/36 = 1/6$. Thus, X is uniformly distributed over $\{1, \dots, 6\}$. Likewise, Y is uniformly distributed over $\{1, \dots, 6\}$, and the random variable (X, Y) is uniformly distributed over Ω. We can also define the random variable $Z := X + Y$, which formally is the function on the sample space defined by

$$Z(s, t) := s + t \text{ for } (s, t) \in \Omega.$$

The image of Z is $\{2, \dots, 12\}$, and its distribution is given by the following table:

u	2	3	4	5	6	7	8	9	10	11	12
$P[Z = u]$	1/36	2/36	3/36	4/36	5/36	6/36	5/36	4/36	3/36	2/36	1/36

. $\square$

Example 8.17. If A is an event, we may define a random variable X as follows: $X := 1$ if the event A occurs, and $X := 0$ otherwise. The variable X is called the **indicator variable for** A. Formally, X is the function that maps $\omega \in A$ to 1, and $\omega \in \Omega \setminus A$ to 0; that is, X is simply the characteristic function of A. The distribution of X is that of a Bernoulli trial: $P[X = 1] = P[A]$ and $P[X = 0] = 1 - P[A]$.

It is not hard to see that $1 - X$ is the indicator variable for $\overline{A}$. Now suppose B is another event, with indicator variable Y. Then it is also not hard to see that XY is the indicator variable for $A \cap B$, and that $X + Y - XY$ is the indicator variable for $A \cup B$; in particular, if $A \cap B = \emptyset$, then $X + Y$ is the indicator variable for $A \cup B$. $\square$

Example 8.18. Consider again Example 8.8, where we have a coin that comes up *heads* with probability p, and *tails* with probability $q := 1 - p$, and we toss it n times. For each $i = 1, \dots, n$, let A_i be the event that the ith toss comes up *heads*, and let X_i be the corresponding indicator variable. Let us also define $X := X_1 + \cdots + X_n$, which represents the total number of tosses that come up *heads*. The image of X is $\{0, \dots, n\}$. By the calculations made in Example 8.8, for each $k = 0, \dots, n$, we have

$$P[X = k] = \binom{n}{k} p^k q^{n-k}.$$

The distribution of the random variable X is called a **binomial distribution**. Such a distribution is parameterized by the success probability p of the underlying Bernoulli trial, and by the number of times n the trial is repeated. $\square$

Uniform distributions are very nice, simple distributions. It is therefore good to have simple criteria that ensure that certain random variables have uniform distributions. The next theorem provides one such criterion. We need a definition: if S and T are finite sets, then we say that a given function $f : S \to T$ is a **regular function** if every element in the image of f has the same number of pre-images under f.

Theorem 8.4. *Suppose* $f : S \to T$ *is a surjective, regular function, and that* X *is a random variable that is uniformly distributed over* S. *Then* $f(X)$ *is uniformly distributed over* T.

Proof. The assumption that f is surjective and regular implies that for every $t \in T$, the set $S_t := f^{-1}(\{t\})$ has size $|S|/|T|$. So, for each $t \in T$, working directly from the definitions, we have

$$P[f(X) = t] = \sum_{\omega \in X^{-1}(S_t)} P(\omega) = \sum_{s \in S_t} \sum_{\omega \in X^{-1}(\{s\})} P(\omega) = \sum_{s \in S_t} P[X = s]$$

$$= \sum_{s \in S_t} 1/|S| = (|S|/|T|)/|S| = 1/|T|. \quad \square$$

As a corollary, we have:

Theorem 8.5. *Suppose that* $\rho : G \to G'$ *is a surjective homomorphism of finite abelian groups* G *and* G', *and that* X *is a random variable that is uniformly distributed over* G. *Then* $\rho(X)$ *is uniformly distributed over* G'.

Proof. It suffices to show that ρ is regular. Recall that the kernel K of ρ is a subgroup of G, and that for every $g' \in G'$, the set $\rho^{-1}(\{g'\})$ is a coset of K (see Theorem 6.19); moreover, every coset of K has the same size (see Theorem 6.14). These facts imply that ρ is regular. $\square$

Example 8.19. Let us continue with Example 8.16. Recall that for a given integer a, and positive integer n, $[a]_n \in \mathbb{Z}_n$ denotes the residue class of a modulo n. Let us define $X' := [X]_6$ and $Y' := [Y]_6$. It is not hard to see that both X' and Y' are uniformly distributed over $\mathbb{Z}_6$, while (X', Y') is uniformly distributed over $\mathbb{Z}_6 \times \mathbb{Z}_6$. Let us define $Z' :- X' + Y'$ (where addition here is in $\mathbb{Z}_6$). We claim that Z' is uniformly distributed over $\mathbb{Z}_6$. This follows immediately from the fact that the map that sends $(a, b) \in \mathbb{Z}_6 \times \mathbb{Z}_6$ to $a + b \in \mathbb{Z}_6$ is a surjective group homomorphism (see Example 6.45). Further, we claim that (X', Z') is uniformly distributed over $\mathbb{Z}_6 \times \mathbb{Z}_6$. This follows immediately from the fact that the map that sends $(a, b) \in \mathbb{Z}_6 \times \mathbb{Z}_6$ to $(a, a + b) \in \mathbb{Z}_6 \times \mathbb{Z}_6$ is a surjective group homomorphism (indeed, it is a group isomorphism). $\square$

Let X be a random variable whose image is S. Let B be an event with $P[B] \neq 0$. The **conditional distribution of** X **given** B is defined to be the distribution of X *relative to the conditional distribution* $P(\cdot|B)$, that is, the distribution $P_{X|B} : S \to [0, 1]$ defined by $P_{X|B}(s) := P[X = s \mid B]$ for $s \in S$.

Suppose X and Y are random variables, with images S and T, respectively. We say X and Y are **independent** if for all $s \in S$ and all $t \in T$, the events $X = s$ and $Y = t$ are independent, which is to say,

$$P[(X = s) \cap (Y = t)] = P[X = s] \, P[Y = t].$$

Equivalently, X and Y are independent if and only if the distribution of (X, Y) is essentially equal to the product of the distribution of X and the distribution of Y. As a special case, if X is uniformly distributed over S, and Y is uniformly distributed over T, then X and Y are independent if and only if (X, Y) is uniformly distributed over $S \times T$.

Independence can also be characterized in terms of conditional probabilities. From the definitions, it is immediate that X and Y are independent if and only if for all values t taken by Y with non-zero probability, we have

$$P[X = s \mid Y = t] = P[X = s]$$

for all $s \in S$; that is, the conditional distribution of X given $Y = t$ is the same as the distribution of X. From this point of view, an intuitive interpretation of independence is that information about the value of one random variable does not reveal any information about the value of the other.

Example 8.20. Let us continue with Examples 8.16 and 8.19. The random variables X and Y are independent: each is uniformly distributed over $\{1, \ldots, 6\}$, and (X, Y) is uniformly distributed over $\{1, \ldots, 6\} \times \{1, \ldots, 6\}$. Let us calculate the conditional distribution of X given $Z = 4$. We have $P[X = s \mid Z = 4] = 1/3$ for $s = 1, 2, 3$, and $P[X = s \mid Z = 4] = 0$ for $s = 4, 5, 6$. Thus, the conditional distribution of X given $Z = 4$ is essentially the uniform distribution on $\{1, 2, 3\}$. Let us calculate the conditional distribution of Z given $X = 1$. We have $P[Z = u \mid X = 1] = 1/6$ for $u = 2, \ldots, 7$, and $P[Z = u \mid X = 1] = 0$ for $u = 8, \ldots, 12$. Thus, the conditional distribution of Z given $X = 1$ is essentially the uniform distribution on $\{2, \ldots, 7\}$. In particular, it is clear that X and Z are *not* independent. The random variables X' and Y' are independent, as are X' and Z': each of X', Y', and Z' is uniformly distributed over $\mathbb{Z}_6$, and each of (X', Y') and (X', Z') is uniformly distributed over $\mathbb{Z}_6 \times \mathbb{Z}_6$. $\square$

We now generalize the notion of independence to families of random variables. Let $\{X_i\}_{i \in I}$ be a finite family of random variables. Let us call a corresponding family of values $\{s_i\}_{i \in I}$ an **assignment** to $\{X_i\}_{i \in I}$ if s_i is in the image of X_i for each $i \in I$. For a given positive integer k, we say that the family $\{X_i\}_{i \in I}$ is **k-wise independent** if for every assignment $\{s_i\}_{i \in I}$ to $\{X_i\}_{i \in I}$, the family of events $\{X_i = s_i\}_{i \in I}$ is k-wise independent.

The notions of pairwise and mutual independence for random variables are defined following the same pattern that was used for events. The family $\{X_i\}_{i \in I}$ is called **pairwise independent** if it is 2-wise independent, which means that for all $i, j \in I$ with $i \neq j$, the variables X_i and X_j are independent. The family $\{X_i\}_{i \in I}$ is called **mutually independent** if it is k-wise independent for all positive integers k. Equivalently, and more explicitly, mutual independence means that for every assignment $\{s_i\}_{i \in I}$ to $\{X_i\}_{i \in I}$, we have

$$P\left[\bigcap_{j \in J}(X_j = s_j)\right] = \prod_{j \in J} P[X_j = s_j] \text{ for all } J \subseteq I. \tag{8.13}$$

If $n := |I| > 0$, mutual independence is equivalent to n-wise independence; moreover, if $0 < k \leq n$, then $\{X_i\}_{i \in I}$ is k-wise independent if and only if $\{X_j\}_{j \in J}$ is mutually independent for every $J \subseteq I$ with $|J| = k$.

Example 8.21. Returning again to Examples 8.16, 8.19, and 8.20, we see that the family of random variables X', Y', Z' is pairwise independent, but not mutually independent; for example,

$$P\left[(X' = [0]_6) \cap (Y' = [0]_6) \cap (Z' = [0]_6)\right] = 1/6^2,$$

but

$$P\left[X' = [0]_6\right] \cdot P\left[Y' = [0]_6\right] \cdot P\left[Z' = [0]_6\right] = 1/6^3. \quad \square$$

Example 8.22. Suppose $\{A_i\}_{i \in I}$ is a finite family of events. Let $\{X_i\}_{i \in I}$ be the corresponding family of indicator variables, so that for each $i \in I$, $X_i = 1$ if A_i occurs, and $X_i = 0$, otherwise. Theorem 8.3 immediately implies that for every positive integer k, $\{A_i\}_{i \in I}$ is k-wise independent if and only if $\{X_i\}_{i \in I}$ is k-wise independent. $\square$

Example 8.23. Consider again Example 8.15, where we toss a fair coin 3 times. For $i = 1, 2, 3$, let X_i be the indicator variable for the event A_i that the ith toss comes up *heads*. Then $\{X_i\}_{i=1}^{3}$ is a mutually independent family of random variables. Let Y_{12} be the indicator variable for the event B_{12} that tosses 1 and 2 agree;

similarly, let Y_{13} be the indicator variable for the event B_{13}, and Y_{23} the indicator variable for B_{23}. Then the family of random variables Y_{12}, Y_{13}, Y_{23} is pairwise independent, but not mutually independent. $\square$

We next present a number of useful tools for establishing independence.

Theorem 8.6. *Let X be a random variable with image S, and Y be a random variable with image T. Further, suppose that $f : S \to [0, 1]$ and $g : T \to [0, 1]$ are functions such that*

$$\sum_{s \in S} f(s) = \sum_{t \in T} g(t) = 1, \tag{8.14}$$

and that for all $s \in S$ and $t \in T$, we have

$$P[(X = s) \cap (Y = t)] = f(s)g(t). \tag{8.15}$$

Then X and Y are independent, the distribution of X is f, and the distribution of Y is g.

Proof. Since $\{Y = t\}_{t \in T}$ is a partition of the sample space, making use of total probability (8.9), along with (8.15) and (8.14), we see that for all $s \in S$, we have

$$P[X = s] = \sum_{t \in T} P[(X = s) \cap (Y = t)] = \sum_{t \in T} f(s)g(t) = f(s) \sum_{t \in T} g(t) = f(s).$$

Thus, the distribution of X is indeed f. Exchanging the roles of X and Y in the above argument, we see that the distribution of Y is g. Combining this with (8.15), we see that X and Y are independent. $\square$

The generalization of Theorem 8.6 to families of random variables is a bit messy, but the basic idea is the same:

Theorem 8.7. *Let $\{X_i\}_{i \in I}$ be a finite family of random variables, where each X_i has image S_i. Also, let $\{f_i\}_{i \in I}$ be a family of functions, where for each $i \in I$, $f_i : S_i \to [0, 1]$ and $\sum_{s_i \in S_i} f_i(s_i) = 1$. Further, suppose that*

$$P\left[\bigcap_{i \in I}(X_i = s_i)\right] = \prod_{i \in I} f_i(s_i)$$

for each assignment $\{s_i\}_{i \in I}$ to $\{X_i\}_{i \in I}$. Then the family $\{X_i\}_{i \in I}$ is mutually independent, and for each $i \in I$, the distribution of X_i is f_i.

Proof. To prove the theorem, it suffices to prove the following statement: for every subset J of I, and every assignment $\{s_j\}_{j \in J}$ to $\{X_j\}_{j \in J}$, we have

$$P\left[\bigcap_{j \in J}(X_j = s_j)\right] = \prod_{j \in J} f_j(s_j).$$

Moreover, it suffices to prove this statement for the case where $J = I \setminus \{d\}$, for an arbitrary $d \in I$: this allows us to eliminate any one variable from the family, without affecting the hypotheses, and by repeating this procedure, we can eliminate any number of variables.

Thus, let $d \in I$ be fixed, let $J := I \setminus \{d\}$, and let $\{s_j\}_{j \in J}$ be a fixed assignment to $\{X_j\}_{j \in J}$. Then, since $\{X_d = s_d\}_{s_d \in S_d}$ is a partition of the sample space, we have

$$P\left[\bigcap_{j \in J}(X_j = s_j)\right] = P\left[\bigcup_{s_d \in S_d}\left(\bigcap_{i \in I}(X_i = s_i)\right)\right] = \sum_{s_d \in S_d} P\left[\bigcap_{i \in I}(X_i = s_i)\right]$$

$$= \sum_{s_d \in S_d}\prod_{i \in I} f_i(s_i) = \prod_{j \in J} f_j(s_j) \cdot \sum_{s_d \in S_d} f_d(s_d) = \prod_{j \in J} f_j(s_j). \quad \square$$

This theorem has several immediate consequences. First of all, mutual independence may be more simply characterized:

Theorem 8.8. *Let $\{X_i\}_{i \in I}$ be a finite family of random variables. Suppose that for every assignment $\{s_i\}_{i \in I}$ to $\{X_i\}_{i \in I}$, we have*

$$P\left[\bigcap_{i \in I}(X_i = s_i)\right] = \prod_{i \in I} P[X_i = s_i].$$

Then $\{X_i\}_{i \in I}$ is mutually independent.

Theorem 8.8 says that to check for mutual independence, we only have to consider the index set $J = I$ in (8.13). Put another way, it says that a family of random variables $\{X_i\}_{i=1}^n$ is mutually independent if and only if the distribution of $(X_1, \ldots, X_n)$ is essentially equal to the product of the distributions of the individual X_i's.

Based on the definition of mutual independence, and its characterization in Theorem 8.8, the following is also immediate:

Theorem 8.9. *Suppose $\{X_i\}_{i=1}^n$ is a family of random variables, and that m is an integer with $0 < m < n$. Then the following are equivalent:*

(i) *$\{X_i\}_{i=1}^n$ is mutually independent;*

(ii) *$\{X_i\}_{i=1}^m$ is mutually independent, $\{X_i\}_{i=m+1}^n$ is mutually independent, and the two variables $(X_1, \ldots, X_m)$ and $(X_{m+1}, \ldots, X_n)$ are independent.*

The following is also an immediate consequence of Theorem 8.7 (it also follows easily from Theorem 8.4).

Theorem 8.10. *Suppose that $X_1, \ldots, X_n$ are random variables, and that $S_1, \ldots, S_n$ are finite sets. Then the following are equivalent:*

(i) *$(X_1, \ldots, X_n)$ is uniformly distributed over $S_1 \times \cdots \times S_n$;*

(ii) $\{X_i\}_{i=1}^n$ is mutually independent, with each X_i uniformly distributed over S_i.

Another immediate consequence of Theorem 8.7 is the following:

Theorem 8.11. *Suppose* P *is the product distribution* $P_1 \cdots P_n$, *where each* P_i *is a probability distribution on a sample space* Ω_i, *so that the sample space of* P *is* $\Omega = \Omega_1 \times \cdots \times \Omega_n$. *For each* $i = 1, \ldots, n$, *let* X_i *be the random variable that projects on the ith coordinate, so that* $X_i(\omega_1, \ldots, \omega_n) = \omega_i$. *Then* $\{X_i\}_{i=1}^n$ *is mutually independent, and for each* $i = 1, \ldots, n$, *the distribution of* X_i *is* P_i.

Theorem 8.11 is often used to synthesize independent random variables "out of thin air," by taking the product of appropriate probability distributions. Other arguments may then be used to prove the independence of variables derived from these.

Example 8.24. Theorem 8.11 immediately implies that in Example 8.18, the family of indicator variables $\{X_i\}_{i=1}^n$ is mutually independent. □

The following theorem gives us yet another way to establish independence.

Theorem 8.12. *Suppose* $\{X_i\}_{i=1}^n$ *is a mutually independent family of random variables. Further, suppose that for* $i = 1, \ldots, n$, $Y_i := g_i(X_i)$ *for some function* g_i. *Then* $\{Y_i\}_{i=1}^n$ *is mutually independent.*

Proof. It suffices to prove the theorem for $n = 2$. The general case follows easily by induction, using Theorem 8.9. For $i = 1, 2$, let t_i be any value in the image of Y_i, and let $S_i' := g_i^{-1}(\{t_i\})$. We have

$$P[(Y_1 = t_1) \cap (Y_2 = t_2)] = P\left[\left(\bigcup_{s_1 \in S_1'} (X_1 = s_1)\right) \cap \left(\bigcup_{s_2 \in S_2'} (X_2 = s_2)\right)\right]$$

$$= P\left[\bigcup_{s_1 \in S_1'} \bigcup_{s_2 \in S_2'} \left((X_1 = s_1) \cap (X_2 = s_2)\right)\right]$$

$$= \sum_{s_1 \in S_1'} \sum_{s_2 \in S_2'} P[(X_1 = s_1) \cap (X_2 = s_2)]$$

$$= \sum_{s_1 \in S_1'} \sum_{s_2 \in S_2'} P[X_1 = s_1] P[X_2 = s_2]$$

$$= \left(\sum_{s_1 \in S_1'} P[X_1 = s_1]\right)\left(\sum_{s_2 \in S_2'} P[X_2 = s_2]\right)$$

$$= P\left[\bigcup_{s_1 \in S_1'} (X_1 = s_1)\right] P\left[\bigcup_{s_2 \in S_2'} (X_2 = s_2)\right] = P[Y_1 = t_1] P[Y_2 = t_2]. \quad \square$$

As a special case of the above theorem, if each g_i is the characteristic function for some subset S_i' of the image of X_i, then $X_1 \in S_1', \ldots, X_n \in S_n'$ form a mutually independent family of events.

The next theorem is quite handy in proving the independence of random variables in a variety of algebraic settings.

Theorem 8.13. *Suppose that G is a finite abelian group, and that W is a random variable uniformly distributed over G. Let Z be another random variable, taking values in some finite set U, and suppose that W and Z are independent. Let $\sigma : U \to G$ be some function, and define $Y := W + \sigma(Z)$. Then Y is uniformly distributed over G, and Y and Z are independent.*

Proof. Consider any fixed values $t \in G$ and $u \in U$. Evidently, the events $(Y = t) \cap (Z = u)$ and $(W = t - \sigma(u)) \cap (Z = u)$ are the same, and therefore, because W and Z are independent, we have

$$P[(Y = t) \cap (Z = u)] = P[W = t - \sigma(u)] \, P[Z = u] = \frac{1}{|G|} P[Z = u]. \qquad (8.16)$$

Since this holds for every $u \in U$, making use of total probability (8.9), we have

$$P[Y = t] = \sum_{u \in U} P[(Y = t) \cap (Z = u)] = \frac{1}{|G|} \sum_{u \in U} P[Z = u] = \frac{1}{|G|}.$$

Thus, Y is uniformly distributed over G, and by (8.16), Y and Z are independent. (This conclusion could also have been deduced directly from (8.16) using Theorem 8.6 — we have repeated the argument here.) $\square$

Note that in the above theorem, we make no assumption about the distribution of Z, or any properties of the function σ.

Example 8.25. Theorem 8.13 may be used to justify the security of the **one-time pad** encryption scheme. Here, the variable W represents a random, secret key — the "pad" — that is shared between Alice and Bob; U represents a space of possible messages; Z represents a "message source," from which Alice draws her message according to some distribution; finally, the function $\sigma : U \to G$ represents some invertible "encoding transformation" that maps messages into group elements.

To encrypt a message drawn from the message source, Alice encodes the message as a group element, and then adds the pad. The variable $Y := W + \sigma(Z)$ represents the resulting ciphertext. Since $Z = \sigma^{-1}(Y - W)$, when Bob receives the ciphertext, he decrypts it by subtracting the pad, and converting the resulting group element back into a message. Because the message source Z and ciphertext Y are independent, an eavesdropping adversary who learns the value of Y does not learn anything about Alice's message: for any particular ciphertext t, the conditional distribution of Z given $Y = t$ is the same as the distribution of Z.

The term "one time" comes from the fact that a given encryption key should be used only once; otherwise, security may be compromised. Indeed, suppose the key is used a second time, encrypting a message drawn from a second source Z'. The second ciphertext is represented by the random variable $Y' := W + \sigma(Z')$. In general, the random variables (Z, Z') and (Y, Y') will not be independent, since $Y - Y' = \sigma(Z) - \sigma(Z')$. To illustrate this more concretely, suppose Z is uniformly distributed over a set of 1000 messages, Z' is uniformly distributed over a set of two messages, say, $\{u'_1, u'_2\}$, and that Z and Z' are independent. Now, without any further information about Z, an adversary would have at best a 1-in-a-1000 chance of guessing its value. However, if he sees that $Y = t$ and $Y' = t'$, for particular values $t, t' \in G$, then he has a 1-in-2-chance, since the value of Z is equally likely to be one of just two messages, namely, $u_1 := \sigma^{-1}(t - t' + \sigma(u'_1))$ and $u_2 := \sigma^{-1}(t - t' + \sigma(u'_2))$; more formally, the conditional distribution of Z given $(Y = t) \cap (Y' = t')$ is essentially the uniform distribution on $\{u_1, u_2\}$.

In practice, it is convenient to define the group G to be the group of all bit strings of some fixed length, with bit-wise exclusive-or as the group operation. The encoding function σ simply "serializes" a message as a bit string. $\square$

Example 8.26. Theorem 8.13 may also be used to justify a very simple type of **secret sharing**. A colorful, if militaristic, motivating scenario is the following. To launch a nuclear missile, two officers who carry special keys must insert their keys simultaneously into the "authorization device" (at least, that is how it works in Hollywood). In the digital version of this scenario, an authorization device contains a secret, digital "launch code," and each officer holds a digital "share" of this code, so that (i) individually, each share reveals no information about the launch code, but (ii) collectively, the two shares may be combined in a simple way to derive the launch code. Thus, to launch the missile, both officers must input their shares into the authorization device; hardware in the authorization device combines the two shares, and compares the resulting code against the launch code it stores — if they match, the missile flies.

In the language of Theorem 8.13, the launch code is represented by the random variable Z, and the two shares by W and $Y := W + \sigma(Z)$, where (as in the previous example) $\sigma : U \to G$ is some simple, invertible encoding function. Because W and Z are independent, information about the share W leaks no information about the launch code Z; likewise, since Y and Z are independent, information about Y leaks no information about Z. However, by combining both shares, the launch code is easily constructed as $Z = \sigma^{-1}(Y - W)$. $\square$

Example 8.27. Let k be a positive integer. This example shows how we can take a mutually independent family of k random variables, and, from it, construct a much larger, k-wise independent family of random variables.

Let p be a prime, with $p \geq k$. Let $\{H_i\}_{i=0}^{k-1}$ be a mutually independent family of random variables, each of which is uniformly distributed over $\mathbb{Z}_p$. Let us set $H := (H_0, \ldots, H_{k-1})$, which, by assumption, is uniformly distributed over $\mathbb{Z}_p^{\times k}$. For each $s \in \mathbb{Z}_p$, we define the function $\rho_s : \mathbb{Z}_p^{\times k} \to \mathbb{Z}_p$ as follows: for $r = (r_0, \ldots, r_{k-1}) \in \mathbb{Z}_p^{\times k}$, $\rho_s(r) := \sum_{i=0}^{k-1} r_i s^i$; that is, $\rho_s(r)$ is the value obtained by evaluating the polynomial $r_0 + r_1 X + \cdots + r_{k-1} X^{k-1} \in \mathbb{Z}_p[X]$ at the point s.

Each $s \in \mathbb{Z}_p$ defines a random variable $\rho_s(H) = H_0 + H_1 s + \cdots + H_{k-1} s^{k-1}$. We claim that the family of random variables $\{\rho_s(H)\}_{s \in \mathbb{Z}_p}$ is k-wise independent, with each individual $\rho_s(H)$ uniformly distributed over $\mathbb{Z}_p$. By Theorem 8.10, it suffices to show the following: for all distinct points $s_1, \ldots, s_k \in \mathbb{Z}_p$, the random variable $W := (\rho_{s_1}(H), \ldots, \rho_{s_k}(H))$ is uniformly distributed over $\mathbb{Z}_p^{\times k}$. So let $s_1, \ldots, s_k$ be fixed, distinct elements of $\mathbb{Z}_p$, and define the function

$$\begin{aligned} \rho : \quad & \mathbb{Z}_p^{\times k} \to \mathbb{Z}_p^{\times k} \\ & r \mapsto (\rho_{s_1}(r), \ldots, \rho_{s_k}(r)). \end{aligned} \tag{8.17}$$

Thus, $W = \rho(H)$, and by Lagrange interpolation (Theorem 7.15), the function ρ is a bijection; moreover, since H is uniformly distributed over $\mathbb{Z}_p^{\times k}$, so is W.

Of course, the field $\mathbb{Z}_p$ may be replaced by an arbitrary finite field. $\square$

Example 8.28. Consider again the secret sharing scenario of Example 8.26. Suppose at the critical moment, one of the officers is missing in action. The military planners would perhaps like a more flexible secret sharing scheme; for example, perhaps shares of the launch code should be distributed to three officers, in such a way that no single officer can authorize a launch, but any two can. More generally, for positive integers k and ℓ, with $\ell \geq k + 1$, the scheme should distribute shares among ℓ officers, so that no coalition of k (or fewer) officers can authorize a launch, yet any coalition of $k + 1$ officers can. Using the construction of the previous example, this is easily achieved, as follows.

Let us model the secret launch code as a random variable Z, taking values in a finite set U. Assume that p is prime, with $p \geq \ell$, and that $\sigma : U \to \mathbb{Z}_p$ is a simple, invertible encoding function. To construct the shares, we make use of random variables $H_0, \ldots, H_{k-1}$, where each H_i is uniformly distributed over $\mathbb{Z}_p$, and the family of random variables $H_0, \ldots, H_{k-1}, Z$ is mutually independent. For each $s \in \mathbb{Z}_p$, we define the random variable

$$Y_s := H_0 + H_1 s + \cdots + H_{k-1} s^{k-1} + \sigma(Z) s^k.$$

We can pick any subset $S \subseteq \mathbb{Z}_p$ of size ℓ that we wish, so that for each $s \in S$, an officer gets the secret share Y_s (along with the public value s).

First, we show how any coalition of $k+1$ officers can reconstruct the launch code from their collection of shares, say, $Y_{s_1}, \ldots, Y_{s_{k+1}}$. This is easily done by means of the Lagrange interpolation formula (again, Theorem 7.15). Indeed, we only need to recover the high-order coefficient, $\sigma(Z)$, which we can obtain via the formula

$$\sigma(Z) = \sum_{i=1}^{k+1} \frac{Y_{s_i}}{\prod_{j \neq i}(s_i - s_j)}.$$

Second, we show that no coalition of k officers learn anything about the launch code, even if they pool their shares. Formally, this means that if $s_1, \ldots, s_k$ are fixed, distinct points, then $Y_{s_1}, \ldots, Y_{s_k}, Z$ form a mutually independent family of random variables. This is easily seen, as follows. Define $H := (H_0, \ldots, H_{k-1})$, and $W := \rho(H)$, where $\rho : \mathbb{Z}_p^{\times k} \to \mathbb{Z}_p^{\times k}$ is as defined in (8.17), and set $Y := (Y_{s_1}, \ldots, Y_{s_k})$. Now, by hypothesis, H and Z are independent, and H is uniformly distributed over $\mathbb{Z}_p^{\times k}$. As we noted in Example 8.27, ρ is a bijection, and hence, W is uniformly distributed over $\mathbb{Z}_p^{\times k}$; moreover (by Theorem 8.12), W and Z are independent. Observe that $Y = W + \sigma'(Z)$, where σ' maps $u \in U$ to $(\sigma(u)s_1^k, \ldots, \sigma(u)s_k^k) \in \mathbb{Z}_p^{\times k}$, and so applying Theorem 8.13 (with the group $\mathbb{Z}_p^{\times k}$, the random variables W and Z, and the function σ'), we see that Y and Z are independent, where Y is uniformly distributed over $\mathbb{Z}_p^{\times k}$. From this, it follows (using Theorems 8.9 and 8.10) that the family of random variables $Y_{s_1}, \ldots, Y_{s_k}, Z$ is mutually independent, with each Y_{s_i} uniformly distributed over $\mathbb{Z}_p$.

Finally, we note that when $k = 1$, $\ell = 2$, and $S = \{0, 1\}$, this construction degenerates to the construction in Example 8.26 (with the additive group $\mathbb{Z}_p$). $\square$

EXERCISE 8.12. Suppose X and X' are random variables that take values in a set S and that have *essentially* the same distribution. Show that if $f : S \to T$ is a function, then $f(X)$ and $f(X')$ have essentially the same distribution.

EXERCISE 8.13. Let $\{X_i\}_{i=1}^n$ be a family of random variables, and let S_i be the image of X_i for $i = 1, \ldots, n$. Show that $\{X_i\}_{i=1}^n$ is mutually independent if and only if for each $i = 2, \ldots, n$, and for all $s_1 \in S_1, \ldots, s_i \in S_i$, we have

$$P[X_i = s_i \mid (X_1 = s_1) \cap \cdots \cap (X_{i-1} = s_{i-1})] = P[X_i = s_i].$$

EXERCISE 8.14. Suppose that $\rho : G \to G'$ is a surjective group homomorphism, where G and G' are finite abelian groups. Show that if $g', h' \in G'$, and X and Y are independent random variables, where X is uniformly distributed over $\rho^{-1}(\{g'\})$, and Y takes values in $\rho^{-1}(\{h'\})$, then $X + Y$ is uniformly distributed over $\rho^{-1}(\{g' + h'\})$.

EXERCISE 8.15. Suppose X and Y are random variables, where X takes values in S, and Y takes values in T. Further suppose that Y' is uniformly distributed over T, and that (X, Y) and Y' are independent. Let ϕ be a predicate on $S \times T$. Show that $P[\phi(X, Y) \cap (Y = Y')] = P[\phi(X, Y)]/|T|$.

EXERCISE 8.16. Let X and Y be independent random variables, where X is uniformly distributed over a set S, and Y is uniformly distributed over a set $T \subseteq S$. Define a third random variable Z as follows: if $X \in T$, then $Z := X$; otherwise, $Z := Y$. Show that Z is uniformly distributed over T.

EXERCISE 8.17. Let n be a positive integer, and let X be a random variable, uniformly distributed over $\{0, \ldots, n-1\}$. For each positive divisor d of n, let us define the random variable $X_d := X \bmod d$. Show that:

    (a) if d is a divisor of n, then the variable X_d is uniformly distributed over $\{0, \ldots, d-1\}$;

    (b) if $d_1, \ldots, d_k$ are divisors of n, then $\{X_{d_i}\}_{i=1}^{k}$ is mutually independent if and only if $\{d_i\}_{i=1}^{k}$ is pairwise relatively prime.

EXERCISE 8.18. Suppose X and Y are random variables, each uniformly distributed over $\mathbb{Z}_2$, but not necessarily independent. Show that the distribution of (X, Y) is the same as the distribution of $(X + 1, Y + 1)$.

EXERCISE 8.19. Let $I := \{1, \ldots, n\}$, where $n \geq 2$, let $B := \{0, 1\}$, and let G be a finite abelian group, with $|G| > 1$. Suppose that $\{X_{ib}\}_{(i,b) \in I \times B}$ is a mutually independent family of random variables, each uniformly distributed over G. For each $\beta = (b_1, \ldots, b_n) \in B^{\times n}$, let us define the random variable $Y_\beta := X_{1b_1} + \cdots + X_{nb_n}$. Show that each Y_β is uniformly distributed over G, and that $\{Y_\beta\}_{\beta \in B^{\times n}}$ is 3-wise independent, but not 4-wise independent.

8.4 Expectation and variance

Let P be a probability distribution on a sample space Ω. If X is a real-valued random variable, then its **expected value**, or **expectation**, is

$$E[X] := \sum_{\omega \in \Omega} X(\omega) \, P(\omega). \tag{8.18}$$

If S is the image of X, and if for each $s \in S$ we group together the terms in (8.18) with $X(\omega) = s$, then we see that

$$E[X] = \sum_{s \in S} s \, P[X = s]. \tag{8.19}$$

From (8.19), it is clear that $E[X]$ depends only on the distribution of X: if X' is another random variable with the same (or essentially the same) distribution as X, then $E[X] = E[X']$.

More generally, suppose X is an arbitrary random variable (not necessarily real valued) whose image is S, and f is a real-valued function on S. Then again, if for each $s \in S$ we group together the terms in (8.18) with $X(\omega) = s$, we see that

$$E[f(X)] = \sum_{s \in S} f(s) P[X = s]. \tag{8.20}$$

We make a few trivial observations about expectation, which the reader may easily verify. First, if X is equal to a constant c (i.e., $X(\omega) = c$ for every $\omega \in \Omega$), then $E[X] = E[c] = c$. Second, if X and Y are random variables such that $X \geq Y$ (i.e., $X(\omega) \geq Y(\omega)$ for every $\omega \in \Omega$), then $E[X] \geq E[Y]$. Similarly, if $X > Y$, then $E[X] > E[Y]$.

In calculating expectations, one rarely makes direct use of (8.18), (8.19), or (8.20), except in rather trivial situations. The next two theorems develop tools that are often quite effective in calculating expectations.

Theorem 8.14 (Linearity of expectation). *If X and Y are real-valued random variables, and a is a real number, then*

$$E[X + Y] = E[X] + E[Y] \quad \text{and} \quad E[aX] = a\,E[X].$$

Proof. It is easiest to prove this using the defining equation (8.18) for expectation. For $\omega \in \Omega$, the value of the random variable $X+Y$ at ω is by definition $X(\omega)+Y(\omega)$, and so we have

$$E[X + Y] = \sum_{\omega} (X(\omega) + Y(\omega)) P(\omega)$$

$$= \sum_{\omega} X(\omega) P(\omega) + \sum_{\omega} Y(\omega) P(\omega)$$

$$= E[X] + E[Y].$$

For the second part of the theorem, by a similar calculation, we have

$$E[aX] = \sum_{\omega} (aX(\omega)) P(\omega) = a \sum_{\omega} X(\omega) P(\omega) = a\,E[X]. \quad \square$$

More generally, the above theorem implies (using a simple induction argument) that if $\{X_i\}_{i \in I}$ is a finite family of real-valued random variables, then we have

$$E\left[\sum_{i \in I} X_i\right] = \sum_{i \in I} E[X_i]. \tag{8.21}$$

So we see that expectation is linear; however, expectation is not in general multiplicative, except in the case of independent random variables:

Theorem 8.15. *If X and Y are independent, real-valued random variables, then* $E[XY] = E[X]E[Y]$.

Proof. It is easiest to prove this using (8.20), with the function $f(s,t) := st$ applied to the random variable (X,Y). We have

$$E[XY] = \sum_{s,t} st\, P[(X = s) \cap (Y = t)]$$

$$= \sum_{s,t} st\, P[X = s]P[Y = t]$$

$$= \left(\sum_s s\, P[X = s]\right)\left(\sum_t t\, P[Y = t]\right)$$

$$= E[X]E[Y]. \quad \square$$

More generally, the above theorem implies (using a simple induction argument) that if $\{X_i\}_{i \in I}$ is a finite, mutually independent family of real-valued random variables, then

$$E\left[\prod_{i \in I} X_i\right] = \prod_{i \in I} E[X_i]. \tag{8.22}$$

The following simple facts are also sometimes quite useful in calculating expectations:

Theorem 8.16. *Let X be a $0/1$-valued random variable. Then* $E[X] = P[X = 1]$.

Proof. $E[X] = 0 \cdot P[X = 0] + 1 \cdot P[X = 1] = P[X = 1]$. $\square$

Theorem 8.17. *If X is a random variable that takes only non-negative integer values, then*

$$E[X] = \sum_{i \geq 1} P[X \geq i].$$

Note that since X has a finite image, the sum appearing above is finite.

Proof. Suppose that the image of X is contained in $\{0, \ldots, n\}$, and for $i = 1, \ldots, n$, let X_i be the indicator variable for the event $X \geq i$. Then $X = X_1 + \cdots + X_n$, and by linearity of expectation and Theorem 8.16, we have

$$E[X] = \sum_{i=1}^{n} E[X_i] = \sum_{i=1}^{n} P[X \geq i]. \quad \square$$

Let X be a real-valued random variable with $\mu := E[X]$. The **variance** of X is $\text{Var}[X] := E[(X - \mu)^2]$. The variance provides a measure of the spread or dispersion of the distribution of X around its expected value. Note that since $(X - \mu)^2$ takes only non-negative values, variance is always non-negative.

Theorem 8.18. *Let X be a real-valued random variable, with $\mu := E[X]$, and let a and b be real numbers. Then we have*

(i) $\text{Var}[X] = E[X^2] - \mu^2$,

(ii) $\text{Var}[aX] = a^2 \text{Var}[X]$, *and*

(iii) $\text{Var}[X + b] = \text{Var}[X]$.

Proof. For part (i), observe that

$$\text{Var}[X] = E[(X - \mu)^2] = E[X^2 - 2\mu X + \mu^2]$$
$$= E[X^2] - 2\mu E[X] + E[\mu^2] = E[X^2] - 2\mu^2 + \mu^2$$
$$= E[X^2] - \mu^2,$$

where in the third equality, we used the fact that expectation is linear, and in the fourth equality, we used the fact that $E[c] = c$ for constant c (in this case, $c = \mu^2$).

For part (ii), observe that

$$\text{Var}[aX] = E[a^2 X^2] - E[aX]^2 = a^2 E[X^2] - (a\mu)^2$$
$$= a^2(E[X^2] - \mu^2) = a^2 \text{Var}[X],$$

where we used part (i) in the first and fourth equality, and the linearity of expectation in the second.

Part (iii) follows by a similar calculation:

$$\text{Var}[X + b] = E[(X + b)^2] - (\mu + b)^2$$
$$= (E[X^2] + 2b\mu + b^2) - (\mu^2 + 2b\mu + b^2)$$
$$= E[X^2] - \mu^2 = \text{Var}[X]. \quad \square$$

The following is an immediate consequence of part (i) of Theorem 8.18, and the fact that variance is always non-negative:

Theorem 8.19. *If X is a real-valued random variable, then $E[X^2] \geq E[X]^2$.*

Unlike expectation, the variance of a sum of random variables is not equal to the sum of the variances, unless the variables are *pairwise independent*:

Theorem 8.20. *If $\{X_i\}_{i \in I}$ is a finite, pairwise independent family of real-valued random variables, then*

$$\text{Var}\left[\sum_{i \in I} X_i\right] = \sum_{i \in I} \text{Var}[X_i].$$

Proof. We have

$$\text{Var}\left[\sum_{i\in I} X_i\right] = E\left[\left(\sum_{i\in I} X_i\right)^2\right] - \left(E\left[\sum_{i\in I} X_i\right]\right)^2$$

$$= \sum_{i\in I} E[X_i^2] + \sum_{\substack{i,j\in I \\ i\neq j}} \left(E[X_iX_j] - E[X_i]E[X_j]\right) - \sum_{i\in I} E[X_i]^2$$

(by linearity of expectation and rearranging terms)

$$= \sum_{i\in I} E[X_i^2] - \sum_{i\in I} E[X_i]^2$$

(by pairwise independence and Theorem 8.15)

$$= \sum_{i\in I} \text{Var}[X_i]. \quad \square$$

Corresponding to Theorem 8.16, we have:

Theorem 8.21. *Let X be a $0/1$-valued random variable, with $p := P[X = 1]$ and $q := P[X = 0] = 1 - p$. Then $\text{Var}[X] = pq$.*

Proof. We have $E[X] = p$ and $E[X^2] = P[X^2 = 1] = P[X = 1] = p$. Therefore,

$$\text{Var}[X] = E[X^2] - E[X]^2 = p - p^2 = p(1 - p) = pq. \quad \square$$

Let B be an event with $P[B] \neq 0$, and let X be a real-valued random variable. We define the **conditional expectation of X given B**, denoted $E[X \mid B]$, to be the expected value of the X *relative to the conditional distribution* $P(\cdot \mid B)$, so that

$$E[X \mid B] = \sum_{\omega\in\Omega} X(\omega)\,P(\omega \mid B) = P[B]^{-1} \sum_{\omega\in B} X(\omega)\,P(\omega).$$

Analogous to (8.19), if S is the image of X, we have

$$E[X \mid B] = \sum_{s\in S} s\,P[X = s \mid B]. \tag{8.23}$$

Furthermore, suppose I is a finite index set, and $\{B_i\}_{i\in I}$ is a partition of the sample space, where each B_i occurs with non-zero probability. If for each $i \in I$ we group together the terms in (8.18) with $\omega \in B_i$, we obtain the **law of total expectation**:

$$E[X] = \sum_{i\in I} E[X \mid B_i]\,P[B_i]. \tag{8.24}$$

Example 8.29. Let X be uniformly distributed over $\{1, \ldots, m\}$. Let us compute $E[X]$ and $\text{Var}[X]$. We have

$$E[X] = \sum_{s=1}^{m} s \cdot \frac{1}{m} = \frac{m(m+1)}{2} \cdot \frac{1}{m} = \frac{m+1}{2}.$$

We also have

$$E[X^2] = \sum_{s=1}^{m} s^2 \cdot \frac{1}{m} = \frac{m(m+1)(2m+1)}{6} \cdot \frac{1}{m} = \frac{(m+1)(2m+1)}{6}.$$

Therefore,

$$Var[X] = E[X^2] - E[X]^2 = \frac{m^2 - 1}{12}. \quad \square$$

Example 8.30. Let X denote the value of a roll of a die. Let $\mathcal{A}$ be the event that X is even. Then the conditional distribution of X given $\mathcal{A}$ is essentially the uniform distribution on $\{2, 4, 6\}$, and hence

$$E[X \mid \mathcal{A}] = \frac{2 + 4 + 6}{3} = 4.$$

Similarly, the conditional distribution of X given $\overline{\mathcal{A}}$ is essentially the uniform distribution on $\{1, 3, 5\}$, and so

$$E[X \mid \overline{\mathcal{A}}] = \frac{1 + 3 + 5}{3} = 3.$$

Using the law of total expectation, we can compute the expected value of X as follows:

$$E[X] = E[X \mid \mathcal{A}] P[\mathcal{A}] + E[X \mid \overline{\mathcal{A}}] P[\overline{\mathcal{A}}] = 4 \cdot \frac{1}{2} + 3 \cdot \frac{1}{2} = \frac{7}{2},$$

which agrees with the calculation in the previous example. $\square$

Example 8.31. Let X be a random variable with a binomial distribution, as in Example 8.18, that counts the number of successes among n Bernoulli trials, each of which succeeds with probability p. Let us compute $E[X]$ and $Var[X]$. We can write X as the sum of indicator variables, $X = \sum_{i=1}^{n} X_i$, where X_i is the indicator variable for the event that the ith trial succeeds; each X_i takes the value 1 with probability p and 0 with probability $q := 1 - p$, and the family of random variables $\{X_i\}_{i=1}^{n}$ is mutually independent (see Example 8.24). By Theorems 8.16 and 8.21, we have $E[X_i] = p$ and $Var[X_i] = pq$ for $i = 1, \ldots, n$. By linearity of expectation, we have

$$E[X] = \sum_{i=1}^{n} E[X_i] = np.$$

By Theorem 8.20, and the fact that $\{X_i\}_{i=1}^{n}$ is mutually independent (and hence pairwise independent), we have

$$Var[X] = \sum_{i=1}^{n} Var[X_i] = npq. \quad \square$$

Example 8.32. Our proof of Theorem 8.1 could be elegantly recast in terms of indicator variables. For $B \subseteq \Omega$, let X_B be the indicator variable for B, so that $X_B(\omega) = \delta_\omega[B]$ for each $\omega \in \Omega$. Equation (8.8) then becomes

$$X_A = \sum_{\emptyset \subsetneq J \subseteq I} (-1)^{|J|-1} X_{A_J},$$

and by Theorem 8.16 and linearity of expectation, we have

$$P[A] = E[X_A] = \sum_{\emptyset \subsetneq J \subseteq I} (-1)^{|J|-1} E[X_{A_J}] = \sum_{\emptyset \subsetneq J \subseteq I} (-1)^{|J|-1} P[X_{A_J}]. \quad \square$$

EXERCISE 8.20. Suppose X is a real-valued random variable. Show that $|E[X]| \leq E[|X|] \leq E[X^2]^{1/2}$.

EXERCISE 8.21. Suppose X and Y take non-negative real values, and that $Y \leq c$ for some constant c. Show that $E[XY] \leq c\, E[X]$

EXERCISE 8.22. Let X be a 0/1-valued random variable. Show that $\mathrm{Var}[X] \leq 1/4$.

EXERCISE 8.23. Let B be an event with $P[B] \neq 0$, and let $\{B_i\}_{i \in I}$ be a finite, pairwise disjoint family of events whose union is B. Generalizing the law of total expectation (8.24), show that for every real-valued random variable X, if $I^* := \{i \in I : P[B_i] \neq 0\}$, then we have

$$E[X \mid B]\, P[B] = \sum_{i \in I^*} E[X \mid B_i]\, P[B_i].$$

Also show that if $E[X \mid B_i] \leq \alpha$ for each $i \in I^*$, then $E[X \mid B] \leq \alpha$.

EXERCISE 8.24. Let B be an event with $P[B] \neq 0$, and let $\{C_i\}_{i \in I}$ be a finite, pairwise disjoint family of events whose union contains B. Again, generalizing the law of total expectation, show that for every real-valued random variable X, if $I^* := \{i \in I : P[B \cap C_i] \neq 0\}$, then we have

$$E[X \mid B] = \sum_{i \in I^*} E[X \mid B \cap C_i]\, P[C_i \mid B].$$

EXERCISE 8.25. This exercise makes use of the notion of *convexity* (see §A8).

(a) Prove **Jensen's inequality**: if f is convex on an interval, and X is a random variable taking values in that interval, then $E[f(X)] \geq f(E[X])$. Hint: use induction on the size of the image of X. (Note that Theorem 8.19 is a special case of this, with $f(s) := s^2$.)

(b) Using part (a), show that if X takes non-negative real values, and α is a positive number, then $E[X^\alpha] \geq E[X]^\alpha$ if $\alpha \geq 1$, and $E[X^\alpha] \leq E[X]^\alpha$ if $\alpha \leq 1$.

(c) Using part (a), show that if X takes positive real values, then $E[X] \geq e^{E[\log X]}$.

(d) Using part (c), derive the **arithmetic/geometric mean inequality**: for all positive numbers $x_1, \ldots, x_n$, we have

$$(x_1 + \cdots + x_n)/n \geq (x_1 \cdots x_n)^{1/n}.$$

EXERCISE 8.26. For real-valued random variables X and Y, their **covariance** is defined as $\mathrm{Cov}[X, Y] := E[XY] - E[X]E[Y]$. Show that:

(a) if X, Y, and Z are real-valued random variables, and a is a real number, then $\mathrm{Cov}[X + Y, Z] = \mathrm{Cov}[X, Z] + \mathrm{Cov}[Y, Z]$ and $\mathrm{Cov}[aX, Z] = a\,\mathrm{Cov}[X, Z]$;

(b) if $\{X_i\}_{i \in I}$ is a finite family of real-valued random variables, then

$$\mathrm{Var}\left[\sum_{i \in I} X_i\right] = \sum_{i \in I} \mathrm{Var}[X_i] + \sum_{\substack{i,j \in I \\ i \neq j}} \mathrm{Cov}[X_i, X_j].$$

EXERCISE 8.27. Let $f : [0, 1] \to \mathbb{R}$ be a function that is "nice" in the following sense: for some constant c, we have $|f(s) - f(t)| \leq c|s - t|$ for all $s, t \in [0, 1]$. This condition is implied, for example, by the assumption that f has a derivative that is bounded in absolute value by c on the interval $[0, 1]$. For each positive integer n, define the polynomial $B_{n,f} := \sum_{k=0}^{n} \binom{n}{k} f(k/n) T^k (1 - T)^{n-k} \in \mathbb{R}[T]$. Show that $|B_{n,f}(p) - f(p)| \leq c/2\sqrt{n}$ for all positive integers n and all $p \in [0, 1]$. Hint: let X be a random variable with a binomial distribution that counts the number of successes among n Bernoulli trials, each of which succeeds with probability p, and begin by observing that $B_{n,f}(p) = E[f(X/n)]$. The polynomial $B_{n,f}$ is called the nth **Bernstein approximation** to f, and this result proves a classical result that any "nice" function can approximated to arbitrary precision by a polynomial of sufficiently high degree.

EXERCISE 8.28. Consider again the game played between Alice and Bob in Example 8.11. Suppose that to play the game, Bob must place a one dollar bet. However, after Alice reveals the sum of the two dice, Bob may elect to double his bet. If Bob's guess is correct, Alice pays him his bet, and otherwise Bob pays Alice his bet. Describe an optimal playing strategy for Bob, and calculate his expected winnings.

EXERCISE 8.29. A die is rolled repeatedly until it comes up "1," or until it is rolled n times (whichever comes first). What is the expected number of rolls of the die?

8.5 Some useful bounds

In this section, we present several theorems that can be used to bound the probability that a random variable deviates from its expected value by some specified amount.

Theorem 8.22 (Markov's inequality). *Let X be a random variable that takes only non-negative real values. Then for every $\alpha > 0$, we have*

$$P[X \geq \alpha] \leq E[X]/\alpha.$$

Proof. We have

$$E[X] = \sum_s s\, P[X = s] = \sum_{s < \alpha} s\, P[X = s] + \sum_{s \geq \alpha} s\, P[X = s],$$

where the summations are over elements s in the image of X. Since X takes only non-negative values, all of the terms are non-negative. Therefore,

$$E[X] \geq \sum_{s \geq \alpha} s\, P[X = s] \geq \sum_{s \geq \alpha} \alpha\, P[X = s] = \alpha\, P[X \geq \alpha]. \quad \square$$

Markov's inequality may be the only game in town when nothing more about the distribution of X is known besides its expected value. However, if the variance of X is also known, then one can get a better bound.

Theorem 8.23 (Chebyshev's inequality). *Let X be a real-valued random variable, with $\mu := E[X]$ and $v := \mathrm{Var}[X]$. Then for every $\alpha > 0$, we have*

$$P[|X - \mu| \geq \alpha] \leq v/\alpha^2.$$

Proof. Let $Y := (X - \mu)^2$. Then Y is always non-negative, and $E[Y] = v$. Applying Markov's inequality to Y, we have

$$P[|X - \mu| \geq \alpha] = P[Y \geq \alpha^2] \leq v/\alpha^2. \quad \square$$

An important special case of Chebyshev's inequality is the following. Suppose that $\{X_i\}_{i \in I}$ is a finite, non-empty, pairwise independent family of real-valued random variables, each with the same distribution. Let μ be the common value of $E[X_i]$, v be the common value of $\mathrm{Var}[X_i]$, and $n := |I|$. Set

$$\bar{X} := \frac{1}{n} \sum_{i \in I} X_i.$$

The variable $\bar{X}$ is called the **sample mean** of $\{X_i\}_{i \in I}$. By the linearity of expectation, we have $E[\bar{X}] = \mu$, and since $\{X_i\}_{i \in I}$ is pairwise independent, it follows from

Theorem 8.20 (along with part (ii) of Theorem 8.18) that $\text{Var}[\bar{X}] = v/n$. Applying Chebyshev's inequality, for every $\varepsilon > 0$, we have

$$P[|\bar{X} - \mu| \geq \varepsilon] \leq \frac{v}{n\varepsilon^2}. \tag{8.25}$$

The inequality (8.25) says that for all $\varepsilon > 0$, and for all $\delta > 0$, there exists n_0 (depending on ε and δ, as well as the variance v) such that $n \geq n_0$ implies

$$P[|\bar{X} - \mu| \geq \varepsilon] \leq \delta. \tag{8.26}$$

In words:

> As n gets large, the sample mean closely approximates the expected
> value μ with high probability.

This fact, known as the **law of large numbers**, justifies the usual intuitive interpretation given to expectation.

Let us now examine an even more specialized case of the above situation, where each X_i is a 0/1-valued random variable, taking the value 1 with probability p, and 0 with probability $q := 1 - p$. By Theorems 8.16 and 8.21, the X_i's have a common expected value p and variance pq. Therefore, by (8.25), for every $\varepsilon > 0$, we have

$$P[|\bar{X} - p| \geq \varepsilon] \leq \frac{pq}{n\varepsilon^2}. \tag{8.27}$$

The bound on the right-hand side of (8.27) decreases linearly in n. If one makes the stronger assumption that the family $\{X_i\}_{i \in I}$ is *mutually independent* (so that $X := \sum_i X_i$ has a binomial distribution), one can obtain a much better bound that decreases *exponentially* in n:

Theorem 8.24 (Chernoff bound). *Let $\{X_i\}_{i \in I}$ be a finite, non-empty, and mutually independent family of random variables, such that each X_i is 1 with probability p and 0 with probability $q := 1 - p$. Assume that $0 < p < 1$. Also, let $n := |I|$ and $\bar{X}$ be the sample mean of $\{X_i\}_{i \in I}$. Then for every $\varepsilon > 0$, we have:*

(i) $P[\bar{X} - p \geq \varepsilon] \leq e^{-n\varepsilon^2/2q}$;

(ii) $P[\bar{X} - p \leq -\varepsilon] \leq e^{-n\varepsilon^2/2p}$;

(iii) $P[|\bar{X} - p| \geq \varepsilon] \leq 2e^{-n\varepsilon^2/2}$.

Proof. First, we observe that (ii) follows directly from (i) by replacing X_i by $1 - X_i$ and exchanging the roles of p and q. Second, we observe that (iii) follows directly from (i) and (ii). Thus, it suffices to prove (i).

Let $\alpha > 0$ be a parameter, whose value will be determined later. Define the random variable $Z := e^{\alpha n(\bar{X} - p)}$. Since the function $x \mapsto e^{\alpha n x}$ is strictly increasing, we have $\bar{X} - p \geq \varepsilon$ if and only if $Z \geq e^{\alpha n\varepsilon}$. By Markov's inequality, it follows that

$$P[\bar{X} - p \geq \varepsilon] = P[Z \geq e^{\alpha n\varepsilon}] \leq E[Z]e^{-\alpha n\varepsilon}. \tag{8.28}$$

So our goal is to bound $E[Z]$ from above.

For each $i \in I$, define the random variable $Z_i := e^{\alpha(X_i-p)}$. Observe that $Z = \prod_{i \in I} Z_i$, that $\{Z_i\}_{i \in I}$ is a mutually independent family of random variables (see Theorem 8.12), and that for each $i \in I$, we have

$$E[Z_i] = e^{\alpha(1-p)}p + e^{\alpha(0-p)}q = pe^{\alpha q} + qe^{-\alpha p}.$$

It follows that

$$E[Z] = E\left[\prod_{i \in I} Z_i\right] = \prod_{i \in I} E[Z_i] = (pe^{\alpha q} + qe^{-\alpha p})^n.$$

We will prove below that

$$pe^{\alpha q} + qe^{-\alpha p} \leq e^{\alpha^2 q/2}. \tag{8.29}$$

From this, it follows that

$$E[Z] \leq e^{\alpha^2 qn/2}. \tag{8.30}$$

Combining (8.30) with (8.28), we obtain

$$P[\bar{X} - p \geq \varepsilon] \leq e^{\alpha^2 qn/2 - \alpha n\varepsilon}. \tag{8.31}$$

Now we choose the parameter α so as to minimize the quantity $\alpha^2 qn/2 - \alpha n\varepsilon$. The optimal value of α is easily seen to be $\alpha = \varepsilon/q$, and substituting this value of α into (8.31) yields (i).

To finish the proof of the theorem, it remains to prove the inequality (8.29). Let

$$\beta := pe^{\alpha q} + qe^{-\alpha p}.$$

We want to show that $\beta \leq e^{\alpha^2 q/2}$, or equivalently, that $\log \beta \leq \alpha^2 q/2$. We have

$$\beta = e^{\alpha q}(p + qe^{-\alpha}) = e^{\alpha q}(1 - q(1 - e^{-\alpha})),$$

and taking logarithms and applying parts (i) and (ii) of §A1, we obtain

$$\log \beta = \alpha q + \log(1 - q(1 - e^{-\alpha})) \leq \alpha q - q(1 - e^{-\alpha}) = q(e^{-\alpha} + \alpha - 1) \leq q\alpha^2/2.$$

This establishes (8.29) and completes the proof of the theorem. □

Thus, the Chernoff bound is a quantitatively superior version of the law of large numbers, although its range of application is clearly more limited.

Example 8.33. Suppose we toss a fair coin 10,000 times. The expected number of *heads* is 5,000. What is an upper bound on the probability α that we get 6,000 or more *heads*? Using Markov's inequality, we get $\alpha \leq 5/6$. Using Chebyshev's inequality, and in particular, the inequality (8.27), we get

$$\alpha \leq \frac{1/4}{10^4 10^{-2}} = \frac{1}{400}.$$

Finally, using the Chernoff bound, we obtain

$$\alpha \le e^{-10^4 10^{-2}/2(0.5)} = e^{-100} \approx 10^{-43.4}. \quad \square$$

EXERCISE 8.30. With notation and assumptions as in Theorem 8.24, and with $p := q := 1/2$, show that there exist constants c_1 and c_2 such that

$$P[|\bar{X} - 1/2| \ge c_1/\sqrt{n}] \le 1/2 \quad \text{and} \quad P[|\bar{X} - 1/2| \ge c_2/\sqrt{n}] \ge 1/2.$$

Hint: for the second inequality, use Exercise 5.16.

EXERCISE 8.31. In each step of a **random walk**, we toss a coin, and move either one unit to the right, or one unit to the left, depending on the outcome of the coin toss. The question is, after n steps, what is our expected distance from the starting point? Let us model this using a mutually independent family of random variables $\{Y_i\}_{i=1}^n$, with each Y_i uniformly distributed over $\{-1, 1\}$, and define $Y := Y_1 + \cdots + Y_n$. Show that the $c_1\sqrt{n} \le E[|Y|] \le c_2\sqrt{n}$, for some constants c_1 and c_2.

EXERCISE 8.32. The goal of this exercise is to prove that with probability very close to 1, a random number between 1 and m has very close to $\log\log m$ prime factors. To prove this result, you will need to use appropriate theorems from Chapter 5. Suppose N is a random variable that is uniformly distributed over $\{1, \ldots, m\}$, where $m \ge 3$. For $i = 1, \ldots, m$, let D_i be the indicator variable for the event that i divides N. Also, define $X := \sum_{p \le m} D_p$, where the sum is over all primes $p \le m$, so that X counts the number of distinct primes dividing N. Show that:

  (a) $1/i - 1/m < E[D_i] \le 1/i$, for each $i = 1, \ldots, m$;

  (b) $|E[X] - \log\log m| \le c_1$ for some constant c_1;

  (c) for all primes p, q, where $p \le m$, $q \le m$, and $p \ne q$, we have

$$\text{Cov}[D_p, D_q] \le \frac{1}{m}\left(\frac{1}{p} + \frac{1}{q}\right),$$

  where Cov is the covariance, as defined in Exercise 8.26;

  (d) $\text{Var}[X] \le \log\log m + c_2$ for some constant c_2;

  (e) for some constant c_3, and for every $\alpha \ge 1$, we have

$$P\left[|X - \log\log m| \ge \alpha(\log\log m)^{1/2}\right] \le \alpha^{-2}\left(1 + c_3(\log\log m)^{-1/2}\right).$$

EXERCISE 8.33. For each positive integer n, let $\tau(n)$ denote the number of positive divisors of n. Suppose that N is uniformly distributed over $\{1, \ldots, m\}$. Show that $E[\tau(N)] = \log m + O(1)$.

EXERCISE 8.34. You are given three biased coins, where for $i = 1, 2, 3$, coin i comes up *heads* with probability p_i. The coins look identical, and all you know is the following: (1) $|p_1 - p_2| > 0.01$ and (2) either $p_3 = p_1$ or $p_3 = p_2$. Your goal is to determine whether p_3 is equal to p_1, or to p_2. Design a random experiment to determine this. The experiment may produce an incorrect result, but this should happen with probability at most 10^{-12}. Try to use a reasonable number of coin tosses.

EXERCISE 8.35. Consider the following game, parameterized by a positive integer n. One rolls a pair of dice, and records the value of their sum. This is repeated until some value ℓ is recorded n times, and this value ℓ is declared the "winner." It is intuitively clear that 7 is the most likely winner. Let α_n be the probability that 7 does not win. Give a careful argument that $\alpha_n \to 0$ as $n \to \infty$. Assume that the rolls of the dice are mutually independent.

8.6 Balls and bins

This section and the next discuss applications of the theory developed so far.

Our first application is a brief study of "balls and bins." Suppose you throw n balls into m bins. A number of questions naturally arise, such as:

- What is the probability that a *collision* occurs, that is, two balls land in the same bin?

- What is the expected value of the maximum number of balls that land in any one bin?

To formalize these questions, we introduce some notation that will be used throughout this section. Let I be a finite set of size $n > 0$, and S a finite set of size $m > 0$. Let $\{X_i\}_{i \in I}$ be a family of random variables, where each X_i is uniformly distributed over the set S. The idea is that I represents a set of labels for our n balls, S represents the set of m bins, and X_i represents the bin into which ball i lands.

We define C to be the event that a collision occurs; formally, this is the event that $X_i = X_j$ for some $i, j \in I$ with $i \neq j$. We also define M to be the random variable that measures that maximum number of balls in any one bin; formally,

$$M := \max\{N_s : s \in S\},$$

where for each $s \in S$, N_s is the number of balls that land in bin s; that is,

$$N_s := |\{i \in I : X_i = s\}|.$$

The questions posed above can now be stated as the problems of estimating $P[C]$

and $E[M]$. However, to estimate these quantities, we have to make some assumptions about the independence of the X_i's. While it is natural to assume that the family of random variables $\{X_i\}_{i \in I}$ is mutually independent, it is also interesting and useful to estimate these quantities under weaker independence assumptions. We shall therefore begin with an analysis under the weaker assumption that $\{X_i\}_{i \in I}$ is *pairwise* independent. We start with a simple observation:

Theorem 8.25. *Suppose* $\{X_i\}_{i \in I}$ *is pairwise independent. Then for all* $i, j \in I$ *with* $i \neq j$, *we have* $P[X_i = X_j] = 1/m$.

Proof. The event $X_i = X_j$ occurs if and only if $X_i = s$ and $X_j = s$ for some $s \in S$. Therefore,

$$P[X_i = X_j] = \sum_{s \in S} P[(X_i = s) \cap (X_j = s)] \quad \text{(by Boole's equality (8.7))}$$

$$= \sum_{s \in S} 1/m^2 \quad \text{(by pairwise independence)}$$

$$= 1/m. \quad \square$$

Theorem 8.26. *Suppose* $\{X_i\}_{i \in I}$ *is pairwise independent. Then*

$$P[C] \leq \frac{n(n-1)}{2m}.$$

Proof. Let $I^{(2)} := \{J \subseteq I : |J| = 2\}$. Then using Boole's inequality (8.6) and Theorem 8.25, we have

$$P[C] \leq \sum_{\{i,j\} \in I^{(2)}} P[X_i = X_j] = \sum_{\{i,j\} \in I^{(2)}} \frac{1}{m} = \frac{|I^{(2)}|}{m} = \frac{n(n-1)}{2m}. \quad \square$$

Theorem 8.27. *Suppose* $\{X_i\}_{i \in I}$ *is pairwise independent. Then*

$$E[M] \leq \sqrt{n^2/m + n}.$$

Proof. To prove this, we use the fact that $E[M]^2 \leq E[M^2]$ (see Theorem 8.19), and that $M^2 \leq Z := \sum_{s \in S} N_s^2$. It will therefore suffice to show that

$$E[Z] \leq n^2/m + n. \tag{8.32}$$

To this end, for $i \in I$ and $s \in S$, let L_{is} be the indicator variable for the event that ball i lands in bin s (i.e., $X_i = s$), and for $i, j \in I$, let C_{ij} be the indicator variable for the event that balls i and j land in the same bin (i.e., $X_i = X_j$). Observing that

$C_{ij} = \sum_{s \in S} L_{is} L_{js}$, we have

$$Z = \sum_{s \in S} N_s^2 = \sum_{s \in S} \left(\sum_{i \in I} L_{is} \right)^2 = \sum_{s \in S} \left(\sum_{i \in I} L_{is} \right) \left(\sum_{j \in I} L_{js} \right) = \sum_{i,j \in I} \sum_{s \in S} L_{is} L_{js}$$

$$= \sum_{i,j \in I} C_{ij}.$$

For $i, j \in I$, we have $E[C_{ij}] = P[X_i = X_j]$ (see Theorem 8.16), and so by Theorem 8.25, we have $E[C_{ij}] = 1/m$ if $i \neq j$, and clearly, $E[C_{ij}] = 1$ if $i = j$. By linearity of expectation, we have

$$E[Z] = \sum_{\substack{i,j \in I}} E[C_{ij}] = \sum_{\substack{i,j \in I \\ i \neq j}} E[C_{ij}] + \sum_{i \in I} E[C_{ii}] = \frac{n(n-1)}{m} + n \leq n^2/m + n,$$

which proves (8.32). $\square$

We next consider the situation where $\{X_i\}_{i \in I}$ is mutually independent. Of course, Theorem 8.26 is still valid in this case, but with our stronger assumption, we can derive a *lower* bound on $P[C]$.

Theorem 8.28. *Suppose $\{X_i\}_{i \in I}$ is mutually independent. Then*

$$P[C] \geq 1 - e^{-n(n-1)/2m}.$$

Proof. Let $\alpha := P[\bar{C}]$. We want to show $\alpha \leq e^{-n(n-1)/2m}$. We may assume that $I = \{1, \ldots, n\}$ (the labels make no difference) and that $n \leq m$ (otherwise, $\alpha = 0$). Under the hypothesis of the theorem, the random variable $(X_1, \ldots, X_n)$ is uniformly distributed over $S^{\times n}$. Among all m^n sequences $(s_1, \ldots, s_n) \in S^{\times n}$, there are a total of $m(m-1) \cdots (m - n + 1)$ that contain no repetitions: there are m choices for s_1, and for any fixed value of s_1, there are $m - 1$ choices for s_2, and so on. Therefore

$$\alpha = m(m-1) \cdots (m - n + 1)/m^n = \left(1 - \frac{1}{m}\right)\left(1 - \frac{2}{m}\right) \cdots \left(1 - \frac{n-1}{m}\right).$$

Using part (i) of §A1, we obtain

$$\alpha \leq e^{-\sum_{i=1}^{n-1} i/m} = e^{-n(n-1)/2m}. \quad \square$$

Theorem 8.26 implies that if $n(n-1) \leq m$, then the probability of a collision is *at most* $1/2$; moreover, Theorem 8.28 implies that if $n(n-1) \geq (2\log 2)m$, then the probability of a collision is *at least* $1/2$. Thus, for n near $\sqrt{m}$, the probability of a collision is roughly $1/2$. A colorful illustration of this is the following fact: in a room with 23 or more people, the odds are better than even that two people in the room have birthdays on the same day of the year. This follows by setting $n = 23$ and $m = 365$ in Theorem 8.28. Here, we are ignoring leap years, and the fact that birthdays are not uniformly distributed over the calendar year (however, any skew in the birthday distribution only increases the odds that two people share the same birthday — see Exercise 8.40 below). Because of this fact, Theorem 8.28 is often called the **birthday paradox** (the "paradox" being the perhaps surprisingly small number of people in the room).

The hypothesis that $\{X_i\}_{i \in I}$ is mutually independent is crucial in Theorem 8.28. Indeed, assuming just pairwise independence, we may have $P[C] = 1/m$, even when $n = m$ (see Exercise 8.42 below). However, useful, non-trivial lower bounds on $P[C]$ can still be obtained under assumptions weaker than mutual independence (see Exercise 8.43 below).

Assuming $\{X_i\}_{i \in I}$ is mutually independent, we can get a much sharper upper bound on $E[M]$ than that provided by Theorem 8.27. For simplicity, we only consider the case where $m = n$; in this case, Theorem 8.27 gives us the bound $E[M] \leq \sqrt{2n}$ (which cannot be substantially improved assuming only pairwise independence — see Exercise 8.44 below).

Theorem 8.29. *Suppose $\{X_i\}_{i \in I}$ is mutually independent and that $m = n$. Then*

$$E[M] \leq (1 + o(1))\frac{\log n}{\log \log n}.$$

Proof. We use Theorem 8.17, which says that $E[M] = \sum_{k \geq 1} P[M \geq k]$.

Claim 1. For $k \geq 1$, we have $P[M \geq k] \leq n/k!$.

To prove Claim 1, we may assume that $k \leq n$ (as otherwise, $P[M \geq k] = 0$). Let $I^{(k)} := \{J \subseteq I : |J| = k\}$. Now, $M \geq k$ if and only if there is an $s \in S$ and a subset $J \in I^{(k)}$, such that $X_j = s$ for all $j \in J$. Therefore,

$$P[M \geq k] \leq \sum_{s \in S} \sum_{J \in I^{(k)}} P\left[\bigcap_{j \in J}(X_j = s)\right] \quad \text{(by Boole's inequality (8.6))}$$

$$= \sum_{s \in S} \sum_{J \in I^{(k)}} \prod_{j \in J} P[X_j = s] \quad \text{(by mutual independence)}$$

$$= n\binom{n}{k}n^{-k} \leq n/k!.$$

That proves Claim 1.

Of course, Claim 1 is only interesting when $n/k! \leq 1$, since $P[M \geq k]$ is always at most 1. Define $F(n)$ to be the smallest positive integer k such that $k! \geq n$.

Claim 2. $F(n) \sim \log n / \log \log n$.

To prove this, let us set $k := F(n)$. It is clear that $n \leq k! \leq nk$, and taking logarithms, $\log n \leq \log k! \leq \log n + \log k$. Moreover, we have

$$\log k! = \sum_{\ell=1}^{k} \log \ell = \int_{1}^{k} \log x \, dx + O(\log k) = k \log k - k + O(\log k) \sim k \log k,$$

where we have estimated the sum by an integral (see §A5). Thus,

$$\log n = \log k! + O(\log k) \sim k \log k.$$

Taking logarithms again, we see that

$$\log \log n = \log k + \log \log k + o(1) \sim \log k,$$

and so $\log n \sim k \log k \sim k \log \log n$, from which Claim 2 follows.

Finally, observe that each term in the sequence $\{n/k!\}_{k=1}^{\infty}$ is at most half the previous term. Combining this observation with Claims 1 and 2, and the fact that $P[M \geq k]$ is always at most 1, we have

$$E[M] = \sum_{k \geq 1} P[M \geq k] = \sum_{k \leq F(n)} P[M \geq k] + \sum_{k > F(n)} P[M \geq k]$$

$$\leq F(n) + \sum_{\ell \geq 1} 2^{-\ell} = F(n) + 1 \sim \log n / \log \log n. \quad \square$$

EXERCISE 8.36. Let $\alpha_1, \ldots, \alpha_m$ be real numbers that sum to 1. Show that $0 \leq \sum_{s=1}^{m} (\alpha_s - 1/m)^2 = \sum_{s=1}^{m} \alpha_s^2 - 1/m$, and in particular, $\sum_{s=1}^{m} \alpha_s^2 \geq 1/m$.

EXERCISE 8.37. Let X and X' be independent random variables, both having the same distribution on a set S of size m. Show that $P[X = X'] = \sum_{s \in S} P[X = s]^2 \geq 1/m$.

EXERCISE 8.38. Suppose that the family of random variables X, Y, Y' is mutually independent, where X has image S, and where Y and Y' have the same distribution on a set T. Let ϕ be a predicate on $S \times T$, and let $\alpha := P[\phi(X, Y)]$. Show that $P[\phi(X, Y) \cap \phi(X, Y')] \geq \alpha^2$. In addition, show that if Y and Y' are both uniformly distributed over T, then $P[\phi(X, Y) \cap \phi(X, Y') \cap (Y \neq Y')] \geq \alpha^2 - \alpha/|T|$.

EXERCISE 8.39. Let $\alpha_1, \ldots, \alpha_m$ be non-negative real numbers that sum to 1. Let $S := \{1, \ldots, m\}$, and for $n = 1, \ldots, m$, let $S^{(n)} := \{T \subseteq S : |T| = n\}$, and define

$$P_n(\alpha_1, \ldots, \alpha_m) := \sum_{T \in S^{(n)}} \prod_{t \in T} \alpha_t.$$

Show that $P_n(\alpha_1, \ldots, \alpha_m)$ is maximized when $\alpha_1 = \cdots = \alpha_m = 1/m$. Hint: first argue that if $\alpha_s < \alpha_t$, then for every $\varepsilon \in [0, \alpha_t - \alpha_s]$, replacing the pair (α_s, α_t) by $(\alpha_s + \varepsilon, \alpha_t - \varepsilon)$ does not decrease the value of $P_n(\alpha_1, \ldots, \alpha_m)$.

EXERCISE 8.40. Suppose that $\{X_i\}_{i \in I}$ is a finite, non-empty, mutually independent family of random variables, where each X_i is uniformly distributed over a finite set S. Suppose that $\{Y_i\}_{i \in I}$ is another finite, non-empty, mutually independent family of random variables, where each Y_i has the same distribution and takes values in the set S. Let α be the probability that the X_i's are distinct, and β be the probability that the Y_i's are distinct. Using the previous exercise, show that $\beta \le \alpha$.

EXERCISE 8.41. Suppose n balls are thrown into m bins. Let $\mathcal{A}$ be the event that there is some bin that is empty. Assuming that the throws are mutually independent, and that $n \ge m(\log m + t)$ for some $t \ge 0$, show that $P[\mathcal{A}] \le e^{-t}$.

EXERCISE 8.42. Show that for every prime p, there exists a pairwise independent family of random variables $\{X_i\}_{i \in \mathbb{Z}_p}$, where each X_i is uniformly distributed over $\mathbb{Z}_p$, and yet the probability that all the X_i's are distinct is $1 - 1/p$.

EXERCISE 8.43. Let $\{X_i\}_{i=1}^n$ be a finite, non-empty, *4-wise independent* family of random variables, each uniformly distributed over a set S. Let α be the probability that the X_i's are distinct. For $i, j = 1, \ldots, n$, let C_{ij} be the indicator variable for the event that $X_i = X_j$, and define $K := \{(i, j) : 1 \le i \le n - 1, \ i + 1 \le j \le n\}$ and $Z := \sum_{(i,j) \in K} C_{ij}$. Show that:

   (a) $\{C_{ij}\}_{(i,j) \in K}$ is pairwise independent;

   (b) $E[Z] = n(n - 1)/2m$ and $Var[Z] = (1 - 1/m) E[Z]$;

   (c) $\alpha \le 1/E[Z]$;

   (d) $\alpha \le 1/2$, provided $n(n - 1) \ge 2m$ (hint: Exercise 8.4).

EXERCISE 8.44. Let k be a positive integer, let $n := k^2 - k + 1$, let I and S be sets of size n, and let s_0 be a fixed element of S. Also, let $I^{(k)} := \{J \subseteq I : |J| = k\}$, and let Π be the set of all permutations on S. For each $J \in I^{(k)}$, let f_J be some function that maps J to s_0, and maps $I \setminus J$ injectively into $S \setminus \{s_0\}$. For $\pi \in \Pi$, $J \in I^{(k)}$, and $i \in I$, define $\rho_i(\pi, J) := \pi(f_J(i))$. Finally, let Y be uniformly distributed over $\Pi \times I^{(k)}$, and for $i \in I$, define $X_i := \rho_i(Y)$. Show that $\{X_i\}_{i \in I}$ is pairwise independent, with each X_i uniformly distributed over S, and yet the number of X_i's with the same value is always at least $\sqrt{n}$.

EXERCISE 8.45. Let S be a set of size $m \ge 1$, and let s_0 be an arbitrary, fixed element of S. Let F be a random variable that is uniformly distributed over the set of all m^m functions from S into S. Let us define random variables X_i, for $i = 0, 1, 2, \ldots$, as follows:

$$X_0 := s_0, \quad X_{i+1} := F(X_i) \ (i = 0, 1, 2, \ldots).$$

Thus, the value of X_i is obtained by applying the function F a total of i times to the starting value s_0. Since S has size m, the sequence $\{X_i\}_{i=0}^{\infty}$ must repeat at some point; that is, there exists a positive integer n (with $n \leq m$) such that $X_n = X_i$ for some $i = 0, \ldots, n - 1$. Define the random variable Y to be the smallest such value n.

(a) Show that for every $i \geq 0$ and for all $s_1, \ldots, s_i \in S$ such that $s_0, s_1, \ldots, s_i$ are distinct, the conditional distribution of X_{i+1} given the event $(X_1 = s_1) \cap \cdots \cap (X_i = s_i)$ is the uniform distribution on S.

(b) Show that for every integer $n \geq 1$, we have $Y \geq n$ if and only if the random variables $X_0, X_1, \ldots, X_{n-1}$ take on distinct values.

(c) From parts (a) and (b), show that for each $n = 1, \ldots, m$, we have

$$P[Y \geq n \mid Y \geq n - 1] = 1 - (n - 1)/m,$$

and conclude that

$$P[Y \geq n] = \prod_{i=1}^{n-1}(1 - i/m) \leq e^{-n(n-1)/2m}.$$

(d) Using part (c), show that

$$E[Y] = \sum_{n \geq 1} P[Y \geq n] \leq \sum_{n \geq 1} e^{-n(n-1)/2m} = O(m^{1/2}).$$

(e) Modify the above argument to show that $E[Y] = \Omega(m^{1/2})$.

EXERCISE 8.46. The setup for this exercise is identical to that of the previous exercise, except that now, F is uniformly distributed over the set of all $m!$ *permutations* of S.

(a) Show that if $Y = n$, then $X_n = X_0$.

(b) Show that for every $i \geq 0$ and all $s_1, \ldots, s_i \in S$ such that $s_0, s_1, \ldots, s_i$ are distinct, the conditional distribution of X_{i+1} given $(X_1 = s_1) \cap \cdots \cap (X_i = s_i)$ is essentially the uniform distribution on $S \setminus \{s_1, \ldots, s_i\}$.

(c) Show that for each $n = 2, \ldots, m$, we have

$$P[Y \geq n \mid Y \geq n - 1] = 1 - \frac{1}{m - n + 2},$$

and conclude that for all $n = 1, \ldots, m$, we have

$$P[Y \geq n] = \prod_{i=0}^{n-2}\left(1 - \frac{1}{m - i}\right) = 1 - \frac{n - 1}{m}.$$

(d) From part (c), show that Y is uniformly distributed over $\{1, \ldots, m\}$, and in particular, $E[Y] = (m + 1)/2$.

8.7 Hash functions

In this section, we apply the tools we have developed thus far to a particularly important area of computer science: the theory and practice of hashing.

Let R, S, and T be finite, non-empty sets. Suppose that for each $r \in R$, we have a function $\Phi_r : S \to T$. We call Φ_r a **hash function** (**from** S **to** T). Elements of R are called **keys**, and if $\Phi_r(s) = t$, we say that s **hashes to** t **under** r.

In applications of hash functions, we are typically interested in what happens when various inputs are hashed under a randomly chosen key. To model such situations, let H be a random variable that is uniformly distributed over R, and for each $s \in S$, let us define the random variable $\Phi_H(s)$, which takes the value $\Phi_r(s)$ when $H = r$.

- We say that the family of hash functions $\{\Phi_r\}_{r \in R}$ is **pairwise independent** if the family of random variables $\{\Phi_H(s)\}_{s \in S}$ is pairwise independent, with each $\Phi_H(s)$ uniformly distributed over T.

- We say that $\{\Phi_r\}_{r \in R}$ is **universal** if

$$P[\Phi_H(s) = \Phi_H(s')] \leq 1/|T|$$

  for all $s, s' \in S$ with $s \neq s'$.

We make a couple of simple observations. First, by Theorem 8.25, if the family of hash functions $\{\Phi_r\}_{r \in R}$ is pairwise independent, then it is universal. Second, by Theorem 8.10, if $|S| > 1$, then $\{\Phi_r\}_{r \in R}$ is pairwise independent if and only if the following condition holds:

the random variable $(\Phi_H(s), \Phi_H(s'))$ is uniformly distributed over $T \times T$, for all $s, s' \in S$ with $s \neq s'$;

or equivalently,

$P[\Phi_H(s) = t \cap \Phi_H(s') = t'] = 1/|T|^2$ for all $s, s' \in S$ with $s \neq s'$, and for all $t, t' \in T$.

Before looking at constructions of pairwise independent and universal families of hash functions, we briefly discuss two important applications.

Example 8.34. Suppose $\{\Phi_r\}_{r \in R}$ is a *universal* family of hash functions from S to T. One can implement a "dictionary" using a so-called **hash table**, which is basically an array A indexed by T, where each entry in A is a list. Entries in the dictionary are drawn from the set S. To insert a word $s \in S$ into the dictionary, s is first hashed to an index t, and then s is appended to the list $A[t]$; likewise, to see if an arbitrary word $s \in S$ is in the dictionary, s is first hashed to an index t, and then the list $A[t]$ is searched for s.

Usually, the set of entries in the dictionary is much smaller than the set S. For example, S may consist of all bit strings of length up to, say 2048, but the dictionary may contain just a few thousand, or a few million, entries. Also, to be practical, the set T should not be too large.

Of course, all entries in the dictionary could end up hashing to the same index, in which case, looking up a word in the dictionary degenerates into linear search. However, we hope that this does not happen, and that entries hash to indices that are nicely spread out over T. As we will now see, in order to ensure reasonable performance (in an expected sense), T needs to be of size roughly equal to the number of entries in the dictionary,

Suppose we create a dictionary containing n entries. Let $m := |T|$, and let $I \subseteq S$ be the set of entries (so $n = |I|$). These n entries are inserted into the hash table using a randomly chosen hash key, which we model as a random variable H that is uniformly distributed over R. For each $s \in S$, we define the random variable L_s to be the number of entries in I that hash to the same index as s under the key H; that is, $L_s := |\{i \in I : \Phi_H(s) = \Phi_H(i)\}|$. Intuitively, L_s measures the cost of looking up the particular word s in the dictionary. We want to bound $E[L_s]$. To this end, we write L_s as a sum of indicator variables: $L_s = \sum_{i \in I} C_{si}$, where C_{si} is the indicator variable for the event that $\Phi_H(s) = \Phi_H(i)$. By Theorem 8.16, we have $E[C_{si}] = P[\Phi_H(s) = \Phi_H(i)]$; moreover, by the universal property, $E[C_{si}] \le 1/m$ if $s \ne i$, and clearly, $E[C_{si}] = 1$ if $s = i$. By linearity of expectation, we have

$$E[L_s] = \sum_{i \in I} E[C_{si}].$$

If $s \notin I$, then each term in the sum is $\le 1/m$, and so $E[L_s] \le n/m$. If $s \in I$, then one term in the sum is 1, and the other $n - 1$ terms are $\le 1/m$, and so $E[L_s] \le 1 + (n-1)/m$. In any case, we have

$$E[L_s] \le 1 + n/m.$$

In particular, this means that if $m \ge n$, then the expected cost of looking up any particular word in the dictionary is bounded by a constant. $\square$

Example 8.35. Suppose Alice wants to send a message to Bob in such a way that Bob can be reasonably sure that the message he receives really came from Alice, and was not modified in transit by some malicious adversary. We present a solution to this problem here that works assuming that Alice and Bob share a randomly generated secret key, and that this key is used to authenticate just a single message (multiple messages can be authenticated using multiple keys).

Suppose that $\{\Phi_r\}_{r \in R}$ is a *pairwise independent* family of hash functions from S to T. We model the shared random key as a random variable H, uniformly distributed over R. We also model Alice's message as a random variable X, taking values in the set S. We make no assumption about the distribution of X, but we do assume that X and H are independent. When Alice sends the message X to Bob, she also sends the "authentication tag" $Y := \Phi_H(X)$. Now, when Bob receives a message X' and tag Y', he checks that $\Phi_H(X') = Y'$; if this holds, he accepts the message X' as authentic; otherwise, he rejects it. Here, X' and Y' are also random variables; however, they may have been created by a malicious adversary who may have even created them after seeing X and Y. We can model such an adversary as a pair of functions $f : S \times T \to S$ and $g : S \times T \to T$, so that $X' := f(X, Y)$ and $Y' := g(X, Y)$. The idea is that after seeing X and Y, the adversary computes X' and Y' and sends X' and Y' to Bob instead of X and Y. Let us say that the adversary *fools* Bob if $\Phi_H(X') = Y'$ and $X' \neq X$. We will show that $P[\mathcal{F}] \leq 1/m$, where $\mathcal{F}$ is the event that the adversary fools Bob, and $m := |T|$. Intuitively, this bound holds because the pairwise independence property guarantees that after seeing the value of Φ_H at one input, the value of Φ_H at any other input is completely unpredictable, and cannot be guessed with probability any better than $1/m$. If m is chosen to be suitably large, the probability that Bob gets fooled can be made acceptably small. For example, S may consist of all bit strings of length up to, say, 2048, while the set T may be encoded using much shorter bit strings, of length, say, 64. This is nice, as it means that the authentication tags consume very little additional bandwidth.

A straightforward calculation justifies the claim that $P[\mathcal{F}] \leq 1/m$:

$$P[\mathcal{F}] = \sum_{s \in S} \sum_{t \in T} P\Big[(X = s) \cap (Y = t) \cap \mathcal{F}\Big] \quad \text{(law of total probability (8.9))}$$

$$= \sum_{s \in S} \sum_{t \in T} P\Big[(X = s) \cap (\Phi_H(s) = t) \cap (\Phi_H(f(s,t)) = g(s,t)) \cap (f(s,t) \neq s)\Big]$$

$$= \sum_{s \in S} \sum_{t \in T} P[X = s] \, P\Big[(\Phi_H(s) = t) \cap (\Phi_H(f(s,t)) = g(s,t)) \cap (f(s,t) \neq s)\Big] \quad \text{(since X and H are independent)}$$

$$\leq \sum_{s \in S} \sum_{t \in T} P[X = s] \cdot (1/m^2) \quad \text{(since $\{\Phi_r\}_{r \in R}$ is pairwise independent)}$$

$$= (1/m) \sum_{s \in S} P[X = s] = 1/m. \quad \square$$

We now present several constructions of pairwise independent and universal families of hash functions.

Example 8.36. By setting $k := 2$ in Example 8.27, for each prime p, we immediately get a pairwise independent family of hash functions $\{\Phi_r\}_{r \in R}$ from $\mathbb{Z}_p$ to $\mathbb{Z}_p$, where $R = \mathbb{Z}_p \times \mathbb{Z}_p$, and for $r = (r_0, r_1) \in R$, the hash function Φ_r is given by

$$\Phi_r : \quad \mathbb{Z}_p \to \mathbb{Z}_p$$
$$s \mapsto r_0 + r_1 s. \quad \Box$$

While very simple and elegant, the family of hash functions in Example 8.36 is not very useful in practice. As we saw in Examples 8.34 and 8.35, what we would really like are families of hash functions that hash *long* inputs to *short* outputs. The next example provides us with a pairwise independent family of hash functions that satisfies this requirement.

Example 8.37. Let p be a prime, and let ℓ be a positive integer. Let $S := \mathbb{Z}_p^{\times \ell}$ and $R := \mathbb{Z}_p^{\times(\ell+1)}$. For each $r = (r_0, r_1, \ldots, r_\ell) \in R$, we define the hash function

$$\Phi_r : \quad\quad\quad S \to \mathbb{Z}_p$$
$$(s_1, \ldots, s_\ell) \mapsto r_0 + r_1 s_1 + \cdots + r_\ell s_\ell.$$

We will show that $\{\Phi_r\}_{r \in R}$ is a pairwise independent family of hash functions from S to $\mathbb{Z}_p$. To this end, let H be a random variable uniformly distributed over R. We want to show that for each $s, s' \in S$ with $s \neq s'$, the random variable $(\Phi_H(s), \Phi_H(s'))$ is uniformly distributed over $\mathbb{Z}_p \times \mathbb{Z}_p$. So let $s \neq s'$ be fixed, and define the function

$$\rho : \quad R \to \mathbb{Z}_p \times \mathbb{Z}_p$$
$$r \mapsto (\Phi_r(s), \Phi_r(s')).$$

Because ρ is a group homomorphism, it will suffice to show that ρ is surjective (see Theorem 8.5). Suppose $s = (s_1, \ldots, s_\ell)$ and $s' = (s'_1, \ldots, s'_\ell)$. Since $s \neq s'$, we must have $s_j \neq s'_j$ for some $j = 1, \ldots, \ell$. For this j, consider the function

$$\rho' : \quad\quad\quad R \to \mathbb{Z}_p \times \mathbb{Z}_p$$
$$(r_0, r_1, \ldots, r_\ell) \mapsto (r_0 + r_j s_j, r_0 + r_j s'_j).$$

Evidently, the image of ρ includes the image of ρ', and by Example 8.36, the function ρ' is surjective. $\Box$

To use the construction in Example 8.37 in applications where the set of inputs consists of bit strings of a given length, one can naturally split such a bit string up into short bit strings which, when viewed as integers, lie in the set $\{0, \ldots, p-1\}$, and which can in turn be viewed as elements of $\mathbb{Z}_p$. This gives us a natural, injective map from bit strings to elements of $\mathbb{Z}_p^{\times \ell}$. The appropriate choice of the prime p depends on the application. Of course, the requirement that p is prime limits our choice in the size of the output set; however, this is usually not a severe restriction, as Bertrand's postulate (Theorem 5.8) tells us that we can always choose p

to within a factor of 2 of any desired value of the output set size. Nevertheless, the construction in the following example gives us a *universal* (but not pairwise independent) family of hash functions with an output set of any size we wish.

Example 8.38. Let p be a prime, and let m be an arbitrary positive integer. Let us introduce some convenient notation: for $\alpha \in \mathbb{Z}_p$, let $[\![\alpha]\!]_m := [\text{rep}(\alpha)]_m \in \mathbb{Z}_m$ (recall that $\text{rep}(\alpha)$ denotes the unique integer $a \in \{0, \ldots, p-1\}$ such that $\alpha = [a]_p$). Let $R := \mathbb{Z}_p \times \mathbb{Z}_p^*$, and for each $r = (r_0, r_1) \in R$, define the hash function

$$\Phi_r : \quad \mathbb{Z}_p \to \mathbb{Z}_m$$
$$s \mapsto [\![r_0 + r_1 s]\!]_m.$$

Our goal is to show that $\{\Phi_r\}_{r \in R}$ is a universal family of hash functions from $\mathbb{Z}_p$ to $\mathbb{Z}_m$. So let $s, s' \in \mathbb{Z}_p$ with $s \neq s'$, let H_0 and H_1 be independent random variables, with H_0 uniformly distributed over $\mathbb{Z}_p$ and H_1 uniformly distributed over $\mathbb{Z}_p^*$, and let $H := (H_0, H_1)$. Also, let C be the event that $\Phi_H(s) = \Phi_H(s')$. We want to show that $P[C] \leq 1/m$. Let us define random variables $Y := H_0 + H_1 s$ and $Y' := H_0 + H_1 s'$. Also, let $\hat{s} := s' - s \neq 0$. Then we have

$$P[C] = P\left[[\![Y]\!]_m = [\![Y']\!]_m\right]$$

$$= P\left[[\![Y]\!]_m = [\![Y + H_1\hat{s}]\!]_m\right] \text{ (since } Y' = Y + H_1\hat{s})$$

$$= \sum_{\alpha \in \mathbb{Z}_p} P\left[([\![Y]\!]_m = [\![Y + H_1\hat{s}]\!]_m) \cap (Y = \alpha)\right] \text{ (law of total probability (8.9))}$$

$$= \sum_{\alpha \in \mathbb{Z}_p} P\left[([\![\alpha]\!]_m = [\![\alpha + H_1\hat{s}]\!]_m) \cap (Y = \alpha)\right]$$

$$= \sum_{\alpha \in \mathbb{Z}_p} P\left[[\![\alpha]\!]_m = [\![\alpha + H_1\hat{s}]\!]_m\right] P[Y = \alpha]$$

(by Theorem 8.13, Y and H_1 are independent).

It will suffice to show that

$$P\left[[\![\alpha]\!]_m = [\![\alpha + H_1\hat{s}]\!]_m\right] \leq 1/m \tag{8.33}$$

for each $\alpha \in \mathbb{Z}_p$, since then

$$P[C] \leq \sum_{\alpha \in \mathbb{Z}_p} (1/m) P[Y = \alpha] = (1/m) \sum_{\alpha \in \mathbb{Z}_p} P[Y = \alpha] = 1/m.$$

So consider a fixed $\alpha \in \mathbb{Z}_p$. As $\hat{s} \neq 0$ and H_1 is uniformly distributed over $\mathbb{Z}_p^*$, it follows that $H_1\hat{s}$ is uniformly distributed over $\mathbb{Z}_p^*$, and hence $\alpha + H_1\hat{s}$ is uniformly distributed over the set $\mathbb{Z}_p \setminus \{\alpha\}$. Let $M_\alpha := \{\beta \in \mathbb{Z}_p : [\![\alpha]\!]_m = [\![\beta]\!]_m\}$. To prove

(8.33), we need to show that $|M_\alpha \setminus \{\alpha\}| \le (p-1)/m$. But it is easy to see that $|M_\alpha| \le \lceil p/m \rceil$, and since M_α certainly contains α, we have

$$|M_\alpha \setminus \{\alpha\}| \le \left\lceil \frac{p}{m} \right\rceil - 1 \le \frac{p}{m} + \frac{m-1}{m} - 1 = \frac{p-1}{m}. \quad \Box$$

One drawback of the family of hash functions in the previous example is that the prime p may need to be quite large (at least as large as the size of the set of inputs) and so to evaluate a hash function, we have to perform modular multiplication of large integers. In contrast, in Example 8.37, the prime p can be much smaller (only as large as the size of the set of outputs), and so these hash functions can be evaluated much more quickly.

Another consideration in designing families of hash functions is the size of key set. The following example gives a variant of the family in Example 8.37 that uses somewhat a smaller key set (relative to the size of the input), but is only a universal family, and not a pairwise independent family.

Example 8.39. Let p be a prime, and let ℓ be a positive integer. Let $S := \mathbb{Z}_p^{\times(\ell+1)}$ and $R := \mathbb{Z}_p^{\times\ell}$. For each $r = (r_1, \ldots, r_\ell) \in R$, we define the hash function

$$\Phi_r : \qquad\qquad S \to \mathbb{Z}_p$$
$$(s_0, s_1, \ldots, s_\ell) \mapsto s_0 + r_1 s_1 + \cdots + r_\ell s_\ell.$$

Our goal is to show that $\{\Phi_r\}_{r\in R}$ is a universal family of hash functions from S to $\mathbb{Z}_p$. So let $s, s' \in S$ with $s \ne s'$, and let H be a random variable that is uniformly distributed over R. We want to show that $\mathsf{P}[\Phi_H(s) = \Phi_H(s')] \le 1/p$. Let $s = (s_0, s_1, \ldots, s_\ell)$ and $s' = (s_0', s_1', \ldots, s_\ell')$, and set $\hat{s}_i := s_i' - s_i$ for $i = 0, 1, \ldots, \ell$. Let us define the function

$$\rho : \qquad\qquad R \to \mathbb{Z}_p$$
$$(r_1, \ldots, r_\ell) \mapsto r_1 \hat{s}_1 + \cdots + r_\ell \hat{s}_\ell.$$

Clearly, $\Phi_H(s) = \Phi_H(s')$ if and only if $\rho(H) = -\hat{s}_0$. Moreover, ρ is a group homomorphism. There are two cases to consider. In the first case, $\hat{s}_i = 0$ for all $i = 1, \ldots, \ell$; in this case, the image of ρ is $\{0\}$, but $\hat{s}_0 \ne 0$ (since $s \ne s'$), and so $\mathsf{P}[\rho(H) = -\hat{s}_0] = 0$. In the second case, $\hat{s}_i \ne 0$ for some $i = 1, \ldots, \ell$; in this case, the image of ρ is $\mathbb{Z}_p$, and so $\rho(H)$ is uniformly distributed over $\mathbb{Z}_p$ (see Theorem 8.5); thus, $\mathsf{P}[\rho(H) = -\hat{s}_0] = 1/p$. $\Box$

One can get significantly smaller key sets, if one is willing to relax the definitions of universal and pairwise independence. Let $\{\Phi_r\}_{r\in R}$ be a family of hash functions from S to T, where $m := |T|$. Let H be a random variable that is uniformly distributed over R. We say that $\{\Phi_r\}_{r\in R}$ is ε-**almost universal** if for all $s, s' \in S$ with $s \ne s'$, we have $\mathsf{P}[\Phi_{||}(s) - \Phi_H(s')] \le c$. Thus, $\{\Phi_r\}_{r\in R}$ is universal if and only if it is $1/m$-almost universal. We say that $\{\Phi_r\}_{r \in R}$ is ε-**almost strongly universal** if $\Phi_H(s)$ is uniformly distributed over T for each $s \in S$, and $P[(\Phi_H(s) = t) \cap (\Phi_H(s') = t')] \le \varepsilon/m$ for all $s, s' \in S$ with $s \ne s'$ and all $t, t' \in T$. Constructions, properties, and applications of these types of hash functions are developed in some of the exercises below.

EXERCISE 8.47. For each positive integer n, let I_n denote $\{0, \ldots, n-1\}$. Let m be a power of a prime, ℓ be a positive integer, $S := I_m^{\times \ell}$, and $R := I_{m^2}^{\times(\ell+1)}$. For each $r = (r_0, r_1, \ldots, r_\ell) \in R$, define the hash function

$$\Phi_r : \qquad\qquad S \to I_m$$

$$(s_1, \ldots, s_\ell) \mapsto \left\lfloor \left((r_0 + r_1 s_1 + \cdots + r_\ell s_\ell) \bmod m^2\right) \middle/ m \right\rfloor.$$

Using the result from Exercise 2.13, show that $\{\Phi_r\}_{r \in R}$ is a pairwise independent family of hash functions from S to I_m. Note that on a typical computer, if m is a suitable power of 2, then it is very easy to evaluate these hash functions, using just multiplications, additions, shifts, and masks (no divisions).

EXERCISE 8.48. Let $\{\Phi_r\}_{r \in R}$ be an ε-almost universal family of hash functions from S to T. Also, let H, X, X' be random variables, where H is uniformly distributed over R, and both X and X' take values in S. Moreover, assume H and (X, X') are independent. Show that $P[\Phi_H(X) = \Phi_H(X')] \le P[X = X'] + \varepsilon$.

EXERCISE 8.49. Let $\{\Phi_r\}_{r \in R}$ be an ε-almost universal a family of hash functions from S to T, and let H be a random variable that is uniformly distributed over R. Let I be a subset of S of size $n > 0$. Let C be the event that $\Phi_H(i) = \Phi_H(j)$ for some $i, j \in I$ with $i \ne j$. We define several random variables: for each $t \in T$, $N_t := |\{i \in I : \Phi_H(i) = t\}|$; $M := \max\{N_t : t \in T\}$; for each $s \in S$, $L_s := |\{i \in I : \Phi_H(s) = \Phi_H(i)\}|$. Show that:

(a) $P[C] \le \varepsilon n(n-1)/2$;

(b) $E[M] \le \sqrt{\varepsilon n^2 + n}$;

(c) for each $s \in S$, $E[L_s] \le 1 + \varepsilon n$.

The results of the previous exercise show that for many applications, the ε-almost universal property is good enough, provided ε is suitably small. The next three exercises develop ε-almost universal families of hash functions with very small sets of keys, even when ε is quite small.

EXERCISE 8.50. Let p be a prime, and let ℓ be a positive integer. Let $S := \mathbb{Z}_p^{\times(\ell+1)}$.

For each $r \in \mathbb{Z}_p$, define the hash function

$$\Phi_r : \qquad S \to \mathbb{Z}_p$$
$$(s_0, s_1, \ldots, s_\ell) \mapsto s_0 + s_1 r + \cdots + s_\ell r^\ell.$$

Show that $\{\Phi_r\}_{r \in \mathbb{Z}_p}$ is an ℓ/p-almost universal family of hash functions from S to $\mathbb{Z}_p$.

EXERCISE 8.51. Let $\{\Phi_r\}_{r \in R}$ be an ε-almost universal family of hash functions from S to T. Let $\{\Phi'_{r'}\}_{r' \in R'}$ be an ε'-almost universal family of hash functions from S' to T', where $T \subseteq S'$. Show that

$$\{\Phi'_{r'} \circ \Phi_r\}_{(r,r') \in R \times R'}$$

is an $(\varepsilon + \varepsilon')$-almost universal family of hash functions from S to T' (here, "$\circ$" denotes function composition).

EXERCISE 8.52. Let m and ℓ be positive integers, and let $0 < \alpha < 1$. Given these parameters, show how to construct an ε-almost universal family of hash functions $\{\Phi_r\}_{r \in R}$ from $\mathbb{Z}_m^{\times \ell}$ to $\mathbb{Z}_m$, such that

$$\varepsilon \leq (1 + \alpha)/m \quad \text{and} \quad \log|R| = O(\log m + \log \ell + \log(1/\alpha)).$$

Hint: use the previous two exercises, and Example 8.38.

EXERCISE 8.53. Let $\{\Phi_r\}_{r \in R}$ be an ε-almost universal family of hash functions from S to T. Show that $\varepsilon \geq 1/|T| - 1/|S|$.

EXERCISE 8.54. Let $\{\Phi_r\}_{r \in R}$ be a family of hash functions from S to T, with $m := |T|$. Show that:

(a) if $\{\Phi_r\}_{r \in R}$ is ε-almost strongly universal, then it is ε-almost universal;

(b) if $\{\Phi_r\}_{r \in R}$ is pairwise independent, then it is $1/m$-almost strongly universal;

(c) if $\{\Phi_r\}_{r \in R}$ is ε-almost universal, and $\{\Phi'_{r'}\}_{r' \in R'}$ is an ε'-almost strongly universal family of hash functions from S' to T', where $T \subseteq S'$, then $\{\Phi'_{r'} \circ \Phi_r\}_{(r,r') \in R \times R'}$ is an $(\varepsilon + \varepsilon')$-almost strongly universal family of hash functions from S to T'.

EXERCISE 8.55. Show that if an ε-almost strongly universal family of hash functions is used in Example 8.35, then Bob gets fooled with probability at most ε.

EXERCISE 8.56. Show how to construct an ε-almost strongly universal family of hash functions satisfying the same bounds as in Exercise 8.52, under the restriction that m is a prime power.

EXERCISE 8.57. Let p be a prime, and let ℓ be a positive integer. Let $S := \mathbb{Z}_p^{\times \ell}$ and $R := \mathbb{Z}_p \times \mathbb{Z}_p$. For each $(r_0, r_1) \in R$, define the hash function

$$\Phi_r : \qquad\qquad S \to \mathbb{Z}_p$$
$$(s_1, \ldots, s_\ell) \mapsto r_0 + s_1 r_1 + \cdots + s_\ell r_1^\ell.$$

Show that $\{\Phi_r\}_{r \in R}$ is an ℓ/p-almost strongly universal family of hash functions from S to $\mathbb{Z}_p$.

8.8 Statistical distance

This section discusses a useful measure of "distance" between two random variables. Although important in many applications, the results of this section (and the next) will play only a very minor role in the remainder of the text.

Let X and Y be random variables which both take values in a finite set S. We define the **statistical distance between X and Y** as

$$\Delta[X; Y] := \frac{1}{2} \sum_{s \in S} \big| P[X = s] - P[Y = s] \big|.$$

Theorem 8.30. *For random variables X, Y, Z, we have*

> *(i)* $0 \le \Delta[X; Y] \le 1$,
>
> *(ii)* $\Delta[X; X] = 0$,
>
> *(iii)* $\Delta[X; Y] = \Delta[Y; X]$, *and*
>
> *(iv)* $\Delta[X; Z] \le \Delta[X; Y] + \Delta[Y; Z]$.

Proof. Exercise. $\square$

It is also clear from the definition that $\Delta[X; Y]$ depends only on the distributions of X and Y, and not on any other properties. As such, we may sometimes speak of the statistical distance between two distributions, rather than between two random variables.

Example 8.40. Suppose X has the uniform distribution on $\{1, \ldots, m\}$, and Y has the uniform distribution on $\{1, \ldots, m - \delta\}$, where $\delta \in \{0, \ldots, m - 1\}$. Let us compute $\Delta[X; Y]$. We could apply the definition directly; however, consider the following graph of the distributions of X and Y:

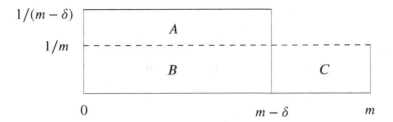

The statistical distance between X and Y is just $1/2$ times the area of regions A and C in the diagram. Moreover, because probability distributions sum to 1, we must have

$$\text{area of } B + \text{area of } A = 1 = \text{area of } B + \text{area of } C,$$

and hence, the areas of region A and region C are the same. Therefore,

$$\Delta[X;Y] = \text{area of } A = \text{area of } C = \delta/m. \quad \square$$

The following characterization of statistical distance is quite useful:

Theorem 8.31. *Let X and Y be random variables taking values in a set S. For every $S' \subseteq S$, we have*

$$\Delta[X;Y] \geq |P[X \in S'] - P[Y \in S']|,$$

and equality holds for some $S' \subseteq S$, and in particular, for the set

$$S' := \{s \in S : P[X = s] < P[Y = s]\},$$

as well as its complement.

Proof. Suppose we split the set S into two disjoint subsets: the set S_0 consisting of those $s \in S$ such that $P[X = s] < P[Y = s]$, and the set S_1 consisting of those $s \in S$ such that $P[X = s] \geq P[Y = s]$. Consider the following rough graph of the distributions of X and Y, where the elements of S_0 are placed to the left of the elements of S_1:

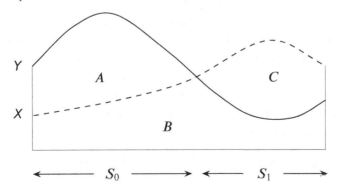

Now, as in Example 8.40,

$$\Delta[X; Y] = \text{area of } A = \text{area of } C.$$

Now consider any subset S' of S, and observe that

$$P[X \in S'] - P[Y \in S'] = \text{area of } C' - \text{area of } A',$$

where C' is the subregion of C that lies above S', and A' is the subregion of A that lies above S'. It follows that $|P[X \in S'] - P[Y \in S']|$ is maximized when $S' = S_0$ or $S' = S_1$, in which case it is equal to $\Delta[X; Y]$. $\square$

We can restate Theorem 8.31 as follows:

$$\Delta[X; Y] = \max\{|P[\phi(X)] - P[\phi(Y)]| : \phi \text{ is a predicate on } S\}.$$

This implies that when $\Delta[X; Y]$ is very small, then for *every* predicate ϕ, the events $\phi(X)$ and $\phi(Y)$ occur with almost the same probability. Put another way, there is no "statistical test" that can effectively distinguish between the distributions of X and Y. For many applications, this means that the distribution of X is "for all practical purposes" equivalent to that of Y, and hence in analyzing the behavior of X, we can instead analyze the behavior of Y, if that is more convenient.

Theorem 8.32. *If S and T are finite sets, X and Y are random variables taking values in S, and $f : S \to T$ is a function, then $\Delta[f(X); f(Y)] \leq \Delta[X; Y]$.*

Proof. We have

$$\Delta[f(X); f(Y)] = |P[f(X) \in T'] - P[f(Y) \in T']| \text{ for some } T' \subseteq T$$
$$\text{(by Theorem 8.31)}$$
$$= |P[X \in f^{-1}(T')] - P[Y \in f^{-1}(T')]|$$
$$\leq \Delta[X; Y] \text{ (again by Theorem 8.31). } \square$$

Example 8.41. Let X be uniformly distributed over the set $\{0, \ldots, m-1\}$, and let Y be uniformly distributed over the set $\{0, \ldots, n-1\}$, for $n \geq m$. Let $f(t) := t \bmod m$. We want to compute an upper bound on the statistical distance between X and $f(Y)$. We can do this as follows. Let $n = qm - r$, where $0 \leq r < m$, so that $q = \lceil n/m \rceil$. Also, let Z be uniformly distributed over $\{0, \ldots, qm - 1\}$. Then $f(Z)$ is uniformly distributed over $\{0, \ldots, m-1\}$, since every element of $\{0, \ldots, m-1\}$ has the same number (namely, q) of pre-images under f which lie in the set $\{0, \ldots, qm - 1\}$. Since statistical distance depends only on the distributions of the random variables, by the previous theorem, we have

$$\Delta[X; f(Y)] = \Delta[f(Z); f(Y)] \leq \Delta[Z; Y],$$

and as we saw in Example 8.40,

$$\Delta[Z; Y] = r/qm < 1/q \le m/n.$$

Therefore,

$$\Delta[X; f(Y)] < m/n. \quad \Box$$

We close this section with two useful theorems.

Theorem 8.33. *Suppose X, Y, and Z are random variables, where X and Z are independent, and Y and Z are independent. Then $\Delta[X, Z; Y, Z] = \Delta[X, Y]$.*

Note that $\Delta[X, Z; Y, Z]$ is shorthand for $\Delta[(X, Z); (Y, Z)]$.

Proof. Suppose X and Y take values in a finite set S, and Z takes values in a finite set T. From the definition of statistical distance,

$$2\Delta[X, Z; Y, Z] = \sum_{s,t} \left| P[(X = s) \cap (Z = t)] - P[(Y = s) \cap (Z = t)] \right|$$

$$= \sum_{s,t} \left| P[X = s] P[Z = t] - P[Y = s] P[Z = t] \right|$$

(by independence)

$$= \sum_{s,t} P[Z = t] \left| P[X = s] - P[Y = s] \right|$$

$$= \left(\sum_t P[Z = t] \right) \left(\sum_s \left| P[X = s] - P[Y = s] \right| \right)$$

$$= 1 \cdot 2\Delta[X; Y]. \quad \Box$$

Theorem 8.34. *Let $X_1, \ldots, X_n, Y_1, \ldots, Y_n$ be random variables, where $\{X_i\}_{i=1}^n$ is mutually independent, and $\{Y_i\}_{i=1}^n$ is mutually independent. Then we have*

$$\Delta[X_1, \ldots, X_n; Y_1, \ldots, Y_n] \le \sum_{i=1}^n \Delta[X_i; Y_i].$$

Proof. Since $\Delta[X_1, \ldots, X_n; Y_1, \ldots, Y_n]$ depends only on the individual distributions of the random variables $(X_1, \ldots, X_n)$ and $(Y_1, \ldots, Y_n)$, without loss of generality, we may assume that $(X_1, \ldots, X_n)$ and $(Y_1, \ldots, Y_n)$ are independent, so that $X_1, \ldots, X_n, Y_1, \ldots, Y_n$ form a mutually independent family of random variables. We introduce random variables $Z_0, \ldots, Z_n$, defined as follows:

$$Z_0 := (X_1, \ldots, X_n),$$
$$Z_i := (Y_1, \ldots, Y_i, X_{i+1}, \ldots, X_n) \quad \text{for } i = 1, \ldots, n-1, \text{ and}$$
$$Z_n := (Y_1, \ldots, Y_n).$$

By definition, $\Delta[X_1, \ldots, X_n; Y_1, \ldots, Y_n] = \Delta[Z_0; Z_n]$. Moreover, by part (iv) of Theorem 8.30, we have $\Delta[Z_0; Z_n] \leq \sum_{i=1}^{n} \Delta[Z_{i-1}; Z_i]$. Now consider any fixed index $i = 1, \ldots, n$. By Theorem 8.33, we have

$$\Delta[Z_{i-1}; Z_i] = \Delta[\, X_i, \, (Y_1, \ldots, Y_{i-1}, X_{i+1}, \ldots, X_n);$$
$$Y_i, \, (Y_1, \ldots, Y_{i-1}, X_{i+1}, \ldots, X_n)]$$
$$= \Delta[X_i; Y_i].$$

The theorem now follows immediately. $\square$

The technique used in the proof of the previous theorem is sometimes called a **hybrid argument**, as one considers the sequence of "hybrid" random variables $Z_0, Z_1, \ldots, Z_n$, and shows that the distance between each consecutive pair of variables is small.

EXERCISE 8.58. Let X and Y be independent random variables, each uniformly distributed over $\mathbb{Z}_p$, where p is prime. Calculate $\Delta[X, Y; X, XY]$.

EXERCISE 8.59. Let n be an integer that is the product of two distinct primes of the same bit length. Let X be uniformly distributed over $\mathbb{Z}_n$, and let Y be uniformly distributed over $\mathbb{Z}_n^*$. Show that $\Delta[X; Y] \leq 3n^{-1/2}$.

EXERCISE 8.60. Let X and Y be 0/1-valued random variables. Show that

$$\Delta[X; Y] = |P[X = 1] - P[Y = 1]|.$$

EXERCISE 8.61. Let S be a finite set, and consider any function $\phi : S \to \{0, 1\}$. Let B be a random variable uniformly distributed over $\{0, 1\}$, and for $b = 0, 1$, let X_b be a random variable taking values in S, and assume that X_b and B are independent. Show that

$$|P[\phi(X_B) = B] - \tfrac{1}{2}| = \tfrac{1}{2}|P[\phi(X_0) = 1] - P[\phi(X_1) = 1]| \leq \tfrac{1}{2}\Delta[X_0; X_1].$$

EXERCISE 8.62. Let X, Y be random variables taking values in a finite set S. For an event B that occurs with non-zero probability, define the **conditional statistical distance**

$$\Delta[X; Y \mid B] := \frac{1}{2} \sum_{s \in S} |P[X = s \mid B] - P[Y = s \mid B]|.$$

Let $\{B_i\}_{i \in I}$ be a finite, pairwise disjoint family of events whose union is B. Show that

$$\Delta[X; Y \mid B]\, P[B] \leq \sum_{\substack{P[B_i] \neq 0}} \Delta[X; Y \mid B_i]\, P[B_i].$$

EXERCISE 8.63. Let $\{\Phi_r\}_{r \in R}$ be a family of hash functions from S to T, with $m := |T|$. We say $\{\Phi_r\}_{r \in R}$ is ε-**variationally universal** if $\Phi_H(s)$ is uniformly distributed over T for each $s \in S$, and $\Delta[\Phi_H(s'); Y \mid \Phi_H(s) = t] \le \varepsilon$ for each $s, s' \in S$ with $s \ne s'$ and each $t \in T$; here, H and Y are independent random variables, with H uniformly distributed over R, and Y uniformly distributed over T. Show that:

(a) if $\{\Phi_r\}_{r \in R}$ is pairwise independent, then it is 0-variationally universal;

(b) if $\{\Phi_r\}_{r \in R}$ is ε-variationally universal, then it is $(1/m + \varepsilon)$-almost strongly universal;

(c) if $\{\Phi_r\}_{r \in R}$ is ε-almost universal, and $\{\Phi'_{r'}\}_{r' \in R'}$ is an ε'-variationally universal family of hash functions from S' to T', where $T \subseteq S'$, then $\{\Phi'_{r'} \circ \Phi_r\}_{(r,r') \in R \times R'}$ is an $(\varepsilon + \varepsilon')$-variationally universal family of hash functions from S to T'.

EXERCISE 8.64. Let $\{\Phi_r\}_{r \in R}$ be a family hash functions from S to T such that (i) each Φ_r maps S injectively into T, and (ii) there exists $\varepsilon \in [0, 1]$ such that $\Delta[\Phi_H(s); \Phi_H(s')] \le \varepsilon$ for all $s, s' \in S$, where H is uniformly distributed over R. Show that $|R| \ge (1 - \varepsilon)|S|$.

EXERCISE 8.65. Let X and Y be random variables that take the same value unless a certain event $\mathcal{F}$ occurs (i.e., $X(\omega) = Y(\omega)$ for all $\omega \in \bar{\mathcal{F}}$). Show that $\Delta[X; Y] \le \mathsf{P}[\mathcal{F}]$.

EXERCISE 8.66. Let X and Y be random variables taking values in the interval $[0, t]$. Show that $|\mathsf{E}[X] - \mathsf{E}[Y]| \le t \cdot \Delta[X; Y]$.

EXERCISE 8.67. Show that Theorem 8.33 is not true if we drop the independence assumptions.

EXERCISE 8.68. Let S be a set of size $m \ge 1$. Let F be a random variable that is uniformly distributed over the set of all functions from S into S. Let G be a random variable that is uniformly distributed over the set of all permutations of S. Let $s_1, \ldots, s_n$ be distinct, fixed elements of S. Show that

$$\Delta[F(s_1), \ldots, F(s_n); G(s_1), \ldots, G(s_n)] \le \frac{n(n-1)}{2m}.$$

EXERCISE 8.69. Let m be a large integer. Consider three random experiments. In the first, we generate a random integer X_1 between 1 and m, and then a random integer Y_1 between 1 and X_1. In the second, we generate a random integer X_2 between 2 and m, and then generate a random integer Y_2 between 1 and X_2. In the third, we generate a random integer X_3 between 2 and m, and then a random integer Y_3

between 2 and X_3. Show that $\Delta[X_1, Y_1; X_2, Y_2] = O(1/m)$ and $\Delta[X_2, Y_2; X_3, Y_3] = O(\log m/m)$, and conclude that $\Delta[X_1, Y_1; X_3, Y_3] = O(\log m/m)$.

8.9 Measures of randomness and the leftover hash lemma (∗)

In this section, we discuss different ways to measure "how random" the distribution of a random variable is, and relations among them.

Let X be a random variable taking values in a finite set S of size m. We define three measures of randomness:

1. the **collision probability of** X is $\sum_{s \in S} P[X = s]^2$;

2. the **guessing probability of** X is $\max\{P[X = s] : s \in S\}$;

3. the **distance of** X **from uniform on** S is $\frac{1}{2} \sum_{s \in S} |P[X = s] - 1/m|$.

Suppose X has collision probability β, guessing probability γ, and distance δ from uniform on S. If X' is another random variable with the same distribution as X, where X and X' independent, then $\beta = P[X = X']$ (see Exercise 8.37). If Y is a random variable that is uniformly distributed over S, then $\delta = \Delta[X; Y]$. If X itself is uniformly distributed over S, then $\beta = \gamma = 1/m$, and $\delta = 0$. The quantity $\log_2(1/\gamma)$ is sometimes called the **min entropy** of X, and the quantity $\log_2(1/\beta)$ is sometimes called the **Renyi entropy** of X.

We first state some easy inequalities:

Theorem 8.35. *Suppose X is a random variable that takes values in a finite set S of size m. If X has collision probability β, guessing probability γ, and distance δ from uniform on S, then:*

(i) $\beta \geq 1/m$;

(ii) $\gamma^2 \leq \beta \leq \gamma \leq 1/m + \delta$.

Proof. Part (i) is immediate from Exercise 8.37. The other inequalities are left as easy exercises. □

This theorem implies that the collision and guessing probabilities are minimal for the uniform distribution, which perhaps agrees with one's intuition.

While the above theorem implies that β and γ are close to $1/m$ when δ is small, the following theorem provides a converse:

Theorem 8.36. *Suppose X is a random variable that takes values in a finite set S of size m. If X has collision probability β, and distance δ from uniform on S, then $\delta \leq \frac{1}{2}\sqrt{m\beta - 1}$.*

Proof. We may assume that $\delta > 0$, since otherwise the theorem is already true, simply from the fact that $\beta \geq 1/m$.

For $s \in S$, let $p_s := P[X = s]$. We have $\delta = \frac{1}{2}\sum_s |p_s - 1/m|$, and hence $1 = \sum_s q_s$, where $q_s := |p_s - 1/m|/2\delta$. So we have

$$\frac{1}{m} \leq \sum_s q_s^2 \quad \text{(by Exercise 8.36)}$$

$$= \frac{1}{4\delta^2} \sum_s (p_s - 1/m)^2$$

$$= \frac{1}{4\delta^2}\left(\sum_s p_s^2 - 1/m\right) \quad \text{(again by Exercise 8.36)}$$

$$= \frac{1}{4\delta^2}(\beta - 1/m),$$

from which the theorem follows immediately. □

We are now in a position to state and prove a very useful result which, intuitively, allows us to convert a "low quality" source of randomness into a "high quality" source of randomness, making use of an almost universal family of hash functions (see end of §8.7).

Theorem 8.37 (Leftover hash lemma). *Let $\{\Phi_r\}_{r \in R}$ be a $(1 + \alpha)/m$-almost universal family of hash functions from S to T, where $m := |T|$. Let H and X be independent random variables, where H is uniformly distributed over R, and X takes values in S. If β is the collision probability of X, and δ' is the distance of $(H, \Phi_H(X))$ from uniform on $R \times T$, then $\delta' \leq \frac{1}{2}\sqrt{m\beta + \alpha}$.*

Proof. Let β' be the collision probability of $(H, \Phi_H(X))$. Our goal is to bound β' from above, and then apply Theorem 8.36 to the random variable $(H, \Phi_H(X))$. To this end, let $\ell := |R|$, and suppose H' and X' are random variables, where H' has the same distribution as H, X' has the same distribution as X, and H, H', X, X' form a mutually independent family of random variables. Then we have

$$\beta' = P[(H = H') \cap (\Phi_H(X) = \Phi_{H'}(X'))]$$
$$= P[(H = H') \cap (\Phi_H(X) = \Phi_H(X'))]$$
$$= \frac{1}{\ell} P[\Phi_H(X) = \Phi_H(X')] \quad \text{(a special case of Exercise 8.15)}$$
$$\leq \frac{1}{\ell}(P[X = X'] + (1 + \alpha)/m) \quad \text{(by Exercise 8.48)}$$
$$= \frac{1}{\ell m}(m\beta + 1 + \alpha).$$

The theorem now follows immediately from Theorem 8.36. □

In the previous theorem, if $\{\Phi_r\}_{r \in R}$ is a universal family of hash functions, then we can take $\alpha = 0$. However, it is convenient to allow $\alpha > 0$, as this allows for the use of families with a smaller key set (see Exercise 8.52).

Example 8.42. Suppose $S := \{0,1\}^{\times 1000}$, $T := \{0,1\}^{\times 64}$, and that $\{\Phi_r\}_{r \in R}$ is a universal family of hash functions from S to T. Suppose X and H are independent random variables, where X is uniformly distributed over some subset S' of S of size $\geq 2^{160}$, and H is uniformly distributed over R. Then the collision and guessing probabilities of X are at most 2^{-160}, and so the leftover hash lemma (with $\alpha = 0$) says that the distance of $(H, \Phi_H(X))$ from uniform on $R \times T$ is δ', where $\delta' \leq \frac{1}{2}\sqrt{2^{64}2^{-160}} = 2^{-49}$. By Theorem 8.32, it follows that the distance of $\Phi_H(X)$ from uniform on T is at most $\delta' \leq 2^{-49}$. $\square$

The leftover hash lemma allows one to convert "low quality" sources of randomness into "high quality" sources of randomness. Suppose that to conduct an experiment, we need to sample a random variable Y whose distribution is uniform on a set T of size m, or at least, its distance from uniform on T is sufficiently small. However, we may not have direct access to a source of "real" randomness whose distribution looks anything like that of the desired uniform distribution, but rather, only to a "low quality" source of randomness. For example, one could model various characteristics of a person's typing at the keyboard, or perhaps various characteristics of the internal state of a computer (both its software and hardware) as a random process. We cannot say very much about the probability distributions associated with such processes, but perhaps we can conservatively estimate the collision or guessing probabilities associated with these distributions. Using the leftover hash lemma, we can hash the output of this random process, using a suitably generated random hash function. The hash function acts like a "magnifying glass": it "focuses" the randomness inherent in the "low quality" source distribution onto the set T, obtaining a "high quality," nearly uniform, distribution on T.

Of course, this approach requires a random hash function, which may be just as difficult to generate as a random element of T. The following theorem shows, however, that we can at least use the same "magnifying glass" many times over, with the statistical distance from uniform of the output distribution increasing linearly in the number of applications of the hash function.

Theorem 8.38. *Let $\{\Phi_r\}_{r \in R}$ be a $(1 + \alpha)/m$-almost universal family of hash functions from S to T, where $m := |T|$. Let $H, X_1, \ldots, X_n$ be random variables, where H is uniformly distributed over R, each X_i takes values in S, and $H, X_1, \ldots, X_n$ form a mutually independent family of random variables. If β is an upper bound on the collision probability of each X_i, and δ' is the distance of $(H, \Phi_H(X_1), \ldots, \Phi_H(X_n))$ from uniform on $R \times T^{\times n}$, then $\delta' \leq \frac{1}{2}n\sqrt{m\beta + \alpha}$.*

Proof. Let $Y_1, \ldots, Y_n$ be random variables, each uniformly distributed over T, and assume that $H, X_1, \ldots, X_n, Y_1, \ldots, Y_n$ form a mutually independent family of random variables. We shall make a hybrid argument (as in the proof of Theorem 8.34). Define random variables $Z_0, Z_1, \ldots, Z_n$ as follows:

$$Z_0 := (H, \Phi_H(X_1), \ldots, \Phi_H(X_n)),$$
$$Z_i := (H, Y_1, \ldots, Y_i, \Phi_H(X_{i+1}), \ldots, \Phi_H(X_n)) \quad \text{for } i = 1, \ldots, n-1, \text{ and}$$
$$Z_n := (H, Y_1, \ldots, Y_n).$$

We have

$$\delta' = \Delta[Z_0; Z_n]$$

$$\leq \sum_{i=1}^n \Delta[Z_{i-1}; Z_i] \quad \text{(by part (iv) of Theorem 8.30)}$$

$$\leq \sum_{i=1}^n \Delta[\, H, Y_1, \ldots, Y_{i-1}, \Phi_H(X_i), X_{i+1}, \ldots, X_n;$$
$$\qquad\quad H, Y_1, \ldots, Y_{i-1}, \quad Y_i, \quad X_{i+1}, \ldots, X_n \,]$$
$$\text{(by Theorem 8.32)}$$

$$= \sum_{i=1}^n \Delta[H, \Phi_H(X_i); H, Y_i] \quad \text{(by Theorem 8.33)}$$

$$\leq \tfrac{1}{2} n \sqrt{m\beta + \alpha} \quad \text{(by Theorem 8.37).} \quad \square$$

Another source of "low quality" randomness arises in certain cryptographic applications, where we have a "secret value" X, which is a random variable that takes values in a set S, and which has small collision or guessing probability. We want to derive from X a "secret key" whose distance from uniform on some specified "key space" T is small. Typically, T is the set of all bit strings of some given length, as in Example 8.25. Theorem 8.38 allows us to do this using a "public" hash function—generated at random once and for all, published for all to see, and used over and over to derive secret keys as needed. However, to apply this theorem, it is crucial that the secret values (and the hash key) are mutually independent.

EXERCISE 8.70. Consider again the situation in Theorem 8.37. Suppose that $T = \{0, \ldots, m-1\}$, but that we would rather have a nearly uniform distribution on $T' = \{0, \ldots, m'-1\}$, for some $m' < m$. While it may be possible to work with a different family of hash functions, we do not have to if m is large enough with respect to m', in which case we can just use the value $Y' := \Phi_H(X) \bmod m'$. Show that the distance of (H, Y') from uniform on $R \times T'$ is at most $\tfrac{1}{2}\sqrt{m\beta + \alpha} + m'/m$.

EXERCISE 8.71. Let $\{\Phi_r\}_{r \in R}$ be a $(1 + \alpha)/m$-almost universal family of hash functions from S to T, where $m := |T|$. Suppose H, X, Y, Z are random variables, where H is uniformly distributed over R, X takes values in S, Y is uniformly distributed over T, and U is the set of values taken by Z with non-zero probability. Assume that the family of random variables $H, Y, (X, Z)$ is mutually independent.

(a) For $u \in U$, define $\beta(u) := \sum_{s \in S} P[X = s \mid Z = u]^2$. Also, let $\beta' := \sum_{u \in U} \beta(u) P[Z = u]$. Show that $\Delta[H, \Phi_H(X), Z; H, Y, Z] \leq \frac{1}{2}\sqrt{m\beta' + \alpha}$.

(b) Suppose that X is uniformly distributed over a subset S' of S, and that $Z = f(X)$ for some function $f : S \to U$. Show that $\Delta[H, \Phi_H(X), Z; H, Y, Z] \leq \frac{1}{2}\sqrt{m|U|/|S'| + \alpha}$.

8.10 Discrete probability distributions

In addition to working with probability distributions over finite sample spaces, one can also work with distributions over infinite sample spaces. If the sample space is *countable*, that is, either finite or *countably infinite* (see §A3), then the distribution is called a **discrete probability distribution**. We shall not consider any other types of probability distributions in this text. The theory developed in §§8.1–8.5 extends fairly easily to the countably infinite setting, and in this section, we discuss how this is done.

8.10.1 Basic definitions

To say that the sample space Ω is countably infinite simply means that there is a bijection f from the set of positive integers onto Ω; thus, we can enumerate the elements of Ω as $\omega_1, \omega_2, \omega_3, \ldots$, where $\omega_i := f(i)$.

As in the finite case, a **probability distribution on** Ω is a function $P : \Omega \to [0, 1]$, where all the probabilities sum to 1, which means that the infinite series $\sum_{i=1}^{\infty} P(\omega_i)$ converges to one. Luckily, the convergence properties of an infinite series whose terms are all non-negative is invariant under a reordering of terms (see §A6), so it does not matter how we enumerate the elements of Ω.

Example 8.43. Suppose we toss a fair coin repeatedly until it comes up *heads*, and let k be the total number of tosses. We can model this experiment as a discrete probability distribution P, where the sample space consists of the set of all positive integers: for each positive integer k, $P(k) := 2^{-k}$. We can check that indeed $\sum_{k=1}^{\infty} 2^{-k} = 1$, as required.

One may be tempted to model this experiment by setting up a probability distribution on the sample space of all infinite sequences of coin tosses; however, this sample space is not countably infinite, and so we cannot construct a discrete probability distribution on this space. While it is possible to extend the notion of a probability distribution to such spaces, this would take us too far afield. □

Example 8.44. More generally, suppose we repeatedly execute a Bernoulli trial until it succeeds, where each execution succeeds with probability $p > 0$ independently of the previous trials, and let k be the total number of trials executed. Then we associate the probability $P(k) := q^{k-1}p$ with each positive integer k, where $q := 1 - p$, since we have $k - 1$ failures before the one success. One can easily check that these probabilities sum to 1. Such a distribution is called a **geometric distribution**. □

Example 8.45. The series $\sum_{k=1}^{\infty} 1/k^3$ converges to some positive number c. Therefore, we can define a probability distribution on the set of positive integers, where we associate with each $k \geq 1$ the probability $1/ck^3$. □

As in the finite case, an event is an arbitrary subset $\mathcal{A}$ of Ω. The probability $P[\mathcal{A}]$ of $\mathcal{A}$ is defined as the sum of the probabilities associated with the elements of $\mathcal{A}$. This sum is treated as an infinite series when $\mathcal{A}$ is infinite. This series is guaranteed to converge, and its value does not depend on the particular enumeration of the elements of $\mathcal{A}$.

Example 8.46. Consider the geometric distribution discussed in Example 8.44, where p is the success probability of each Bernoulli trial, and $q := 1 - p$. For a given integer $i \geq 1$, consider the event $\mathcal{A}$ that the number of trials executed is at least i. Formally, $\mathcal{A}$ is the set of all integers greater than or equal to i. Intuitively, $P[\mathcal{A}]$ should be q^{i-1}, since we perform at least i trials if and only if the first $i - 1$ trials fail. Just to be sure, we can compute

$$P[\mathcal{A}] = \sum_{k \geq i} P(k) = \sum_{k \geq i} q^{k-1}p = q^{i-1}p \sum_{k \geq 0} q^k = q^{i-1}p \cdot \frac{1}{1-q} = q^{i-1}. \quad \square$$

It is an easy matter to check that all the statements and theorems in §8.1 carry over *verbatim* to the case of countably infinite sample spaces. Moreover, Boole's inequality (8.6) and equality (8.7) are also valid for countably infinite families of events:

Theorem 8.39. *Suppose $\mathcal{A} := \bigcup_{i=1}^{\infty} \mathcal{A}_i$, where $\{\mathcal{A}_i\}_{i=1}^{\infty}$ is an infinite sequence of events. Then*

(i) $P[\mathcal{A}] \leq \sum_{i=1}^{\infty} P[\mathcal{A}_i]$, and

(ii) $P[\mathcal{A}] = \sum_{i=1}^{\infty} P[\mathcal{A}_i]$ if $\{\mathcal{A}_i\}_{i=1}^{\infty}$ is pairwise disjoint.

Proof. As in the proof of Theorem 8.1, for $\omega \in \Omega$ and $\mathcal{B} \subseteq \Omega$, define $\delta_\omega[\mathcal{B}] := 1$ if $\omega \in \mathcal{B}$, and $\delta_\omega[\mathcal{B}] := 0$ if $\omega \notin \mathcal{B}$. First, suppose that $\{\mathcal{A}_i\}_{i=1}^{\infty}$ is pairwise disjoint.

Evidently, $\delta_\omega[\mathcal{A}] = \sum_{i=1}^\infty \delta_\omega[\mathcal{A}_i]$ for each $\omega \in \Omega$, and so

$$P[\mathcal{A}] = \sum_{\omega \in \Omega} P(\omega)\delta_\omega[\mathcal{A}] = \sum_{\omega \in \Omega} P(\omega) \sum_{i=1}^\infty \delta_\omega[\mathcal{A}_i]$$

$$= \sum_{i=1}^\infty \sum_{\omega \in \Omega} P(\omega)\delta_\omega[\mathcal{A}_i] = \sum_{i=1}^\infty P[\mathcal{A}_i],$$

where we use the fact that we may reverse the order of summation in an infinite double summation of non-negative terms (see §A7). That proves (ii), and (i) follows from (ii), applied to the sequence $\{\mathcal{A}'_i\}_{i=1}^\infty$, where $\mathcal{A}'_i := \mathcal{A}_i \setminus \bigcup_{j=1}^{i-1} \mathcal{A}_i$, as $P[\mathcal{A}] = \sum_{i=1}^\infty P[\mathcal{A}'_i] \le \sum_{i=1}^\infty P[\mathcal{A}_i]$. $\square$

8.10.2 Conditional probability and independence

All of the definitions and results in §8.2 carry over *verbatim* to the countably infinite case. The law of total probability (equations (8.9) and (8.10)), as well as Bayes' theorem (8.11), extend to families of events $\{\mathcal{B}_i\}_{i \in I}$ indexed by any countably infinite set I. The definitions of independent families of events (k-wise and mutually) extend *verbatim* to infinite families.

8.10.3 Random variables

All of the definitions and results in §8.3 carry over *verbatim* to the countably infinite case. Note that the image of a random variable may be either finite or countably infinite. The definitions of independent families of random variables (k-wise and mutually) extend *verbatim* to infinite families.

8.10.4 Expectation and variance

We define the expected value of a real-valued random variable X exactly as in (8.18); that is, $E[X] := \sum_\omega X(\omega) P(\omega)$, but where this sum is now an infinite series. If this series converges absolutely (see §A6), then we say that X has **finite expectation**, or that $E[X]$ is **finite**. In this case, the series defining $E[X]$ converges to the same finite limit, regardless of the ordering of the terms.

If $E[X]$ is not finite, then under the right conditions, $E[X]$ may still exist, although its value will be $\pm\infty$. Consider first the case where X takes only non-negative values. In this case, if $E[X]$ is not finite, then we naturally define $E[X] := \infty$, as the series defining $E[X]$ diverges to ∞, regardless of the ordering of the terms. In the general case, we may define random variables X^+ and X^-, where

$$X^+(\omega) := \max\{0, X(\omega)\} \quad \text{and} \quad X^-(\omega) := \max\{0, -X(\omega)\},$$

so that $X = X^+ - X^-$, and both X^+ and X^- take only non-negative values. Clearly, X has finite expectation if and only if both X^+ and X^- have finite expectation. Now suppose that $E[X]$ is not finite, so that one of $E[X^+]$ or $E[X^-]$ is infinite. If $E[X^+] = E[X^-] = \infty$, then we say that $E[X]$ **does not exist**; otherwise, we define $E[X] := E[X^+] - E[X^-]$, which is $\pm\infty$; in this case, the series defining $E[X]$ diverges to $\pm\infty$, regardless of the ordering of the terms.

Example 8.47. Let X be a random variable whose distribution is as in Example 8.45. Since the series $\sum_{k=1}^{\infty} 1/k^2$ converges and the series $\sum_{k=1}^{\infty} 1/k$ diverges, the expectation $E[X]$ is finite, while $E[X^2] = \infty$. One may also verify that the random variable $(-1)^X X^2$ has no expectation. $\square$

All of the results in §8.4 carry over essentially unchanged, although one must pay some attention to "convergence issues."

If $E[X]$ exists, then we can regroup the terms in the series $\sum_\omega X(\omega) P(\omega)$, without affecting its value. In particular, equation (8.19) holds provided $E[X]$ exists, and equation (8.20) holds provided $E[f(X)]$ exists.

Theorem 8.14 still holds, under the additional hypothesis that $E[X]$ and $E[Y]$ are finite. Equation (8.21) also holds, provided the individual expectations $E[X_i]$ are finite. More generally, if $E[X]$ and $E[Y]$ exist, then $E[X + Y] = E[X] + E[Y]$, unless $E[X] = \infty$ and $E[Y] = -\infty$, or $E[X] = -\infty$ and $E[Y] = \infty$. Also, if $E[X]$ exists, then $E[aX] = a E[X]$, unless $a = 0$ and $E[X] = \pm\infty$.

One might consider generalizing (8.21) to countably infinite families of random variables. To this end, suppose $\{X_i\}_{i=1}^{\infty}$ is an infinite sequence of real-valued random variables. The random variable $X := \sum_{i=1}^{\infty} X_i$ is well defined, provided the series $\sum_{i=1}^{\infty} X_i(\omega)$ converges for each $\omega \in \Omega$. One might hope that $E[X] = \sum_{i=1}^{\infty} E[X_i]$; however, this is not in general true, even if the individual expectations, $E[X_i]$, are non-negative, and even if the series defining X converges absolutely for each ω; nevertheless, it is true when the X_i's are non-negative:

Theorem 8.40. *Let $\{X_i\}_{i=1}^{\infty}$ be an infinite sequence of random variables. Suppose that for each $i \geq 1$, X_i takes non-negative values only, and has finite expectation. Also suppose that $\sum_{i=1}^{\infty} X_i(\omega)$ converges for each $\omega \in \Omega$, and define $X := \sum_{i=1}^{\infty} X_i$. Then we have*

$$E[X] = \sum_{i=1}^{\infty} E[X_i].$$

Proof. This is a calculation just like the one made in the proof of Theorem 8.39, where, again, we use the fact that we may reverse the order of summation in an infinite double summation of non-negative terms:

$$E[X] = \sum_{\omega \in \Omega} P(\omega)X(\omega) = \sum_{\omega \in \Omega} P(\omega) \sum_{i=1}^{\infty} X_i(\omega)$$

$$= \sum_{i=1}^{\infty} \sum_{\omega \in \Omega} P(\omega)X_i(\omega) = \sum_{i=1}^{\infty} E[X_i]. \quad \square$$

Theorem 8.15 holds under the additional hypothesis that $E[X]$ and $E[Y]$ are finite. Equation (8.22) also holds, provided the individual expectations $E[X_i]$ are finite. Theorem 8.16 still holds, of course. Theorem 8.17 also holds, but where now the sum may be infinite; it can be proved using essentially the same argument as in the finite case, combined with Theorem 8.40.

Example 8.48. Suppose X is a random variable with a geometric distribution, as in Example 8.44, with an associated success probability p and failure probability $q := 1 - p$. As we saw in Example 8.46, for every integer $i \geq 1$, we have $P[X \geq i] = q^{i-1}$. We may therefore apply the infinite version of Theorem 8.17 to easily compute the expected value of X:

$$E[X] = \sum_{i=1}^{\infty} P[X \geq i] = \sum_{i=1}^{\infty} q^{i-1} = \frac{1}{1-q} = \frac{1}{p}. \quad \square$$

Example 8.49. To illustrate that Theorem 8.40 does not hold in general, consider the geometric distribution on the positive integers, where $P(j) = 2^{-j}$ for $j \geq 1$. For $i \geq 1$, define the random variable X_i so that $X_i(i) = 2^i$, $X_i(i+1) = -2^{i+1}$, and $X_i(j) = 0$ for all $j \notin \{i, i+1\}$. Then $E[X_i] = 0$ for all $i \geq 1$, and so $\sum_{i \geq 1} E[X_i] = 0$. Now define $X := \sum_{i \geq 1} X_i$. This is well defined, and in fact $X(1) = 2$, while $X(j) = 0$ for all $j > 1$. Hence $E[X] = 1$. $\square$

The variance $\mathrm{Var}[X]$ of X exists only when $\mu := E[X]$ is finite, in which case it is defined as usual as $E[(X - \mu)^2]$, which may be either finite or infinite. Theorems 8.18, 8.19, and 8.20 hold provided all the relevant expectations and variances are finite.

The definition of conditional expectation carries over verbatim. Equation (8.23) holds, provided $E[X \mid B]$ exists, and the law of total expectation (8.24) holds, provided $E[X]$ exists. The law of total expectation also holds for a countably infinite partition $\{B_i\}_{i \in I}$, provided $E[X]$ exists, and each of the conditional expectations $E[X \mid B_i]$ is finite.

8.10.5 Some useful bounds

All of the results in this section hold, provided the relevant expectations and variances are finite.

EXERCISE 8.72. Let $\{A_i\}_{i=1}^{\infty}$ be a family of events, such that $A_i \subseteq A_{i+1}$ for each $i \geq 1$, and let $A := \bigcup_{i=1}^{\infty} A_i$. Show that $P[A] = \lim_{i \to \infty} P[A_i]$.

EXERCISE 8.73. Generalize Exercises 8.6, 8.7, 8.23, and 8.24 to the discrete setting, allowing a countably infinite index set I.

EXERCISE 8.74. Suppose X is a random variable taking positive integer values, and that for some real number q, with $0 \leq q \leq 1$, and for all integers $i \geq 1$, we have $P[X \geq i] = q^{i-1}$. Show that X has a geometric distribution with associated success probability $p := 1 - q$.

EXERCISE 8.75. This exercise extends Jensen's inequality (see Exercise 8.25) to the discrete setting. Suppose that f is a convex function on an interval I. Let X be a random variable whose image is a countably infinite subset of I, and assume that both $E[X]$ and $E[f(X)]$ are finite. Show that $E[f(X)] \geq f(E[X])$. Hint: use continuity.

EXERCISE 8.76. A gambler plays a simple game in a casino: with each play of the game, the gambler may bet any number m of dollars; a fair coin is tossed, and if it comes up *heads*, the casino pays m dollars to the gambler, and otherwise, the gambler pays m dollars to the casino. The gambler plays the game repeatedly, using the following strategy: he initially bets a dollar, and with each subsequent play, he doubles his bet; if he ever wins, he quits and goes home; if he runs out of money, he also goes home; otherwise, he plays again. Show that if the gambler has an infinite amount of money, then his expected winnings are one dollar, and if he has a finite amount of money, his expected winnings are zero.

8.11 Notes

The idea of sharing a secret via polynomial evaluation and interpolation (see Example 8.28) is due to Shamir [90].

Our Chernoff bound (Theorem 8.24) is one of a number of different types of bounds that appear in the literature under the rubric of "Chernoff bound."

Universal and pairwise independent hash functions, with applications to hash tables and message authentication codes, were introduced by Carter and Wegman [25, 105]. The notions of ε-almost universal and ε-almost strongly universal hashing were developed by Stinson [101]. The notion of ε-variationally universal hashing (see Exercise 8.63) is from Krovetz and Rogaway [57].

The leftover hash lemma (Theorem 8.37) was originally stated and proved by Impagliazzo, Levin, and Luby [48], who use it to obtain an important result in the theory of cryptography. Our proof of the leftover hash lemma is loosely based on one by Impagliazzo and Zuckermann [49], who also present further applications.

9

Probabilistic algorithms

It is sometimes useful to endow our algorithms with the ability to generate random numbers. In fact, we have already seen two examples of how such probabilistic algorithms may be useful:

- at the end of §3.4, we saw how a probabilistic algorithm might be used to build a simple and efficient primality test; however, this test might incorrectly assert that a composite number is prime; in the next chapter, we will see how a small modification to this algorithm will ensure that the probability of making such a mistake is extremely small;

- in §4.5, we saw how a probabilistic algorithm could be used to make Fermat's two squares theorem constructive; in this case, the use of randomization never leads to incorrect results, but the running time of the algorithm was only bounded "in expectation."

We will see a number of other probabilistic algorithms in this text, and it is high time that we place them on a firm theoretical foundation. To simplify matters, we only consider algorithms that generate random bits. Where such random bits actually come from will not be of great concern to us here. In a practical implementation, one would use a pseudo-random bit generator, which should produce bits that "for all practical purposes" are "as good as random." While there is a well-developed theory of pseudo-random bit generation (some of which builds on the ideas in §8.9), we will not delve into this here. Moreover, the pseudo-random bit generators used in practice are not based on this general theory, and are much more ad hoc in design. So, although we will present a rigorous formal theory of probabilistic algorithms, the application of this theory to practice is ultimately a bit heuristic; nevertheless, experience with these algorithms has shown that the theory is a very good predictor of the real-world behavior of these algorithms.

9.1 Basic definitions

Formally speaking, we will add a new type of instruction to our random access machine (described in §3.2):

random bit This type of instruction is of the form $\gamma \leftarrow \text{RAND}$, where γ takes the same form as in arithmetic instructions. Execution of this type of instruction assigns to γ a value sampled from the uniform distribution on $\{0, 1\}$, independently from the execution of all other random-bit instructions.

Algorithms that use random-bit instructions are called **probabilistic** (or **randomized**), while those that do not are called **deterministic**.

In describing probabilistic algorithms at a high level, we shall write "$y \xleftarrow{\mathscr{R}} \{0, 1\}$" to denote the assignment of a random bit to the variable y, and "$y \xleftarrow{\mathscr{R}} \{0, 1\}^{\times \ell}$" to denote the assignment of a random bit string of length ℓ to the variable y.

To analyze the behavior of a probabilistic algorithm, we first need a probability distribution that appropriately models its execution. Once we have done this, we shall define the running time and output to be *random variables* associated with this distribution.

9.1.1 Defining the distribution

It would be desirable to define a probability distribution that could be used for all algorithms and all inputs. While this can be done in principle, it would require notions from the theory of probability more advanced than those we developed in the previous chapter. Instead, for a given probabilistic algorithm A and input x, we shall define a discrete probability distribution that models A's execution on input x. Thus, every algorithm/input pair yields a different distribution.

To motivate our definition, consider Example 8.43. We could view the sample space in that example to be the set of all bit strings consisting of zero or more 0 bits, followed by a single 1 bit, and to each such bit string ω of this special form, we assign the probability $2^{-|\omega|}$, where $|\omega|$ denotes the length of ω. The "random experiment" we have in mind is to generate random bits one at a time until one of these special "halting" strings is generated. In developing the definition of the probability distribution for a probabilistic algorithm, we simply consider more general sets of "halting" strings, as determined by the algorithm and its input.

So consider a fixed algorithm A and input x. Let λ be a finite bit string of length, say, ℓ. We can use λ to "drive" the execution of A on input x for up to ℓ execution steps, as follows: for each step $i = 1, \ldots, \ell$, if the ith instruction executed by A is $\gamma \leftarrow \text{RAND}$, the ith bit of λ is assigned to γ. In this context, we shall refer to λ as an **execution path**. The reader may wish to visualize λ as a finite path in an infinite binary tree, where we start at the root, branching to the left if the next bit in λ is a 0 bit, and branching to the right if the next bit in λ is a 1 bit.

After using λ to drive A on input x for up to ℓ steps, we might find that the algorithm executed a *halt* instruction at some point during the execution, in which case we call λ a **complete** execution path; moreover, if this *halt* instruction was the ℓth instruction executed by A, then we call λ an **exact** execution path.

Our intent is to define the probability distribution associated with A on input x to be $P : \Omega \to [0, 1]$, where the sample space Ω is the set of all *exact* execution paths, and $P(\omega) := 2^{-|\omega|}$ for each $\omega \in \Omega$. However, for this to work, all the probabilities must sum to 1. The next theorem at least guarantees that these probabilities sum to at most 1. The only property of Ω that really matters in the proof of this theorem is that it is **prefix free**, which means that no exact execution path is a proper prefix of any other.

Theorem 9.1. *Let Ω be the set of all exact execution paths for A on input x. Then $\sum_{\omega \in \Omega} 2^{-|\omega|} \leq 1$.*

Proof. Let k be a non-negative integer. Let $\Omega_k \subseteq \Omega$ be the set of all exact execution paths of length at most k, and let $\alpha_k := \sum_{\omega \in \Omega_k} 2^{-|\omega|}$. We shall show below that

$$\alpha_k \leq 1. \tag{9.1}$$

From this, it will follow that

$$\sum_{\omega \in \Omega} 2^{-|\omega|} = \lim_{k \to \infty} \alpha_k \leq 1.$$

To prove the inequality (9.1), consider the set C_k of all *complete* execution paths of length *equal* to k. We claim that

$$\alpha_k = 2^{-k} |C_k|, \tag{9.2}$$

from which (9.1) follows, since clearly, $|C_k| \leq 2^k$. So now we are left to prove (9.2). Observe that by definition, each $\lambda \in C_k$ extends some $\omega \in \Omega_k$; that is, ω is a prefix of λ; moreover, ω is uniquely determined by λ, since no exact execution path is a proper prefix of any other exact execution path. Also observe that for each $\omega \in \Omega_k$, if $C_k(\omega)$ is the set of execution paths $\lambda \in C_k$ that extend ω, then $|C_k(\omega)| = 2^{k-|\omega|}$, and by the previous observation, $\{C_k(\omega)\}_{\omega \in \Omega_k}$ is a partition of C_k. Thus, we have

$$\alpha_k = \sum_{\omega \in \Omega_k} 2^{-|\omega|} = \sum_{\omega \in \Omega_k} 2^{-|\omega|} \sum_{\lambda \in C_k(\omega)} 2^{-k+|\omega|} = 2^{-k} \sum_{\omega \in \Omega_k} \sum_{\lambda \in C_k(\omega)} 1 = 2^{-k} |C_k|,$$

which proves (9.2). $\square$

From the above theorem, if Ω is the set of all exact execution paths for A on input x, then

$$\alpha := \sum_{\omega \in \Omega} 2^{-|\omega|} \leq 1,$$

and we say that A **halts with probability** α **on input** x. If $\alpha = 1$, we define the distribution $\mathsf{P} : \Omega \to [0, 1]$ associated with A on input x, where $\mathsf{P}(\omega) := 2^{-|\omega|}$ for each $\omega \in \Omega$.

We shall mainly be interested in algorithms that halt with probability 1 on all inputs. The following four examples provide some simple criteria that guarantee this.

Example 9.1. Suppose that on input x, A always halts within a finite number of steps, regardless of its random choices. More precisely, this means that there is a bound ℓ (depending on A and x), such that all execution paths of length ℓ are complete. In this case, we say that A's running time on input x is **strictly bounded by** ℓ, and it is clear that A halts with probability 1 on input x. Moreover, one can much more simply model A's computation on input x by working with the uniform distribution on execution paths of length ℓ. $\square$

Example 9.2. Suppose A and B are probabilistic algorithms that both halt with probability 1 on all inputs. Using A and B as subroutines, we can form their **serial composition**; that is, we can construct the algorithm

$C(x) :$ output $B(A(x))$,

which on input x, first runs A on input x, obtaining a value y, then runs B on input y, obtaining a value z, and finally, outputs z. We claim that C halts with probability 1 on all inputs.

For simplicity, we may assume that A places its output y in a location in memory where B expects to find its input, and that B places its output in a location in memory where C's output should go. With these assumptions, the program for C is obtained by simply concatenating the programs for A and B, making the following adjustments: every *halt* instruction in A's program is translated into an instruction that branches to the first instruction of B's program, and every target in a branch instruction in B's program is increased by the length of A's program.

Let Ω be the sample space representing A's execution on an input x. Each $\omega \in \Omega$ determines an output y, and a corresponding sample space Ω'_ω representing B's execution on input y. The sample space representing C's execution on input x is

$$\Omega'' = \{\omega\omega' : \omega \in \Omega, \omega' \in \Omega'_\omega\},$$

where $\omega\omega'$ is the concatenation of ω and ω'. We have

$$\sum_{\omega\omega'\in\Omega''} 2^{-|\omega\omega'|} = \sum_{\omega\in\Omega} 2^{-|\omega|} \sum_{\omega'\in\Omega'_\omega} 2^{-|\omega'|} = \sum_{\omega\in\Omega} 2^{-|\omega|} \cdot 1 = 1,$$

which shows that C halts with probability 1 on input x. $\Box$

Example 9.3. Suppose A, B, and C are probabilistic algorithms that halt with probability 1 on all inputs, and that A always outputs either *true* or *false*. Then we can form the **conditional construct**

$D(x)$:   if $A(x)$ then output $B(x)$ else output $C(x)$.

By a calculation similar to that in the previous example, it is easy to see that D halts with probability 1 on all inputs. $\Box$

Example 9.4. Suppose A and B are probabilistic algorithms that halt with probability 1 on all inputs, and that A always outputs either *true* or *false*. We can form the **iterative construct**

$C(x)$:   while $A(x)$ do $x \leftarrow B(x)$
output x.

Algorithm C may or may not halt with probability 1. To analyze C, we define an infinite sequence of algorithms $\{C_n\}_{n=0}^\infty$; namely, we define C_0 as

$C_0(x)$:   halt,

and for $n > 0$, we define C_n as

$C_n(x)$:   if $A(x)$ then $C_{n-1}(B(x))$.

Essentially, C_n drives C for up to n loop iterations before halting, if necessary, in C_0. By the previous three examples, it follows by induction on n that each C_n halts with probability 1 on all inputs. Therefore, we have a well-defined probability distribution for each C_n and each input x.

Consider a fixed input x. For each $n \geq 0$, let β_n be the probability that on input x, C_n terminates by executing algorithm C_0. Intuitively, β_n is the probability that C executes at least n loop iterations; however, this probability is defined with respect to the probability distribution associated with algorithm C_n on input x. It is not hard to see that the sequence $\{\beta_n\}_{n=0}^\infty$ is non-increasing, and so the limit $\beta := \lim_{n\to\infty} \beta_n$ exists; moreover, C halts with probability $1 - \beta$ on input x.

On the one hand, if the loop in algorithm C is guaranteed to terminate after a finite number of iterations (as in a "for loop"), then C certainly halts with probability 1. Indeed, if on input x, there is a bound ℓ (depending on x) such that the number of loop iterations is always at most ℓ, then $\beta_{\ell+1} = \beta_{\ell+2} = \cdots = 0$. On the other hand, if on input x, C enters into a good, old-fashioned infinite loop, then C

certainly does not halt with probability 1, as $\beta_0 = \beta_1 = \cdots = 1$. Of course, there may be in-between cases, which require further analysis. $\square$

We now illustrate the above criteria with a couple of some simple, concrete examples.

Example 9.5. Consider the following algorithm, which models an experiment in which we toss a fair coin repeatedly until it comes up *heads*:

> repeat
>> $b \stackrel{\text{\cent}}{\leftarrow} \{0, 1\}$
>
> until $b = 1$

For each positive integer n, let β_n be the probability that the algorithm executes at least n loop iterations, in the sense of Example 9.4. It is not hard to see that $\beta_n = 2^{-n+1}$, and since $\beta_n \to 0$ as $n \to \infty$, the algorithm halts with probability 1, even though the loop is not guaranteed to terminate after any particular, finite number of steps. $\square$

Example 9.6. Consider the following algorithm:

> $i \leftarrow 0$
> repeat
>> $i \leftarrow i + 1$
>> $\sigma \stackrel{\text{\cent}}{\leftarrow} \{0, 1\}^{\times i}$
>
> until $\sigma = 0^{\times i}$

For each positive integer n, let β_n be the probability that the algorithm executes at least n loop iterations, in the sense of Example 9.4. It is not hard to see that

$$\beta_n = \prod_{i=1}^{n-1}(1 - 2^{-i}) \geq \prod_{i=1}^{n-1} e^{-2^{-i+1}} = e^{-\sum_{i=0}^{n-2} 2^{-i}} \geq e^{-2},$$

where we have made use of the estimate (iii) in §A1. Therefore,

$$\lim_{n \to \infty} \beta_n \geq e^{-2} > 0,$$

and so the algorithm does not halt with probability 1, even though it never falls into an infinite loop. $\square$

9.1.2 Defining the running time and output

Let A be a probabilistic algorithm that halts with probability 1 on a fixed input x. We may define the random variable Z that represents A's running time on input x, and the random variable Y that represents A's output on input x.

Formally, Z and Y are defined using the probability distribution on the sample space Ω, defined in §9.1.2. The sample space Ω consists of all exact execution paths for A on input x. For each $\omega \in \Omega$, $Z(\omega) := |\omega|$, and $Y(\omega)$ is the output produced by A on input x, using ω to drive its execution.

The **expected running time of A on input** x is defined to be $E[Z]$. Note that in defining the expected running time, we view the input as fixed, rather than drawn from some probability distribution. Also note that the expected running time may be infinite.

We say that A runs in **expected polynomial time** if there exist constants a, b, and c, such that for all n, and for all inputs x of size n, the expected running time of A on input x is at most $an^b + c$. We say that A runs in **strict polynomial time** if there exist constants a, b, and c, such that for all n, and for all inputs x of size n, A's running time on input x is strictly bounded by $an^b + c$ (as in Example 9.1).

Example 9.7. Consider again the algorithm in Example 9.5. Let L be the random variable that represents the number of loop iterations executed by the algorithm. The distribution of L is a geometric distribution, with associated success probability $1/2$ (see Example 8.44). Therefore, $E[L] = 2$ (see Example 8.46). Let Z be the random variable that represents the running time of the algorithm. We have $Z \leq cL$, for some implementation-dependent constant c. Therefore, $E[Z] \leq c\,E[L] = 2c$. $\square$

Example 9.8. Consider the following probabilistic algorithm that takes as input a positive integer m. It models an experiment in which we toss a fair coin repeatedly until it comes up *heads* m times.

$$k \leftarrow 0$$
$$\text{repeat}$$
$$\qquad b \xleftarrow{\ell} \{0, 1\}$$
$$\qquad \text{if } b = 1 \text{ then } k \leftarrow k + 1$$
$$\text{until } k = m$$

Let L be the random variable that represents the number of loop iterations executed the algorithm on a fixed input m. We claim that $E[L] = 2m$. To see this, define random variables $L_1, \ldots, L_m$, where L_1 is the number of loop iterations needed to get $b = 1$ for the first time, L_2 is the number of additional loop iterations needed to get $b = 1$ for the second time, and so on. Clearly, we have $L = L_1 + \cdots + L_m$, and moreover, $E[L_i] = 2$ for $i = 1, \ldots, m$; therefore, by linearity of expectation, we have $E[L] = E[L_1] + \cdots + E[L_m] = 2m$. It follows that the expected running time of this algorithm on input m is $O(m)$. $\square$

Example 9.9. Consider the following algorithm:

$$n \leftarrow 0$$

repeat $n \leftarrow n + 1, \ b \stackrel{\text{\tiny ¢}}{\leftarrow} \{0, 1\}$ until $b = 1$

repeat $\sigma \stackrel{\text{\tiny ¢}}{\leftarrow} \{0, 1\}^{\times n}$ until $\sigma = 0^{\times n}$

The expected running time is infinite (even though it does halt with probability 1). To see this, define random variables L_1 and L_2, where L_1 is the number of iterations of the first loop, and L_2 is the number of iterations of the second. As in Example 9.7, the distribution of L_1 is a geometric distribution with associated success probability $1/2$, and $E[L_1] = 2$. For each $k \geq 1$, the conditional distribution of L_2 given $L_1 = k$ is a geometric distribution with associated success probability $1/2^k$, and so $E[L_2 \mid L_1 = k] = 2^k$. Therefore,

$$E[L_2] = \sum_{k \geq 1} E[L_2 \mid L_1 = k] \, P[L_1 = k] = \sum_{k \geq 1} 2^k \cdot 2^{-k} = \sum_{k \geq 1} 1 = \infty. \ \square$$

We have presented a fairly rigorous definitional framework for probabilistic algorithms, but from now on, we shall generally reason about such algorithms at a higher, and more intuitive, level. Nevertheless, all of our arguments can be translated into this rigorous framework, the details of which we leave to the interested reader. Moreover, all of the algorithms we shall present halt with probability 1 on all inputs, but we shall not go into the details of proving this (but the criteria in Examples 9.1–9.4 can be used to easily verify this).

EXERCISE 9.1. Suppose A is a probabilistic algorithm that halts with probability 1 on input x, and let $P : \Omega \to [0, 1]$ be the corresponding probability distribution. Let λ be an execution path of length ℓ, and assume that no proper prefix of λ is exact. Let $\mathcal{E}_\lambda := \{\omega \in \Omega : \omega \text{ extends } \lambda\}$. Show that $P[\mathcal{E}_\lambda] = 2^{-\ell}$.

EXERCISE 9.2. Let A be a probabilistic algorithm that on a given input x, halts with probability 1, and produces an output in the set T. Let P be the corresponding probability distribution, and let Y and Z be random variables representing the output and running time, respectively. For each $k \geq 0$, let P_k be the uniform distribution on all execution paths λ of length k. We define random variables Y_k and Z_k, associated with P_k, as follows: if λ is complete, we define $Y_k(\lambda)$ to be the output produced by A, and $Z_k(\lambda)$ to be the actual number of steps executed by A; otherwise, we define $Y_k(\lambda)$ to be the special value "$\perp$" and $Z_k(\lambda)$ to be k. For each $t \in T$, let p_{tk} be the probability (relative to P_k) that $Y_k = t$, and let μ_k be the expected value (relative to P_k) of Z_k. Show that:

(a) for each $t \in T$, $P[Y = t] = \lim\limits_{k \to \infty} p_{tk}$;

(b) $E[Z] = \lim_{k \to \infty} \mu_k$.

EXERCISE 9.3. Let A_1 and A_2 be probabilistic algorithms. Let B be any probabilistic algorithm that always outputs 0 or 1. For $i = 1, 2$, let A_i' be the algorithm
that on input x computes and outputs $B(A_i(x))$. Fix an input x, and let Y_1 and Y_2
be random variables representing the outputs of A_1 and A_2, respectively, on input
x, and let Y_1' and Y_2' be random variables representing the outputs of A_1' and A_2',
respectively, on input x. Assume that the images of Y_1 and Y_2 are finite, and let
$\delta := \Delta[Y_1; Y_2]$ be their statistical distance. Show that $|P[Y_1' = 1] - P[Y_2' = 1]| \le \delta$.

9.2 Generating a random number from a given interval

Suppose we want to generate a number, uniformly at random from the interval
$\{0, \ldots, m - 1\}$, for a given positive integer m.

If m is a power of 2, say $m = 2^\ell$, then we can do this directly as follows: generate
a random ℓ-bit string σ, and convert σ to the integer $I(\sigma)$ whose base-2 representation is σ; that is, if $\sigma = b_{\ell-1} b_{\ell-2} \cdots b_0$, where the b_i's are bits, then

$$I(\sigma) := \sum_{i=0}^{\ell-1} b_i 2^i.$$

In the general case, we do not have a direct way to do this, since we can only
directly generate random bits. But the following algorithm does the job:

Algorithm RN. On input m, where m is a positive integer, do the following, where
$\ell := \lceil \log_2 m \rceil$:

> repeat
> > $\sigma \xleftarrow{\$} \{0, 1\}^{\times \ell}$
> > $y \leftarrow I(\sigma)$
> until $y < m$
> output y

Theorem 9.2. *The expected running time of Algorithm RN is $O(\mathrm{len}(m))$, and its
output is uniformly distributed over $\{0, \ldots, m - 1\}$.*

Proof. Note that $m \le 2^\ell < 2m$. Let L denote the number of loop iterations of this
algorithm, and Z its running time. With every loop iteration, the algorithm halts
with probability $m/2^\ell$, and so the distribution of L is a geometric distribution with
associated success probability $m/2^\ell > 1/2$. Therefore, $E[L] = 2^\ell/m < 2$. Since
$Z \le c \, \mathrm{len}(m) \cdot L$ for some constant c, it follows that $E[Z] = O(\mathrm{len}(m))$.

Next, we analyze the output distribution. Let Y denote the output of the algorithm. We want to show that Y is uniformly distributed over $\{0, \ldots, m - 1\}$. This

is perhaps intuitively obvious, but let us give a rigorous justification of this claim. To do this, for $i = 1, 2, \ldots$, let Y_i denote the value of y in the ith loop iteration; for completeness, if the ith loop iteration is not executed, then we define $Y_i := \bot$. Also, for $i = 1, 2 \ldots$, let $\mathcal{H}_i$ be the event that the algorithm halts in the ith loop iteration (i.e., $\mathcal{H}_i$ is the event that $L = i$). Let $t \in \{0, \ldots, m - 1\}$ be fixed.

First, by total probability (specifically, the infinite version of (8.9), discussed in §8.10.2), we have

$$P[Y = t] = \sum_{i \geq 1} P[(Y = t) \cap \mathcal{H}_i] = \sum_{i \geq 1} P[(Y_i = t) \cap \mathcal{H}_i]. \tag{9.3}$$

Next, observe that as each loop iteration works the same as any other, it follows that for each $i \geq 1$, we have

$$P[(Y_i = t) \cap \mathcal{H}_i \mid L \geq i] = P[(Y_1 = t) \cap \mathcal{H}_1] = P[Y_1 = t] = 2^{-\ell}.$$

Moreover, since $\mathcal{H}_i$ implies $L \geq i$, we have

$$P[(Y_i = t) \cap \mathcal{H}_i] = P[(Y_i = t) \cap \mathcal{H}_i \cap (L \geq i)]$$
$$= P[(Y_i = t) \cap \mathcal{H}_i \mid L \geq i] \, P[L \geq i] = 2^{-\ell} \, P[L \geq i],$$

and so using (9.3) and the infinite version of Theorem 8.17 (discussed in §8.10.4), we have

$$P[Y = t] = \sum_{i \geq 1} P[(Y_i = t) \cap \mathcal{H}_i] = \sum_{i \geq 1} 2^{-\ell} \, P[L \geq i] = 2^{-\ell} \sum_{i \geq 1} P[L \geq i]$$
$$= 2^{-\ell} \cdot E[L] = 2^{-\ell} \cdot 2^\ell / m = 1/m.$$

This shows that Y is uniformly distributed over $\{0, \ldots, m - 1\}$. $\square$

Of course, by adding an appropriate value to the output of Algorithm RN, we can generate random numbers uniformly in the interval $\{m_1, \ldots, m_2\}$, for any given m_1 and m_2. In what follows, we shall denote the execution of this algorithm as

$$y \xleftarrow{\$} \{m_1, \ldots, m_2\}.$$

More generally, if T is any finite, non-empty set for which we have an efficient algorithm whose output is uniformly distributed over T, we shall denote the execution of this algorithm as

$$y \xleftarrow{\$} T.$$

For example, we may write

$$y \xleftarrow{\$} \mathbb{Z}_m$$

to denote assignment to y of a randomly chosen element of $\mathbb{Z}_m$. Of course, this is done by running Algorithm RN on input m, and viewing its output as a residue class modulo m.

We also mention the following alternative algorithm for generating an almost-random number from an interval.

Algorithm RN′. On input m, k, where both m and k are positive integers, do the following, where $\ell := \lceil \log_2 m \rceil$:

$$\sigma \xleftarrow{\mathscr{L}} \{0, 1\}^{\times(\ell+k)}$$
$$y \leftarrow I(\sigma) \bmod m$$
$$\text{output } y$$

Compared with Algorithm RN, Algorithm RN′ has the advantage that there are no loops—it always halts in a bounded number of steps; however, it has the disadvantage that its output is *not* uniformly distributed over the interval $\{0, \dots, m-1\}$. Nevertheless, the statistical distance between its output distribution and the uniform distribution on $\{0, \dots, m-1\}$ is at most 2^{-k} (see Example 8.41 in §8.8). Thus, by choosing k suitably large, we can make the output distribution "as good as uniform" for most practical purposes.

EXERCISE 9.4. Prove that if m is not a power of 2, there is no probabilistic algorithm whose running time is strictly bounded and whose output distribution is uniform on $\{0, \dots, m-1\}$.

EXERCISE 9.5. You are to design and analyze an efficient probabilistic algorithm B that takes as input two integers n and y, with $n > 0$ and $0 \le y \le n$, and always outputs 0 or 1. Your algorithm should satisfy the following property. Suppose A is a probabilistic algorithm that takes two inputs, n and x, and always outputs an integer between 0 and n. Let Y be a random variable representing A's output on input n, x. Then for all inputs n, x, we should have $\mathsf{P}[B(n, A(n, x)) \text{ outputs } 1] = \mathsf{E}[Y]/n$.

9.3 The generate and test paradigm

Algorithm RN, which was discussed in §9.2, is a specific instance of a very general type of construction that may be called the "generate and test" paradigm.

Suppose we have two probabilistic algorithms, A and B, and we combine them to form a new algorithm

$$C(x): \quad \text{repeat} \quad y \leftarrow A(x) \quad \text{until } B(x, y)$$
$$\text{output } y.$$

Here, we assume that $B(x, y)$ always outputs either *true* or *false*.

Our goal is to answer the following questions about C for a fixed input x:

1. Does C halt with probability 1?

2. What is the expected running time of C?

3. What is the output distribution of C?

The answer to the first question is "yes," provided (i) A halts with probability 1 on input x, (ii) for all possible outputs y of $A(x)$, B halts with probability 1 on input (x, y), and (iii) for some possible output y of $A(x)$, $B(x, y)$ outputs *true* with non-zero probability. We shall assume this from now on.

To address the second and third questions, let us define random variables L, Z, and Y, where L is the total number of loop iterations of C, Z is the total running time of C, and Y is the output of C. We can reduce the study of L, Z, and Y to the study of a single iteration of the main loop. Instead of working with a new probability distribution that directly models a single iteration of the loop, it is more convenient to simply study the *first* iteration of the loop in C. To this end, we define random variables Z_1 and Y_1, where Z_1 is the running time of the first loop iteration of C, and Y_1 is the value assigned to y in the first loop iteration of C. Also, let $\mathcal{H}_1$ be the event that the algorithm halts in the first loop iteration, and let T be the set of possible outputs of $A(x)$. Note that by the assumption in the previous paragraph, $P[\mathcal{H}_1] > 0$.

Theorem 9.3. *Under the assumptions above,*

(i) *L has a geometric distribution with associated success probability $P[\mathcal{H}_1]$, and in particular, $E[L] = 1/P[\mathcal{H}_1]$;*

(ii) *$E[Z] = E[Z_1]E[L] = E[Z_1]/P[\mathcal{H}_1]$;*

(iii) *for every $t \in T$, $P[Y = t] = P[Y_1 = t \mid \mathcal{H}_1]$.*

Proof. (i) is clear.

To prove (ii), for $i \geq 1$, let Z_i be the time spent by the algorithm in the ith loop iteration, so that $Z = \sum_{i \geq 1} Z_i$. Now, the conditional distribution of Z_i given $L \geq i$ is (essentially) the same as the distribution of Z_1; moreover, $Z_i = 0$ when $L < i$. Therefore, by the law of total expectation (8.24), for each $i \geq 1$, we have

$$E[Z_i] = E[Z_i \mid L \geq i]\,P[L \geq i] + E[Z_i \mid L < i]\,P[L < i] = E[Z_1]\,P[L \geq i].$$

We may assume that $E[Z_1]$ is finite, as otherwise (ii) is trivially true. By Theorem 8.40 and the infinite version of Theorem 8.17 (discussed in §8.10.4), we have

$$E[Z] = \sum_{i \geq 1} E[Z_i] = \sum_{i \geq 1} E[Z_1]\,P[L \geq i] = E[Z_1] \sum_{i \geq 1} P[L \geq i] = E[Z_1]\,E[L].$$

To prove (iii), for $i \geq 1$, let Y_i be the value assigned to y in loop iteration i, with $Y_i := \perp$ if $L < i$, and let $\mathcal{H}_i$ be the event that the algorithm halts in loop iteration i

(i.e., $\mathcal{H}_i$ is the event that $L = i$). By a calculation similar to that made in the proof of Theorem 9.2, for each $t \in T$, we have

$$
P[Y = t] = \sum_{i \geq 1} P[(Y = t) \cap \mathcal{H}_i] = \sum_{i \geq 1} P[(Y_i = t) \cap \mathcal{H}_i \mid L \geq i] P[L \geq i]
$$

$$
= P[(Y_1 = t) \cap \mathcal{H}_1] \sum_{i \geq 1} P[L \geq i] = P[(Y_1 = t) \cap \mathcal{H}_1] \cdot E[L]
$$

$$
= P[(Y_1 = t) \cap \mathcal{H}_1] / P[\mathcal{H}_1] = P[Y_1 = t \mid \mathcal{H}_1]. \quad \square
$$

Example 9.10. Suppose T is a finite set, and T' is a non-empty, finite subset of T. Consider the following generalization of Algorithm RN:

> repeat
>> $y \xleftarrow{\mathscr{L}} T$
>
> until $y \in T'$
> output y

Here, we assume that we have an algorithm to generate a random element of T (i.e., uniformly distributed over T), and an efficient algorithm to test for membership in T'. Let L denote the number of loop iterations, and Y the output. Also, let Y_1 be the value of y in the first iteration, and $\mathcal{H}_1$ the event that the algorithm halts in the first iteration. Since Y_1 is uniformly distributed over T, and $\mathcal{H}_1$ is the event that $Y_1 \in T'$, we have $P[\mathcal{H}_1] = |T'|/|T|$. It follows that $E[L] = |T|/|T'|$. As for the output, for every $t \in T$, we have

$$
P[Y = t] = P[Y_1 = t \mid \mathcal{H}_1] = P[Y_1 = t \mid Y_1 \in T'],
$$

which is 0 if $t \notin T'$ and is $1/|T'|$ if $t \in T'$. It follows that Y is uniformly distributed over T'. $\square$

Example 9.11. Let us analyze the following algorithm:

> repeat
>> $y \xleftarrow{\mathscr{L}} \{1, 2, 3, 4\}$
>> $z \xleftarrow{\mathscr{L}} \{1, \ldots, y\}$
>
> until $z = 1$
> output y

With each loop iteration, the algorithm chooses y uniformly at random, and then decides to halt with probability $1/y$. Let L denote the number of loop iterations, and Y the output. Also, let Y_1 be the value of y in the first iteration, and $\mathcal{H}_1$ the event that the algorithm halts in the first iteration. Y_1 is uniformly distributed over

$\{1, \dots, 4\}$, and for $t = 1, \dots, 4$, $P[\mathcal{H}_1 \mid Y_1 = t] = 1/t$. Therefore,

$$P[\mathcal{H}_1] = \sum_{t=1}^{4} P[\mathcal{H}_1 \mid Y_1 = t] P[Y_1 = t] = \sum_{t=1}^{4} (1/t)(1/4) = 25/48.$$

Thus, $E[L] = 48/25$. For the output distribution, for $t = 1, \dots, 4$, we have

$$P[Y = t] = P[Y_1 = t \mid \mathcal{H}_1] = P[(Y_1 = t) \cap \mathcal{H}_1] / P[\mathcal{H}_1]$$

$$= P[\mathcal{H}_1 \mid Y_1 = t] P[Y_1 = t] / P[\mathcal{H}_1] = (1/t)(1/4)(48/25) = \frac{12}{25t}.$$

This example illustrates how a probabilistic test can be used to create a biased output distribution. $\square$

EXERCISE 9.6. Design and analyze an efficient probabilistic algorithm that takes as input an integer $n \geq 2$, and outputs a random element of $\mathbb{Z}_n^*$.

EXERCISE 9.7. Consider the following probabilistic algorithm that takes as input a positive integer m:

$S \leftarrow \emptyset$
repeat
    $n \xleftarrow{\text{\textcent}} \{1, \dots, m\}, \ S \leftarrow S \cup \{n\}$
until $|S| = m$

Show that the expected number of iterations of the main loop is $\sim m \log m$.

EXERCISE 9.8. Consider the following algorithm (which takes no input):

$j \leftarrow 1$
repeat
    $j \leftarrow j + 1, \ n \xleftarrow{\text{\textcent}} \{0, \dots, j-1\}$
until $n = 0$

Show that the expected running time of this algorithm is infinite (even though it does halt with probability 1).

EXERCISE 9.9. Now consider the following modification to the algorithm in the previous exercise:

$j \leftarrow 2$
repeat
    $j \leftarrow j + 1, \ n \xleftarrow{\text{\textcent}} \{0, \dots, j-1\}$
until $n = 0$ or $n = 1$

Show that the expected running time of this algorithm is finite.

EXERCISE 9.10. Consider again Algorithm RN in §9.2. On input m, this algorithm may use up to $\approx 2\ell$ random bits on average, where $\ell := \lceil \log_2 m \rceil$. Indeed, each loop iteration generates ℓ random bits, and the expected number of loop iterations will be ≈ 2 when $m \approx 2^{\ell-1}$. This exercise asks you to analyze an alternative algorithm that uses just $\ell + O(1)$ random bits on average, which may be useful in settings where random bits are a scarce resource. This algorithm runs as follows:

> repeat
> $\qquad y \leftarrow 0, \ i \leftarrow 1$
> $\qquad$ while $y < m$ and $i \le \ell$ do
> (∗) $\qquad\qquad b \xleftarrow{\text{\tiny{ℓ}}} \{0, 1\}, \ y \leftarrow y + 2^{\ell-i} b, \ i \leftarrow i + 1$
> $\qquad$ until $y < m$
> $\qquad$ output y

Define random variables K and Y, where K is the number of times the line marked (∗) is executed, and Y is the output. Show that $E[K] = \ell + O(1)$ and that Y is uniformly distributed over $\{0, \ldots, m-1\}$.

EXERCISE 9.11. Let S and T be finite, non-empty sets, and let $f : S \times T \to \{-1, 0, 1\}$ be a function. Consider the following probabilistic algorithm:

> $x \xleftarrow{\text{\tiny{ℓ}}} S, y \xleftarrow{\text{\tiny{ℓ}}} T$
> if $f(x, y) = 0$ then
> $\qquad y' \leftarrow y$
> else
> $\qquad y' \xleftarrow{\text{\tiny{ℓ}}} T$
> (∗) $\qquad$ while $f(x, y') = 0$ do $y' \xleftarrow{\text{\tiny{ℓ}}} T$

Here, we assume we have algorithms to generate random elements in S and T, and a deterministic algorithm to evaluate f. Define random variables X, Y, Y', and L, where X is the value assigned to x, Y is the value assigned to y, Y' is the *final* value assigned to y', and L is the number of times that f is evaluated at the line marked (∗).

(a) Show that (X, Y') has the same distribution as (X, Y).

(b) Show that $E[L] \le 1$.

(c) Give an explicit example of S, T, and f, such that if the line marked (∗) is deleted, then $E[f(X, Y)] > E[f(X, Y')] = 0$.

9.4 Generating a random prime

Suppose we are given an integer $m \geq 2$, and want to generate a random prime between 2 and m. One way to proceed is simply to generate random numbers until we get a prime. This idea will work, assuming the existence of an efficient, deterministic algorithm *IsPrime* that determines whether or not a given integer is prime. We will present such an algorithm later, in Chapter 21. For the moment, we shall just assume we have such an algorithm, and use it as a "black box." Let us assume that on inputs of bit length at most ℓ, *IsPrime* runs in time at most $\tau(\ell)$. Let us also assume (quite reasonably) that $\tau(\ell) = \Omega(\ell)$.

Algorithm RP. On input m, where m is an integer ≥ 2, do the following:

> repeat
>> $n \xleftarrow{\text{\$}} \{2, \ldots, m\}$
> until *IsPrime*(n)
> output n

We now wish to analyze the running time and output distribution of Algorithm RP on an input m, where $\ell := \text{len}(m)$. This is easily done, using the results of §9.3, and more specifically, by Example 9.10. The expected number of loop iterations performed by Algorithm RP is $(m - 1)/\pi(m)$, where $\pi(m)$ is the number of primes up to m. By Chebyshev's theorem (Theorem 5.1), $\pi(m) = \Theta(m/\ell)$. It follows that the expected number of loop iterations is $\Theta(\ell)$. Furthermore, the expected running time of any one loop iteration is $O(\tau(\ell))$ (the expected running time for generating n is $O(\ell)$, and this is where we use the assumption that $\tau(\ell) = \Omega(\ell)$). It follows that the expected total running time is $O(\ell\tau(\ell))$. As for the output, it is clear that it is uniformly distributed over the set of primes up to m.

9.4.1 Using a probabilistic primality test

In the above analysis, we assumed that *IsPrime* was an efficient, deterministic algorithm. While such an algorithm exists, there are in fact simpler and far more efficient primality tests that are probabilistic. We shall discuss such an algorithm in detail in the next chapter. This algorithm (like several other probabilistic primality tests) has *one-sided error*, in the following sense: if the input n is prime, then the algorithm always outputs *true*; otherwise, if n is composite, the output may be *true* or *false*, but the probability that the output is *true* is at most ε, where ε is a very small number (the algorithm may be easily tuned to make ε quite small, e.g., 2^{-100}).

Let us analyze the behavior of Algorithm RP under the assumption that *IsPrime* is implemented by a probabilistic algorithm with an error probability for composite inputs bounded by ε, as discussed in the previous paragraph. Let $\bar{\tau}(\ell)$ be a bound on the expected running time of this algorithm for all inputs of bit length at most ℓ. Again, we assume that $\bar{\tau}(\ell) = \Omega(\ell)$.

We use the technique developed in §9.3. Consider a fixed input m, and let $\ell := \text{len}(m)$. Let L, Z, and N be random variables representing, respectively, the number of loop iterations, the total running time, and output of Algorithm RP on input m. Also, let Z_1 be the random variable representing the running time of the first loop iteration, and let N_1 be the random variable representing the value assigned to n in the first loop iteration. Let $\mathcal{H}_1$ be the event that the algorithm halts in the first loop iteration, and let C_1 be the event that N_1 is composite.

Clearly, N_1 is uniformly distributed over $\{2, \dots, m\}$. Also, by our assumptions about *IsPrime*, we have

$$E[Z_1] = O(\bar{\tau}(\ell)),$$

and moreover, for each $j \in \{2, \dots, m\}$, we have

$$P[\mathcal{H}_1 \mid N_1 = j] \le \varepsilon \text{ if } j \text{ is composite},$$

and

$$P[\mathcal{H}_1 \mid N_1 = j] = 1 \text{ if } j \text{ is prime}.$$

In particular,

$$P[\mathcal{H}_1 \mid C_1] \le \varepsilon \text{ and } P[\mathcal{H}_1 \mid \bar{C}_1] = 1.$$

It follows that

$$P[\mathcal{H}_1] = P[\mathcal{H}_1 \mid C_1]P[C_1] + P[\mathcal{H}_1 \mid \bar{C}_1]P[\bar{C}_1] \ge P[\mathcal{H}_1 \mid \bar{C}_1]P[\bar{C}_1]$$
$$= \pi(m)/(m-1).$$

Therefore,

$$E[L] \le (m-1)/\pi(m) = O(\ell)$$

and

$$E[Z] = E[L]E[Z_1] = O(\ell\bar{\tau}(\ell)).$$

That takes care of the running time. Now consider the output. For every $j \in \{2, \dots, m\}$, we have

$$P[N = j] = P[N_1 = j \mid \mathcal{H}_1].$$

If j is prime, then

$$P[N = j] = P[N_1 = j \mid \mathcal{H}_1] = \frac{P[(N_1 = j) \cap \mathcal{H}_1]}{P[\mathcal{H}_1]}$$

$$= \frac{P[\mathcal{H}_1 \mid N_1 = j] P[N_1 = j]}{P[\mathcal{H}_1]} = \frac{1}{(m-1)P[\mathcal{H}_1]}.$$

Thus, every prime is output with equal probability; however, the algorithm may also output a number that is not prime. Let us bound the probability of this event. One might be tempted to say that this happens with probability at most ε; however, in drawing such a conclusion, one would be committing the fallacy of Example 8.13—to correctly analyze the probability that Algorithm RP mistakenly outputs a composite, one must take into account the rate of incidence of the "primality disease," as well as the error rate of the test for this disease. Indeed, if C is the event that N is composite, then we have

$$P[C] = P[C_1 \mid \mathcal{H}_1] = \frac{P[C_1 \cap \mathcal{H}_1]}{P[\mathcal{H}_1]} = \frac{P[\mathcal{H}_1 \mid C_1] P[C_1]}{P[\mathcal{H}_1]}$$

$$\leq \frac{\varepsilon}{P[\mathcal{H}_1]} \leq \frac{\varepsilon}{\pi(m)/(m-1)} = O(\ell\varepsilon).$$

Another way of analyzing the output distribution of Algorithm RP is to consider its statistical distance Δ from the uniform distribution on the set of primes between 2 and m. As we have already argued, every prime between 2 and m is equally likely to be output, and in particular, any fixed prime is output with probability at most $1/\pi(m)$. It follows from Theorem 8.31 that $\Delta = P[C] = O(\ell\varepsilon)$.

9.4.2 Generating a random ℓ-bit prime

Instead of generating a random prime between 2 and m, we may instead want to generate a random ℓ-bit prime, that is, a prime between $2^{\ell-1}$ and $2^{\ell} - 1$. Bertrand's postulate (Theorem 5.8) tells us that there exist such primes for every $\ell \geq 2$, and that in fact, there are $\Omega(2^{\ell}/\ell)$ such primes. Because of this, we can modify Algorithm RP, so that each candidate n is chosen at random from the interval $\{2^{\ell-1}, \ldots, 2^{\ell} - 1\}$, and all of the results for that algorithm carry over essentially without change. In particular, the expected number of trials until the algorithm halts is $O(\ell)$, and if a probabilistic primality test as in §9.4.1 is used, with an error probability of ε, the probability that the output is not prime is $O(\ell\varepsilon)$.

EXERCISE 9.12. Suppose Algorithm RP is implemented using an imperfect random number generator, so that the statistical distance between the output distribution of the random number generator and the uniform distribution on $\{2, \ldots, m\}$ is equal to δ (e.g., Algorithm RN′ in §9.2). Assume that $2\delta < \pi(m)/(m-1)$. Also, let μ denote the expected number of iterations of the main loop of Algorithm RP, let Δ denote the statistical distance between its output distribution and the uniform distribution on the primes up to m, and let $\ell := \mathrm{len}(m)$.

(a) Assuming the primality test is deterministic, show that $\mu = O(\ell)$ and $\Delta = O(\delta\ell)$.

(b) Assuming the primality test is probabilistic, with one-sided error ε, as in §9.4.1, show that $\mu = O(\ell)$ and $\Delta = O((\delta + \varepsilon)\ell)$.

9.5 Generating a random non-increasing sequence

The following algorithm will be used in the next section as a fundamental subroutine in a beautiful algorithm (Algorithm RFN) that generates random numbers in *factored form*.

Algorithm RS. On input m, where m is an integer ≥ 2, do the following:

$n_0 \leftarrow m$
$k \leftarrow 0$
repeat
$\quad k \leftarrow k + 1$
$\quad n_k \xleftarrow{\mathscr{R}} \{1, \ldots, n_{k-1}\}$
until $n_k = 1$
output $(n_1, \ldots, n_k)$

We analyze first the output distribution, and then the running time.

9.5.1 Analysis of the output distribution

Let $N_1, N_2, \ldots$ be random variables denoting the choices of $n_1, n_2, \ldots$ (for completeness, define $N_i := 1$ if loop i is never entered).

A particular output of the algorithm is a non-increasing sequence $(j_1, \ldots, j_h)$, where $j_1 \geq j_2 \geq \cdots \geq j_{h-1} > j_h = 1$. For any such sequence, we have

$$P\left[\bigcap_{v=1}^{h}(N_v = j_v)\right] = P[N_1 = j_1] \cdot \prod_{v=2}^{h} P\left[N_v = j_v \mid \bigcap_{w<v}(N_w = j_w)\right]$$

$$= \frac{1}{m} \cdot \frac{1}{j_1} \cdots \frac{1}{j_{h-1}}. \tag{9.4}$$

This completely describes the output distribution, in the sense that we have determined the probability with which each non-increasing sequence appears as an output. However, there is another way to characterize the output distribution that is significantly more useful. For $j = 2, \ldots, m$, define the random variable O_j to be the number of occurrences of the integer j in the output sequence. The O_j's determine the N_i's, and *vice versa*. Indeed, $O_m = e_m, \ldots, O_2 = e_2$ if and only if the output of the algorithm is the sequence

$$(\underbrace{m, \ldots, m}_{e_m \text{ times}}, \underbrace{m - 1, \ldots, m - 1}_{e_{m-1} \text{ times}}, \ldots, \underbrace{2, \ldots, 2}_{e_2 \text{ times}}, 1).$$

From (9.4), we can therefore directly compute

$$P\left[\bigcap_{j=2}^{m}(O_j = e_j)\right] = \frac{1}{m} \prod_{j=2}^{m} \frac{1}{j^{e_j}}. \tag{9.5}$$

Moreover, we can write $1/m$ as a telescoping product,

$$\frac{1}{m} = \frac{m-1}{m} \cdot \frac{m-2}{m-1} \cdots \cdot \frac{2}{3} \cdot \frac{1}{2} = \prod_{j=2}^{m}(1 - 1/j),$$

and so re-write (9.5) as

$$P\left[\bigcap_{j=2}^{m}(O_j = e_j)\right] = \prod_{j=2}^{m} j^{-e_j}(1 - 1/j). \tag{9.6}$$

Notice that for $j = 2, \ldots, m$,

$$\sum_{e_j \geq 0} j^{-e_j}(1 - 1/j) = 1,$$

and so by (a discrete version of) Theorem 8.7, the family of random variables $\{O_j\}_{j=2}^{m}$ is mutually independent, and for each $j = 2, \ldots, m$ and each integer $e_j \geq 0$, we have

$$P[O_j = e_j] = j^{-e_j}(1 - 1/j). \tag{9.7}$$

In summary, we have shown:

that the family $\{O_j\}_{j=2}^{m}$ is mutually independent, where for each $j = 2, \ldots, m$, the variable $O_j + 1$ has a geometric distribution with an associated success probability of $1 - 1/j$.

Another, perhaps more intuitive, analysis of the distribution of the O_j's runs as follows. Conditioning on the event $O_m = e_m, \ldots, O_{j+1} = e_{j+1}$, one sees that the value of O_j is the number of times the value j appears in the sequence $N_i, N_{i+1}, \ldots$, where $i = e_m + \cdots + e_{j+1} + 1$; moreover, in this conditional probability distribution, it is not too hard to convince oneself that N_i is uniformly distributed over $\{1, \ldots, j\}$. Hence the probability that $O_j = e_j$ in this conditional probability distribution is the probability of getting a run of exactly e_j copies of the value j in an experiment in which we successively choose numbers between 1 and j at random, and this latter probability is clearly $j^{-e_j}(1 - 1/j)$.

9.5.2 Analysis of the running time

Let $\ell := \text{len}(m)$, and let K be the random variable that represents the number of loop iterations performed by the algorithm. With the random variables $O_2, \ldots, O_m$ defined as above, we can write $K = 1 + \sum_{j=2}^{m} O_j$. Moreover, for each j, $O_j + 1$ has a geometric distribution with associated success probability $1 - 1/j$, and hence

$$E[O_j] = \frac{1}{1 - 1/j} - 1 = \frac{1}{j - 1}.$$

Thus,

$$E[K] = 1 + \sum_{j=2}^{m} E[O_j] = 1 + \sum_{j=1}^{m-1} \frac{1}{j} \leq 2 + \int_1^m \frac{dy}{y} = \log m + 2,$$

where we have estimated the sum by an integral (see §A5).

Intuitively, this is roughly as we would expect, since with probability $1/2$, each successive n_i is at most one half as large as its predecessor, and so after $O(\ell)$ steps, we expect to reach 1.

Let Z be the total running time of the algorithm. We may bound $E[Z]$ using essentially the same argument that was used in the proof of Theorem 9.3. First, write $Z = \sum_{i \geq 1} Z_i$, where Z_i is the time spent in the ith loop iteration. Each loop iteration, if executed at all, runs in expected time $O(\ell)$. That is, there exists a constant c, such that for each $i \geq 1$,

$$E[Z_i \mid K \geq i] \leq c\ell \quad \text{and} \quad E[Z_i \mid K < i] = 0.$$

Thus,

$$E[Z_i] = E[Z_i \mid K \geq i] P[K \geq i] + E[Z_i \mid K < i] P[K < i] \leq c\ell\, P[K \geq i],$$

and so

$$E[Z] = \sum_{i \geq 1} E[Z_i] \leq c\ell \sum_{i \geq 1} P[K \geq i] = c\ell\, E[K] = O(\ell^2).$$

In summary, we have shown:

the expected running time of Algorithm RS on ℓ-bit inputs is $O(\ell^2)$.

EXERCISE 9.13. Show that when Algorithm RS runs on input m, the expected number of (not necessarily distinct) primes in the output sequence is $\sim \log \log m$.

9.6 Generating a random factored number

We now present an efficient algorithm that generates a random factored number. That is, on input $m \geq 2$, the algorithm generates a number y uniformly distributed over the interval $\{1, \ldots, m\}$, but instead of the usual output format for such a number y, the output consists of the prime factorization of y.

As far as anyone knows, there are no efficient algorithms for factoring large numbers, despite years of active research in search of such an algorithm. So our algorithm to generate a random factored number will *not* work by generating a random number and then factoring it.

Our algorithm will use Algorithm RS in §9.5 as a subroutine. In addition, as we did in §9.4, we shall assume the existence of an efficient, deterministic primality test *IsPrime*. In the analysis of the algorithm, we shall make use of Mertens' theorem, which we proved in Chapter 5 (Theorem 5.13).

Algorithm RFN. On input m, where m is an integer ≥ 2, do the following:

> repeat
>> run Algorithm RS on input m, obtaining $(n_1, \ldots, n_k)$
>
> (∗)   let $(p_1, \ldots, p_r)$ be the subsequence of primes in $(n_1, \ldots, n_k)$
>
> (∗∗) $y \leftarrow p_1 \cdots p_r$
>
>> if $y \leq m$ then
>>> $x \xleftarrow{\ell} \{1, \ldots, m\}$
>>>
>>> if $x \leq y$ then output $(p_1, \ldots, p_r)$ and halt
>
> forever

Notes:

(∗) For $i = 1, \ldots, k - 1$, the number n_i is tested for primality using algorithm *IsPrime*. The sequence $(n_1, \ldots, n_k)$ may contain duplicates, and if these are prime, they are appear in $(p_1, \ldots, p_r)$ with the same multiplicity.

(∗∗) We assume that the product is computed by a simple iterative procedure that halts as soon as the partial product exceeds m. This ensures that the time spent forming the product is always $O(\text{len}(m)^2)$, which simplifies the analysis.

We now analyze the running time and output distribution of Algorithm RFN on input m, using the generate-and-test paradigm discussed in §9.3; here, the "generate" part consists of the first two lines of the loop body, which generates the sequence $(p_1, \ldots, p_r)$, while the "test" part consists of the last four lines of the loop body.

Let $\ell := \text{len}(m)$. We assume that each call to *IsPrime* takes time at most $\tau(\ell)$, and for simplicity, we assume $\tau(\ell) = \Omega(\ell)$.

Let K_1 be the value of k in the first loop iteration, Z_1 be the running time of the first loop iteration, Y_1 be the value of y in the first loop iteration, and $\mathcal{H}_1$ be the event that the algorithm halts in the first loop iteration. Also, let Z be the total running time of the algorithm, and let Y be the value of y in the last loop iteration (i.e., the number whose factorization is output).

We begin with three preliminary calculations.

First, let $t = 1, \ldots, m$ be a fixed integer, and let us calculate the probability that $Y_1 = t$. Suppose $t = \prod_{p \leq m} p^{e_p}$ is the prime factorization of t. Let $O_2, \ldots, O_m$ be random variables as defined in §9.5, so that O_j represents the number of occurrences of j in the output sequence of the first invocation of Algorithm RS. Then $Y_1 = t$ if and only if $O_p = e_p$ for all primes $p \leq m$, and so by the analysis in §9.5, we have

$$P[Y_1 = t] = \prod_{p \leq m} p^{-e_p}(1 - 1/p) = \frac{g(m)}{t},$$

where

$$g(m) := \prod_{p \leq m}(1 - 1/p).$$

Second, we calculate $P[\mathcal{H}_1]$. Observe that for $t = 1, \ldots, m$, we have

$$P[\mathcal{H}_1 \mid Y_1 = t] = t/m,$$

and so

$$P[\mathcal{H}_1] = \sum_{t=1}^{m} P[\mathcal{H}_1 \mid Y_1 = t] P[Y_1 = t] = \sum_{t=1}^{m} \frac{t}{m}\frac{g(m)}{t} = g(m).$$

Third, let $t = 1, \ldots, m$ be a fixed integer, and let us calculate the conditional probability that $Y_1 = t$ given $\mathcal{H}_1$. We have

$$P[Y_1 = t \mid \mathcal{H}_1] = \frac{P[(Y_1 = t) \cap \mathcal{H}_1]}{P[\mathcal{H}_1]} = \frac{P[\mathcal{H}_1 \mid Y_1 = t] P[Y_1 = t]}{P[\mathcal{H}_1]}$$
$$= \frac{(t/m)(g(m)/t)}{g(m)} = \frac{1}{m}.$$

We may now easily analyze the output distribution of Algorithm RFN. By Theorem 9.3, for each $t = 1, \ldots, m$, we have

$$P[Y = t] = P[Y_1 = t \mid \mathcal{H}_1] = \frac{1}{m},$$

which shows that the output is indeed uniformly distributed over all integers in $\{1, \ldots, m\}$, represented in factored form.

Finally, we analyze the expected running time of Algorithm RFN. It is easy to see that $E[Z_1] = O(E[K_1]\tau(\ell) + \ell^2)$, and by the analysis in §9.5, we know that $E[K_1] = O(\ell)$, and hence $E[Z_1] = O(\ell\tau(\ell))$. By Theorem 9.3, we have

$$E[Z] = E[Z_1]/P[\mathcal{H}_1] = E[Z_1]g(m)^{-1}.$$

By Mertens' theorem, $g(m)^{-1} = O(\ell)$. We conclude that

$$E[Z] = O(\ell^2\tau(\ell)).$$

That is, the expected running time of Algorithm RFN is $O(\ell^2\tau(\ell))$.

9.6.1 Using a probabilistic primality test (∗)

Analogous to the discussion in §9.4.1, we can analyze the behavior of Algorithm RFN under the assumption that *IsPrime* is a probabilistic algorithm which may erroneously indicate that a composite number is prime with probability at most ε. Let $\ell := \text{len}(m)$, and as we did in §9.4.1, let $\bar{\tau}(\ell)$ be a bound on the expected running time of *IsPrime* for all inputs of bit length at most ℓ (and again, assume $\bar{\tau}(\ell) = \Omega(\ell)$).

The random variables K_1, Z_1, Y_1, Z, Y and the event $\mathcal{H}_1$ are defined as above. Let us also define $\mathcal{F}_1$ to be the event that the primality test makes a mistake in the first loop iteration, and $\mathcal{F}$ to be the event that the output of the algorithm is not a list of primes. Let $\delta := P[\mathcal{F}_1]$.

Again, we begin with three preliminary calculations.

First, let $t = 1, \ldots, m$ be fixed and let us calculate $P[(Y_1 = t) \cap \bar{\mathcal{F}}_1]$. To do this, define the random variable Y_1' to be the product of the actual primes among the output of the first invocation of Algorithm RS (because the primality test may err, Y_1 may contain additional factors). Evidently, the events $(Y_1 = t) \cap \bar{\mathcal{F}}_1$ and $(Y_1' = t) \cap \bar{\mathcal{F}}_1$ are the same. Moreover, we claim that the events $Y_1' = t$ and $\bar{\mathcal{F}}_1$ are independent. To see this, recall that the family $\{O_j\}_{j=2}^{m}$ is mutually independent, and also observe that the event $Y_1' = t$ depends only on the random variables O_j, where j is prime, while the event $\bar{\mathcal{F}}_1$ depends only on the random variables O_j, where j is composite, along with the execution paths of *IsPrime* on corresponding inputs. Thus, by a calculation analogous to one we made above,

$$P[(Y_1 = t) \cap \bar{\mathcal{F}}_1] = P[Y_1' = t]\,P[\bar{\mathcal{F}}_1] = \frac{g(m)}{t}(1 - \delta).$$

Second, we calculate $P[\mathcal{H}_1 \cap \bar{\mathcal{F}}_1]$. Observe that for $t = 1, \ldots, m$, we have

$$P[\mathcal{H}_1 \mid (Y_1 = t) \cap \bar{\mathcal{F}}_1] = t/m,$$

and so

$$P[\mathcal{H}_1 \cap \overline{F}_1] = \sum_{t=1}^{m} P[\mathcal{H}_1 \cap (Y_1 = t) \cap \overline{F}_1]$$

$$= \sum_{t=1}^{m} P[\mathcal{H}_1 \mid (Y_1 = t) \cap \overline{F}_1] P[(Y_1 = t) \cap \overline{F}_1]$$

$$= \sum_{t=1}^{m} \frac{t}{m} \frac{g(m)}{t} (1 - \delta) = g(m)(1 - \delta).$$

Third, let $t = 1, \ldots, m$ be a fixed integer, and let us calculate the conditional probability that $(Y_1 = t) \cap \overline{F}_1$ given $\mathcal{H}_1$. We have

$$P[(Y_1 = t) \cap \overline{F}_1 \mid \mathcal{H}_1] = \frac{P[(Y_1 = t) \cap \overline{F}_1 \cap \mathcal{H}_1]}{P[\mathcal{H}_1]}$$

$$= \frac{P[\mathcal{H}_1 \mid (Y_1 = t) \cap \overline{F}_1] P[(Y_1 = t) \cap \overline{F}_1]}{P[\mathcal{H}_1]}$$

$$= \frac{(t/m)((1-\delta)g(m)/t)}{P[\mathcal{H}_1]} = \frac{g(m)(1-\delta)}{m \, P[\mathcal{H}_1]}.$$

We may now easily analyze the output distribution of Algorithm RFN. By Theorem 9.3, for each $t = 1, \ldots, m$, we have

$$P[(Y = t) \cap \overline{F}] = P[(Y_1 = t) \cap \overline{F}_1 \mid \mathcal{H}_1] = \frac{g(m)(1-\delta)}{m \, P[\mathcal{H}_1]}.$$

Thus, every integer between 1 and m is equally likely to be output by Algorithm RFN in correct factored form.

Let us also calculate an upper bound on the probability $P[F]$ that Algorithm RFN outputs an integer that is not in correct factored form. Making use of Exercise 8.1, we have

$$P[F_1 \mid \mathcal{H}_1] = \frac{P[F_1 \cap \mathcal{H}_1]}{P[\mathcal{H}_1]} \leq \frac{P[F_1]}{P[F_1 \cup \mathcal{H}_1]}.$$

Moreover,

$$P[F_1 \cup \mathcal{H}_1] = P[F_1] + P[\mathcal{H}_1 \cap \overline{F}_1] = \delta + g(m)(1 - \delta)$$
$$\geq g(m)\delta + g(m)(1 - \delta) = g(m).$$

By Theorem 9.3, it follows that

$$P[F] = P[F_1 \mid \mathcal{H}_1] \leq \delta/g(m).$$

Now, the reader may verify that

$$\delta \leq \epsilon \cdot (F[K_1] - 1),$$

and by our calculations in §9.5, $E[K_1] \leq \log m + 2$. Thus,

$$\delta \leq \varepsilon \cdot (\log m + 1),$$

and so by Mertens' theorem,

$$P[\mathcal{F}] = O(\ell^2 \varepsilon).$$

We may also analyze the statistical distance Δ between the output distribution of Algorithm RFN and the uniform distribution on $\{1, \ldots, m\}$ (in factored form). It follows from Theorem 8.31 that $\Delta = P[\mathcal{F}] \leq \delta/g(m) = O(\ell^2 \varepsilon)$.

Finally, we analyze the expected running time of Algorithm RFN. We have

$$P[\mathcal{H}_1] \geq P[\mathcal{H}_1 \cap \overline{\mathcal{F}}_1] = g(m)(1 - \delta).$$

We leave it to the reader to verify that $E[Z_1] = O(\ell \bar{\tau}(\ell))$, from which it follows by Theorem 9.3 that

$$E[Z] = E[Z_1]/P[\mathcal{H}_1] = O(\ell^2 \bar{\tau}(\ell)/(1 - \delta)).$$

If ε is moderately small, so that $\varepsilon(\log m + 1) \leq 1/2$, and hence $\delta \leq 1/2$, then

$$E[Z] = O(\ell^2 \bar{\tau}(\ell)).$$

9.7 Some complexity theory

We close this chapter with a few observations about probabilistic algorithms from a more "complexity theoretic" point of view.

Suppose f is a function mapping bit strings to bit strings. We may have an algorithm A that **approximately computes** f in the following sense: there exists a constant ε, with $0 \leq \varepsilon < 1/2$, such that for all inputs x, $A(x)$ outputs $f(x)$ with probability at least $1 - \varepsilon$. The value ε is a bound on the **error probability**, which is defined as the probability that $A(x)$ does not output $f(x)$.

9.7.1 Reducing the error probability

There is a standard "trick" by which one can make the error probability very small; namely, run A on input x some number, say k, times, and take the majority output as the answer. Suppose $\varepsilon < 1/2$ is a bound on the error probability. Using the Chernoff bound (Theorem 8.24), the error probability for the iterated version of A is bounded by

$$\exp[-(1/2 - \varepsilon)^2 k/2], \tag{9.8}$$

and so the error probability decreases exponentially with the number of iterations. This bound is derived as follows. For $i = 1, \ldots, k$, let X_i be the indicator variable for the event that the ith iteration of $A(x)$ does not output $f(x)$. The expected value of the sample mean $\bar{X} := \frac{1}{k} \sum_{i=1}^{k} X_i$ is at most ε, and if the majority output of the iterated algorithm is wrong (or indeed, if there is no majority), then $\bar{X}$ exceeds its expectation by at least $1/2 - \varepsilon$. The bound (9.8) follows immediately from part (i) of Theorem 8.24.

9.7.2 Strict polynomial time

If we have an algorithm A that runs in expected polynomial time, and which approximately computes a function f, then we can easily turn it into a new algorithm A' that runs in *strict* polynomial time, and also approximates f, as follows. Suppose that $\varepsilon < 1/2$ is a bound on the error probability, and $Q(n)$ is a polynomial bound on the expected running time for inputs of size n. Then A' simply runs A for at most $kQ(n)$ steps, where k is any constant chosen so that $\varepsilon + 1/k < 1/2$—if A does not halt within this time bound, then A' simply halts with an arbitrary output. The probability that A' errs is at most the probability that A errs plus the probability that A runs for more than $kQ(n)$ steps. By Markov's inequality (Theorem 8.22), the latter probability is at most $1/k$, and hence A' approximates f as well, but with an error probability bounded by $\varepsilon + 1/k$.

9.7.3 Language recognition

An important special case of approximately computing a function is when the output of the function f is either 0 or 1 (or equivalently, *false* or *true*). In this case, f may be viewed as the characteristic function of the language $L := \{x : f(x) = 1\}$. (It is the tradition of computational complexity theory to call sets of bit strings "languages.") There are several "flavors" of probabilistic algorithms for approximately computing the characteristic function f of a language L that are traditionally considered—for the purposes of these definitions, we may restrict ourselves to algorithms that output either 0 or 1:

- We call a probabilistic, expected polynomial-time algorithm an **Atlantic City algorithm** for recognizing L if it approximately computes f with error probability bounded by a constant $\varepsilon < 1/2$.

- We call a probabilistic, expected polynomial-time algorithm A a **Monte Carlo algorithm** for recognizing L if for some constant $\delta > 0$, we have:
  - $P[A(x) \text{ outputs } 1] \geq \delta$ for all $x \in L$;
  - $P[A(x) \text{ outputs } 1] = 0$ for all $x \notin L$.

- We call a probabilistic, expected polynomial-time algorithm a **Las Vegas algorithm** for recognizing L if it computes f correctly on all inputs x.

One also says an Atlantic City algorithm has **two-sided error**, a Monte Carlo algorithm has **one-sided error**, and a Las Vegas algorithm has **zero-sided error**.

EXERCISE 9.14. Show that every language recognized by a Las Vegas algorithm is also recognized by a Monte Carlo algorithm, and that every language recognized by a Monte Carlo algorithm is also recognized by an Atlantic City algorithm.

EXERCISE 9.15. Show that if L is recognized by an Atlantic City algorithm that runs in expected polynomial time, then it is recognized by an Atlantic City algorithm that runs in strict polynomial time, and whose error probability is at most 2^{-n} on inputs of size n.

EXERCISE 9.16. Show that if L is recognized by a Monte Carlo algorithm that runs in expected polynomial time, then it is recognized by a Monte Carlo algorithm that runs in strict polynomial time, and whose error probability is at most 2^{-n} on inputs of size n.

EXERCISE 9.17. Show that a language is recognized by a Las Vegas algorithm if and only if the language and its complement are recognized by Monte Carlo algorithms.

EXERCISE 9.18. Show that if L is recognized by a Las Vegas algorithm that runs in strict polynomial time, then L may be recognized in deterministic polynomial time.

EXERCISE 9.19. Suppose that for a given language L, there exists a probabilistic algorithm A that runs in expected polynomial time, and always outputs either 0 or 1. Further suppose that for some constants α and c, where

- α is a rational number with $0 \leq \alpha < 1$, and
- c is a positive integer,

and for all sufficiently large n, and all inputs x of size n, we have

- if $x \notin L$, then $P[A(x)$ outputs $1] \leq \alpha$, and
- if $x \in L$, then $P[A(x)$ outputs $1] \geq \alpha + 1/n^c$.

(a) Show that there exists an Atlantic City algorithm for L.

(b) Show that if $\alpha = 0$, then there exists a Monte Carlo algorithm for L.

9.8 Notes

Our approach in §9.1 to defining the probability distribution associated with the execution of a probabilistic algorithm is not the only possible one. For example, one could define the output distribution and expected running time of an algorithm on a given input directly, using the identities in Exercise 9.2, and avoid the construction of an underlying probability distribution altogether; however, we would then have very few tools at our disposal to analyze the behavior of an algorithm. Yet another approach is to define a distribution that models an infinite random bit string. This can be done, but requires more advanced notions from probability theory than those that have been covered in this text.

The algorithm presented in §9.6 for generating a random factored number is due to Kalai [52], although the analysis presented here is a bit different, and our analysis using a probabilistic primality test is new. Kalai's algorithm is significantly simpler, though less efficient, than an earlier algorithm due to Bach [9], which uses an expected number of $O(\ell)$ primality tests, as opposed to the $O(\ell^2)$ primality tests used by Kalai's algorithm.

See Luby [63] for an exposition of the theory of pseudo-random bit generation.

10

Probabilistic primality testing

In this chapter, we discuss some simple and efficient probabilistic algorithms for testing whether a given integer is prime.

10.1 Trial division

Suppose we are given an integer $n > 1$, and we want to determine whether n is prime or composite. The simplest algorithm to describe and to program is **trial division**. We simply divide n by 2, 3, and so on, testing if any of these numbers evenly divide n. Of course, we only need to divide by primes up to $\sqrt{n}$, since if n is composite, it must have a prime factor no greater than $\sqrt{n}$ (see Exercise 1.2). Not only does this algorithm determine whether n is prime or composite, it also produces a non-trivial factor of n in case n is composite.

Of course, the drawback of this algorithm is that it is terribly inefficient: it requires $\Theta(\sqrt{n})$ arithmetic operations, which is exponential in the bit length of n. Thus, for practical purposes, this algorithm is limited to quite small n. Suppose, for example, that n has 100 decimal digits, and that a computer can perform 1 billion divisions per second (this is much faster than any computer existing today). Then it would take on the order of 10^{33} *years* to perform $\sqrt{n}$ divisions.

In this chapter, we discuss a much faster primality test that allows 100-decimal-digit numbers to be tested for primality in less than a second. Unlike the above test, however, this test does not find a factor of n when n is composite. Moreover, the algorithm is probabilistic, and may in fact make a mistake. However, the probability that it makes a mistake can be made so small as to be irrelevant for all practical purposes. Indeed, we can easily make the probability of error as small as 2^{-100} — should one really care about an event that happens with such a miniscule probability?

10.2 The Miller–Rabin test

We describe in this section a fast (polynomial time) test for primality, known as the **Miller–Rabin test**. As discussed above, the algorithm is probabilistic, and may (with small probability) make a mistake.

We assume for the remainder of this section that the number n we are testing for primality is an odd integer greater than 1.

We recall some basic algebraic facts that will play a critical role in this section (see §7.5). Suppose $n = p_1^{e_1} \cdots p_r^{e_r}$ is the prime factorization of n (since n is odd, each p_i is odd). The Chinese remainder theorem gives us a ring isomorphism

$$\theta : \quad \mathbb{Z}_n \to \mathbb{Z}_{p_1^{e_1}} \times \cdots \times \mathbb{Z}_{p_r^{e_r}}$$

$$[a]_n \mapsto ([a]_{p_1^{e_1}}, \ldots, [a]_{p_r^{e_r}}),$$

and restricting θ to $\mathbb{Z}_n^*$ yields a group isomorphism

$$\mathbb{Z}_n^* \cong \mathbb{Z}_{p_1^{e_1}}^* \times \cdots \times \mathbb{Z}_{p_r^{e_r}}^*.$$

Moreover, Theorem 7.28 says that each $\mathbb{Z}_{p_i^{e_i}}^*$ is a cyclic group, whose order, of course, is $\varphi(p_i^{e_i}) = p_i^{e_i-1}(p_i - 1)$, where φ is Euler's phi function.

Several probabilistic primality tests, including the Miller–Rabin test, have the following general structure. Define $\mathbb{Z}_n^+$ to be the set of non-zero elements of $\mathbb{Z}_n$; thus, $|\mathbb{Z}_n^+| = n - 1$, and if n is prime, $\mathbb{Z}_n^+ = \mathbb{Z}_n^*$. Suppose also that we define a set $L_n \subseteq \mathbb{Z}_n^+$ such that:

- there is an efficient algorithm that on input n and $\alpha \in \mathbb{Z}_n^+$, determines if $\alpha \in L_n$;
- if n is prime, then $L_n = \mathbb{Z}_n^*$;
- if n is composite, $|L_n| \le c(n - 1)$ for some constant $c < 1$.

To test n for primality, we set a "repetition parameter" k, and choose random elements $\alpha_1, \ldots, \alpha_k \in \mathbb{Z}_n^+$. If $\alpha_i \in L_n$ for all $i = 1, \ldots, k$, then we output *true*; otherwise, we output *false*.

It is easy to see that if n is prime, this algorithm always outputs *true*, and if n is composite this algorithm outputs *true* with probability at most c^k. If $c = 1/2$ and k is chosen large enough, say $k = 100$, then the probability that the output is wrong is so small that for all practical purposes, it is "just as good as zero."

We now make a first attempt at defining a suitable set L_n. Let us define

$$L_n := \{\alpha \in \mathbb{Z}_n^+ : \alpha^{n-1} = 1\}.$$

Note that $L_n \subseteq \mathbb{Z}_n^*$, since if $\alpha^{n-1} = 1$, then α has a multiplicative inverse, namely,

α^{n-2}. We can test if $\alpha \in L_n$ in time $O(\text{len}(n)^3)$, using a repeated-squaring algorithm.

Theorem 10.1. *If n is prime, then $L_n = \mathbb{Z}_n^*$. If n is composite and $L_n \subsetneq \mathbb{Z}_n^*$, then $|L_n| \leq (n-1)/2$.*

Proof. Note that L_n is the kernel of the $(n-1)$-power map on $\mathbb{Z}_n^*$, and hence is a subgroup of $\mathbb{Z}_n^*$.

If n is prime, then we know that $\mathbb{Z}_n^*$ is a group of order $n-1$. Since the order of a group element divides the order of the group, we have $\alpha^{n-1} = 1$ for all $\alpha \in \mathbb{Z}_n^*$. That is, $L_n = \mathbb{Z}_n^*$.

Suppose that n is composite and $L_n \subsetneq \mathbb{Z}_n^*$. Since the order of a subgroup divides the order of the group, we have $|\mathbb{Z}_n^*| = t|L_n|$ for some integer $t > 1$. From this, we conclude that

$$|L_n| = \frac{1}{t}|\mathbb{Z}_n^*| \leq \frac{1}{2}|\mathbb{Z}_n^*| \leq \frac{n-1}{2}. \quad \Box$$

Unfortunately, there are odd composite numbers n such that $L_n = \mathbb{Z}_n^*$. Such numbers are called **Carmichael numbers**. The smallest Carmichael number is

$$561 = 3 \cdot 11 \cdot 17.$$

Carmichael numbers are extremely rare, but it is known that there are infinitely many of them, so we cannot ignore them. The following theorem puts some constraints on Carmichael numbers.

Theorem 10.2. *Every Carmichael number n is of the form $n = p_1 \cdots p_r$, where the p_i's are distinct primes, $r \geq 3$, and $(p_i - 1) \mid (n-1)$ for $i = 1, \ldots, r$.*

Proof. Let $n = p_1^{e_1} \cdots p_r^{e_r}$ be a Carmichael number. By the Chinese remainder theorem, we have an isomorphism of $\mathbb{Z}_n^*$ with the group

$$\mathbb{Z}_{p_1^{e_1}}^* \times \cdots \times \mathbb{Z}_{p_r^{e_r}}^*,$$

and we know that each group $\mathbb{Z}_{p_i^{e_i}}^*$ is cyclic of order $p_i^{e_i-1}(p_i - 1)$. Thus, the power $n - 1$ kills the group $\mathbb{Z}_n^*$ if and only if it kills all the groups $\mathbb{Z}_{p_i^{e_i}}^*$, which happens if and only if $p_i^{e_i-1}(p_i - 1) \mid (n-1)$. Now, on the one hand, $n \equiv 0 \pmod{p_i}$. On the other hand, if $e_i > 1$, we would have $n \equiv 1 \pmod{p_i}$, which is clearly impossible. Thus, we must have $e_i = 1$.

It remains to show that $r \geq 3$. Suppose $r = 2$, so that $n = p_1 p_2$. We have

$$n - 1 = p_1 p_2 - 1 = (p_1 - 1)p_2 + (p_2 - 1).$$

Since $(p_1 - 1) \mid (n-1)$, we must have $(p_1 - 1) \mid (p_2 - 1)$. By a symmetric argument, $(p_2 - 1) \mid (p_1 - 1)$. Hence, $p_1 = p_2$, a contradiction. $\quad \Box$

To obtain a good primality test, we need to define a different set L'_n, which we do as follows. Let $n - 1 = t2^h$, where t is odd (and $h \geq 1$ since n is assumed odd), and define

$$L'_n := \{\alpha \in \mathbb{Z}_n^+ : \quad \alpha^{t2^h} = 1 \text{ and}$$
$$\alpha^{t2^{j+1}} = 1 \implies \alpha^{t2^j} = \pm 1 \text{ for } j = 0, \dots, h - 1\}.$$

The Miller–Rabin test uses this set L'_n, in place of the set L_n defined above. It is clear from the definition that $L'_n \subseteq L_n$.

Testing whether a given $\alpha \in \mathbb{Z}_n^+$ belongs to L'_n can be done using the following procedure:

$\beta \leftarrow \alpha^t$
if $\beta = 1$ then return *true*
for $j \leftarrow 0$ to $h - 1$ do
       if $\beta = -1$ then return *true*
       if $\beta = +1$ then return *false*
       $\beta \leftarrow \beta^2$
return *false*

It is clear that using a repeated-squaring algorithm, this procedure runs in time $O(\text{len}(n)^3)$. We leave it to the reader to verify that this procedure correctly determines membership in L'_n.

Theorem 10.3. *If n is prime, then $L'_n = \mathbb{Z}_n^*$. If n is composite, then $|L'_n| \leq (n-1)/4$.*

Proof. Let $n - 1 = t2^h$, where t is odd.

Case 1: n is prime. Let $\alpha \in \mathbb{Z}_n^*$. Since $\mathbb{Z}_n^*$ is a group of order $n - 1$, and the order of a group element divides the order of the group, we know that $\alpha^{t2^h} = \alpha^{n-1} = 1$. Now consider any index $j = 0, \dots, h-1$ such that $\alpha^{t2^{j+1}} = 1$, and consider the value $\beta := \alpha^{t2^j}$. Then since $\beta^2 = \alpha^{t2^{j+1}} = 1$, the only possible choices for β are ± 1—this is because $\mathbb{Z}_n^*$ is cyclic of even order and so there are exactly two elements of $\mathbb{Z}_n^*$ whose multiplicative order divides 2, namely ± 1. So we have shown that $\alpha \in L'_n$.

Case 2: $n = p^e$, where p is prime and $e > 1$. Certainly, L'_n is contained in the kernel K of the $(n-1)$-power map on $\mathbb{Z}_n^*$. By Theorem 6.32, $|K| = \gcd(\varphi(n), n-1)$. Since $n = p^e$, we have $\varphi(n) = p^{e-1}(p - 1)$, and so

$$|L'_n| \leq |K| = \gcd(p^{e-1}(p - 1), p^e - 1) = p - 1 = \frac{p^e - 1}{p^{e-1} + \cdots + 1} \leq \frac{n - 1}{4}.$$

Case 3: $n = p_1^{e_1} \cdots p_r^{e_r}$ is the prime factorization of n, and $r > 1$. Let

$$\theta : \mathbb{Z}_n \to \mathbb{Z}_{p_1^{e_1}} \times \cdots \times \mathbb{Z}_{p_r^{e_r}}$$

be the ring isomorphism provided by the Chinese remainder theorem. Also, let

$\varphi(p_i^{e_i}) = t_i 2^{h_i}$, with t_i odd, for $i = 1, \ldots, r$, and let $g := \min\{h, h_1, \ldots, h_r\}$. Note that $g \geq 1$, and that each $\mathbb{Z}_{p_i^{e_i}}^*$ is a cyclic group of order $t_i 2^{h_i}$.

We first claim that for every $\alpha \in L'_n$, we have $\alpha^{t2^g} = 1$. To prove this, first note that if $g = h$, then by definition, $\alpha^{t2^g} = 1$, so suppose that $g < h$. By way of contradiction, suppose that $\alpha^{t2^g} \neq 1$, and let j be the smallest index in the range $g, \ldots, h - 1$ such that $\alpha^{t2^{j+1}} = 1$. By the definition of L'_n, we must have $\alpha^{t2^j} = -1$. Since $g < h$, we must have $g = h_i$ for some particular index $i = 1, \ldots, r$. Writing $\theta(\alpha) = (\alpha_1, \ldots, \alpha_r)$, we have $\alpha_i^{t2^j} = -1$. This implies that the multiplicative order of α_i^t is equal to 2^{j+1} (see Theorem 6.37). However, since $j \geq g = h_i$, this contradicts the fact that the order of a group element (in this case, α_i^t) must divide the order of the group (in this case, $\mathbb{Z}_{p_i^{e_i}}^*$).

For $j = 0, \ldots, h$, let us define ρ_j to be the $(t2^j)$-power map on $\mathbb{Z}_n^*$. From the claim in the previous paragraph, and the definition of L'_n, it follows that each $\alpha \in L'_n$ satisfies $\alpha^{t2^{g-1}} = \pm 1$. In other words, $L'_n \subseteq \rho_{g-1}^{-1}(\{\pm 1\})$, and hence

$$|L'_n| \leq 2|\text{Ker } \rho_{g-1}|. \tag{10.1}$$

From the group isomorphism $\mathbb{Z}_n^* \cong \mathbb{Z}_{p_1^{e_1}}^* \times \cdots \times \mathbb{Z}_{p_r^{e_r}}^*$, and Theorem 6.32, we have

$$|\text{Ker } \rho_j| = \prod_{i=1}^{r} \gcd(t_i 2^{h_i}, t2^j) \tag{10.2}$$

for each $j = 0, \ldots, h$. Since $g \leq h$, and $g \leq h_i$ for $i = 1, \ldots, r$, it follows immediately from (10.2) that

$$2^r |\text{Ker } \rho_{g-1}| = |\text{Ker } \rho_g| \leq |\text{Ker } \rho_h|. \tag{10.3}$$

Combining (10.3) with (10.1), we obtain

$$|L'_n| \leq 2^{-r+1} |\text{Ker } \rho_h|. \tag{10.4}$$

If $r \geq 3$, then (10.4) directly implies that $|L'_n| \leq |\mathbb{Z}_n^*|/4 \leq (n - 1)/4$, and we are done. So suppose that $r = 2$. In this case, Theorem 10.2 implies that n is not a Carmichael number, which implies that $|\text{Ker } \rho_h| \leq |\mathbb{Z}_n^*|/2$, and so again, (10.4) implies $|L'_n| \leq |\mathbb{Z}_n^*|/4 \leq (n - 1)/4$. $\square$

EXERCISE 10.1. Show that an integer $n > 1$ is prime if and only if there exists an element in $\mathbb{Z}_n^*$ of multiplicative order $n - 1$.

EXERCISE 10.2. Show that Carmichael numbers satisfy Fermat's little theorem; that is, if n is a Carmichael number, then $\alpha^n = \alpha$ for all $\alpha \in \mathbb{Z}_n$.

EXERCISE 10.3. Let p be a prime. Show that $n := 2p + 1$ is a prime if and only if $2^{n-1} \equiv 1 \pmod{n}$.

EXERCISE 10.4. Here is another primality test that takes as input an odd integer $n > 1$, and a positive integer parameter k. The algorithm chooses $\alpha_1, \ldots, \alpha_k \in \mathbb{Z}_n^+$ at random, and computes

$$\beta_i := \alpha_i^{(n-1)/2} \quad (i = 1, \ldots, k).$$

If $(\beta_1, \ldots, \beta_k)$ is of the form $(\pm 1, \pm 1, \ldots, \pm 1)$, but is not equal to $(1, 1, \ldots, 1)$, the algorithm outputs *true*; otherwise, the algorithm outputs *false*. Show that if n is prime, then the algorithm outputs *false* with probability at most 2^{-k}, and if n is composite, the algorithm outputs *true* with probability at most 2^{-k}.

In the terminology of §9.7, the algorithm in the above exercise is an example of an "Atlantic City" algorithm for the language of prime numbers (or equivalently, the language of composite numbers), while the Miller–Rabin test is an example of a "Monte Carlo" algorithm for the language of *composite* numbers.

10.3 Generating random primes using the Miller–Rabin test

The Miller–Rabin test is the most practical algorithm known for testing primality, and because of this, it is widely used in many applications, especially cryptographic applications where one needs to generate large, random primes (as we saw in §4.7). In this section, we discuss how one uses the Miller–Rabin test in several practically relevant scenarios where one must generate large primes.

10.3.1 Generating a random prime between 2 and m

Suppose we are given an integer $m \geq 2$, and want to generate a random prime between 2 and m. We can do this by simply picking numbers at random until one of them passes a primality test. We discussed this problem in some detail in §9.4, where we assumed that we had a primality test *IsPrime*. The reader should review §9.4, and §9.4.1 in particular. In this section, we discuss aspects of this problem that are specific to the situation where the Miller–Rabin test is used to implement *IsPrime*. To be more precise, let us define the following algorithm:

Algorithm MR. On input n, k, where n and k are integers with $n > 1$ and $k \geq 1$, do the following:

if $n = 2$ then return *true*

if n is even then return *false*

repeat k times

$\qquad \alpha \xleftarrow{\text{\$}} \mathbb{Z}_n^+$

$\qquad$ if $\alpha \notin L'_n$ return *false*

return *true*

So we shall implement *IsPrime*$(\cdot)$ as MR$(\cdot, k)$, where k is an auxiliary parameter. By Theorem 10.3, if n is prime, the output of MR(n, k) is always *true*, while if n is composite, the output is *true* with probability at most 4^{-k}. Thus, this implementation of *IsPrime* satisfies the assumptions in §9.4.1, with $\varepsilon = 4^{-k}$.

Let $\gamma(m, k)$ be the probability that the output of Algorithm RP in §9.4—using this implementation of *IsPrime*—is composite. Then as we discussed in §9.4.1,

$$\gamma(m, k) \leq 4^{-k} \cdot \frac{m - 1}{\pi(m)} = O(4^{-k}\ell), \tag{10.5}$$

where $\ell := \text{len}(m)$, and $\pi(m)$ is the number of primes up to m. Furthermore, if the output of Algorithm RP is prime, then every prime is equally likely; that is, the conditional distribution of the output, given that the output is prime, is (essentially) the uniform distribution on the set of primes up to m.

Let us now consider the expected running time of Algorithm RP. As discussed in §9.4.1, the expected number of iterations of the main loop in Algorithm RP is $O(\ell)$. Clearly, the expected running time of a single loop iteration is $O(k\ell^3)$, since MR(n, k) executes at most k iterations of the Miller–Rabin test, and each such test takes time $O(\ell^3)$. This leads to a bound on the expected total running time of Algorithm RP of $O(k\ell^4)$. However, this estimate is overly pessimistic, because when n is composite, we expect to perform very few Miller–Rabin tests — only when n is prime do we actually perform all k of them.

To make a rigorous argument, let us define random variables measuring various quantities during the *first* iteration of the main loop in Algorithm RP: N_1 is the value of n; K_1 is the number of Miller–Rabin tests actually performed; Z_1 is the running time. Of course, N_1 is uniformly distributed over $\{2, \ldots, m\}$. Let C_1 be the event that N_1 is composite. Consider the conditional distribution of K_1 given C_1. This is not exactly a geometric distribution, since K_1 never takes on values greater than k; nevertheless, using Theorem 8.17, we can easily calculate

$$\mathsf{E}[K_1 \mid C_1] = \sum_{i \geq 1} \mathsf{P}[K_1 \geq i \mid C_1] \leq \sum_{i \geq 1} (1/4)^{i-1} = 4/3.$$

Using the law of total expectation (8.24), it follows that

$$E[K_1] = E[K_1 \mid C_1] P[C_1] + E[K_1 \mid \bar{C}_1] P[\bar{C}_1]$$
$$\leq 4/3 + k\pi(m)/(m-1).$$

Thus, $E[K_1] \leq 4/3 + O(k/\ell)$, and hence $E[Z_1] = O(\ell^3 E[K_1]) = O(\ell^3 + k\ell^2)$. Therefore, if Z is the total running time of Algorithm RP, then $E[Z] = O(\ell E[Z_1])$, and so

$$E[Z] = O(\ell^4 + k\ell^3). \qquad (10.6)$$

Note that the above estimate (10.5) for $\gamma(m, k)$ is actually quite pessimistic. This is because the error probability 4^{-k} is a worst-case estimate; in fact, for "most" composite integers n, the probability that $MR(n, k)$ outputs *true* is much smaller than this. In fact, $\gamma(m, 1)$ is *very* small for large m. For example, the following is known:

Theorem 10.4. *We have*

$$\gamma(m, 1) \leq \exp[-(1 + o(1)) \log(m) \log(\log(\log(m)))/\log(\log(m))].$$

Proof. Literature—see §10.5. $\square$

The bound in the above theorem goes to zero quite quickly: faster than $(\log m)^{-c}$ for every positive constant c. While the above theorem is asymptotically very good, in practice, one needs explicit bounds. For example, the following *lower* bounds for $-\log_2(\gamma(2^\ell, 1))$ are known:

ℓ	200	300	400	500	600
	3	19	37	55	74

Given an upper bound on $\gamma(m, 1)$, we can bound $\gamma(m, k)$ for $k \geq 2$ using the following inequality:

$$\gamma(m, k) \leq \frac{\gamma(m, 1)}{1 - \gamma(m, 1)} 4^{-k+1}. \qquad (10.7)$$

To prove (10.7), it is not hard to see that on input m, the output distribution of Algorithm RP is the same as that of the following algorithm:

repeat
    repeat
        $n' \xleftarrow{\ell} \{2, \ldots, m\}$
    until $MR(n', 1)$
    $n \leftarrow n'$
until $MR(n, k-1)$
output n

Let N_1 be the random variable representing the value of n in the first iteration of the main loop in this algorithm, let C_1 be the event that N_1 is composite, and let $\mathcal{H}_1$ be the event that this algorithm halts at the end of the first iteration of the main loop. Using Theorem 9.3, we see that

$$\gamma(m, k) = \mathsf{P}[C_1 \mid \mathcal{H}_1] = \frac{\mathsf{P}[C_1 \cap \mathcal{H}_1]}{\mathsf{P}[\mathcal{H}_1]} \le \frac{\mathsf{P}[C_1 \cap \mathcal{H}_1]}{\mathsf{P}[\bar{C}_1]} = \frac{\mathsf{P}[\mathcal{H}_1 \mid C_1] \, \mathsf{P}[C_1]}{\mathsf{P}[\bar{C}_1]}$$

$$\le \frac{4^{-k+1} \gamma(m, 1)}{1 - \gamma(m, 1)},$$

which proves (10.7).

Given that $\gamma(m, 1)$ is so small, for large m, Algorithm RP actually exhibits the following behavior in practice: it generates a random value $n \in \{2, \ldots, m\}$; if n is odd and composite, then the very *first* iteration of the Miller–Rabin test will detect this with overwhelming probability, and no more iterations of the test are performed on this n; otherwise, if n is prime, the algorithm will perform $k - 1$ more iterations of the Miller–Rabin test, "just to make sure."

EXERCISE 10.5. Consider the problem of generating a random Sophie Germain prime between 2 and m (see §5.5.5). One algorithm to do this is as follows:

    repeat
        $n \stackrel{\text{\$}}{\leftarrow} \{2, \ldots, m\}$
        if $MR(n, k)$ then
            if $MR(2n + 1, k)$ then
                output n and halt
    forever

Assuming Conjecture 5.24, show that this algorithm runs in expected time $O(\ell^5 + k\ell^4)$, and outputs a number that is not a Sophie Germain prime with probability $O(4^{-k}\ell^2)$. As usual, $\ell := \text{len}(m)$.

EXERCISE 10.6. Improve the algorithm in the previous exercise, so that under the same assumptions, it runs in expected time $O(\ell^5 + k\ell^3)$, and outputs a number that is not a Sophie Germain prime with probability $O(4^{-k}\ell^2)$, or even better, show that this probability is at most $\gamma(m, k)\pi^*(m)/\pi(m) = O(\gamma(m, k)\ell)$, where $\pi^*(m)$ is defined as in §5.5.5.

EXERCISE 10.7. Suppose in Algorithm RFN in §9.6 we implement algorithm *IsPrime*$(\cdot)$ as $MR(\cdot, k)$, where k is a parameter satisfying $4^{-k}(\log m + 1) \le 1/2$, and m is the input to RFN. Show that the expected running time of Algorithm RFN in this case is $O(\ell^5 + k\ell^4 \, \text{len}(\ell))$. Hint: use Exercise 9.13.

10.3.2 Trial division up to a small bound

In generating a random prime, most candidates will in fact be composite, and so it makes sense to cast these out as quickly as possible. Significant efficiency gains can be achieved by testing if a given candidate n is divisible by any prime up to a given bound s, before we subject n to a Miller–Rabin test. This strategy makes sense, since for a small, "single precision" prime p, we can test if $p \mid n$ essentially in time $O(\text{len}(n))$, while a single iteration of the Miller–Rabin test takes time $O(\text{len}(n)^3)$.

To be more precise, let us define the following algorithm:

Algorithm MRS. On input n, k, s, where $n, k, s \in \mathbb{Z}$, and $n > 1$, $k \geq 1$, and $s > 1$, do the following:

> for each prime $p \leq s$ do
> > if $p \mid n$ then
> > > if $p = n$ then return *true* else return *false*
>
> repeat k times
> > $\alpha \xleftarrow{\$} \mathbb{Z}_n^+$
> > if $\alpha \notin L'_n$ return *false*
>
> return *true*

In an implementation of the above algorithm, one would most likely use the sieve of Eratosthenes (see §5.4) to generate the small primes.

Note that $\text{MRS}(n, k, 2)$ is equivalent to $\text{MR}(n, k)$. Also, it is clear that the probability that $\text{MRS}(n, k, s)$ makes a mistake is no more than the probability that $\text{MR}(n, k)$ makes a mistake. Therefore, using MRS in place of MR will not increase the probability that the output of Algorithm RP is a composite—indeed, it is likely that this probability decreases significantly.

Let us now analyze the impact on the running time Algorithm RP. To do this, we need to estimate the probability $\sigma(m, s)$ that a randomly chosen integer between 2 and m is not divisible by any prime up to s. If m is sufficiently large with respect to s, the following heuristic argument can be made rigorous, as we will discuss below. The probability that a random integer is divisible by a prime p is about $1/p$, so the probability that it is not divisible by p is about $1 - 1/p$. Assuming that these events are essentially independent for different values of p (this is the heuristic part), we estimate

$$\sigma(m, s) \approx \prod_{p \leq s}(1 - 1/p). \tag{10.8}$$

Assuming for the time being that the approximation in (10.8) is sufficiently accurate, then using Mertens' theorem (Theorem 5.13), we may deduce that

$$\sigma(m, s) = O(1/\log s). \tag{10.9}$$

Later, when we make this argument more rigorous, we shall see that (10.9) holds provided s is not too large relative to m, and in particular, if $s = O((\log m)^c)$ for some constant c.

The estimate (10.9) gives us a bound on the probability that a random integer passes the trial division phase, and so must be subjected to Miller–Rabin; however, performing the trial division takes some time, so we also need to estimate the expected number $\kappa(m, s)$ of trial divisions performed on a random integer between 2 and m. Of course, in the worst case, we divide by all primes up to s, and so $\kappa(m, s) \leq \pi(s) = O(s/\log s)$, but we can get a better bound, as follows. Let $p_1, p_2, \ldots, p_r$ be the primes up to s, and for $i = 1, \ldots, r$, let q_i be the probability that we perform at least i trial divisions. By Theorem 8.17, we have

$$\kappa(m, s) = \sum_{i=1}^{r} q_i.$$

Moreover, $q_1 = 1$, and $q_i = \sigma(m, p_{i-1})$ for $i = 2, \ldots, r$. From this, and (10.9), it follows that

$$\kappa(m, s) = 1 + \sum_{i=2}^{r} \sigma(m, p_{i-1}) = O\left(\sum_{p \leq s} 1/\log p\right).$$

As a simple consequence of Chebyshev's theorem (in particular, see Exercise 5.3), we obtain

$$\kappa(m, s) = O(s/(\log s)^2). \tag{10.10}$$

We now derive a bound on the running time of Algorithm RP, assuming that *IsPrime*($\cdot$) is implemented using MRS($\cdot, k, s$). Let $\ell := \text{len}(m)$. Our argument follows the same lines as was used to derive the estimate (10.6). Let us define random variables measuring various quantities during the *first* iteration of the main loop in Algorithm RP: N_1 is the value of n; K_1 is the number of Miller–Rabin tests actually performed; Z_1 is the running time. Also, let C_1 be the event that N_1 is composite, and let D_1 be the event that N_1 passes the trial division check. Then we have

$$\begin{aligned}
E[K_1] &= E[K_1 \mid C_1 \cap D_1] P[C_1 \cap D_1] + E[K_1 \mid C_1 \cap \bar{D}_1] P[C_1 \cap \bar{D}_1] \\
&\quad + E[K_1 \mid \bar{C}_1] P[\bar{C}_1] \\
&\leq 4/3 \cdot P[C_1 \cap D_1] + 0 \cdot P[C_1 \cap \bar{D}_1] + k \cdot P[\bar{C}_1] \\
&\leq 4/3 \cdot P[D_1] + k \cdot P[\bar{C}_1].
\end{aligned}$$

By (10.9) and Chebyshev's theorem, it follows that

$$E[K_1] = O(1/\text{len}(s) + k/\ell). \tag{10.11}$$

Let us write $Z_1 = Z_1' + Z_1''$, where Z_1' is the amount of time spent performing the Miller–Rabin test, and Z_1'' is the amount of time spent performing trial division. By (10.11), we have $E[Z_1'] = O(\ell^3/\operatorname{len}(s) + k\ell^2)$. Further, assuming that each individual trial division step takes time $O(\ell)$, then by (10.10) we have $E[Z_1''] = O(\ell s/\operatorname{len}(s)^2)$. Hence,

$$E[Z_1] = O(\ell^3/\operatorname{len}(s) + k\ell^2 + \ell s/\operatorname{len}(s)^2).$$

It follows that if Z is the total running time of Algorithm RP, then

$$E[Z] = O(\ell^4/\operatorname{len}(s) + k\ell^3 + \ell^2 s/\operatorname{len}(s)^2).$$

Clearly, we want to choose the parameter s so that the time spent performing trial division is dominated by the time spent performing the Miller–Rabin test. To this end, let us assume that $\ell \leq s \leq \ell^2$. Then we have

$$E[Z] = O(\ell^4/\operatorname{len}(\ell) + k\ell^3). \tag{10.12}$$

This estimate does not take into account the time to generate the small primes using the sieve of Eratosthenes. These values might be pre-computed, in which case this time is zero, but even if we compute them on the fly, this takes time $O(s\operatorname{len}(\operatorname{len}(s)))$, which is dominated by the running time of the rest of the algorithm for the values of s under consideration.

Thus, by sieving up to a bound s, where $\ell \leq s \leq \ell^2$, then compared to (10.6), we effectively reduce the running time by a factor proportional to $\operatorname{len}(\ell)$, which is a very real and noticeable improvement in practice.

As we already mentioned, the above analysis is heuristic, but the results are correct. We shall now discuss how this analysis can be made rigorous; however, we should remark that any such rigorous analysis is mainly of theoretical interest only — in any practical implementation, the optimal choice of the parameter s is best determined by experiment, with the analysis being used only as a rough guide. Now, to make the analysis rigorous, we need prove that the estimate (10.8) is sufficiently accurate. Proving such estimates takes us into the realm of "sieve theory." The larger m is with respect to s, the easier it is to prove such estimates. We shall prove only the simplest and most naive such estimate, but it is still good enough for our purposes.

Before stating any results, let us restate the problem slightly differently. For a given real number $y \geq 0$, let us call a positive integer "y-rough" if it is not divisible by any prime p up to y. For all real numbers $x \geq 0$ and $y \geq 0$, let us define $R(x, y)$ to be the number of y-rough positive integers up to x. Thus, since $\sigma(m, s)$ is the probability that a random integer between 2 and m is s-rough, and 1 is by definition s-rough, we have $\sigma(m, s) = (R(m, s) - 1)/(m - 1)$.

Theorem 10.5. *For all real $x \geq 0$ and $y \geq 0$, we have*

$$\left| R(x, y) - x \prod_{p \leq y}(1 - 1/p) \right| \leq 2^{\pi(y)}.$$

Proof. To simplify the notation, we shall use the Möbius function μ (see §2.9). Also, for a real number u, let us write $u = \lfloor u \rfloor + \{u\}$, where $0 \leq \{u\} < 1$. Let Q be the product of the primes up to the bound y.

Now, there are $\lfloor x \rfloor$ positive integers up to x, and of these, for each prime p dividing Q, precisely $\lfloor x/p \rfloor$ are divisible by p, for each pair p, p' of distinct primes dividing Q, precisely $\lfloor x/pp' \rfloor$ are divisible by pp', and so on. By inclusion/exclusion (see Theorem 8.1), we have

$$R(x, y) = \sum_{d|Q} \mu(d)\lfloor x/d \rfloor = \sum_{d|Q} \mu(d)(x/d) - \sum_{d|Q} \mu(d)\{x/d\}.$$

Moreover,

$$\sum_{d|Q} \mu(d)(x/d) = x \sum_{d|Q} \mu(d)/d = x \prod_{p \leq y}(1 - 1/p),$$

and

$$\left| \sum_{d|Q} \mu(d)\{x/d\} \right| \leq \sum_{d|Q} 1 = 2^{\pi(y)}.$$

That proves the theorem. $\square$

This theorem says something non-trivial only when y is quite small. Nevertheless, using Chebyshev's theorem on the density of primes, along with Mertens' theorem, it is not hard to see that this theorem implies that (10.9) holds when $s = O((\log m)^c)$ for some constant c (see Exercise 10.8), which implies the estimate (10.12) above, when $\ell \leq s \leq \ell^2$.

EXERCISE 10.8. Suppose that s is a function of m such that $s = O((\log m)^c)$ for some positive constant c. Show that $\sigma(m, s) = O(1/\log s)$.

EXERCISE 10.9. Let f be a polynomial with integer coefficients. For real $x \geq 0$ and $y \geq 0$, define $R_f(x, y)$ to be the number of positive integers t up to x such that $f(t)$ is y-rough. For each positive integer m, define $\omega_f(m)$ to be the number of integers $t \in \{0, \ldots, m - 1\}$ such that $f(t) \equiv 0 \pmod{m}$. Show that

$$\left| R_f(x, y) - x \prod_{p \leq y}(1 - \omega_f(p)/p) \right| \leq \prod_{p \leq y}(1 + \omega_f(p)).$$

EXERCISE 10.10. Consider again the problem of generating a random Sophie Germain prime, as discussed in Exercises 10.5 and 10.6. A useful idea is to first test if *either n or 2n + 1* are divisible by any small primes up to some bound *s*, before performing any more expensive tests. Using this idea, design and analyze an algorithm that improves the running time of the algorithm in Exercise 10.6 to $O(\ell^5/\operatorname{len}(\ell)^2 + k\ell^3)$—under the same assumptions, and achieving the same error probability bound as in that exercise. Hint: first show that the previous exercise implies that the number of positive integers *t* up to *x* such that both *t* and *2t + 1* are *y*-rough is at most

$$x \cdot \frac{1}{2} \prod_{2 < p \leq y} (1 - 2/p) + 3^{\pi(y)}.$$

EXERCISE 10.11. Design an algorithm that takes as input a prime *q* and a bound *m*, and outputs a random prime *p* between 2 and *m* such that $p \equiv 1 \pmod{q}$. Clearly, we need to assume that *m* is sufficiently large with respect to *q*. Analyze your algorithm assuming Conjecture 5.22. State how large *m* must be with respect to *q*, and under these assumptions, show that your algorithm runs in time $O(\ell^4/\operatorname{len}(\ell) + k\ell^3)$, and that its output is incorrect with probability $O(4^{-k}\ell)$. As usual, $\ell := \operatorname{len}(m)$.

10.3.3 Generating a random ℓ-bit prime

In some applications, we want to generate a random prime of fixed size—a random 1024-bit prime, for example. More generally, let us consider the following problem: given an integer $\ell \geq 2$, generate a random ℓ-bit prime, that is, a prime in the interval $[2^{\ell-1}, 2^{\ell})$.

Bertrand's postulate (Theorem 5.8) implies that there exists a constant $c > 0$ such that $\pi(2^{\ell}) - \pi(2^{\ell-1}) \geq c2^{\ell-1}/\ell$ for all $\ell \geq 2$.

Now let us modify Algorithm RP so that it takes as input an integer $\ell \geq 2$, and repeatedly generates a random *n* in the interval $\{2^{\ell-1}, \ldots, 2^{\ell} - 1\}$ until *IsPrime(n)* returns *true*. Let us call this variant Algorithm RP'. Further, let us implement *IsPrime(·)* as MR(·, *k*), for some auxiliary parameter *k*, and define $\gamma'(\ell, k)$ to be the probability that the output of Algorithm RP' —with this implementation of *IsPrime*—is composite.

Then using exactly the same reasoning as in §10.3.1, we have

$$\gamma'(\ell, k) \leq 4^{-k} \frac{2^{\ell-1}}{\pi(2^{\ell}) - \pi(2^{\ell-1})} = O(4^{-k}\ell);$$

moreover, if the output of Algorithm RP' is prime, then every ℓ-bit prime is equally likely, and the expected running time is $O(\ell^4 + k\ell^3)$. By doing some trial division as in §10.3.2, this can be reduced to $O(\ell^4 / \operatorname{len}(\ell) + k\ell^3)$.

The function $\gamma'(\ell, k)$ has been studied a good deal; for example, the following explicit bound is known:

Theorem 10.6. *For all $\ell \geq 2$, we have*

$$\gamma'(\ell, 1) \leq \ell^2 4^{2-\sqrt{\ell}}.$$

Proof. Literature—see §10.5. $\square$

Upper bounds for $\gamma'(\ell, k)$ for specific values of ℓ and k have been computed. The following table lists some known *lower* bounds for $-\log_2(\gamma'(\ell, k))$ for various values of ℓ and k:

$k \backslash \ell$	200	300	400	500	600
1	11	19	37	56	75
2	25	33	46	63	82
3	34	44	55	70	88
4	41	53	63	78	95
5	47	60	72	85	102

Using exactly the same reasoning as the derivation of (10.7), one sees that

$$\gamma'(\ell, k) \leq \frac{\gamma'(\ell, 1)}{1 - \gamma'(\ell, 1)} 4^{-k+1}.$$

10.4 Factoring and computing Euler's phi function

In this section, we use some of the ideas developed to analyze the Miller–Rabin test to prove that the problem of factoring n and the problem of computing $\varphi(n)$ are equivalent. By equivalent, we mean that given an efficient algorithm to solve one problem, we can efficiently solve the other, and *vice versa*.

Clearly, one direction is easy: if we can factor n into primes, so

$$n = p_1^{e_1} \cdots p_r^{e_r}, \tag{10.13}$$

then we can simply compute $\varphi(n)$ using the formula

$$\varphi(n) = p_1^{e_1 - 1}(p_1 - 1) \cdots p_r^{e_r - 1}(p_r - 1).$$

For the other direction, first consider the special case where $n = pq$, for distinct primes p and q. Suppose we are given n and $\varphi(n)$, so that we have two equations in the unknowns p and q:

$$n = pq \quad \text{and} \quad \varphi(n) = (p - 1)(q - 1).$$

Substituting n/p for q in the second equation, and simplifying, we obtain

$$p^2 + (\varphi(n) - n - 1)p + n = 0,$$

which can be solved using the quadratic formula.

For the general case, it is just as easy to prove a stronger result: given any non-zero multiple of the *exponent* of $\mathbb{Z}_n^*$, we can efficiently factor n. In particular, this will show that we can efficiently factor Carmichael numbers.

Before stating the algorithm in its full generality, we can convey the main idea by considering the special case where $n = pq$, where p and q are distinct primes, with $p \equiv q \equiv 3 \pmod{4}$. Suppose we are given such an n, along with a non-zero multiple f of the exponent of $\mathbb{Z}_n^*$. Now, $\mathbb{Z}_n^* \cong \mathbb{Z}_p^* \times \mathbb{Z}_q^*$, and since $\mathbb{Z}_p^*$ is a cyclic group of order $p - 1$ and $\mathbb{Z}_q^*$ is a cyclic group of order $q - 1$, this means that f is a non-zero common multiple of $p - 1$ and $q - 1$. Let $f = t2^h$, where t is odd, and consider the following probabilistic algorithm:

$$\alpha \xleftarrow{\mathscr{R}} \mathbb{Z}_n^+$$
$$d \leftarrow \gcd(\mathrm{rep}(\alpha), n)$$
if $d \neq 1$ then output d and halt
$$\beta \leftarrow \alpha^t$$
$$d' \leftarrow \gcd(\mathrm{rep}(\beta) + 1, n)$$
if $d' \notin \{1, n\}$ then output d' and halt
output "failure"

Recall that $\mathrm{rep}(\alpha)$ denotes the canonical representative of α, that is, the unique integer a such that $[a]_n = \alpha$ and $0 \leq a < n$. We shall prove that this algorithm outputs a non-trivial divisor of n with probability at least $1/2$.

Let ρ be the t-power map on $\mathbb{Z}_n^*$, and let $G := \rho^{-1}(\{\pm 1\})$. We shall show that

- $G \subsetneq \mathbb{Z}_n^*$, and
- if the algorithm chooses $\alpha \notin G$, then it splits n.

Since G is a subgroup of $\mathbb{Z}_n^*$, it follows that $|G|/|\mathbb{Z}_n^+| \leq |G|/|\mathbb{Z}_n^*| \leq 1/2$, and this implies the algorithm succeeds with probability at least $1/2$.

Let $\theta : \mathbb{Z}_n \to \mathbb{Z}_p \times \mathbb{Z}_q$ be the ring isomorphism from the Chinese remainder theorem. The assumption that $p \equiv 3 \pmod 4$ means that $(p - 1)/2$ is an odd integer, and since f is a multiple of $p - 1$, it follows that $\gcd(t, p - 1) = (p - 1)/2$, and hence the image of $\mathbb{Z}_p^*$ under the t-power map is the subgroup of $\mathbb{Z}_p^*$ of order 2, which is $\{\pm 1\}$. Likewise, the image of $\mathbb{Z}_q^*$ under the t-power map is $\{\pm 1\}$. Thus,

$$\theta(\mathrm{Im}\,\rho) = \theta((\mathbb{Z}_n^*)^t) = (\theta(\mathbb{Z}_n^*))^t = (\mathbb{Z}_p^*)^t \times (\mathbb{Z}_q^*)^t = \{\pm 1\} \times \{\pm 1\},$$

and so $\mathrm{Im}\,\rho$ consists of the four elements:

$$1 = \theta^{-1}(1, 1), \; -1 = \theta^{-1}(-1, -1), \; \theta^{-1}(-1, 1), \; \theta^{-1}(1, -1).$$

By the observations in the previous paragraph, not all elements of $\mathbb{Z}_n^*$ map to ± 1 under ρ, which means that $G \subsetneq \mathbb{Z}_n^*$. Suppose that the algorithm chooses $\alpha \in \mathbb{Z}_n^* \setminus G$. We want to show that n gets split. If $\alpha \notin \mathbb{Z}_n^*$, then $\gcd(\text{rep}(\alpha), n)$ is a non-trivial divisor of n, and the algorithm splits n. So let us assume that $\alpha \in \mathbb{Z}_n^* \setminus G$. Consider the value $\beta = \alpha^t = \rho(\alpha)$ computed by the algorithm. Since $\alpha \notin G$, we have $\beta \neq \pm 1$, and by the observations in the previous paragraph, we have $\theta(\beta) = (-1, 1)$ or $\theta(\beta) = (1, -1)$. In the first case, $\theta(\beta + 1) = (0, 2)$, and so $\gcd(\text{rep}(\beta) + 1, n) = p$, while in the second case, $\theta(\beta + 1) = (2, 0)$, and so $\gcd(\text{rep}(\beta) + 1, n) = q$. In either case, the algorithm splits n.

We now consider the general case, where n is an arbitrary positive integer. Let $\lambda(n)$ denote the exponent of $\mathbb{Z}_n^*$. If the prime factorization of n is as in (10.13), then by the Chinese remainder theorem, we have

$$\lambda(n) = \text{lcm}(\lambda(p_1^{e_1}), \ldots, \lambda(p_r^{e_r})).$$

Moreover, for every prime power p^e, by Theorem 7.28, we have

$$\lambda(p^e) = \begin{cases} p^{e-1}(p-1) & \text{if } p \neq 2 \text{ or } e \leq 2, \\ 2^{e-2} & \text{if } p = 2 \text{ and } e \geq 3. \end{cases}$$

In particular, if $d \mid n$, then $\lambda(d) \mid \lambda(n)$.

Now, assume we are given n, along with a non-zero multiple f of $\lambda(n)$. We would like to calculate the complete prime factorization of n. We may proceed recursively: first, if $n = 1$, we may obviously halt; otherwise, we test if n is prime, using an efficient primality test, and if so, halt (if we are using the Miller–Rabin test, then we may erroneously halt even when n is composite, but we can ensure that this happens with negligible probability); otherwise, we split n as $n = d_1 d_2$, using an algorithm to be described below, and then recursively factor both d_1 and d_2; since $\lambda(d_1) \mid f$ and $\lambda(d_2) \mid f$, we may use the same value f in the recursion.

So let us assume that $n > 1$ and n is not prime, and our goal now is to use f to obtain a non-trivial factorization of n. If n is even, then we can certainly do this. Moreover, if n is a perfect power—that is, if $n = a^b$ for some integers $a > 1$ and $b > 1$—we can also obtain a non-trivial factorization of n (see Exercise 3.31).

So let us assume not only that $n > 1$ and n is not prime, but also that n is odd, and n is not a perfect power. Let $f = t2^h$, where t is odd. Consider the following probabilistic algorithm:

$\alpha \overset{\$}{\leftarrow} \mathbb{Z}_n^+$

$d \leftarrow \gcd(\mathrm{rep}(\alpha), n)$

if $d \neq 1$ then output d and halt

$\beta \leftarrow \alpha^t$

for $j \leftarrow 0$ to $h - 1$ do

        $d' \leftarrow \gcd(\mathrm{rep}(\beta) + 1, n)$

        if $d' \notin \{1, n\}$ then output d' and halt

        $\beta \leftarrow \beta^2$

output "failure"

We want to show that this algorithm outputs a non-trivial factor of n with probability at least $1/2$. To do this, suppose the prime factorization of n is as in (10.13). Then by our assumptions about n, we have $r \geq 2$ and each p_i is odd. Let $\lambda(p_i^{e_i}) = t_i 2^{h_i}$, where t_i is odd, for $i = 1, \ldots, r$, and let $g := \max\{h_1, \ldots, h_r\}$. Note that since $\lambda(n) \mid f$, we have $1 \leq g \leq h$.

Let ρ be the $(t2^{g-1})$-power map on $\mathbb{Z}_n^*$, and let $G := \rho^{-1}(\{\pm 1\})$. As above, we shall show that

- $G \subsetneq \mathbb{Z}_n^*$, and
- if the algorithm chooses $\alpha \notin G$, then it splits n,

which will prove that the algorithm splits n with probability at least $1/2$.

Let

$$\theta : \mathbb{Z}_n \rightarrow \mathbb{Z}_{p_1^{e_1}} \times \cdots \times \mathbb{Z}_{p_r^{e_r}}$$

be the ring isomorphism of the Chinese remainder theorem. We have

$$\theta(\mathrm{Im}\,\rho) = G_1 \times \cdots \times G_r,$$

where

$$G_i := \left(\mathbb{Z}_{p_i^{e_i}}^* \right)^{t2^{g-1}} \quad \text{for } i = 1, \ldots, r.$$

Let us assume the p_i's are ordered so that $h_i = g$ for $i = 1, \ldots, r'$, and $h_i < g$ for $i = r' + 1, \ldots, r$, where we have $1 \leq r' \leq r$. Then we have $G_i = \{\pm 1\}$ for $i = 1, \ldots, r'$, and $G_i = \{1\}$ for $i = r' + 1, \ldots, r$.

By the observations in the previous paragraph, and the fact that $r \geq 2$, the image of ρ contains elements other than ± 1; for example, $\theta^{-1}(-1, 1, \ldots, 1)$ is such an element. This means that $G \subsetneq \mathbb{Z}_n^*$. Suppose the algorithm chooses $\alpha \in \mathbb{Z}_n^+ \setminus G$. We want to show that n gets split. If $\alpha \notin \mathbb{Z}_n^*$, then $\gcd(\mathrm{rep}(\alpha), n)$ is a non-trivial divisor of n, and so the algorithm certainly splits n. So assume $\alpha \in \mathbb{Z}_n^* \setminus G$. In loop iteration $j = g - 1$, the value of β is equal to $\rho(\alpha)$, and writing $\theta(\beta) = (\beta_1, \ldots, \beta_r)$, we have $\beta_i = \pm 1$ for $i = 1, \ldots, r$. Let S be the set of indices i such that $\beta_i = -1$.

As $\alpha \notin G$, we know that $\beta \neq \pm 1$, and so $\emptyset \subsetneq S \subsetneq \{1, \ldots, r\}$. Thus,

$$\gcd(\mathrm{rep}(\beta) + 1, n) = \prod_{i \in S} p_i^{e_i}$$

is a non-trivial factor of n. This means that the algorithm splits n in loop iteration $j = g - 1$ (if not in some earlier loop iteration).

So we have shown that the above algorithm splits n with probability at least $1/2$. If we iterate the algorithm until n gets split, the expected number of loop iterations required will be at most 2. Combining this with the above recursive algorithm, we get an algorithm that completely factors an arbitrary n in expected polynomial time.

EXERCISE 10.12. Suppose you are given an integer n of the form $n = pq$, where p and q are distinct, ℓ-bit primes, with $p = 2p' + 1$ and $q = 2q' + 1$, where p' and q' are themselves prime. Suppose that you are also given an integer t such that $\gcd(t, p'q') \neq 1$. Show how to efficiently factor n.

EXERCISE 10.13. Suppose there is a probabilistic algorithm A that takes as input an integer n of the form $n = pq$, where p and q are distinct, ℓ-bit primes, with $p = 2p' + 1$ and $q = 2q' + 1$, where p' and q' are prime. The algorithm also takes as input $\alpha, \beta \in (\mathbb{Z}_n^*)^2$. It outputs either "failure," or integers x, y, not both zero, such that $\alpha^x \beta^y = 1$. Furthermore, assume that A runs in expected polynomial time, and that for all n of the above form, and for randomly chosen $\alpha, \beta \in (\mathbb{Z}_n^*)^2$, A succeeds in finding x, y as above with probability $\varepsilon(n)$. Here, the probability is taken over the random choice of α and β, as well as the random choices made during the execution of A on input (n, α, β). Show how to use A to construct another probabilistic algorithm A' that takes as input n as above, runs in expected polynomial time, and that satisfies the following property:

if $\varepsilon(n) \geq 0.001$, then A' factors n with probability at least 0.999.

10.5 Notes

The Miller–Rabin test is due to Miller [67] and Rabin [79]. The paper by Miller defined the set L'_n, but did not give a probabilistic analysis. Rather, Miller showed that under a generalization of the Riemann hypothesis, for composite n, the least positive integer a such that $[a]_n \in \mathbb{Z}_n \setminus L'_n$ is at most $O((\log n)^2)$, thus giving rise to a deterministic primality test whose correctness depends on the above unproved hypothesis. The later paper by Rabin re-interprets Miller's result in the context of probabilistic algorithms.

Bach [10] gives an explicit version of Miller's result, showing that under the same assumptions, the least positive integer a such that $[a]_n \in \mathbb{Z}_n \setminus L'_n$ is at most $2(\log n)^2$; more generally, Bach shows that the following holds under a generalization of the Riemann hypothesis:

> For every positive integer n, and every subgroup $G \subsetneq \mathbb{Z}_n^*$, the least positive integer a such that $[a]_n \in \mathbb{Z}_n \setminus G$ is at most $2(\log n)^2$, and the least positive integer b such that $[b]_n \in \mathbb{Z}_n^* \setminus G$ is at most $3(\log n)^2$.

The first efficient probabilistic primality test was invented by Solovay and Strassen [99] (their paper was actually submitted for publication in 1974). Later, in Chapter 21, we shall discuss a recently discovered, deterministic, polynomial-time (though not very practical) primality test, whose analysis does not rely on any unproved hypothesis.

Carmichael numbers are named after R. D. Carmichael, who was the first to discuss them, in work published in the early 20th century. Alford, Granville, and Pomerance [7] proved that there are infinitely many Carmichael numbers.

Exercise 10.4 is based on Lehmann [58].

Theorem 10.4, as well as the table of values just below it, are from Kim and Pomerance [55]. In fact, these bounds hold for the weaker test based on L_n.

Our analysis in §10.3.2 is loosely based on a similar analysis in §4.1 of Maurer [65]. Theorem 10.5 and its generalization in Exercise 10.9 are certainly not the best results possible in this area. The general goal of "sieve theory" is to prove useful upper and lower bounds for quantities like $R_f(x, y)$ that hold when y is as large as possible with respect to x. For example, using a technique known as Brun's pure sieve, one can show that for $\log y < \sqrt{\log x}$, there exist β and β', both of absolute value at most 1, such that

$$R_f(x, y) = (1 + \beta e^{-\sqrt{\log x}})x \prod_{p \leq y}(1 - \omega_f(p)/p) + \beta' \sqrt{x}.$$

Thus, this gives us very sharp estimates for $R_f(x, y)$ when x tends to infinity, and y is bounded by any fixed polynomial in $\log x$. For a proof of this result, see §2.2 of Halberstam and Richert [44] (the result itself is stated as equation 2.16). Brun's pure sieve is really just the first non-trivial sieve result, developed in the early 20th century; even stronger results, extending the useful range of y (but with larger error terms), have subsequently been proved.

Theorem 10.6, as well as the table of values immediately below it, are from Damgård, Landrock, and Pomerance [32].

The algorithm presented in §10.4 for factoring an integer given a multiple of $\varphi(n)$ (or, for that matter, $\lambda(n)$) is essentially due to Miller [67]. However, just as for his primality test, Miller presents his algorithm as a deterministic algorithm, which he analyzes under a generalization of the Riemann hypothesis. The probabilistic version of Miller's factoring algorithm appears to be "folklore."

11

Finding generators and discrete logarithms in $\mathbb{Z}_p^*$

As we have seen in Theorem 7.28, for a prime p, $\mathbb{Z}_p^*$ is a cyclic group of order $p - 1$. This means that there exists a generator $\gamma \in \mathbb{Z}_p^*$, such that each $\alpha \in \mathbb{Z}_p^*$ can be written uniquely as $\alpha = \gamma^x$, where x is an integer with $0 \leq x < p-1$; the integer x is called the **discrete logarithm** of α to the base γ, and is denoted $\log_\gamma \alpha$.

This chapter discusses some computational problems in this setting; namely, how to efficiently find a generator γ, and given γ and α, how to compute $\log_\gamma \alpha$.

More generally, if γ generates a subgroup G of $\mathbb{Z}_p^*$ of order q, where $q \mid (p - 1)$, and $\alpha \in G$, then $\log_\gamma \alpha$ is defined to be the unique integer x with $0 \leq x < q$ and $\alpha = \gamma^x$. In some situations it is more convenient to view $\log_\gamma \alpha$ as an element of $\mathbb{Z}_q$. Also for $x \in \mathbb{Z}_q$, with $x = [a]_q$, one may write γ^x to denote γ^a. There can be no confusion, since if $x = [a']_q$, then $\gamma^{a'} = \gamma^a$. However, in this chapter, we shall view $\log_\gamma \alpha$ as an integer.

Although we work in the group $\mathbb{Z}_p^*$, all of the algorithms discussed in this chapter trivially generalize to any finite cyclic group that has a suitably compact representation of group elements and an efficient algorithm for performing the group operation on these representations.

11.1 Finding a generator for $\mathbb{Z}_p^*$

In this section, we consider the problem of how to find a generator for $\mathbb{Z}_p^*$. There is no efficient algorithm known for this problem, unless the prime factorization of $p-1$ is given, and even then, we must resort to the use of a probabilistic algorithm. Of course, factoring in general is believed to be a very difficult problem, so it may not be easy to get the prime factorization of $p - 1$. However, if our goal is to construct a large prime p, together with a generator for $\mathbb{Z}_p^*$, then we may use Algorithm RFN in §9.6 to generate a random factored number n in some range, test $n + 1$ for primality, and then repeat until we get a factored number n such that

$p = n + 1$ is prime. In this way, we can generate a random prime p in a given range along with the factorization of $p - 1$.

We now present an efficient probabilistic algorithm that takes as input an odd prime p, along with the prime factorization

$$p - 1 = \prod_{i=1}^{r} q_i^{e_i},$$

and outputs a generator for $\mathbb{Z}_p^*$. It runs as follows:

> for $i \leftarrow 1$ to r do
>> repeat
>>> choose $\alpha \in \mathbb{Z}_p^*$ at random
>>> compute $\beta \leftarrow \alpha^{(p-1)/q_i}$
>> until $\beta \neq 1$
>> $\gamma_i \leftarrow \alpha^{(p-1)/q_i^{e_i}}$
> $\gamma \leftarrow \prod_{i=1}^{r} \gamma_i$
> output γ

First, let us analyze the correctness of this algorithm. When the ith loop iteration terminates, by construction, we have

$$\gamma_i^{q_i^{e_i}} = 1 \quad \text{but} \quad \gamma_i^{q_i^{e_i-1}} \neq 1.$$

It follows (see Theorem 6.37) that γ_i has multiplicative order $q_i^{e_i}$. From this, it follows (see Theorem 6.38) that γ has multiplicative order $p - 1$. Thus, we have shown that if the algorithm terminates, its output is always correct.

Let us now analyze the running time of this algorithm. Fix $i = 1, \ldots, r$, and consider the repeat/until loop in the ith iteration of the outer loop. Let L_i be the random variable whose value is the number of iterations of this repeat/until loop. Since α is chosen at random from $\mathbb{Z}_p^*$, the value of β is uniformly distributed over the image of the $(p-1)/q_i$-power map (see Theorem 8.5), and since the latter is a subgroup of $\mathbb{Z}_p^*$ of order q_i (see Example 7.61), we see that $\beta = 1$ with probability $1/q_i$. Thus, L_i has a geometric distribution with associated success probability $1 - 1/q_i$, and $\mathsf{E}[L_i] = 1/(1 - 1/q_i) \leq 2$ (see Theorem 9.3).

Now set $L := L_1 + \cdots + L_r$. By linearity of expectation (Theorem 8.14), we have $\mathsf{E}[L] = \mathsf{E}[L_1] + \cdots + \mathsf{E}[L_r] \leq 2r$. The running time Z of the entire algorithm is $O(L \cdot \text{len}(p)^3)$, and hence the expected running time is $\mathsf{E}[Z] = O(r\,\text{len}(p)^3)$, and since $r \leq \log_2 p$, we have $\mathsf{E}[Z] = O(\text{len}(p)^4)$.

Although this algorithm is quite practical, there are asymptotically faster algorithms for this problem (see Exercise 11.2).

EXERCISE 11.1. Suppose we are not given the prime factorization of $p - 1$, but rather, just a prime q dividing $p-1$, and we want to find an element of multiplicative order q in $\mathbb{Z}_p^*$. Design and analyze an efficient algorithm to do this.

EXERCISE 11.2. Suppose we are given a prime p, along with the prime factorization $p - 1 = \prod_{i=1}^r q_i^{e_i}$.

(a) If, in addition, we are given $\alpha \in \mathbb{Z}_p^*$, show how to compute the multiplicative order of α in time $O(r \operatorname{len}(p)^3)$. Hint: use Exercise 6.40.

(b) Improve the running time bound to $O(\operatorname{len}(r) \operatorname{len}(p)^3)$. Hint: use Exercise 3.39.

(c) Modifying the algorithm you developed for part (b), show how to construct a generator for $\mathbb{Z}_p^*$ in expected time $O(\operatorname{len}(r) \operatorname{len}(p)^3)$.

EXERCISE 11.3. Suppose we are given a positive integer n, along with its prime factorization $n = p_1^{e_1} \cdots p_r^{e_r}$, and that for each $i = 1, \ldots, r$, we are also given the prime factorization of $p_i - 1$. Show how to efficiently compute the multiplicative order of any element $\alpha \in \mathbb{Z}_n^*$.

EXERCISE 11.4. Suppose there is an efficient algorithm that takes as input a positive integer n and an element $\alpha \in \mathbb{Z}_n^*$, and computes the multiplicative order of α. Show how to use this algorithm to build an efficient integer factoring algorithm.

11.2 Computing discrete logarithms in $\mathbb{Z}_p^*$

In this section, we consider algorithms for computing the discrete logarithm of $\alpha \in \mathbb{Z}_p^*$ to a given base γ. The algorithms we present here are, in the worst case, exponential-time algorithms, and are by no means the best possible; however, in some special cases, these algorithms are not so bad.

11.2.1 Brute-force search

Suppose that $\gamma \in \mathbb{Z}_p^*$ generates a subgroup G of $\mathbb{Z}_p^*$ of order $q > 1$ (not necessarily prime), and we are given p, q, γ, and $\alpha \in G$, and wish to compute $\log_\gamma \alpha$.

The simplest algorithm to solve this problem is **brute-force search**:

$$\beta \leftarrow 1$$
$$i \leftarrow 0$$
while $\beta \neq \alpha$ do
$$\quad \beta \leftarrow \beta \cdot \gamma$$
$$\quad i \leftarrow i + 1$$
output i

This algorithm is clearly correct, and the main loop will always halt after at most q iterations (assuming, as we are, that $\alpha \in G$). So the total running time is $O(q \operatorname{len}(p)^2)$.

11.2.2 Baby step/giant step method

As above, suppose that $\gamma \in \mathbb{Z}_p^*$ generates a subgroup G of $\mathbb{Z}_p^*$ of order $q > 1$ (not necessarily prime), and we are given p, q, γ, and $\alpha \in G$, and wish to compute $\log_\gamma \alpha$.

A faster algorithm than brute-force search is the **baby step/giant step method**. It works as follows.

Let us choose an approximation m to $q^{1/2}$. It does not have to be a very good approximation — we just need $m = \Theta(q^{1/2})$. Also, let $m' = \lfloor q/m \rfloor$, so that $m' = \Theta(q^{1/2})$ as well.

The idea is to compute all the values γ^i for $i = 0, \ldots, m - 1$ (the "baby steps") and to build an "associative array" (or "lookup table") T that maps the key γ^i to the value i. For $\beta \in \mathbb{Z}_p^*$, we shall write $T[\beta]$ to denote the value associated with the key β, writing $T[\beta] = \bot$ if there is no such value. We shall assume that T is implemented so that accessing $T[\beta]$ is fast. Using an appropriate data structure, T can be implemented so that accessing individual elements takes time $O(\operatorname{len}(p))$. One such data structure is a *radix tree* (also called a *search trie*). Other data structures may be used (for example, a *hash table* or a *binary search tree*), but these may have somewhat different access times.

We can build the associative array T using the following algorithm:

> initialize T // $T[\beta] = \bot$ *for all* $\beta \in \mathbb{Z}_p^*$
> $\beta \leftarrow 1$
> for $i \leftarrow 0$ to $m - 1$ do
> $T[\beta] \leftarrow i$
> $\beta \leftarrow \beta \cdot \gamma$

Clearly, this algorithm takes time $O(q^{1/2} \operatorname{len}(p)^2)$.

After building the lookup table, we execute the following procedure (the "giant steps"):

> $\gamma' \leftarrow \gamma^{-m}$
> $\beta \leftarrow \alpha, \ j \leftarrow 0, \ i \leftarrow T[\beta]$
> while $i = \bot$ do
> $\beta \leftarrow \beta \cdot \gamma', \ j \leftarrow j + 1, \ i \leftarrow T[\beta]$
> $x \leftarrow jm + i$
> output x

To analyze this procedure, suppose that $\alpha = \gamma^x$ with $0 \le x < q$. Now, x can be written in a unique way as $x = vm + u$, where u and v are integers with $0 \le u < m$ and $0 \le v \le m'$. In the jth loop iteration, for $j = 0, 1, \ldots$, we have

$$\beta = \alpha\gamma^{-mj} = \gamma^{(v-j)m+u}.$$

So we will detect $i \ne \perp$ precisely when $j = v$, in which case $i = u$. Thus, the output will be correct, and the total running time of the algorithm (for both the "baby steps" and "giant steps" parts) is easily seen to be $O(q^{1/2} \operatorname{len}(p)^2)$.

While this algorithm is much faster than brute-force search, it has the drawback that it requires space for about $q^{1/2}$ elements of $\mathbb{Z}_p$. Of course, there is a "time/space trade-off" here: by choosing m smaller, we get a table of size $O(m)$, but the running time will be proportional to $O(q/m)$. In §11.2.5 below, we discuss an algorithm that runs (at least heuristically) in time $O(q^{1/2} \operatorname{len}(q) \operatorname{len}(p)^2)$, but which requires space for only a constant number of elements of $\mathbb{Z}_p$.

11.2.3 Groups of order q^e

Suppose that $\gamma \in \mathbb{Z}_p^*$ generates a subgroup G of $\mathbb{Z}_p^*$ of order q^e, where $q > 1$ and $e \ge 1$, and we are given p, q, e, γ, and $\alpha \in G$, and wish to compute $\log_\gamma \alpha$.

There is a simple algorithm that allows one to reduce this problem to the problem of computing discrete logarithms in the subgroup of $\mathbb{Z}_p^*$ of order q.

It is perhaps easiest to describe the algorithm recursively. The base case is when $e = 1$, in which case, we use an algorithm for the subgroup of $\mathbb{Z}_p^*$ of order q. For this, we might employ the algorithm in §11.2.2, or if q is *very* small, the algorithm in §11.2.1.

Suppose now that $e > 1$. We choose an integer f with $0 < f < e$. Different strategies for choosing f yield different algorithms—we discuss this below. Suppose $\alpha = \gamma^x$, where $0 \le x < q^e$. Then we can write $x = q^f v + u$, where u and v are integers with $0 \le u < q^f$ and $0 \le v < q^{e-f}$. Therefore,

$$\alpha^{q^{e-f}} = \gamma^{q^{e-f}u}.$$

Note that $\gamma^{q^{e-f}}$ has multiplicative order q^f, and so if we recursively compute the discrete logarithm of $\alpha^{q^{e-f}}$ to the base $\gamma^{q^{e-f}}$, we obtain u.

Having obtained u, observe that

$$\alpha/\gamma^u = \gamma^{q^f v}.$$

Note also that γ^{q^f} has multiplicative order q^{e-f}, and so if we recursively compute the discrete logarithm of α/γ^u to the base γ^{q^f}, we obtain v, from which we then compute $x = q^f v + u$.

Let us put together the above ideas succinctly in a recursive procedure:

Algorithm RDL. On input p, q, e, γ, α as above, do the following:

> if $e = 1$ then
>> return $\log_\gamma \alpha$ // *base case: use a different algorithm*
>
> else
>> select $f \in \{1, \ldots, e - 1\}$
>>
>> $u \leftarrow \text{RDL}(p, q, f, \gamma^{q^{e-f}}, \alpha^{q^{e-f}})$ // $0 \leq u < q^f$
>>
>> $v \leftarrow \text{RDL}(p, q, e - f, \gamma^{q^f}, \alpha/\gamma^u)$ // $0 \leq v < q^{e-f}$
>>
>> return $q^f v + u$

To analyze the running time of this recursive algorithm, note that the running time of the body of one recursive invocation (not counting the running time of the recursive calls it makes) is $O(e \, \text{len}(q) \, \text{len}(p)^2)$. To calculate the total running time, we have to sum up the running times of all the recursive calls plus the running times of all the base cases.

Regardless of the strategy for choosing f, the total number of base case invocations is e. Note that all the base cases compute discrete logarithms to the base $\gamma^{q^{e-1}}$. Assuming we implement the base case using the baby step/giant step algorithm in §11.2.2, the total running time for all the base cases is therefore $O(e q^{1/2} \, \text{len}(p)^2)$.

The total running time for the recursion (not including the base case computations) depends on the strategy used to choose the split f. It is helpful to represent the behavior of the algorithm using a **recursion tree**. This is a binary tree, where every node represents one recursive invocation of the algorithm; the root of the tree represents the initial invocation of the algorithm; for every node N in the tree, if N represents the recursive invocation I, then N's children (if any) represent the recursive invocations made by I. We can naturally organize the nodes of the recursion tree by **levels**: the root of the recursion tree is at level 0, its children are at level 1, its grandchildren at level 2, and so on. The **depth** of the recursion tree is defined to be the maximum level of any node.

We consider two different strategies for choosing the split f:

- If we always choose $f = 1$ or $f = e - 1$, then the depth of the recursion tree is $O(e)$. The running time contributed by each level of the recursion tree is $O(e \, \text{len}(q) \, \text{len}(p)^2)$, and so the total running time for the recursion is $O(e^2 \, \text{len}(q) \, \text{len}(p)^2)$. Note that if $f = 1$, then the algorithm is essentially tail recursive, and so may be easily converted to an iterative algorithm without the need for a stack.

- If we use a "balanced" divide-and-conquer strategy, choosing $f \approx e/2$, then the depth of the recursion tree is $O(\text{len}(e))$, while the running time contributed by each level of the recursion tree is still $O(e \operatorname{len}(q) \operatorname{len}(p)^2)$. Thus, the total running time for the recursion is $O(e \operatorname{len}(e) \operatorname{len}(q) \operatorname{len}(p)^2)$.

Assuming we use the faster, balanced recursion strategy, and that we use the baby step/giant step algorithm for the base case, the total running time of Algorithm RDL is:

$$O((eq^{1/2} + e \operatorname{len}(e) \operatorname{len}(q)) \cdot \operatorname{len}(p)^2).$$

11.2.4 Discrete logarithms in $\mathbb{Z}_p^*$

Suppose that we are given a prime p, along with the prime factorization

$$p - 1 = \prod_{i=1}^{r} q_i^{e_i},$$

a generator γ for $\mathbb{Z}_p^*$, and $\alpha \in \mathbb{Z}_p^*$. We wish to compute $\log_\gamma \alpha$.

Suppose that $\alpha = \gamma^x$, where $0 \le x < p - 1$. Then for $i = 1, \ldots, r$, we have

$$\alpha^{(p-1)/q_i^{e_i}} = \left(\gamma^{(p-1)/q_i^{e_i}} \right)^x.$$

Note that $\gamma^{(p-1)/q_i^{e_i}}$ has multiplicative order $q_i^{e_i}$, and if x_i is the discrete logarithm of $\alpha^{(p-1)/q_i^{e_i}}$ to the base $\gamma^{(p-1)/q_i^{e_i}}$, then we have $0 \le x_i < q_i^{e_i}$ and $x \equiv x_i \pmod{q_i^{e_i}}$.

Thus, if we compute the values $x_1, \ldots, x_r$, using Algorithm RDL in §11.2.3, we can obtain x using the algorithm of the Chinese remainder theorem (see Theorem 4.6). If we define $q := \max\{q_1, \ldots, q_r\}$, then the running time of this algorithm will be bounded by $q^{1/2} \operatorname{len}(p)^{O(1)}$.

We conclude that

> *the difficulty of computing discrete logarithms in* $\mathbb{Z}_p^*$ *is determined by the size of the largest prime dividing* $p - 1$.

11.2.5 A space-efficient square-root time algorithm

We present a more space-efficient alternative to the algorithm in §11.2.2, the analysis of which we leave as a series of exercises for the reader.

The algorithm makes a somewhat heuristic assumption that we have a function that "behaves" for all practical purposes like a random function. Such functions can indeed be constructed using cryptographic techniques under reasonable intractability assumptions; however, for the particular application here, one can get by in practice with much simpler constructions.

Let p be a prime, q a prime dividing $p - 1$, γ an element of $\mathbb{Z}_p^*$ that generates a subgroup G of $\mathbb{Z}_p^*$ of order q, and $\alpha \in G$. Let F be a function mapping elements of G to $\{0, \ldots, q - 1\}$. Define $H : G \to G$ to be the function that sends β to $\beta \alpha \gamma^{F(\beta)}$.

The algorithm runs as follows:

$i \leftarrow 1$
$x \leftarrow 0, \beta \leftarrow \alpha,$
$x' \leftarrow F(\beta), \beta' \leftarrow H(\beta)$
while $\beta \neq \beta'$ do
    $x \leftarrow (x + F(\beta)) \bmod q, \beta \leftarrow H(\beta)$
    repeat 2 times
        $x' \leftarrow (x' + F(\beta')) \bmod q, \beta' \leftarrow H(\beta')$
    $i \leftarrow i + 1$
if $i < q$ then
    output $(x - x')i^{-1} \bmod q$
else
    output "fail"

To analyze this algorithm, let us define $\beta_1, \beta_2, \ldots,$ as follows: $\beta_1 := \alpha$ and for $i > 1$, $\beta_i := H(\beta_{i-1})$.

EXERCISE 11.5. Show that each time the main loop of the algorithm is entered, we have $\beta = \beta_i = \gamma^x \alpha^i$, and $\beta' = \beta_{2i} = \gamma^{x'} \alpha^{2i}$.

EXERCISE 11.6. Show that if the loop terminates with $i < q$, the value output is equal to $\log_\gamma \alpha$.

EXERCISE 11.7. Let j be the smallest index such that $\beta_j = \beta_k$ for some index $k < j$. Show that $j \leq q + 1$ and that the loop terminates with $i < j$ (and in particular, $i \leq q$).

EXERCISE 11.8. Assume that F is a random function, meaning that it is chosen at random, uniformly from among all functions from G into $\{0, \ldots, q-1\}$. Show that this implies that H is a random function, meaning that it is uniformly distributed over all functions from G into G.

EXERCISE 11.9. Assuming that F is a random function as in the previous exercise, apply the result of Exercise 8.45 to conclude that the expected running time of the algorithm is $O(q^{1/2} \operatorname{len}(q) \operatorname{len}(p)^2)$, and that the probability that the algorithm fails is exponentially small in q.

11.3 The Diffie–Hellman key establishment protocol

One of the main motivations for studying algorithms for computing discrete logarithms is the relation between this problem and the problem of breaking a protocol called the **Diffie–Hellman key establishment protocol**, named after its inventors.

In this protocol, Alice and Bob need never to have talked to each other before, but nevertheless, can establish a shared secret key that nobody else can easily compute. To use this protocol, a third party must provide a "telephone book," which contains the following information:

- p, q, and γ, where p and q are primes with $q \mid (p - 1)$, and γ is an element generating a subgroup G of $\mathbb{Z}_p^*$ of order q;
- an entry for each user, such as Alice or Bob, that contains the user's name, along with a "public key" for that user, which is an element of the group G.

To use this system, Alice posts her public key in the telephone book, which is of the form $\alpha = \gamma^x$, where $x \in \{0, \ldots, q - 1\}$ is chosen by Alice at random. The value x is Alice's "secret key," which Alice never divulges to anybody. Likewise, Bob posts his public key, which is of the form $\beta = \gamma^y$, where $y \in \{0, \ldots, q - 1\}$ is chosen by Bob at random, and is his secret key.

To establish a shared key known only between them, Alice retrieves Bob's public key β from the telephone book, and computes $\kappa_A := \beta^x$. Likewise, Bob retrieves Alice's public key α, and computes $\kappa_B := \alpha^y$. It is easy to see that

$$\kappa_A = \beta^x = (\gamma^y)^x = \gamma^{xy} = (\gamma^x)^y = \alpha^y = \kappa_B,$$

and hence Alice and Bob share the same secret key $\kappa := \kappa_A = \kappa_B$.

Using this shared secret key, they can then use standard methods for encryption and message authentication to hold a secure conversation. We shall not go any further into how this is done; rather, we briefly (and only superficially) discuss some aspects of the security of the key establishment protocol itself. Clearly, if an attacker obtains α and β from the telephone book, and computes $x = \log_\gamma \alpha$, then he can compute Alice and Bob's shared key as $\kappa = \beta^x$ — in fact, given x, an attacker can efficiently compute *any* key shared between Alice and another user.

Thus, if this system is to be secure, it should be very difficult to compute discrete logarithms. However, the assumption that computing discrete logarithms is hard is not enough to guarantee security. Indeed, it is not entirely inconceivable that the discrete logarithm problem is hard, and yet the problem of computing κ from α and β is easy. The latter problem — computing κ from α and β — is called the **Diffie–Hellman problem**.

As in the discussion of the RSA cryptosystem in §4.7, the reader is warned that the above discussion about security is a bit of an oversimplification. A complete discussion of all the security issues related to the above protocol is beyond the scope of this text.

Note that in our presentation of the Diffie–Hellman protocol, we work with a generator of a subgroup G of $\mathbb{Z}_p^*$ of prime order, rather than a generator for $\mathbb{Z}_p^*$. There are several reasons for doing this: one is that there are no known discrete logarithm algorithms that are any more practical in G than in $\mathbb{Z}_p^*$, provided the order q of G is sufficiently large; another is that by working in G, the protocol becomes substantially more efficient. In typical implementations, p is 1024 bits long, so as to protect against subexponential-time algorithms such as those discussed later in §15.2, while q is 160 bits long, which is enough to protect against the square-root-time algorithms discussed in §11.2.2 and §11.2.5. The modular exponentiations in the protocol will run several times faster using "short," 160-bit exponents rather than "long," 1024-bit exponents.

For the following exercise, we need the following notions from complexity theory.

- We say problem A is **deterministic poly-time reducible** to problem B if there exists a deterministic algorithm R for solving problem A that makes calls to a subroutine for problem B, where the running time of R (not including the running time for the subroutine for B) is polynomial in the input length.

- We say that problems A and B are **deterministic poly-time equivalent** if A is deterministic poly-time reducible to B and B is deterministic poly-time reducible to A.

EXERCISE 11.10. Consider the following problems.

(a) Given a prime p, a prime q that divides $p - 1$, an element $\gamma \in \mathbb{Z}_p^*$ generating a subgroup G of $\mathbb{Z}_p^*$ of order q, and two elements $\alpha, \beta \in G$, compute γ^{xy}, where $x := \log_\gamma \alpha$ and $y := \log_\gamma \beta$. (This is just the Diffie–Hellman problem.)

(b) Given a prime p, a prime q that divides $p - 1$, an element $\gamma \in \mathbb{Z}_p^*$ generating a subgroup G of $\mathbb{Z}_p^*$ of order q, and an element $\alpha \in G$, compute γ^{x^2}, where $x := \log_\gamma \alpha$.

(c) Given a prime p, a prime q that divides $p - 1$, an element $\gamma \in \mathbb{Z}_p^*$ generating a subgroup G of $\mathbb{Z}_p^*$ of order q, and two elements $\alpha, \beta \in G$, with $\beta \neq 1$, compute $\gamma^{xy'}$, where $x := \log_\gamma \alpha$, $y' := y^{-1} \bmod q$, and $y := \log_\gamma \beta$.

(d) Given a prime p, a prime q that divides $p - 1$, an element $\gamma \in \mathbb{Z}_p^*$ generating a subgroup G of $\mathbb{Z}_p^*$ of order q, and an element $\alpha \in G$, with $\alpha \neq 1$, compute $\gamma^{x'}$, where $x' := x^{-1} \bmod q$ and $x := \log_\gamma \alpha$.

Show that these problems are deterministic poly-time equivalent. Moreover, your reductions should preserve the values of p, q, and γ; that is, if the algorithm that reduces one problem to another takes as input an instance of the former problem of the form $(p, q, \gamma, \ldots)$, it should invoke the subroutine for the latter problem with inputs of the form $(p, q, \gamma, \ldots)$.

EXERCISE 11.11. Suppose there is a probabilistic algorithm A that takes as input a prime p, a prime q that divides $p - 1$, and an element $\gamma \in \mathbb{Z}_p^*$ generating a subgroup G of $\mathbb{Z}_p^*$ of order q. The algorithm also takes as input $\alpha \in G$. It outputs either "failure," or $\log_\gamma \alpha$. Furthermore, assume that A runs in expected polynomial time, and that for all p, q, and γ of the above form, and for randomly chosen $\alpha \in G$, A succeeds in computing $\log_\gamma \alpha$ with probability $\varepsilon(p, q, \gamma)$. Here, the probability is taken over the random choice of α, as well as the random choices made during the execution of A. Show how to use A to construct another probabilistic algorithm A' that takes as input p, q, and γ as above, as well as $\alpha \in G$, runs in expected polynomial time, and that satisfies the following property:

if $\varepsilon(p, q, \gamma) \geq 0.001$, then *for all $\alpha \in G$, A'* computes $\log_\gamma \alpha$ with probability at least 0.999.

The algorithm A' in the previous exercise is an example of a **random self-reduction**, which means an algorithm that reduces the task of solving an *arbitrary* instance of a given problem to that of solving a *random* instance of the *same* problem. Intuitively, the existence of such a reduction means that the problem is no harder in the worst case than on average.

EXERCISE 11.12. Let p be a prime, q a prime that divides $p - 1$, $\gamma \in \mathbb{Z}_p^*$ an element that generates a subgroup G of $\mathbb{Z}_p^*$ of order q, and $\alpha \in G$. For $\delta \in G$, a **representation of δ with respect to γ and α** is a pair of integers (r, s), with $0 \leq r < q$ and $0 \leq s < q$, such that $\gamma^r \alpha^s = \delta$.

(a) Show that for every $\delta \in G$, there are precisely q representations (r, s) of δ with respect to γ and α, and among these, there is precisely one with $s = 0$.

(b) Show that given a representation (r, s) of 1 with respect to γ and α such that $s \neq 0$, we can efficiently compute $\log_\gamma \alpha$.

(c) Show that given any $\delta \in G$, along with any two distinct representations of δ with respect to γ and α, we can efficiently compute $\log_\gamma \alpha$.

(d) Suppose we are given access to an "oracle" that, when presented with any $\delta \in G$, tells us some representation of δ with respect to γ and α. Show how to use this oracle to efficiently compute $\log_\gamma \alpha$.

The following two exercises examine the danger of the use of "short" exponents in discrete logarithm based cryptographic schemes that *do not* work with a group of prime order.

EXERCISE 11.13. Let p be a prime and let $p - 1 = q_1^{e_1} \cdots q_r^{e_r}$ be the prime factorization of $p - 1$. Let γ be a generator for $\mathbb{Z}_p^*$. Let y be a positive number, and let $Q_p(y)$ be the product of all the prime powers $q_i^{e_i}$ with $q_i \leq y$. Suppose you are given p, y, the primes q_i dividing $p - 1$ with $q_i \leq y$, along with γ, an element α of $\mathbb{Z}_p^*$, and a bound $\hat{x}$, where $x := \log_\gamma \alpha < \hat{x}$. Show how to compute x in time

$$(y^{1/2} + (\hat{x}/Q_p(y))^{1/2}) \cdot \text{len}(p)^{O(1)}.$$

EXERCISE 11.14. Continuing with the previous, let $Q_p'(y)$ denote the product of all the primes q_i dividing $p - 1$ with $q_i \leq y$. Note that $Q_p'(y) \mid Q_p(y)$. The goal of this exercise is to estimate the expected value of $\log Q_p'(y)$, assuming p is a large, random prime. To this end, let R be a random variable that is uniformly distributed over all ℓ-bit primes, and assume that $y \leq 2^{\ell/3}$. Assuming Conjecture 5.22, show that asymptotically (as $\ell \to \infty$), we have $\mathsf{E}[\log Q_R'(y)] = \log y + O(1)$.

The results of the previous two exercises caution against the use of "short" exponents in cryptographic schemes based on the discrete logarithm problem for $\mathbb{Z}_p^*$. For example, suppose that p is a random 1024-bit prime, and that for reasons of efficiency, one chooses $\hat{x} \approx 2^{160}$, thinking that a method such as the baby step/giant step method would require $\approx 2^{80}$ steps to recover x. However, if we choose $y \approx 2^{80}$, then the above analysis implies that $Q_p(y)$ is at least $\approx 2^{80}$ with a reasonable probability, in which case $\hat{x}/Q_p(y)$ is at most $\approx 2^{80}$, and so we can in fact recover x in $\approx 2^{40}$ steps (there are known methods to find the primes up to y that divide $p - 1$ quickly enough). While 2^{80} may not be a feasible number of steps, 2^{40} may very well be. Of course, none of these issues arise if one works in a subgroup of $\mathbb{Z}_p^*$ of large prime order, which is the recommended practice.

An interesting fact about the Diffie–Hellman problem is that there is no known efficient algorithm to recognize a solution to the problem. Some cryptographic protocols actually rely on the apparent difficulty of this decision problem, which is called the **decisional Diffie–Hellman problem**. The following three exercises develop a random self-reducibility property for this decision problem.

EXERCISE 11.15. Let p be a prime, q a prime dividing $p - 1$, and γ an element of $\mathbb{Z}_p^*$ that generates a subgroup G of order q. Let $\alpha \in G$, and let H be the subgroup of $G \times G$ generated by (γ, α). Let $\tilde{\gamma}, \tilde{\alpha}$ be arbitrary elements of G, and define the map

$$\rho: \quad \mathbb{Z}_q \times \mathbb{Z}_q \to G \times G$$
$$([r]_q, [s]_q) \mapsto (\gamma^r \tilde{\gamma}^s, \alpha^r \tilde{\alpha}^s).$$

Show that the definition of ρ is unambiguous, that ρ is a group homomorphism, and that

- if $(\tilde{\gamma}, \tilde{\alpha}) \in H$, then $\text{Im } \rho = H$, and
- if $(\tilde{\gamma}, \tilde{\alpha}) \notin H$, then $\text{Im } \rho = G \times G$.

EXERCISE 11.16. For p, q, γ as in the previous exercise, let $\mathcal{D}_{p,q,\gamma}$ be the set of all triples of the form $(\gamma^x, \gamma^y, \gamma^{xy})$, and let $\mathcal{R}_{p,q,\gamma}$ be the set of all triples of the form $(\gamma^x, \gamma^y, \gamma^z)$. Using the result from the previous exercise, design a probabilistic algorithm that runs in expected polynomial time, and that on input p, q, γ, along with a triple $\Gamma \in \mathcal{R}_{p,q,\gamma}$, outputs a triple $\Gamma^* \in \mathcal{R}_{p,q,\gamma}$ such that

- if $\Gamma \in \mathcal{D}_{p,q,\gamma}$, then Γ^* is uniformly distributed over $\mathcal{D}_{p,q,\gamma}$, and
- if $\Gamma \notin \mathcal{D}_{p,q,\gamma}$, then Γ^* is uniformly distributed over $\mathcal{R}_{p,q,\gamma}$.

EXERCISE 11.17. Suppose that A is a probabilistic algorithm that takes as input p, q, γ as in the previous exercise, along with a triple $\Gamma^* \in \mathcal{R}_{p,q,\gamma}$, and outputs either 0 or 1. Furthermore, assume that A runs in expected polynomial time. Define two random variables, $X_{p,q,\gamma}$ and $Y_{p,q,\gamma}$, as follows:

- $X_{p,q,\gamma}$ is defined to be the output of A on input p, q, γ, and Γ^*, where Γ^* is uniformly distributed over $\mathcal{D}_{p,q,\gamma}$, and
- $Y_{p,q,\gamma}$ is defined to be the output of A on input p, q, γ, and Γ^*, where Γ^* is uniformly distributed over $\mathcal{R}_{p,q,\gamma}$.

In both cases, the value of the random variable is determined by the random choice of Γ^*, as well as the random choices made by the algorithm. Define

$$\varepsilon(p, q, \gamma) := \left| P[X_{p,q,\gamma} = 1] - P[Y_{p,q,\gamma} = 1] \right|.$$

Using the result of the previous exercise, show how to use A to design a probabilistic, expected polynomial-time algorithm that takes as input p, q, γ as above, along with $\Gamma \in \mathcal{R}_{p,q,\gamma}$, and outputs either "yes" or "no," so that

if $\varepsilon(p, q, \gamma) \geq 0.001$, then for *all* $\Gamma \in \mathcal{R}_{p,q,\gamma}$, the probability that A' correctly determines whether $\Gamma \in \mathcal{D}_{p,q,\gamma}$ is at least 0.999.

Hint: use the Chernoff bound.

The following exercise demonstrates that the problem of distinguishing "Diffie–Hellman triples" from "random triples" is hard only if the order of the underlying group is not divisible by any small primes, which is another reason we have chosen to work with groups of large prime order.

EXERCISE 11.18. Assume the notation of the previous exercise, but let us drop the restriction that q is prime. Design and analyze a deterministic algorithm A that takes inputs p, q, γ and $\Gamma^* \in \mathcal{R}_{p,q,\gamma}$, that outputs 0 or 1, and that satisfies the following property: if t is the *smallest* prime dividing q, then A runs in time $(t + \text{len}(p))^{O(1)}$, and the "distinguishing advantage" $\varepsilon(p, q, \gamma)$ for A on inputs p, q, γ is at least $1/t$.

11.4 Notes

The probabilistic algorithm in §11.1 for finding a generator for $\mathbb{Z}_p^*$ can be made deterministic under a generalization of the Riemann hypothesis. Indeed, as discussed in §10.5, under such a hypothesis, Bach's result [10] implies that for each prime $q \mid (p-1)$, the least positive integer a such that $[a]_p \in \mathbb{Z}_p^* \setminus (\mathbb{Z}_p^*)^q$ is at most $2 \log p$.

Related to the problem of constructing a generator for $\mathbb{Z}_p^*$ is the question of how big is the smallest positive integer g such that $[g]_p$ is a generator for $\mathbb{Z}_p^*$; that is, how big is the smallest (positive) primitive root modulo p. The best bounds on the least primitive root are also obtained using the same generalization of the Riemann hypothesis mentioned above. Under this hypothesis, Wang [104] showed that the least primitive root modulo p is $O(r^6 \operatorname{len}(p)^2)$, where r is the number of distinct prime divisors of $p-1$. Shoup [95] improved Wang's bound to $O(r^4 \operatorname{len}(r)^4 \operatorname{len}(p)^2)$ by adapting a result of Iwaniec [50, 51] and applying it to Wang's proof. The best unconditional bound on the smallest primitive root modulo p is $p^{1/4+o(1)}$ (this bound is also in Wang [104]). Of course, even if there exists a small primitive root, there is no known way to efficiently recognize a primitive root modulo p without knowing the prime factorization of $p-1$.

As we already mentioned, all of the algorithms presented in this chapter are completely "generic," in the sense that they work in *any* finite cyclic group—we really did not exploit any properties of $\mathbb{Z}_p^*$ other than the fact that it is a cyclic group. In fact, as far as such "generic" algorithms go, the algorithms presented here for discrete logarithms are optimal [71, 98]. However, there are faster, "non-generic" algorithms (though still not polynomial time) for discrete logarithms in $\mathbb{Z}_p^*$. We shall examine one such algorithm later, in Chapter 15.

The "baby step/giant step" algorithm in §11.2.2 is due to Shanks [91]. See, for example, the book by Cormen, Leiserson, Rivest, and Stein [29] for appropriate data structures to implement the lookup table used in that algorithm. In particular, see Problem 12-2 in [29] for a brief introduction to radix trees, which is the data structure that yields the best running time (at least in principle) for our application.

The algorithms in §11.2.3 and §11.2.4 are variants of an algorithm published by Pohlig and Hellman [75]. See Chapter 4 of [29] for details on how one analyzes recursive algorithms, such as the one presented in §11.2.3; in particular, Section 4.2 in [29] discusses in detail the notion of a recursion tree.

The algorithm in §11.2.5 is a variant of an algorithm of Pollard [76]; in fact, Pollard's algorithm is a bit more efficient than the one presented here, but the analysis of its running time depends on stronger heuristics. Pollard's paper also describes an algorithm for computing discrete logarithms that lie in a restricted interval—if the interval has width w, this algorithm runs (heuristically) in time

$w^{1/2} \operatorname{len}(p)^{O(1)}$, and requires space for $O(\operatorname{len}(w))$ elements of $\mathbb{Z}_p$. This algorithm is useful in reducing the space requirement for the algorithm of Exercise 11.13.

The key establishment protocol in §11.3 is from Diffie and Hellman [34]. That paper initiated the study of **public key cryptography**, which has proved to be a very rich field of research. Exercises 11.13 and 11.14 are based on van Oorschot and Wiener [74]. For more on the decisional Diffie–Hellman assumption, see Boneh [18].

12

Quadratic reciprocity and computing modular square roots

In §2.8, we initiated an investigation of quadratic residues. This chapter continues this investigation. Recall that an integer a is called a quadratic residue modulo a positive integer n if $\gcd(a, n) = 1$ and $a \equiv b^2 \pmod{n}$ for some integer b.

First, we derive the famous law of quadratic reciprocity. This law, while historically important for reasons of pure mathematical interest, also has important computational applications, including a fast algorithm for testing if an integer is a quadratic residue modulo a prime.

Second, we investigate the problem of computing modular square roots: given a quadratic residue a modulo n, compute an integer b such that $a \equiv b^2 \pmod{n}$. As we will see, there are efficient probabilistic algorithms for this problem when n is prime, and more generally, when the factorization of n into primes is known.

12.1 The Legendre symbol

For an odd prime p and an integer a with $\gcd(a, p) = 1$, the **Legendre symbol** $(a \mid p)$ is defined to be 1 if a is a quadratic residue modulo p, and -1 otherwise. For completeness, one defines $(a \mid p) = 0$ if $p \mid a$. The following theorem summarizes the essential properties of the Legendre symbol.

Theorem 12.1. *Let p be an odd prime, and let $a, b \in \mathbb{Z}$. Then we have:*

(i) $(a \mid p) \equiv a^{(p-1)/2} \pmod{p}$; *in particular,* $(-1 \mid p) = (-1)^{(p-1)/2}$;

(ii) $(a \mid p)(b \mid p) = (ab \mid p)$;

(iii) $a \equiv b \pmod{p}$ *implies* $(a \mid p) = (b \mid p)$;

(iv) $(2 \mid p) = (-1)^{(p^2-1)/8}$;

(v) *if q is an odd prime, then* $(p \mid q) = (-1)^{\frac{p-1}{2}\frac{q-1}{2}}(q \mid p)$.

Part (i) of the theorem is just a restatement of Euler's criterion (Theorem 2.21).

As was observed in Theorem 2.31, this implies that -1 is a quadratic residue modulo p if and only if $p \equiv 1 \pmod 4$. Thus, the quadratic residuosity of -1 modulo p is determined by the residue class of p modulo 4.

Part (ii) of the theorem follows immediately from part (i), and part (iii) is an immediate consequence of the definition of the Legendre symbol.

Part (iv), which we will prove below, can also be recast as saying that 2 is a quadratic residue modulo p if and only if $p \equiv \pm 1 \pmod 8$. Thus, the quadratic residuosity of 2 modulo p is determined by the residue class of p modulo 8.

Part (v), which we will also prove below, is the **law of quadratic reciprocity**. Note that when $p = q$, both $(p \mid q)$ and $(q \mid p)$ are zero, and so the statement of part (v) is trivially true — the interesting case is when $p \neq q$, and in this case, part (v) is equivalent to saying that

$$(p \mid q)(q \mid p) = (-1)^{\frac{p-1}{2}\frac{q-1}{2}}.$$

Thus, the Legendre symbols $(p \mid q)$ and $(q \mid p)$ have the same values if and only if either $p \equiv 1 \pmod 4$ or $q \equiv 1 \pmod 4$. As the following examples illustrate, this result also shows that for a given odd prime q, the quadratic residuosity of q modulo another odd prime p is determined by the residue class of p modulo either q or $4q$.

Example 12.1. Let us characterize those primes p modulo which 5 is a quadratic residue. Since $5 \equiv 1 \pmod 4$, the law of quadratic reciprocity tells us that $(5 \mid p) = (p \mid 5)$. Now, among the numbers $\pm 1, \pm 2$, the quadratic residues modulo 5 are ± 1. It follows that 5 is a quadratic residue modulo p if and only if $p \equiv \pm 1 \pmod 5$. This example obviously generalizes, replacing 5 by any prime $q \equiv 1 \pmod 4$, and replacing the above congruences modulo 5 by appropriate congruences modulo q. □

Example 12.2. Let us characterize those primes p modulo which 3 is a quadratic residue. Since $3 \not\equiv 1 \pmod 4$, we must be careful in our application of the law of quadratic reciprocity. First, suppose that $p \equiv 1 \pmod 4$. Then $(3 \mid p) = (p \mid 3)$, and so 3 is a quadratic residue modulo p if and only if $p \equiv 1 \pmod 3$. Second, suppose that $p \not\equiv 1 \pmod 4$. Then $(3 \mid p) = -(p \mid 3)$, and so 3 is a quadratic residue modulo p if and only if $p \equiv -1 \pmod 3$. Putting this all together, we see that 3 is quadratic residue modulo p if and only if

$$p \equiv 1 \pmod 4 \text{ and } p \equiv 1 \pmod 3$$

or

$$p \equiv -1 \pmod 4 \text{ and } p \equiv -1 \pmod 3.$$

Using the Chinese remainder theorem, we can restate this criterion in terms of residue classes modulo 12: 3 is quadratic residue modulo p if and only if $p \equiv \pm 1 \pmod{12}$. This example obviously generalizes, replacing 3 by any prime $q \equiv -1 \pmod{4}$, and replacing the above congruences modulo 12 by appropriate congruences modulo $4q$. $\square$

The rest of this section is devoted to a proof of parts (iv) and (v) of Theorem 12.1. The proof is completely elementary, although a bit technical.

Theorem 12.2 (Gauss' lemma). *Let p be an odd prime and let a be an integer not divisible by p. Define $\alpha_j := ja \bmod p$ for $j = 1, \ldots, (p-1)/2$, and let n be the number of indices j for which $\alpha_j > p/2$. Then $(a \mid p) = (-1)^n$.*

Proof. Let $r_1, \ldots, r_n$ denote the values α_j that exceed $p/2$, and let $s_1, \ldots, s_k$ denote the remaining values α_j. The r_i's and s_i's are all distinct and non-zero. We have $0 < p - r_i < p/2$ for $i = 1, \ldots, n$, and no $p - r_i$ is an s_j; indeed, if $p - r_i = s_j$, then $s_j \equiv -r_i \pmod{p}$, and writing $s_j = ua \bmod p$ and $r_i = va \bmod p$, for some $u, v = 1, \ldots, (p-1)/2$, we have $ua \equiv -va \pmod{p}$, which implies $u \equiv -v \pmod{p}$, which is impossible.

It follows that the sequence of numbers $s_1, \ldots, s_k, p - r_1, \ldots, p - r_n$ is just a reordering of $1, \ldots, (p-1)/2$. Then we have

$$((p-1)/2)! \equiv s_1 \cdots s_k (-r_1) \cdots (-r_n)$$
$$\equiv (-1)^n s_1 \cdots s_k r_1 \cdots r_n$$
$$\equiv (-1)^n ((p-1)/2)! \, a^{(p-1)/2} \pmod{p},$$

and canceling the factor $((p-1)/2)!$, we obtain $a^{(p-1)/2} \equiv (-1)^n \pmod{p}$, and the result follows from the fact that $(a \mid p) \equiv a^{(p-1)/2} \pmod{p}$. $\square$

Theorem 12.3. *If p is an odd prime and $\gcd(a, 2p) = 1$, then $(a \mid p) = (-1)^t$ where $t = \sum_{j=1}^{(p-1)/2} \lfloor ja/p \rfloor$. Also, $(2 \mid p) = (-1)^{(p^2-1)/8}$.*

Proof. Let a be an integer not divisible by p, but which may be even, and let us adopt the same notation as in the statement and proof of Theorem 12.2; in particular, $\alpha_1, \ldots, \alpha_{(p-1)/2}$, $r_1, \ldots, r_n$, and $s_1, \ldots, s_k$ are as defined there. Note that $ja = p \lfloor ja/p \rfloor + \alpha_j$, for $j = 1, \ldots, (p-1)/2$, so we have

$$\sum_{j=1}^{(p-1)/2} ja = \sum_{j=1}^{(p-1)/2} p \lfloor ja/p \rfloor + \sum_{j=1}^{n} r_j + \sum_{j=1}^{k} s_j. \tag{12.1}$$

Moreover, as we saw in the proof of Theorem 12.2, the sequence of numbers

$s_1, \ldots, s_k, p - r_1, \ldots, p - r_n$ is a reordering of $1, \ldots, (p-1)/2$, and hence

$$\sum_{j=1}^{(p-1)/2} j = \sum_{j=1}^{n}(p - r_j) + \sum_{j=1}^{k} s_j = np - \sum_{j=1}^{n} r_j + \sum_{j=1}^{k} s_j. \qquad (12.2)$$

Subtracting (12.2) from (12.1), we get

$$(a - 1) \sum_{j=1}^{(p-1)/2} j = p \left(\sum_{j=1}^{(p-1)/2} \lfloor ja/p \rfloor - n \right) + 2 \sum_{j=1}^{n} r_j. \qquad (12.3)$$

Note that

$$\sum_{j=1}^{(p-1)/2} j = \frac{p^2 - 1}{8}, \qquad (12.4)$$

which together with (12.3) implies

$$(a - 1)\frac{p^2 - 1}{8} \equiv \sum_{j=1}^{(p-1)/2} \lfloor ja/p \rfloor - n \ (\text{mod } 2). \qquad (12.5)$$

If a is odd, (12.5) implies

$$n \equiv \sum_{j=1}^{(p-1)/2} \lfloor ja/p \rfloor \ (\text{mod } 2). \qquad (12.6)$$

If $a = 2$, then $\lfloor 2j/p \rfloor = 0$ for $j = 1, \ldots, (p-1)/2$, and (12.5) implies

$$n \equiv \frac{p^2 - 1}{8} \ (\text{mod } 2). \qquad (12.7)$$

The theorem now follows from (12.6) and (12.7), together with Theorem 12.2. $\square$

Note that this last theorem proves part (iv) of Theorem 12.1. The next theorem proves part (v).

Theorem 12.4. *If p and q are distinct odd primes, then*

$$(p \mid q)(q \mid p) = (-1)^{\frac{p-1}{2} \frac{q-1}{2}}.$$

Proof. Let S be the set of pairs of integers (x, y) with $1 \leq x \leq (p-1)/2$ and $1 \leq y \leq (q-1)/2$. Note that S contains no pair (x, y) with $qx = py$, so let us partition S into two subsets: S_1 contains all pairs (x, y) with $qx > py$, and S_2 contains all pairs (x, y) with $qx < py$. Note that $(x, y) \in S_1$ if and only if

$1 \le x \le (p-1)/2$ and $1 \le y \le \lfloor qx/p \rfloor$. So $|S_1| = \sum_{x=1}^{(p-1)/2} \lfloor qx/p \rfloor$. Similarly, $|S_2| = \sum_{y=1}^{(q-1)/2} \lfloor py/q \rfloor$. So we have

$$\frac{p-1}{2}\frac{q-1}{2} = |S| = |S_1| + |S_2| = \sum_{x=1}^{(p-1)/2} \lfloor qx/p \rfloor + \sum_{y=1}^{(q-1)/2} \lfloor py/q \rfloor,$$

and Theorem 12.3 implies

$$(p \mid q)(q \mid p) = (-1)^{\frac{p-1}{2}\frac{q-1}{2}}. \quad \square$$

EXERCISE 12.1. Characterize those odd primes p for which $(15 \mid p) = 1$, in terms of the residue class of p modulo 60.

EXERCISE 12.2. Let p be an odd prime. Show that the following are equivalent:

(a) $(-2 \mid p) = 1$;

(b) $p \equiv 1$ or $3 \pmod 8$;

(c) $p = r^2 + 2t^2$ for some $r, t \in \mathbb{Z}$.

12.2 The Jacobi symbol

Let a, n be integers, where n is positive and odd, so that $n = q_1 \cdots q_k$, where the q_i's are odd primes, not necessarily distinct. Then the **Jacobi symbol** $(a \mid n)$ is defined as

$$(a \mid n) := (a \mid q_1) \cdots (a \mid q_k),$$

where $(a \mid q_i)$ is the Legendre symbol. By definition, $(a \mid 1) = 1$ for all $a \in \mathbb{Z}$. Thus, the Jacobi symbol essentially extends the domain of definition of the Legendre symbol. Note that $(a \mid n) \in \{0, \pm 1\}$, and that $(a \mid n) = 0$ if and only if $\gcd(a, n) > 1$. The following theorem summarizes the essential properties of the Jacobi symbol.

Theorem 12.5. *Let m, n be odd, positive integers, and let $a, b \in \mathbb{Z}$. Then we have:*

(i) $(ab \mid n) = (a \mid n)(b \mid n)$;

(ii) $(a \mid mn) = (a \mid m)(a \mid n)$;

(iii) $a \equiv b \pmod n$ *implies* $(a \mid n) = (b \mid n)$;

(iv) $(-1 \mid n) = (-1)^{(n-1)/2}$;

(v) $(2 \mid n) = (-1)^{(n^2-1)/8}$;

(vi) $(m \mid n) = (-1)^{\frac{m-1}{2}\frac{n-1}{2}}(n \mid m)$.

Proof. Parts (i)–(iii) follow directly from the definition (exercise).

For parts (iv) and (vi), one can easily verify (exercise) that for all odd integers $n_1, \ldots, n_k$,

$$\sum_{i=1}^{k} (n_i - 1)/2 \equiv (n_1 \cdots n_k - 1)/2 \pmod{2}.$$

Part (iv) easily follows from this fact, along with part (ii) of this theorem and part (i) of Theorem 12.1 (exercise). Part (vi) easily follows from this fact, along with parts (i) and (ii) of this theorem, and part (v) of Theorem 12.1 (exercise).

For part (v), one can easily verify (exercise) that for odd integers $n_1, \ldots, n_k$,

$$\sum_{i=1}^{k} (n_i^2 - 1)/8 \equiv (n_1^2 \cdots n_k^2 - 1)/8 \pmod{2}.$$

Part (v) easily follows from this fact, along with part (ii) of this theorem, and part (iv) of Theorem 12.1 (exercise). □

As we shall see later, this theorem is extremely useful from a computational point of view — with it, one can efficiently compute $(a \mid n)$, without having to know the prime factorization of either a or n. Also, in applying this theorem it is useful to observe that for all odd integers m, n,

- $(-1)^{(n-1)/2} = 1 \iff n \equiv 1 \pmod{4}$;
- $(-1)^{(n^2-1)/8} = 1 \iff n \equiv \pm 1 \pmod{8}$;
- $(-1)^{((m-1)/2)((n-1)/2)} = 1 \iff m \equiv 1 \pmod{4}$ or $n \equiv 1 \pmod{4}$.

Suppose a is a quadratic residue modulo n, so that $a \equiv b^2 \pmod{n}$, where $\gcd(a, n) = 1 = \gcd(b, n)$. Then by parts (iii) and (i) of Theorem 12.5, we have $(a \mid n) = (b^2 \mid n) = (b \mid n)^2 = 1$. Thus, if a is a quadratic residue modulo n, then $(a \mid n) = 1$. The converse, however, does not hold: $(a \mid n) = 1$ does *not* imply that a is a quadratic residue modulo n (see Exercise 12.3 below).

It is sometimes useful to view the Jacobi symbol as a group homomorphism. Let n be an odd, positive integer. Define the **Jacobi map**

$$J_n : \quad \mathbb{Z}_n^* \to \{\pm 1\}$$
$$[a]_n \mapsto (a \mid n).$$

First, we note that by part (iii) of Theorem 12.5, this definition is unambiguous. Second, we note that since $\gcd(a, n) = 1$ implies $(a \mid n) = \pm 1$, the image of J_n is indeed contained in $\{\pm 1\}$. Third, we note that by part (i) of Theorem 12.5, J_n is a group homomorphism. Since J_n is a group homomorphism, it follows that its kernel, Ker J_n, is a subgroup of $\mathbb{Z}_n^*$.

EXERCISE 12.3. Let n be an odd, positive integer, and consider the Jacobi map J_n.

    (a) Show that $(\mathbb{Z}_n^*)^2 \subseteq \operatorname{Ker} J_n$.

    (b) Show that if n is the square of an integer, then $\operatorname{Ker} J_n = \mathbb{Z}_n^*$.

    (c) Show that if n is not the square of an integer, then $[\mathbb{Z}_n^* : \operatorname{Ker} J_n] = 2$ and $[\operatorname{Ker} J_n : (\mathbb{Z}_n^*)^2] = 2^{r-1}$, where r is the number of distinct prime divisors of n.

EXERCISE 12.4. Let p and q be distinct primes, with $p \equiv q \equiv 3 \pmod 4$, and let $n := pq$.

    (a) Show that $[-1]_n \in \operatorname{Ker} J_n \setminus (\mathbb{Z}_n^*)^2$, and from this, conclude that the cosets of $(\mathbb{Z}_n^*)^2$ in $\operatorname{Ker} J_n$ are the two distinct cosets $(\mathbb{Z}_n^*)^2$ and $[-1]_n (\mathbb{Z}_n^*)^2$.

    (b) Let $\delta \in \mathbb{Z}_n^* \setminus \operatorname{Ker} J_n$. Show that the map from $\{0, 1\} \times \{0, 1\} \times (\mathbb{Z}_n^*)^2$ to $\mathbb{Z}_n^*$ that sends (a, b, γ) to $\delta^a (-1)^b \gamma$ is a bijection.

12.3 Computing the Jacobi symbol

Suppose we are given an odd, positive integer n, along with an integer a, and we want to compute the Jacobi symbol $(a \mid n)$. Theorem 12.5 suggests the following algorithm:

$\sigma \leftarrow 1$
repeat
    // loop invariant: n is odd and positive

    $a \leftarrow a \bmod n$
    if $a = 0$ then
        if $n = 1$ then return σ else return 0

    compute a', h such that $a = 2^h a'$ and a' is odd
    if $h \not\equiv 0 \pmod 2$ and $n \not\equiv \pm 1 \pmod 8$ then $\sigma \leftarrow -\sigma$
    if $a' \not\equiv 1 \pmod 4$ and $n \not\equiv 1 \pmod 4$ then $\sigma \leftarrow -\sigma$
    $(a, n) \leftarrow (n, a')$
forever

That this algorithm correctly computes the Jacobi symbol $(a \mid n)$ follows directly from Theorem 12.5. Using an analysis similar to that of Euclid's algorithm, one easily sees that the running time of this algorithm is $O(\operatorname{len}(a) \operatorname{len}(n))$.

EXERCISE 12.5. Develop a "binary" Jacobi symbol algorithm, that is, one that uses only addition, subtractions, and "shift" operations, analogous to the binary gcd algorithm in Exercise 4.6.

EXERCISE 12.6. This exercise develops a probabilistic primality test based on the Jacobi symbol. For odd integer $n > 1$, define

$$G_n := \{\alpha \in \mathbb{Z}_n^* : \alpha^{(n-1)/2} = J_n(\alpha)\},$$

where $J_n : \mathbb{Z}_n^* \to \{\pm 1\}$ is the Jacobi map.

(a) Show that G_n is a subgroup of $\mathbb{Z}_n^*$.

(b) Show that if n is prime, then $G_n = \mathbb{Z}_n^*$.

(c) Show that if n is composite, then $G_n \subsetneq \mathbb{Z}_n^*$.

(d) Based on parts (a)–(c), design and analyze an efficient probabilistic primality test that works by choosing a random, non-zero element $\alpha \in \mathbb{Z}_n$, and testing if $\alpha \in G_n$.

12.4 Testing quadratic residuosity

In this section, we consider the problem of testing whether a is a quadratic residue modulo n, for given integers a and n, from a computational perspective.

12.4.1 Prime modulus

For an odd prime p, we can test if an integer a is a quadratic residue modulo p by either performing the exponentiation $a^{(p-1)/2} \bmod p$ or by computing the Legendre symbol $(a \mid p)$. Assume that $0 \le a < p$. Using a standard repeated squaring algorithm, the former method takes time $O(\mathrm{len}(p)^3)$, while using the Euclidean-like algorithm of the previous section, the latter method takes time $O(\mathrm{len}(p)^2)$. So clearly, the latter method is to be preferred.

12.4.2 Prime-power modulus

For an odd prime p, we know that a is a quadratic residue modulo p^e if and only if a is a quadratic residue modulo p (see Theorem 2.30). So this case immediately reduces to the previous one.

12.4.3 Composite modulus

For odd, composite n, if we know the factorization of n, then we can also determine if a is a quadratic residue modulo n by determining if it is a quadratic residue modulo each prime divisor p of n (see Exercise 2.39). However, without knowledge of this factorization (which is in general believed to be hard to compute), there is no efficient algorithm known. We can compute the Jacobi symbol $(a \mid n)$; if this is -1 or 0, we can conclude that a is not a quadratic residue; otherwise, we cannot conclude much of anything.

12.5 Computing modular square roots

In this section, we consider the problem of computing a square root of a modulo n, given integers a and n, where a is a quadratic residue modulo n.

12.5.1 Prime modulus

Let p be an odd prime, and let a be an integer such that $0 < a < p$ and $(a \mid p) = 1$. We would like to compute a square root of a modulo p. Let $\alpha := [a]_p \in \mathbb{Z}_p^*$, so that we can restate our problem as that of finding $\beta \in \mathbb{Z}_p^*$ such that $\beta^2 = \alpha$, given $\alpha \in (\mathbb{Z}_p^*)^2$.

We first consider the special case where $p \equiv 3 \pmod 4$, in which it turns out that this problem can be solved very easily. Indeed, we claim that in this case

$$\beta := \alpha^{(p+1)/4}$$

is a square root of α—note that since $p \equiv 3 \pmod 4$, the number $(p + 1)/4$ is an integer. To show that $\beta^2 = \alpha$, suppose $\alpha = \tilde{\beta}^2$ for some $\tilde{\beta} \in \mathbb{Z}_p^*$. We know that there is such a $\tilde{\beta}$, since we are assuming that $\alpha \in (\mathbb{Z}_p^*)^2$. Then we have

$$\beta^2 = \alpha^{(p+1)/2} = \tilde{\beta}^{p+1} = \tilde{\beta}^2 = \alpha,$$

where we used Fermat's little theorem for the third equality. Using a repeated-squaring algorithm, we can compute β in time $O(\mathrm{len}(p)^3)$.

Now we consider the general case, where we may have $p \not\equiv 3 \pmod 4$. Here is one way to efficiently compute a square root of α, assuming we are given, in addition to α, an auxiliary input $\gamma \in \mathbb{Z}_p^* \setminus (\mathbb{Z}_p^*)^2$ (how one obtains such a γ is discussed below).

Let us write $p - 1 = 2^h m$, where m is odd. For every $\delta \in \mathbb{Z}_p^*$, δ^m has multiplicative order dividing 2^h. Since $\alpha^{2^{h-1}m} = 1$, α^m has multiplicative order dividing 2^{h-1}. Since $\gamma^{2^{h-1}m} = -1$, γ^m has multiplicative order precisely 2^h. Since there is only one subgroup of $\mathbb{Z}_p^*$ of order 2^h, it follows that γ^m generates this subgroup, and that $\alpha^m = \gamma^{mx}$ for some integer x, where $0 \le x < 2^h$ and x is even. We can find x by computing the discrete logarithm of α^m to the base γ^m, using the algorithm in §11.2.3. Setting $\kappa = \gamma^{mx/2}$, we have

$$\kappa^2 = \alpha^m.$$

We are not quite done, since we now have a square root of α^m, and not of α.

Since m is odd, we may write $m = 2t + 1$ for some non-negative integer t. It then follows that

$$(\kappa\alpha^{-t})^2 = \kappa^2\alpha^{-2t} = \alpha^m\alpha^{-2t} = \alpha^{m-2t} = \alpha.$$

Thus, $\kappa\alpha^{-t}$ is a square root of α.

Let us summarize the above algorithm for computing a square root of $\alpha \in (\mathbb{Z}_p^*)^2$, assuming we are given $\gamma \in \mathbb{Z}_p^* \setminus (\mathbb{Z}_p^*)^2$, in addition to α:

> compute positive integers m, h such that $p - 1 = 2^h m$ with m odd
> $\gamma' \leftarrow \gamma^m, \alpha' \leftarrow \alpha^m$
> compute $x \leftarrow \log_{\gamma'} \alpha'$ // *note that $0 \le x < 2^h$ and x is even*
> $\beta \leftarrow (\gamma')^{x/2}\alpha^{-\lfloor m/2 \rfloor}$
> output β

The work done outside the discrete logarithm calculation amounts to just a handful of exponentiations modulo p, and so takes time $O(\text{len}(p)^3)$. The time to compute the discrete logarithm is $O(h\,\text{len}(h)\,\text{len}(p)^2)$. So the total running time of this procedure is

$$O(\text{len}(p)^3 + h\,\text{len}(h)\,\text{len}(p)^2).$$

The above procedure assumed we had at hand a non-square γ. If $h = 1$, which means that $p \equiv 3 \pmod 4$, then $(-1 \mid p) = -1$, and so we are done. However, we have already seen how to efficiently compute a square root in this case.

If $h > 1$, we can find a non-square γ using a probabilistic search algorithm. Simply choose γ at random, test if it is a square, and if so, repeat. The probability that a random element of $\mathbb{Z}_p^*$ is a square is $1/2$; thus, the expected number of trials until we find a non-square is 2; moreover, the running time per trial is $O(\text{len}(p)^2)$, and hence the expected running time of this probabilistic search algorithm is $O(\text{len}(p)^2)$.

12.5.2 Prime-power modulus

Let p be an odd prime, let a be an integer relatively prime to p, and let $e > 1$ be an integer. We know that a is a quadratic residue modulo p^e if and only if a is a quadratic residue modulo p. Suppose that a is a quadratic residue modulo p, and that we have found an integer b such that $b^2 \equiv a \pmod p$, using, say, one of the procedures described in §12.5.1. From this, we can easily compute a square root of a modulo p^e using the following technique, which is known as **Hensel lifting**.

More generally, suppose that for some $f \ge 1$, we have computed an integer b satisfying the congruence $b^2 \equiv a \pmod{p^f}$, and we want to find an integer c satisfying the congruence $c^2 \equiv a \pmod{p^{f+1}}$. Clearly, if $c^2 \equiv a \pmod{p^{f+1}}$, then

$c^2 \equiv a \pmod{p^f}$, and so $c \equiv \pm b \pmod{p^f}$. So let us set $c = b + p^f h$, and solve for h. We have

$$c^2 \equiv (b + p^f h)^2 \equiv b^2 + 2bp^f h + p^{2f} h^2 \equiv b^2 + 2bp^f h \pmod{p^{f+1}}.$$

So we want to find an integer h satisfying the linear congruence

$$2bp^f h \equiv a - b^2 \pmod{p^{f+1}}. \tag{12.8}$$

Since $p \nmid 2b$, we have $\gcd(2bp^f, p^{f+1}) = p^f$. Furthermore, since $b^2 \equiv a \pmod{p^f}$, we have $p^f \mid (a - b^2)$. Therefore, Theorem 2.5 implies that (12.8) has a unique solution h modulo p, which we can efficiently compute as in Example 4.3.

By iterating the above procedure, starting with a square root of a modulo p, we can quickly find a square root of a modulo p^e. We leave a detailed analysis of the running time of this procedure to the reader.

12.5.3 Composite modulus

To find square roots modulo n, where n is an odd composite modulus, if we know the prime factorization of n, then we can use the above procedures for finding square roots modulo primes and prime powers, and then use the algorithm of the Chinese remainder theorem to get a square root modulo n.

However, if the factorization of n is not known, then there is no efficient algorithm known for computing square roots modulo n. In fact, one can show that the problem of finding square roots modulo n is at least as hard as the problem of factoring n, in the sense that if there is an efficient algorithm for computing square roots modulo n, then there is an efficient (probabilistic) algorithm for factoring n.

We now present an algorithm to factor n, using a modular square-root algorithm A as a subroutine. For simplicity, we assume that A is deterministic, and that for all n and for all $\alpha \in (\mathbb{Z}_n^*)^2$, $A(n, \alpha)$ outputs a square root of α. Also for simplicity, we shall assume that n is of the form $n = pq$, where p and q are distinct, odd primes. In Exercise 12.15 below, you are asked to relax these restrictions. Our algorithm runs as follows:

$$\beta \xleftarrow{\text{\$}} \mathbb{Z}_n^+, d \leftarrow \gcd(\text{rep}(\beta), n)$$
if $d > 1$ then
      output d
else
      $\alpha \leftarrow \beta^2, \beta' \leftarrow A(n, \alpha)$
      if $\beta = \pm\beta'$
            then output "failure"
            else output $\gcd(\text{rep}(\beta - \beta'), n)$

Here, $\mathbb{Z}_n^+$ denotes the set of non-zero elements of $\mathbb{Z}_n$. Also, recall that $\operatorname{rep}(\beta)$ denotes the canonical representative of β.

First, we argue that the algorithm outputs either "failure" or a non-trivial factor of n. Clearly, if $\beta \notin \mathbb{Z}_n^*$, then the value d computed by the algorithm is a non-trivial factor. So suppose $\beta \in \mathbb{Z}_n^*$. In this case, the algorithm invokes A on inputs n and $\alpha := \beta^2$, obtaining a square root β' of α. Suppose that $\beta \neq \pm\beta'$, and set $\gamma := \beta - \beta'$. What we need to show is that $\gcd(\operatorname{rep}(\gamma), n)$ is a non-trivial factor of n. To see this, consider the ring isomorphism of the Chinese remainder theorem

$$\theta : \quad \mathbb{Z}_n \to \mathbb{Z}_p \times \mathbb{Z}_q$$
$$[a]_n \mapsto ([a]_p, [a]_q).$$

Suppose $\theta(\beta') = (\beta_1', \beta_2')$. Then the four square roots of α are

$$\beta' = \theta^{-1}(\beta_1', \beta_2'), \ -\beta' = \theta^{-1}(-\beta_1', -\beta_2'), \ \theta^{-1}(-\beta_1', \beta_2'), \ \theta^{-1}(\beta_1', -\beta_2').$$

The assumption that $\beta \neq \pm\beta'$ implies that $\theta(\beta) = (-\beta_1', \beta_2')$ or $\theta(\beta) = (\beta_1', -\beta_2')$. In the first case, $\theta(\gamma) = (-2\beta_1', 0)$, which implies $\gcd(\operatorname{rep}(\gamma), n) = q$. In the second case, $\theta(\gamma) = (0, -2\beta_2')$, which implies $\gcd(\operatorname{rep}(\gamma), n) = p$.

Second, we argue that $\mathsf{P}[\mathcal{F}] \leq 1/2$, where $\mathcal{F}$ is the event that the algorithm outputs "failure." Viewed as a random variable, β is uniformly distributed over $\mathbb{Z}_n^+$. Clearly, $\mathsf{P}[\mathcal{F} \mid \beta \notin \mathbb{Z}_n^*] = 0$. Now consider any fixed $\alpha' \in (\mathbb{Z}_n^*)^2$. Observe that the conditional distribution of β given that $\beta^2 = \alpha'$ is (essentially) the uniform distribution on the set of four square roots of α'. Also observe that the output of A depends only on n and β^2, and so with respect to the conditional distribution given that $\beta^2 = \alpha'$, the output β' of A is fixed. Thus,

$$\mathsf{P}[\mathcal{F} \mid \beta^2 = \alpha'] = \mathsf{P}[\beta = \pm\beta' \mid \beta^2 = \alpha'] = 1/2.$$

Putting everything together, using total probability, we have

$$\mathsf{P}[\mathcal{F}] = \mathsf{P}[\mathcal{F} \mid \beta \notin \mathbb{Z}_n^*] \mathsf{P}[\beta \notin \mathbb{Z}_n^*] + \sum_{\alpha' \in (\mathbb{Z}_n^*)^2} \mathsf{P}[\mathcal{F} \mid \beta^2 = \alpha'] \mathsf{P}[\beta^2 = \alpha']$$

$$= 0 \cdot \mathsf{P}[\beta \notin \mathbb{Z}_n^*] + \sum_{\alpha' \in (\mathbb{Z}_n^*)^2} \frac{1}{2} \cdot \mathsf{P}[\beta^2 = \alpha'] \leq \frac{1}{2}.$$

Thus, the above algorithm fails to split n with probability at most $1/2$. If we like, we can repeat the algorithm until it succeeds. The expected number of iterations performed will be at most 2.

EXERCISE 12.7. Let p be an odd prime, and let $f \in \mathbb{Z}_p[X]$ be a polynomial with $0 \leq \deg(f) \leq 2$. Design and analyze an efficient, deterministic algorithm that takes as input p, f, and an element of $\mathbb{Z}_p^* \setminus (\mathbb{Z}_p^*)^2$, and which determines if f has any roots in $\mathbb{Z}_p$, and if so, finds all of the roots. Hint: see Exercise 7.17.

EXERCISE 12.8. Show how to deterministically compute square roots modulo primes $p \equiv 5 \pmod 8$ in time $O(\text{len}(p)^3)$.

EXERCISE 12.9. This exercise develops an alternative algorithm for computing square roots modulo a prime. Let p be an odd prime, let $\beta \in \mathbb{Z}_p^*$, and set $\alpha := \beta^2$. Define $B_\alpha := \{\gamma \in \mathbb{Z}_p : \gamma^2 - \alpha \in (\mathbb{Z}_p^*)^2\}$.

(a) Show that $B_\alpha = \{\gamma \in \mathbb{Z}_p : g(\gamma) = 0\}$, where

$$g := (X - \beta)^{(p-1)/2} - (X + \beta)^{(p-1)/2} \in \mathbb{Z}_p[X].$$

(b) Let $\gamma \in \mathbb{Z}_p \setminus B_\alpha$, and suppose $\gamma^2 \neq \alpha$. Let μ, v be the uniquely determined elements of $\mathbb{Z}_p$ satisfying the polynomial congruence

$$\mu + vX \equiv (\gamma - X)^{(p-1)/2} \pmod{X^2 - \alpha}.$$

Show that $\mu = 0$ and $v^{-2} = \alpha$.

(c) Using parts (a) and (b), design and analyze a probabilistic algorithm that computes a square root of a given $\alpha \in (\mathbb{Z}_p^*)^2$ in expected time $O(\text{len}(p)^3)$.

Note that when $p - 1 = 2^h m$ (m odd), and h is large (e.g., $h \approx \text{len}(p)/2$), the algorithm in the previous exercise is asymptotically faster than the one in §12.5.1; however, the latter algorithm is likely to be faster in practice for the typical case where h is small.

EXERCISE 12.10. Show that the following two problems are deterministic, poly-time equivalent (see discussion just above Exercise 11.10 in §11.3):

(a) Given an odd prime p and $\alpha \in (\mathbb{Z}_p^*)^2$, find $\beta \in \mathbb{Z}_p^*$ such that $\beta^2 = \alpha$.

(b) Given an odd prime p, find an element of $\mathbb{Z}_p^* \setminus (\mathbb{Z}_p^*)^2$.

EXERCISE 12.11. Design and analyze an efficient, deterministic algorithm that takes as input primes p and q, such that $q \mid (p - 1)$, along with an element $\alpha \in \mathbb{Z}_p^*$, and determines whether or not $\alpha \in (\mathbb{Z}_p^*)^q$.

EXERCISE 12.12. Design and analyze an efficient, deterministic algorithm that takes as input primes p and q, such that $q \mid (p - 1)$ but $q^2 \nmid (p - 1)$, along with an element $\alpha \in (\mathbb{Z}_p^*)^q$, and computes a qth root of α, that is, an element $\beta \in \mathbb{Z}_p^*$ such that $\beta^q = \alpha$.

EXERCISE 12.13. Design and analyze an algorithm that takes as input primes p and q, such that $q \mid (p - 1)$, along with an element $\alpha \in (\mathbb{Z}_p^*)^q$, and computes a qth root of α. (Unlike Exercise 12.12, we now allow $q^2 \mid (p-1)$.) Your algorithm may be probabilistic, and should have an expected running time that is bounded by $q^{1/2}$ times a polynomial in $\text{len}(p)$. Hint: Exercise 4.13 may be useful.

EXERCISE 12.14. Let p be an odd prime, γ be a generator for $\mathbb{Z}_p^*$, and α be any element of $\mathbb{Z}_p^*$. Define

$$B(p, \gamma, \alpha) := \begin{cases} 1 & \text{if } \log_\gamma \alpha \geq (p-1)/2; \\ 0 & \text{if } \log_\gamma \alpha < (p-1)/2. \end{cases}$$

Suppose that there is an algorithm that efficiently computes $B(p, \gamma, \alpha)$ for all p, γ, α as above. Show how to use this algorithm as a subroutine in an efficient, probabilistic algorithm that computes $\log_\gamma \alpha$ for all p, γ, α as above. Hint: in addition to the algorithm that computes B, use algorithms for testing quadratic residuosity and computing square roots modulo p, and "read off" the bits of $\log_\gamma \alpha$ one at a time.

EXERCISE 12.15. Suppose there is a probabilistic algorithm A that takes as input a positive integer n, and an element $\alpha \in (\mathbb{Z}_n^*)^2$. Assume that for all n, and for a randomly chosen $\alpha \in (\mathbb{Z}_n^*)^2$, A computes a square root of α with probability at least 0.001. Here, the probability is taken over the random choice of α and the random choices of A. Show how to use A to construct another probabilistic algorithm A' that takes n as input, runs in expected polynomial time, and that satisfies the following property:

for all n, A' outputs the complete factorization of n into primes with probability at least 0.999.

EXERCISE 12.16. Suppose there is a probabilistic algorithm A that takes as input positive integers n and m, and an element $\alpha \in (\mathbb{Z}_n^*)^m$. It outputs either "failure," or an mth root of α. Furthermore, assume that A runs in expected polynomial time, and that for all n and m, and for randomly chosen $\alpha \in (\mathbb{Z}_n^*)^m$, A succeeds in computing an mth root of α with probability $\varepsilon(n, m)$. Here, the probability is taken over the random choice of α, as well as the random choices made during the execution of A. Show how to use A to construct another probabilistic algorithm A' that takes as input n, m, and $\alpha \in (\mathbb{Z}_n^*)^m$, runs in expected polynomial time, and that satisfies the following property:

if $\varepsilon(n, m) \geq 0.001$, then *for all* $\alpha \in (\mathbb{Z}_n^*)^m$, A' computes an mth root of α with probability at least 0.999.

12.6 The quadratic residuosity assumption

Loosely speaking, the **quadratic residuosity (QR)** assumption is the assumption that it is hard to distinguish squares from non-squares in $\mathbb{Z}_n^*$, where n is of the form

$n = pq$, and p and q are distinct primes. This assumption plays an important role in cryptography. Of course, since the Jacobi symbol is easy to compute, for this assumption to make sense, we have to restrict our attention to elements of $\mathrm{Ker}\,J_n$, where $J_n : \mathbb{Z}_n^* \to \{\pm 1\}$ is the Jacobi map. We know that $(\mathbb{Z}_n^*)^2 \subseteq \mathrm{Ker}\,J_n$ (see Exercise 12.3). Somewhat more precisely, the QR assumption is the assumption that it is hard to distinguish a random element in $\mathrm{Ker}\,J_n \setminus (\mathbb{Z}_n^*)^2$ from a random element in $(\mathbb{Z}_n^*)^2$, given n (but not its factorization!).

To give a rough idea as to how this assumption may be used in cryptography, assume that $p \equiv q \equiv 3 \pmod 4$, so that $[-1]_n \in \mathrm{Ker}\,J_n \setminus (\mathbb{Z}_n^*)^2$, and moreover, $\mathrm{Ker}\,J_n \setminus (\mathbb{Z}_n^*)^2 = [-1]_n(\mathbb{Z}_n^*)^2$ (see Exercise 12.4). The value n can be used as a public key in a public-key cryptosystem (see §4.7). Alice, knowing the public key, can encrypt a single bit $b \in \{0, 1\}$ as $\beta := (-1)^b \alpha^2$, where Alice chooses $\alpha \in \mathbb{Z}_n^*$ at random. The point is, if $b = 0$, then β is uniformly distributed over $(\mathbb{Z}_n^*)^2$, and if $b = 1$, then β is uniformly distributed over $\mathrm{Ker}\,J_n \setminus (\mathbb{Z}_n^*)^2$. Now Bob, knowing the secret key, which is the factorization of n, can easily determine if $\beta \in (\mathbb{Z}_n^*)^2$ or not, and hence deduce the value of the encrypted bit b. However, under the QR assumption, an eavesdropper, seeing just n and β, cannot effectively figure out what b is.

Of course, the above scheme is much less efficient than the RSA cryptosystem presented in §4.7, but nevertheless, has attractive properties; in particular, its security is very closely tied to the QR assumption, whereas the security of RSA is a bit less well understood.

EXERCISE 12.17. Suppose that A is a probabilistic algorithm that takes as input n of the form $n = pq$, where p and q are distinct primes such that $p \equiv q \equiv 3 \pmod 4$. The algorithm also takes as input $\alpha \in \mathrm{Ker}\,J_n$, and outputs either 0 or 1. Furthermore, assume that A runs in expected polynomial time. Define two random variables, X_n and Y_n, as follows: X_n is defined to be the output of A on input n and a value α chosen at random from $\mathrm{Ker}\,J_n \setminus (\mathbb{Z}_n^*)^2$, and Y_n is defined to be the output of A on input n and a value α chosen at random from $(\mathbb{Z}_n^*)^2$. In both cases, the value of the random variable is determined by the random choice of α, as well as the random choices made by the algorithm. Define $\varepsilon(n) := |P[X_n = 1] - P[Y_n = 1]|$. Show how to use A to design a probabilistic, expected polynomial time algorithm A' that takes as input n as above and $\alpha \in \mathrm{Ker}\,J_n$, and outputs either "square" or "non-square," with the following property:

> if $\varepsilon(n) \geq 0.001$, then *for all* $\alpha \in \mathrm{Ker}\,J_n$, the probability that A' correctly identifies whether $\alpha \in (\mathbb{Z}_n^*)^2$ is at least 0.999.

Hint: use the Chernoff bound.

EXERCISE 12.18. Assume the same notation as in the previous exercise. Define the random variable X'_n to be the output of A on input n and a value α chosen at random from $\operatorname{Ker} J_n$. Show that $|P[X'_n = 1] - P[Y_n = 1]| = \varepsilon(n)/2$. Thus, the problem of distinguishing $\operatorname{Ker} J_n$ from $(\mathbb{Z}_n^*)^2$ is essentially equivalent to the problem of distinguishing $\operatorname{Ker} J_n \setminus (\mathbb{Z}_n^*)^2$ from $(\mathbb{Z}_n^*)^2$.

12.7 Notes

The proof we present here of Theorem 12.1 is essentially the one from Niven and Zuckerman [72]. Our proof of Theorem 12.5 follows closely the one found in Bach and Shallit [11].

Exercise 12.6 is based on Solovay and Strassen [99].

The probabilistic algorithm in §12.5.1 can be made deterministic under a generalization of the Riemann hypothesis. Indeed, as discussed in §10.5, under such a hypothesis, Bach's result [10] implies that the least positive integer that is not a quadratic residue modulo p is at most $2 \log p$ (this follows by applying Bach's result with the subgroup $(\mathbb{Z}_p^*)^2$ of $\mathbb{Z}_p^*$). Thus, we may find the required element $\gamma \in \mathbb{Z}_p^* \setminus (\mathbb{Z}_n^*)^2$ in deterministic polynomial time, just by brute-force search. The best *unconditional* bound on the smallest positive integer that is not a quadratic residue modulo p is due to Burgess [22], who gives a bound of $p^{\alpha+o(1)}$, where $\alpha := 1/(4\sqrt{e}) \approx 0.15163$.

Goldwasser and Micali [41] introduced the quadratic residuosity assumption to cryptography (as discussed in §12.6). This assumption has subsequently been used as the basis for numerous cryptographic schemes.

13

Modules and vector spaces

In this chapter, we introduce the basic definitions and results concerning modules over a ring R and vector spaces over a field F. The reader may have seen some of these notions before, but perhaps only in the context of vector spaces over a specific field, such as the real or complex numbers, and not in the context of, say, finite fields like $\mathbb{Z}_p$.

13.1 Definitions, basic properties, and examples

Throughout this section, R denotes a ring (i.e., a commutative ring with unity).

Definition 13.1. *An **R-module** is a set M together with an addition operation on M and a function $\mu : R \times M \rightarrow M$, such that the set M under addition forms an abelian group, and moreover, for all $c, d \in R$ and $\alpha, \beta \in M$, we have:*

(i) $\mu(c, \mu(d, \alpha)) = \mu(cd, \alpha)$;

(ii) $\mu(c + d, \alpha) = \mu(c, \alpha) + \mu(d, \alpha)$;

(iii) $\mu(c, \alpha + \beta) = \mu(c, \alpha) + \mu(c, \beta)$;

(iv) $\mu(1_R, \alpha) = \alpha$.

One may also call an R-module M a **module over** R, and elements of R are sometimes called **scalars**. The function μ in the definition is called a **scalar multiplication map**, and the value $\mu(c, \alpha)$ is called the **scalar product** of c and α. Usually, we shall simply write $c\alpha$ (or $c \cdot \alpha$) instead of $\mu(c, \alpha)$. When we do this, properties (i)–(iv) of the definition may be written as follows:

$$c(d\alpha) = (cd)\alpha, \quad (c + d)\alpha = c\alpha + d\alpha, \quad c(\alpha + \beta) = c\alpha + c\beta, \quad 1_R\alpha = \alpha.$$

Note that there are two addition operations at play here: addition in R (such as $c + d$) and addition in M (such as $\alpha + \beta$). Likewise, there are two multiplication operations at play: multiplication in R (such as cd) and scalar multiplication (such as $c\alpha$). Note that by property (i), we may write $cd\alpha$ without any ambiguity, as both possible interpretations, $c(d\alpha)$ and $(cd)\alpha$, yield the same value.

For fixed $c \in R$, the map that sends $\alpha \in M$ to $c\alpha \in M$ is a group homomorphism with respect to the additive group operation of M (by property (iii) of the definition); likewise, for fixed $\alpha \in M$, the map that sends $c \in R$ to $c\alpha \in M$ is a group homomorphism from the additive group of R into the additive group of M (by property (ii)). Combining these observations with basic facts about group homomorphisms (see Theorem 6.19), we may easily derive the following basic facts about R-modules:

Theorem 13.2. *If M is a module over R, then for all $c \in R$, $\alpha \in M$, and $k \in \mathbb{Z}$, we have:*

(i) $0_R \cdot \alpha = 0_M$;

(ii) $c \cdot 0_M = 0_M$;

(iii) $(-c)\alpha = -(c\alpha) = c(-\alpha)$;

(iv) $(kc)\alpha = k(c\alpha) = c(k\alpha)$.

Proof. Exercise. □

An R-module M may be **trivial**, consisting of just the zero element 0_M. If R is the trivial ring, then any R-module M is trivial, since for every $\alpha \in M$, we have $\alpha = 1_R\alpha = 0_R\alpha = 0_M$.

Example 13.1. The ring R itself can be viewed as an R-module in the obvious way, with addition and scalar multiplication defined in terms of the addition and multiplication operations of R. □

Example 13.2. The set $R^{\times n}$, which consists of all of n-tuples of elements of R, forms an R-module, with addition and scalar multiplication defined component-wise: for $\alpha = (a_1, \ldots, a_n) \in R^{\times n}$, $\beta = (b_1, \ldots, b_n) \in R^{\times n}$, and $c \in R$, we define

$$\alpha + \beta := (a_1 + b_1, \ldots, a_n + b_n) \text{ and } c\alpha := (ca_1, \ldots, ca_n). \square$$

Example 13.3. The ring of polynomials $R[X]$ over R forms an R-module in the natural way, with addition and scalar multiplication defined in terms of the addition and multiplication operations of the polynomial ring. □

Example 13.4. As in Example 7.39, let f be a non-zero polynomial over R with $\mathrm{lc}(f) \in R^*$, and consider the quotient ring $E := R[X]/(f)$. Then E is a module over R, with addition defined in terms of the addition operation of E, and scalar multiplication defined by $c[g]_f := [c]_f \cdot [g]_f = [cg]_f$, for $c \in R$ and $g \in R[X]$. □

Example 13.5. Generalizing Example 13.3, if E is any ring containing R as a subring (i.e., E is an extension ring of R), then E is a module over R, with addition and scalar multiplication defined in terms of the addition and multiplication operations of E. □

Example 13.6. Any abelian group G, written additively, can be viewed as a $\mathbb{Z}$-module, with scalar multiplication defined in terms of the usual integer multiplication map (see Theorem 6.4). □

Example 13.7. Let G be any group, written additively, whose exponent divides n. Then we may define a scalar multiplication that maps $[k]_n \in \mathbb{Z}_n$ and $\alpha \in G$ to $k\alpha$. That this map is unambiguously defined follows from the fact that G has exponent dividing n, so that if $k \equiv k' \pmod{n}$, we have $k\alpha - k'\alpha = (k - k')\alpha = 0_G$, since $n \mid (k - k')$. It is easy to check that this scalar multiplication map indeed makes G into a $\mathbb{Z}_n$-module. □

Example 13.8. Of course, viewing a group as a module does not depend on whether or not we happen to use additive notation for the group operation. If we specialize the previous example to the group $G = \mathbb{Z}_p^*$, where p is prime, then we may view G as a $\mathbb{Z}_{p-1}$-module. However, since the group operation itself is written multiplicatively, the "scalar product" of $[k]_{p-1} \in \mathbb{Z}_{p-1}$ and $\alpha \in \mathbb{Z}_p^*$ is the power α^k. □

Example 13.9. If $M_1, \ldots, M_k$ are R-modules, then so is their direct product $M_1 \times \cdots \times M_k$, where addition and scalar product are defined component-wise. If $M = M_1 = \cdots = M_k$, we write this as $M^{\times k}$. □

Example 13.10. If I is an arbitrary set, and M is an R-module, then $\text{Map}(I, M)$, which is the set of all functions $f : I \to M$, may be naturally viewed as an R-module, with point-wise addition and scalar multiplication: for $f, g \in \text{Map}(I, M)$ and $c \in R$, we define

$$(f + g)(i) := f(i) + g(i) \text{ and } (cf)(i) := cf(i) \text{ for all } i \in I. \square$$

13.2 Submodules and quotient modules

Again, throughout this section, R denotes a ring. The notions of subgroups and quotient groups extend in the obvious way to R-modules.

Definition 13.3. *Let M be an R-module. A subset N of M is a* **submodule (over R) of** *M if*

    (i) N is a subgroup of the additive group M, and

    (ii) $c\alpha \in N$ for all $c \in R$ and $\alpha \in N$ (i.e., N is closed under scalar multiplication).

It is easy to see that a submodule N of an R-module M is also an R-module in its own right, with addition and scalar multiplication operations inherited from M.

Expanding the above definition, we see that a non-empty subset N of M is a submodule if and only if for all $c \in R$ and all $\alpha, \beta \in N$, we have

$$\alpha + \beta \in N, \quad -\alpha \in N, \quad \text{and} \quad c\alpha \in N.$$

Observe that the condition $-\alpha \in N$ is redundant, as it is implied by the condition $c\alpha \in N$ with $c = -1_R$.

Clearly, $\{0_M\}$ and M are submodules of M. For $k \in \mathbb{Z}$, it is easy to see that not only are kM and $M\{k\}$ subgroups of M (see Theorems 6.7 and 6.8), they are also submodules of M. Moreover, for $c \in R$,

$$cM := \{c\alpha : \alpha \in M\} \quad \text{and} \quad M\{c\} := \{\alpha \in M : c\alpha = 0_M\}$$

are also submodules of M. Further, for $\alpha \in M$,

$$R\alpha := \{c\alpha : c \in R\}$$

is a submodule of M. Finally, if N_1 and N_2 are submodules of M, then $N_1 + N_2$ and $N_1 \cap N_2$ are not only subgroups of M, they are also submodules of M. We leave it to the reader to verify all these facts: they are quite straightforward.

Let $\alpha_1, \ldots, \alpha_k \in M$. The submodule

$$R\alpha_1 + \cdots + R\alpha_k$$

is called the **submodule (over R) generated by** $\alpha_1, \ldots, \alpha_k$. It consists of all **$R$-linear combinations**

$$c_1\alpha_1 + \cdots + c_k\alpha_k,$$

where the c_i's are elements of R, and is the smallest submodule of M that contains the elements $\alpha_1, \ldots, \alpha_k$. We shall also write this submodule as $\langle \alpha_1, \ldots, \alpha_k \rangle_R$. As a matter of definition, we allow $k = 0$, in which case this submodule is $\{0_M\}$. We say that M is **finitely generated (over R)** if $M = \langle \alpha_1, \ldots, \alpha_k \rangle_R$ for some $\alpha_1, \ldots, \alpha_k \in M$.

Example 13.11. For a given integer $\ell \geq 0$, define $R[X]_{<\ell}$ to be the set of polynomials of degree less than ℓ. The reader may verify that $R[X]_{<\ell}$ is a submodule of the R-module $R[X]$, and indeed, is the submodule generated by $1, X, \ldots, X^{\ell-1}$. If $\ell = 0$, then this submodule is the trivial submodule $\{0_R\}$. $\square$

Example 13.12. Let G be an abelian group. As in Example 13.6, we can view G as a $\mathbb{Z}$-module in a natural way. Subgroups of G are just the same thing as submodules of G, and for $a_1, \ldots, a_k \in G$, the subgroup $\langle a_1, \ldots, a_k \rangle$ is the same as the submodule $\langle a_1, \ldots, a_k \rangle_\mathbb{Z}$. $\square$

Example 13.13. As in Example 13.1, we may view the ring R itself as an R-module. With respect to this module structure, ideals of R are just the same thing as submodules of R, and for $a_1, \ldots, a_k \in R$, the ideal $(a_1, \ldots, a_k)$ is the same as the submodule $\langle a_1, \ldots, a_k \rangle_R$. Note that for $a \in R$, the ideal generated by a may be written either as aR, using the notation introduced in §7.3, or as Ra, using the notation introduced in this section. □

Example 13.14. If E is an extension ring of R, then we may view E as an R-module, as in Example 13.5. It is easy to see that every ideal of E is a submodule; however, the converse is not true in general. Indeed, the submodule $R[X]_{<\ell}$ of $R[X]$ discussed in Example 13.11 is not an ideal of the ring $R[X]$. □

If N is a submodule of M, then in particular, it is also a subgroup of M, and we can form the quotient group M/N in the usual way (see §6.3), which consists of all cosets $[\alpha]_N$, where $\alpha \in M$. Moreover, because N is closed under scalar multiplication, we can also define a scalar multiplication on M/N in a natural way. Namely, for $c \in R$ and $\alpha \in M$, we define

$$c \cdot [\alpha]_N := [c\alpha]_N.$$

As usual, one must check that this definition is unambiguous, which means that $c\alpha \equiv c\alpha' \pmod{N}$ whenever $\alpha \equiv \alpha' \pmod{N}$. But this follows (as the reader may verify) from the fact that N is closed under scalar multiplication. One can also easily check that with scalar multiplication defined in this way, M/N is an R-module; it is called the **quotient module (over R) of M modulo N**.

Example 13.15. Suppose E is an extension ring of R, and I is an ideal of E. Viewing E as an R-module, I is a submodule of E, and hence the quotient ring E/I may naturally be viewed as an R-module, with scalar multiplication defined by $c \cdot [\alpha]_I := [c\alpha]_I$ for $c \in R$ and $\alpha \in E$. Example 13.4 is a special case of this, applied to the extension ring $R[X]$ and the ideal (f). □

EXERCISE 13.1. Show that if N is a submodule of an R-module M, then a set $P \subseteq N$ is a submodule of M if and only if P is a submodule of N.

EXERCISE 13.2. Let M_1 and M_2 be R-modules, and let N_1 be a submodule of M_1 and N_2 a submodule of M_2. Show that $N_1 \times N_2$ is a submodule of $M_1 \times M_2$.

EXERCISE 13.3. Show that if R is non-trivial, then the R-module $R[X]$ is not finitely generated.

13.3 Module homomorphisms and isomorphisms

Again, throughout this section, R is a ring. The notion of a group homomorphism extends in the obvious way to R-modules.

Definition 13.4. *Let M and M' be modules over R. An **R-module homomorphism** from M to M' is a function $\rho : M \to M'$, such that*

   (i) *ρ is a group homomorphism from M to M', and*

   (ii) *$\rho(c\alpha) = c\rho(\alpha)$ for all $c \in R$ and $\alpha \in M$.*

An R-module homomorphism is also called an **R-linear map**. We shall use this terminology from now on. Expanding the definition, we see that a map $\rho : M \to M'$ is an R-linear map if and only if $\rho(\alpha + \beta) = \rho(\alpha) + \rho(\beta)$ and $\rho(c\alpha) = c\rho(\alpha)$ for all $\alpha, \beta \in M$ and all $c \in R$.

Example 13.16. If N is a submodule of an R-module M, then the inclusion map $i : N \to M$ is obviously an R-linear map. $\square$

Example 13.17. Suppose N is a submodule of an R-module M. Then the *natural map* (see Example 6.36)

$$\rho : \quad M \to M/N$$
$$\alpha \mapsto [\alpha]_N$$

is not just a group homomorphism, it is also easily seen to be an R-linear map. $\square$

Example 13.18. Let M be an R-module, and let k be an integer. Then the k-multiplication map on M (see Example 6.38) is not only a group homomorphism, but it is also easily seen to be an R-linear map. Its image is the submodule kM, and its kernel the submodule $M\{k\}$. $\square$

Example 13.19. Let M be an R-module, and let c be an element of R. The map

$$\rho : \quad M \to M$$
$$\alpha \mapsto c\alpha$$

is called c-**multiplication map on** M, and is easily seen to be an R-linear map whose image is the submodule cM, and whose kernel is the submodule $M\{c\}$. The set of all $c \in R$ for which $cM = \{0_M\}$ is called the **R-exponent of** M, and is easily seen to be an ideal of R. $\square$

Example 13.20. Let M be an R-module, and let α be an element of M. The map

$$\rho : \quad R \to M$$
$$c \mapsto c\alpha$$

is easily seen to be an R-linear map whose image is the submodule $R\alpha$ (i.e., the submodule generated by α). The kernel of this map is called the **R-order of** α, and is easily seen to be an ideal of R. $\square$

Example 13.21. Generalizing the previous example, let M be an R-module, and let $\alpha_1, \ldots, \alpha_k$ be elements of M. The map

$$\rho: \qquad R^{\times k} \to M$$
$$(c_1, \ldots, c_k) \mapsto c_1\alpha_1 + \cdots + c_k\alpha_k$$

is easily seen to be an R-linear map whose image is the submodule $R\alpha_1 + \cdots + R\alpha_k$ (i.e., the submodule generated by $\alpha_1, \ldots, \alpha_k$). $\square$

Example 13.22. Suppose that $M_1, \ldots, M_k$ are submodules of an R-module M. Then the map

$$\rho: \quad M_1 \times \cdots \times M_k \to M$$
$$(\alpha_1, \ldots, \alpha_k) \mapsto \alpha_1 + \cdots + \alpha_k$$

is easily seen to be an R-linear map whose image is the submodule $M_1 + \cdots + M_k$. $\square$

Example 13.23. Let E be an extension ring of R. As we saw in Example 13.5, E may be viewed as an R-module in a natural way. Let $\alpha \in E$, and consider the α-multiplication map on E, which sends $\beta \in E$ to $\alpha\beta \in E$. Then it is easy to see that this is an R-linear map. $\square$

Example 13.24. Let E and E' be extension rings of R, which may be viewed as R-modules as in Example 13.5. Suppose that $\rho: E \to E'$ is a ring homomorphism whose restriction to R is the identity map (i.e., $\rho(c) = c$ for all $c \in R$). Then ρ is an R-linear map. Indeed, for every $c \in R$ and $\alpha, \beta \in E$, we have $\rho(\alpha+\beta) = \rho(\alpha)+\rho(\beta)$ and $\rho(c\alpha) = \rho(c)\rho(\alpha) = c\rho(\alpha)$. $\square$

Example 13.25. Let G and G' be abelian groups. As we saw in Example 13.6, G and G' may be viewed as $\mathbb{Z}$-modules. In addition, every group homomorphism $\rho: G \to G'$ is also a $\mathbb{Z}$-linear map. $\square$

Since an R-module homomorphism is also a group homomorphism on the underlying additive groups, all of the statements in Theorem 6.19 apply. In particular, an R-linear map is injective if and only if the kernel is trivial (i.e., contains only the zero element). However, in the case of R-module homomorphisms, we can extend Theorem 6.19, as follows:

Theorem 13.5. *Let $\rho: M \to M'$ be an R-linear map. Then:*

(i) *for every submodule N of M, $\rho(N)$ is a submodule of M'; in particular (setting $N := M$), $\operatorname{Im}\rho$ is a submodule of M';*

(ii) *for every submodule N' of M', $\rho^{-1}(N')$ is a submodule of M'; in particular (setting $N' := \{0_{M'}\}$), $\operatorname{Ker} \rho$ is a submodule of M.*

Proof. Exercise. □

Theorems 6.20 and 6.21 have natural R-module analogs, which the reader may easily verify:

Theorem 13.6. *If $\rho : M \to M'$ and $\rho' : M' \to M''$ are R-linear maps, then so is their composition $\rho' \circ \rho : M \to M''$.*

Theorem 13.7. *Let $\rho_i : M \to M'_i$, for $i = 1, \ldots, k$, be R-linear maps. Then the map*

$$\rho : \quad M \to M'_1 \times \cdots \times M'_k$$
$$\alpha \mapsto (\rho_1(\alpha), \ldots, \rho_k(\alpha))$$

is an R-linear map.

If an R-linear map $\rho : M \to M'$ is bijective, then it is called an **R-module isomorphism** of M with M'. If such an R-module isomorphism ρ exists, we say that M **is isomorphic to** M', and write $M \cong M'$. Moreover, if $M = M'$, then ρ is called an **R-module automorphism** on M.

Theorems 6.22–6.26 also have natural R-module analogs, which the reader may easily verify:

Theorem 13.8. *If ρ is an R-module isomorphism of M with M', then the inverse function ρ^{-1} is an R-module isomorphism of M' with M.*

Theorem 13.9 (First isomorphism theorem). *Let $\rho : M \to M'$ be an R-linear map with kernel K and image N'. Then we have an R-module isomorphism*

$$M/K \cong N'.$$

Specifically, the map

$$\bar{\rho} : \quad M/K \to M'$$
$$[\alpha]_K \mapsto \rho(\alpha)$$

is an injective R-linear map whose image is N'.

Theorem 13.10. *Let $\rho : M \to M'$ be an R-linear map. Then for every submodule N of M with $N \subseteq \operatorname{Ker} \rho$, we may define an R-linear map*

$$\bar{\rho} : \quad M/N \to M'$$
$$[\alpha]_N \mapsto \rho(\alpha).$$

Moreover, $\operatorname{Im} \bar{\rho} = \operatorname{Im} \rho$, and ρ is injective if and only if $N = \operatorname{Ker} \rho$.

Theorem 13.11 (Internal direct product). *Let M be an R-module with submodules N_1, N_2, where $N_1 \cap N_2 = \{0_M\}$. Then we have an R-module isomorphism*

$$N_1 \times N_2 \cong N_1 + N_2$$

given by the map

$$\rho : \quad N_1 \times N_2 \to N_1 + N_2$$
$$(\alpha_1, \alpha_2) \mapsto \alpha_1 + \alpha_2.$$

Theorem 13.12. *Let M and M' be R-modules, and consider the R-module of functions $\mathrm{Map}(M, M')$ (see Example 13.10). Then*

$$\mathrm{Hom}_R(M, M') := \{\sigma \in \mathrm{Map}(M, M') : \sigma \text{ is an } R\text{-linear map}\}$$

is a submodule of $\mathrm{Map}(M, M')$.

Example 13.26. Consider again the R-module $R[X]/(f)$ discussed in Example 13.4, where $f \in R[X]$ is of degree $\ell \geq 0$ and $\mathrm{lc}(f) \in R^*$. As an R-module, $R[X]/(f)$ is isomorphic to $R[X]_{<\ell}$ (see Example 13.11). Indeed, based on the observations in Example 7.39, the map $\rho : R[X]_{<\ell} \to R[X]/(f)$ that sends a polynomial $g \in R[X]$ of degree less than ℓ to $[g]_f \in R[X]/(f)$ is an isomorphism of $R[X]_{<\ell}$ with $R[X]/(f)$. Furthermore, $R[X]_{<\ell}$ is isomorphic as an R-module to $R^{\times \ell}$. Indeed, the map $\rho' : R[X]_{<\ell} \to R^{\times \ell}$ that sends $g = \sum_{i=0}^{\ell-1} a_i X^i \in R[X]_{<\ell}$ to $(a_0, \ldots, a_{\ell-1}) \in R^{\times \ell}$ is an isomorphism of $R[X]_{<\ell}$ with $R^{\times \ell}$. $\square$

EXERCISE 13.4. Verify that the "is isomorphic to" relation on R-modules is an equivalence relation; that is, for all R-modules M_1, M_2, M_3, we have:

    (a) $M_1 \cong M_1$;

    (b) $M_1 \cong M_2$ implies $M_2 \cong M_1$;

    (c) $M_1 \cong M_2$ and $M_2 \cong M_3$ implies $M_1 \cong M_3$.

EXERCISE 13.5. Let $\rho_i : M_i \to M_i'$, for $i = 1, \ldots, k$, be R-linear maps. Show that the map

$$\rho : \quad M_1 \times \cdots \times M_k \to M_1' \times \cdots \times M_k'$$
$$(\alpha_1, \ldots, \alpha_k) \mapsto (\rho_1(\alpha_1), \ldots, \rho_k(\alpha_k))$$

is an R-linear map.

EXERCISE 13.6. Let $\rho : M \to M'$ be an R-linear map, and let $c \in R$. Show that $\rho(cM) = c\rho(M)$.

EXERCISE 13.7. Let $\rho : M \to M'$ be an R-linear map. Let N be a submodule of

M, and let $\tau : N \to M'$ be the restriction of ρ to N. Show that τ is an R-linear map and that $\operatorname{Ker} \tau = \operatorname{Ker} \rho \cap N$.

EXERCISE 13.8. Suppose $M_1, \ldots, M_k$ are R-modules. Show that for each $i = 1, \ldots, k$, the projection map $\pi_i : M_1 \times \cdots \times M_k \to M_i$ that sends $(\alpha_1, \ldots, \alpha_k)$ to α_i is a surjective R-linear map.

EXERCISE 13.9. Show that if $M = M_1 \times M_2$ for R-modules M_1 and M_2, and N_1 is a subgroup of M_1 and N_2 is a subgroup of M_2, then we have an R-module isomorphism $M/(N_1 \times N_2) \cong M_1/N_1 \times M_2/N_2$.

EXERCISE 13.10. Let M be an R-module with submodules N_1 and N_2. Show that we have an R-module isomorphism $(N_1 + N_2)/N_2 \cong N_1/(N_1 \cap N_2)$.

EXERCISE 13.11. Let M be an R-module with submodules N_1, N_2, and A, where $N_2 \subseteq N_1$. Show that $(N_1 \cap A)/(N_2 \cap A)$ is isomorphic to a submodule of N_1/N_2.

EXERCISE 13.12. Let $\rho : M \to M'$ be an R-linear map with kernel K. Let N be a submodule of M. Show that we have an R-module isomorphism $M/(N + K) \cong \rho(M)/\rho(N)$.

EXERCISE 13.13. Let $\rho : M \to M'$ be a surjective R-linear map. Let S be the set of all submodules of M that contain $\operatorname{Ker} \rho$, and let S' be the set of all submodules of M'. Show that the sets S and S' are in one-to-one correspondence, via the map that sends $N \in S$ to $\rho(N) \in S'$.

13.4 Linear independence and bases

Throughout this section, R denotes a ring.

Definition 13.13. *Let M be an R-module, and let $\{\alpha_i\}_{i=1}^n$ be a family of elements of M. We say that $\{\alpha_i\}_{i=1}^n$*

    (i) *is **linearly dependent (over R)** if there exist $c_1, \ldots, c_n \in R$, not all zero, such that $c_1\alpha_1 + \cdots + c_n\alpha_n = 0_M$;*

   (ii) *is **linearly independent (over R)** if it is not linearly dependent;*

  (iii) *****spans M (over R)** if for every $\alpha \in M$, there exist $c_1, \ldots, c_n \in R$ such that $c_1\alpha_1 + \cdots + c_n\alpha_n = \alpha$;*

  (iv) *is a **basis for M (over R)** if it is linearly independent and spans M.*

The family $\{\alpha_i\}_{i=1}^n$ always spans some submodule of M, namely, the submodule N generated by $\alpha_1, \ldots, \alpha_n$. In this case, we may also call N the **submodule (over R) spanned by** $\{\alpha_i\}_{i=1}^n$.

The family $\{\alpha_i\}_{i=1}^n$ may contain duplicates, in which case it is linearly dependent (unless R is trivial). Indeed, if, say, $\alpha_1 = \alpha_2$, then setting $c_1 := 1$, $c_2 := -1$, and $c_3 := \cdots := c_n := 0$, we have the linear relation $\sum_{i=1}^{n} c_i \alpha_i = 0_M$.

If the family $\{\alpha_i\}_{i=1}^{n}$ contains 0_M, then it is also linear dependent (unless R is trivial). Indeed, if, say, $\alpha_1 = 0_M$, then setting $c_1 := 1$ and $c_2 := \cdots := c_n := 0$, we have the linear relation $\sum_{i=1}^{n} c_i \alpha_i = 0_M$.

The family $\{\alpha_i\}_{i=1}^{n}$ may also be empty (i.e., $n = 0$), in which case it is linearly independent, and spans the submodule $\{0_M\}$.

In the above definition, the ordering of the elements $\alpha_1, \ldots, \alpha_n$ makes no difference. As such, when convenient, we may apply the terminology in the definition to any family $\{\alpha_i\}_{i \in I}$, where I is an arbitrary, finite index set.

Example 13.27. Consider the R-module $R^{\times n}$. Define $\alpha_1, \ldots, \alpha_n \in R^{\times n}$ as follows:

$$\alpha_1 := (1, 0, \ldots, 0), \ \alpha_2 := (0, 1, 0, \ldots, 0), \ldots, \ \alpha_n := (0, \ldots, 0, 1);$$

that is, α_i has a 1 in position i and is zero everywhere else. It is easy to see that $\{\alpha_i\}_{i=1}^{n}$ is a basis for $R^{\times n}$. Indeed, for all $c_1, \ldots, c_n \in R$, we have

$$c_1 \alpha_1 + \cdots + c_n \alpha_n = (c_1, \ldots, c_n),$$

from which it is clear that $\{\alpha_i\}_{i=1}^{n}$ spans $R^{\times n}$ and is linearly independent. The family $\{\alpha_i\}_{i=1}^{n}$ is called the **standard basis** for $R^{\times n}$. $\square$

Example 13.28. Consider the $\mathbb{Z}$-module $\mathbb{Z}^{\times 3}$. In addition to the standard basis, which consists of the tuples

$$(1, 0, 0), (0, 1, 0), (0, 0, 1),$$

the tuples

$$\alpha_1 := (1, 1, 1), \ \alpha_2 := (0, 1, 0), \ \alpha_3 := (2, 0, 1)$$

also form a basis. To see this, first observe that for all $c_1, c_2, c_3, d_1, d_2, d_3 \in \mathbb{Z}$, we have

$$(d_1, d_2, d_3) = c_1 \alpha_1 + c_2 \alpha_2 + c_3 \alpha_3$$

if and only if

$$d_1 = c_1 + 2c_3, \ d_2 = c_1 + c_2, \ \text{and} \ d_3 = c_1 + c_3. \tag{13.1}$$

If (13.1) holds with $d_1 = d_2 = d_3 = 0$, then subtracting the equation $c_1 + c_3 = 0$ from $c_1 + 2c_3 = 0$, we see that $c_3 = 0$, from which it easily follows that $c_1 = c_2 = 0$. This shows that the family $\{\alpha_i\}_{i=1}^{3}$ is linearly independent. To show that it spans $\mathbb{Z}^{\times 3}$, the reader may verify that for any given $d_1, d_2, d_3 \in \mathbb{Z}$, the values

$$c_1 := -d_1 + 2d_3, \ c_2 := d_1 + d_2 - 2d_3, \ c_3 := d_1 - d_3$$

satisfy (13.1).

The family of tuples $(1, 1, 1), (0, 1, 0), (1, 0, 1)$ is not a basis, as it is linearly dependent: the third tuple is equal to the first minus the second.

The family of tuples $(1, 0, 12), (0, 1, 30), (0, 0, 18)$ is linearly independent, but does not span $\mathbb{Z}^{\times 3}$: the last component of any $\mathbb{Z}$-linear combination of these tuples must be divisible by $\gcd(12, 30, 18) = 6$. However, this family of tuples is a basis for the $\mathbb{Q}$-module $\mathbb{Q}^{\times 3}$. $\square$

Example 13.29. Consider again the submodule $R[X]_{<\ell}$ of $R[X]$, where $\ell \geq 0$, consisting of all polynomials of degree less than ℓ (see Example 13.11). Then $\{X^{i-1}\}_{i=1}^{\ell}$ is a basis for $R[X]_{<\ell}$ over R. $\square$

Example 13.30. Consider again the ring $E = R[X]/(f)$, where $f \in R[X]$ with $\deg(f) = \ell \geq 0$ and $\mathrm{lc}(f) \in R^*$. As in Example 13.4, we may naturally view E as a module over R. From the observations in Example 7.39, it is clear that $\{\xi^{i-1}\}_{i=1}^{\ell}$ is a basis for E over R, where $\xi := [X]_f \in E$. $\square$

The next theorem highlights a critical property of bases:

Theorem 13.14. *If $\{\alpha_i\}_{i=1}^{n}$ is a basis for an R-module M, then the map*

$$\varepsilon : \qquad R^{\times n} \to M$$
$$(c_1, \ldots, c_n) \mapsto c_1 \alpha_1 + \cdots + c_n \alpha_n$$

is an R-module isomorphism. In particular, every element of M can be expressed in a unique way as $c_1 \alpha_1 + \cdots + c_n \alpha_n$, for $c_1, \ldots, c_n \in R$.

Proof. We already saw that ε is an R-linear map in Example 13.21. Since $\{\alpha_i\}_{i=1}^{n}$ is linearly independent, it follows that the kernel of ε is trivial, so that ε is injective. That ε is surjective follows immediately from the fact that $\{\alpha_i\}_{i=1}^{n}$ spans M. $\square$

The following is an immediate corollary of this theorem:

Theorem 13.15. *Any two R-modules with bases of the same size are isomorphic.*

The following theorem develops an important connection between bases and linear maps.

Theorem 13.16. *Let $\{\alpha_i\}_{i=1}^{n}$ be a basis for an R-module M, and let $\rho : M \to M'$ be an R-linear map. Then:*

(i) *ρ is surjective if and only if $\{\rho(\alpha_i)\}_{i=1}^{n}$ spans M';*

(ii) *ρ is injective if and only if $\{\rho(\alpha_i)\}_{i=1}^{n}$ is linearly independent;*

(iii) *ρ is an isomorphism if and only if $\{\rho(\alpha_i)\}_{i=1}^{n}$ is a basis for M'.*

Proof. By the previous theorem, we know that every element of M can be written uniquely as $\sum_i c_i \alpha_i$, where the c_i's are in R. Therefore, every element in $\operatorname{Im} \rho$ can be expressed as $\rho(\sum_i c_i \alpha_i) = \sum_i c_i \rho(\alpha_i)$. It follows that $\operatorname{Im} \rho$ is equal to the subspace of M' spanned by $\{\rho(\alpha_i)\}_{i=1}^n$. From this, (i) is clear.

For (ii), consider a non-zero element $\sum_i c_i \alpha_i$ of M, so that not all c_i's are zero. Now, $\sum_i c_i \alpha_i \in \operatorname{Ker} \rho$ if and only if $\sum_i c_i \rho(\alpha_i) = 0_{M'}$, and thus, $\operatorname{Ker} \rho$ is non-trivial if and only if $\{\rho(\alpha_i)\}_{i=1}^n$ is linearly dependent. That proves (ii).

(iii) follows from (i) and (ii). $\square$

EXERCISE 13.14. Let M be an R-module. Suppose $\{\alpha_i\}_{i=1}^n$ is a linearly independent family of elements of M. Show that for every $J \subseteq \{1, \ldots, n\}$, the subfamily $\{\alpha_j\}_{j \in J}$ is also linearly independent.

EXERCISE 13.15. Suppose $\rho : M \to M'$ is an R-linear map. Show that if $\{\alpha_i\}_{i=1}^n$ is a linearly dependent family of elements of M, then $\{\rho(\alpha_i)\}_{i=1}^n$ is also linearly dependent.

EXERCISE 13.16. Suppose $\rho : M \to M'$ is an injective R-linear map and that $\{\alpha_i\}_{i=1}^n$ is a linearly independent family of elements of M. Show that $\{\rho(\alpha_i)\}_{i=1}^n$ is linearly independent.

EXERCISE 13.17. Suppose that $\{\alpha_i\}_{i=1}^n$ spans an R-module M and that $\rho : M \to M'$ is an R-linear map. Show that:

(a) ρ is surjective if and only if $\{\rho(\alpha_i)\}_{i=1}^n$ spans M';

(b) if $\{\rho(\alpha_i)\}_{i=1}^n$ is linearly independent, then ρ is injective.

13.5 Vector spaces and dimension

Throughout this section, F denotes a field.

A module over a field is also called a **vector space**. In particular, an F-module is called an **F-vector space**, or a **vector space over F**.

For vector spaces over F, one typically uses the terms **subspace** and **quotient space**, instead of (respectively) submodule and quotient module; likewise, one usually uses the terms **F-vector space homomorphism**, **isomorphism** and **automorphism**, as appropriate.

We now develop the basic theory of dimension for *finitely generated* vector spaces. Recall that a vector space V over F is finitely generated if we have $V = \langle \alpha_1, \ldots, \alpha_n \rangle_F$ for some $\alpha_1, \ldots, \alpha_n$ of V. The main results here are that

• every finitely generated vector space has a basis, and

• all such bases have the same number of elements.

Throughout the rest of this section, V denotes a vector space over F. We begin with a technical fact that will be used several times throughout this section:

Theorem 13.17. *Suppose that $\{\alpha_i\}_{i=1}^{n}$ is a linearly independent family of elements that spans a subspace $W \subsetneq V$, and that $\alpha_{n+1} \in V \setminus W$. Then $\{\alpha_i\}_{i=1}^{n+1}$ is also linearly independent.*

Proof. Suppose we have a linear relation

$$0_V = c_1\alpha_1 + \cdots + c_n\alpha_n + c_{n+1}\alpha_{n+1},$$

where the c_i's are in F. We want to show that all the c_i's are zero. If $c_{n+1} \neq 0$, then we have

$$\alpha_{n+1} = -c_{n+1}^{-1}(c_1\alpha_1 + \cdots + c_n\alpha_n) \in W,$$

contradicting the assumption that $\alpha_{n+1} \notin W$. Therefore, we must have $c_{n+1} = 0$, and the linear independence of $\{\alpha_i\}_{i=1}^{n}$ implies that $c_1 = \cdots = c_n = 0$. $\square$

The next theorem says that every finitely generated vector space has a basis, and in fact, any family that spans a vector space contains a subfamily that is a basis for the vector space.

Theorem 13.18. *Suppose $\{\alpha_i\}_{i=1}^{n}$ is a family of elements that spans V. Then for some subset $J \subseteq \{1,\ldots,n\}$, the subfamily $\{\alpha_j\}_{j\in J}$ is a basis for V.*

Proof. We prove this by induction on n. If $n = 0$, the theorem is clear, so assume $n > 0$. Consider the subspace W of V spanned by $\{\alpha_i\}_{i=1}^{n-1}$. By the induction hypothesis, for some $K \subseteq \{1,\ldots,n-1\}$, the subfamily $\{\alpha_k\}_{k\in K}$ is a basis for W. There are two cases to consider.

Case 1: $\alpha_n \in W$. In this case, $W = V$, and the theorem clearly holds with $J := K$.

Case 2: $\alpha_n \notin W$. We claim that setting $J := K \cup \{n\}$, the subfamily $\{\alpha_j\}_{j\in J}$ is a basis for V. Indeed, since $\{\alpha_k\}_{k\in K}$ is linearly independent, and $\alpha_n \notin W$, Theorem 13.17 immediately implies that $\{\alpha_j\}_{j\in J}$ is linearly independent. Also, since $\{\alpha_k\}_{k\in K}$ spans W, it is clear that $\{\alpha_j\}_{j\in J}$ spans $W + \langle\alpha_n\rangle_F = V$. $\square$

Theorem 13.19. *If V is spanned by some family of n elements of V, then every family of $n + 1$ elements of V is linearly dependent.*

Proof. We prove this by induction on n. If $n = 0$, the theorem is clear, so assume that $n > 0$. Let $\{\alpha_i\}_{i=1}^{n}$ be a family that spans V, and let $\{\beta_i\}_{i=1}^{n+1}$ be an arbitrary family of elements of V. We wish to show that $\{\beta_i\}_{i=1}^{n+1}$ is linearly dependent.

We know that β_{n+1} is a linear combination of the α_i's, say,

$$\beta_{n+1} = c_1\alpha_1 + \cdots + c_n\alpha_n. \tag{13.2}$$

If all the c_i's were zero, then we would have $\beta_{n+1} = 0_V$, and so trivially, $\{\beta_i\}_{i=1}^{n+1}$ is linearly dependent. So assume that some c_i is non-zero, and for concreteness, say $c_n \neq 0$. Dividing equation (13.2) through by c_n, it follows that α_n is an F-linear combination of $\alpha_1, \ldots, \alpha_{n-1}, \beta_{n+1}$. Therefore,

$$\langle \alpha_1, \ldots, \alpha_{n-1}, \beta_{n+1} \rangle_F \supseteq \langle \alpha_1, \ldots, \alpha_{n-1} \rangle_F + \langle \alpha_n \rangle_F = V.$$

Now consider the subspace $W := \langle \beta_{n+1} \rangle_F$ and the quotient space V/W. Since the family of elements $\alpha_1, \ldots, \alpha_{n-1}, \beta_{n+1}$ spans V, it is easy to see that $\{[\alpha_i]_W\}_{i=1}^{n-1}$ spans V/W; therefore, by induction, $\{[\beta_i]_W\}_{i=1}^{n}$ is linearly dependent. This means that there exist $d_1, \ldots, d_n \in F$, not all zero, such that $d_1 \beta_1 + \cdots + d_n \beta_n \equiv 0 \pmod{W}$, which means that for some $d_{n+1} \in F$, we have $d_1 \beta_1 + \cdots + d_n \beta_n = d_{n+1} \beta_{n+1}$. That proves that $\{\beta_i\}_{i=1}^{n+1}$ is linearly dependent. $\square$

An important corollary of Theorem 13.19 is the following:

Theorem 13.20. *If V is finitely generated, then any two bases for V have the same size.*

Proof. If one basis had more elements than another, then Theorem 13.19 would imply that the first basis was linearly dependent, which contradicts the definition of a basis. $\square$

Theorem 13.20 allows us to make the following definition:

Definition 13.21. *If V is finitely generated, the common size of any basis is called the **dimension** of V, and is denoted $\dim_F(V)$.*

Note that from the definitions, we have $\dim_F(V) = 0$ if and only if V is the trivial vector space (i.e., $V = \{0_V\}$). We also note that one often refers to a finitely generated vector space as a **finite dimensional** vector space. We shall give preference to this terminology from now on.

To summarize the main results in this section up to this point: if V is finite dimensional, it has a basis, and any two bases have the same size, which is called the dimension of V.

Theorem 13.22. *Suppose that $\dim_F(V) = n$, and that $\{\alpha_i\}_{i=1}^{n}$ is a family of n elements of V. The following are equivalent:*

    (i) $\{\alpha_i\}_{i=1}^{n}$ is linearly independent;

    (ii) $\{\alpha_i\}_{i=1}^{n}$ spans V;

    (iii) $\{\alpha_i\}_{i=1}^{n}$ is a basis for V.

Proof. Let W be the subspace of V spanned by $\{\alpha_i\}_{i=1}^{n}$.

First, let us show that (i) implies (ii). Suppose $\{\alpha_i\}_{i=1}^{n}$ is linearly independent.

Also, by way of contradiction, suppose that $W \subsetneq V$, and choose $\alpha_{n+1} \in V \setminus W$. Then Theorem 13.17 implies that $\{\alpha_i\}_{i=1}^{n+1}$ is linearly independent. But then we have a linearly independent family of $n + 1$ elements of V, which is impossible by Theorem 13.19.

Second, let us prove that (ii) implies (i). Let us assume that $\{\alpha_i\}_{i=1}^{n}$ is linearly dependent, and prove that $W \subsetneq V$. By Theorem 13.18, we can find a basis for W among the α_i's, and since $\{\alpha_i\}_{i=1}^{n}$ is linearly dependent, this basis must contain strictly fewer than n elements. Hence, $\dim_F(W) < \dim_F(V)$, and therefore, $W \subsetneq V$.

The theorem now follows from the above arguments, and the fact that, by definition, (iii) holds if and only if both (i) and (ii) hold. $\square$

We next examine the dimension of subspaces of finite dimensional vector spaces.

Theorem 13.23. *Suppose that V is finite dimensional and W is a subspace of V. Then W is also finite dimensional, with $\dim_F(W) \leq \dim_F(V)$. Moreover, $\dim_F(W) = \dim_F(V)$ if and only if $W = V$.*

Proof. Suppose $\dim_F(V) = n$. Consider the set S of all linearly independent families of the form $\{\alpha_i\}_{i=1}^{m}$, where $m \geq 0$ and each α_i is in W. The set S is certainly non-empty, as it contains the empty family. Moreover, by Theorem 13.19, every member of S must have at most n elements. Therefore, we may choose some particular element $\{\alpha_i\}_{i=1}^{m}$ of S, where m is as large as possible. We claim that this family $\{\alpha_i\}_{i=1}^{m}$ is a basis for W. By definition, $\{\alpha_i\}_{i=1}^{m}$ is linearly independent and spans some subspace W' of W. If $W' \subsetneq W$, we can choose an element $\alpha_{m+1} \in W \setminus W'$, and by Theorem 13.17, the family $\{\alpha_i\}_{i=1}^{m+1}$ is linearly independent, and therefore, this family also belongs to S, contradicting the assumption that m is as large as possible.

That proves that W is finite dimensional with $\dim_F(W) \leq \dim_F(V)$. It remains to show that these dimensions are equal if and only if $W = V$. Now, if $W = V$, then clearly $\dim_F(W) = \dim_F(V)$. Conversely, if $\dim_F(W) = \dim_F(V)$, then by Theorem 13.22, any basis for W must already span V. $\square$

Theorem 13.24. *If V is finite dimensional, and W is a subspace of V, then the quotient space V/W is also finite dimensional, and*

$$\dim_F(V/W) = \dim_F(V) - \dim_F(W).$$

Proof. Suppose that $\{\alpha_i\}_{i=1}^{n}$ spans V. Then it is clear that $\{[\alpha_i]_W\}_{i=1}^{n}$ spans V/W. By Theorem 13.18, we know that V/W has a basis of the form $\{[\alpha_i]_W\}_{i=1}^{\ell}$, where $\ell \leq n$ (renumbering the α_i's as necessary). By Theorem 13.23, we know that W has a basis, say $\{\beta_j\}_{j=1}^{m}$. The theorem will follow immediately from the following:

Claim. The elements

$$\alpha_1, \ldots, \alpha_\ell, \ \beta_1, \ldots, \beta_m \tag{13.3}$$

form a basis for V.

To see that this family spans V, consider any element γ of V. Then since $\{[\alpha_i]_W\}_{i=1}^\ell$ spans V/W, we have $\gamma \equiv \sum_i c_i\alpha_i \pmod{W}$ for some $c_1, \ldots, c_\ell \in F$. If we set $\beta := \gamma - \sum_i c_i\alpha_i \in W$, then since $\{\beta_j\}_{j=1}^m$ spans W, we have $\beta = \sum_j d_j\beta_j$ for some $d_1, \ldots, d_m \in F$, and hence $\gamma = \sum_i c_i\alpha_i + \sum_j d_j\beta_j$. That proves that the family of elements (13.3) spans V. To prove this family is linearly independent, suppose we have a relation of the form $\sum_i c_i\alpha_i + \sum_j d_j\beta_j = 0_V$, where $c_1, \ldots, c_\ell \in F$ and $d_1, \ldots, d_m \in F$. If any of the c_i's were non-zero, this would contradict the assumption that $\{[\alpha_i]_W\}_{i=1}^\ell$ is linearly independent. So assume that all the c_i's are zero. If any of the d_j's were non-zero, this would contradict the assumption that $\{\beta_j\}_{j=1}^m$ is linearly independent. Thus, all the c_i's and d_j's must be zero, which proves that the family of elements (13.3) is linearly independent. That proves the claim. $\square$

Theorem 13.25. *If V is finite dimensional, then every linearly independent family of elements of V can be extended to form a basis for V.*

Proof. One can prove this by generalizing the proof of Theorem 13.18. Alternatively, we can adapt the proof of the previous theorem. Let $\{\beta_j\}_{j=1}^m$ be a linearly independent family of elements that spans a subspace W of V. As in the proof of the previous theorem, if $\{[\alpha_i]_W\}_{i=1}^\ell$ is a basis for the quotient space V/W, then the elements

$$\alpha_1, \ldots, \alpha_\ell, \ \beta_1, \ldots, \beta_m$$

form a basis for V. $\square$

Example 13.31. Suppose that F is finite, say $|F| = q$, and that V is finite dimensional, say $\dim_F(V) = n$. Then clearly $|V| = q^n$. If W is a subspace with $\dim_F(W) = m$, then $|W| = q^m$, and by Theorem 13.24, $\dim_F(V/W) = n - m$, and hence $|V/W| = q^{n-m}$. Just viewing V and W as additive groups, we know that the index of W in V is $[V : W] = |V/W| = |V|/|W| = q^{n-m}$, which agrees with the above calculations. $\square$

We next consider the relation between the notion of dimension and linear maps. First, observe that by Theorem 13.15, if two finite dimensional vector spaces have the same dimension, then they are isomorphic. The following theorem is the converse:

Theorem 13.26. *If V is of finite dimension n, and V is isomorphic to V', then V' is also of finite dimension n.*

Proof. If $\{\alpha_i\}_{i=1}^n$ is a basis for V, then by Theorem 13.16, $\{\rho(\alpha_i)\}_{i=1}^n$ is a basis for V'. $\square$

Thus, two finite dimensional vector spaces are isomorphic if and only if they have the same dimension.

We next illustrate one way in which the notion of dimension is particularly useful. In general, if we have a function $f : A \to B$, injectivity does not imply surjectivity, nor does surjectivity imply injectivity. If A and B are finite sets of equal size, then these implications do indeed hold. The following theorem gives us another important setting where these implications hold, with finite dimensionality playing the role corresponding to finite cardinality:

Theorem 13.27. *If $\rho : V \to V'$ is an F-linear map, and if V and V' are finite dimensional with $\dim_F(V) = \dim_F(V')$, then we have:*

ρ is injective if and only if ρ is surjective.

Proof. Let $\{\alpha_i\}_{i=1}^n$ be a basis for V. Then

ρ is injective $\iff$ $\{\rho(\alpha_i)\}_{i=1}^n$ is linearly independent (by Theorem 13.16)

$\iff$ $\{\rho(\alpha_i)\}_{i=1}^n$ spans V' (by Theorem 13.22)

$\iff$ ρ is surjective (again by Theorem 13.16). $\square$

This theorem may be generalized as follows:

Theorem 13.28. *If V is finite dimensional, and $\rho : V \to V'$ is an F-linear map, then $\operatorname{Im} \rho$ is a finite dimensional vector space, and*

$$\dim_F(V) = \dim_F(\operatorname{Im} \rho) + \dim_F(\operatorname{Ker} \rho).$$

Proof. As the reader may verify, this follows immediately from Theorem 13.24, together with Theorems 13.26 and 13.9. $\square$

Intuitively, one way to think of Theorem 13.28 is as a "law of conservation" for dimension: any "dimensionality" going into ρ that is not "lost" to the kernel of ρ must show up in the image of ρ.

EXERCISE 13.18. Show that if $V_1, \ldots, V_n$ are finite dimensional vector spaces over F, then $V_1 \times \cdots \times V_n$ has dimension $\sum_{i=1}^n \dim_F(V_i)$.

EXERCISE 13.19. Show that if V is a finite dimensional vector space over F with subspaces W_1 and W_2, then

$$\dim_F(W_1 + W_2) = \dim_F(W_1) + \dim_F(W_2) - \dim_F(W_1 \cap W_2).$$

EXERCISE 13.20. From the previous exercise, one might be tempted to think that a more general "inclusion/exclusion principle" for dimension holds. Determine if the following statement is true or false: if V is a finite dimensional vector space over F with subspaces W_1, W_2, and W_3, then

$$\dim_F(W_1 + W_2 + W_3) = \dim_F(W_1) + \dim_F(W_2) + \dim_F(W_3)$$
$$- \dim_F(W_1 \cap W_2) - \dim_F(W_1 \cap W_3) - \dim_F(W_2 \cap W_3)$$
$$+ \dim_F(W_1 \cap W_2 \cap W_3).$$

EXERCISE 13.21. Suppose that V and W are vector spaces over F, V is finite dimensional, and $\{\alpha_i\}_{i=1}^k$ is a linearly independent family of elements of V. In addition, let $\beta_1, \ldots, \beta_k$ be arbitrary elements of W. Show that there exists an F-linear map $\rho : V \to W$ such that $\rho(\alpha_i) = \beta_i$ for $i = 1, \ldots, k$.

EXERCISE 13.22. Let V be a vector space over F with basis $\{\alpha_i\}_{i=1}^n$. Let S be a finite, non-empty subset of F, and define

$$B := \left\{ \sum_{i=1}^n c_i \alpha_i : c_1, \ldots, c_n \in S \right\}.$$

Show that if W is a subspace of V, with $W \subsetneq V$, then $|B \cap W| \leq |S|^{n-1}$.

EXERCISE 13.23. The theory of dimension for finitely generated vector spaces is quite elegant and powerful. There is a theory of dimension (of sorts) for modules over an arbitrary, non-trivial ring R, but it is much more awkward and limited. This exercise develops a proof of one aspect of this theory: if an R-module M has a basis at all, then any two bases have the same size. To prove this, we need the fact that any non-trivial ring has a maximal ideal (this was proved in Exercise 7.40 for countable rings). Let n, m be positive integers, let $\alpha_1, \ldots, \alpha_m$ be elements of $R^{\times n}$, and let I be an ideal of R.

(a) Show that if $\{\alpha_i\}_{i=1}^m$ spans $R^{\times n}$, then every element of $I^{\times n}$ can be expressed as $c_1 \alpha_1 + \cdots + c_m \alpha_m$, where $c_1, \ldots, c_m$ belong to I.

(b) Show that if $m > n$ and I is a maximal ideal, then there exist $c_1, \ldots, c_m \in R$, not all in I, such that $c_1 \alpha_1 + \cdots + c_m \alpha_m \in I^{\times n}$.

(c) From (a) and (b), deduce that if $m > n$, then $\{\alpha_i\}_{i=1}^m$ cannot be a basis for $R^{\times n}$.

(d) From (c), conclude that any two bases for a given R-module M must have the same size.

14

Matrices

In this chapter, we discuss basic definitions and results concerning matrices. We shall start out with a very general point of view, discussing matrices whose entries lie in an arbitrary ring R. Then we shall specialize to the case where the entries lie in a field F, where much more can be said.

One of the main goals of this chapter is to discuss "Gaussian elimination," which is an algorithm that allows us to efficiently compute bases for the image and kernel of an F-linear map.

In discussing the complexity of algorithms for matrices over a ring R, we shall treat a ring R as an "abstract data type," so that the running times of algorithms will be stated in terms of the number of arithmetic operations in R. If R is a finite ring, such as $\mathbb{Z}_m$, we can immediately translate this into a running time on a RAM (in later chapters, we will discuss other finite rings and efficient algorithms for doing arithmetic in them).

If R is, say, the field of rational numbers, a complete running time analysis would require an additional analysis of the sizes of the numbers that appear in the execution of the algorithm. We shall not attempt such an analysis here—however, we note that all the algorithms discussed in this chapter do in fact run in polynomial time when $R = \mathbb{Q}$, assuming we represent rational numbers as fractions in lowest terms. Another possible approach for dealing with rational numbers is to use floating point approximations. While this approach eliminates the size problem, it creates many new problems because of round-off errors. We shall not address any of these issues here.

14.1 Basic definitions and properties

Throughout this section, R denotes a ring.

For positive integers m and n, an $m \times n$ **matrix** A over a ring R is a rectangular array

$$A = \begin{pmatrix} a_{11} & a_{12} & \cdots & a_{1n} \\ a_{21} & a_{22} & \cdots & a_{2n} \\ \vdots & \vdots & & \vdots \\ a_{m1} & a_{m2} & \cdots & a_{mn} \end{pmatrix},$$

where each entry a_{ij} in the array is an element of R; the element a_{ij} is called the (i, j) **entry** of A, which we denote by $A(i, j)$. For $i = 1, \ldots, m$, the i**th row** of A is

$$(a_{i1}, \ldots, a_{in}),$$

which we denote by $\mathrm{Row}_i(A)$, and for $j = 1, \ldots, n$, the j**th column** of A is

$$\begin{pmatrix} a_{1j} \\ a_{2j} \\ \vdots \\ a_{mj} \end{pmatrix},$$

which we denote by $\mathrm{Col}_j(A)$. We regard a row of A as a $1 \times n$ matrix, and a column of A as an $m \times 1$ matrix.

The set of all $m \times n$ matrices over R is denoted by $R^{m \times n}$. Elements of $R^{1 \times n}$ are called **row vectors (of dimension** n**)** and elements of $R^{m \times 1}$ are called **column vectors (of dimension** m**)**. Elements of $R^{n \times n}$ are called **square matrices (of dimension** n**)**. We do not make a distinction between $R^{1 \times n}$ and $R^{\times n}$; that is, we view standard n-tuples as row vectors.

We can define the familiar operations of **matrix addition** and **scalar multiplication**:

- If $A, B \in R^{m \times n}$, then $A + B$ is the $m \times n$ matrix whose (i, j) entry is $A(i, j) + B(i, j)$.
- If $c \in R$ and $A \in R^{m \times n}$, then cA is the $m \times n$ matrix whose (i, j) entry is $cA(i, j)$.

The $m \times n$ **zero matrix** is the $m \times n$ matrix, all of whose entries are 0_R; we denote this matrix by $0_R^{m \times n}$ (or just 0, when the context is clear).

Theorem 14.1. *With addition and scalar multiplication as defined above, $R^{m \times n}$ is an R-module. The matrix $0_R^{m \times n}$ is the additive identity, and the additive inverse of a matrix $A \in R^{m \times n}$ is the $m \times n$ matrix whose (i, j) entry is $-A(i, j)$.*

Proof. To prove this, one first verifies that matrix addition is associative and commutative, which follows from the associativity and commutativity of addition in R. The claims made about the additive identity and additive inverses are also easily verified. These observations establish that $R^{m \times n}$ is an abelian group. One also has to check that all of the properties in Definition 13.1 hold. We leave this to the reader. □

We can also define the familiar operation of **matrix multiplication**:

- If $A \in R^{m \times n}$ and $B \in R^{n \times p}$, then AB is the $m \times p$ matrix whose (i, k) entry is

$$\sum_{j=1}^{n} A(i, j) B(j, k).$$

The $n \times n$ **identity matrix** is the matrix $I \in R^{n \times n}$, where $I(i, i) := 1_R$ and $I(i, j) := 0_R$ for $i \neq j$. That is, I has 1_R's on the diagonal that runs from the upper left corner to the lower right corner, and 0_R's everywhere else.

Theorem 14.2.

 (i) *Matrix multiplication is associative; that is, $A(BC) = (AB)C$ for all $A \in R^{m \times n}$, $B \in R^{n \times p}$, and $C \in R^{p \times q}$.*

 (ii) *Matrix multiplication distributes over matrix addition; that is, $A(C + D) = AC + AD$ and $(A + B)C = AC + BC$ for all $A, B \in R^{m \times n}$ and $C, D \in R^{n \times p}$.*

 (iii) *The $n \times n$ identity matrix $I \in R^{n \times n}$ acts as a multiplicative identity; that is, $AI = A$ and $IB = B$ for all $A \in R^{m \times n}$ and $B \in R^{n \times m}$; in particular, $CI = C = IC$ for all $C \in R^{n \times n}$.*

 (iv) *Scalar multiplication and matrix multiplication associate; that is, $c(AB) = (cA)B = A(cB)$ for all $c \in R$, $A \in R^{m \times n}$, and $B \in R^{n \times p}$.*

Proof. All of these are trivial, except for (i), which requires just a bit of computation to show that the (i, ℓ) entry of both $A(BC)$ and $(AB)C$ is equal to (as the reader may verify)

$$\sum_{\substack{1 \leq j \leq n \\ 1 \leq k \leq p}} A(i, j) B(j, k) C(k, \ell). \quad \Box$$

Note that while matrix addition is commutative, matrix multiplication in general is not. Indeed, Theorems 14.1 and 14.2 imply that $R^{n \times n}$ satisfies all the properties of a ring except for commutativity of multiplication.

Some simple but useful facts to keep in mind are the following:

- If $A \in R^{m \times n}$ and $B \in R^{n \times p}$, then the ith row of AB is equal to vB, where $v = \text{Row}_i(A)$; also, the kth column of AB is equal to Aw, where $w = \text{Col}_k(B)$.

- If $A \in R^{m \times n}$ and $v = (c_1, \ldots, c_m) \in R^{1 \times m}$, then

$$vA = \sum_{i=1}^{m} c_i \operatorname{Row}_i(A).$$

  In words: vA is a linear combination of the rows of A, with coefficients taken from the corresponding entries of v.

- If $A \in R^{m \times n}$ and

$$w = \begin{pmatrix} d_1 \\ \vdots \\ d_n \end{pmatrix} \in R^{n \times 1},$$

  then

$$Aw = \sum_{j=1}^{n} d_j \operatorname{Col}_j(A).$$

  In words: Aw is a linear combination of the columns of A, with coefficients taken from the corresponding entries of w.

If $A \in R^{m \times n}$, the **transpose** of A, denoted by $A^\top$, is defined to be the $n \times m$ matrix whose (j, i) entry is $A(i, j)$.

Theorem 14.3. *If $A, B \in R^{m \times n}$, $C \in R^{n \times p}$, and $c \in R$, then:*

  (i) $(A + B)^\top = A^\top + B^\top$;

  (ii) $(cA)^\top = cA^\top$;

  (iii) $(A^\top)^\top = A$;

  (iv) $(AC)^\top = C^\top A^\top$.

Proof. Exercise. □

If A_i is an $n_i \times n_{i+1}$ matrix, for $i = 1, \ldots, k$, then by associativity of matrix multiplication, we may write the product matrix $A_1 \cdots A_k$, which is an $n_1 \times n_{k+1}$ matrix, without any ambiguity.

For an $n \times n$ matrix A, and a positive integer k, we write A^k to denote the product $A \cdots A$, where there are k terms in the product. Note that $A^1 = A$. We may extend this notation to $k = 0$, defining A^0 to be the $n \times n$ identity matrix. One may readily verify the usual rules of exponent arithmetic: for all non-negative integers k, ℓ, we have

$$(A^\ell)^k = A^{k\ell} = (A^k)^\ell \quad \text{and} \quad A^k A^\ell = A^{k+\ell}.$$

It is easy also to see that part (iv) of Theorem 14.3 implies that for all non-negative integers k, we have

$$(A^k)^\top = (A^\top)^k.$$

Algorithmic issues

For computational purposes, matrices are represented in the obvious way as arrays of elements of R. As remarked at the beginning of this chapter, we shall treat R as an "abstract data type," and not worry about how elements of R are actually represented; in discussing the complexity of algorithms, we shall simply count "operations in R," by which we mean additions, subtractions, and multiplications; we shall sometimes also include equality testing and computing multiplicative inverses as "operations in R." In any real implementation, there will be other costs, such as incrementing counters, and so on, which we may safely ignore, as long as their number is at most proportional to the number of operations in R.

The following statements are easy to verify:

- We can multiply an $m \times n$ matrix by a scalar using mn operations in R.
- We can add two $m \times n$ matrices using mn operations in R.
- We can compute the product of an $m \times n$ matrix and an $n \times p$ matrix using $O(mnp)$ operations in R.

It is also easy to see that given an $n \times n$ matrix A, and a non-negative integer e, we can adapt the repeated squaring algorithm discussed in §3.4 so as to compute A^e using $O(\text{len}(e))$ multiplications of $n \times n$ matrices, and hence $O(\text{len}(e)n^3)$ operations in R.

EXERCISE 14.1. Let $A \in R^{m \times n}$. Show that if $vA = 0_R^{1 \times n}$ for all $v \in R^{1 \times m}$, then $A = 0_R^{m \times n}$.

14.2 Matrices and linear maps

Let R be a ring.

For positive integers m and n, consider the R-modules $R^{1 \times m}$ and $R^{1 \times n}$. If A is an $m \times n$ matrix over R, then the map

$$\lambda_A : \quad R^{1 \times m} \to R^{1 \times n}$$
$$v \mapsto vA$$

is easily seen to be an R-linear map—this follows immediately from parts (ii) and (iv) of Theorem 14.2. We call λ_A the **linear map corresponding to** A.

If $v = (c_1, \ldots, c_m) \in R^{1 \times m}$, then

$$\lambda_A(v) = vA = \sum_{i=1}^{m} c_i \operatorname{Row}_i(A).$$

From this, it is clear that

- the image of λ_A is the submodule of $R^{1 \times n}$ spanned by $\{\operatorname{Row}_i(A)\}_{i=1}^m$; in particular, λ_A is surjective if and only if $\{\operatorname{Row}_i(A)\}_{i=1}^m$ spans $R^{1 \times n}$;

- λ_A is injective if and only if $\{\operatorname{Row}_i(A)\}_{i=1}^m$ is linearly independent.

There is a close connection between matrix multiplication and composition of corresponding linear maps. Specifically, let $A \in R^{m \times n}$ and $B \in R^{n \times p}$, and consider the corresponding linear maps $\lambda_A : R^{1 \times m} \to R^{1 \times n}$ and $\lambda_B : R^{1 \times n} \to R^{1 \times p}$. Then we have

$$\lambda_B \circ \lambda_A = \lambda_{AB}. \tag{14.1}$$

This follows immediately from the associativity of matrix multiplication.

We have seen how vector/matrix multiplication defines a linear map. Conversely, we shall now see that the action of any R-linear map can be viewed as a vector/matrix multiplication, provided the R-modules involved have bases (which will always be the case for finite dimensional vector spaces).

Let M be an R-module, and suppose that $S = \{\alpha_i\}_{i=1}^m$ is a basis for M, where $m > 0$. As we know (see Theorem 13.14), every element $\alpha \in M$ can be written uniquely as $c_1 \alpha_1 + \cdots + c_m \alpha_m$, where the c_i's are in R. Let us define

$$\operatorname{Vec}_S(\alpha) := (c_1, \ldots, c_m) \in R^{1 \times m}.$$

We call $\operatorname{Vec}_S(\alpha)$ the **coordinate vector of α relative to** S. The function

$$\operatorname{Vec}_S : M \to R^{1 \times m}$$

is an R-module isomorphism (it is the inverse of the isomorphism ε in Theorem 13.14).

Let N be another R-module, and suppose that $\mathcal{T} = \{\beta_j\}_{j=1}^n$ is a basis for N, where $n > 0$. Just as in the previous paragraph, every element $\beta \in N$ has a unique coordinate vector $\operatorname{Vec}_{\mathcal{T}}(\beta) \in R^{1 \times n}$ relative to $\mathcal{T}$.

Now let $\rho : M \to N$ be an arbitrary R-linear map. Our goal is to define a matrix $A \in R^{m \times n}$ with the following property:

$$\operatorname{Vec}_{\mathcal{T}}(\rho(\alpha)) = \operatorname{Vec}_S(\alpha)A \quad \text{for all } \alpha \in M. \tag{14.2}$$

In words: if we multiply the coordinate vector of α on the right by A, we get the coordinate vector of $\rho(\alpha)$.

Constructing such a matrix A is easy: we define A to be the matrix whose ith row, for $i = 1, \dots, m$, is the coordinate vector of $\rho(\alpha_i)$ relative to $\mathcal{T}$. That is,

$$\mathrm{Row}_i(A) = \mathrm{Vec}_{\mathcal{T}}(\rho(\alpha_i)) \quad \text{for } i = 1, \dots, m.$$

Then for an arbitrary $\alpha \in M$, if $(c_1, \dots, c_m)$ is the coordinate vector of α relative to S, we have

$$\rho(\alpha) = \rho\left(\sum_i c_i \alpha_i\right) = \sum_i c_i \rho(\alpha_i)$$

and so

$$\mathrm{Vec}_{\mathcal{T}}(\rho(\alpha)) = \sum_i c_i \, \mathrm{Vec}_{\mathcal{T}}(\rho(\alpha_i)) = \sum_i c_i \, \mathrm{Row}_i(A) = \mathrm{Vec}_S(\alpha) A.$$

Furthermore, A is the only matrix satisfying (14.2). Indeed, if A' also satisfies (14.2), then subtracting, we obtain

$$\mathrm{Vec}_S(\alpha)(A - A') = 0_R^{1 \times n}$$

for all $\alpha \in M$. Since the map $\mathrm{Vec}_S : M \to R^{1 \times m}$ is surjective, this means that $v(A - A')$ is zero for all $v \in R^{1 \times m}$, and from this, it is clear (see Exercise 14.1) that $A - A'$ is the zero matrix, and so $A = A'$.

We call the unique matrix A satisfying (14.2) the **matrix of ρ relative to S and $\mathcal{T}$**, and denote it by $\mathrm{Mat}_{S,\mathcal{T}}(\rho)$.

Recall that $\mathrm{Hom}_R(M, N)$ is the R-module consisting of all R-linear maps from M to N (see Theorem 13.12). We can view $\mathrm{Mat}_{S,\mathcal{T}}$ as a function mapping elements of $\mathrm{Hom}_R(M, N)$ to elements of $R^{m \times n}$.

Theorem 14.4. *The function $\mathrm{Mat}_{S,\mathcal{T}} : \mathrm{Hom}_R(M, N) \to R^{m \times n}$ is an R-module isomorphism. In particular, for every $A \in R^{m \times n}$, the pre-image of A under $\mathrm{Mat}_{S,\mathcal{T}}$ is $\mathrm{Vec}_{\mathcal{T}}^{-1} \circ \lambda_A \circ \mathrm{Vec}_S$, where $\lambda_A : R^{1 \times m} \to R^{1 \times n}$ is the linear map corresponding to A.*

Proof. To show that $\mathrm{Mat}_{S,\mathcal{T}}$ is an R-linear map, let $\rho, \rho' \in \mathrm{Hom}_R(M, N)$, and let $c \in R$. Also, let $A := \mathrm{Mat}_{S,\mathcal{T}}(\rho)$ and $A' := \mathrm{Mat}_{S,\mathcal{T}}(\rho')$. Then for all $\alpha \in M$, we have

$$\mathrm{Vec}_{\mathcal{T}}((\rho + \rho')(\alpha)) = \mathrm{Vec}_{\mathcal{T}}(\rho(\alpha) + \rho'(\alpha)) = \mathrm{Vec}_{\mathcal{T}}(\rho(\alpha)) + \mathrm{Vec}_{\mathcal{T}}(\rho'(\alpha))$$
$$= \mathrm{Vec}_S(\alpha) A + \mathrm{Vec}_S(\alpha) A' = \mathrm{Vec}_S(\alpha)(A + A').$$

As this holds for all $\alpha \in M$, and since the matrix of a linear map is uniquely determined, we must have $\mathrm{Mat}_{S,\mathcal{T}}(\rho + \rho') = A + A'$. A similar argument shows that $\mathrm{Mat}_{S,\mathcal{T}}(c\rho) = cA$. This shows that $\mathrm{Mat}_{S,\mathcal{T}}$ is an R-linear map.

To show that the map $\mathrm{Mat}_{S,\mathcal{T}}$ is injective, it suffices to show that its kernel is trivial. If ρ is in the kernel of this map, then setting $A := 0_R^{m \times n}$ in (14.2), we see that $\text{Vec}_T(\rho(\alpha))$ is zero for all $\alpha \in M$. But since the map $\text{Vec}_T : N \to R^{1 \times n}$ is injective, this implies $\rho(\alpha)$ is zero for all $\alpha \in M$. Thus, ρ must be the zero map.

To show surjectivity, we show that every $A \in R^{m \times n}$ has a pre-image under $\text{Mat}_{S,T}$ as described in the statement of the theorem. So let A be an $m \times n$ matrix, and let $\rho := \text{Vec}_T^{-1} \circ \lambda_A \circ \text{Vec}_S$. Again, since the matrix of a linear map is uniquely determined, it suffices to show that (14.2) holds for this particular A and ρ. For every $\alpha \in M$, we have

$$\text{Vec}_T(\rho(\alpha)) = \text{Vec}_T(\text{Vec}_T^{-1}(\lambda_A(\text{Vec}_S(\alpha)))) = \lambda_A(\text{Vec}_S(\alpha))$$
$$= \text{Vec}_S(\alpha)A.$$

That proves the theorem. $\square$

As a special case of the above, suppose that $M = R^{1 \times m}$ and $N = R^{1 \times n}$, and S and T are the standard bases for M and N (see Example 13.27). In this case, the functions Vec_S and Vec_T are the identity maps, and the previous theorem implies that the function

$$\Lambda : \quad R^{m \times n} \to \text{Hom}_R(R^{1 \times m}, R^{1 \times n})$$

$$A \mapsto \lambda_A$$

is the inverse of the function $\text{Mat}_{S,T} : \text{Hom}_R(R^{1 \times m}, R^{1 \times n}) \to R^{m \times n}$. Thus, the function Λ is also an R-module isomorphism.

To summarize, we see that an R-linear map ρ from M to N, together with particular bases for M and N, uniquely determine a matrix A such that the action of multiplication on the right by A implements the action of ρ with respect to the given bases. There may be many bases for M and N to choose from, and different choices will in general lead to different matrices. Also, note that in general, a basis may be indexed by an arbitrary finite set; however, in defining coordinate vectors and matrices of linear maps, the index set must be ordered in some way. In any case, from a computational perspective, the matrix A gives us an efficient way to compute the map ρ, assuming elements of M and N are represented as coordinate vectors with respect to the given bases.

We have taken a "row-centric" point of view. Of course, if one prefers, by simply transposing everything, one can equally well take a "column-centric" point of view, where the action of ρ corresponds to multiplication of a column vector on the left by a matrix.

Example 14.1. Consider the quotient ring $E = R[X]/(f)$, where $f \in R[X]$ with $\deg(f) = \ell > 0$ and $\text{lc}(f) \in R^*$. Let $\xi := [X]_f \in E$. As an R-module, E has a basis $S := \{\xi^{i-1}\}_{i=1}^{\ell}$ (see Example 13.30). Let $\rho : E \to E$ be the ξ-multiplication map, which sends $\alpha \in E$ to $\xi\alpha \in E$. This is an R-linear map. If

$f = c_0 + c_1 X + \cdots + c_{\ell-1} X^{\ell-1} + c_\ell X^\ell$, then the matrix of ρ relative to S is the $\ell \times \ell$ matrix

$$A = \begin{pmatrix} 0 & 1 & 0 & \cdots & 0 \\ 0 & 0 & 1 & \cdots & 0 \\ & & & \ddots & \\ 0 & 0 & 0 & \cdots & 1 \\ -c_0/c_\ell & -c_1/c_\ell & -c_2/c_\ell & \cdots & -c_{\ell-1}/c_\ell \end{pmatrix},$$

where for $i = 1, \ldots, \ell - 1$, the ith row of A contains a 1 in position $i + 1$, and is zero everywhere else. The matrix A is called the **companion matrix of** f. $\square$

Example 14.2. Let $x_1, \ldots, x_k \in R$. Let $R[X]_{<k}$ be the set of polynomials $g \in R[X]$ with $\deg(g) < k$, which is an R-module with a basis $S := \{X^{i-1}\}_{i=1}^k$ (see Example 13.29). The multi-point evaluation map

$$\rho : \quad R[X]_{<k} \to R^{1 \times k}$$
$$g \mapsto (g(x_1), \ldots, g(x_k))$$

is an R-linear map. Let $\mathcal{T}$ be the standard basis for $R^{1 \times k}$. Then the matrix of ρ relative to S and $\mathcal{T}$ is the $k \times k$ matrix

$$A = \begin{pmatrix} 1 & 1 & \cdots & 1 \\ x_1 & x_2 & \cdots & x_k \\ x_1^2 & x_2^2 & \cdots & x_k^2 \\ \vdots & \vdots & & \vdots \\ x_1^{k-1} & x_2^{k-1} & \cdots & x_k^{k-1} \end{pmatrix}.$$

The matrix A is called a **Vandermonde matrix.** $\square$

EXERCISE 14.2. Let $\sigma : M \to N$ and $\tau : N \to P$ be R-linear maps, and suppose that M, N, and P have bases S, $\mathcal{T}$, and $\mathcal{U}$, respectively. Show that

$$\text{Mat}_{S,\mathcal{U}}(\tau \circ \sigma) = \text{Mat}_{S,\mathcal{T}}(\sigma) \cdot \text{Mat}_{\mathcal{T},\mathcal{U}}(\tau).$$

EXERCISE 14.3. Let V be a vector space over a field F with basis $S = \{\alpha_i\}_{i=1}^m$. Suppose that U is a subspace of V of dimension $\ell < m$. Show that there exists a matrix $A \in F^{m \times (m-\ell)}$ such that for all $\alpha \in V$, we have $\alpha \in U$ if and only if $\text{Vec}_S(\alpha)A$ is zero. Such a matrix A is called a **parity check matrix** for U.

EXERCISE 14.4. Let F be a finite field, and let A be a non-zero $m \times n$ matrix over F. Suppose one chooses a vector $v \in F^{1 \times m}$ at random. Show that the probability that vA is the zero vector is at most $1/|F|$.

EXERCISE 14.5. Design and analyze a probabilistic algorithm that takes as input matrices $A, B, C \in \mathbb{Z}_p^{m \times m}$, where p is a prime. The algorithm should run in time $O(m^2 \operatorname{len}(p)^2)$ and should output either "yes" or "no" so that the following holds:

- if $C = AB$, then the algorithm should always output "yes";
- if $C \neq AB$, then the algorithm should output "no" with probability at least 0.999.

14.3 The inverse of a matrix

Let R be a ring.

For a square matrix $A \in R^{n \times n}$, we call a matrix $B \in R^{n \times n}$ an **inverse** of A if $BA = AB = I$, where I is the $n \times n$ identity matrix. It is easy to see that if A has an inverse, then the inverse is unique: if B and C are inverses of A, then

$$B = BI = B(AC) = (BA)C = IC = C.$$

Because the inverse of A is uniquely determined, we denote it by A^{-1}. If A has an inverse, we say that A is **invertible**, or **non-singular**. If A is not invertible, it is sometimes called **singular**. We will use the terms "invertible" and "not invertible." Observe that A is the inverse of A^{-1}; that is, $(A^{-1})^{-1} = A$.

If A and B are invertible $n \times n$ matrices, then so is their product: in fact, it is easily verified that $(AB)^{-1} = B^{-1}A^{-1}$. It follows that if A is an invertible matrix, and k is a non-negative integer, then A^k is invertible with inverse $(A^{-1})^k$, which we also denote by A^{-k}.

It is also easy to see that A is invertible if and only if the transposed matrix $A^\top$ is invertible, in which case $(A^\top)^{-1} = (A^{-1})^\top$. Indeed, $AB = I = BA$ holds if and only if $B^\top A^\top = I = A^\top B^\top$.

We now develop a connection between invertible matrices and R-module isomorphisms. Recall from the previous section the R-module isomorphism

$$\Lambda : \quad R^{n \times n} \to \operatorname{Hom}_R(R^{1 \times n}, R^{1 \times n})$$

$$A \mapsto \lambda_A,$$

where for each $A \in R^{n \times n}$, λ_A is the corresponding R-linear map

$$\lambda_A : \quad R^{1 \times n} \to R^{1 \times n}$$

$$v \mapsto vA.$$

Evidently, λ_I is the identity map.

Theorem 14.5. *Let $A \in R^{n \times n}$, and let $\lambda_A : R^{1 \times n} \to R^{1 \times n}$ be the corresponding R-linear map. Then A is invertible if and only if λ_A is bijective, in which case $\lambda_{A^{-1}} = \lambda_A^{-1}$.*

Proof. Suppose A is invertible, and that B is its inverse. We have $AB = BA = I$, and hence $\lambda_{AB} = \lambda_{BA} = \lambda_I$, from which it follows (see (14.1)) that $\lambda_B \circ \lambda_A = \lambda_A \circ \lambda_B = \lambda_I$. Since λ_I is the identity map, this implies λ_A is bijective.

Suppose λ_A is bijective. We know that the inverse map λ_A^{-1} is also an R-linear map, and since the mapping Λ above is surjective, we have $\lambda_A^{-1} = \lambda_B$ for some $B \in R^{n \times n}$. Therefore, we have $\lambda_B \circ \lambda_A = \lambda_A \circ \lambda_B = \lambda_I$, and hence (again, see (14.1)) $\lambda_{AB} = \lambda_{BA} = \lambda_I$. Since the mapping Λ is injective, it follows that $AB = BA = I$. This implies A is invertible, with $A^{-1} = B$. $\square$

We also have:

Theorem 14.6. *Let $A \in R^{n \times n}$. The following are equivalent:*

(i) *A is invertible;*

(ii) *$\{\text{Row}_i(A)\}_{i=1}^n$ is a basis for $R^{1 \times n}$;*

(iii) *$\{\text{Col}_j(A)\}_{j=1}^n$ is a basis for $R^{n \times 1}$.*

Proof. We first prove the equivalence of (i) and (ii). By the previous theorem, A is invertible if and only if λ_A is bijective. Also, in the previous section, we observed that λ_A is surjective if and only if $\{\text{Row}_i(A)\}_{i=1}^n$ spans $R^{1 \times n}$, and that λ_A is injective if and only if $\{\text{Row}_i(A)\}_{i=1}^n$ is linearly independent.

The equivalence of (i) and (iii) follows by considering the transpose of A. $\square$

EXERCISE 14.6. Let R be a ring, and let A be a square matrix over R. Let us call B a **left inverse** of A if $BA = I$, and let us call C a **right inverse** of A if $AC = I$.

(a) Show that if A has both a left inverse B and a right inverse C, then $B = C$ and hence A is invertible.

(b) Assume that R is a field. Show that if A has either a left inverse or a right inverse, then A is invertible.

Note that part (b) of the previous exercise holds for arbitrary rings, but the proof of this is non-trivial, and requires the development of the theory of determinants, which we do not cover in this text.

EXERCISE 14.7. Show that if A and B are two square matrices over a field such that their product AB is invertible, then both A and B themselves must be invertible.

EXERCISE 14.8. Show that if A is a square matrix over an arbitrary ring, and A^k is invertible for some $k > 0$, then A is invertible.

EXERCISE 14.9. With notation as in Example 14.1, show that the matrix A is invertible if and only if $c_0 \in R^*$.

EXERCISE 14.10. With notation as in Example 14.2, show that the matrix A is invertible if and only if $x_i - x_j \in R^*$ for all $i \neq j$.

14.4 Gaussian elimination

Throughout this section, F denotes a field.

A matrix $B \in F^{m \times n}$ is said to be in **reduced row echelon form** if there exists a sequence of integers $(p_1, \ldots, p_r)$, with $0 \leq r \leq m$ and $1 \leq p_1 < p_2 < \cdots < p_r \leq n$, such that the following holds:

- for $i = 1, \ldots, r$, all of the entries in row i of B to the left of entry (i, p_i) are zero; that is, $B(i, j) = 0_F$ for $j = 1, \ldots, p_i - 1$;

- for $i = 1, \ldots, r$, all of the entries in column p_i of B above entry (i, p_i) are zero; that is, $B(i', p_i) = 0_F$ for $i' = 1, \ldots, i - 1$;

- for $i = 1, \ldots, r$, we have $B(i, p_i) = 1_F$;

- all entries in rows $r + 1, \ldots, m$ of B are zero; that is, $B(i, j) = 0_F$ for $i = r + 1, \ldots, m$ and $j = 1, \ldots, n$.

It is easy to see that if B is in reduced row echelon form, then the sequence $(p_1, \ldots, p_r)$ above is uniquely determined, and we call it the **pivot sequence** of B. Several further remarks are in order:

- All of the entries of B are completely determined by the pivot sequence, except for the entries (i, j) with $1 \leq i \leq r$ and $j > p_i$ with $j \notin \{p_{i+1}, \ldots, p_r\}$, which may be arbitrary.

- If B is an $n \times n$ matrix in reduced row echelon form whose pivot sequence is of length n, then B must be the $n \times n$ identity matrix.

- We allow for an empty pivot sequence (i.e., $r = 0$), which will be the case precisely when $B = 0_F^{m \times n}$.

Example 14.3. The following 4×6 matrix B over the rational numbers is in reduced row echelon form:

$$B = \begin{pmatrix} 0 & 1 & -2 & 0 & 0 & 3 \\ 0 & 0 & 0 & 1 & 0 & 2 \\ 0 & 0 & 0 & 0 & 1 & -4 \\ 0 & 0 & 0 & 0 & 0 & 0 \end{pmatrix}.$$

The pivot sequence of B is $(2, 4, 5)$. Notice that the first three rows of B form a linearly independent family of vectors, that columns 2, 4, and 5 form a linearly independent family of vectors, and that all of other columns of B are linear combinations of columns 2, 4, and 5. Indeed, if we truncate the pivot columns to their first three rows, we get the 3×3 identity matrix. □

Generalizing the previous example, if a matrix is in reduced row echelon form, it is easy to deduce the following properties, which turn out to be quite useful:

Theorem 14.7. *If B is a matrix in reduced row echelon form with pivot sequence $(p_1, \ldots, p_r)$, then:*

(i) *rows $1, 2, \ldots, r$ of B form a linearly independent family of vectors;*

(ii) *columns $p_1, \ldots, p_r$ of B form a linearly independent family of vectors, and all other columns of B can be expressed as linear combinations of columns $p_1, \ldots, p_r$.*

Proof. Exercise — just look at the matrix! $\square$

Gaussian elimination is an algorithm that transforms a given matrix $A \in F^{m \times n}$ into a matrix $B \in F^{m \times n}$, where B is in reduced row echelon form, and is obtained from A by a sequence of **elementary row operations**. There are three types of elementary row operations:

Type I: swap two rows;

Type II: multiply a row by a non-zero scalar;

Type III: add a scalar multiple of one row to a different row.

The application of any specific elementary row operation to an $m \times n$ matrix C can be affected by multiplying C on the left by a suitable $m \times m$ matrix X. Indeed, the matrix X corresponding to a particular elementary row operation is simply the matrix obtained by applying the same elementary row operation to the $m \times m$ identity matrix. It is easy to see that for every elementary row operation, the corresponding matrix X is invertible.

We now describe the basic version of Gaussian elimination. The input is an $m \times n$ matrix A, and the algorithm is described in Fig. 14.1.

The algorithm works as follows. First, it makes a copy B of A (this is not necessary if the original matrix A is not needed afterwards). The algorithm proceeds column by column, starting with the left-most column, so that after processing column j, the first j columns of B are in reduced row echelon form, and the current value of r represents the length of the pivot sequence. To process column j, in steps 3–6 the algorithm first searches for a non-zero element among $B(r+1, j), \ldots, B(m, j)$; if none is found, then the first $j + 1$ columns of B are already in reduced row echelon form. Otherwise, one of these non-zero elements is selected as the **pivot element** (the choice is arbitrary), which is then used in steps 8–13 to bring column j into the required form. After incrementing r, the pivot element is brought into position (r, j), using a Type I operation in step 9. Then the entry (r, j) is set to 1_F, using a Type II operation in step 10. Finally, all the entries above and below entry (r, j) are set to 0_F, using Type III operations in steps 11–13. Note that because

1. $B \leftarrow A, r \leftarrow 0$
2.   for $j \leftarrow 1$ to n do
3. $\ell \leftarrow 0, i \leftarrow r$
4.   while $\ell = 0$ and $i \leq m$ do
5. $i \leftarrow i + 1$
6.   if $B(i, j) \neq 0_F$ then $\ell \leftarrow i$
7.   if $\ell \neq 0$ then
8. $r \leftarrow r + 1$
9.   swap rows r and ℓ of B
10. $\mathrm{Row}_r(B) \leftarrow B(r, j)^{-1} \mathrm{Row}_r(B)$
11.   for $i \leftarrow 1$ to m do
12.   if $i \neq r$ then
13. $\mathrm{Row}_i(B) \leftarrow \mathrm{Row}_i(B) - B(i, j) \mathrm{Row}_r(B)$
14. output B

Fig. 14.1. Gaussian elimination

columns $1, \ldots, j - 1$ of B were already in reduced row echelon form, none of these operations changes any values in these columns.

As for the complexity of the algorithm, it is easy to see that it performs $O(mn)$ elementary row operations, each of which takes $O(n)$ operations in F, so a total of $O(mn^2)$ operations in F.

Example 14.4. Consider the execution of the Gaussian elimination algorithm on input

$$A = \begin{pmatrix} [0] & [1] & [1] \\ [2] & [1] & [2] \\ [2] & [2] & [0] \end{pmatrix} \in \mathbb{Z}_3^{3 \times 3}.$$

After copying A into B, the algorithm transforms B as follows:

$$\begin{pmatrix} [0] & [1] & [1] \\ [2] & [1] & [2] \\ [2] & [2] & [0] \end{pmatrix} \xrightarrow{\mathrm{Row}_1 \leftrightarrow \mathrm{Row}_2} \begin{pmatrix} [2] & [1] & [2] \\ [0] & [1] & [1] \\ [2] & [2] & [0] \end{pmatrix} \xrightarrow{\mathrm{Row}_1 \leftarrow [2]\,\mathrm{Row}_1} \begin{pmatrix} [1] & [2] & [1] \\ [0] & [1] & [1] \\ [2] & [2] & [0] \end{pmatrix}$$

$$\xrightarrow{\mathrm{Row}_3 \leftarrow \mathrm{Row}_3 - [2]\,\mathrm{Row}_1} \begin{pmatrix} [1] & [2] & [1] \\ [0] & [1] & [1] \\ [0] & [1] & [1] \end{pmatrix} \xrightarrow{\mathrm{Row}_1 \leftarrow \mathrm{Row}_1 - [2]\,\mathrm{Row}_2} \begin{pmatrix} [1] & [0] & [2] \\ [0] & [1] & [1] \\ [0] & [1] & [1] \end{pmatrix}$$

$$\xrightarrow{\text{Row}_3 \leftarrow \text{Row}_3 - \text{Row}_2} \begin{pmatrix} [1] & [0] & [2] \\ [0] & [1] & [1] \\ [0] & [0] & [0] \end{pmatrix} . \quad \square$$

Suppose the Gaussian elimination algorithm performs a total of t elementary row operations. Then as discussed above, the application of the eth elementary row operation, for $e = 1, \dots, t$, amounts to multiplying the current value of the matrix B on the left by a particular invertible $m \times m$ matrix X_e. Therefore, the final output value of B satisfies the equation

$$B = XA \quad \text{where} \quad X = X_t X_{t-1} \cdots X_1.$$

Since the product of invertible matrices is also invertible, we see that X itself is invertible.

Although the algorithm as presented does not compute the matrix X, it can be easily modified to do so. The resulting algorithm, which we call **extended Gaussian elimination**, is the same as plain Gaussian elimination, except that we initialize the matrix X to be the $m \times m$ identity matrix, and we add the following steps:

- just before step 9: swap rows r and ℓ of X;
- just before step 10: $\text{Row}_r(X) \leftarrow B(r, j)^{-1} \text{Row}_r(X)$;
- just before step 13: $\text{Row}_i(X) \leftarrow \text{Row}_i(X) - B(i, j) \text{Row}_r(X)$.

At the end of the algorithm we output X in addition to B.

So we simply perform the same elementary row operations on X that we perform on B. The reader may verify that the above algorithm is correct, and that it uses $O(mn(m + n))$ operations in F.

Example 14.5. Continuing with Example 14.4, the execution of the extended Gaussian elimination algorithm initializes X to the identity matrix, and then transforms X as follows:

$$\begin{pmatrix} [1] & [0] & [0] \\ [0] & [1] & [0] \\ [0] & [0] & [1] \end{pmatrix} \xrightarrow{\text{Row}_1 \leftrightarrow \text{Row}_2} \begin{pmatrix} [0] & [1] & [0] \\ [1] & [0] & [0] \\ [0] & [0] & [1] \end{pmatrix} \xrightarrow{\text{Row}_1 \leftarrow [2] \text{Row}_1} \begin{pmatrix} [0] & [2] & [0] \\ [1] & [0] & [0] \\ [0] & [0] & [1] \end{pmatrix}$$

$$\xrightarrow{\text{Row}_3 \leftarrow \text{Row}_3 - [2] \text{Row}_1} \begin{pmatrix} [0] & [2] & [0] \\ [1] & [0] & [0] \\ [0] & [2] & [1] \end{pmatrix} \xrightarrow{\text{Row}_1 \leftarrow \text{Row}_1 - [2] \text{Row}_2} \begin{pmatrix} [1] & [2] & [0] \\ [1] & [0] & [0] \\ [0] & [2] & [1] \end{pmatrix}$$

$$\xrightarrow{\text{Row}_3 \leftarrow \text{Row}_3 - \text{Row}_2} \begin{pmatrix} [1] & [2] & [0] \\ [1] & [0] & [0] \\ [2] & [2] & [1] \end{pmatrix}. \quad \square$$

EXERCISE 14.11. For each type of elementary row operation, describe the matrix X which corresponds to it, as well as X^{-1}.

EXERCISE 14.12. Given a matrix $B \in F^{m \times n}$ in reduced row echelon form, show how to compute its pivot sequence using $O(n)$ operations in F.

EXERCISE 14.13. In §4.4, we saw how to speed up matrix multiplication over $\mathbb{Z}$ using the Chinese remainder theorem. In this exercise, you are to do the same, but for performing Gaussian elimination over $\mathbb{Z}_p$, where p is a large prime. Suppose you are given an $m \times m$ matrix A over $\mathbb{Z}_p$, where $\text{len}(p) = \Theta(m)$. Straightforward application of Gaussian elimination would require $O(m^3)$ operations in $\mathbb{Z}_p$, each of which takes time $O(m^2)$, leading to a total running time of $O(m^5)$. Show how to use the techniques of §4.4 to reduce the running time of Gaussian elimination to $O(m^4)$.

14.5 Applications of Gaussian elimination

Throughout this section, A is an arbitrary $m \times n$ matrix over a field F, and $XA = B$, where X is an invertible $m \times m$ matrix, and B is an $m \times n$ matrix in reduced row echelon form with pivot sequence $(p_1, \ldots, p_r)$. This is precisely the information produced by the extended Gaussian elimination algorithm, given A as input (the pivot sequence can easily be "read" directly from B—see Exercise 14.12). Also, let

$$\lambda_A : \quad F^{1 \times m} \to F^{1 \times n}$$

$$v \mapsto vA$$

be the linear map corresponding to A.

Computing the image and kernel

Consider first the **row space** of A, by which we mean the subspace of $F^{1 \times n}$ spanned by $\{\text{Row}_i(A)\}_{i=1}^{m}$, or equivalently, the image of λ_A.

We claim that the row space of A is the same as the row space of B. To see this, note that since $B = XA$, for every $v \in F^{1 \times m}$, we have $vB = v(XA) = (vX)A$, and so the row space of B is contained in the row space of A. For the other containment, note that since X is invertible, we can write $A = X^{-1}B$, and apply the same argument.

Further, note that the row space of B, and hence that of A, clearly has dimension r. Indeed, as stated in Theorem 14.7, rows $1, \ldots, r$ of B form a basis for the row space of B.

Consider next the kernel K of λ_A, or what we might call the **row null space** of A. We claim that $\{\text{Row}_i(X)\}_{i=r+1}^m$ is a basis for K. Clearly, just from the fact that $XA = B$ and the fact that rows $r + 1, \ldots, m$ of B are zero, it follows that rows $r + 1, \ldots, m$ of X are contained in K. Furthermore, as X is invertible, $\{\text{Row}_i(X)\}_{i=1}^m$ is a basis for $F^{1 \times m}$ (see Theorem 14.6). Thus, the family of vectors $\{\text{Row}_i(X)\}_{i=r+1}^m$ is linearly independent and spans a subspace K' of K. It suffices to show that $K' = K$. Suppose to the contrary that $K' \subsetneq K$, and let $v \in K \setminus K'$. As $\{\text{Row}_i(X)\}_{i=1}^m$ spans $F^{1 \times m}$, we may write $v = \sum_{i=1}^m c_i \text{Row}_i(X)$; moreover, as $v \notin K'$, we must have $c_i \neq 0_F$ for some $i = 1, \ldots, r$. Setting $\tilde{v} := (c_1, \ldots, c_m)$, we see that $v = \tilde{v} X$, and so

$$\lambda_A(v) = vA = (\tilde{v} X)A = \tilde{v}(X A) = \tilde{v} B.$$

Furthermore, since $\{\text{Row}_i(B)\}_{i=1}^r$ is linearly independent, rows $r + 1, \ldots, m$ of B are zero, and $\tilde{v}$ has a non-zero entry in one of its first r positions, we see that $\tilde{v} B$ is not the zero vector. We have derived a contradiction, and hence may conclude that $K' = K$.

Finally, note that if $m = n$, then A is invertible if and only if its row space has dimension m, which holds if and only if $r = m$, and in the latter case, B is the identity matrix, and hence X is the inverse of A.

Let us summarize the above discussion:

- The first r rows of B form a basis for the row space of A (i.e., the image of λ_A).
- The last $m - r$ rows of X form a basis for the row null space of A (i.e., the kernel of λ_A).
- If $m = n$, then A is invertible (i.e., λ_A is an isomorphism) if and only if $r = m$, in which case X is the inverse of A (i.e., the matrix of λ_A^{-1} relative to the standard basis).

So we see that from the output of the extended Gaussian elimination algorithm, we can simply "read off" bases for both the image and the kernel, as well as the inverse (if it exists), of a linear map represented as a matrix with respect to given bases. Also note that this procedure provides a "constructive" version of Theorem 13.28.

Example 14.6. Continuing with Examples 14.4 and 14.5, we see that the vectors $([1], [0], [2])$ and $([0], [1], [1])$ form a basis for the row space of A, while the vector $([2], [2], [1])$ is a basis for the row null space of A. $\square$

Solving systems of linear equations

Suppose that in addition to the matrix A, we are given $w \in F^{1 \times n}$, and want to find a solution $v \in F^{1 \times m}$ (or perhaps describe all solutions), to the equation

$$vA = w. \tag{14.3}$$

Equivalently, we can phrase the problem as finding an element (or describing all elements) of the set $\lambda_A^{-1}(\{w\})$.

Now, if there exists a solution at all, say $v \in F^{1 \times m}$, then $\lambda_A(v) = \lambda_A(v')$ if and only if $v \equiv v' \pmod{K}$, where K is the kernel of λ_A. It follows that the set of all solutions to (14.3) is $v + K = \{v + v_0 : v_0 \in K\}$. Thus, given a basis for K and any solution v to (14.3), we have a complete and concise description of the set of solutions to (14.3).

As we have discussed above, the last $m - r$ rows of X form a basis for K, so it suffices to determine if $w \in \operatorname{Im} \lambda_A$, and if so, determine a single pre-image v of w.

Also as we discussed, $\operatorname{Im} \lambda_A$, that is, the row space of A, is equal to the row space of B, and because of the special form of B, we can quickly and easily determine if the given w is in the row space of B, as follows. By definition, w is in the row space of B if and only if there exists a vector $\bar{v} \in F^{1 \times m}$ such that $\bar{v}B = w$. We may as well assume that all but the first r entries of $\bar{v}$ are zero. Moreover, $\bar{v}B = w$ implies that for $i = 1, \dots, r$, the ith entry of $\bar{v}$ is equal to the p_ith entry of w. Thus, the vector $\bar{v}$, if it exists, is completely determined by the entries of w at positions $p_1, \dots, p_r$. We can construct $\bar{v}$ satisfying these conditions, and then test if $\bar{v}B = w$. If not, then we may conclude that (14.3) has no solutions; otherwise, setting $v := \bar{v}X$, we see that $vA = (\bar{v}X)A = \bar{v}(XA) = \bar{v}B = w$, and so v is a solution to (14.3).

One easily verifies that if we implement the above procedure as an algorithm, the work done in addition to running the extended Gaussian elimination algorithm amounts to $O(m(n + m))$ operations in F.

A special case of the above procedure is when $m = n$ and A is invertible, in which case (14.3) has a unique solution, namely, $v := wX$, since in this case, $X = A^{-1}$.

The rank of a matrix

We define the **row rank** of A to be the dimension of its row space, which is equal to $\dim_F(\operatorname{Im} \lambda_A)$. The **column space** of A is defined as the subspace of $F^{m \times 1}$ spanned by $\{\operatorname{Col}_j(A)\}_{j=1}^n$; that is, the column space of A is $\{Az : z \in F^{n \times 1}\}$. The **column rank** of A is the dimension of its column space.

Now, the column space of A need not be the same as the column space of B, but from the identity $B = XA$, and the fact that X is invertible, it easily follows that these two subspaces are isomorphic (via the map that sends $y \in F^{m \times 1}$ to Xy), and hence have the same dimension. Moreover, by Theorem 14.7, the column rank of B is r, which is the same as the row rank of A.

So we may conclude: *The column rank and row rank of A are the same.*

Because of this, we may define the **rank** of a matrix to be the common value of its row and column rank.

The orthogonal complement of a subspace

So as to give equal treatment to rows and columns, one can also define the **column null space** of A to be the kernel of the linear map defined by multiplication on the left by A; that is, the column null space of A is $\{z \in F^{n \times 1} : Az = 0_F^{m \times 1}\}$. By applying the results above to the transpose of A, we see that the column null space of A has dimension $n - r$, where r is the rank of A.

Let $U \subseteq F^{1 \times n}$ be the row space of A, and let $U^\perp \subseteq F^{1 \times n}$ denote the set of all vectors $\bar{u} \in F^{1 \times n}$ whose transpose $\bar{u}^\top$ belongs to the column null space of A. Now, U is a subspace of $F^{1 \times n}$ of dimension r and $U^\perp$ is a subspace of $F^{1 \times n}$ of dimension $n - r$. The space $U^\perp$ consists precisely of all vectors $\bar{u} \in F^{1 \times n}$ that are "orthogonal" to all vectors $u \in U$, in the sense that the "inner product" $u\bar{u}^\top$ is zero. For this reason, $U^\perp$ is sometimes called the "orthogonal complement of U."

Clearly, $U^\perp$ is determined by the subspace U itself, and does not depend on the particular choice of matrix A. It is also easy to see that the orthogonal complement of $U^\perp$ is U; that is, $(U^\perp)^\perp = U$. This follows immediately from the fact that $U \subseteq (U^\perp)^\perp$ and $\dim_F((U^\perp)^\perp) = n - \dim_F(U^\perp) = \dim_F(U)$.

Now suppose that $U \cap U^\perp = \{0\}$. Then by Theorem 13.11, we have an isomorphism of $U \times U^\perp$ with $U + U^\perp$, and since $U \times U^\perp$ has dimension n, it must be the case that $U + U^\perp = F^{1 \times n}$. It follows that every element of $F^{1 \times n}$ can be expressed uniquely as $u + \bar{u}$, where $u \in U$ and $\bar{u} \in U^\perp$.

We emphasize that the observations in the previous paragraph hinged on the assumption that $U \cap U^\perp = \{0\}$, which itself holds provided U contains no non-zero "self-orthogonal vectors" u such that $uu^\top$ is zero. If F is the field of real numbers, then of course there are no non-zero self-orthogonal vectors, since $uu^\top$ is the sum of the squares of the entries of u. However, for other fields, there may very well be non-zero self-orthogonal vectors. As an example, if $F = \mathbb{Z}_2$, then any vector u with an even number of 1-entries is self orthogonal.

So we see that while much of the theory of vector spaces and matrices carries over without change from familiar ground fields, like the real numbers, to arbitrary ground fields F, not everything does. In particular, the usual decomposition of a vector space into a subspace and its orthogonal complement breaks down, as does any other procedure that relies on properties specific to "inner product spaces."

For the following three exercises, as above, A is an arbitrary $m \times n$ matrix over a field F, and $XA = B$, where X is an invertible $m \times m$ matrix, and B is in reduced row echelon form.

EXERCISE 14.14. Show that the column null space of A is the same as the column null space of B.

EXERCISE 14.15. Show how to compute a basis for the column null space of A using $O(r(n-r))$ operations in F, given A and B.

EXERCISE 14.16. Show that the matrix B is uniquely determined by A; more precisely, show that if $X'A = B'$, where X' is an invertible $m \times m$ matrix, and B' is in reduced row echelon form, then $B' = B$.

In the following two exercises, the theory of determinants could be used; however, they can all be solved directly, without too much difficulty, using just the ideas developed so far in the text.

EXERCISE 14.17. Let p be a prime. A matrix $A \in \mathbb{Z}^{m \times m}$ is called **invertible modulo** p if there exists a matrix $B \in \mathbb{Z}^{m \times m}$ such that $AB \equiv BA \equiv I \pmod{p}$, where I is the $m \times m$ integer identity matrix. Here, two matrices are considered congruent with respect to a given modulus if their corresponding entries are congruent. Show that A is invertible modulo p if and only if A is invertible over $\mathbb{Q}$, and the entries of A^{-1} lie in $\mathbb{Q}^{(p)}$ (see Example 7.26).

EXERCISE 14.18. You are given a matrix $A \in \mathbb{Z}^{m \times m}$ and a prime p such that A is invertible modulo p (see previous exercise). Suppose that you are also given $w \in \mathbb{Z}^{1 \times m}$.

(a) Show how to efficiently compute a vector $v \in \mathbb{Z}^{1 \times m}$ such that $vA \equiv w \pmod{p}$, and that v is uniquely determined modulo p.

(b) Given a vector v as in part (a), along with an integer $e \geq 1$, show how to efficiently compute $\hat{v} \in \mathbb{Z}^{1 \times m}$ such that $\hat{v}A \equiv w \pmod{p^e}$, and that $\hat{v}$ is uniquely determined modulo p^e. Hint: mimic the "lifting" procedure discussed in §12.5.2.

(c) Using parts (a) and (b), design and analyze an efficient algorithm that takes the matrix A and the prime p as input, together with a bound H on the absolute value of the numerator and denominator of the entries of the vector v' that is the unique (rational) solution to the equation $v'A = w$. Your algorithm should run in time polynomial in the length of H, the length of p, and the sum of the lengths of the entries of A and w. Hint: use rational reconstruction, but be sure to fully justify its application.

Note that in the previous exercise, one can use the theory of determinants to derive good bounds, in terms of the lengths of the entries of A and w, on the size of the least prime p such that A is invertible modulo p (assuming A is invertible over the rationals), and on the length of the numerator and denominator of the entries of rational solution v' to the equation $v'A = w$. The interested reader who is familiar with the basic theory of determinants is encouraged to establish such bounds.

The next two exercises illustrate how Gaussian elimination can be adapted, in certain cases, to work in rings that are not necessarily fields. Let R be an arbitrary ring. A matrix $B \in R^{m \times n}$ is said to be in **row echelon form** if there exists a pivot sequence $(p_1, \ldots, p_r)$, with $0 \le r \le m$ and $1 \le p_1 < p_2 < \cdots < p_r \le n$, such that the following holds:

- for $i = 1, \ldots, r$, all of the entries in row i of B to the left of entry (i, p_i) are zero;
- for $i = 1, \ldots, r$, we have $B(i, p_i) \ne 0_R$;
- all entries in rows $r + 1, \ldots, m$ of B are zero.

EXERCISE 14.19. Let R be the ring $\mathbb{Z}_{p^e}$, where p is prime and $e > 1$. Let $\pi := [p] \in R$. The goal of this exercise is to develop an efficient algorithm for the following problem: given a matrix $A \in R^{m \times n}$, with $m > n$, find a vector $v \in R^{1 \times m}$ such that $vA = 0_R^{1 \times n}$ but $v \notin \pi R^{1 \times m}$.

(a) Show how to modify the extended Gaussian elimination algorithm to solve the following problem: given a matrix $A \in R^{m \times n}$, compute $X \in R^{m \times m}$ and $B \in R^{m \times n}$, such that $XA = B$, X is invertible, and B is in row echelon form. Your algorithm should run in time $O(mn(m + n)e^2 \operatorname{len}(p)^2)$. Assume that the input includes the values p and e. Hint: when choosing a pivot element, select one divisible by a minimal power of π; as in ordinary Gaussian elimination, your algorithm should only use elementary row operations to transform the input matrix.

(b) Using the fact that the matrix X computed in part (a) is invertible, argue that none of its rows belong to $\pi R^{1 \times m}$.

(c) Argue that if $m > n$ and the matrix B computed in part (a) has pivot sequence $(p_1, \ldots, p_r)$, then $m - r > 0$ and if v is any one of the last $m - r$ rows of X, then $vA = 0_R^{1 \times n}$.

(d) Give an example that shows that $\{\operatorname{Row}_i(B)\}_{i=1}^r$ need not be linearly independent, and that $\{\operatorname{Row}_i(X)\}_{i=r+1}^m$ need not span the kernel of the linear map λ_A corresponding to A.

EXERCISE 14.20. Let R be the ring $\mathbb{Z}_\ell$, where $\ell > 1$ is an integer. You are given a matrix $A \subset R^{m \times n}$. Show how to efficiently compute $X \in R^{m \times m}$ and $B \in R^{m \times n}$

such that $XA = B$, X is invertible, and B is in row echelon form. Your algorithm should run in time $O(mn(m + n) \operatorname{len}(\ell)^2)$. Hint: to zero-out entries, you should use "rotations"—for integers a, b, d, s, t with

$$d = \gcd(a, b) \neq 0 \quad \text{and} \quad as + bt = d,$$

and for row indices r, i, a rotation simultaneously updates rows r and i of a matrix C as follows:

$$(\operatorname{Row}_r(C), \operatorname{Row}_i(C)) \leftarrow (s \operatorname{Row}_r(C) + t \operatorname{Row}_i(C), -\frac{b}{d} \operatorname{Row}_r(C) + \frac{a}{d} \operatorname{Row}_i(C));$$

observe that if $C(r, j) = [a]_\ell$ and $C(i, j) = [b]_\ell$ before applying the rotation, then $C(r, j) = [d]_\ell$ and $C(i, j) = [0]_\ell$ after the rotation.

EXERCISE 14.21. Consider again the setting in Exercise 14.3. Show that $A \in F^{m \times (m-\ell)}$ is a parity check matrix for U if and only if $\{\operatorname{Col}_j(A)^\top\}_{i=1}^{m-\ell}$ is a basis for the orthogonal complement of $\operatorname{Vec}_S(U) \subseteq F^{1 \times m}$.

EXERCISE 14.22. Let $\{v_i\}_{i=1}^n$ be a family of vectors, where $v_i \in \mathbb{R}^{1 \times \ell}$ for each $i = 1, \ldots, n$. We say that $\{v_i\}_{i=1}^n$ is **pairwise orthogonal** if $v_i v_j^\top = 0$ for all $i \neq j$. Show that every pairwise orthogonal family of non-zero vectors over $\mathbb{R}$ is linearly independent.

EXERCISE 14.23. The purpose of this exercise is to use linear algebra to prove that any pairwise independent family of hash functions (see §8.7) must contain a large number of hash functions. More precisely, let $\{\Phi_r\}_{r \in R}$ be a pairwise independent family of hash functions from S to T, with $|T| \geq 2$. Our goal is to show that $|R| \geq |S|$. Let $n := |S|$, and $m := |T|$, and $\ell := |R|$. Write $R = \{r_1, \ldots, r_\ell\}$ and $S = \{s_1, \ldots, s_n\}$. Without loss of generality, we may assume that T is a set of non-zero real numbers that sum to zero (e.g., $T = \{1, \ldots, m - 1, -m(m - 1)/2\}$). Now define the matrix $A \in \mathbb{R}^{n \times \ell}$ with $A(i, j) := \Phi_{r_j}(s_i)$. Show that $\{\operatorname{Row}_i(A)\}_{i=1}^n$ is a pairwise orthogonal family of non-zero vectors (see previous exercise). From this, deduce that $\ell \geq n$.

14.6 Notes

While a trivial application of the defining formulas yields a simple algorithm for multiplying two $n \times n$ matrices over a ring R that uses $O(n^3)$ operations in R, this algorithm is not the best, asymptotically speaking. The currently fastest algorithm for this problem, due to Coppersmith and Winograd [28], uses $O(n^\omega)$ operations in R, where $\omega < 2.376$. We note, however, that the good old $O(n^3)$ algorithm is still the only one used in almost any practical setting.

15

Subexponential-time discrete logarithms and factoring

This chapter presents subexponential-time algorithms for computing discrete logarithms and for factoring integers. These algorithms share a common technique, which makes essential use of the notion of a **smooth number**.

15.1 Smooth numbers

If y is a non-negative real number and m is a positive integer, then we say that m is y-**smooth** if all prime divisors of m are at most y.

For $0 \le y \le x$, let us define $\Psi(y, x)$ to be the number of y-smooth integers up to x. The following theorem gives us a lower bound on $\Psi(y, x)$, which will be crucial in the analysis of our discrete logarithm and factoring algorithms.

Theorem 15.1. *Let y be a function of x such that*

$$\frac{y}{\log x} \to \infty \text{ and } u := \frac{\log x}{\log y} \to \infty$$

as $x \to \infty$. Then

$$\Psi(y, x) \ge x \cdot \exp[(-1 + o(1))u \log \log x].$$

Proof. Let us write $u = \lfloor u \rfloor + \delta$, where $0 \le \delta < 1$. Let us split the primes up to y into two sets: the set V of "very small" primes that are at most $y^\delta/2$, and the set W of other primes that are greater than $y^\delta/2$ but at most y. To simplify matters, let us also include the integer 1 in the set V.

By Bertrand's postulate (Theorem 5.8), there exists a constant $C > 0$ such that $|W| \ge Cy/\log y$ for sufficiently large y. By the assumption that $y/\log x \to \infty$ as $x \to \infty$, we also have $|W| \ge 2\lfloor u \rfloor$ for sufficiently large x.

To derive the lower bound, we shall count those integers that can be built up by multiplying together $\lfloor u \rfloor$ distinct elements of W, together with one element of V.

399

These products are clearly distinct, y-smooth numbers, and each is bounded by x, since each is at most $y^{\lfloor u \rfloor} y^{\delta} = y^u = x$.

If S denotes the set of all of these products, then for x sufficiently large, we have

$$
\begin{aligned}
|S| &= \binom{|W|}{\lfloor u \rfloor} \cdot |V| \\
&= \frac{|W|(|W| - 1) \cdots (|W| - \lfloor u \rfloor + 1)}{\lfloor u \rfloor !} \cdot |V| \\
&\geq \left(\frac{|W|}{2u} \right)^{\lfloor u \rfloor} \cdot |V| \\
&\geq \left(\frac{Cy}{2u \log y} \right)^{\lfloor u \rfloor} \cdot |V| \\
&= \left(\frac{Cy}{2 \log x} \right)^{u - \delta} \cdot |V|.
\end{aligned}
$$

Taking logarithms, we have

$$
\begin{aligned}
\log|S| &\geq (u - \delta)(\log y - \log \log x + \log(C/2)) + \log|V| \\
&= \log x - u \log \log x + (\log|V| - \delta \log y) + \\
&\quad O(u + \log \log x).
\end{aligned}
\tag{15.1}
$$

To prove the theorem, it suffices to show that

$$
\log|S| \geq \log x - (1 + o(1))u \log \log x.
$$

Under our assumption that $u \to \infty$, the term $O(u + \log \log x)$ in (15.1) is clearly $o(u \log \log x)$, and so it will suffice to show that the term $(\log|V| - \delta \log y)$ is also $o(u \log \log x)$. But by Chebyshev's theorem (Theorem 5.1), for some positive constant D, we have

$$
Dy^{\delta} / \log y \leq |V| \leq y^{\delta},
$$

and taking logarithms, and again using the fact that $u \to \infty$, we have

$$
\log|V| - \delta \log y = O(\log \log y) = o(u \log \log x). \quad \square
$$

15.2 An algorithm for discrete logarithms

We now present a probabilistic, subexponential-time algorithm for computing discrete logarithms. The input to the algorithm is p, q, γ, α, where p and q are primes, with $q \mid (p - 1)$, γ is an element of $\mathbb{Z}_p^*$ generating a subgroup G of $\mathbb{Z}_p^*$ of order q, and $\alpha \in G$.

We shall make the simplifying assumption that $q^2 \nmid (p - 1)$, which is equivalent to saying that $q \nmid m := (p - 1)/q$. Although not strictly necessary, this assumption simplifies the design and analysis of the algorithm, and moreover, for cryptographic applications, this assumption is almost always satisfied. Exercises 15.1–15.3 below explore how this assumption may be lifted, as well as other generalizations.

At a high level, the main goal of our discrete logarithm algorithm is to find a random representation of 1 with respect to γ and α — as discussed in Exercise 11.12, this allows us to compute $\log_\gamma \alpha$ (with high probability). More precisely, our main goal is to compute integers r and s in a probabilistic fashion, such that $\gamma^r \alpha^s = 1$ and $[s]_q$ is uniformly distributed over $\mathbb{Z}_q$. Having accomplished this, then with probability $1 - 1/q$, we shall have $s \not\equiv 0 \pmod{q}$, which allows us to compute $\log_\gamma \alpha$ as $-rs^{-1} \bmod q$.

Let H be the subgroup of $\mathbb{Z}_p^*$ of order m. Our assumption that $q \nmid m$ implies that $G \cap H = \{1\}$, since the multiplicative order of any element in the intersection must divide both q and m, and so the only possibility is that the multiplicative order is 1. Therefore, the map $\rho : G \times H \to \mathbb{Z}_p^*$ that sends (β, δ) to $\beta\delta$ is injective (Theorem 6.25), and since $|\mathbb{Z}_p^*| = qm$, it must be surjective as well.

We shall use this fact in the following way: if β is chosen uniformly at random from G, and δ is chosen uniformly at random from H (and independent of β), then $\beta\delta$ is uniformly distributed over $\mathbb{Z}_p^*$. Furthermore, since H is the image of the q-power map on $\mathbb{Z}_p^*$, we may generate a random $\delta \in H$ simply by choosing $\hat{\delta} \in \mathbb{Z}_p^*$ at random, and setting $\delta := \hat{\delta}^q$.

The discrete logarithm algorithm uses a "smoothness parameter" y. We will discuss choice of y below, when we analyze the running time of the algorithm; for now, we only assume that $y < p$. Let $p_1, \ldots, p_k$ be an enumeration of the primes up to y. Let $\pi_i := [p_i]_p \in \mathbb{Z}_p^*$ for $i = 1, \ldots, k$.

The algorithm has two stages.

In the first stage, we find relations of the form

$$\gamma^{r_i} \alpha^{s_i} \delta_i = \pi_1^{e_{i1}} \ldots \pi_k^{e_{ik}}, \tag{15.2}$$

for $i = 1, \ldots, k + 1$, where $r_i, s_i, e_{i1}, \ldots, e_{ik} \in \mathbb{Z}$ and $\delta_i \in H$ for each i.

We obtain each such relation by a randomized search, as follows: we choose $r_i, s_i \in \{0, \ldots, q - 1\}$ at random, as well as $\hat{\delta}_i \in \mathbb{Z}_p^*$ at random; we then compute $\delta_i := \hat{\delta}_i^q$, $\beta_i := \gamma^{r_i} \alpha^{s_i}$, and $m_i := \mathrm{rep}(\beta_i \delta_i)$. Now, the value β_i is uniformly distributed over G, while δ_i is uniformly distributed over H; therefore, the product $\beta_i \delta_i$ is uniformly distributed over $\mathbb{Z}_p^*$, and hence m_i is uniformly distributed over $\{1, \ldots, p - 1\}$. Next, we simply try to factor m_i by trial division, trying all the primes $p_1, \ldots, p_k$ up to y. If we are lucky, we completely factor m_i in this way, obtaining a factorization

$$m_i = p_1^{e_{i1}} \cdots p_k^{e_{ik}},$$

for some exponents $e_{i1}, \ldots, e_{ik}$, and we get the relation (15.2). If we are unlucky, then we simply keep trying until we are lucky.

For $i = 1, \ldots, k+1$, let $v_i := (e_{i1}, \ldots, e_{ik}) \in \mathbb{Z}^{\times k}$, and let $\bar{v}_i$ denote the image of v_i in $\mathbb{Z}_q^{\times k}$ (i.e., $\bar{v}_i := ([e_{i1}]_q, \ldots, [e_{ik}]_q)$). Since $\mathbb{Z}_q^{\times k}$ is a vector space over the field $\mathbb{Z}_q$ of dimension k, the family of vectors $\bar{v}_1, \ldots, \bar{v}_{k+1}$ must be linearly dependent. The second stage of the algorithm uses Gaussian elimination over $\mathbb{Z}_q$ (see §14.4) to find a linear dependence among the vectors $\bar{v}_1, \ldots, \bar{v}_{k+1}$, that is, to find integers $c_1, \ldots, c_{k+1} \in \{0, \ldots, q-1\}$, not all zero, such that

$$(e_1, \ldots, e_k) := c_1 v_1 + \cdots + c_{k+1} v_{k+1} \in q\mathbb{Z}^{\times k}.$$

Raising each equation (15.2) to the corresponding power c_i, and multiplying them all together, we obtain

$$\gamma^r \alpha^s \delta = \pi_1^{e_1} \cdots \pi_k^{e_k},$$

where

$$r := \sum_{i=1}^{k+1} c_i r_i, \quad s := \sum_{i=1}^{k+1} c_i s_i, \quad \text{and } \delta := \prod_{i=1}^{k+1} \delta_i^{c_i}.$$

Now, $\delta \in H$, and since each e_j is a multiple of q, we also have $\pi_j^{e_j} \in H$ for $j = 1, \ldots, k$. It follows that $\gamma^r \alpha^s \in H$. But since $\gamma^r \alpha^s \in G$ as well, and $G \cap H = \{1\}$, it follows that $\gamma^r \alpha^s = 1$. If we are lucky (and we will be with overwhelming probability, as we discuss below), we will have $s \not\equiv 0 \pmod{q}$, in which case, we can compute $s' := s^{-1} \bmod q$, obtaining

$$\alpha = \gamma^{-rs'},$$

and hence $-rs' \bmod q$ is the discrete logarithm of α to the base γ. If we are very unlucky, we will have $s \equiv 0 \pmod{q}$, at which point the algorithm simply quits, reporting "failure."

The entire algorithm, called Algorithm SEDL, is presented in Fig. 15.1.

As already argued above, if Algorithm SEDL does not output "failure," then its output is indeed the discrete logarithm of α to the base γ. There remain three questions to answer:

1. What is the expected running time of Algorithm SEDL?

2. How should the smoothness parameter y be chosen so as to minimize the expected running time?

3. What is the probability that Algorithm SEDL outputs "failure"?

Let us address these questions in turn. As for the expected running time, let σ be the probability that a random element of $\{1, \ldots, p-1\}$ is y-smooth. Then

$i \leftarrow 0$

repeat

    $i \leftarrow i + 1$

    repeat

        choose $r_i, s_i \in \{0, \ldots, q-1\}$ at random

        choose $\hat{\delta}_i \in \mathbb{Z}_p^*$ at random

        $\beta_i \leftarrow \gamma^{r_i}\alpha^{s_i}, \quad \delta_i \leftarrow \hat{\delta}_i^q, \quad m_i \leftarrow \mathrm{rep}(\beta_i\delta_i)$

        test if m_i is y-smooth (trial division)

    until $m_i = p_1^{e_{i1}} \cdots p_k^{e_{ik}}$ for some integers $e_{i1}, \ldots, e_{ik}$

    until $i = k + 1$

set $v_i \leftarrow (e_{i1}, \ldots, e_{ik}) \in \mathbb{Z}^{\times k}$ for $i = 1, \ldots, k + 1$

apply Gaussian elimination over $\mathbb{Z}_q$ to find integers $c_1, \ldots, c_{k+1} \in$
    $\{0, \ldots, q-1\}$, not all zero, such that
    $c_1 v_1 + \cdots + c_{k+1} v_{k+1} \in q\mathbb{Z}^{\times k}$.

$r \leftarrow \sum_{i=1}^{k+1} c_i r_i, \quad s \leftarrow \sum_{i=1}^{k+1} c_i s_i$

if $s \equiv 0 \pmod{q}$

    then output "failure"

    else output $-rs^{-1} \bmod q$

Fig. 15.1. Algorithm SEDL

the expected number of attempts needed to produce a single relation is σ^{-1}, and so the expected number of attempts to produce $k + 1$ relations is $(k + 1)\sigma^{-1}$. In each attempt, we perform trial division using $p_1, \ldots, p_k$, along with a few other minor computations, leading to a total expected running time in stage 1 of $k^2\sigma^{-1} \cdot \mathrm{len}(p)^{O(1)}$. The running time in stage 2 is dominated by the Gaussian elimination step, which takes time $k^3 \cdot \mathrm{len}(p)^{O(1)}$. Thus, if Z is the total running time of the algorithm, then we have

$$E[Z] \leq (k^2\sigma^{-1} + k^3) \cdot \mathrm{len}(p)^{O(1)}. \tag{15.3}$$

Let us assume for the moment that

$$y = \exp[(\log p)^{\lambda + o(1)}] \tag{15.4}$$

for some constant λ with $0 < \lambda < 1$. Our final choice of y will indeed satisfy this assumption. Consider the probability σ. We have

$$\sigma = \Psi(y, p - 1)/(p - 1) = \Psi(y, p)/(p - 1) \geq \Psi(y, p)/p,$$

where for the second equality we use the assumption that $y < p$, so p is not y-smooth. With our assumption (15.4), we may apply Theorem 15.1 (with the given value of y and $x := p$), obtaining

$$\sigma \geq \exp[(-1 + o(1))(\log p / \log y) \log \log p].$$

By Chebyshev's theorem (Theorem 5.1), we know that $k = \Theta(y / \log y)$, and so $\log k = (1 + o(1)) \log y$. Moreover, assumption (15.4) implies that the factor $\text{len}(p)^{O(1)}$ in (15.3) is of the form $\exp[o(\min(\log y, \log p / \log y))]$, and so we have

$$E[Z] \leq \exp[(1 + o(1)) \max\{(\log p / \log y) \log \log p + 2 \log y, \ 3 \log y\}]. \quad (15.5)$$

Let us find the value of y that minimizes the right-hand side of (15.5), ignoring the "$o(1)$" terms. Let $\mu := \log y$, $A := \log p \log \log p$, $S_1 := A / \mu + 2\mu$, and $S_2 := 3\mu$. We want to find μ that minimizes $\max\{S_1, S_2\}$. Using a little calculus, one sees that S_1 is minimized at $\mu = (A/2)^{1/2}$. With this choice of μ, we have $S_1 = (2\sqrt{2})A^{1/2}$ and $S_2 = (3/\sqrt{2})A^{1/2} < S_1$. Thus, choosing

$$y = \exp[(1/\sqrt{2})(\log p \log \log p)^{1/2}],$$

we obtain

$$E[Z] \leq \exp[(2\sqrt{2} + o(1))(\log p \log \log p)^{1/2}].$$

That takes care of the first two questions, although strictly speaking, we have only obtained an upper bound for the expected running time, and we have not shown that the choice of y is actually optimal, but we shall nevertheless content ourselves (for now) with these results. Finally, we deal with the third question, on the probability that the algorithm outputs "failure."

Lemma 15.2. *The probability that Algorithm* SEDL *outputs "failure" is* $1/q$.

Proof. Let F be the event that the algorithm outputs "failure." For $i = 1, \ldots, k+1$, we may view the *final* values assigned to r_i, s_i, δ_i, and m_i as random variables, which we shall denote by these same names (to avoid additional notation). Similarly, we may view s as a random variable.

Let $m'_1, \ldots, m'_{k+1}$ be arbitrary, fixed y-smooth numbers, and let B be the event that $m_1 = m'_1, \ldots, m_{k+1} = m'_{k+1}$. We shall show that $P[F \mid B] = 1/q$, and since this holds for all relevant B, it follows by total probability that $P[F] = 1/q$.

For the rest of the argument, we focus on the conditional distribution given B. With respect to this conditional distribution, the distribution of each random variable (r_i, s_i, δ_i) is (essentially) the uniform distribution on the set

$$P_i := \{(r', s', \delta') \in I_q \times I_q \times H : \gamma^{r'} \alpha^{s'} \delta' = [m'_i]_p\},$$

where $I_q := \{0, \ldots, q-1\}$; also, the family of random variables $\{(r_i, s_i, \delta_i)\}_{i=1}^{k+1}$

is mutually independent. It is easy to see that for $i = 1, \ldots, k + 1$, and for each $s' \in I_q$, there exist unique values $r' \in I_q$ and $\delta' \in H$ such that $(r', s', \delta') \in P_i$. From this, it easily follows that each s_i is uniformly distributed over I_q, and the family of random variables $\{s_i\}_{i=1}^{k+1}$ is mutually independent. Also, the values $c_1, \ldots, c_{k+1}$ computed by the algorithm are *fixed* (as they are determined by $m'_1, \ldots, m'_{k+1}$), and since $s = c_1 s_1 + \cdots + c_{k+1} s_{k+1}$, and not all the c_i's are zero modulo q, it follows that $s \bmod q$ is uniformly distributed over I_q, and so is equal to zero with probability $1/q$. $\square$

Let us summarize the above discussion in the following theorem.

Theorem 15.3. *With the smoothness parameter set as*

$$y := \exp[(1/\sqrt{2})(\log p \log \log p)^{1/2}],$$

the expected running time of Algorithm SEDL is at most

$$\exp[(2\sqrt{2} + o(1))(\log p \log \log p)^{1/2}].$$

The probability that Algorithm SEDL outputs "failure" is $1/q$.

In the description and analysis of Algorithm SEDL, we have assumed that the primes $p_1, \ldots, p_k$ were pre-computed. Of course, we can construct this list of primes using, for example, the sieve of Eratosthenes (see §5.4), and the running time of this pre-computation will be dominated by the running time of Algorithm SEDL.

In the analysis of Algorithm SEDL, we relied crucially on the fact that in generating a relation, each candidate element $\gamma^{r_i} \alpha^{s_i} \delta_i$ was uniformly distributed over $\mathbb{Z}_p^*$. If we simply left out the δ_i's, then the candidate element would be uniformly distributed over the subgroup G, and Theorem 15.1 simply would not apply. Although the algorithm might anyway work as expected, we would not be able to prove this.

EXERCISE 15.1. Using the result of Exercise 14.19, show how to modify Algorithm SEDL to work in the case where $p - 1 = q^e m$, $e > 1$, $q \nmid m$, γ generates the subgroup G of $\mathbb{Z}_p^*$ of order q^e, and $\alpha \in G$. Your algorithm should compute $\log_\gamma \alpha$ with roughly the same expected running time and success probability as Algorithm SEDL.

EXERCISE 15.2. Using the algorithm of the previous exercise as a subroutine, design and analyze an algorithm for the following problem. The input is p, q, γ, α, where p is a prime, q is a prime dividing $p - 1$, γ generates the subgroup G of $\mathbb{Z}_p^*$ of order q, and $\alpha \in G$; note that we may have $q^2 \mid (p - 1)$. The output is $\log_\gamma \alpha$. Your algorithm should always succeed in computing this discrete logarithm, and its expected running time should be bounded by a constant times the expected running time of the algorithm of the previous exercise.

EXERCISE 15.3. Using the result of Exercise 14.20, show how to modify Algorithm SEDL to solve the following problem: given a prime p, a generator γ for $\mathbb{Z}_p^*$, and an element $\alpha \in \mathbb{Z}_p^*$, compute $\log_\gamma \alpha$. Your algorithm should work without knowledge of the factorization of $p-1$; its expected running time should be roughly the same as that of Algorithm SEDL, but its success probability may be lower. In addition, explain how the success probability may be significantly increased at almost no cost by collecting a few extra relations.

EXERCISE 15.4. Let $n = pq$, where p and q are distinct, large primes. Let e be a prime, with $e < n$ and $e \nmid (p-1)(q-1)$. Let x be a positive integer, with $x < n$. Suppose you are given n (but not its factorization!) along with e and x. In addition, you are given access to two "oracles," which you may invoke as often as you like.

- The first oracle is a "challenge oracle": each invocation of the oracle produces a "challenge" $a \in \{1, \ldots, x\}$—distributed uniformly, and independent of all other challenges.

- The second oracle is a "solution oracle": you invoke this oracle with the index of a previous challenge oracle; if the corresponding challenge was a, the solution oracle returns the eth root of a modulo n; that is, the solution oracle returns $b \in \{1, \ldots, n-1\}$ such that $b^e \equiv a \pmod{n}$—note that b always exists and is uniquely determined.

Let us say that you "win" if you are able to compute the eth root modulo n of any challenge, but *without* invoking the solution oracle with the corresponding index of the challenge (otherwise, winning would be trivial, of course).

(a) Design a probabilistic algorithm that wins the above game, using an expected number of

$$\exp[(c + o(1))(\log x \log \log x)^{1/2}] \cdot \operatorname{len}(n)^{O(1)}$$

steps, for some constant c, where a "step" is either a computation step or an oracle invocation (either challenge or solution). Hint: Gaussian elimination over the field $\mathbb{Z}_e$.

(b) Suppose invocations of the challenge oracle are "cheap," while invocations of the solution oracle are relatively "expensive." How would you modify your strategy in part (a)?

Exercise 15.4 has implications in cryptography. A popular way of implementing a public-key primitive known as a "digital signature" works as follows: to digitally sign a message M (which may be an arbitrarily long bit string), first apply a "hash function" or "message digest" H to M, obtaining an integer a in some fixed range $\{1, \ldots, x\}$, and then compute the signature of M as the eth root b of a modulo n. Anyone can verify that such a signature b is correct by checking that $b^e \equiv H(M) \pmod{n}$; however, it would appear to be difficult to "forge" a signature without knowing the factorization of n. Indeed, one can prove the security of this signature scheme by assuming that it is hard to compute the eth root of a random number modulo n, and by making the heuristic assumption that H is a random function (see §15.5). However, for this proof to work, the value of x must be close to n; otherwise, if x is significantly smaller than n, as the result of this exercise, one can break the signature scheme at a cost that is roughly the same as the cost of factoring numbers around the size of x, rather than the size of n.

15.3 An algorithm for factoring integers

We now present a probabilistic, subexponential-time algorithm for factoring integers. The algorithm uses techniques very similar to those used in Algorithm SEDL in §15.2.

Let $n > 1$ be the integer we want to factor. We make a few simplifying assumptions. First, we assume that n is odd — this is not a real restriction, since we can always pull out any factors of 2 in a pre-processing step. Second, we assume that n is not a perfect power, that is, not of the form a^b for integers $a > 1$ and $b > 1$ — this is also not a real restriction, since we can always partially factor n using the algorithm from Exercise 3.31 if n is a perfect power. Third, we assume that n is not prime — this may be efficiently checked using, say, the Miller–Rabin test (see §10.2). Fourth, we assume that n is not divisible by any primes up to a "smoothness parameter" y — we can ensure this using trial division, and it will be clear that the running time of this pre-computation is dominated by that of the algorithm itself.

With these assumptions, the prime factorization of n is of the form

$$n = q_1^{f_1} \cdots q_w^{f_w},$$

where $w > 1$, the q_i's are distinct, odd primes, each greater than y, and the f_i's are positive integers.

The main goal of our factoring algorithm is to find a random square root of 1 in $\mathbb{Z}_n^*$. Let

$$\theta: \quad \mathbb{Z}_n \to \mathbb{Z}_{q_1^{f_1}} \times \cdots \times \mathbb{Z}_{q_w^{f_w}}$$

$$[a]_n \mapsto ([a]_{q_1^{f_1}}, \ldots, [a]_{q_w^{f_w}})$$

be the ring isomorphism of the Chinese remainder theorem. The square roots of 1 in $\mathbb{Z}_n^*$ are precisely those elements $\gamma \in \mathbb{Z}_n^*$ such that $\theta(\gamma) = (\pm 1, \ldots, \pm 1)$. If γ is a random square root of 1, then with probability $1 - 2^{-w+1} \geq 1/2$, we have

$\theta(\gamma) = (\gamma_1, \ldots, \gamma_w)$, where the γ_i's are neither all 1 nor all -1 (i.e., $\gamma \neq \pm 1$). If this happens, then $\theta(\gamma - 1) = (\gamma_1 - 1, \ldots, \gamma_w - 1)$, and so we see that some, but not all, of the values $\gamma_i - 1$ will be zero. The value of $\gcd(\mathrm{rep}(\gamma - 1), n)$ is precisely the product of the prime powers $q_i^{f_i}$ such that $\gamma_i - 1 = 0$, and hence this gcd will yield a non-trivial factorization of n, unless $\gamma = \pm 1$.

Let $p_1, \ldots, p_k$ be the primes up to the smoothness parameter y mentioned above. Let $\pi_i := [p_i]_n \in \mathbb{Z}_n^*$ for $i = 1, \ldots, k$.

We first describe a simplified version of the algorithm, after which we modify the algorithm slightly to deal with a technical problem. Like Algorithm SEDL, this algorithm proceeds in two stages. In the first stage, we find relations of the form

$$\alpha_i^2 = \pi_1^{e_{i1}} \cdots \pi_k^{e_{ik}}, \tag{15.6}$$

for $i = 1, \ldots, k+1$, where $e_{i1}, \ldots, e_{ik} \in \mathbb{Z}$ and $\alpha_i \in \mathbb{Z}_n^*$ for each i.

We can obtain each such relation by randomized search, as follows: we select $\alpha_i \in \mathbb{Z}_n^*$ at random, square it, and try to factor $m_i := \mathrm{rep}(\alpha_i^2)$ by trial division, trying all the primes $p_1, \ldots, p_k$ up to y. If we are lucky, we obtain a factorization

$$m_i = p_1^{e_{i1}} \cdots p_k^{e_{ik}},$$

for some exponents $e_{i1}, \ldots, e_{ik}$, yielding the relation (15.6); if not, we just keep trying.

For $i = 1, \ldots, k+1$, let $v_i := (e_{i1}, \ldots, e_{ik}) \in \mathbb{Z}^{\times k}$, and let $\bar{v}_i$ denote the image of v_i in $\mathbb{Z}_2^{\times k}$ (i.e., $\bar{v}_i := ([e_{i1}]_2, \ldots, [e_{ik}]_2)$). Since $\mathbb{Z}_2^{\times k}$ is a vector space over the field $\mathbb{Z}_2$ of dimension k, the family of vectors $\bar{v}_1, \ldots, \bar{v}_{k+1}$ must be linearly dependent. The second stage of the algorithm uses Gaussian elimination over $\mathbb{Z}_2$ to find a linear dependence among the vectors $\bar{v}_1, \ldots, \bar{v}_{k+1}$, that is, to find integers $c_1, \ldots, c_{k+1} \in \{0, 1\}$, not all zero, such that

$$(e_1, \ldots, e_k) := c_1 v_1 + \cdots + c_{k+1} v_{k+1} \in 2\mathbb{Z}^{\times k}.$$

Raising each equation (15.6) to the corresponding power c_i, and multiplying them all together, we obtain

$$\alpha^2 = \pi_1^{e_1} \cdots \pi_k^{e_k},$$

where

$$\alpha := \prod_{i=1}^{k+1} \alpha_i^{c_i}.$$

Since each e_i is even, we can compute

$$\beta := \pi_1^{e_1/2} \cdots \pi_k^{e_k/2},$$

and we see that $\alpha^2 = \beta^2$, and hence $(\alpha/\beta)^2 = 1$. Thus, $\gamma := \alpha/\beta$ is a square root of 1 in $\mathbb{Z}_n^*$. A more careful analysis (see below) shows that in fact, γ is uniformly distributed over all square roots of 1, and hence, with probability at least $1/2$, if we compute $\gcd(\text{rep}(\gamma - 1), n)$, we get a non-trivial factor of n.

That is the basic idea of the algorithm. There is, however, a technical problem. Namely, in the method outlined above for generating a relation, we attempt to factor $m_i := \text{rep}(\alpha_i^2)$. Thus, the running time of the algorithm will depend in a crucial way on the probability that a random square modulo n is y-smooth. Unfortunately for us, Theorem 15.1 does not say anything about this situation—it only applies to the situation where a number is chosen at random from an interval $[1, x]$. There are (at least) three different ways to address this problem:

1. Ignore it, and just assume that the bounds in Theorem 15.1 apply to random squares modulo n (taking $x := n$ in the theorem).

2. Prove a version of Theorem 15.1 that applies to random squares modulo n.

3. Modify the factoring algorithm, so that Theorem 15.1 applies.

The first choice, while not unreasonable from a practical point of view, is not very satisfying mathematically. It turns out that the second choice is indeed a viable option (i.e., the theorem is true and is not so difficult to prove), but we opt for the third choice, as it is somewhat easier to carry out, and illustrates a probabilistic technique that is more generally useful.

So here is how we modify the basic algorithm. Instead of generating relations of the form (15.6), we generate relations of the form

$$\alpha_i^2 \delta = \pi_1^{e_{i1}} \cdots \pi_k^{e_{ik}}, \tag{15.7}$$

for $i = 1, \ldots, k + 2$, where $e_{i1}, \ldots, e_{ik} \in \mathbb{Z}$ and $\alpha_i \in \mathbb{Z}_n^*$ for each i, and $\delta \in \mathbb{Z}_n^*$. Note that the value δ is the same in all relations.

We generate these relations as follows. For the very first relation (i.e., $i = 1$), we repeatedly choose α_1 and δ in $\mathbb{Z}_n^*$ at random, until $\text{rep}(\alpha_1^2 \delta)$ is y-smooth. Then, after having found the first relation, we find each subsequent relation (i.e., for $i > 1$) by repeatedly choosing α_i in $\mathbb{Z}_n^*$ at random until $\text{rep}(\alpha_i^2 \delta)$ is y-smooth, where δ is the same value that was used in the first relation. Now, Theorem 15.1 will apply directly to determine the success probability of each attempt to generate the first relation. When we have found this relation, the value $\alpha_1^2 \delta$ will be uniformly distributed over all y-smooth elements of $\mathbb{Z}_n^*$ (i.e., elements whose integer representations are y-smooth). Consider the various cosets of $(\mathbb{Z}_n^*)^2$ in $\mathbb{Z}_n^*$. Intuitively, it is much more likely that a random y-smooth element of $\mathbb{Z}_n^*$ lies in a coset that contains many y-smooth elements than in a coset with very few, and indeed, it is reasonably likely that the fraction of y-smooth elements in the coset containing δ is not much less than the overall fraction of y-smooth elements in $\mathbb{Z}_n^*$. Therefore, for $i > 1$, each attempt to find a relation should succeed with reasonably high probability. This intuitive argument will be made rigorous in the analysis to follow.

The second stage is then modified as follows. For $i = 1, \ldots, k + 2$, let $v_i := (e_{i1}, \ldots, e_{ik}, 1) \in \mathbb{Z}^{\times(k+1)}$, and let $\bar{v}_i$ denote the image of v_i in $\mathbb{Z}_2^{\times(k+1)}$. Since $\mathbb{Z}_2^{\times(k+1)}$ is a vector space over the field $\mathbb{Z}_2$ of dimension $k + 1$, the family of vectors $\bar{v}_1, \ldots, \bar{v}_{k+2}$ must be linearly dependent. Therefore, we use Gaussian elimination over $\mathbb{Z}_2$ to find a linear dependence among the vectors $\bar{v}_1, \ldots, \bar{v}_{k+2}$, that is, to find integers $c_1, \ldots, c_{k+2} \in \{0, 1\}$, not all zero, such that

$$(e_1, \ldots, e_{k+1}) := c_1 v_1 + \cdots + c_{k+2} v_{k+2} \in 2\mathbb{Z}^{\times(k+1)}.$$

Raising each equation (15.7) to the corresponding power c_i, and multiplying them all together, we obtain

$$\alpha^2 \delta^{e_{k+1}} = \pi_1^{e_1} \cdots \pi_k^{e_k},$$

where

$$\alpha := \prod_{i=1}^{k+2} \alpha_i^{c_i}.$$

Since each e_i is even, we can compute

$$\beta := \pi_1^{e_1/2} \cdots \pi_k^{e_k/2} \delta^{-e_{k+1}/2},$$

so that $\alpha^2 = \beta^2$ and $\gamma := \alpha/\beta$ is a square root of 1 in $\mathbb{Z}_n^*$.

The entire algorithm, called Algorithm SEF, is presented in Fig. 15.2.

Now the analysis. From the discussion above, it is clear that Algorithm SEF either outputs "failure," or outputs a non-trivial factor of n. So we have the same three questions to answer as we did in the analysis of Algorithm SEDL:

1. What is the expected running time of Algorithm SEF?

2. How should the smoothness parameter y be chosen so as to minimize the expected running time?

3. What is the probability that Algorithm SEF outputs "failure"?

To answer the first question, let σ denote the probability that (the canonical representative of) a random element of $\mathbb{Z}_n^*$ is y-smooth. For $i = 1, \ldots, k + 2$, let L_i denote the number of iterations of the inner loop in the ith iteration of the main loop in stage 1; that is, L_i is the number of attempts made in finding the ith relation.

Lemma 15.4. *For $i = 1, \ldots, k + 2$, we have* $\mathsf{E}[L_i] \leq \sigma^{-1}$.

Proof. We first compute $\mathsf{E}[L_1]$. As δ is chosen uniformly from $\mathbb{Z}_n^*$ and independent of α_1, at each attempt to find a relation, $\alpha_1^2 \delta$ is uniformly distributed over $\mathbb{Z}_n^*$,

$i \leftarrow 0$
repeat
    $i \leftarrow i + 1$
    repeat
        choose $\alpha_i \in \mathbb{Z}_n^*$ at random
        if $i = 1$ then choose $\delta \in \mathbb{Z}_n^*$ at random
        $m_i \leftarrow \mathrm{rep}(\alpha_i^2 \delta)$
        test if m_i is y-smooth (trial division)
    until $m_i = p_1^{e_{i1}} \cdots p_k^{e_{ik}}$ for some integers $e_{i1}, \ldots, e_{ik}$
  until $i = k + 2$

set $v_i \leftarrow (e_{i1}, \ldots, e_{ik}, 1) \in \mathbb{Z}^{\times(k+1)}$ for $i = 1, \ldots, k + 2$

apply Gaussian elimination over $\mathbb{Z}_2$ to find integers $c_1, \ldots, c_{k+2} \in$
    $\{0, 1\}$, not all zero, such that
    $(e_1, \ldots, e_{k+1}) := c_1 v_1 + \cdots + c_{k+2} v_{k+2} \in 2\mathbb{Z}^{\times(k+1)}$.

$\alpha \leftarrow \prod_{i=1}^{k+2} \alpha_i^{c_i}$, $\beta \leftarrow \pi_1^{e_1/2} \cdots \pi_k^{e_k/2} \delta^{-e_{k+1}/2}$, $\gamma \leftarrow \alpha/\beta$

if $\gamma = \pm 1$
    then output "failure"
    else output $\gcd(\mathrm{rep}(\gamma - 1), n)$

Fig. 15.2. Algorithm SEF

and hence the probability that the attempt succeeds is precisely σ. This means
$\mathsf{E}[L_1] = \sigma^{-1}$.

We next compute $\mathsf{E}[L_i]$ for $i > 1$. To this end, let us denote the cosets of $(\mathbb{Z}_n^*)^2$
by $\mathbb{Z}_n^*$ as $C_1, \ldots, C_t$. As it happens, $t = 2^w$, but this fact plays no role in the
analysis. For $j = 1, \ldots, t$, let σ_j denote the probability that a random element of
C_j is y-smooth, and let τ_j denote the probability that the final value of δ belongs to
C_j.

We claim that for $j = 1, \ldots, t$, we have $\tau_j = \sigma_j \sigma^{-1} t^{-1}$. To see this, note that each
coset C_j has the same number of elements, namely, $|\mathbb{Z}_n^*| t^{-1}$, and so the number of
y-smooth elements in C_j is equal to $\sigma_j |\mathbb{Z}_n^*| t^{-1}$. Moreover, the final value of $\alpha_1^2 \delta$
is equally likely to be any one of the y-smooth numbers in $\mathbb{Z}_n^*$, of which there are
$\sigma |\mathbb{Z}_n^*|$, and hence

$$\tau_j = \frac{\sigma_j |\mathbb{Z}_n^*| t^{-1}}{\sigma |\mathbb{Z}_n^*|} = \sigma_j \sigma^{-1} t^{-1},$$

which proves the claim.

Now, for a fixed value of δ and a random choice of $\alpha_i \in \mathbb{Z}_n^*$, one sees that $\alpha_i^2 \delta$ is uniformly distributed over the coset containing δ. Therefore, for $j = 1, \ldots, t$, if $\tau_j > 0$, we have

$$E[L_i \mid \delta \in C_j] = \sigma_j^{-1}.$$

Summing over all $j = 1, \ldots, t$ with $\tau_j > 0$, it follows that

$$E[L_i] = \sum_{\tau_j > 0} E[L_i \mid \delta \in C_j] \cdot P[\delta \in C_j]$$

$$= \sum_{\tau_j > 0} \sigma_j^{-1} \cdot \tau_j = \sum_{\tau_j > 0} \sigma_j^{-1} \cdot \sigma_j \sigma^{-1} t^{-1} \le \sigma^{-1},$$

which proves the lemma. $\square$

So in stage 1, the expected number of attempts made in generating a single relation is σ^{-1}, each such attempt takes time $k \cdot \text{len}(n)^{O(1)}$, and we have to generate $k+2$ relations, leading to a total expected running time in stage 1 of $\sigma^{-1} k^2 \cdot \text{len}(n)^{O(1)}$. Stage 2 is dominated by the cost of performing Gaussian elimination, which takes time $k^3 \cdot \text{len}(n)^{O(1)}$. Thus, if Z is the total running time of the algorithm, we have

$$E[Z] \le (\sigma^{-1} k^2 + k^3) \cdot \text{len}(n)^{O(1)}.$$

By our assumption that n is not divisible by any primes up to y, all y-smooth integers up to $n - 1$ are in fact relatively prime to n. Therefore, the number of y-smooth elements of $\mathbb{Z}_n^*$ is equal to $\Psi(y, n - 1)$, and since n itself is not y-smooth, this is equal to $\Psi(y, n)$. From this, it follows that

$$\sigma = \Psi(y, n) / |\mathbb{Z}_n^*| \ge \Psi(y, n) / n.$$

The rest of the running time analysis is essentially the same as in the analysis of Algorithm SEDL; that is, assuming $y = \exp[(\log n)^{\lambda + o(1)}]$ for some constant $0 < \lambda < 1$, we obtain

$$E[Z] \le \exp[(1 + o(1)) \max\{(\log n / \log y) \log \log n + 2 \log y, \ 3 \log y\}]. \quad (15.8)$$

Setting $y = \exp[(1/\sqrt{2})(\log n \log \log n)^{1/2}]$, we obtain

$$E[Z] \le \exp[(2\sqrt{2} + o(1))(\log n \log \log n)^{1/2}].$$

That basically takes care of the first two questions. As for the third, we have:

Lemma 15.5. *Algorithm SEF outputs "failure" with probability $2^{-w+1} \le 1/2$.*

Proof. Let $\mathcal{F}$ be the event that the algorithm outputs "failure." We may view the *final* values assigned to δ and $\alpha_1, \ldots, \alpha_{k+2}$ as random variables, which we shall denote by these same names. Let $\delta' \in \mathbb{Z}_n^*$ and $\alpha_1', \ldots, \alpha_{k+2}' \in (\mathbb{Z}_n^*)^2$ be arbitrary, fixed values such that $\text{rep}(\alpha_i'\delta')$ is y-smooth for $i = 1, \ldots, k + 2$. Let B be the event that $\delta = \delta'$ and $\alpha_i^2 = \alpha_i'$ for $i = 1, \ldots, k + 2$. We shall show that $P[\mathcal{F} \mid B] = 2^{-w+1}$, and since this holds for all relevant B, it follows by total probability that $P[\mathcal{F}] = 2^{-w+1}$.

For the rest of the argument, we focus on the conditional distribution given B. With respect to this conditional distribution, the distribution of each random variable α_i is (essentially) the uniform distribution on $\rho^{-1}(\{\alpha_i'\})$, where ρ is the squaring map on $\mathbb{Z}_n^*$. Moreover, the family of random variables $\{\alpha_i\}_{i=1}^{k+2}$ is mutually independent. Also, the values β and $c_1, \ldots, c_{k+2}$ computed by the algorithm are *fixed*. It follows (see Exercise 8.14) that the distribution of α is (essentially) the uniform distribution on $\rho^{-1}(\{\beta^2\})$, and hence $\gamma := \alpha/\beta$ is a random square root of 1 in $\mathbb{Z}_n^*$. Thus, $\gamma = \pm 1$ with probability 2^{-w+1}. $\square$

Let us summarize the above discussion in the following theorem.

Theorem 15.6. *With the smoothness parameter set as*

$$y := \exp[(1/\sqrt{2})(\log n \log \log n)^{1/2}],$$

the expected running time of Algorithm SEF is at most

$$\exp[(2\sqrt{2} + o(1))(\log n \log \log n)^{1/2}].$$

The probability that Algorithm SEF outputs "failure" is at most $1/2$.

EXERCISE 15.5. It is perhaps a bit depressing that after all that work, Algorithm SEF only succeeds (in the worst case) with probability $1/2$. Of course, to reduce the failure probability, we can simply repeat the entire computation—with ℓ repetitions, the failure probability drops to $2^{-\ell}$. However, there is a better way to reduce the failure probability. Suppose that in stage 1, instead of collecting $k + 2$ relations, we collect $k + 1 + \ell$ relations, where $\ell \geq 1$ is an integer parameter.

(a) Show that in stage 2, we can use Gaussian elimination over $\mathbb{Z}_2$ to find integer vectors

$$c^{(j)} = (c_1^{(j)}, \ldots, c_{k+1+\ell}^{(j)}) \in \{0, 1\}^{\times(k+1+\ell)} \quad (j = 1, \ldots, \ell)$$

such that

- over the field $\mathbb{Z}_2$, the images of the vectors $c^{(1)}, \ldots, c^{(\ell)}$ in $\mathbb{Z}_2^{\times(k+1+\ell)}$ form a linearly independent family of vectors, and
- for $j = 1, \ldots, \ell$, we have

$$c_1^{(j)} v_1 + \cdots + c_{k+1+\ell}^{(j)} v_{k+1+\ell} \in 2\mathbb{Z}^{\times(k+2)}.$$

(b) Show that given vectors $c^{(1)}, \ldots, c^{(\ell)}$ as in part (a), if for $j = 1, \ldots, \ell$, we set

$$(e_1^{(j)}, \ldots, e_{k+1}^{(j)}) \leftarrow c_1^{(j)} v_1 + \cdots + c_{k+1+\ell}^{(j)} v_{k+1+\ell},$$

$$\alpha^{(j)} \leftarrow \prod_{i=1}^{k+1+\ell} \alpha_i^{c_i^{(j)}}, \quad \beta^{(j)} \leftarrow \pi_1^{e_1^{(j)}/2} \cdots \pi_k^{e_k^{(j)}/2} \delta^{-e_{k+1}^{(j)}/2}, \quad \gamma^{(j)} \leftarrow \alpha^{(j)}/\beta^{(j)},$$

then the family of random variables $\gamma^{(1)}, \ldots, \gamma^{(\ell)}$ is mutually independent, with each $\gamma^{(j)}$ uniformly distributed over the set of all square roots of 1 in $\mathbb{Z}_n^*$, and hence at least one of $\gcd(\mathrm{rep}(\gamma^{(j)} - 1), n)$ splits n with probability at least $1 - 2^{-\ell}$.

So, for example, if we set $\ell = 20$, then the failure probability is reduced to less than one in a million, while the increase in running time over Algorithm SEF will hardly be noticeable.

15.4 Practical improvements

Our presentation and analysis of algorithms for discrete logarithms and factoring were geared towards simplicity and mathematical rigor. However, if one really wants to compute discrete logarithms or factor numbers, then a number of important practical improvements should be considered. In this section, we briefly sketch some of these improvements, focusing our attention on algorithms for factoring numbers (although some of the techniques apply to discrete logarithms as well).

15.4.1 Better smoothness density estimates

From an algorithmic point of view, the simplest way to improve the running times of both Algorithms SEDL and SEF is to use a more accurate smoothness density estimate, which dictates a different choice of the smoothness bound y in those algorithms, speeding them up significantly. While our Theorem 15.1 is a valid *lower bound* on the density of smooth numbers, it is not "tight," in the sense that the actual density of smooth numbers is somewhat higher. We quote from the literature the following result:

Theorem 15.7. *Let y be a function of x such that for some $\varepsilon > 0$, we have*

$$y = \Omega((\log x)^{1+\varepsilon}) \quad \text{and} \quad u := \frac{\log x}{\log y} \to \infty$$

as $x \to \infty$. Then

$$\Psi(y, x) = x \cdot \exp[(-1 + o(1))u \log u].$$

Proof. See §15.5. □

Let us apply this result to the analysis of Algorithm SEF. Assume that

$$y = \exp[(\log n)^{1/2+o(1)}].$$

Our choice of y will in fact be of this form. With this assumption, we have $\log \log y = (1/2 + o(1)) \log \log n$, and using Theorem 15.7, we can improve the inequality (15.8), obtaining instead (as the reader may verify)

$$E[Z] \le \exp[(1 + o(1)) \max \{ \tfrac{1}{2}(\log n / \log y) \log \log n + 2 \log y, 3 \log y \}].$$

From this, if we set

$$y := \exp[\tfrac{1}{2}(\log n \log \log n)^{1/2})],$$

we obtain

$$E[Z] \le \exp[(2 + o(1))(\log n \log \log n)^{1/2}].$$

An analogous improvement can be obtained for Algorithm SEDL.

Although this improvement only reduces the constant $2\sqrt{2} \approx 2.828$ to 2, the constant is in the exponent, and so this improvement is not to be scoffed at!

15.4.2 The quadratic sieve algorithm

We now describe a practical improvement to Algorithm SEF. This algorithm, known as the **quadratic sieve**, is faster in practice than Algorithm SEF; however, its analysis is somewhat heuristic.

First, let us return to the simplified version of Algorithm SEF, where we collect relations of the form (15.6). Furthermore, instead of choosing the values α_i at random, we will choose them in a special way, as we now describe. Let

$$\tilde{n} := \lfloor \sqrt{n} \rfloor,$$

and define the polynomial

$$F := (X + \tilde{n})^2 - n \in \mathbb{Z}[X].$$

In addition to the usual "smoothness parameter" y, we need a "sieving parameter" z, whose choice will be discussed below. We shall assume that both y and z are of the form $\exp[(\log n)^{1/2+o(1)}]$, and our ultimate choices of y and z will indeed satisfy this assumption.

For all $s = 1, 2, \ldots, \lfloor z \rfloor$, we shall determine which values of s are "good," in the sense that the corresponding value $F(s)$ is y-smooth. For each good s, since we have $F(s) \equiv (s + \tilde{n})^2 \pmod{n}$, we obtain one relation of the form (15.6), with $\alpha_i := [s + \tilde{n}]_n$. If we find at least $k + 1$ good values of s, then we can apply

Gaussian elimination as usual to find a square root γ of 1 in $\mathbb{Z}_n^*$. Hopefully, we will have $\gamma \neq \pm 1$, allowing us to split n.

Observe that for $1 \leq s \leq z$, we have

$$1 \leq F(s) \leq z^2 + 2zn^{1/2} \leq n^{1/2+o(1)}.$$

Now, although the values $F(s)$ are not at all random, we might expect heuristically that the number of good s up to z is roughly equal to $\hat{\sigma}z$, where $\hat{\sigma}$ is the probability that a random integer in the interval $[1, n^{1/2}]$ is y-smooth, and by Theorem 15.7, we have

$$\hat{\sigma} = \exp[(-\tfrac{1}{4} + o(1))(\log n / \log y) \log \log n].$$

If our heuristics are valid, this already yields an improvement over Algorithm SEF, since now we are looking for y-smooth numbers near $n^{1/2}$, which are much more common than y-smooth numbers near n. But there is another improvement possible; namely, instead of testing each individual number $F(s)$ for smoothness using trial division, we can test them all at once using the following "sieving procedure."

The sieving procedure works as follows. First, we create an array $v[1 \ldots \lfloor z \rfloor]$, and initialize $v[s]$ to $F(s)$, for $1 \leq s \leq z$. Then, for each prime p up to y, we do the following:

1. Compute the roots of the polynomial F modulo p.

   This can be done quite efficiently, as follows. For $p = 2$, F has exactly one root modulo p, which is determined by the parity of $\tilde{n}$. For $p > 2$, we may use the familiar quadratic formula together with an algorithm for computing square roots modulo p, as discussed in Exercise 12.7. A quick calculation shows that the discriminant of F is $4n$, and thus, F has a root modulo p if and only if n is a quadratic residue modulo p, in which case it will have two roots (under our usual assumptions, we cannot have $p \mid n$).

2. Assume that F has v_p distinct roots modulo p lying in the interval $[1, p]$; call them $r_1, \ldots, r_{v_p}$.

   Note that $v_p = 1$ for $p = 2$ and $v_p \in \{0, 2\}$ for $p > 2$. Also note that $F(s) \equiv 0 \pmod{p}$ if and only if $s \equiv r_i \pmod{p}$ for some $i = 1, \ldots, v_p$.

   For $i = 1, \ldots, v_p$, do the following:

   $s \leftarrow r_i$
   while $s \leq z$ do
       repeat $v[s] \leftarrow v[s]/p$ until $p \nmid v[s]$
       $s \leftarrow s + p$

At the end of this sieving procedure, the good values of s may be identified as precisely those such that $v[s] = 1$. The running time of this sieving procedure is at most $\text{len}(n)^{O(1)}$ times

$$\sum_{p \leq y} \frac{z}{p} = z \sum_{p \leq y} \frac{1}{p} = O(z \log \log y) = z^{1+o(1)}.$$

Here, we have made use of Theorem 5.10, although this is not really necessary — for our purposes, the bound $\sum_{p \leq y} 1/p = O(\log y)$ would suffice. Note that this sieving procedure is a factor of $k^{1+o(1)}$ faster than the method for finding smooth numbers based on trial division. With just a little extra book-keeping, we can not only identify the good values of s but also compute the factorization of $F(s)$ into primes, at essentially no extra cost.

Now, let us put together all the pieces. We have to choose z just large enough so as to find at least $k + 1$ good values of s up to z. So we should choose z so that $z \approx k/\hat{\sigma}$ — in practice, we could choose an initial estimate for z, and if this choice of z does not yield enough relations, we could keep doubling z until we do get enough relations. Assuming that $z \approx k/\hat{\sigma}$, the cost of sieving is $(k/\hat{\sigma})^{1+o(1)}$, or

$$\exp[(1 + o(1))(\tfrac{1}{4}(\log n / \log y) \log \log n + \log y)].$$

The cost of Gaussian elimination is still $O(k^3)$, or

$$\exp[(3 + o(1)) \log y].$$

Thus, the total running time is bounded by

$$\exp[(1 + o(1)) \max \{\tfrac{1}{4}(\log n / \log y) \log \log n + \log y, \ 3 \log y\}].$$

Let $\mu := \log y$, $A := (1/4) \log n \log \log n$, $S_1 := A/\mu + \mu$ and $S_2 := 3\mu$, and let us find the value of μ that minimizes $\max\{S_1, S_2\}$. Using a little calculus, one finds that S_1 is minimized at $\mu = A^{1/2}$. For this value of μ, we have $S_1 = 2A^{1/2}$ and $S_2 = 3A^{1/2} > S_1$, and so this choice of μ is a bit larger than optimal. For $\mu < A^{1/2}$, S_1 is decreasing (as a function of μ), while S_2 is always increasing. It follows that the optimal value of μ is obtained by setting

$$A/\mu + \mu = 3\mu,$$

and solving for μ. This yields $\mu = (A/2)^{1/2}$. So setting

$$y := \exp[(1/2\sqrt{2})(\log n \log \log n)^{1/2}],$$

the total running time of the quadratic sieve factoring algorithm is bounded by

$$\exp[(3/2\sqrt{2} + o(1))(\log n \log \log n)^{1/2}].$$

Thus, we have reduced the constant in the exponent from 2 (for Algorithm SEF with the more accurate smoothness density estimates) to $3/2\sqrt{2} \approx 1.061$.

We mention one final improvement. The matrix to which we apply Gaussian elimination in stage 2 is "sparse"; indeed, since any integer less than n has $O(\log n)$ prime factors, the total number of non-zero entries in the matrix is $k^{1+o(1)}$. There are special algorithms for working with such sparse matrices, which allow us to perform stage 2 of the factoring algorithm in time $k^{2+o(1)}$, or

$$\exp[(2 + o(1)) \log y].$$

Setting

$$y := \exp[\tfrac{1}{2}(\log n \log \log n)^{1/2}],$$

the total running time is bounded by

$$\exp[(1 + o(1))(\log n \log \log n)^{1/2}].$$

Thus, this improvement reduces the constant in the exponent from $3/2\sqrt{2} \approx 1.061$ to 1. Moreover, the special algorithms designed to work with sparse matrices typically use much less space than ordinary Gaussian elimination (even if the input to Gaussian elimination is sparse, the intermediate matrices will not be). We shall discuss in detail later, in §18.4, one such algorithm for solving sparse systems of linear equations.

The quadratic sieve may fail to factor n, for one of two reasons: first, it may fail to find $k + 1$ relations; second, it may find these relations, but in stage 2, it finds only a trivial square root of 1. There is no rigorous theory to say why the algorithm should not fail for one of these two reasons, but experience shows that the algorithm does indeed work as expected.

15.5 Notes

Many of the algorithmic ideas in this chapter were first developed for the problem of factoring integers, and then later adapted to the discrete logarithm problem. The first (heuristic) subexponential-time algorithm for factoring integers, called the **continued fraction method** (not discussed here), was introduced by Lehmer and Powers [59], and later refined and implemented by Morrison and Brillhart [70]. The first rigorously analyzed subexponential-time algorithm for factoring integers was introduced by Dixon [35]. Algorithm SEF is a variation of Dixon's algorithm, which works the same way as Algorithm SEF, except that it generates relations of the form (15.6) directly (and indeed, it is possible to prove a variant of Theorem 15.1, and for that matter, Theorem 15.7, for random squares modulo n). Algorithm SEF is based on an idea suggested by Rackoff (personal communication).

Theorem 15.7 was proved by Canfield, Erdős, and Pomerance [23].

The quadratic sieve was introduced by Pomerance [78]. Recall that the quadratic sieve has a heuristic running time of

$$\exp[(1 + o(1))(\log n \log \log n)^{1/2}].$$

This running time bound can also be achieved *rigorously* by a result of Lenstra and Pomerance [61], and to date, this is the best rigorous running time bound for factoring algorithms. We should stress, however, that most practitioners in this field are not so much interested in rigorous running time analyses as they are in actually factoring integers, and, for such purposes, heuristic running time estimates are quite acceptable. Indeed, the quadratic sieve is much more practical than the algorithm in [61], which is mainly of theoretical interest.

There are two other factoring algorithms not discussed here, but that should anyway at least be mentioned. The first is the **elliptic curve method**, introduced by Lenstra [60]. Unlike all of the other known subexponential-time algorithms, the running time of this algorithm is sensitive to the sizes of the factors of n; in particular, if p is the smallest prime dividing n, the algorithm will find p (heuristically) in expected time

$$\exp[(\sqrt{2} + o(1))(\log p \log \log p)^{1/2}] \cdot \mathrm{len}(n)^{O(1)}.$$

This algorithm is quite practical, and is the method of choice when it is known (or suspected) that n has some small factors. It also has the advantage that it uses only polynomial space (unlike all of the other known subexponential-time factoring algorithms).

The second is the **number field sieve**, the basic idea of which was introduced by Pollard [77], and later generalized and refined by Buhler, Lenstra, and Pomerance [21], as well as by others. The number field sieve will split n (heuristically) in expected time

$$\exp[(c + o(1))(\log n)^{1/3}(\log \log n)^{2/3}],$$

where c is a constant (currently, the smallest value of c is 1.902, a result due to Coppersmith [27]). The number field sieve is currently the asymptotically fastest known factoring algorithm (at least, heuristically), and it is also practical, having been used to set the latest factoring record — the factorization of a 200-decimal-digit integer that is the product of two primes of about the same size. See the web page www.crypto-world.com/FactorRecords.html for more details (as well as for announcements of new records).

As for subexponential-time algorithms for discrete logarithms, Adleman [1] adapted the ideas used for factoring to the discrete logarithm problem, although it seems that some of the basic ideas were known much earlier. Algorithm SEDL is a variation on this algorithm, and the basic technique is usually referred to as the

index calculus method. The basic idea of the number field sieve was adapted to the discrete logarithm problem by Gordon [42]; see also Adleman [2] and Schirokauer, Weber, and Denny [84].

For many more details and references for subexponential-time algorithms for factoring and discrete logarithms, see Chapter 6 of Crandall and Pomerance [30]. Also, see the web page `www.crypto-world.com/FactorWorld.html` for links to research papers and implementation reports.

For more details regarding the security of signature schemes, as discussed following Exercise 15.4, see the paper by Bellare and Rogaway [13].

Last, but not least, we should mention the fact that there are in fact *polynomial-time* algorithms for factoring and for computing discrete logarithms; however, these algorithms require special hardware, namely, a **quantum computer**. Shor [92, 93] showed that these problems could be solved in polynomial time on such a device; however, at the present time, it is unclear when and if such machines will ever be built. Much, indeed most, of modern-day cryptography will crumble if this happens, or if efficient "classical" algorithms for these problems are discovered (which is still a real possibility).

16

More rings

This chapter develops a number of more advanced concepts concerning rings. These concepts will play important roles later in the text, and we prefer to discuss them now, so as to avoid too many interruptions of the flow of subsequent discussions.

16.1 Algebras

Throughout this section, R denotes a ring (i.e., a commutative ring with unity).

Sometimes, a ring may also be naturally viewed as an R-module, in which case, both the theory of rings and the theory of modules may be brought to bear to study its properties.

Definition 16.1. *An **R-algebra** is a set E, together with addition and multiplication operations on E, and a function $\mu : R \times E \to E$, such that*

   (i) with respect to addition and multiplication, E forms a ring;

  (ii) with respect to addition and the scalar multiplication map μ, E forms an R-module;

 (iii) for all $c \in R$, and $\alpha, \beta \in E$, we have

$$\mu(c, \alpha)\beta = \mu(c, \alpha\beta) = \alpha\mu(c, \beta).$$

An R-algebra E may also be called an **algebra over** R. As we usually do for R-modules, we shall write $c\alpha$ (or $c \cdot \alpha$) instead of $\mu(c, \alpha)$. When we do this, part (iii) of the definition states that

$$(c\alpha)\beta = c(\alpha\beta) = \alpha(c\beta)$$

for all $c \in R$ and $\alpha, \beta \in E$. In particular, we may write $c\alpha\beta$ without any ambiguity. Note that there are two multiplication operations at play here: scalar multiplication (such as $c\alpha$), and ring multiplication (such as $\alpha\beta$). Also note that since we are assuming E is commutative, the second equality in part (iii) is already implied by the first. A simple consequence of the definition is that for all $c, d \in R$ and $\alpha, \beta \in E$, we have $(c\alpha)(d\beta) = (cd)(\alpha\beta)$. From this, it follows that for all $c \in R$, $\alpha \in E$, and $k \geq 0$, we have $(c\alpha)^k = c^k\alpha^k$.

Example 16.1. Suppose E is a ring and $\tau : R \to E$ is a ring homomorphism. With scalar multiplication defined by $c\alpha := \tau(c)\alpha$ for $c \in R$ and $\alpha \in E$, one may easily check that E is indeed an R-algebra. In this case, we say that E is an R-algebra **via the map** τ. $\square$

Example 16.2. If R is a subring of E, then with $\tau : R \to E$ being the inclusion map, we can view E as an R-algebra as in the previous example. In this case, we say that E is an R-algebra **via inclusion**. $\square$

Example 16.3. If $\tau : R \to E$ is a natural embedding of rings, then by a slight abuse of terminology, just as we sometimes say that R is a subring of E, we shall also say that E is an R-algebra via inclusion. $\square$

In fact, all R-algebras can be viewed as special cases of Example 16.1:

Theorem 16.2. *If E is an R-algebra, then the map*

$$\tau : \quad R \to E$$
$$c \mapsto c \cdot 1_E,$$

is a ring homomorphism, and $c\alpha = \tau(c)\alpha$ for all $c \in R$ and $\alpha \in E$.

Proof. Exercise. $\square$

In the special situation where R is a *field*, we can say even more. In this situation, and with τ as in the above theorem, then either E is trivial or τ is injective (see Exercise 7.47). In the latter case, E contains an isomorphic copy of R as a subring. To summarize:

Theorem 16.3. *If R is a field, then an R-algebra is either the trivial ring or contains an isomorphic copy of R as a subring.*

The following examples give further important constructions of R-algebras.

Example 16.4. If $E_1, \ldots, E_k$ are R-algebras, then their direct product $E_1 \times \cdots \times E_k$ is an R-algebra as well, where addition, multiplication, and scalar multiplication are defined component-wise. As usual, if $E = E_1 = \cdots = E_k$, we write this as $E^{\times k}$. $\square$

Example 16.5. If I is an arbitrary set, and E is an R-algebra, then $\mathrm{Map}(I, E)$, which is the set of all functions $f : I \to E$, may be naturally viewed as an R-algebra, with addition, multiplication, and scalar multiplication defined point-wise. $\square$

Example 16.6. Let E be an R-algebra and let I be an ideal of E. Then it is easily verified that I is also a submodule of E. This means that the quotient ring E/I may also be viewed as an R-module, and indeed, it is an R-algebra, called the **quotient algebra (over R) of E modulo I**. For $\alpha, \beta \in E$ and $c \in R$, addition, multiplication, and scalar multiplication in E are defined as follows:

$$[\alpha]_I + [\beta]_I := [\alpha + \beta]_I, \quad [\alpha]_I \cdot [\beta]_I := [\alpha \cdot \beta]_I, \quad c \cdot [\alpha]_I := [c \cdot \alpha]_I. \; \square$$

Example 16.7. The ring of polynomials $R[X]$ is an R-algebra via inclusion. Let $f \in R[X]$ be a non-zero polynomial with $\mathrm{lc}(f) \in R^*$. We may form the quotient ring $E := R[X]/(f)$, which may naturally be viewed as an R-algebra, as in the previous example. If $\deg(f) = 0$, then E is trivial; so assume $\deg(f) > 0$, and consider the map

$$\tau : \quad R \to E$$
$$c \mapsto c \cdot 1_E$$

from Theorem 16.2. By definition, $\tau(c) = [c]_f$. As discussed in Example 7.55, the map τ is a natural embedding of rings, and so by identifying R with its image in E under τ, we can view R as a subring of E; therefore, we can also view E as an R-algebra via inclusion. $\square$

Subalgebras

Let E be an R-algebra. A subset S of E is called a **subalgebra (over R) of** E if it is both a subring of E and a submodule of E. This means that S contains 1_E, and is closed under addition, multiplication, and scalar multiplication; restricting these operations to S, we may view S as an R-algebra in its own right.

The following theorem gives a simple but useful characterization of subalgebras, in relation to subrings:

Theorem 16.4. *If E is an R-algebra via inclusion, and S is a subring of E, then S is a subalgebra if and only if S contains R. More generally, if E is an arbitrary R-algebra, and S is a subring of E, then S is a subalgebra of E if and only if S contains $c \cdot 1_E$ for all $c \in R$.*

Proof. Exercise. $\square$

R-algebra homomorphisms

Let E and E' be R-algebras. A function $\rho : E \to E'$ is called an **R-algebra homomorphism** if ρ is both a ring homomorphism and an R-linear map. This means that $\rho(1_E) = 1_{E'}$, and

$$\rho(\alpha + \beta) = \rho(\alpha) + \rho(\beta), \quad \rho(\alpha\beta) = \rho(\alpha)\rho(\beta), \quad \text{and } \rho(c\alpha) = c\rho(\alpha)$$

for all $\alpha, \beta \in E$ and all $c \in R$. As usual, if ρ is bijective, then it is called an **R-algebra isomorphism**, and if, in addition, $E = E'$, it is called an **R-algebra automorphism**.

The following theorem gives a simple but useful characterization of R-algebra homomorphisms, in relation to ring homomorphisms:

Theorem 16.5. *If E and E' are R-algebras via inclusion, and $\rho : E \to E'$ is a ring homomorphism, then ρ is an R-algebra homomorphism if and only if the restriction of ρ to R is the identity map. More generally, if E and E' are arbitrary R-algebras and $\rho : E \to E'$ is a ring homomorphism, then ρ is an R-algebra homomorphism if and only if $\rho(c \cdot 1_E) = c \cdot 1_{E'}$ for all $c \in R$.*

Proof. Exercise. $\square$

Example 16.8. If E is an R-algebra and I is an ideal of E, then as observed in Example 16.6, I is also a submodule of E, and we may form the quotient algebra E/I. The natural map

$$\rho : \quad E \to E/I$$
$$\alpha \mapsto [\alpha]_I$$

is both a ring homomorphism and an R-linear map, and hence is an R-algebra homomorphism. $\square$

Example 16.9. Since $\mathbb{C}$ contains $\mathbb{R}$ as a subring, we may naturally view $\mathbb{C}$ as an $\mathbb{R}$-algebra via inclusion. The complex conjugation map on $\mathbb{C}$ that sends $a + bi$ to $a - bi$, for $a, b \in \mathbb{R}$, is an $\mathbb{R}$-algebra automorphism on $\mathbb{C}$ (see Example 7.5). $\square$

Many simple facts about R-algebra homomorphisms can be obtained by combining corresponding facts for ring and R-module homomorphisms. For example, the composition of two R-algebra homomorphisms is again an R-algebra homomorphism, since the composition is both a ring homomorphism and an R-linear map (Theorems 7.22 and 13.6). As another example, if $\rho : E \to E'$ is an R-algebra homomorphism, then its image S' is both a subring and a submodule of E', and hence, S' is a subalgebra of E'. The kernel K of ρ is an ideal of E, and we may form the quotient algebra E/K. The first isomorphism theorems for rings and modules (Theorems 7.26 and 13.9) tell us that E/K and S' are isomorphic both as rings and as R-modules, and hence, they are isomorphic as R-algebras. Specifically, the map

$$\bar{\rho}: \quad E/K \to E'$$
$$[\alpha]_K \mapsto \rho(\alpha)$$

is an injective R-algebra homomorphism whose image is S'.

The following theorem isolates an important subalgebra associated with any R-algebra homomorphism $\rho : E \to E$.

Theorem 16.6. *Let E be an R-algebra, and let $\rho : E \to E$ be an R-algebra homomorphism. Then the set $S := \{\alpha \in E : \rho(\alpha) = \alpha\}$ is a subalgebra of E, called the* **subalgebra of E fixed by** *ρ. Moreover, if E is a field, then so is S.*

Proof. Let us verify that S is closed under addition. If $\alpha, \beta \in S$, then we have

$$\rho(\alpha + \beta) = \rho(\alpha) + \rho(\beta) \quad \text{(since } \rho \text{ is a group homomorphism)}$$
$$= \alpha + \beta \quad \text{(since } \alpha, \beta \in S).$$

Using the fact that ρ is a ring homomorphism, one can similarly show that S is closed under multiplication, and that $1_E \in S$. Likewise, using the fact that ρ is an R-linear map, one can also show that S is closed under scalar multiplication.

This shows that S is a subalgebra, proving the first statement. For the second statement, suppose that E is a field. Let α be a non-zero element of S, and suppose $\beta \in E$ is its multiplicative inverse, so that $\alpha\beta = 1_E$. We want to show that β lies in S. Again, using the fact that ρ is a ring homomorphism, we have

$$\alpha\beta = 1_E = \rho(1_E) = \rho(\alpha\beta) = \rho(\alpha)\rho(\beta) = \alpha\rho(\beta),$$

and hence $\alpha\beta = \alpha\rho(\beta)$; canceling α, we obtain $\beta = \rho(\beta)$, and so $\beta \in S$. $\square$

Example 16.10. The subalgebra of $\mathbb{C}$ fixed by the complex conjugation map is $\mathbb{R}$. $\square$

Polynomial evaluation

Let E be an R-algebra. Consider the ring of polynomials $R[X]$ (which is an R-algebra via inclusion). Any polynomial $g \in R[X]$ naturally defines a function on E: if $g = \sum_i a_i X^i$, with each $a_i \in R$, and $\alpha \in E$, then

$$g(\alpha) := \sum_i a_i \alpha^i.$$

Just as for rings, we say that α is a **root** of g if $g(\alpha) = 0_E$.

For fixed $\alpha \in E$, the **polynomial evaluation map**

$$\rho: \quad R[X] \to E$$
$$g \mapsto g(\alpha)$$

is easily seen to be an R-algebra homomorphism. The image of ρ is denoted $R[\alpha]$, and is a subalgebra of E. Indeed, $R[\alpha]$ is the smallest subalgebra of E containing α, and is called the **subalgebra (over R) generated by** α. Note that if E is an R-algebra via inclusion, then the notation $R[\alpha]$ has the same meaning as that introduced in Example 7.44.

We next state a very simple, but extremely useful, fact:

Theorem 16.7. *Let $\rho : E \to E'$ be an R-algebra homomorphism. Then for all $g \in R[X]$ and $\alpha \in E$, we have*

$$\rho(g(\alpha)) = g(\rho(\alpha)).$$

Proof. Let $g = \sum_i a_i X^i \in R[X]$. Then we have

$$\rho(g(\alpha)) = \rho\left(\sum_i a_i \alpha^i\right) = \sum_i \rho(a_i \alpha^i) = \sum_i a_i \rho(\alpha^i) = \sum_i a_i \rho(\alpha)^i$$

$$= g(\rho(\alpha)). \quad \square$$

As a special case of Theorem 16.7, if $E = R[\alpha]$ for some $\alpha \in E$, then every element of E can be expressed as $g(\alpha)$ for some $g \in R[X]$, and $\rho(g(\alpha)) = g(\rho(\alpha))$; hence, the action of ρ is completely determined by its action on α.

Example 16.11. Let $f \in R[X]$ be a non-zero polynomial with $\mathrm{lc}(f) \in R^*$. As in Example 16.7, we may form the quotient algebra $E := R[X]/(f)$.

Let $\xi := [X]_f \in E$. Then $E = R[\xi]$, and moreover, every element of E can be expressed uniquely as $g(\xi)$, where $g \in R[X]$ and $\deg(g) < \deg(f)$. In addition, ξ is a root of f. If $\deg(f) > 0$, these facts were already observed in Example 7.55, and otherwise, they are trivial.

Now let E' be any R-algebra, and suppose that $\rho : E \to E'$ is an R-algebra homomorphism, and let $\xi' := \rho(\xi)$. By the previous theorem, ρ sends $g(\xi)$ to $g(\xi')$, for each $g \in R[X]$. Thus, the image of ρ is $R[\xi']$. Also, we have $f(\xi') = f(\rho(\xi)) = \rho(f(\xi)) = \rho(0_E) = 0_{E'}$. Therefore, ξ' must be a root of f.

Conversely, suppose that $\xi' \in E'$ is a root of f. Then the polynomial evaluation map from $R[X]$ to E' that sends $g \in R[X]$ to $g(\xi') \in E'$ is an R-algebra homomorphism whose kernel contains f. Using the generalized versions of the first isomorphism theorems for rings and R-modules (Theorems 7.27 and 13.10), we obtain the R-algebra homomorphism

$$\rho: \quad E \to E'$$
$$g(\xi) \mapsto g(\xi').$$

One sees that complex conjugation is just a special case of this construction (see Example 7.57). $\square$

EXERCISE 16.1. Let E be an R-algebra. For $\alpha \in E$, consider the α-multiplication map on E, which sends $\beta \in E$ to $\alpha\beta \in E$. Show that this map is an R-linear map.

EXERCISE 16.2. Show that every ring may be viewed in a unique way as a $\mathbb{Z}$-algebra, and that subrings are subalgebras, and ring homomorphisms are $\mathbb{Z}$-algebra homomorphisms.

EXERCISE 16.3. Show that the only $\mathbb{R}$-algebra homomorphisms from $\mathbb{C}$ into itself are the identity map and the complex conjugation map.

16.2 The field of fractions of an integral domain

Let D be an integral domain. Just as we can construct the field of rational numbers by forming fractions involving integers, we can construct a field consisting of fractions whose numerators and denominators are elements of D. This construction is quite straightforward, though a bit tedious.

To begin with, let S be the set of all pairs of the form (a, b), with $a, b \in D$ and $b \neq 0_D$. Intuitively, such a pair (a, b) is a "formal fraction," with numerator a and denominator b. We define a binary relation $\sim$ on S as follows: for $(a_1, b_1), (a_2, b_2) \in S$, we say $(a_1, b_1) \sim (a_2, b_2)$ if and only if $a_1 b_2 = a_2 b_1$. Our first task is to show that this is an equivalence relation:

Lemma 16.8. *For all* $(a_1, b_1), (a_2, b_2), (a_3, b_3) \in S$, *we have*

    (i) $(a_1, b_1) \sim (a_1, b_1)$;

    (ii) $(a_1, b_1) \sim (a_2, b_2)$ *implies* $(a_2, b_2) \sim (a_1, b_1)$;

    (iii) $(a_1, b_1) \sim (a_2, b_2)$ *and* $(a_2, b_2) \sim (a_3, b_3)$ *implies* $(a_1, b_1) \sim (a_3, b_3)$.

Proof. (i) and (ii) are rather trivial, and we do not comment on these any further. As for (iii), assume that $a_1 b_2 = a_2 b_1$ and $a_2 b_3 = a_3 b_2$. Multiplying the first equation by b_3, we obtain $a_1 b_2 b_3 = a_2 b_1 b_3$ and substituting $a_3 b_2$ for $a_2 b_3$ on the right-hand side of this last equation, we obtain $a_1 b_2 b_3 = a_3 b_2 b_1$. Now, using the fact that b_2 is non-zero and that D is an integral domain, we may cancel b_2 from both sides, obtaining $a_1 b_3 = a_3 b_1$. $\square$

Since $\sim$ is an equivalence relation, it partitions S into equivalence classes, and for $(a, b) \in S$, we denote by $[a, b]$ the equivalence class containing (a, b), and we denote by K the set of all such equivalence classes. Our next task is to define addition and multiplication operations on equivalence classes, mimicking the usual rules of arithmetic with fractions. We want to define the sum of $[a_1, b_1]$ and $[a_2, b_2]$ to be $[a_1 b_2 + a_2 b_1, b_1 b_2]$, and the product of $[a_1, b_1]$ and $[a_2, b_2]$ to be $[a_1 a_2, b_1 b_2]$. Note that since D is an integral domain, if b_1 and b_2 are non-zero, then so is the product $b_1 b_2$, and therefore $[a_1 b_2 + a_2 b_1, b_1 b_2]$ and $[a_1 a_2, b_1 b_2]$ are indeed equivalence classes. However, to ensure that this definition is unambiguous, and does not depend on the particular choice of representatives of the equivalence classes $[a_1, b_1]$ and $[a_2, b_2]$, we need the following lemma.

Lemma 16.9. *Let* $(a_1, b_1), (a_1', b_1'), (a_2, b_2), (a_2', b_2') \in S$, *where* $(a_1, b_1) \sim (a_1', b_1')$ *and* $(a_2, b_2) \sim (a_2', b_2')$. *Then we have*

$$(a_1 b_2 + a_2 b_1, b_1 b_2) \sim (a_1' b_2' + a_2' b_1', b_1' b_2')$$

and

$$(a_1 a_2, b_1 b_2) \sim (a_1' a_2', b_1' b_2').$$

Proof. This is a straightforward calculation. Since $a_1 b_1' = a_1' b_1$ and $a_2 b_2' = a_2' b_2$, we have

$$(a_1 b_2 + a_2 b_1) b_1' b_2' = a_1 b_2 b_1' b_2' + a_2 b_1 b_1' b_2' = a_1' b_2 b_1 b_2' + a_2' b_1 b_1' b_2$$
$$= (a_1' b_2' + a_2' b_1') b_1 b_2$$

and

$$a_1 a_2 b_1' b_2' = a_1' a_2 b_1 b_2' = a_1' a_2' b_1 b_2. \quad \square$$

In light of this lemma, we may unambiguously define addition and multiplication on K as follows: for $[a_1, b_1], [a_2, b_2] \in K$, we define

$$[a_1, b_1] + [a_2, b_2] := [a_1 b_2 + a_2 b_1, b_1 b_2]$$

and

$$[a_1, b_1] \cdot [a_2, b_2] := [a_1 a_2, b_1 b_2].$$

The next task is to show that K is a ring—we leave the details of this (which are quite straightforward) to the reader.

Lemma 16.10. *With addition and multiplication as defined above, K is a ring, with additive identity $[0_D, 1_D]$ and multiplicative identity $[1_D, 1_D]$.*

Proof. Exercise. $\square$

Finally, we observe that K is in fact a field: it is clear that $[a, b]$ is a non-zero element of K if and only if $a \neq 0_D$, and hence any non-zero element $[a, b]$ of K has a multiplicative inverse, namely, $[b, a]$.

The field K is called the **field of fractions of** D. Consider the map $\tau : D \to K$ that sends $a \in D$ to $[a, 1_D] \in K$. It is easy to see that this map is a ring homomorphism, and one can also easily verify that it is injective. So, starting from D, we can synthesize "out of thin air" its field of fractions K, which essentially contains D as a subring, via the natural embedding $\tau : D \to K$.

Now suppose that we are given a field L that contains D as a subring. Consider the set K' consisting of all elements of L of the form ab^{-1}, where $a, b \in D$ and $b \neq 0_D$ — note that here, the arithmetic operations are performed using the rules for arithmetic in L. One may easily verify that K' is a subfield of L that contains D, and it is easy to see that this is the smallest subfield of L that contains D. The subfield K' of L may be referred to as the **field of fractions of** D **within** L. One may easily verify that the map $\rho : K \to L$ that sends $[a, b] \in K$ to $ab^{-1} \in L$ is an unambiguously defined ring homomorphism that maps K injectively onto K'. If we view K and L as D-algebras via inclusion, and we see that the map ρ is in fact a D-algebra homomorphism. Thus, K and K' are isomorphic as D-algebras. It is in this sense that the field of fractions K is the smallest field that contains D as a subring.

From now on, we shall simply write an element $[a, b]$ of K as the fraction a/b. In this notation, the above rules for addition, multiplication, and testing equality in K now look quite familiar:

$$\frac{a_1}{b_1} + \frac{a_2}{b_2} = \frac{a_1 b_2 + a_2 b_1}{b_1 b_2}, \quad \frac{a_1}{b_1} \cdot \frac{a_2}{b_2} = \frac{a_1 a_2}{b_1 b_2}, \quad \frac{a_1}{b_1} = \frac{a_2}{b_2} \iff a_1 b_2 = a_2 b_1.$$

Function fields

An important special case of the above construction for the field of fractions of D is when $D = F[X]$, where F is a field. In this case, the field of fractions is denoted $F(X)$, and is called the **field of rational functions (over** F**)**. This terminology is a bit unfortunate, since just as with polynomials, although the elements of $F(X)$ define functions, they are not (in general) in one-to-one correspondence with these functions.

Since $F[X]$ is a subring of $F(X)$, and since F is a subring of $F[X]$, we see that F is a subfield of $F(X)$.

More generally, we may apply the above construction to $D = F[X_1, \ldots, X_n]$, the ring of multi-variate polynomials over the field F, in which case the field of fractions is denoted $F(X_1, \ldots, X_n)$, and is also called the field of rational functions (over F, in the variables $X_1, \ldots, X_n$).

EXERCISE 16.4. Let F be a field of characteristic zero. Show that F contains an isomorphic copy of $\mathbb{Q}$.

EXERCISE 16.5. Show that the field of fractions of $\mathbb{Z}[i]$ within $\mathbb{C}$ is $\mathbb{Q}[i]$. (See Example 7.25 and Exercise 7.14.)

16.3 Unique factorization of polynomials

Throughout this section, F denotes a field.

Like the ring $\mathbb{Z}$, the ring $F[X]$ of polynomials is an integral domain, and because of the division with remainder property for polynomials, $F[X]$ has many other properties in common with $\mathbb{Z}$. Indeed, essentially all the ideas and results from Chapter 1 can be carried over almost verbatim from $\mathbb{Z}$ to $F[X]$, and in this section, we shall do just that.

Recall that the units of $F[X]$ are precisely the units F^* of F, that is, the non-zero constants. We call two polynomials $g, h \in F[X]$ **associate** if $g = ch$ for some $c \in F^*$. It is easy to see that g and h are associate if and only if $g \mid h$ and $h \mid g$ — indeed, this follows as a special case of part (i) of Theorem 7.4. Clearly, any non-zero polynomial g is associate to a unique monic polynomial (i.e., a polynomial with leading coefficient 1), called the **monic associate** of g; indeed, the monic associate of g is $\mathrm{lc}(g)^{-1} \cdot g$ (where, as usual, $\mathrm{lc}(g)$ denotes the leading coefficient of g).

We call a polynomial $f \in F[X]$ **irreducible** if it is non-constant and all divisors of f are associate to 1 or f. Conversely, we call f **reducible** if it is non-constant and is not irreducible. Equivalently, a non-constant polynomial f is reducible if and only if there exist polynomials $g, h \in F[X]$ of degree strictly less than that of f such that $f = gh$.

Clearly, if g and h are associate polynomials, then g is irreducible if and only if h is irreducible.

The irreducible polynomials play a role similar to that of the prime numbers. Just as it is convenient to work with only *positive* prime numbers, it is also convenient to restrict attention to *monic* irreducible polynomials.

Corresponding to Theorem 1.3, every non-zero polynomial can be expressed as a unit times a product of monic irreducibles in an essentially unique way:

Theorem 16.11. *Every non-zero polynomial $f \in F[X]$ can be expressed as*

$$f = c \cdot p_1^{e_1} \cdots p_r^{e_r},$$

where $c \in F^*$, $p_1, \ldots, p_r$ are distinct monic irreducible polynomials, and $e_1, \ldots, e_r$ are positive integers. Moreover, this expression is unique, up to a reordering of the irreducible polynomials.

To prove this theorem, we may assume that f is monic, since the non-monic case trivially reduces to the monic case.

The proof of the existence part of Theorem 16.11 is just as for Theorem 1.3. If f is 1 or a monic irreducible, we are done. Otherwise, there exist $g, h \in F[X]$ of degree strictly less than that of f such that $f = gh$, and again, we may assume that g and h are monic. By induction on degree, both g and h can be expressed as a product of monic irreducible polynomials, and hence, so can f.

The proof of the uniqueness part of Theorem 16.11 is almost identical to that of Theorem 1.3. The key to the proof is the division with remainder property, Theorem 7.10, from which we can easily derive the following analog of Theorem 1.6:

Theorem 16.12. *Let I be an ideal of $F[X]$. Then there exists a unique polynomial $d \in F[X]$ such that $I = dF[X]$ and d is either zero or monic.*

Proof. We first prove the existence part of the theorem. If $I = \{0\}$, then $d = 0$ does the job, so let us assume that $I \neq \{0\}$. Since I contains non-zero polynomials, it must contain monic polynomials, since if g is a non-zero polynomial in I, then its monic associate $\mathrm{lc}(g)^{-1}g$ is also in I. Let d be a monic polynomial of minimal degree in I. We want to show that $I = dF[X]$.

We first show that $I \subseteq dF[X]$. To this end, let g be any element in I. It suffices to show that $d \mid g$. Using Theorem 7.10, we may write $g = dq + r$, where $\deg(r) < \deg(d)$. Then by the closure properties of ideals, one sees that $r = g - dq$ is also an element of I, and by the minimality of the degree of d, we must have $r = 0$. Thus, $d \mid g$.

We next show that $dF[X] \subseteq I$. This follows immediately from the fact that $d \in I$ and the closure properties of ideals.

That proves the existence part of the theorem. As for uniqueness, note that if $dF[X] = eF[X]$, we have $d \mid e$ and $e \mid d$, from which it follows that d and e are associate, and so if d and e are both either monic or zero, they must be equal. $\square$

For $g, h \in F[X]$, we call $d \in F[X]$ a **common divisor** of g and h if $d \mid g$ and $d \mid h$; moreover, we call such a d a **greatest common divisor** of g and h if d is monic or zero, and all other common divisors of g and h divide d. Analogous to Theorem 1.7, we have:

Theorem 16.13. *For all $g, h \in F[X]$, there exists a unique greatest common divisor d of g and h, and moreover, $gF[X] + hF[X] = dF[X]$.*

Proof. We apply the previous theorem to the ideal $I := gF[X] + hF[X]$. Let

$d \in F[X]$ with $I = dF[X]$, as in that theorem. Note that $g, h, d \in I$ and d is monic or zero.

It is clear that d is a common divisor of g and h. Moreover, there exist $s, t \in F[X]$ such that $gs + ht = d$. If $d' \mid g$ and $d' \mid h$, then clearly $d' \mid (gs + ht)$, and hence $d' \mid d$.

Finally, for uniqueness, if e is a greatest common divisor of g and h, then $d \mid e$ and $e \mid d$, and hence e is associate to d, and the requirement that e is monic or zero implies that $e = d$. $\square$

For $g, h \in F[X]$, we denote by $\gcd(g, h)$ the greatest common divisor of g and h. Note that as we have defined it, $\mathrm{lc}(g) \gcd(g, 0) = g$. Also note that when at least one of g or h are non-zero, $\gcd(g, h)$ is the unique monic polynomial of maximal degree that divides both g and h.

An immediate consequence of Theorem 16.13 is that for all $g, h \in F[X]$, there exist $s, t \in F[X]$ such that $gs + ht = \gcd(g, h)$, and that when at least one of g or h are non-zero, $\gcd(g, h)$ is the unique monic polynomial of minimal degree that can be expressed as $gs + ht$ for some $s, t \in F[X]$.

We say that $g, h \in F[X]$ are **relatively prime** if $\gcd(g, h) = 1$, which is the same as saying that the only common divisors of g and h are units. It is immediate from Theorem 16.13 that g and h are relatively prime if and only if $gF[X] + hF[X] = F[X]$, which holds if and only if there exist $s, t \in F[X]$ such that $gs + ht = 1$.

Analogous to Theorem 1.9, we have:

Theorem 16.14. *For $f, g, h \in F[X]$ such that $f \mid gh$ and $\gcd(f, g) = 1$, we have $f \mid h$.*

Proof. Suppose that $f \mid gh$ and $\gcd(f, g) = 1$. Then since $\gcd(f, g) = 1$, by Theorem 16.13 we have $fs + gt = 1$ for some $s, t \in F[X]$. Multiplying this equation by h, we obtain $fhs + ght = h$. Since $f \mid f$ by definition, and $f \mid gh$ by hypothesis, it follows that $f \mid h$. $\square$

Analogous to Theorem 1.10, we have:

Theorem 16.15. *Let $p \in F[X]$ be irreducible, and let $g, h \in F[X]$. Then $p \mid gh$ implies that $p \mid g$ or $p \mid h$.*

Proof. Assume that $p \mid gh$. The only divisors of p are associate to 1 or p. Thus, $\gcd(p, g)$ is either 1 or the monic associate of p. If $p \mid g$, we are done; otherwise, if $p \nmid g$, we must have $\gcd(p, g) = 1$, and by the previous theorem, we conclude that $p \mid h$. $\square$

Now to prove the uniqueness part of Theorem 16.11. Suppose we have

$$p_1 \cdots p_r = q_1 \cdots q_s,$$

where $p_1, \ldots, p_r$ and $q_1, \ldots, q_s$ are monic irreducible polynomials (with duplicates allowed among the p_i's and among the q_j's). If $r = 0$, we must have $s = 0$ and we are done. Otherwise, as p_1 divides the right-hand side, by inductively applying Theorem 16.15, one sees that p_1 is equal to q_j for some j. We can cancel these terms and proceed inductively (on r).

That completes the proof of Theorem 16.11.

Analogous to Theorem 1.11, we have:

Theorem 16.16. *There are infinitely many monic irreducible polynomials in $F[X]$.*

If F is infinite, then this theorem is true simply because there are infinitely many monic, linear polynomials; in any case, one can easily prove this theorem by mimicking the proof of Theorem 1.11 (as the reader may verify).

For a monic irreducible polynomial p, we may define the function v_p, mapping non-zero polynomials to non-negative integers, as follows: for every polynomial $f \neq 0$, if $f = p^e g$, where $p \nmid g$, then $v_p(f) := e$. We may then write the factorization of f into irreducibles as

$$f = c \prod_p p^{v_p(f)},$$

where the product is over all monic irreducible polynomials p, with all but finitely many of the terms in the product equal to 1.

Just as for integers, we may extend the domain of definition of v_p to include 0, defining $v_p(0) := \infty$. For all polynomials g, h, we have

$$v_p(g \cdot h) = v_p(g) + v_p(h) \quad \text{for all } p. \tag{16.1}$$

From this, it follows that for all polynomials g, h, we have

$$h \mid g \iff v_p(h) \leq v_p(g) \quad \text{for all } p, \tag{16.2}$$

and

$$v_p(\gcd(g, h)) = \min(v_p(g), v_p(h)) \quad \text{for all } p. \tag{16.3}$$

For $g, h \in F[X]$, a **common multiple** of g and h is a polynomial m such that $g \mid m$ and $h \mid m$; moreover, such an m is the **least common multiple** of g and h if m is monic or zero, and m divides all common multiples of g and h. In light of Theorem 16.11, it is clear that the least common multiple exists and is unique, and we denote the least common multiple of g and h by $\text{lcm}(a, b)$. Note that as we have defined it, $\mathrm{lcm}(g, 0) = 0$, and that when both g and h are non-zero, $\mathrm{lcm}(g, h)$ is the unique monic polynomial of minimal degree that is divisible by both g and h. Also, for all $g, h \in F[X]$, we have

$$v_p(\mathrm{lcm}(g, h)) = \max(v_p(g), v_p(h)) \text{ for all } p. \tag{16.4}$$

Just as in §1.3, the notions of greatest common divisor and least common multiple generalize naturally from two to any number of polynomials. We also say that a family of polynomials $\{g_i\}_{i=1}^k$ is **pairwise relatively prime** if $\gcd(g_i, g_j) = 1$ for all indices i, j with $i \neq j$.

Also just as in §1.3, any rational function $g/h \in F(X)$ can be expressed as a fraction g_0/h_0 in **lowest terms**—that is, $g/h = g_0/h_0$ and $\gcd(g_0, h_0) = 1$—and this representation is unique up to multiplication by units.

Many of the exercises in Chapter 1 carry over naturally to polynomials — the reader is encouraged to look over all of the exercises in that chapter, determining which have natural polynomial analogs, and work some of these out.

Example 16.12. Let $f \in F[X]$ be a polynomial of degree 2 or 3. Then it is easy to see that f is irreducible if and only if f has no roots in F. Indeed, if f is reducible, then it must have a factor of degree 1, which we can assume is monic; thus, we can write $f = (X - x)g$, where $x \in F$ and $g \in F[X]$, and so $f(x) = (x - x)g(x) = 0$. Conversely, if $x \in F$ is a root of f, then $X - x$ divides f (see Theorem 7.12), and so f is reducible. □

Example 16.13. As a special case of the previous example, consider the polynomials $f := X^2 - 2 \in \mathbb{Q}[X]$ and $g := X^3 - 2 \in \mathbb{Q}[X]$. We claim that as polynomials over $\mathbb{Q}$, f and g are irreducible. Indeed, neither of them have integer roots, and so neither of them have rational roots (see Exercise 1.26); therefore, they are irreducible. □

Example 16.14. In discussing the factorization of polynomials, one must be clear about the coefficient domain. Indeed, if we view f and g in the previous example as polynomials over $\mathbb{R}$, then they factor into irreducibles as

$$f = (X - \sqrt{2})(X + \sqrt{2}), \quad g = (X - \sqrt[3]{2})(X^2 + \sqrt[3]{2} X + \sqrt[3]{4}),$$

and over $\mathbb{C}$, g factors even further, as

$$g = (X - \sqrt[3]{2})(X - \sqrt[3]{2}(1 + i\sqrt{3})/2)(X - \sqrt[3]{2}(1 - i\sqrt{3})/2). □$$

EXERCISE 16.6. Suppose $f = \sum_{i=0}^{\ell} c_i X^i$ is an irreducible polynomial over F, where $c_0 \neq 0$ and $c_\ell \neq 0$. Show that the "reverse" polynomial $\tilde{f} := \sum_{i=0}^{\ell} c_{\ell-i} X^i$ is also irreducible.

16.4 Polynomial congruences

Throughout this section, F denotes a field.

Many of the results from Chapter 2 on congruences modulo a positive integer n carry over almost verbatim to congruences modulo a non-zero polynomial $f \in F[X]$. We state these results here—the proofs of these results are essentially the same as in the integer case, and as such, are omitted for the most part.

Because of the division with remainder property for polynomials, we have the analog of Theorem 2.4:

Theorem 16.17. *Let* $g, f \in F[X]$, *where* $f \neq 0$. *Then there exists a unique* $z \in F[X]$ *such that* $z \equiv g \pmod{f}$ *and* $\deg(z) < \deg(f)$, *namely,* $z := g \bmod f$.

Corresponding to Theorem 2.5, we have:

Theorem 16.18. *Let* $g, f \in F[X]$ *with* $f \neq 0$, *and let* $d := \gcd(g, f)$.

   (i) For every $h \in F[X]$, *the congruence* $gz \equiv h \pmod{f}$ *has a solution* $z \in F[X]$ *if and only if* $d \mid h$.

   (ii) For every $z \in F[X]$, *we have* $gz \equiv 0 \pmod{f}$ *if and only if* $z \equiv 0$ $\pmod{f/d}$.

   (iii) For all $z, z' \in F[X]$, *we have* $gz \equiv gz' \pmod{f}$ *if and only if* $z \equiv z'$ $\pmod{f/d}$.

Let $g, f \in F[X]$ with $f \neq 0$. Part (iii) of Theorem 16.18 gives us a **cancellation law** for polynomial congruences:

$$\text{if } \gcd(g, f) = 1 \text{ and } gz \equiv gz' \pmod{f}, \text{ then } z \equiv z' \pmod{f}.$$

We say that $z \in F[X]$ is a **multiplicative inverse of g modulo f** if $gz \equiv 1 \pmod{f}$. Part (i) of Theorem 16.18 says that g has a multiplicative inverse modulo f if and only if $\gcd(g, f) = 1$. Moreover, part (iii) of Theorem 16.18 says that the multiplicative inverse of g, if it exists, is uniquely determined modulo f.

As for integers, we may generalize the "mod" operation as follows. Suppose $g, h, f \in F[X]$, with $f \neq 0$, $g \neq 0$, and $\gcd(g, f) = 1$. If s is the rational function $h/g \in F(X)$, then we define $s \bmod f$ to be the unique polynomial $z \in F[X]$ satisfying

$$gz \equiv h \pmod{f} \quad \text{and} \quad \deg(z) < \deg(f).$$

With this notation, we can simply write $g^{-1} \bmod f$ to denote the unique multiplicative inverse of g modulo f of degree less than $\deg(f)$.

Corresponding to Theorem 2.6, we have:

Theorem 16.19 (Chinese remainder theorem). *Let $\{f_i\}_{i=1}^k$ be a pairwise rela-*
tively prime family of non-zero polynomials in $F[X]$, and let $g_1, \ldots, g_k$ be arbi-
trary polynomials in $F[X]$. Then there exists a solution $g \in F[X]$ to the system of
congruences

$$g \equiv g_i \pmod{f_i} \quad (i = 1, \ldots, k).$$

Moreover, any $g' \in F[X]$ is a solution to this system of congruences if and only if
$g \equiv g' \pmod{f}$, where $f := \prod_{i=1}^k f_i$.

Let us recall the formula for the solution g (see proof of Theorem 2.6). We have

$$g := \sum_{i=1}^k g_i e_i,$$

where

$$e_i := f_i^* t_i, \quad f_i^* := f/f_i, \quad t_i := (f_i^*)^{-1} \bmod f_i \quad (i = 1, \ldots, k).$$

Now, let us consider the special case of the Chinese remainder theorem where
$f_i = X - x_i$ with $x_i \in F$, and $g_i = y_i \in F$, for $i = 1, \ldots, k$. The condition that
$\{f_i\}_{i=1}^k$ is pairwise relatively prime is equivalent to the condition that the x_i's are
distinct. Observe that a polynomial $g \in F[X]$ satisfies the system of congruences

$$g \equiv g_i \pmod{f_i} \quad (i = 1, \ldots, k)$$

if and only if

$$g(x_i) = y_i \quad (i = 1, \ldots, k).$$

Moreover, we have $f_i^* = \prod_{j \neq i}(X - x_j)$ and $t_i = 1/\prod_{j \neq i}(x_i - x_j) \in F$. So we get

$$g = \sum_{i=1}^k y_i \frac{\prod_{j \neq i}(X - x_j)}{\prod_{j \neq i}(x_i - x_j)}.$$

The reader will recognize this as the usual *Lagrange interpolation formula* (see
Theorem 7.15). Thus, the Chinese remainder theorem for polynomials includes
Lagrange interpolation as a special case.

Polynomial quotient algebras. Let $f \in F[X]$ be a polynomial of degree $\ell \geq 0$,
and consider the quotient ring $E := F[X]/(f)$. As discussed in Example 16.7, we
may naturally view E as an F-algebra. Moreover, if we set $\xi := [X]_f \in E$, then
$E = F[\xi]$, and viewing E as a vector space over F, we see that $\{\xi^{i-1}\}_{i=1}^\ell$ is a basis
for E.

Now suppose $\alpha \in E$. We have $\alpha = [g]_f = g(\xi)$ for some $g \in F[X]$, and from the above discussion about polynomial congruences, we see that α is a unit if and only if $\gcd(g, f) = 1$.

If $\ell = 0$, then E is trivial. If f is irreducible, then E is a field, since $g \not\equiv 0 \pmod{f}$ implies $\gcd(g, f) = 1$. If f is reducible, then E is not a field, and indeed, not even an integral domain: for any non-trivial factor $g \in F[X]$ of f, $[g]_f \in E$ is a zero divisor.

The Chinese remainder theorem for polynomials also has a more algebraic interpretation. Namely, if $\{f_i\}_{i=1}^k$ is a pairwise relatively prime family of non-zero polynomials in $F[X]$, and $f := \prod_{i=1}^k f_i$, then the map

$$\theta : \quad F[X]/(f) \rightarrow F[X]/(f_1) \times \cdots \times F[X]/(f_k)$$
$$[g]_f \mapsto ([g]_{f_1}, \ldots, [g]_{f_k})$$

is unambiguously defined, and is in fact an F-algebra isomorphism. This map may be seen as a generalization of the ring isomorphism $\bar{\rho}$ discussed in Example 7.54.

Example 16.15. The polynomial $X^2 + 1$ is irreducible over $\mathbb{R}$, since if it were not, it would have a root in $\mathbb{R}$ (see Example 16.12), which is clearly impossible, since -1 is not the square of any real number. It follows immediately that $\mathbb{C} = \mathbb{R}[X]/(X^2+1)$ is a field, without having to explicitly calculate a formula for the inverse of a non-zero complex number. $\square$

Example 16.16. Consider the polynomial $f := X^4 + X^3 + 1$ over $\mathbb{Z}_2$. We claim that f is irreducible. It suffices to show that f has no irreducible factors of degree 1 or 2.

If f had a factor of degree 1, then it would have a root; however, $f(0) = 0 + 0 + 1 = 1$ and $f(1) = 1 + 1 + 1 = 1$. So f has no factors of degree 1.

Does f have a factor of degree 2? The polynomials of degree 2 are X^2, $X^2 + X$, $X^2 + 1$, and $X^2 + X + 1$. The first and second of these polynomials are divisible by X, and hence not irreducible, while the third has a 1 as a root, and hence is also not irreducible. The last polynomial, $X^2 + X + 1$, has no roots, and hence is the only irreducible polynomial of degree 2 over $\mathbb{Z}_2$. So now we may conclude that if f were not irreducible, it would have to be equal to

$$(X^2 + X + 1)^2 = X^4 + 2X^3 + 3X^2 + 2X + 1 = X^4 + X^2 + 1,$$

which it is not.

Thus, $E := \mathbb{Z}_2[X]/(f)$ is a field with $2^4 = 16$ elements. We may think of elements E as bit strings of length 4, where the rule for addition is bit-wise "exclusive-or." The rule for multiplication is more complicated: to multiply two given bit strings, we interpret the bits as coefficients of polynomials (with the left-most bit the coefficient of X^3), multiply the polynomials, reduce the product modulo f, and write down the bit string corresponding to the reduced product polynomial. For example, to multiply 1001 and 0011, we compute

$$(X^3 + 1)(X + 1) = X^4 + X^3 + X + 1,$$

and

$$(X^4 + X^3 + X + 1) \bmod (X^4 + X^3 + 1) = X.$$

Hence, the product of 1001 and 0011 is 0010.

Theorem 7.29 says that E^* is a cyclic group. Indeed, the element $\xi := 0010$ (i.e., $\xi = [X]_f$) is a generator for E^*, as the following table of powers shows:

i	ξ^i	i	ξ^i
1	0010	8	1110
2	0100	9	0101
3	1000	10	1010
4	1001	11	1101
5	1011	12	0011
6	1111	13	0110
7	0111	14	1100
		15	0001

Such a table of powers is sometimes useful for computations in small finite fields such as this one. Given $\alpha, \beta \in E^*$, we can compute $\alpha\beta$ by obtaining (by table lookup) i, j such that $\alpha = \xi^i$ and $\beta = \xi^j$, computing $k := (i + j) \bmod 15$, and then obtaining $\alpha\beta = \xi^k$ (again by table lookup). $\square$

16.5 Minimal polynomials

Throughout this section, F denotes a field.

Suppose that E is an arbitrary F-algebra, and let α be an element of E. Consider the polynomial evaluation map

$$\rho : \quad F[X] \to E$$

$$g \mapsto g(\alpha),$$

which is an F-algebra homomorphism. By definition, the image of ρ is $F[\alpha]$. The kernel of ρ is an ideal of $F[X]$, and since every ideal of $F[X]$ is principal, it follows that $\operatorname{Ker} \rho = \phi F[X]$ for some polynomial $\phi \in F[X]$; moreover, we can make the choice of ϕ unique by insisting that it is monic or zero. The polynomial ϕ is called the **minimal polynomial of α (over F)**.

On the one hand, suppose $\phi \neq 0$. Since any polynomial that is zero at α is a polynomial multiple of ϕ, we see that ϕ is the unique monic polynomial of smallest degree that vanishes at α. Moreover, the first isomorphism theorems for rings and modules tell us that $F[\alpha]$ is isomorphic (as an F-algebra) to $F[X]/(\phi)$, via the isomorphism

$$\bar{\rho}: \quad F[X]/(\phi) \to F[\alpha]$$
$$[g]_\phi \mapsto g(\alpha).$$

Under this isomorphism, $[X]_\phi \in F[X]/(\phi)$ corresponds to $\alpha \in F[\alpha]$, and we see that $\{\alpha^{i-1}\}_{i=1}^m$ is a basis for $F[\alpha]$ over F, where $m = \deg(\phi)$. In particular, every element of $F[\alpha]$ can be written uniquely as $\sum_{i=1}^m c_i\alpha^{i-1}$, where $c_1, \ldots, c_m \in F$.

On the other hand, suppose $\phi = 0$. This means that no non-zero polynomial vanishes at α. Also, it means that the map ρ is injective, and hence $F[\alpha]$ is isomorphic (as an F-algebra) to $F[X]$; in particular, $F[\alpha]$ is not finitely generated as a vector space over F.

Note that if $\alpha \in E$ has a minimal polynomial $\phi \neq 0$, then $\deg(\phi) > 0$, unless E is trivial (i.e., $1_E = 0_E$), in which case $\phi = 1$.

Example 16.17. Consider the real numbers $\sqrt{2}$ and $\sqrt[3]{2}$.

We claim that $X^2 - 2$ is the minimal polynomial of $\sqrt{2}$ over $\mathbb{Q}$. To see this, first observe that $\sqrt{2}$ is a root of $X^2 - 2$. Thus, the minimal polynomial of $\sqrt{2}$ divides $X^2 - 2$. However, as we saw in Example 16.13, the polynomial $X^2 - 2$ is irreducible over $\mathbb{Q}$, and hence must be equal to the minimal polynomial of $\sqrt{2}$ over $\mathbb{Q}$.

A similar argument shows that $X^3 - 2$ is the minimal polynomial of $\sqrt[3]{2}$ over $\mathbb{Q}$.

We also see that $\mathbb{Q}[\sqrt{2}]$ is isomorphic (as a $\mathbb{Q}$-algebra) to $\mathbb{Q}[X]/(X^2 - 2)$, and since $X^2 - 2$ is irreducible, it follows that the ring $\mathbb{Q}[\sqrt{2}]$ is actually a field. As a vector space over $\mathbb{Q}$, $\mathbb{Q}[\sqrt{2}]$ has dimension 2, and every element of $\mathbb{Q}[\sqrt{2}]$ may be written uniquely as $a + b\sqrt{2}$ for $a, b \in \mathbb{Q}$. Indeed, for all $a, b \in \mathbb{Q}$, not both zero, the multiplicative inverse of $a + b\sqrt{2}$ is $(a/c) + (b/c)\sqrt{2}$, where $c := a^2 - 2b^2$.

Similarly, $\mathbb{Q}[\sqrt[3]{2}]$ is a field and has dimension 3 as a vector space over $\mathbb{Q}$, and every element of $\mathbb{Q}[\sqrt[3]{2}]$ may be written uniquely as $a + b\sqrt[3]{2} + c\sqrt[3]{4}$ for $a, b, c \in \mathbb{Q}$. $\square$

A simple but important fact is the following:

Theorem 16.20. *Suppose E is an F-algebra, and that as an F-vector space, E has finite dimension n. Then every $\alpha \in E$ has a non-zero minimal polynomial of degree at most n.*

Proof. Indeed, the family of elements

$$1_E, \alpha, \ldots, \alpha^n$$

must be linearly dependent (as must any family of $n + 1$ elements of a vector space of dimension n), and hence there exist $c_0, \ldots, c_n \in F$, not all zero, such that

$$c_0 1_E + c_1 \alpha + \cdots + c_n \alpha^n = 0_E,$$

and therefore, the non-zero polynomial $f := \sum_i c_i X^i$ vanishes at α. $\square$

Example 16.18. Let $f \in F[X]$ be a monic polynomial of degree ℓ, and consider the F-algebra $E := F[X]/(f) = F[\xi]$, where $\xi := [X]_f \in E$. Clearly, the minimal polynomial of ξ over F is f. Moreover, as a vector space over F, E has dimension ℓ, with $\{\xi^{i-1}\}_{i=1}^{\ell}$ being a basis. Therefore, every $\alpha \in E$ has a non-zero minimal polynomial of degree at most ℓ. $\square$

EXERCISE 16.7. In the field E in Example 16.16, what is the minimal polynomial of 1011 over $\mathbb{Z}_2$?

EXERCISE 16.8. Let $\rho : E \to E'$ be an F-algebra homomorphism, let $\alpha \in E$, let ϕ be the minimal polynomial of α over F, and let ϕ' be the minimal polynomial of $\rho(\alpha)$ over F. Show that $\phi' \mid \phi$, and that $\phi' = \phi$ if ρ is injective.

EXERCISE 16.9. Show that if the factorization of f over $F[X]$ into monic irreducibles is $f = f_1^{e_1} \cdots f_r^{e_r}$, and if $\alpha = [h]_f \in F[X]/(f)$, then the minimal polynomial ϕ of α over F is $\text{lcm}(\phi_1, \ldots, \phi_r)$, where each ϕ_i is the minimal polynomial of $[h]_{f_i^{e_i}} \in F[X]/(f_i^{e_i})$ over F.

16.6 General properties of extension fields

We now discuss a few general notions related to extension fields. These are all quite simple applications of the theory developed so far. Recall that if F and E are fields, with F being a subring of E, then F is called a subfield of E, and E is called an extension field of F. As usual, we shall blur the distinction between a subring and a natural embedding; that is, if $\tau : F \to E$ is a natural embedding, we shall simply identify elements of F with their images in E under τ, and in so doing, we may view E as an extension field of F. Usually, the map τ will be clear from context; for example, if $E = F[X]/(f)$ for some irreducible polynomial $f \in F[X]$, then we shall simply say that E is an extension field of F, although strictly speaking, F is embedded in E via the map that sends $c \in F$ to $[c]_f \in E$.

We start with some definitions. Let E be an extension field of a field F. Then E is an F-algebra via inclusion, and in particular, an F-vector space. If E is a finite dimensional F-vector space, then we say that E is a **finite extension of** F, and $\dim_F(E)$ is called the **degree (over** F**)** of the extension, and is denoted $(E : F)$; otherwise, we say that E is an **infinite extension of** F.

An element $\alpha \in E$ is called **algebraic over** F if there exists a non-zero polynomial $g \in F[X]$ such that $g(\alpha) = 0$, and in this case, we define the **degree of** α (**over** F) to be the degree of its minimal polynomial over F (see §16.5); otherwise, α is called **transcendental over** F. If all elements of E are algebraic over F, then we call E an **algebraic extension of** F.

Suppose E is an extension field of a field F. For $\alpha \in E$, we define

$$F(\alpha) := \{g(\alpha)/h(\alpha) : g, h \in F[X], \ h(\alpha) \neq 0\}.$$

It is easy to see that $F(\alpha)$ is a subfield of E, and indeed, it is the smallest subfield of E containing F and α. Clearly, the ring $F[\alpha] = \{g(\alpha) : g \in F[X]\}$, which is the smallest subring of E containing F and α, is a subring of $F(\alpha)$. We derive some basic properties of $F(\alpha)$ and $F[\alpha]$. The analysis naturally breaks down into two cases, depending on whether α is algebraic or transcendental over F.

On the one hand, suppose α is algebraic over F. Let ϕ be the minimal polynomial of α over F, so that $\deg(\phi) > 0$, and the quotient ring $F[X]/(\phi)$ is isomorphic (as an F-algebra) to the ring $F[\alpha]$ (see §16.5). Since $F[\alpha]$ is a subring of a field, it must be an integral domain, which implies that $F[X]/(\phi)$ is an integral domain, and so ϕ is irreducible. This in turn implies that $F[X]/(\phi)$ is a field, and so $F[\alpha]$ is not just a subring of E, it is a *subfield* of E. Since $F[\alpha]$ is itself already a subfield of E containing F and α, it follows that $F(\alpha) = F[\alpha]$. Moreover, $F[\alpha]$ is a finite extension of F; indeed $(F[\alpha] : F) = \deg(\phi) =$ the degree of α over F, and the elements $1, \alpha, \dots, \alpha^{m-1}$, where $m := \deg(\phi)$, form a basis for $F[\alpha]$ over F.

On the other hand, suppose that α is transcendental over F. In this case, the minimal polynomial of α over F is the zero polynomial, and the ring $F[\alpha]$ is isomorphic (as an F-algebra) to the ring $F[X]$ (see §16.5), which is definitely not a field. But consider the "rational function evaluation map" that sends $g/h \in F(X)$ to $g(\alpha)/h(\alpha) \in F(\alpha)$. Since no non-zero polynomial over F vanishes at α, it is easy to see that this map is well defined, and is in fact an F-algebra isomorphism. Thus, we see that $F(\alpha)$ is isomorphic (as an F-algebra) to $F(X)$. It is also clear that $F(\alpha)$ is an infinite extension of F.

Let us summarize the above discussion in the following theorem:

Theorem 16.21. *Let E be an extension field of a field F.*

(i) *If $\alpha \in E$ is algebraic over F, then $F(\alpha) = F[\alpha]$, and $F[\alpha]$ is isomorphic (as an F-algebra) to $F[X]/(\phi)$, where ϕ is the minimal polynomial of α over F, which is irreducible; moreover, $F[\alpha]$ is a finite extension of F, and $(F[\alpha] : F) = \deg(\phi) =$ the degree of α over F, and the elements $1, \alpha, \dots, \alpha^{m-1}$, where $m := \deg(\phi)$, form a basis for $F[\alpha]$ over F.*

(ii) *If $\alpha \in E$ is transcendental over F, then $F(\alpha)$ is isomorphic (as an F-algebra) to the rational function field $F(X)$, while the subring $F[\alpha]$ is*

isomorphic (as an F-algebra) to the ring of polynomials F[X]; moreover,
F(α) is an infinite extension of F.

Suppose E is an extension field of a field K, which itself is an extension of a field F. Then E is also an extension field of F. The following theorem examines the relation between the degrees of these extensions, in the case where E is a finite extension of K, and K is a finite extension of F. The proof is a simple calculation, which we leave to the reader to verify.

Theorem 16.22. *Suppose E is a finite extension of a field K, with a basis $\{\beta_j\}_{j=1}^m$ over K, and K is a finite extension of F, with a basis $\{\alpha_i\}_{i=1}^n$ over F. Then the elements*

$$\alpha_i \beta_j \quad (i = 1, \ldots, n; \ j = 1, \ldots, m)$$

form a basis for E over F. In particular, E is a finite extension of F and

$$(E : F) = (E : K)(K : F).$$

Now suppose that E is a finite extension of a field F. Let K be an intermediate field, that is, a subfield of E containing F. Then evidently, E is a finite extension of K (since any basis for E over F also spans E over K), and K is a finite extension of F (since as F-vector spaces, K is a subspace of E). The previous theorem then implies that $(E : F) = (E : K)(K : F)$. We have proved:

Theorem 16.23. *If E is a finite extension of a field F, and K is a subfield of E containing F, then E is a finite extension of K, K is a finite extension of F, and $(E : F) = (E : K)(K : F)$.*

Again, suppose that E is a finite extension of a field F. Theorem 16.20 implies that E is algebraic over F, and indeed, that each element of E has degree over F bounded by $(E : F)$. However, we can say a bit more about these degrees. Suppose $\alpha \in E$. Then the degree of α over F is equal to $(F[\alpha] : F)$, and by the previous theorem, applied to $K := F[\alpha]$, we have $(E : F) = (E : F[\alpha])(F[\alpha] : F)$. In particular, the degree of α over F divides $(E : F)$. We have proved:

Theorem 16.24. *If E is a finite extension of a field F, then it is an algebraic extension, and for each $\alpha \in E$, the degree of α over F divides $(E : F)$.*

Example 16.19. Continuing with Example 16.17, we see that the real numbers $\sqrt{2}$ and $\sqrt[3]{2}$ are algebraic over $\mathbb{Q}$. The fields $\mathbb{Q}[\sqrt{2}]$ and $\mathbb{Q}[\sqrt[3]{2}]$ are extension fields of $\mathbb{Q}$, where $(\mathbb{Q}[\sqrt{2}] : \mathbb{Q}) = 2 =$ the degree of $\sqrt{2}$ over $\mathbb{Q}$, and $(\mathbb{Q}[\sqrt[3]{2}] : \mathbb{Q}) = 3 =$ the degree of $\sqrt[3]{2}$ over $\mathbb{Q}$. As both of these fields are finite extensions of $\mathbb{Q}$, they are algebraic extensions as well. Since their degrees over $\mathbb{Q}$ are prime numbers, it follows that they have no subfields other than themselves and $\mathbb{Q}$. In particular, if $\alpha \in \mathbb{Q}[\sqrt{2}] \setminus \mathbb{Q}$, then $\mathbb{Q}[\alpha] = \mathbb{Q}[\sqrt{2}]$. Similarly, if $\alpha \in \mathbb{Q}[\sqrt[3]{2}] \setminus \mathbb{Q}$, then $\mathbb{Q}[\alpha] = \mathbb{Q}[\sqrt[3]{2}]$. □

Example 16.20. Continuing with Example 16.18, suppose $f \in F[X]$ is a monic *irreducible* polynomial of degree ℓ, so that $E := F[X]/(f) = F[\xi]$, where $\xi := [X]_f \in E$, is an extension field of F. The element ξ is algebraic of degree ℓ over F. Moreover, E is a finite extension of F, with $(E : F) = \ell$; in particular, E is an algebraic extension of F, and for each $\alpha \in E$, the degree of α over F divides ℓ. □

As we have seen in Example 16.14, an irreducible polynomial over a field may be reducible when viewed as a polynomial over an extension field. A **splitting field** is a finite extension of the coefficient field in which a given polynomial splits completely into linear factors. As the next theorem shows, splitting fields always exist.

Theorem 16.25. *Let F be a field, and $f \in F[X]$ a non-zero polynomial of degree n. Then there exists a finite extension E of F over which f factors as*

$$f = c(X - \alpha_1)(X - \alpha_2) \cdots (X - \alpha_n),$$

where $c \in F$ and $\alpha_1, \ldots, \alpha_n \in E$.

Proof. We may assume that f is monic. We prove the existence of E by induction on the degree n of f. If $n = 0$, then the theorem is trivially true. Otherwise, let h be an irreducible factor of f, and set $K := F[X]/(h)$, so that $\xi := [X]_h \in K$ is a root of h, and hence of f. So over K, which is a finite extension of F, the polynomial f factors as

$$f = (X - \xi)g,$$

where $g \in K[X]$ is a monic polynomial of degree $n - 1$. Applying the induction hypothesis, there exists a finite extension E of K over which g splits into linear factors. Thus, over E, f splits into linear factors, and by Theorem 16.22, E is a finite extension of F. □

EXERCISE 16.10. In the field E in Example 16.16, find all the elements of degree 2 over $\mathbb{Z}_2$.

EXERCISE 16.11. Let E be an extension field of a field F, and let $\alpha_1, \ldots, \alpha_n \in E$ be algebraic over F. Show that the ring $F[\alpha_1, \ldots, \alpha_n]$ (see Example 7.45) is in fact a field, and that $F[\alpha_1, \ldots, \alpha_n]$ is a finite (and hence algebraic) extension of F.

EXERCISE 16.12. Consider the real numbers $\sqrt{2}$ and $\sqrt[3]{2}$. Show that

$$(\mathbb{Q}[\sqrt{2}, \sqrt[3]{2}] : \mathbb{Q}) = (\mathbb{Q}[\sqrt{2} + \sqrt[3]{2}] : \mathbb{Q}) = 6.$$

EXERCISE 16.13. Consider the real numbers $\sqrt{2}$ and $\sqrt{3}$. Show that

$$(\mathbb{Q}[\sqrt{2}, \sqrt{3}] : \mathbb{Q}) = (\mathbb{Q}[\sqrt{2} + \sqrt{3}] : \mathbb{Q}) = 4.$$

EXERCISE 16.14. Show that if E is an algebraic extension of K, and K is an algebraic extension of F, then E is an algebraic extension of F.

EXERCISE 16.15. Let E be an extension of F. Show that the set of all elements of E that are algebraic over F is a subfield of E containing F.

EXERCISE 16.16. Consider a field F and its field of rational functions $F(X)$. Let $\alpha \in F(X) \setminus F$. Show that X is algebraic over $F(\alpha)$, and that α is transcendental over F.

EXERCISE 16.17. Let E be an extension field of a field F. Suppose $\alpha \in E$ is transcendental over F, and that E is algebraic over $F(\alpha)$. Show that for every $\beta \in E$, β is transcendental over F if and only if E is algebraic over $F(\beta)$.

16.7 Formal derivatives

Throughout this section, R denotes a ring.

Consider a polynomial $g \in R[X]$. If Y is another indeterminate, we may evaluate g at $X + Y$, and collecting monomials of like degree in Y, we may write

$$g(X + Y) = g_0 + g_1 Y + g_2 Y^2 + \cdots \tag{16.5}$$

where $g_i \in R[X]$ for $i = 0, 1, 2, \ldots$. Evidently, $g_0 = g$ (just substitute 0 for Y in (16.5)), and we may write

$$g(X + Y) \equiv g + g_1 Y \pmod{Y^2}. \tag{16.6}$$

We define the **formal derivative** of g, denoted $\mathbf{D}(g)$, to be the unique polynomial $g_1 \in R[X]$ satisfying (16.6). We stress that unlike the "analytical" notion of derivative from calculus, which is defined in terms of limits, this definition is purely "symbolic." Nevertheless, some of the usual rules for derivatives still hold:

Theorem 16.26. *We have:*

  (i) $\mathbf{D}(c) = 0$ *for all $c \in R$;*

  (ii) $\mathbf{D}(X) = 1$;

  (iii) $\mathbf{D}(g + h) = \mathbf{D}(g) + \mathbf{D}(h)$ *for all $g, h \in R[X]$;*

  (iv) $\mathbf{D}(gh) = \mathbf{D}(g)h + g\mathbf{D}(h)$ *for all $g, h \in R[X]$.*

Proof. Parts (i) and (ii) are immediate from the definition. Parts (iii) and (iv) follow from the definition by a simple calculation. Suppose

$$g(X + Y) \equiv g + g_1 Y \pmod{Y^2} \text{ and } h(X + Y) \equiv h + h_1 Y \pmod{Y^2}$$

where $g_1 = \mathbf{D}(g)$ and $h_1 = \mathbf{D}(h)$. Then

$$(g + h)(X + Y) \equiv g(X + Y) + h(X + Y) \equiv (g + h) + (g_1 + h_1)Y \pmod{Y^2},$$

and

$$(gh)(X + Y) \equiv g(X + Y)h(X + Y) \equiv gh + (g_1 h + g h_1)Y \pmod{Y^2}. \quad \Box$$

Combining parts (i) and (iv) of this theorem, we see that $\mathbf{D}(cg) = c\mathbf{D}(g)$ for all $c \in R$ and $g \in R[X]$. This fact can also be easily derived directly from the definition of the derivative.

Combining parts (ii) and (iv) of this theorem, together with a simple induction argument, we see that $\mathbf{D}(X^n) = nX^{n-1}$ for all positive integers n. This fact can also be easily derived directly from the definition of the derivative by considering the binomial expansion of $(X + Y)^n$.

Combining part (iii) of this theorem and the observations in the previous two paragraphs, we see that for any polynomial $g = \sum_{i=0}^k a_i X^i \in R[X]$, we have

$$\mathbf{D}(g) = \sum_{i=1}^{k} i a_i X^{i-1}, \tag{16.7}$$

which agrees with the usual formula for the derivative of a polynomial.

The notion of a formal derivative can be generalized to multi-variate polynomials. Let $g \in R[X_1, \ldots, X_n]$. For any $i = 1, \ldots, n$, we can view g as a polynomial in the variable X_i, whose coefficients are elements of $R[X_1, \ldots, X_{i-1}, X_{i+1}, \ldots, X_n]$. Then if we formally differentiate with respect to the variable X_i, we obtain the formal "partial" derivative $\mathbf{D}_{X_i}(g)$.

EXERCISE 16.18. Show that for $g_1, \ldots, g_n \in R[X]$, we have

$$\mathbf{D}\left(\prod_i g_i\right) = \sum_i \mathbf{D}(g_i) \prod_{j \neq i} g_j$$

and that for $g \in R[X]$, and $n \geq 1$, we have

$$\mathbf{D}(g^n) = ng^{n-1}\mathbf{D}(g).$$

EXERCISE 16.19. Prove the "chain rule" for formal derivatives: if $g, h \in R[X]$

and $f = g(h) \in R[X]$, then $\mathbf{D}(f) = \mathbf{D}(g)(h) \cdot \mathbf{D}(h)$; more generally, if $g \in R[X_1, ..., X_n]$, and $h_1, ..., h_n \in R[X]$, and $f = g(h_1, ..., h_n) \in R[X]$, then

$$\mathbf{D}_X(f) = \sum_{i=1}^{n} \mathbf{D}_{X_i}(g)(h_1, ..., h_n)\mathbf{D}_X(h_i).$$

EXERCISE 16.20. Let $g \in R[X]$, and let $x \in R$ be a root of g. Show that x is a multiple root of g if and only if x is also a root of $\mathbf{D}(g)$ (see Exercise 7.18).

EXERCISE 16.21. Let $g \in R[X]$ with $\deg(g) = k \geq 0$, and let $x \in R$. Show that if we evaluate g at $X + x$, writing

$$g\left(X + x \right) = \sum_{i=0}^{k} b_i X^i,$$

with $b_0, \ldots, b_k \in R$, then we have

$$i! \cdot b_i = (\mathbf{D}^i(g))(x) \text{ for } i = 0, \ldots, k.$$

EXERCISE 16.22. Suppose p is a prime, $g \in \mathbb{Z}[X]$, and $x \in \mathbb{Z}$, such that $g(x) \equiv 0 \pmod{p}$ and $\mathbf{D}(g)(x) \not\equiv 0 \pmod{p}$. Show that for every positive integer e, there exists an integer $\hat{x}$ such that $g(\hat{x}) \equiv 0 \pmod{p^e}$, and give an efficient procedure to compute such an $\hat{x}$, given p, g, x, and e. Hint: mimic the "lifting" procedure discussed in §12.5.2.

16.8 Formal power series and Laurent series

We discuss generalizations of polynomials that allow an infinite number of non-zero coefficients. Although we are mainly interested in the case where the coefficients come from a field F, we develop the basic theory for general rings R.

16.8.1 Formal power series

The ring $R[\![X]\!]$ of **formal power series over** R consists of all formal expressions of the form

$$g = a_0 + a_1 X + a_2 X^2 + \cdots ,$$

where $a_0, a_1, a_2, \ldots \in R$. Unlike ordinary polynomials, we allow an infinite number of non-zero coefficients. We may write such a formal power series as

$$g = \sum_{i=0}^{\infty} a_i X^i.$$

Formally, such a formal power series is an infinite sequence $\{a_i\}_{i=0}^{\infty}$, and the rules for addition and multiplication are *exactly* the same as for polynomials. Indeed, the formulas (7.2) and (7.3) in §7.2 for addition and multiplication may be applied directly — all of the relevant sums are finite, and so everything is well defined. We leave it to the reader to verify that with addition and multiplication so defined, $R[[X]]$ indeed forms a ring. We shall not attempt to interpret a formal power series as a function, and therefore, "convergence" issues shall simply not arise.

Clearly, $R[[X]]$ contains $R[X]$ as a subring. Let us consider the group of units of $R[[X]]$.

Theorem 16.27. *Let* $g = \sum_{i=0}^{\infty} a_i X^i \in R[[X]]$. *Then* $g \in (R[[X]])^*$ *if and only if* $a_0 \in R^*$.

Proof. If a_0 is not a unit, then it is clear that g is not a unit, since the constant term of a product of formal power series is equal to the product of the constant terms.

Conversely, if a_0 is a unit, we show how to define the coefficients of the inverse $h = \sum_{i=0}^{\infty} b_i X^i$ of g. Let $f = gh = \sum_{i=0}^{\infty} c_i X^i$. We want $f = 1$, which means that $c_0 = 1$ and $c_i = 0$ for all $i > 0$. Now, $c_0 = a_0 b_0$, so we set $b_0 := a_0^{-1}$. Next, we have $c_1 = a_0 b_1 + a_1 b_0$, so we set $b_1 := -a_1 b_0 \cdot a_0^{-1}$. Next, we have $c_2 = a_0 b_2 + a_1 b_1 + a_2 b_0$, so we set $b_2 := -(a_1 b_1 + a_2 b_0) \cdot a_0^{-1}$. Continuing in this way, we see that if we define $b_i := -(a_1 b_{i-1} + \cdots + a_i b_0) \cdot a_0^{-1}$ for $i \geq 1$, then $gh = 1$. $\square$

Example 16.21. In the ring $R[[X]]$, the multiplicative inverse of $1 - X$ is $\sum_{i=0}^{\infty} X^i$. $\square$

EXERCISE 16.23. Let F be a field. Show that every non-zero ideal of $F[[X]]$ is of the form (X^m) for some uniquely determined integer $m \geq 0$.

16.8.2 Formal Laurent series

One may generalize formal power series to allow a finite number of negative powers of X. The ring $R((X))$ of **formal Laurent series over** R consists of all formal expressions of the form

$$g = a_m X^m + a_{m+1} X^{m+1} + \cdots,$$

where m is allowed to be any integer (possibly negative), and $a_m, a_{m+1}, \ldots \in R$. Thus, elements of $R((X))$ may have an infinite number of terms involving positive powers of X, but only a finite number of terms involving negative powers of X. We may write such a formal Laurent series as

$$g = \sum_{i=m}^{\infty} a_i X^i.$$

Formally, such a formal Laurent series is a doubly infinite sequence $\{a_i\}_{i=-\infty}^{\infty}$, with the restriction that for some integer m, we have $a_i = 0$ for all $i < m$. We may again use the usual formulas (7.2) and (7.3) to define addition and multiplication (where the indices i, j, and k now range over all integers, not just the non-negative integers). Note that while the sum in (7.3) has an infinite number of terms, only finitely many of them are non-zero.

One may naturally view $R[\![X]\!]$ as a subring of $R(\!(X)\!)$, and of course, $R[X]$ is a subring of $R[\![X]\!]$ and so also a subring of $R(\!(X)\!)$.

Theorem 16.28. *If D is an integral domain, then $D(\!(X)\!)$ is an integral domain.*

Proof. Let $g = \sum_{i=m}^{\infty} a_i X^i$ and $h = \sum_{i=n}^{\infty} b_i X^i$, where $a_m \neq 0$ and $b_n \neq 0$. Then $gh = \sum_{i=m+n}^{\infty} c_i X^i$, where $c_{m+n} = a_m b_n \neq 0$. $\square$

Theorem 16.29. *Let $g \in R(\!(X)\!)$, and suppose that $g \neq 0$ and $g = \sum_{i=m}^{\infty} a_i X^i$ with $a_m \in R^*$. Then g has a multiplicative inverse in $R(\!(X)\!)$.*

Proof. We can write $g = X^m g'$, where g' is a formal power series whose constant term is a unit, and hence there is a formal power series h such that $g'h = 1$. Thus, $X^{-m}h$ is the multiplicative inverse of g in $R(\!(X)\!)$. $\square$

As an immediate corollary, we have:

Theorem 16.30. *If F is a field, then $F(\!(X)\!)$ is a field.*

EXERCISE 16.24. Let F be a field. Show that $F(\!(X)\!)$ is the field of fractions of $F[\![X]\!]$; that is, there is no subfield $E \subsetneq F(\!(X)\!)$ that contains $F[\![X]\!]$.

16.8.3 Reversed Laurent series

While formal Laurent series are useful in some situations, in many others, it is more useful and natural to consider **reversed Laurent series over** R. These are formal expressions of the form

$$g = \sum_{i=-\infty}^{m} a_i X^i,$$

where $a_m, a_{m-1}, \ldots \in R$. Thus, in a reversed Laurent series, we allow an infinite number of terms involving negative powers of X, but only a finite number of terms involving positive powers of X. Formally, such a reversed Laurent series is a doubly infinite sequence $\{a_i\}_{i=-\infty}^{\infty}$, with the restriction that for some integer m, we have $a_i = 0$ for all $i > m$. We may again use the usual formulas (7.2) and (7.3) to define addition and multiplication — and again, the sum in (7.3) has only finitely many non-zero terms.

The ring of all reversed Laurent series is denoted $R((X^{-1}))$, and as the notation suggests, the map that sends X to X^{-1} (and acts as the identity on R) is an R-algebra isomorphism of $R((X))$ with $R((X^{-1}))$. Also, one may naturally view $R[X]$ as a subring of $R((X^{-1}))$.

For $g = \sum_{i=-\infty}^{m} a_i X^i \in R((X^{-1}))$ with $a_m \neq 0$, let us define the **degree of** g, denoted $\deg(g)$, to be the value m, and the **leading coefficient of** g, denoted $\mathrm{lc}(g)$, to be the value a_m. As for ordinary polynomials, we define the degree of 0 to be $-\infty$, and the leading coefficient of 0 to be 0. Note that if g happens to be a polynomial, then these definitions of degree and leading coefficient agree with that for ordinary polynomials.

Theorem 16.31. *For* $g, h \in R((X^{-1}))$, *we have* $\deg(gh) \leq \deg(g) + \deg(h)$, *where equality holds unless both* $\mathrm{lc}(g)$ *and* $\mathrm{lc}(h)$ *are zero divisors. Furthermore, if* $h \neq 0$ *and* $\mathrm{lc}(h)$ *is a unit, then* h *is a unit, and we have* $\deg(gh^{-1}) = \deg(g) - \deg(h)$.

Proof. Exercise. □

It is also natural to define a **floor function** for reversed Laurent series: for $g \in R((X^{-1}))$ with $g = \sum_{i=-\infty}^{m} a_i X^i$, we define

$$\lfloor g \rfloor := \sum_{i=0}^{m} a_i X^i \in R[X];$$

that is, we compute the floor function by simply throwing away all terms involving negative powers of X.

Theorem 16.32. *Let* $g, h \in R[X]$ *with* $h \neq 0$ *and* $\mathrm{lc}(h) \in R^*$, *and using the usual division with remainder property for polynomials, write* $g = hq + r$, *where* $q, r \in R[X]$ *with* $\deg(r) < \deg(h)$. *Let* h^{-1} *denote the multiplicative inverse of* h *in* $R((X^{-1}))$. *Then* $q = \lfloor gh^{-1} \rfloor$.

Proof. Multiplying the equation $g = hq + r$ by h^{-1}, we obtain $gh^{-1} = q + rh^{-1}$, and $\deg(rh^{-1}) < 0$, from which it follows that $\lfloor gh^{-1} \rfloor = q$. □

Let F be a field, so that $F((X^{-1}))$ is also field (this is immediate from Theorem 16.31). Now, $F((X^{-1}))$ contains $F[X]$ as a subring, and hence contains (an isomorphic copy of) the rational function field $F(X)$. Just as $F(X)$ corresponds to the field of rational numbers, $F((X^{-1}))$ corresponds to the field real numbers. Indeed, we can think of real numbers as decimal numbers with a finite number of digits to the left of the decimal point and an infinite number to the right, and reversed Laurent series have a similar "syntactic" structure. In many ways, this syntactic similarity between the real numbers and reversed Laurent series is more than just superficial.

EXERCISE 16.25. Write down the rule for determining the multiplicative inverse of an element of $R((X^{-1}))$ whose leading coefficient is a unit in R.

EXERCISE 16.26. Let F be a field of characteristic other than 2. Show that a non-zero $g \in F((X^{-1}))$ has a square-root in $F((X^{-1}))$ if and only if $\deg(g)$ is even and $\mathrm{lc}(g)$ has a square-root in F.

EXERCISE 16.27. Let R be a ring, and let $a \in R$. Show that the multiplicative inverse of $X - a$ in $R((X^{-1}))$ is $\sum_{j=1}^{\infty} a^{j-1} X^{-j}$.

EXERCISE 16.28. Let R be an arbitrary ring, let $a_1, \ldots, a_\ell \in R$, and let

$$f := (X - a_1)(X - a_2) \cdots (X - a_\ell) \in R[X].$$

For $j \geq 0$, define the "power sum"

$$s_j := \sum_{i=1}^{\ell} a_i^j.$$

Show that in the ring $R((X^{-1}))$, we have

$$\frac{\mathbf{D}(f)}{f} = \sum_{i=1}^{\ell} \frac{1}{(X - a_i)} = \sum_{j=1}^{\infty} s_{j-1} X^{-j},$$

where $\mathbf{D}(f)$ is the formal derivative of f.

EXERCISE 16.29. Continuing with the previous exercise, derive **Newton's identities**, which state that if $f = X^\ell + c_1 X^{\ell-1} + \cdots + c_\ell$, with $c_1, \ldots, c_\ell \in R$, then

$$s_1 + c_1 = 0$$
$$s_2 + c_1 s_1 + 2c_2 = 0$$
$$s_3 + c_1 s_2 + c_2 s_1 + 3c_3 = 0$$
$$\vdots$$
$$s_\ell + c_1 s_{\ell-1} + \cdots + c_{\ell-1} s_1 + \ell c_\ell = 0$$
$$s_{j+\ell} + c_1 s_{j+\ell-1} + \cdots + c_{\ell-1} s_{j+1} + c_\ell s_j = 0 \ (j \geq 1).$$

16.9 Unique factorization domains (∗)

As we have seen, both the ring of integers and the ring of polynomials over a field enjoy a unique factorization property. These are special cases of a more general phenomenon, which we explore here.

Throughout this section, D denotes an integral domain.

We call $a, b \in D$ **associate** if $a = ub$ for some $u \in D^*$. Equivalently, a and b are associate if and only if $a \mid b$ and $b \mid a$ (see part (i) of Theorem 7.4). A non-zero element $p \in D$ is called **irreducible** if it is not a unit, and all divisors of p are associate to 1 or p. Equivalently, a non-zero, non-unit $p \in D$ is irreducible if and only if it cannot be expressed as $p = ab$ where neither a nor b are units.

Definition 16.33. *We call D a **unique factorization domain (UFD)** if*

(i) *every non-zero element of D that is not a unit can be written as a product of irreducibles in D, and*

(ii) *such a factorization into irreducibles is unique up to associates and the order in which the factors appear.*

Another way to state part (ii) of the above definition is that if $p_1 \cdots p_r$ and $p'_1 \cdots p'_s$ are two factorizations of some element as a product of irreducibles, then $r = s$, and there exists a permutation π on the indices $\{1, \ldots, r\}$ such that p_i and $p'_{\pi(i)}$ are associate.

As we have seen, both $\mathbb{Z}$ and $F[X]$ are UFDs. In both of those cases, we chose to single out a distinguished irreducible element among all those associate to any given irreducible: for $\mathbb{Z}$, we always chose positive primes, and for $F[X]$, we chose monic irreducible polynomials. For any specific unique factorization domain D, there may be such a natural choice, but in the general case, there will not be (but see Exercise 16.30 below).

Example 16.22. Having already seen two examples of UFDs, it is perhaps a good idea to look at an example of an integral domain that is not a UFD. Consider the subring $\mathbb{Z}[\sqrt{-3}]$ of the complex numbers, which consists of all complex numbers of the form $a + b\sqrt{-3}$, where $a, b \in \mathbb{Z}$. As this is a subring of the field $\mathbb{C}$, it is an integral domain (one may also view $\mathbb{Z}[\sqrt{-3}]$ as the quotient ring $\mathbb{Z}[X]/(X^2 + 3)$).

Let us first determine the units in $\mathbb{Z}[\sqrt{-3}]$. For $a, b \in \mathbb{Z}$, we have $N(a+b\sqrt{-3}) = a^2 + 3b^2$, where N is the usual norm map on $\mathbb{C}$ (see Example 7.5). If $\alpha \in \mathbb{Z}[\sqrt{-3}]$ is a unit, then there exists $\alpha' \in \mathbb{Z}[\sqrt{-3}]$ such that $\alpha\alpha' = 1$. Taking norms, we obtain

$$1 = N(1) = N(\alpha\alpha') = N(\alpha)N(\alpha').$$

Since the norm of an element of $\mathbb{Z}[\sqrt{-3}]$ is a non-negative integer, this implies that $N(\alpha) = 1$. If $\alpha = a + b\sqrt{-3}$, with $a, b \in \mathbb{Z}$, then $N(\alpha) = a^2 + 3b^2$, and it is clear that $N(\alpha) = 1$ if and only if $\alpha = \pm 1$. We conclude that the only units in $\mathbb{Z}[\sqrt{-3}]$ are ± 1.

Now consider the following two factorizations of 4 in $\mathbb{Z}[\sqrt{-3}]$:

$$4 = 2 \cdot 2 = (1 + \sqrt{-3})(1 - \sqrt{-3}). \tag{16.8}$$

We claim that 2 is irreducible. For suppose, say, that $2 = \alpha\alpha'$, for $\alpha, \alpha' \in \mathbb{Z}[\sqrt{-3}]$, with neither a unit. Taking norms, we have $4 = N(2) = N(\alpha)N(\alpha')$, and therefore, $N(\alpha) = N(\alpha') = 2$—but this is impossible, since there are no integers a and b such that $a^2 + 3b^2 = 2$. By the same reasoning, since $N(1 + \sqrt{-3}) = N(1 - \sqrt{-3}) = 4$, we see that $1 + \sqrt{-3}$ and $1 - \sqrt{-3}$ are both irreducible. Further, it is clear that 2 is not associate to either $1 + \sqrt{-3}$ or $1 - \sqrt{-3}$, and so the two factorizations of 4 in (16.8) are fundamentally different. □

For $a, b \in D$, we call $d \in D$ a **common divisor** of a and b if $d \mid a$ and $d \mid b$; moreover, we call such a d a **greatest common divisor** of a and b if all other common divisors of a and b divide d. We say that a and b are **relatively prime** if the only common divisors of a and b are units. It is immediate from the definition of a greatest common divisor that it is unique, up to multiplication by units, if it exists at all. Unlike in the case of $\mathbb{Z}$ and $F[X]$, in the general setting, greatest common divisors need not exist; moreover, even when they do, we shall not attempt to "normalize" greatest common divisors, and we shall speak only of "a" greatest common divisor, rather than "the" greatest common divisor.

Just as for integers and polynomials, we can generalize the notion of a greatest common divisor in an arbitrary integral domain D from two to any number of elements of D, and we can also define a **least common multiple** of any number of elements as well.

Although these greatest common divisors and least common multiples need not exist in an arbitrary integral domain D, if D is a UFD, they will always exist. The existence question easily reduces to the question of the existence of a greatest common divisor and least common multiple of a and b, where a and b are non-zero elements of D. So assuming that D is a UFD, we may write

$$a = u \prod_{i=1}^{r} p_i^{e_i} \text{ and } b = v \prod_{i=1}^{r} p_i^{f_i},$$

where u and v are units, $p_1, \ldots, p_r$ are non-associate irreducibles, and $e_1, \ldots, e_r$ and $f_1, \ldots, f_r$ are non-negative integers, and it is easily seen that

$$\prod_{i=1}^{r} p_i^{\min(e_i, f_i)}$$

is a greatest common divisor of a and b, while

$$\prod_{i=1}^{r} p_i^{\max(e_i, f_i)}$$

is a least common multiple of a and b.

It is also evident that in a UFD D, if $c \mid ab$ and c and a are relatively prime, then $c \mid b$. In particular, if p is irreducible and $p \mid ab$, then $p \mid a$ or $p \mid b$. This is equivalent to saying that if p is irreducible, then the quotient ring D/pD is an integral domain (and the ideal pD is a prime ideal — see Exercise 7.38). The converse also holds:

Theorem 16.34. *Suppose D satisfies part (i) of Definition 16.33, and that D/pD is an integral domain for every irreducible $p \in D$. Then D is a UFD.*

Proof. Exercise. □

EXERCISE 16.30.   (a) Show that the "is associate to" relation is an equivalence relation.

  (b) Consider an equivalence class C induced by the "is associate to" relation. Show that if C contains an irreducible element, then all elements of C are irreducible.

  (c) Suppose that for every equivalence class C that contains irreducibles, we choose one element of C, and call it a **distinguished irreducible**. Show that D is a UFD if and only if every non-zero element of D can be expressed as $u p_1^{e_1} \cdots p_r^{e_r}$, where u is a unit, $p_1, \ldots, p_r$ are distinguished irreducibles, and this expression is unique up to a reordering of the p_i's.

EXERCISE 16.31. Show that the ring $\mathbb{Z}[\sqrt{-5}]$ is not a UFD.

EXERCISE 16.32. Let D be a UFD and F its field of fractions. Show that

  (a) every element $x \in F$ can be expressed as $x = a/b$, where $a, b \in D$ are relatively prime, and

  (b) that if $x = a/b$ for $a, b \in D$ relatively prime, then for any other $a', b' \in D$ with $x = a'/b'$, we have $a' = ca$ and $b' = cb$ for some $c \in D$.

EXERCISE 16.33. Let D be a UFD and let $p \in D$ be irreducible. Show that there is no prime ideal Q of D with $\{0_D\} \subsetneq Q \subsetneq pD$ (see Exercise 7.38).

16.9.1 Unique factorization in Euclidean and principal ideal domains

Our proofs of the unique factorization property in both $\mathbb{Z}$ and $F[X]$ hinged on the division with remainder property for these rings. This notion can be generalized, as follows.

Definition 16.35. *We say D is a **Euclidean domain** if there is a "size function" S mapping the non-zero elements of D to the set of non-negative integers, such that for all $a, b \in D$ with $b \neq 0$, there exist $q, r \in D$, with the property that $a = bq + r$ and either $r = 0$ or $S(r) < S(b)$.*

Example 16.23. Both $\mathbb{Z}$ and $F[X]$ are Euclidean domains. In $\mathbb{Z}$, we can take the ordinary absolute value function $|\cdot|$ as a size function, and for $F[X]$, the function $\deg(\cdot)$ will do. □

Example 16.24. Recall again the ring

$$\mathbb{Z}[i] = \{a + bi : a, b \in \mathbb{Z}\}$$

of Gaussian integers from Example 7.25. Let us show that this is a Euclidean domain, using the usual norm map N on complex numbers (see Example 7.5) for the size function. Let $\alpha, \beta \in \mathbb{Z}[i]$, with $\beta \neq 0$. We want to show the existence of $\kappa, \rho \in \mathbb{Z}[i]$ such that $\alpha = \beta\kappa + \rho$, where $N(\rho) < N(\beta)$. Suppose that in the field $\mathbb{C}$, we compute $\alpha\beta^{-1} = r + si$, where $r, s \in \mathbb{Q}$. Let m, n be integers such that $|m - r| \leq 1/2$ and $|n - s| \leq 1/2$—such integers m and n always exist, but may not be uniquely determined. Set $\kappa := m + ni \in \mathbb{Z}[i]$ and $\rho := \alpha - \beta\kappa$. Then we have

$$\alpha\beta^{-1} = \kappa + \delta,$$

where $\delta \in \mathbb{C}$ with $N(\delta) \leq 1/4 + 1/4 = 1/2$, and

$$\rho = \alpha - \beta\kappa = \alpha - \beta(\alpha\beta^{-1} - \delta) = \delta\beta,$$

and hence

$$N(\rho) = N(\delta\beta) = N(\delta)N(\beta) \leq \frac{1}{2}N(\beta). \quad \square$$

Theorem 16.36. *If D is a Euclidean domain and I is an ideal of D, then there exists $d \in D$ such that $I = dD$.*

Proof. If $I = \{0\}$, then $d = 0$ does the job, so let us assume that $I \neq \{0\}$. Let d be any non-zero element of I such that $S(d)$ is minimal, where S is a size function that makes D into a Euclidean domain. We claim that $I = dD$.

It will suffice to show that for all $c \in I$, we have $d \mid c$. Now, we know that there exists $q, r \in D$ such that $c = dq + r$, where either $r = 0$ or $S(r) < S(d)$. If $r = 0$, we are done; otherwise, r is a non-zero element of I with $S(r) < S(d)$, contradicting the minimality of $S(d)$. □

Recall that an ideal of the form $I = dD$ is called a principal ideal. If all ideals of D are principal, then D is called a **principal ideal domain (PID)**. Theorem 16.36 says that every Euclidean domain is a PID.

PIDs enjoy many nice properties, including:

Theorem 16.37. *If D is a PID, then D is a UFD.*

For the rings $\mathbb{Z}$ and $F[X]$, the proof of part (i) of Definition 16.33 was a quite straightforward induction argument (as it also would be for any Euclidean domain). For a general PID, however, this requires a different sort of argument. We begin with the following fact:

Theorem 16.38. *If D is a PID, and $I_1 \subseteq I_2 \subseteq \cdots$ are ideals of D, then there exists an integer k such that $I_k = I_{k+1} = \cdots$.*

Proof. Let $I := \bigcup_{i=1}^{\infty} I_i$, which is an ideal of D (see Exercise 7.37). Thus, $I = dD$ for some $d \in D$. But $d \in \bigcup_{i=1}^{\infty} I_i$ implies that $d \in I_k$ for some k, which shows that $I = dD \subseteq I_k$. It follows that $I = I_k = I_{k+1} = \cdots$. $\square$

We can now prove the existence part of Theorem 16.37:

Theorem 16.39. *If D is a PID, then every non-zero, non-unit element of D can be expressed as a product of irreducibles in D.*

Proof. Let $c \in D$, $c \neq 0$, and c not a unit. If c is irreducible, we are done. Otherwise, we can write $c = ab$, where neither a nor b are units. As ideals, we have $cD \subsetneq aD$ and $cD \subsetneq bD$. If we continue this process recursively, building up a "factorization tree" where c is at the root, a and b are the children of c, and so on, then the recursion must stop, since any infinite path in the tree would give rise to ideals

$$cD = I_1 \subsetneq I_2 \subsetneq \cdots,$$

contradicting Theorem 16.38. $\square$

The proof of the uniqueness part of Theorem 16.37 is essentially the same as for proofs we gave for $\mathbb{Z}$ and $F[X]$.

Analogous to Theorems 1.7 and 16.13, we have:

Theorem 16.40. *Let D be a PID. For all $a, b \in D$, there exists a greatest common divisor d of a and b, and moreover, $aD + bD = dD$.*

Proof. Exercise. $\square$

As an immediate consequence of the previous theorem, we see that in a PID D, for all $a, b \in D$ with greatest common divisor d, there exist $s, t \in D$ such that

$as + bt = d$; moreover, $a, b \in D$ are relatively prime if and only if there exist $s, t \in D$ such that $as + bt = 1$.

Analogous to Theorems 1.9 and 16.14, we have:

Theorem 16.41. *Let D be a PID. For all $a, b, c \in D$ such that $c \mid ab$ and a and c are relatively prime, we have $c \mid b$.*

Proof. Exercise. $\square$

Analogous to Theorems 1.10 and 16.15, we have:

Theorem 16.42. *Let D be a PID. Let $p \in D$ be irreducible, and let $a, b \in D$. Then $p \mid ab$ implies that $p \mid a$ or $p \mid b$.*

Proof. Exercise. $\square$

Theorem 16.37 now follows immediately from Theorems 16.39, 16.42, and 16.34.

EXERCISE 16.34. Show that $\mathbb{Z}[\sqrt{-2}]$ is a Euclidean domain.

EXERCISE 16.35. Consider the polynomial

$$X^3 - 1 = (X - 1)(X^2 + X + 1).$$

Over $\mathbb{C}$, the roots of $X^3 - 1$ are $1, (-1 \pm \sqrt{-3})/2$. Let $\omega := (-1 + \sqrt{-3})/2$, and note that $\omega^2 = -1 - \omega = (-1 - \sqrt{-3})/2$, and $\omega^3 = 1$.

(a) Show that the ring $\mathbb{Z}[\omega]$ consists of all elements of the form $a + b\omega$, where $a, b \in \mathbb{Z}$, and is an integral domain. This ring is called the ring of **Eisenstein integers**.

(b) Show that the only units in $\mathbb{Z}[\omega]$ are ± 1, $\pm \omega$, and $\pm \omega^2$.

(c) Show that $\mathbb{Z}[\omega]$ is a Euclidean domain.

EXERCISE 16.36. Show that in a PID, all non-zero prime ideals are maximal (see Exercise 7.38).

Recall that for a complex number $\alpha = a + bi$, with $a, b \in \mathbb{R}$, the norm of α was defined as $N(\alpha) = \alpha\bar{\alpha} = a^2 + b^2$ (see Example 7.5). There are other measures of the "size" of a complex number that are useful. The **absolute value** of α is defined as $|\alpha| := \sqrt{N(\alpha)} = \sqrt{a^2 + b^2}$. The **max norm** of α is defined as $M(\alpha) := \max\{|a|, |b|\}$.

EXERCISE 16.37. Let $\alpha, \beta \in \mathbb{C}$. Prove the following statements:

(a) $|\alpha\beta| = |\alpha||\beta|$;

(b) $|\alpha + \beta| \le |\alpha| + |\beta|$;

(c) $N(\alpha + \beta) \le 2(N(\alpha) + N(\beta))$;

(d) $M(\alpha) \le |\alpha| \le \sqrt{2}M(\alpha)$.

The following exercises develop algorithms for computing with Gaussian integers. For computational purposes, we assume that a Gaussian integer $\alpha = a + bi$, with $a, b \in \mathbb{Z}$, is represented as the pair of integers (a, b).

EXERCISE 16.38. Let $\alpha, \beta \in \mathbb{Z}[i]$.

(a) Show how to compute $M(\alpha)$ in time $O(\text{len}(M(\alpha)))$ and $N(\alpha)$ in time $O(\text{len}(M(\alpha))^2)$.

(b) Show how to compute $\alpha + \beta$ in time $O(\text{len}(M(\alpha)) + \text{len}(M(\beta)))$.

(c) Show how to compute $\alpha \cdot \beta$ in time $O(\text{len}(M(\alpha)) \cdot \text{len}(M(\beta)))$.

(d) Assuming $\beta \ne 0$, show how to compute $\kappa, \rho \in \mathbb{Z}[i]$ such that $\alpha = \beta\kappa + \rho$, $N(\rho) \le \frac{1}{2}N(\beta)$, and $N(\kappa) \le 4N(\alpha)/N(\beta)$. Your algorithm should run in time $O(\text{len}(M(\alpha)) \cdot \text{len}(M(\beta)))$. Hint: see Example 16.24; also, to achieve the stated running time bound, your algorithm should first test if $M(\beta) \ge 2M(\alpha)$.

EXERCISE 16.39. Using the division with remainder algorithm from part (d) of the previous exercise, adapt the Euclidean algorithm for (ordinary) integers to work with Gaussian integers. On inputs $\alpha, \beta \in \mathbb{Z}[i]$, your algorithm should compute a greatest common divisor $\delta \in \mathbb{Z}[i]$ of α and β in time $O(\ell^3)$, where $\ell := \max\{\text{len}(M(\alpha)), \text{len}(M(\beta))\}$.

EXERCISE 16.40. Extend the algorithm of the previous exercise, so that it computes $\sigma, \tau \in \mathbb{Z}[i]$ such that $\alpha\sigma + \beta\tau = \delta$. Your algorithm should run in time $O(\ell^3)$, and it should also be the case that $\text{len}(M(\sigma))$ and $\text{len}(M(\tau))$ are $O(\ell)$.

The algorithms in the previous two exercises for computing greatest common divisors in $\mathbb{Z}[i]$ run in time cubic in the length of their input, whereas the corresponding algorithms for $\mathbb{Z}$ run in time quadratic in the length of their input. This is essentially because the running time of the algorithm for division with remainder discussed in Exercise 16.38 is insensitive to the size of the quotient.

To get a quadratic-time algorithm for computing greatest common divisors in $\mathbb{Z}[i]$, in the following exercises we shall develop an analog of the binary gcd algorithm for $\mathbb{Z}$.

EXERCISE 16.41. Let $\pi := 1 + i \in \mathbb{Z}[i]$.

(a) Show that $2 = \pi\bar{\pi} = -i\pi^2$, that $N(\pi) = 2$, and that π is irreducible in $\mathbb{Z}[i]$.

(b) Let $\alpha \in \mathbb{Z}[i]$, with $\alpha = a + bi$ for $a, b \in \mathbb{Z}$. Show that $\pi \mid \alpha$ if and only if $a - b$ is even, in which case

$$\frac{\alpha}{\pi} = \frac{a+b}{2} + \frac{b-a}{2}i.$$

(c) Show that for all $\alpha \in \mathbb{Z}[i]$, we have $\alpha \equiv 0 \pmod{\pi}$ or $\alpha \equiv 1 \pmod{\pi}$.

(d) Show that the quotient ring $\mathbb{Z}[i]/\pi\mathbb{Z}[i]$ is isomorphic to the ring $\mathbb{Z}_2$.

(e) Show that for all $\alpha \in \mathbb{Z}[i]$ with $\alpha \equiv 1 \pmod{\pi}$, there exists a unique $\varepsilon \in \{\pm 1, \pm i\}$ such that $\alpha \equiv \varepsilon \pmod{2\pi}$.

(f) Show that for all $\alpha, \beta \in \mathbb{Z}[i]$ with $\alpha \equiv \beta \equiv 1 \pmod{\pi}$, there exists a unique $\varepsilon \in \{\pm 1, \pm i\}$ such that $\alpha \equiv \varepsilon\beta \pmod{2\pi}$.

EXERCISE 16.42. We now present a "$(1 + i)$-ary gcd algorithm" for Gaussian integers. Let $\pi := 1 + i \in \mathbb{Z}[i]$. The algorithm takes non-zero $\alpha, \beta \in \mathbb{Z}[i]$ as input, and runs as follows:

$$\rho \leftarrow \alpha, \ \rho' \leftarrow \beta, \ e \leftarrow 0$$
while $\pi \mid \rho$ and $\pi \mid \rho'$ do $\rho \leftarrow \rho/\pi, \ \rho' \leftarrow \rho'/\pi, \ e \leftarrow e + 1$
repeat
       while $\pi \mid \rho$ do $\rho \leftarrow \rho/\pi$
       while $\pi \mid \rho'$ do $\rho' \leftarrow \rho'/\pi$
       if $M(\rho') < M(\rho)$ then $(\rho, \rho') \leftarrow (\rho', \rho)$
       determine $\varepsilon \in \{\pm 1, \pm i\}$ such that $\rho' \equiv \varepsilon\rho \pmod{2\pi}$
(*) $\rho' \leftarrow \rho' - \varepsilon\rho$
until $\rho' = 0$
$\delta \leftarrow \pi^e \cdot \rho$
output δ

Show that this algorithm correctly computes a greatest common divisor of α and β, and that it can be implemented so as to run in time $O(\ell^2)$, where $\ell := \max(\text{len}(M(\alpha)), \text{len}(M(\beta)))$. Hint: to analyze the running time, for $i = 1, 2, \ldots$, let v_i (respectively, v'_i) denote the value of $|\rho\rho'|$ just before (respectively, after) the execution of the line marked (*) in loop iteration i, and show that

$$v'_i \leq (1 + \sqrt{2})v_i \quad \text{and} \quad v_{i+1} \leq v'_i/2\sqrt{2}.$$

EXERCISE 16.43. Extend the algorithm of the previous exercise, so that it computes $\sigma, \tau \in \mathbb{Z}[i]$ such that $\alpha\sigma + \beta\tau = \delta$. Your algorithm should run in time $O(\ell^2)$, and it should also be the case that $\text{len}(M(\sigma))$ and $\text{len}(M(\tau))$ are $O(\ell)$. Hint: adapt the algorithm in Exercise 4.10.

EXERCISE 16.44. In Exercise 16.41, we saw that 2 factors as $-i(1 + i)^2$ in $\mathbb{Z}[i]$, where $1 + i$ is irreducible. This exercise examines the factorization in $\mathbb{Z}[i]$ of prime numbers $p > 2$. Show that:

(a) for every irreducible $\pi \in \mathbb{Z}[i]$, there exists a unique prime number p such that π divides p;

(b) for all prime numbers $p \equiv 1 \pmod{4}$, we have $p = \pi\bar{\pi}$, where $\pi \in \mathbb{Z}[i]$ is irreducible, and the complex conjugate $\bar{\pi}$ of π is also irreducible and not associate to π;

(c) all prime numbers $p \equiv 3 \pmod{4}$ are irreducible in $\mathbb{Z}[i]$.

Hint: for parts (b) and (c), use Theorem 2.34.

16.9.2 Unique factorization in $D[X]$

In this section, we prove the following:

Theorem 16.43. *If D is a UFD, then so is $D[X]$.*

This theorem implies, for example, that $\mathbb{Z}[X]$ is a UFD. Applying the theorem inductively, one also sees that $\mathbb{Z}[X_1, \ldots, X_n]$ is a UFD, as is $F[X_1, \ldots, X_n]$ for every field F.

We begin with some simple observations. First, recall that for an integral domain D, $D[X]$ is an integral domain, and the units in $D[X]$ are precisely the units in D. Second, it is easy to see that an element of D is irreducible in D if and only if it is irreducible in $D[X]$. Third, for $c \in D$ and $f = \sum_i c_i X^i \in D[X]$, we have $c \mid f$ if and only if $c \mid c_i$ for all i.

We call a non-zero polynomial $f \in D[X]$ **primitive** if the only elements of D that divide f are units. If D is a UFD, then given any non-zero polynomial $f \in D[X]$, we can write it as $f = cf'$, where $c \in D$ and $f' \in D[X]$ is a primitive polynomial: just take c to be a greatest common divisor of all the coefficients of f.

Example 16.25. In $\mathbb{Z}[X]$, the polynomial $f = 4X^2 + 6X + 20$ is not primitive, but we can write $f = 2f'$, where $f' = 2X^2 + 3X + 10$ is primitive. □

It is easy to prove the existence part of Theorem 16.43:

Theorem 16.44. *Let D be a UFD. Every non-zero, non-unit element of $D[X]$ can be expressed as a product of irreducibles in $D[X]$.*

Proof. Let f be a non-zero, non-unit polynomial in $D[X]$. If f is a constant, then because D is a UFD, f factors into irreducibles in D. So assume f is not constant. If f is not primitive, we can write $f = cf'$, where c is a non-zero, non-unit in D, and f' is a primitive, non-constant polynomial in $D[X]$. Again, as D is a UFD, c factors into irreducibles in D.

From the above discussion, it suffices to prove the theorem for non-constant, primitive polynomials $f \in D[X]$. If f is itself irreducible, we are done. Otherwise, we can write $f = gh$, where $g, h \in D[X]$ and neither g nor h are units. Further, by the assumption that f is a primitive, non-constant polynomial, both g and h must also be primitive, non-constant polynomials; in particular, both g and h have degree strictly less than $\deg(f)$, and the theorem follows by induction on degree. $\square$

The uniqueness part of Theorem 16.43 is (as usual) more difficult. We begin with the following fact:

Theorem 16.45. *Let D be a UFD, let p be an irreducible in D, and let $g, h \in D[X]$. Then $p \mid gh$ implies $p \mid g$ or $p \mid h$.*

Proof. Consider the quotient ring D/pD, which is an integral domain (because D is a UFD), and the corresponding ring of polynomials $(D/pD)[X]$, which is also an integral domain. Also consider the natural map that sends $a \in D$ to $\bar{a} := [a]_p \in D/pD$, which we can extend coefficient-wise to a ring homomorphism from $D[X]$ to $(D/pD)[X]$ (see Example 7.46). If $p \mid gh$, then we have

$$0 = \overline{gh} = \bar{g}\bar{h},$$

and since $(D/pD)[X]$ is an integral domain, it follows that $\bar{g} = 0$ or $\bar{h} = 0$, which means that $p \mid g$ or $p \mid h$. $\square$

Theorem 16.46. *Let D be a UFD. The product of two primitive polynomials in $D[X]$ is also primitive.*

Proof. Let $g, h \in D[X]$ be primitive polynomials, and let $f := gh$. If f is not primitive, then $c \mid f$ for some non-zero, non-unit $c \in D$, and as D is a UFD, there is some irreducible element $p \in D$ that divides c, and therefore, divides f as well. By Theorem 16.45, it follows that $p \mid g$ or $p \mid h$, which implies that either g is not primitive or h is not primitive. $\square$

Suppose that D is a UFD and that F is its field of fractions. Any non-zero polynomial $f \in F[X]$ can always be written as $f = (c/d)f'$, where $c, d \in D$, with $d \neq 0$, and $f' \in D[X]$ is primitive. To see this, clear the denominators of the coefficients of f, writing $df = f''$, where $0 \neq d \in D$ and $f'' \in D[X]$. Then take c to be a greatest common divisor of the coefficients of f'', so that $f'' = cf'$, where $f' \in D[X]$ is primitive. Then we have $f = (c/d)f'$, as required. Of course, we may assume that c and d are relatively prime—if not, we may divide c and d by a greatest common divisor.

Example 16.26. Let $f = (3/5)X^2 + 9X + 3/2 \in \mathbb{Q}[X]$. Then we can write $f = (3/10)f'$, where $f' = 2X^2 + 30X + 5 \in \mathbb{Z}[X]$ is primitive. $\square$

As a consequence of the previous theorem, we have:

Theorem 16.47. *Let D be a UFD and let F be its field of fractions. Suppose that $f, g \in D[X]$ and $h \in F[X]$ are non-zero polynomials such that $f = gh$ and g is primitive. Then $h \in D[X]$.*

Proof. Write $h = (c/d)h'$, where $c, d \in D$ and $h' \in D[X]$ is primitive. Let us assume that c and d are relatively prime. Then we have

$$d \cdot f = c \cdot gh'. \tag{16.9}$$

We claim that $d \in D^*$. To see this, note that (16.9) implies that $d \mid (c \cdot gh')$, and the assumption that c and d are relatively prime implies that $d \mid gh'$. But by Theorem 16.46, gh' is primitive, from which it follows that d is a unit. That proves the claim.

It follows that $c/d \in D$, and hence $h = (c/d)h' \in D[X]$. $\square$

Theorem 16.48. *Let D be a UFD and F its field of fractions. If $f \in D[X]$ with $\deg(f) > 0$ is irreducible, then f is also irreducible in $F[X]$.*

Proof. Suppose that f is not irreducible in $F[X]$, so that $f = gh$ for non-constant polynomials $g, h \in F[X]$, both of degree strictly less than that of f. We may write $g = (c/d)g'$, where $c, d \in D$ and $g' \in D[X]$ is primitive. Set $h' := (c/d)h$, so that $f = gh = g'h'$. By Theorem 16.47, we have $h' \in D[X]$, and this shows that f is not irreducible in $D[X]$. $\square$

Theorem 16.49. *Let D be a UFD. Let $f \in D[X]$ with $\deg(f) > 0$ be irreducible, and let $g, h \in D[X]$. If f divides gh in $D[X]$, then f divides either g or h in $D[X]$.*

Proof. Suppose that $f \in D[X]$ with $\deg(f) > 0$ is irreducible. This implies that f is a primitive polynomial. By Theorem 16.48, f is irreducible in $F[X]$, where F is the field of fractions of D. Suppose f divides gh in $D[X]$. Then because $F[X]$ is a UFD, f divides either g or h in $F[X]$. But Theorem 16.47 implies that f divides either g or h in $D[X]$. $\square$

Theorem 16.43 now follows immediately from Theorems 16.44, 16.45, and 16.49, together with Theorem 16.34.

In the proof of Theorem 16.43, there is a clear connection between factorization in $D[X]$ and $F[X]$, where F is the field of fractions of D. We should perhaps make this connection more explicit. Let $f \in D[X]$ be a non-zero polynomial. We may write f as

$$f = up_1^{a_1} \cdots p_r^{a_r} f_1^{b_1} \cdots f_s^{b_s}.$$

where $u \in D^*$, the p_i's are non-associate, irreducible elements of D, and the f_j's are non-associate, irreducible, non-constant polynomials over D (and in particular, primitive). For $j = 1, \ldots, s$, let $g_j := \mathrm{lc}(f_j)^{-1} f_j$ be the monic associate of f_j in $F[X]$. Then in $F[X]$, f factors as

$$f = c g_1^{b_1} \cdots g_s^{b_s},$$

where

$$c := u \cdot \prod_i p_i^{a_i} \cdot \prod_j \mathrm{lc}(f_j)^{b_j} \in F,$$

and the g_j's are distinct, irreducible, monic polynomials over F.

Example 16.27. Consider the polynomial $f = 4X^2 + 2X - 2 \in \mathbb{Z}[X]$. Over $\mathbb{Z}[X]$, f factors as $2(2X - 1)(X + 1)$, where each of these three factors is irreducible in $\mathbb{Z}[X]$. However, over $\mathbb{Q}[X]$, f factors as $4(X - 1/2)(X + 1)$, where 4 is a unit, and the other two factors are irreducible. $\square$

The following theorem provides a useful criterion for establishing that a polynomial is irreducible.

Theorem 16.50 (Eisenstein's criterion). *Let D be a UFD and F its field of fractions. Let $f = c_n X^n + c_{n-1} X^{n-1} + \cdots + c_0 \in D[X]$. If there exists an irreducible $p \in D$ such that*

$$p \nmid c_n, \ p \mid c_{n-1}, \ \cdots, \ p \mid c_0, \ p^2 \nmid c_0,$$

then f is irreducible over F.

Proof. Let f be as above, and suppose it were not irreducible in $F[X]$. Then by Theorem 16.48, we could write $f = gh$, where $g, h \in D[X]$, both of degree strictly less than that of f. Let us write

$$g = a_k X^k + \cdots + a_0 \ \text{ and } \ h = b_\ell X^\ell + \cdots + b_0,$$

where $a_k \neq 0$ and $b_\ell \neq 0$, so that $0 < k < n$ and $0 < \ell < n$. Now, since $c_n = a_k b_\ell$, and $p \nmid c_n$, it follows that $p \nmid a_k$ and $p \nmid b_\ell$. Further, since $c_0 = a_0 b_0$, and $p \mid c_0$ but $p^2 \nmid c_0$, it follows that p divides one of a_0 or b_0, but not both—for concreteness, let us assume that $p \mid a_0$ but $p \nmid b_0$. Also, let m be the smallest positive integer such that $p \nmid a_m$—note that $0 < m \leq k < n$.

Now consider the natural map that sends $a \in D$ to $\bar{a} := [a]_p \in D/pD$, which we can extend coefficient-wise to a ring homomorphism from $D[X]$ to $(D/pD)[X]$ (see Example 7.46). Because D is a UFD and p is irreducible, D/pD is an integral domain. Since $f = gh$, we have

$$\bar{c}_n X^n = \bar{f} = \bar{g}\bar{h} = (\bar{a}_k X^k + \cdots + \bar{a}_m X^m)(\bar{b}_\ell X^\ell + \cdots + \bar{b}_0). \tag{16.10}$$

But notice that when we multiply out the two polynomials on the right-hand side of (16.10), the coefficient of X^m is $\bar{a}_m \bar{b}_0 \neq 0$, and as $m < n$, this clearly contradicts the fact that the coefficient of X^m in the polynomial on the left-hand side of (16.10) is zero. ☐

As an application of Eisenstein's criterion, we have:

Theorem 16.51. *For every prime number q, the qth cyclotomic polynomial*

$$\Phi_q := \frac{X^q - 1}{X - 1} = X^{q-1} + X^{q-2} + \cdots + 1$$

is irreducible over $\mathbb{Q}$.

Proof. Let

$$f := \Phi_q(X + 1) = \frac{(X + 1)^q - 1}{(X + 1) - 1}.$$

It is easy to see that

$$f = \sum_{i=0}^{q-1} c_i X^i, \quad \text{where} \quad c_i = \binom{q}{i+1} \quad (i = 0, \ldots, q - 1).$$

Thus, $c_{q-1} = 1$, $c_0 = q$, and for $0 < i < q - 1$, we have $q \mid c_i$ (see Exercise 1.14). Theorem 16.50 therefore applies, and we conclude that f is irreducible over $\mathbb{Q}$. It follows that Φ_q is irreducible over $\mathbb{Q}$, since if $\Phi_q = gh$ were a non-trivial factorization of Φ_q, then $f = \Phi_q(X + 1) = g(X + 1) \cdot h(X + 1)$ would be a non-trivial factorization of f. ☐

EXERCISE 16.45. Show that neither $\mathbb{Z}[X]$ nor $F[X, Y]$ (where F is a field) are PIDs (even though they are UFDs).

EXERCISE 16.46. Let $f \in \mathbb{Z}[X]$ be a monic polynomial. Show that if f has a root $x \in \mathbb{Q}$, then $x \in \mathbb{Z}$, and x divides the constant term of f.

EXERCISE 16.47. Let D be a UFD, let p be an irreducible element of D, and consider the natural map that sends $a \in D$ to $\bar{a} := [a]_p \in D/pD$, which we extend coefficient-wise to a ring homomorphism from $D[X]$ to $(D/pD)[X]$ (see Example 7.46). Show that if $f \in D[X]$ is a primitive polynomial such that $p \nmid \mathrm{lc}(f)$ and $\bar{f} \in (D/pD)[X]$ is irreducible, then f is irreducible.

EXERCISE 16.48. Let a be a non-zero, square-free integer, with $a \notin \{\pm 1\}$, and let n be a positive integer. Show that the polynomial $X^n - a$ is irreducible in $\mathbb{Q}[X]$.

EXERCISE 16.49. Show that the polynomial $X^4 + 1$ is irreducible in $\mathbb{Q}[X]$.

EXERCISE 16.50. Let F be a field, and consider the ring of bivariate polynomials $F[X, Y]$. Show that in this ring, the polynomial $X^2 + Y^2 - 1$ is irreducible, provided F does not have characteristic 2. What happens if F has characteristic 2?

EXERCISE 16.51. Design and analyze an efficient algorithm for the following problem. The input is a pair of polynomials $g, h \in \mathbb{Z}[X]$, along with their greatest common divisor d in the ring $\mathbb{Q}[X]$. The output is the greatest common divisor of g and h in the ring $\mathbb{Z}[X]$.

EXERCISE 16.52. Let $g, h \in \mathbb{Z}[X]$ be non-zero polynomials with $d := \gcd(g, h) \in \mathbb{Z}[X]$. Show that for every prime p not dividing $\mathrm{lc}(g) \, \mathrm{lc}(h)$, we have $\bar{d} \mid \gcd(\bar{g}, \bar{h})$, and except for finitely many primes p, we have $\bar{d} = \gcd(\bar{g}, \bar{h})$. Here, $\bar{d}$, $\bar{g}$, and $\bar{h}$ denote the images of d, g, and h in $\mathbb{Z}_p[X]$ under the coefficient-wise extension of the natural map from $\mathbb{Z}$ to $\mathbb{Z}_p$ (see Example 7.47).

EXERCISE 16.53. Let F be a field, and let $g, h \in F[X, Y]$. Define $V(g, h) := \{(x, y) \in F \times F : g(x, y) = h(x, y) = 0\}$. Show that if g and h are relatively prime, then $V(g, h)$ is a finite set. Hint: consider the rings $F(X)[Y]$ and $F(Y)[X]$.

16.10 Notes

The "$(1 + i)$-ary gcd algorithm" in Exercise 16.42 for computing greatest common divisors of Gaussian integers is based on algorithms in Weilert [106] and Damgård and Frandsen [31]. The latter paper also develops a corresponding algorithm for Eisenstein integers (see Exercise 16.35). Weilert [107] presents an asymptotically fast algorithm that computes the greatest common divisor of Gaussian integers of length at most ℓ in time $O(\ell^{1+o(1)})$.

17

Polynomial arithmetic and applications

In this chapter, we study algorithms for performing arithmetic on polynomials. Initially, we shall adopt a very general point of view, discussing polynomials whose coefficients lie in an arbitrary ring R, and then specialize to the case where the coefficient ring is a field F.

There are many similarities between arithmetic in $\mathbb{Z}$ and in $R[X]$, and the similarities between $\mathbb{Z}$ and $F[X]$ run even deeper. Many of the algorithms we discuss in this chapter are quite similar to the corresponding algorithms for integers.

As we did in Chapter 14 for matrices, we shall treat R as an "abstract data type," and measure the complexity of algorithms for polynomials over a ring R by counting "operations in R."

17.1 Basic arithmetic

Throughout this section, R denotes a non-trivial ring.

For computational purposes, we shall assume that a polynomial $g = \sum_{i=0}^{k-1} a_i X^i \in R[X]$ is represented as a coefficient vector $(a_0, a_1, \ldots, a_{k-1})$. Further, when g is non-zero, the coefficient a_{k-1} should be non-zero.

The basic algorithms for addition, subtraction, multiplication, and division of polynomials are quite straightforward adaptations of the corresponding algorithms for integers. In fact, because of the lack of "carries," these algorithms are actually much simpler in the polynomial case. We briefly discuss these algorithms here — analogous to our treatment of integer arithmetic, we do not discuss the details of "stripping" leading zero coefficients.

For addition and subtraction, all we need to do is to add or subtract coefficient vectors.

For multiplication, let $g = \sum_{i=0}^{k-1} a_i X^i \in R[X]$ and $h = \sum_{i=0}^{\ell-1} b_i X^i \in R[X]$, where $k \geq 1$ and $\ell \geq 1$. The product $f := g \cdot h$ is of the form $f = \sum_{i=0}^{k+\ell-2} c_i X^i$, the coefficients of which can be computed using $O(k\ell)$ operations in R as follows:

for $i \leftarrow 0$ to $k + \ell - 2$ do $c_i \leftarrow 0$
for $i \leftarrow 0$ to $k - 1$ do
    for $j \leftarrow 0$ to $\ell - 1$ do
        $c_{i+j} \leftarrow c_{i+j} + a_i \cdot b_j$

For division, let $g = \sum_{i=0}^{k-1} a_i X^i \in R[X]$ and $h = \sum_{i=0}^{\ell-1} b_i X^i \in R[X]$, where $b_{\ell-1} \in R^*$. We want to compute polynomials $q, r \in R[X]$ such that $g = hq + r$, where $\deg(r) < \ell - 1$. If $k < \ell$, we can simply set $q \leftarrow 0$ and $r \leftarrow g$; otherwise, we can compute q and r using $O(\ell \cdot (k - \ell + 1))$ operations in R using the following algorithm:

$t \leftarrow b_{\ell-1}^{-1} \in R$
for $i \leftarrow 0$ to $k - 1$ do $r_i \leftarrow a_i$
for $i \leftarrow k - \ell$ down to 0 do
    $q_i \leftarrow t \cdot r_{i+\ell-1}$
    for $j \leftarrow 0$ to $\ell - 1$ do
        $r_{i+j} \leftarrow r_{i+j} - q_i \cdot b_j$
$q \leftarrow \sum_{i=0}^{k-\ell} q_i X^i$, $r \leftarrow \sum_{i=0}^{\ell-2} r_i X^i$

With these simple algorithms, we obtain the polynomial analog of Theorem 3.3. Let us define the **length** of $g \in R[X]$, denoted $\text{len}(g)$, to be the length of its coefficient vector; more precisely, we define

$$\text{len}(g) := \begin{cases} \deg(g) + 1 & \text{if } g \neq 0, \\ 1 & \text{if } g = 0. \end{cases}$$

Sometimes (but not always) it is clearer and more convenient to state the running times of algorithms in terms of the length, rather than the degree, of a polynomial (the latter has the inconvenient habit of taking on the value 0, or worse, $-\infty$).

Theorem 17.1. *Let g and h be arbitrary polynomials in $R[X]$.*

   (i) *We can compute $g \pm h$ with $O(\text{len}(g) + \text{len}(h))$ operations in R.*

   (ii) *We can compute $g \cdot h$ with $O(\text{len}(g) \text{len}(h))$ operations in R.*

   (iii) *If $\text{lc}(h) \in R^*$, we can compute $q, r \in R[X]$ such that $g = hq + r$ and $\deg(r) < \deg(h)$ with $O(\text{len}(h) \text{len}(q))$ operations in R.*

Analogous to algorithms for modular integer arithmetic, we can also do arithmetic in the residue class ring $R[X]/(f)$, where $f \in R[X]$ is a polynomial with $\text{lc}(f) \in R^*$. For each $\alpha \in R[X]/(f)$, there exists a unique polynomial $g \in R[X]$ with $\deg(g) < \deg(f)$ and $\alpha = [g]_f$; we call this polynomial g the **canonical representative of** α, and denote it by $\text{rep}(\alpha)$. For computational purposes, we represent elements of $R[X]/(f)$ by their canonical representatives.

With this representation, addition and subtraction in $R[X]/(f)$ can be performed using $O(\mathrm{len}(f))$ operations in R, while multiplication takes $O(\mathrm{len}(f)^2)$ operations in R.

The repeated-squaring algorithm for computing powers works equally well in this setting: given $\alpha \in R[X]/(f)$ and a non-negative exponent e, we can compute α^e using $O(\mathrm{len}(e))$ multiplications in $R[X]/(f)$, for a total of $O(\mathrm{len}(e)\,\mathrm{len}(f)^2)$ operations in R.

EXERCISE 17.1. State and re-work the polynomial analogs of Exercises 3.26–3.28.

EXERCISE 17.2. Given a polynomial $g \in R[X]$ and an element $x \in R$, a particularly elegant and efficient way of computing $g(x)$ is called **Horner's rule**. Suppose $g = \sum_{i=0}^{k-1} a_i X^i$, where $k \geq 0$ and $a_i \in R$ for $i = 0, \ldots, k - 1$. Horner's rule computes $g(x)$ as follows:

$$y \leftarrow 0_R$$
$$\text{for } i \leftarrow k - 1 \text{ down to } 0 \text{ do}$$
$$\quad y \leftarrow yx + a_i$$
$$\text{output } y$$

Show that this algorithm correctly computes $g(x)$ using k multiplications in R and k additions in R.

EXERCISE 17.3. Let $f \in R[X]$ be a polynomial of degree $\ell > 0$ with $\mathrm{lc}(f) \in R^*$, and let $E := R[X]/(f)$. Suppose that in addition to f, we are given a polynomial $g \in R[X]$ of degree less than k and an element $\alpha \in E$, and we want to compute $g(\alpha) \in E$. This is called the **modular composition** problem.

(a) Show that a straightforward application of Horner's rule yields an algorithm that uses $O(k\ell^2)$ operations in R, and requires space for storing $O(\ell)$ elements of R.

(b) Show how to compute $g(\alpha)$ using just $O(k\ell + k^{1/2}\ell^2)$ operations in R, at the expense of requiring space for storing $O(k^{1/2}\ell)$ elements of R. Hint: first compute a table of powers $1, \alpha, \ldots, \alpha^m$, for $m \approx k^{1/2}$.

EXERCISE 17.4. Given polynomials $g, h \in R[X]$, show how to compute their composition $g(h) \in R[X]$ using $O(\mathrm{len}(g)^2\,\mathrm{len}(h)^2)$ operations in R.

EXERCISE 17.5. Suppose you are given three polynomials $f, g, h \in \mathbb{Z}_p[X]$, where p is a large prime, in particular, $p \geq 2\deg(g)\deg(h)$. Design an efficient probabilistic algorithm that tests if $f = g(h)$ (i.e., if f equals g composed with h). Your algorithm should have the following properties: if $f = g(h)$, it should always output "true," and otherwise, it should output "false" with probability at least 0.999. The expected running time of your algorithm should be $O((\text{len}(f) + \text{len}(g) + \text{len}(h))\,\text{len}(p)^2)$.

EXERCISE 17.6. Let $x, a_0, \ldots, a_{\ell-1} \in R$, and let k be an integer with $0 < k \le \ell$. For $i = 0, \ldots, \ell - k$, define $g_i := \sum_{j=i}^{i+k-1} a_j X^j \in R[X]$. Show how to compute the $\ell - k + 1$ values $g_0(x), \ldots, g_{\ell-k}(x)$ using $O(\ell)$ operations in R.

17.2 Computing minimal polynomials in $F[X]/(f)$ (I)

In this section, we shall examine a computational problem to which we shall return on several occasions, as it will serve to illustrate a number of interesting algebraic and algorithmic concepts.

Let F be a field, and let $f \in F[X]$ be a monic polynomial of degree $\ell > 0$. Also, let $E := F[X]/(f)$, which is an F-algebra, and in particular, an F-vector space. As an F-vector space, E has dimension ℓ. Suppose we are given an element $\alpha \in E$, and want to efficiently compute the minimal polynomial of α over F—that is, the monic polynomial $\phi \in F[X]$ of least degree such that $\phi(\alpha) = 0$, which we know has degree at most ℓ (see §16.5).

We can solve this problem using polynomial arithmetic and Gaussian elimination, as follows. Consider the F-linear map $\rho : F[X]_{\le \ell} \to E$ that sends a polynomial $g \in F[X]$ of degree at most ℓ to $g(\alpha)$. To perform the linear algebra, we need to specify bases for $F[X]_{\le \ell}$ and E. For $F[X]_{\le \ell}$, let us work with the basis $S := \{X^{\ell+1-i}\}_{i=1}^{\ell+1}$. With this choice of basis, for $g = \sum_{i=0}^{\ell} a_i X^i \in F[X]_{\le \ell}$, the coordinate vector of g is $\text{Vec}_S(g) = (a_\ell, \ldots, a_0) \in F^{1 \times (\ell+1)}$. For E, let us work with the basis $\mathcal{T} := \{\xi^{i-1}\}_{i=1}^{\ell}$, where $\xi := [X]_f \in E$. Let

$$A := \text{Mat}_{S,\mathcal{T}}(\rho) \in F^{(\ell+1) \times \ell};$$

that is, A is the matrix of ρ relative to S and $\mathcal{T}$ (see §14.2). For $i = 1, \ldots, \ell + 1$, the ith row of A is the coordinate vector $\text{Vec}_{\mathcal{T}}(\alpha^{\ell+1-i}) \in F^{1 \times \ell}$.

We compute the matrix A by computing the powers $1, \alpha, \ldots, \alpha^\ell$, reading off the ith row of A directly from the canonical representative of the $\alpha^{\ell+1-i}$. We then apply Gaussian elimination to A to find row vectors $v_1, \ldots, v_s \in F^{1 \times (\ell+1)}$ that are coordinate vectors corresponding to a basis for the kernel of ρ. Now, the coordinate vector of the minimal polynomial of α is a linear combination of $v_1, \ldots, v_s$. To find it, we form the $s \times (\ell + 1)$ matrix B whose rows consist of $v_1, \ldots, v_s$, and apply Gaussian elimination to B, obtaining an $s \times (\ell+1)$ matrix B' in reduced row echelon form whose row space is the same as that of B. Let ϕ be the polynomial whose coordinate vector is the last row of B'.

Because of the choice of basis for $F[X]_{\le \ell}$, and because B' is in reduced row echelon form, it is clear that no non-zero polynomial in Ker ρ has degree less than that of ϕ. Moreover, as ϕ is already monic (again, by the fact that B' is in reduced row echelon form), it follows that ϕ is in fact the minimal polynomial of α over F.

The total amount of work performed by this algorithm is $O(\ell^3)$ operations in F to build the matrix A (this just amounts to computing ℓ successive powers of α, that is, $O(\ell)$ multiplications in E, each of which takes $O(\ell^2)$ operations in F), and $O(\ell^3)$ operations in F to perform both Gaussian elimination steps.

17.3 Euclid's algorithm

In this section, F denotes a field, and we consider the computation of greatest common divisors in $F[X]$.

The Euclidean algorithm for integers is easily adapted to compute $\gcd(g, h)$ for polynomials $g, h \in F[X]$. Analogous to the integer case, we assume that $\deg(g) \geq \deg(h)$; however, we shall also assume that $g \neq 0$. This is not a serious restriction, of course, as $\gcd(0, 0) = 0$, and making this restriction will simplify the presentation a bit. Recall that we defined $\gcd(g, h)$ to be either zero or monic, and the assumption that $g \neq 0$ means that $\gcd(g, h)$ is non-zero, and hence monic.

The following is the analog of Theorem 4.1, and is based on the division with remainder property for polynomials.

Theorem 17.2. *Let $g, h \in F[X]$, with $\deg(g) \geq \deg(h)$ and $g \neq 0$. Define the polynomials $r_0, r_1, \ldots, r_{\lambda+1} \in F[X]$ and $q_1, \ldots, q_\lambda \in F[X]$, where $\lambda \geq 0$, as follows:*

$$g = r_0,$$
$$h = r_1,$$
$$r_0 = r_1 q_1 + r_2 \quad (0 \leq \deg(r_2) < \deg(r_1)),$$
$$\vdots$$
$$r_{i-1} = r_i q_i + r_{i+1} \quad (0 \leq \deg(r_{i+1}) < \deg(r_i)),$$
$$\vdots$$
$$r_{\lambda-2} = r_{\lambda-1} q_{\lambda-1} + r_\lambda \quad (0 \leq \deg(r_\lambda) < \deg(r_{\lambda-1})),$$
$$r_{\lambda-1} = r_\lambda q_\lambda \quad (r_{\lambda+1} = 0).$$

Note that by definition, $\lambda = 0$ if $h = 0$, and $\lambda > 0$ otherwise. Then we have $r_\lambda / \operatorname{lc}(r_\lambda) = \gcd(g, h)$, and if $h \neq 0$, then $\lambda \leq \deg(h) + 1$.

Proof. Arguing as in the proof of Theorem 4.1, one sees that

$$\gcd(g, h) = \gcd(r_0, r_1) = \cdots = \gcd(r_\lambda, r_{\lambda+1}) = \gcd(r_\lambda, 0) = r_\lambda / \operatorname{lc}(r_\lambda).$$

That proves the first statement.

For the second statement, if $h \neq 0$, then the degree sequence

$$\deg(r_1), \deg(r_2), \ldots, \deg(r_\lambda)$$

is strictly decreasing, with $\deg(r_\lambda) \geq 0$, from which it follows that $\deg(h) = \deg(r_1) \geq \lambda - 1$. $\square$

This gives us the following polynomial version of the Euclidean algorithm:

Euclid's algorithm. On input g, h, where $g, h \in F[X]$ with $\deg(g) \geq \deg(h)$ and $g \neq 0$, compute $d = \gcd(g, h)$ as follows:

$r \leftarrow g, \; r' \leftarrow h$
while $r' \neq 0$ do
$\qquad r'' \leftarrow r \bmod r'$
$\qquad (r, r') \leftarrow (r', r'')$
$d \leftarrow r / \mathrm{lc}(r) \quad // \; make \; monic$
output d

Theorem 17.3. *Euclid's algorithm for polynomials performs* $O(\mathrm{len}(g)\,\mathrm{len}(h))$ *operations in* F.

Proof. The proof is almost identical to that of Theorem 4.2. Details are left to the reader. $\square$

Just as for integers, if $d = \gcd(g, h)$, then $g F[X] + h F[X] = d F[X]$, and so there exist polynomials s and t such that $gs + ht = d$. The procedure for calculating s and t is precisely the same as in the integer case; however, in the polynomial case, we can be much more precise about the relative sizes of the objects involved in the calculation.

Theorem 17.4. *Let* $g, h, r_0, \ldots, r_{\lambda+1}$ *and* $q_1, \ldots, q_\lambda$ *be as in Theorem 17.2. Define polynomials* $s_0, \ldots, s_{\lambda+1} \in F[X]$ *and* $t_0, \ldots, t_{\lambda+1} \in F[X]$ *as follows:*

$$s_0 := 1, \quad t_0 := 0,$$
$$s_1 := 0, \quad t_1 := 1,$$

and for $i = 1, \ldots, \lambda$,

$$s_{i+1} := s_{i-1} - s_i q_i, \quad t_{i+1} := t_{i-1} - t_i q_i.$$

Then:

(i) *for* $i = 0, \ldots, \lambda + 1$, *we have* $g s_i + h t_i = r_i$; *in particular,* $g s_\lambda + h t_\lambda = \mathrm{lc}(r_\lambda) \gcd(g, h)$;

(ii) *for* $i = 0, \ldots, \lambda$, *we have* $s_i t_{i+1} - t_i s_{i+1} = (-1)^i$;

(iii) for $i = 0, \ldots, \lambda + 1$, *we have* $\gcd(s_i, t_i) = 1$;

(iv) for $i = 1, \ldots, \lambda + 1$, *we have*

$$\deg(t_i) = \deg(g) - \deg(r_{i-1}),$$

and for $i = 2, \ldots, \lambda + 1$, *we have*

$$\deg(s_i) = \deg(h) - \deg(r_{i-1});$$

(v) for $i = 1, \ldots, \lambda + 1$, *we have* $\deg(t_i) \leq \deg(g)$ *and* $\deg(s_i) \leq \deg(h)$; *if* $\deg(g) > 0$ *and* $h \neq 0$, *then* $\deg(t_\lambda) < \deg(g)$ *and* $\deg(s_\lambda) < \deg(h)$.

Proof. (i), (ii), and (iii) are proved just as in the corresponding parts of Theorem 4.3.

For (iv), the proof will hinge on the following facts:

- For $i = 1, \ldots, \lambda$, we have $\deg(r_{i-1}) \geq \deg(r_i)$, and since q_i is the quotient in dividing r_{i-1} by r_i, we have $\deg(q_i) = \deg(r_{i-1}) - \deg(r_i)$.
- For $i = 2, \ldots, \lambda$, we have $\deg(r_{i-1}) > \deg(r_i)$.

We prove the statement involving the t_i's by induction on i, and leave the proof of the statement involving the s_i's to the reader.

One can see by inspection that this statement holds for $i = 1$, since $\deg(t_1) = 0$ and $r_0 = g$. If $\lambda = 0$, there is nothing more to prove, so assume that $\lambda > 0$ and $h \neq 0$.

Now, for $i = 2$, we have $t_2 = 0 - 1 \cdot q_1 = -q_1$. Thus, $\deg(t_2) = \deg(q_1) = \deg(r_0) - \deg(r_1) = \deg(g) - \deg(r_1)$.

Now for the induction step. Assume $i \geq 3$. Then we have

$$\begin{aligned}
\deg(t_{i-1} q_{i-1}) &= \deg(t_{i-1}) + \deg(q_{i-1}) \\
&= \deg(g) - \deg(r_{i-2}) + \deg(q_{i-1}) \quad \text{(by induction)} \\
&= \deg(g) - \deg(r_{i-1}) \\
&\quad \text{(since } \deg(q_{i-1}) = \deg(r_{i-2}) - \deg(r_{i-1})) \\
&> \deg(g) - \deg(r_{i-3}) \quad \text{(since } \deg(r_{i-3}) > \deg(r_{i-1})) \\
&= \deg(t_{i-2}) \quad \text{(by induction)}.
\end{aligned}$$

By definition, $t_i = t_{i-2} - t_{i-1} q_{i-1}$, and from the above reasoning, we see that

$$\deg(g) - \deg(r_{i-1}) = \deg(t_{i-1} q_{i-1}) > \deg(t_{i-2}),$$

from which it follows that $\deg(t_i) = \deg(g) - \deg(r_{i-1})$.

(v) follows easily from (iv). $\square$

From this theorem, we obtain the following algorithm:

The extended Euclidean algorithm. On input g, h, where $g, h \in F[X]$ with $\deg(g) \ge \deg(h)$ and $g \ne 0$, compute d, s, and t, where $d, s, t \in F[X]$, $d = \gcd(g, h)$ and $gs + ht = d$, as follows:

$r \leftarrow g,\ r' \leftarrow h$
$s \leftarrow 1,\ s' \leftarrow 0$
$t \leftarrow 0,\ t' \leftarrow 1$
while $r' \ne 0$ do
  compute q, r'' such that $r = r'q + r''$, with $\deg(r'') < \deg(r')$
  $(r, s, t, r', s', t') \leftarrow (r', s', t', r'', s - s'q, t - t'q)$
$c \leftarrow \mathrm{lc}(r)$
$d \leftarrow r/c,\ s \leftarrow s/c,\ t \leftarrow t/c$ // *make monic*
output d, s, t

Theorem 17.5. *The extended Euclidean algorithm for polynomials performs* $O(\mathrm{len}(g)\,\mathrm{len}(h))$ *operations in* F.

Proof. Exercise. □

EXERCISE 17.7. State and re-work the polynomial analogs of Exercises 4.2, 4.3, 4.4, 4.5, and 4.8.

17.4 Computing modular inverses and Chinese remaindering

In this and the remaining sections of this chapter, we explore various applications of Euclid's algorithm for polynomials. Most of these applications are analogous to their integer counterparts, although there are some differences to watch for. Throughout this section, F denotes a field.

We begin with the obvious application of the extended Euclidean algorithm for polynomials to the problem of computing multiplicative inverses in $F[X]/(f)$.

Theorem 17.6. *Suppose we are given polynomials* $f, h \in F[X]$, *where* $\deg(h) < \deg(f)$. *Then using* $O(\mathrm{len}(f)^2)$ *operations in* F, *we can determine if* h *is relatively prime to* f, *and if so, compute* $h^{-1} \bmod f$.

Proof. We may assume $\deg(f) > 0$, since $\deg(f) = 0$ implies $h = 0 = h^{-1} \bmod f$. We run the extended Euclidean algorithm on input f, h, obtaining polynomials d, s, t such that $d = \gcd(f, h)$ and $fs + ht = d$. If $d \ne 1$, then h does not have a multiplicative inverse modulo f. Otherwise, if $d = 1$, then t is a multiplicative inverse of h modulo f. Moreover, by part (v) of Theorem 17.4, we have $\deg(t) < \deg(f)$, and so $t = h^{-1} \bmod f$. Based on Theorem 17.5, it is clear that all the computations can be performed using $O(\mathrm{len}(f)^2)$ operations in F. □

We also observe that the Chinese remainder theorem for polynomials (Theorem 16.19) can be made computationally effective as well:

Theorem 17.7 (Effective Chinese remainder theorem). *Suppose we are given polynomials* $f_1, \ldots, f_k \in F[X]$ *and* $g_1, \ldots, g_k \in F[X]$, *where the family* $\{f_i\}_{i=1}^{k}$ *is pairwise relatively prime, and where* $\deg(f_i) > 0$ *and* $\deg(g_i) < \deg(f_i)$ *for* $i = 1, \ldots, k$. *Let* $f := \prod_{i=1}^{k} f_i$. *Then using* $O(\mathrm{len}(f)^2)$ *operations in* F, *we can compute the unique polynomial* $g \in F[X]$ *satisfying* $\deg(g) < \deg(f)$ *and* $g \equiv g_i \pmod{f_i}$ *for* $i = 1, \ldots, k$.

Proof. Exercise (just use the formulas given after Theorem 16.19). $\square$

Polynomial interpolation

We remind the reader of the discussion following Theorem 16.19, where the point was made that when $f_i = X - x_i$ and $g_i = y_i$, for $i = 1, \ldots, k$, then the Chinese remainder theorem for polynomials reduces to Lagrange interpolation. Thus, Theorem 17.7 says that given distinct elements $x_1, \ldots, x_k \in F$, along with elements $y_1, \ldots, y_k \in F$, we can compute the unique polynomial $g \in F[X]$ of degree less than k such that

$$g(x_i) = y_i \quad (i = 1, \ldots, k),$$

using $O(k^2)$ operations in F.

It is perhaps worth noting that we could also solve the polynomial interpolation problem using Gaussian elimination, by inverting the corresponding Vandermonde matrix (see Example 14.2). However, this algorithm would use $O(k^3)$ operations in F. This is a specific instance of a more general phenomenon: there are many computational problems involving polynomials over fields that can be solved using Gaussian elimination, but which can be solved more efficiently using more specialized algorithmic techniques.

Speeding up algorithms via modular computation

In §4.4, we discussed how the Chinese remainder theorem could be used to speed up certain types of computations involving integers. The example we gave was the multiplication of integer matrices. We can use the same idea to speed up certain types of computations involving polynomials. For example, if one wants to multiply two matrices whose entries are elements of $F[X]$, one can use the Chinese remainder theorem for polynomials to speed things up. This strategy is most easily implemented if F is sufficiently large, so that we can use polynomial evaluation and interpolation directly, and do not have to worry about constructing irreducible polynomials.

EXERCISE 17.8. Adapt the algorithms of Exercises 4.14 and 4.15 to obtain an algorithm for polynomial interpolation. This algorithm is called **Newton interpolation**.

17.5 Rational function reconstruction and applications

Throughout this section, F denotes a field.

We next state and prove the polynomial analog of Theorem 4.9. As we are now "reconstituting" a rational function, rather than a rational number, we call this procedure **rational function reconstruction**. Because of the relative simplicity of polynomials compared to integers, the rational reconstruction theorem for polynomials is a bit "sharper" than the rational reconstruction theorem for integers, and much simpler to prove.

To state the result precisely, let us introduce some notation. For polynomials $g, h \in F[X]$ with $\deg(g) \geq \deg(h)$ and $g \neq 0$, let us define

$$\mathrm{EEA}(g, h) := \left\{ (r_i, s_i, t_i) \right\}_{i=0}^{\lambda+1},$$

where r_i, s_i, and t_i, for $i = 0, \ldots, \lambda + 1$, are defined as in Theorem 17.4.

Theorem 17.8 (Rational function reconstruction). *Let $f, h \in F[X]$ be polynomials, and let r^*, t^* be non-negative integers, such that*

$$\deg(h) < \deg(f) \quad \text{and} \quad r^* + t^* \leq \deg(f).$$

Further, let $\mathrm{EEA}(f, h) = \{(r_i, s_i, t_i)\}_{i=0}^{\lambda+1}$, and let j be the smallest index (among $0, \ldots, \lambda + 1$) such that $\deg(r_j) < r^*$, and set

$$r' := r_j, \quad s' := s_j, \quad \text{and} \quad t' := t_j.$$

Finally, suppose that there exist polynomials $r, s, t \in F[X]$ such that

$$r = fs + ht, \quad \deg(r) < r^*, \quad \text{and} \quad 0 \leq \deg(t) \leq t^*.$$

Then for some non-zero polynomial $q \in F[X]$, we have

$$r = r'q, \quad s = s'q, \quad t = t'q.$$

Proof. Since $\deg(r_0) = \deg(f) \geq r^* > -\infty = \deg(r_{\lambda+1})$, the value of j is well defined, and moreover, $j \geq 1$, $\deg(r_{j-1}) \geq r^*$, and $t_j \neq 0$.

From the equalities $r_j = f s_j + h t_j$ and $r = f s + h t$, we have the two congruences:

$$r_j \equiv h t_j \pmod{f},$$
$$r \equiv h t \pmod{f}.$$

Subtracting t times the first from t_j times the second, we obtain

$$r t_j \equiv r_j t \pmod{f}.$$

This says that f divides $r t_j - r_j t$.

We want to show that, in fact, $r t_j - r_j t = 0$. To this end, first observe that by part (iv) of Theorem 17.4 and the inequality $\deg(r_{j-1}) \geq r^*$, we have

$$\deg(t_j) = \deg(f) - \deg(r_{j-1}) \leq \deg(f) - r^*.$$

Combining this with the inequality $\deg(r) < r^*$, we see that

$$\deg(r t_j) = \deg(r) + \deg(t_j) < \deg(f).$$

Furthermore, using the inequalities

$$\deg(r_j) < r^*, \quad \deg(t) \leq t^*, \quad \text{and} \quad r^* + t^* \leq \deg(f),$$

we see that

$$\deg(r_j t) = \deg(r_j) + \deg(t) < \deg(f),$$

and it immediately follows that

$$\deg(r t_j - r_j t) < \deg(f).$$

Since f divides $r t_j - r_j t$ and $\deg(r t_j - r_j t) < \deg(f)$, the only possibility is that

$$r t_j - r_j t = 0.$$

The rest of the proof follows exactly the same line of reasoning as in the last paragraph in the proof of Theorem 4.9, as the reader may easily verify. $\square$

17.5.1 Application: recovering rational functions from their reversed Laurent series

We now discuss the polynomial analog of the application in §4.6.1. This is an entirely straightforward translation of the results in §4.6.1, but we shall see in the next chapter that this problem has its own interesting applications.

Suppose Alice knows a rational function $z = s/t \in F(X)$, where s and t are polynomials with $\deg(s) < \deg(t)$, and tells Bob some of the high-order coefficients of the reversed Laurent series (see §16.8) representing z in $F((X^{-1}))$. We shall show that if $\deg(t) \leq \ell$ and Bob is given the bound ℓ on $\deg(t)$, along with the high-order 2ℓ coefficients of z, then Bob can determine z, expressed as a rational function in lowest terms.

So suppose that $z = s/t = \sum_{i=1}^{\infty} z_i X^{-i}$, and that Alice tells Bob the coefficients $z_1, \ldots, z_{2\ell}$. Equivalently, Alice gives Bob the polynomial

$$h := z_1 X^{2\ell-1} + \cdots + z_{2\ell-1} X + z_{2\ell}.$$

Also, let us define $f := X^{2\ell}$. Here is Bob's algorithm for recovering z:

1. Run the extended Euclidean algorithm on input f, h to obtain $\mathrm{EEA}(f, h)$, and apply Theorem 17.8 with f, h, $r^* := \ell$, and $t^* := \ell$, to obtain the polynomials r', s', t'.

2. Output s', t'.

We claim that $z = -s'/t'$. To prove this, first observe that $h = \lfloor fz \rfloor = \lfloor fs/t \rfloor$ (see Theorem 16.32). So if we set $r := fs \bmod t$, then we have

$$r = fs - ht, \ \deg(r) < r^*, \ 0 \le \deg(t) \le t^*, \ \text{and} \ r^* + t^* \le \deg(f).$$

It follows that the polynomials s', t' from Theorem 17.8 satisfy $s = s'q$ and $-t = t'q$ for some non-zero polynomial q, and thus, $s'/t' = -s/t$, which proves the claim.

We may further observe that since the extended Euclidean algorithm guarantees that $\gcd(s', t') = 1$, not only do we obtain z, but we obtain z expressed as a fraction in lowest terms.

It is clear that this algorithm takes $O(\ell^2)$ operations in F.

17.5.2 Application: polynomial interpolation with errors

We now discuss the polynomial analog of the application in §4.6.2.

If we "encode" a polynomial $g \in F[X]$, with $\deg(g) < k$, as the sequence $(y_1, \ldots, y_k) \in F^{\times k}$, where $y_i = g(x_i)$, then we can efficiently recover g from this encoding, using an algorithm for polynomial interpolation. Here, of course, the x_i's are distinct elements of F.

Now suppose that Alice encodes g as $(y_1, \ldots, y_k)$, and sends this encoding to Bob, but that some, say at most ℓ, of the y_i's may be corrupted during transmission. Let $(z_1, \ldots, z_k)$ denote the vector actually received by Bob.

Here is how we can use Theorem 17.8 to recover the original value of g from $(z_1, \ldots, z_k)$, assuming:

- the original polynomial g has degree less than m,
- at most ℓ errors occur in transmission, and
- $k \ge 2\ell + m$.

Let us set $f_i := X - x_i$ for $i = 1, \ldots, k$, and $f := f_1 \cdots f_k$. Now, suppose Bob obtains the corrupted encoding $(z_1, \ldots, z_k)$. Here is what Bob does to recover g:

1. Interpolate, obtaining a polynomial h, with $\deg(h) < k$ and $h(x_i) = z_i$ for $i = 1, \ldots, k$.

2. Run the extended Euclidean algorithm on input f, h to obtain $\text{EEA}(f, h)$, and apply Theorem 17.8 with f, h, $r^* := m + \ell$ and $t^* := \ell$, to obtain the polynomials r', s', t'.

3. If $t' \mid r'$, output r'/t'; otherwise, output "error."

We claim that the above procedure outputs g, under the assumptions listed above. To see this, let t be the product of the f_i's for those values of i where an error occurred. Now, assuming at most ℓ errors occurred, we have $\deg(t) \le \ell$. Also, let $r := gt$, and note that $\deg(r) < m + \ell$. We claim that

$$r \equiv ht \pmod{f}. \tag{17.1}$$

To show that (17.1) holds, it suffices to show that

$$gt \equiv ht \pmod{f_i} \tag{17.2}$$

for all $i = 1, \ldots, k$. To show this, consider first an index i at which no error occurred, so that $y_i = z_i$. Then $gt \equiv y_i t \pmod{f_i}$ and $ht \equiv z_i t \equiv y_i t \pmod{f_i}$, and so (17.2) holds for this i. Next, consider an index i for which an error occurred. Then by construction, $gt \equiv 0 \pmod{f_i}$ and $ht \equiv 0 \pmod{f_i}$, and so (17.2) holds for this i. Thus, (17.1) holds, from which it follows that the values r', t' obtained from Theorem 17.8 satisfy

$$\frac{r'}{t'} = \frac{r}{t} = \frac{gt}{t} = g.$$

One easily checks that both the procedures to encode and decode a value g run in time $O(k^2)$. The above scheme is an example of an **error correcting code** called a **Reed–Solomon code**.

17.5.3 Applications to symbolic algebra

Rational function reconstruction has applications in symbolic algebra, analogous to those discussed in §4.6.3. In that section, we discussed the application of solving systems of linear equations over the integers using rational reconstruction. In exactly the same way, one can use rational function reconstruction to solve systems of linear equations over $F[X]$ — the solution to such a system of equations will be a vector whose entries are elements of $F(X)$, the field of rational functions.

EXERCISE 17.9. Consider again the secret sharing problem, as discussed in Example 8.28. There, we presented a scheme that distributes shares of a secret among several parties in such a way that no coalition of k or fewer parties can reconstruct the secret, while every coalition of $k+1$ parties can. Now suppose that some parties may be corrupt: in the protocol to reconstruct the secret, a corrupted party may contribute an incorrect share. Show how to modify the protocol in Example 8.28 so that if shares are distributed among several parties, then

(a) no coalition of k or fewer parties can reconstruct the secret, and

(b) if at most k parties are corrupt, then every coalition of $3k+1$ parties (which may include some of the corrupted parties) can correctly reconstruct the secret.

The following exercises are the polynomial analogs of Exercises 4.20, 4.22, and 4.23.

EXERCISE 17.10. Let F be a field. Show that given polynomials $s, t \in F[X]$ and integer k, with $\deg(s) < \deg(t)$ and $k > 0$, we can compute the kth coefficient in the reversed Laurent series representing s/t using $O(\operatorname{len}(k)\operatorname{len}(t)^2)$ operations in F.

EXERCISE 17.11. Let F be a field. Let $z \in F((X^{-1}))$ be a reversed Laurent series whose coefficient sequence is ultimately periodic. Show that $z \in F(X)$.

EXERCISE 17.12. Let F be a field. Let $z = s/t$, where $s, t \in F[X]$, $\deg(s) < \deg(t)$, and $\gcd(s, t) = 1$.

(a) Show that if F is finite, there exist integers k, k' such that $0 \leq k < k'$ and $sX^k \equiv sX^{k'} \pmod{t}$.

(b) Show that for integers k, k' with $0 \leq k < k'$, the sequence of coefficients of the reversed Laurent series representing z is $(k, k' - k)$-periodic if and only if $sX^k \equiv sX^{k'} \pmod{t}$.

(c) Show that if F is finite and $X \nmid t$, then the reversed Laurent series representing z is purely periodic with period equal to the multiplicative order of $[X]_t \in (F[X]/(t))^*$.

(d) More generally, show that if F is finite and $t = X^k t'$, with $X \nmid t'$, then the reversed Laurent series representing z is ultimately periodic with pre-period k and period equal to the multiplicative order of $[X]_{t'} \in (F[X]/(t'))^*$.

17.6 Faster polynomial arithmetic (∗)

The algorithms discussed in §3.5 for faster integer arithmetic are easily adapted to polynomials over a ring. Throughout this section, R denotes a non-trivial ring.

EXERCISE 17.13. State and re-work the analog of Exercise 3.41 for $R[X]$. Your algorithm should multiply two polynomials over R of length at most ℓ using $O(\ell^{\log_2 3})$ operations in R.

It is in fact possible to multiply polynomials over R of length at most ℓ using $O(\ell \operatorname{len}(\ell) \operatorname{len}(\operatorname{len}(\ell)))$ operations in R—we shall develop some of the ideas that lead to such a result below in Exercises 17.21–17.24 (see also the discussion in §17.7).

In Exercises 17.14–17.19 below, assume that we have an algorithm that multiplies two polynomials over R of length at most ℓ using at most $M(\ell)$ operations in R, where M is a well-behaved complexity function (as defined in §3.5).

EXERCISE 17.14. State and re-work the analog of Exercises 3.46 and 3.47 for $R[X]$.

EXERCISE 17.15. This problem is the analog of Exercise 3.48 for $R[X]$. Let us first define the notion of a "floating point" reversed Laurent series $\hat{z}$, which is represented as a pair (g, e), where $g \in R[X]$ and $e \in \mathbb{Z}$—the value of $\hat{z}$ is $gX^e \in R((X^{-1}))$, and we call $\operatorname{len}(g)$ the **precision** of $\hat{z}$. We say that $\hat{z}$ is a **length k approximation** of $z \in R((X^{-1}))$ if $\hat{z}$ has precision k and $\hat{z} = (1 + \varepsilon)z$ for $\varepsilon \in R((X^{-1}))$ with $\deg(\varepsilon) \le -k$, which is the same as saying that the high-order k coefficients of $\hat{z}$ and z are equal. Show that given $h \in R[X]$ with $\operatorname{lc}(h) \in R^*$, and positive integer k, we can compute a length k approximation of $1/h \in R((X^{-1}))$ using $O(M(k))$ operations in R. Hint: using Newton iteration, show how to go from a length t approximation of $1/h$ to a length $2t$ approximation, making use of just the high-order $2t$ coefficients of h, and using $O(M(t))$ operations in R.

EXERCISE 17.16. State and re-work the analog of Exercise 3.49 for $R[X]$.

EXERCISE 17.17. State and re-work the analog of Exercise 3.50 for $R[X]$. Conclude that a polynomial of length at most k can be evaluated at k points using $O(M(k) \operatorname{len}(k))$ operations in R.

EXERCISE 17.18. State and re-work the analog of Exercise 3.52 for $R[X]$, assuming $2_R \in R^*$.

The next two exercises develop a useful technique known as **Kronecker substitution**.

EXERCISE 17.19. Let $g, h \in R[X, Y]$ with $g = \sum_{i=0}^{m-1} g_i Y^i$ and $h = \sum_{i=0}^{m-1} h_i Y^i$, where each g_i and h_i is a polynomial in X of degree less than k. The product $f := gh \in R[X, Y]$ may be written $f = \sum_{i=0}^{2m-2} f_i Y^i$, where each f_i is a polynomial in X. Show how to compute f, given g and h, using $O(M(km))$ operations in R. Hint: for an appropriately chosen integer $t > 0$, first convert g, h to $\tilde{g}, \tilde{h} \in R[X]$, where $\tilde{g} := \sum_{i=0}^{m-1} g_i X^{ti}$ and $\tilde{h} := \sum_{i=0}^{m-1} h_i X^{ti}$; next, compute $\tilde{f} := \tilde{g}\tilde{h} \in R[X]$; finally, "read off" the f_i's from the coefficients of $\tilde{f}$.

EXERCISE 17.20. Assume that integers of length at most ℓ can be multiplied in time $\overline{M}(\ell)$, where $\overline{M}$ is a well-behaved complexity function. Let $g, h \in \mathbb{Z}[X]$ with $g = \sum_{i=0}^{m-1} a_i X^i$ and $h = \sum_{i=0}^{m-1} b_i X^i$, where each a_i and b_i is a non-negative integer, strictly less than 2^k. The product $f := gh \in \mathbb{Z}[X]$ may be written $f = \sum_{i=0}^{2m-2} c_i X^i$, where each c_i is a non-negative integer. Show how to compute f, given g and h, using $O(\overline{M}((k + \text{len}(m))m))$ operations in R. Hint: for an appropriately chosen integer $t > 0$, first convert g, h to $a, b \in \mathbb{Z}$, where $a := \sum_{i=0}^{m-1} a_i 2^{ti}$ and $b := \sum_{i=0}^{m-1} b_i 2^{ti}$; next, compute $c := ab \in \mathbb{Z}$; finally, "read off" the c_i's from the bits of c.

The following exercises develop an important algorithm for multiplying polynomials in almost-linear time. For an integer $n \geq 0$, let us call $\omega \in R$ a **primitive 2^nth root of unity** if $n \geq 1$ and $\omega^{2^{n-1}} = -1_R$, or $n = 0$ and $\omega = 1_R$; if $2_R \neq 0_R$, then in particular, ω has multiplicative order 2^n. For $n \geq 0$, and $\omega \in R$ a primitive 2^nth root of unity, let us define the R-linear map $\mathcal{E}_{n,\omega} : R^{\times 2^n} \to R^{\times 2^n}$ that sends the vector $(a_0, \ldots, a_{2^n-1})$ to the vector $(g(1_R), g(\omega), \ldots, g(\omega^{2^n-1}))$, where $g := \sum_{i=0}^{2^n-1} a_i X^i \in R[X]$.

EXERCISE 17.21. Suppose $2_R \in R^*$ and $\omega \in R$ is a primitive 2^nth root of unity.

(a) Let k be any integer, and consider $\gcd(k, 2^n)$, which must be of the form 2^m for some $m = 0, \ldots, n$. Show that ω^k is a primitive 2^{n-m}th root of unity.

(b) Show that if $n \geq 1$, then $\omega - 1_R \in R^*$.

(c) Show that $\omega^k - 1_R \in R^*$ for all integers $k \not\equiv 0 \pmod{2^n}$.

(d) Show that for every integer k, we have

$$\sum_{i=0}^{2^n-1} \omega^{ki} = \begin{cases} 2_R^n & \text{if } k \equiv 0 \pmod{2^n}, \\ 0_R & \text{if } k \not\equiv 0 \pmod{2^n}. \end{cases}$$

(e) Let M_2 be the 2-multiplication map on $R^{\times 2^n}$, which is a bijective, R-linear map. Show that

$$\mathcal{E}_{n,\omega} \circ \mathcal{E}_{n,\omega^{-1}} = M_2^n = \mathcal{E}_{n,\omega^{-1}} \circ \mathcal{E}_{n,\omega},$$

and conclude that $\mathcal{E}_{n,\omega}$ is bijective, with $M_2^{-n} \circ \mathcal{E}_{n,\omega^{-1}}$ being its inverse. Hint: write down the matrices representing the maps $\mathcal{E}_{n,\omega}$ and $\mathcal{E}_{n,\omega^{-1}}$.

EXERCISE 17.22. This exercise develops a fast algorithm, called the **fast Fourier transform** or **FFT**, for computing the function $\mathcal{E}_{n,\omega}$. This is a recursive algorithm

$\text{FFT}(n, \omega; a_0, \ldots, a_{2^n-1})$ that takes as input an integer $n \geq 0$, a primitive 2^nth root of unity $\omega \in R$, and elements $a_0, \ldots, a_{2^n-1} \in R$, and runs as follows:

> if $n = 0$ then
>> return a_0
> else
>> $(\alpha_0, \ldots, \alpha_{2^{n-1}-1}) \leftarrow \text{FFT}(n-1, \omega^2; a_0, a_2, \ldots, a_{2^n-2})$
>> $(\beta_0, \ldots, \beta_{2^{n-1}-1}) \leftarrow \text{FFT}(n-1, \omega^2; a_1, a_3, \ldots, a_{2^n-1})$
>> for $i \leftarrow 0$ to $2^{n-1} - 1$ do
>>> $\gamma_i \leftarrow \alpha_i + \beta_i \omega^i, \quad \gamma_{i+2^{n-1}} \leftarrow \alpha_i - \beta_i \omega^i$
>> return $(\gamma_0, \ldots, \gamma_{2^n-1})$

Show that this algorithm correctly computes $\mathcal{E}_{n,\omega}(a_0, \ldots, a_{2^n-1})$ using $O(2^n n)$ operations in R.

EXERCISE 17.23. Assume $2_R \in R^*$. Suppose that we are given two polynomials $g, h \in R[X]$ of length at most ℓ, along with a primitive 2^nth root of unity $\omega \in R$, where $2\ell \leq 2^n < 4\ell$. Let us "pad" g and h, writing $g = \sum_{i=0}^{2^n-1} a_i X^i$ and $h = \sum_{i=0}^{2^n-1} b_i X^i$, where a_i and b_i are zero for $i \geq \ell$. Show that the following algorithm correctly computes the product of g and h using $O(\ell \, \text{len}(\ell))$ operations in R:

> $(\alpha_0, \ldots, \alpha_{2^n-1}) \leftarrow \text{FFT}(n, \omega; a_0, \ldots, a_{2^n-1})$
> $(\beta_0, \ldots, \beta_{2^n-1}) \leftarrow \text{FFT}(n, \omega; b_0, \ldots, b_{2^n-1})$
> $(\gamma_0, \ldots, \gamma_{2^n-1}) \leftarrow (\alpha_0\beta_0, \ldots, \alpha_{2^n-1}\beta_{2^n-1})$
> $(c_0, \ldots, c_{2^n-1}) \leftarrow 2_R^{-n} \text{FFT}(n, \omega^{-1}; \gamma_0, \ldots, \gamma_{2^n-1})$
> output $\sum_{i=0}^{2\ell-2} c_i X^i$

Also, argue more carefully that the algorithm performs $O(\ell \, \text{len}(\ell))$ additions and subtractions in R, $O(\ell \, \text{len}(\ell))$ multiplications in R by powers of ω, and $O(\ell)$ other multiplications in R.

EXERCISE 17.24. Assume $2_R \in R^*$. In this exercise, we use the FFT to develop an algorithm that multiplies polynomials over R of length at most ℓ using $O(\ell \, \text{len}(\ell)^\beta)$ operations in R, where β is a constant. Unlike the previous exercise, we do not assume that R contains any particular primitive roots of unity; rather, the algorithm will create them "out of thin air." Suppose that $g, h \in R[X]$ are of length at most ℓ. Set $k := \lfloor \sqrt{\ell/2} \rfloor$, $m := \lceil \ell/k \rceil$. We may write $g = \sum_{i=0}^{m-1} g_i X^{ki}$ and $h = \sum_{i=0}^{m-1} h_i X^{ki}$, where the g_i's and h_i's are polynomials of length at most k. Let n be the integer determined by $2m \leq 2^n < 4m$. Let $q := X^{2^{n-1}} + 1_R \in R[X]$, $E := R[X]/(q)$, and $\omega := [X]_q \in E$.

　(a) Show that ω is a primitive 2^nth root of unity in E, and that given an element

$\zeta \in E$ and an integer i between 0 and $2^n - 1$, we can compute $\zeta\omega^i \in E$ using $O(\ell^{1/2})$ operations in R.

(b) Let $\bar{g} := \sum_{i=0}^{m-1} [g_i]_q Y^i \in E[Y]$ and $\bar{h} := \sum_{i=0}^{m-1} [h_i]_q Y^i \in E[Y]$. Using the FFT (over E), show how to compute $\bar{f} := \bar{g}\bar{h} \in E[Y]$ by computing $O(\ell^{1/2})$ products in $R[X]$ of polynomials of length $O(\ell^{1/2})$, along with $O(\ell \operatorname{len}(\ell))$ additional operations in R.

(c) Show how to compute the coefficients of $f := gh \in R[X]$ from the value $\bar{f} \in E[Y]$ computed in part (b), using $O(\ell)$ operations in R.

(d) Based on parts (a)–(c), we obtain a recursive multiplication algorithm: on inputs of length at most ℓ, it performs at most $\alpha_0 \ell \operatorname{len}(\ell)$ operations in R, and calls itself recursively on at most $\alpha_1 \ell^{1/2}$ subproblems, each of length at most $\alpha_2 \ell^{1/2}$; here, α_0, α_1 and α_2 are constants. If we just perform one level of recursion, and immediately switch to a quadratic multiplication algorithm, we obtain an algorithm whose operation count is $O(\ell^{1.5})$. If we perform two levels of recursion, this is reduced to $O(\ell^{1.25})$. For practical purposes, this is probably enough; however, to get an asymptotically better complexity bound, we can let the algorithm recurse all the way down to inputs of some (appropriately chosen) constant length. Show that if we do this, the operation count of the recursive algorithm is $O(\ell \operatorname{len}(\ell)^\beta)$ for some constant β (whose value depends on α_1 and α_2).

The approach used in the previous exercise was a bit sloppy. With a bit more care, one can use the same ideas to get an algorithm that multiplies polynomials over R of length at most ℓ using $O(\ell \operatorname{len}(\ell) \operatorname{len}(\operatorname{len}(\ell)))$ operations in R, assuming $2_R \in R^*$. The next exercise applies similar ideas, but with a few twists, to the problem of *integer* multiplication.

EXERCISE 17.25. This exercise uses the FFT to develop a linear-time algorithm for integer multiplication; however, a rigorous analysis depends on an unproven conjecture (which follows from a generalization of the Riemann hypothesis). Suppose we want to multiply two positive integers a and b, each of length at most ℓ (represented internally using the data structure described in §3.3). Throughout this exercise, assume that all computations are done on a RAM, and that arithmetic on integers of length $O(\operatorname{len}(\ell))$ takes time $O(1)$. Let k be an integer parameter with $k = \Theta(\operatorname{len}(\ell))$, and let $m := \lceil \ell/k \rceil$. We may write $a = \sum_{i=0}^{m-1} a_i 2^{ki}$ and $b = \sum_{i=0}^{m-1} b_i 2^{ki}$, where $0 \le a_i < 2^k$ and $0 \le b_i < 2^k$. Let n be the integer determined by $2m \le 2^n < 4m$.

(a) Assuming Conjecture 5.22, and assuming a deterministic, polynomial-time primality test (such as the one to be presented in Chapter 21), show how to efficiently generate a prime $p \equiv 1 \pmod{2^n}$ and an element $\omega \in \mathbb{Z}_p^*$ of multiplicative order 2^n, such that

$$2^{2k}m < p \leq \ell^{O(1)}.$$

Your algorithm should be probabilistic, and run in expected time polynomial in $\text{len}(\ell)$.

(b) Assuming you have computed p and ω as in part (a), let $g := \sum_{i=0}^{m-1}[a_i]_p X^i \in \mathbb{Z}_p[X]$ and $h := \sum_{i=0}^{m-1}[b_i]_p X^i \in \mathbb{Z}_p[X]$, and show how to compute $f := gh \in \mathbb{Z}_p[X]$ in time $O(\ell)$ using the FFT (over $\mathbb{Z}_p$). Here, you may store elements of $\mathbb{Z}_p$ in single memory cells, so that operations in $\mathbb{Z}_p$ take time $O(1)$.

(c) Assuming you have computed $f \in \mathbb{Z}_p[X]$ as in part (b), show how to obtain $c := ab$ in time $O(\ell)$.

(d) Conclude that assuming Conjecture 5.22, we can multiply two integers of length at most ℓ on a RAM in time $O(\ell)$.

Note that even if one objects to our accounting practices, and insists on charging $O(\text{len}(\ell)^2)$ time units for arithmetic on numbers of length $O(\text{len}(\ell))$, the algorithm in the previous exercise runs in time $O(\ell \, \text{len}(\ell)^2)$, which is "almost" linear time.

EXERCISE 17.26. Continuing with the previous exercise:

(a) Show how the algorithm presented there can be implemented on a RAM that has only built-in addition, subtraction, and branching instructions, but no multiplication or division instructions, and still run in time $O(\ell)$. Also, memory cells should store numbers of length at most $\text{len}(\ell) + O(1)$. Hint: represent elements of $\mathbb{Z}_p$ as sequences of base-2^t digits, where $t \approx \alpha \, \text{len}(\ell)$ for some constant $\alpha < 1$; use table lookup to multiply t-bit numbers, and to perform $2t$-by-t-bit divisions—for α sufficiently small, you can build these tables in time $o(\ell)$.

(b) Using Theorem 5.23, show how to make this algorithm fully deterministic and rigorous, assuming that on inputs of length ℓ, it is provided with a certain bit string σ_ℓ of length $O(\text{len}(\ell))$ (this is called a *non-uniform* algorithm).

EXERCISE 17.27. This exercise shows how the algorithm in Exercise 17.25 can be made quite concrete, and fairly practical, as well.

(a) The number $p := 2^{59} 27 + 1$ is a 64-bit prime. Show how to use this value of p in conjunction with the algorithm in Exercise 17.25 with $k = 20$ and any value of ℓ up to 2^{27}.

(b) The numbers $p_1 := 2^{30} 3 + 1$, $p_2 := 2^{28} 13 + 1$, and $p_3 := 2^{27} 29 + 1$ are 32-bit primes. Show how to use the Chinese remainder theorem to modify the algorithm in Exercise 17.25, so that it uses the three primes p_1, p_2, p_3, and so that it works with $k = 32$ and any value of ℓ up to 2^{31}. This variant may be quite practical on a 32-bit machine with built-in instructions for 32-bit multiplication and 64-by-32-bit division.

The previous three exercises indicate that we can multiply integers in essentially linear time, both in theory and in practice. As mentioned in §3.6, there is a different, fully deterministic and rigorously analyzed algorithm that multiplies integers in linear time on a RAM. In fact, that algorithm works on a very restricted type of machine called a "pointer machine," which can be simulated in "real time" on a RAM with a very restricted instruction set (including the type in the previous exercise). That algorithm works with finite approximations to complex roots of unity, rather than roots of unity in a finite field.

We close this section with a cute application of fast polynomial multiplication to the problem of factoring integers.

EXERCISE 17.28. Let n be a large, positive integer. We can factor n using trial division in time $n^{1/2+o(1)}$; however, using fast polynomial arithmetic in $\mathbb{Z}_n[X]$, one can get a simple, deterministic, and rigorous algorithm that factors n in time $n^{1/4+o(1)}$. Note that all of the factoring algorithms discussed in Chapter 15, while faster, are either probabilistic, or deterministic but heuristic. Assume that we can multiply polynomials in $\mathbb{Z}_n[X]$ of length at most ℓ using $M(\ell)$ operations in $\mathbb{Z}_n$, where M is a well-behaved complexity function, and $M(\ell) = \ell^{1+o(1)}$ (the algorithm from Exercise 17.24 would suffice).

(a) Let ℓ be a positive integer, and for $i = 1, \ldots, \ell$, let

$$a_i := \prod_{j=0}^{\ell-1} (i\ell - j) \bmod n.$$

Using fast polynomial arithmetic, show how to compute $(a_1, \ldots, a_\ell)$ in time $\ell^{1+o(1)} \operatorname{len}(n)^{O(1)}$.

(b) Using the result of part (a), show how to factor n in time $n^{1/4+o(1)}$ using a deterministic algorithm.

17.7 Notes

Reed–Solomon codes were first proposed by Reed and Solomon [81], although the decoder presented here was developed later. Theorem 17.8 was proved by Mills [68]. The Reed–Solomon code is just one way of detecting and correcting errors— we have barely scratched the surface of this subject.

Just as in the case of integer arithmetic, the basic "pencil and paper" quadratic-time algorithms discussed in this chapter for polynomial arithmetic are not the best possible. The fastest known algorithms for multiplication of polynomials of length at most ℓ over a ring R take $O(\ell \operatorname{len}(\ell) \operatorname{len}(\operatorname{len}(\ell)))$ operations in R. These algorithms are all variations on the basic FFT algorithm (see Exercise 17.23), but work without assuming that $2_R \in R^*$ or that R contains any particular primitive roots of unity (we developed some of the ideas in Exercise 17.24). The Euclidean and extended Euclidean algorithms for polynomials over a field F can be implemented so as to take $O(\ell \operatorname{len}(\ell)^2 \operatorname{len}(\operatorname{len}(\ell)))$ operations in F, as can the algorithms for Chinese remaindering and rational function reconstruction. See the book by von zur Gathen and Gerhard [39] for details (as well for an analysis of the Euclidean algorithm for polynomials over the field of rational numbers and over function fields). Depending on the setting and many implementation details, such asymptotically fast algorithms for multiplication and division can be significantly faster than the quadratic-time algorithms, even for quite moderately sized inputs of practical interest. However, the fast Euclidean algorithms are only useful for significantly larger inputs.

Exercise 17.3 is based on an algorithm of Brent and Kung [20]. Using fast matrix and polynomial arithmetic, Brent and Kung show how to solve the modular composition problem using $O(\ell^{(\omega+1)/2})$ operations in R, where ω is the exponent for matrix multiplication (see §14.6), and so $(\omega+1)/2 < 1.7$. Modular composition arises as a subproblem in a number of algorithms.†

---

† Very recently, faster algorithms for modular composition have been discovered. See the papers by C. Umans [Fast polynomial factorization and modular composition in small characteristic, to appear in *40th Annual ACM Symposium on Theory of Computing*, 2008] and K. Kedlaya and C. Umans [Fast modular composition in any characteristic, manuscript, April 2008], both of which are available at www.cs.caltech.edu/~umans/research.

18

Linearly generated sequences and applications

In this chapter, we develop some of the theory of linearly generated sequences. As an application, we develop an efficient algorithm for solving sparse systems of linear equations, such as those that arise in the subexponential-time algorithms for discrete logarithms and factoring in Chapter 15. These topics illustrate the beautiful interplay between the arithmetic of polynomials, linear algebra, and the use of randomization in the design of algorithms.

18.1 Basic definitions and properties

Let F be a field, let V be an F-vector space, and consider an infinite sequence

$$\Psi = \{\alpha_i\}_{i=0}^{\infty}$$

where $\alpha_i \in V$ for $i = 0, 1, 2 \ldots$. We say that Ψ is **linearly generated (over F)** if there exist scalars $c_0, \ldots, c_{k-1} \in F$ such that the following recurrence relation holds:

$$\alpha_{k+i} = \sum_{j=0}^{k-1} c_j \alpha_{j+i} \quad (\text{for } i = 0, 1, 2, \ldots).$$

In this case, all of the elements of the sequence Ψ are determined by the initial segment $\alpha_0, \ldots, \alpha_{k-1}$, together with the coefficients $c_0, \ldots, c_{k-1}$ defining the recurrence relation.

The general problem we consider is this: how to determine the coefficients defining such a recurrence relation, given a sufficiently long initial segment of Ψ. To study this problem, it turns out to be very useful to rephrase the problem slightly. Let $g \in F[X]$ be a polynomial of degree, say, k, and write $g = \sum_{j=0}^{k} a_j X^j$. Next, define

$$g \star \Psi := \sum_{j=0}^{k} a_j \alpha_j.$$

Then it is clear that Ψ is linearly generated if and only if there exists a non-zero polynomial g such that

$$(X^i g) \star \Psi = 0 \quad \text{(for } i = 0, 1, 2, \ldots\text{)}. \tag{18.1}$$

Indeed, if there is such a non-zero polynomial g, then we can take

$$c_0 := -(a_0/a_k), \ c_1 := -(a_1/a_k), \ \ldots, \ c_{k-1} := -(a_{k-1}/a_k)$$

as coefficients defining the recurrence relation for Ψ. We call a polynomial g satisfying (18.1) a **generating polynomial** for Ψ. The sequence Ψ will in general have many generating polynomials. Note that the zero polynomial is technically considered a generating polynomial, but is not a very interesting one.

Let $G(\Psi)$ be the set of all generating polynomials for Ψ.

Theorem 18.1. *The set $G(\Psi)$ is an ideal of $F[X]$.*

Proof. First, note that for all $g, h \in F[X]$, we have $(g+h) \star \Psi = (g \star \Psi) + (h \star \Psi)$ — this is clear from the definitions. It is also clear that for all $c \in F$ and $g \in F[X]$, we have $(cg) \star \Psi = c \cdot (g \star \Psi)$. From these two observations, it follows that $G(\Psi)$ is closed under addition and scalar multiplication. It is also easy to see from the definition that $G(\Psi)$ is closed under multiplication by X; indeed, if $(X^i g) \star \Psi = 0$ for all $i \geq 0$, then certainly, $(X^i (Xg)) \star \Psi = (X^{i+1} g) \star \Psi = 0$ for all $i \geq 0$. But any non-empty subset of $F[X]$ that is closed under addition, multiplication by elements of F, and multiplication by X is an ideal of $F[X]$ (see Exercise 7.27). □

Since all ideals of $F[X]$ are principal, it follows that $G(\Psi)$ is the ideal of $F[X]$ generated by some polynomial $\phi \in F[X]$ — we can make this polynomial unique by choosing the monic associate (if it is non-zero), and we call this polynomial the **minimal polynomial of Ψ**. Thus, a polynomial $g \in F[X]$ is a generating polynomial for Ψ if and only if ϕ divides g; in particular, Ψ is linearly generated if and only if $\phi \neq 0$.

We can now restate our main objective as follows: given a sufficiently long initial segment of a linearly generated sequence, determine its minimal polynomial.

Example 18.1. One can always define a linearly generated sequence by simply choosing an initial segment $\alpha_0, \alpha_1, \ldots, \alpha_{k-1}$, along with scalars $c_0, \ldots, c_{k-1} \in F$ defining the recurrence relation. One can enumerate as many elements of the sequence as one wants by using storage for k elements of V, along with storage for the scalars $c_0, \ldots, c_{k-1}$, as follows:

$(\beta_0, \ldots, \beta_{k-1}) \leftarrow (\alpha_0, \ldots, \alpha_{k-1})$
repeat
    output β_0
    $\beta' \leftarrow \sum_{j=0}^{k-1} c_j \beta_j$
    $(\beta_0, \ldots, \beta_{k-1}) \leftarrow (\beta_1, \ldots, \beta_{k-1}, \beta')$
forever

Because of the structure of the above algorithm, linearly generated sequences are sometimes also called **shift register sequences**. Also observe that if F is a finite field, and V is finite dimensional, the value stored in the "register" $(\beta_0, \ldots, \beta_{k-1})$ must repeat at some point. It follows that the linearly generated sequence must be ultimately periodic (see definitions above Exercise 4.21). □

Example 18.2. Linearly generated sequences can also arise in a natural way, as this example and the next illustrate. Let $E := F[X]/(f)$, where $f \in F[X]$ is a monic polynomial of degree $\ell > 0$, and let α be an element of E. Consider the sequence $\Psi := \{\alpha^i\}_{i=0}^{\infty}$ of powers of α. For every polynomial $g = \sum_{j=0}^{k} a_j X^j \in F[X]$, we have

$$g \star \Psi = \sum_{j=0}^{k} a_j \alpha^j = g(\alpha).$$

Now, if $g(\alpha) = 0$, then clearly $(X^i g) \star \Psi = \alpha^i g(\alpha) = 0$ for all $i \geq 0$. Conversely, if $(X^i g) \star \Psi = 0$ for all $i \geq 0$, then in particular, $g(\alpha) = 0$. Thus, g is a generating polynomial for Ψ if and only if $g(\alpha) = 0$. It follows that the minimal polynomial ϕ of Ψ is the same as the minimal polynomial of α over F, as defined in §16.5. Furthermore, $\phi \neq 0$, and the degree m of ϕ may be characterized as the smallest positive integer m such that $\{\alpha^i\}_{i=0}^{m}$ is linearly dependent; moreover, as E has dimension ℓ over F, we must have $m \leq \ell$. □

Example 18.3. Let V be a vector space over F of dimension $\ell > 0$, and let $\tau : V \to V$ be an F-linear map. Let $\beta \in V$, and consider the sequence $\Psi := \{\alpha_i\}_{i=0}^{\infty}$, where $\alpha_i = \tau^i(\beta)$; that is, $\alpha_0 = \beta$, $\alpha_1 = \tau(\beta)$, $\alpha_2 = \tau(\tau(\beta))$, and so on. For every polynomial $g = \sum_{j=0}^{k} a_j X^j \in F[X]$, we have

$$g \star \Psi = \sum_{j=0}^{k} a_j \tau^j(\beta),$$

and for every $i \geq 0$, we have

$$(X^i g) \star \Psi = \sum_{j=0}^{k} a_j \tau^{i+j}(\beta) = \tau^i \left(\sum_{j=0}^{k} a_j \tau^j(\beta) \right) = \tau^i(g \star \Psi).$$

Thus, if $g \star \Psi = 0$, then clearly $(X^i g) \star \Psi = \tau^i(g \star \Psi) = \tau^i(0) = 0$ for all $i \geq 0$. Conversely, if $(X^i g) \star \Psi = 0$ for all $i \geq 0$, then in particular, $g \star \Psi = 0$. Thus, g is a generating polynomial for Ψ if and only if $g \star \Psi = 0$. The minimal polynomial ϕ of Ψ is non-zero and its degree m is at most ℓ; indeed, m may be characterized as the least non-negative integer such that $\{\tau^i(\beta)\}_{i=0}^m$ is linearly dependent, and since V has dimension ℓ over F, we must have $m \leq \ell$.

The previous example can be seen as a special case of this one, by taking V to be E, τ to be the α-multiplication map on E, and setting β to 1. $\square$

The problem of computing the minimal polynomial of a linearly generated sequence can always be solved by means of Gaussian elimination. For example, the minimal polynomial of the sequence discussed in Example 18.2 can be computed using the algorithm described in §17.2. The minimal polynomial of the sequence discussed in Example 18.3 can be computed in a similar manner. Also, Exercise 18.3 below shows how one can reformulate another special case of the problem so that it is easily solved by Gaussian elimination. However, in the following sections, we will present algorithms for computing minimal polynomials for certain types of linearly generated sequences that are much more efficient than any algorithm based on Gaussian elimination.

EXERCISE 18.1. Show that the only sequence for which 1 is a generating polynomial is the "all zero" sequence.

EXERCISE 18.2. Let $\Psi = \{\alpha_i\}_{i=0}^{\infty}$ be a sequence of elements of an F-vector space V. Further, suppose that Ψ has non-zero minimal polynomial ϕ.

    (a) Show that for all polynomials $g, h \in F[X]$, if $g \equiv h \pmod{\phi}$, then $g \star \Psi = h \star \Psi$.

    (b) Let $m := \deg(\phi)$. Show that if $g \in F[X]$ and $(X^i g) \star \Psi = 0$ for all $i = 0, \ldots, m - 1$, then g is a generating polynomial for Ψ.

EXERCISE 18.3. This exercise develops an alternative characterization of linearly generated sequences. Let $\Psi = \{z_i\}_{i=0}^{\infty}$ be a sequence of elements of F. Further, suppose that Ψ has minimal polynomial $\phi = \sum_{j=0}^{m} c_j X^j$ with $m > 0$ and $c_m = 1$. Define the matrix

$$A := \begin{pmatrix} z_0 & z_1 & \cdots & z_{m-1} \\ z_1 & z_2 & \cdots & z_m \\ \vdots & \vdots & \ddots & \vdots \\ z_{m-1} & z_m & \cdots & z_{2m-2} \end{pmatrix} \in F^{m \times m}$$

and the vector

$$w := (z_m, \ldots, z_{2m-1}) \in F^{1 \times m}.$$

Show that

$$v = (-c_0, \ldots, -c_{m-1}) \in F^{1 \times m}$$

is the *unique* solution to the equation

$$vA = w.$$

Hint: show that the rows of A form a linearly independent family of vectors by making use of Exercise 18.2 and the fact that no polynomial of degree less than m is a generating polynomial for Ψ.

EXERCISE 18.4. Let $c_0, \ldots, c_{k-1} \in F$ and $z_0, \ldots, z_{k-1} \in F$. For each $i \geq 0$, let

$$z_{k+i} := \sum_{j=0}^{k-1} c_j z_{j+i}.$$

Given $n \geq 0$, along with $c_0, \ldots, c_{k-1}$ and $z_0, \ldots, z_{k-1}$, show how to compute z_n using $O(\text{len}(n)k^2)$ operations in F.

EXERCISE 18.5. Let V be a vector space over F, and consider the set $V^{\times \infty}$ of all infinite sequences $\{\alpha_i\}_{i=0}^{\infty}$, where the α_i's are in V. Let us define the scalar product of $g \in F[X]$ and $\Psi \in V^{\times \infty}$ as

$$g \cdot \Psi = \{(X^i g) \star \Psi\}_{i=0}^{\infty} \in V^{\times \infty}.$$

Show that with this scalar product, and addition defined component-wise, $V^{\times \infty}$ is an $F[X]$-module, and that a polynomial $g \in F[X]$ is a generating polynomial for $\Psi \in V^{\times \infty}$ if and only if $g \cdot \Psi = 0$.

18.2 Computing minimal polynomials: a special case

We now tackle the problem of efficiently computing the minimal polynomial of a linearly generated sequence from a sufficiently long initial segment.

We shall first address a special case of this problem, namely, the case where the vector space V is just the field F. In this case, we have

$$\Psi = \{z_i\}_{i=0}^{\infty},$$

where $z_i \in F$ for $i = 0, 1, 2, \ldots$.

Suppose that we do not know the minimal polynomial ϕ of Ψ, but we know an upper bound $M > 0$ on its degree. Then it turns out that the initial segment $z_0, z_1, \ldots z_{2M-1}$ completely determines ϕ, and moreover, we can very efficiently compute ϕ given this initial segment. The following theorem provides the essential ingredient.

Theorem 18.2. *Let* $\Psi = \{z_i\}_{i=0}^{\infty}$ *be a sequence of elements of* F, *and define the reversed Laurent series*

$$z := \sum_{i=0}^{\infty} z_i X^{-(i+1)} \in F((X^{-1})),$$

whose coefficients are the elements of the sequence Ψ. *Then for every* $g \in F[X]$, *we have* $g \in G(\Psi)$ *if and only if* $gz \in F[X]$. *In particular,* Ψ *is linearly generated if and only if* z *is a rational function, in which case, its minimal polynomial is the denominator of* z *when expressed as a fraction in lowest terms.*

Proof. Observe that for every polynomial $g \in F[X]$ and every integer $i \geq 0$, the coefficient of $X^{-(i+1)}$ in the product gz is equal to $X^i g \star \Psi$—just look at the formulas defining these expressions! It follows that g is a generating polynomial for Ψ if and only if the coefficients of the negative powers of X in gz are all zero, which is the same as saying that $gz \in F[X]$. Further, if $g \neq 0$ and $h := gz \in F[X]$, then $\deg(h) < \deg(g)$—this follows simply from the fact that $\deg(z) < 0$ (together with the fact that $\deg(h) = \deg(g) + \deg(z)$). All the statements in the theorem follow immediately from these observations. $\square$

By virtue of Theorem 18.2, we can compute the minimal polynomial ϕ of Ψ using the algorithm in §17.5.1 for computing the numerator and denominator of a rational function from its reversed Laurent series expansion. More precisely, we can compute ϕ given the bound M on its degree, along with the first $2M$ elements $z_0, \ldots, z_{2M-1}$ of Ψ, using $O(M^2)$ operations in F. Just for completeness, we write down this algorithm:

1. Run the extended Euclidean algorithm on inputs

$$f := X^{2M} \quad \text{and} \quad h := z_0 X^{2M-1} + z_1 X^{2M-2} + \cdots + z_{2M-1},$$

and apply Theorem 17.8 with f, h, $r^* := M$, and $t^* := M$, to obtain the polynomials r', s', t'.

2. Output $\phi := t' / \mathrm{lc}(t')$.

EXERCISE 18.6. Suppose F is a finite field and that $\Psi := \{z_i\}_{i=0}^{\infty}$ is linearly generated, with minimal polynomial ϕ. Further, suppose $X \nmid \phi$. Show that Ψ is purely periodic with period equal to the multiplicative order of $[X]_\phi \in (F[X]/(\phi))^*$. Hint: use Exercise 17.12 and Theorem 18.2.

18.3 Computing minimal polynomials: a more general case

Having dealt with the problem of finding the minimal polynomial of a linearly generated sequence Ψ, whose elements lie in F, we address the more general problem, where the elements of Ψ lie in a vector space V over F. We shall only deal with a special case of this problem, but it is one which has useful applications:

- First, we shall assume that V has finite dimension $\ell > 0$ over F.

- Second, we shall assume that the sequence $\Psi = \{\alpha_i\}_{i=0}^{\infty}$ has **full rank**, by which we mean the following: if the minimal polynomial ϕ of Ψ over F has degree m, then $\{\alpha_i\}_{i=0}^{m-1}$ is linearly independent. This property implies that the minimal polynomial of Ψ is the monic polynomial $\phi \in F[X]$ of least degree such that $\phi \star \Psi = 0$. The sequences considered in Examples 18.2 and 18.3 are of this type.

- Third, we shall assume that F is a finite field.

The dual space. Before presenting our algorithm for computing minimal polynomials, we need to discuss the **dual space** $\mathcal{D}_F(V)$ of V (over F), which consists of all F-linear maps from V into F. Thus, $\mathcal{D}_F(V) = \mathrm{Hom}_F(V, F)$, and is a vector space over F, with addition and scalar multiplication defined point-wise (see Theorem 13.12). We shall call elements of $\mathcal{D}_F(V)$ **projections**.

Now, fix a basis $S = \{\gamma_i\}_{i=1}^{\ell}$ for V. As was discussed in §14.2, every element $\delta \in V$ has a unique coordinate vector $\mathrm{Vec}_S(\delta) = (c_1, \ldots, c_\ell) \in F^{1 \times \ell}$, where $\delta = \sum_i c_i \gamma_i$. Moreover, the map $\mathrm{Vec}_S : V \to F^{1 \times \ell}$ is a vector space isomorphism.

To each projection $\pi \in \mathcal{D}_F(V)$ we may also associate the **coordinate vector** $(\pi(\gamma_1), \ldots, \pi(\gamma_\ell))^{\top} \in F^{\ell \times 1}$. If $\mathcal{U}$ is the basis for F consisting of the single element 1_F, then the coordinate vector of π is $\mathrm{Mat}_{S, \mathcal{U}}(\pi)$, that is, the matrix of π relative to the bases S and $\mathcal{U}$. By Theorem 14.4, the map $\mathrm{Mat}_{S, \mathcal{U}} : \mathcal{D}_F(V) \to F^{\ell \times 1}$ is a vector space isomorphism.

In working with algorithms that compute with elements of V and $\mathcal{D}_F(V)$, we shall assume that such elements are represented using coordinate vectors relative to some convenient, fixed basis for V. If $\delta \in V$ has coordinate vector $(c_1, \ldots, c_\ell) \in F^{1 \times \ell}$, and $\pi \in \mathcal{D}_F(V)$ has coordinate vector $(d_1, \ldots, d_\ell)^{\top} \in F^{\ell \times 1}$, then $\pi(\delta)$ is easily computed, using $O(\ell)$ operations in F, as $\sum_{i=1}^{\ell} c_i d_i$.

We now return to the problem of computing the minimal polynomial ϕ of the linearly generated sequence $\Psi = \{\alpha_i\}_{i=0}^{\infty}$. Assume we have a bound $M > 0$ on the degree of ϕ. Since Ψ has full rank and $\dim_F(V) = \ell$, we may assume that $M \leq \ell$.

For each $\pi \in \mathcal{D}_F(V)$, we may consider the *projected* sequence $\Psi_\pi := \{\pi(\alpha_i)\}_{i=0}^{\infty}$. Observe that ϕ is a generating polynomial for Ψ_π; indeed, for every polynomial $g \in F[X]$, we have $g \star \Psi_\pi = \pi(g \star \Psi)$, and hence, for all $i \geq 0$, we have $(X^i \phi) \star \Psi_\pi = \pi((X^i \phi) \star \Psi) = \pi(0) = 0$. Let $\phi_\pi \in F[X]$ denote the minimal polynomial of Ψ_π. Since ϕ_π divides every generating polynomial of Ψ_π, and since ϕ is a generating polynomial for Ψ_π, it follows that ϕ_π divides ϕ.

This suggests the following algorithm for efficiently computing the minimal polynomial of Ψ, using the first $2M$ terms of Ψ:

Algorithm MP. Given the first $2M$ terms of the sequence $\Psi = \{\alpha_i\}_{i=0}^{\infty}$, do the following:

> $g \leftarrow 1 \in F[X]$
> repeat
>> choose $\pi \in \mathcal{D}_F(V)$ at random
>> compute the first $2M$ terms of the projected sequence Ψ_π
>> use the algorithm in §18.2 to compute the minimal polynomial
>>> ϕ_π of Ψ_π
>> $g \leftarrow \mathrm{lcm}(g, \phi_\pi)$
> until $g \star \Psi = 0$
> output g

A few remarks on the above procedure are in order:

- in every iteration of the main loop, g is the least common multiple of a number of divisors of ϕ, and hence is itself a divisor of ϕ; in particular, $\deg(g) \leq M$;
- under our assumption that Ψ has full rank, and since g is a monic divisor of ϕ, if $g \star \Psi = 0$, we may safely conclude that $g = \phi$;
- under our assumption that F is finite, choosing a random element π of $\mathcal{D}_F(V)$ amounts to simply choosing at random the entries of the coordinate vector of π, relative to some basis for V;
- we also assume that elements of V are represented as coordinate vectors, so that applying a projection $\pi \in \mathcal{D}_F(V)$ to an element of V takes $O(\ell)$ operations in F; in particular, in each loop iteration, we can compute the first $2M$ terms of the projected sequence Ψ_π using $O(M\ell)$ operations in F;
- similarly, adding two elements of V, or multiplying an element of V by a scalar, takes $O(\ell)$ operations in F; in particular, in each loop iteration, we can compute $g \star \Psi$ using $O(M\ell)$ operations in F (and using the first $M + 1 \leq 2M$ terms of Ψ).

Based on the above observations, it follows that when the algorithm halts, its output is correct, and that the cost of each loop iteration is $O(M\ell)$ operations in F. The remaining question to be answered is this: what is the expected number of iterations of the main loop? The answer to this question is $O(1)$, which leads to a total expected cost of Algorithm MP of $O(M\ell)$ operations in F.

The key to establishing that the expected number of iterations of the main loop is constant is provided by the following theorem.

Theorem 18.3. *Let* $\Psi = \{\alpha_i\}_{i=0}^{\infty}$ *be a linearly generated sequence over the field* F, *where the* α_i's *are elements of a vector space* V *of finite dimension* $\ell > 0$. *Let* ϕ *be the minimal polynomial of* Ψ *over* F, *let* $m := \deg(\phi)$, *and assume that* Ψ *has full rank (i.e.,* $\{\alpha_i\}_{i=0}^{m-1}$ *is linearly independent). Finally, let* $F[X]_{<m}$ *denote the vector space over* F *consisting of all polynomials in* $F[X]$ *of degree less than* m.

Under the above assumptions, there exists a surjective F-*linear map*

$$\sigma : D_F(V) \to F[X]_{<m}$$

such that for all $\pi \in D_F(V)$, *the minimal polynomial* ϕ_π *of the projected sequence* $\Psi_\pi := \{\pi(\alpha_i)\}_{i=0}^{\infty}$ *satisfies*

$$\phi_\pi = \frac{\phi}{\gcd(\sigma(\pi), \phi)}.$$

Proof. While the statement of this theorem looks a bit complicated, its proof is quite straightforward, given our characterization of linearly generated sequences in Theorem 18.2 in terms of rational functions. We build the linear map σ as the composition of two linear maps, σ_0 and σ_1.

Let us define the map

$$\sigma_0 : \quad D_F(V) \to F((X^{-1}))$$

$$\pi \mapsto \sum_{i=0}^{\infty} \pi(\alpha_i) X^{-(i+1)}.$$

We also define the map σ_1 to be the ϕ-multiplication map on $F((X^{-1}))$—that is, the map that sends $z \in F((X^{-1}))$ to $\phi \cdot z \in F((X^{-1}))$. The map σ is just the composition $\sigma = \sigma_1 \circ \sigma_0$. It is clear that both σ_0 and σ_1 are F-linear maps, and hence, so is σ.

First, observe that for $\pi \in D_F(V)$, the series $z := \sigma_0(\pi)$ is the series associated with the projected sequence Ψ_π, as in Theorem 18.2. Let ϕ_π be the minimal polynomial of Ψ_π. Since ϕ is a generating polynomial for Ψ, it is also a generating polynomial for Ψ_π. Therefore, Theorem 18.2 tells us that

$$h := \sigma(\pi) = \phi \cdot z \in F[X]_{<m},$$

and that ϕ_π is the denominator of z when expressed as a fraction in lowest terms. Now, we have $z = h/\phi$, and it follows that $\phi_\pi = \phi/\gcd(h, \phi)$ is this denominator.

Second, the hypothesis that $\{\alpha_i\}_{i=0}^{m-1}$ is linearly independent implies that $\dim_F(\operatorname{Im} \sigma_0) \geq m$ (see Exercise 13.21). Also, observe that σ_1 is an injective map. Therefore, $\dim_F(\operatorname{Im} \sigma) \geq m$. In the previous paragraph, we observed that $\text{Im}\,\sigma \subseteq F[X]_{<m}$, and since $\dim_F(F[X]_{<m}) = m$, we may conclude that $\text{Im}\,\sigma = F[X]_{<m}$. That proves the theorem. $\square$

Given the above theorem, we can analyze the expected number of iterations of the main loop of Algorithm MP.

First of all, we may as well assume that the degree m of ϕ is greater than 0, as otherwise, we are sure to get ϕ in the very first iteration. Let $\pi_1, \ldots, \pi_s$ be the random projections chosen in the first s iterations of Algorithm MP. By Theorem 18.3, each $\sigma(\pi_i)$ is uniformly distributed over $F[X]_{<m}$, and we have $g = \phi$ at the end of loop iteration s if and only if $\gcd(\phi, \sigma(\pi_1), \ldots, \sigma(\pi_s)) = 1$.

Let us define $\Lambda_F^\phi(s)$ to be the probability that $\gcd(\phi, f_1, \ldots, f_s) = 1$, where $f_1, \ldots, f_s$ are randomly chosen from $F[X]_{<m}$. Thus, the probability that we have $g = \phi$ at the end of loop iteration s is equal to $\Lambda_F^\phi(s)$. While one can analyze the quantity $\Lambda_F^\phi(s)$, it turns out to be easier, and sufficient for our purposes, to analyze a different quantity. Let us define $\Lambda_F^m(s)$ to be the probability that $\gcd(f_1, \ldots, f_s) = 1$, where $f_1, \ldots, f_s$ are randomly chosen from $F[X]_{<m}$. Clearly, $\Lambda_F^\phi(s) \geq \Lambda_F^m(s)$.

Theorem 18.4. *If F is a finite field of cardinality q, and m and s are positive integers, then we have*

$$\Lambda_F^m(s) = 1 - 1/q^{s-1} + (q-1)/q^{sm}.$$

Proof. For each positive integer n, let U_n denote the set of all tuples of polynomials $(f_1, \ldots, f_s) \in F[X]_{<n}^{\times s}$ with $\gcd(f_1, \ldots, f_s) = 1$, and let $u_n := |U_n|$. Also, for each monic polynomial $h \in F[X]$ of degree less that n, let $U_{n,h}$ denote the set of all s-tuples of polynomials of degree less than n whose gcd is h. Observe that the set $U_{n,h}$ is in one-to-one correspondence with U_{n-k}, where $k := \deg(h)$, via the map that sends $(f_1, \ldots, f_s) \in U_{n,h}$ to $(f_1/h, \ldots, f_s/h) \in U_{n-k}$. As there are q^k possible choices for h of degree k, if we define $V_{n,k}$ to be the set of tuples $(f_1, \ldots, f_s) \in F[X]_{<n}^{\times s}$ with $\deg(\gcd(f_1, \ldots, f_s)) = k$, we see that $|V_{n,k}| = q^k u_{n-k}$. Every non-zero tuple in $F[X]_{<n}^{\times s}$ appears in exactly one of the sets $V_{n,k}$, for $k = 0, \ldots, n-1$. Taking into account the zero tuple, it follows that

$$q^{sn} = 1 + \sum_{k=0}^{n-1} q^k u_{n-k}, \qquad (18.2)$$

which holds for all $n \geq 1$. Replacing n by $n-1$ in (18.2), we obtain

$$q^{s(n-1)} = 1 + \sum_{k=0}^{n-2} q^k u_{n-1-k}, \qquad (18.3)$$

which holds for all $n \geq 2$, and indeed, holds for $n = 1$ as well. Subtracting q times (18.3) from (18.2), we deduce that for all $n \geq 1$,

$$q^{sn} - q^{sn-s+1} = 1 + u_n - q,$$

and rearranging terms:

$$u_n = q^{sn} - q^{sn-s+1} + q - 1.$$

Therefore,

$$\Lambda_F^m(s) = u_m/q^{sm} = 1 - 1/q^{s-1} + (q-1)/q^{sm}. \quad \Box$$

From the above theorem, it follows that for $s \geq 1$, the probability P_s that Algorithm MP runs for more than s loop iterations is at most $1/q^{s-1}$. If L is the total number of loop iterations, then

$$\mathsf{E}[L] = \sum_{i \geq 1} \mathsf{P}[L \geq i] = 1 + \sum_{s \geq 1} P_s \leq 1 + \sum_{s \geq 1} 1/q^{s-1} = 1 + \frac{q}{q-1} \leq 3.$$

Let us summarize all of the above analysis with the following:

Theorem 18.5. *Let Ψ be a sequence of elements of an F-vector space V of finite dimension $\ell > 0$ over F, where F is a finite field. Assume that Ψ is linearly generated over F with minimal polynomial $\phi \in F[X]$ of degree m, and that Ψ has full rank (i.e., the first m terms of Ψ form a linearly independent family of elements). Then given an upper bound $M > 0$ on m, along with the first $2M$ elements of Ψ, Algorithm MP correctly computes ϕ using an expected number of $O(M\ell)$ operations in F.*

We close this section with the following observation. Suppose the sequence Ψ is of the form $\{\tau^i(\beta)\}_{i=0}^{\infty}$, where $\beta \in V$ and $\tau : V \to V$ is an F-linear map. Suppose that with respect to some basis S for V, elements of V are represented by their coordinate vectors (which are elements of $F^{1 \times \ell}$), and elements of $D_F(V)$ are represented by their coordinate vectors (which are elements of $F^{\ell \times 1}$). The linear map τ also has a corresponding matrix $A = \mathrm{Mat}_{S,S}(V,V) \in F^{\ell \times \ell}$, so that evaluating τ at a point α in V corresponds to multiplying the coordinate vector of α on the right by A. Now, suppose $\beta \in V$ has coordinate vector $v \in F^{1 \times \ell}$ and that $\pi \in D_F(V)$ has coordinate vector $w \in F^{\ell \times 1}$. Then if Ψ' is the sequence of coordinate vectors of the elements of Ψ, we have

$$\Psi' = \{vA^i\}_{i=0}^{\infty} \quad \text{and} \quad \Psi_\pi = \{vA^i w\}_{i=0}^{\infty}.$$

This more concrete, matrix-oriented point of view is sometimes useful; in particular, it makes quite transparent the symmetry of the roles played by β and π in forming the projected sequence.

EXERCISE 18.7. If $|F| = q$ and $\phi \in F[X]$ is monic and factors into monic irreducible polynomials in $F[X]$ as $\phi = \phi_1^{e_1} \cdots \phi_r^{e_r}$, show that

$$\Lambda_F^\phi(1) = \prod_{i=1}^r (1 - q^{-\deg(\phi_i)}) \geq 1 - \sum_{i=1}^r q^{-\deg(\phi_i)}.$$

From this, conclude that the probability that Algorithm MP terminates after just one loop iteration is $1 - O(m/q)$, where $m = \deg(\phi)$. Thus, if q is very large relative to m, it is highly likely that Algorithm MP terminates after just one iteration of the main loop.

18.4 Solving sparse linear systems

Let V be a vector space of finite dimension $\ell > 0$ over a finite field F, and let $\tau : V \to V$ be an F-linear map. The goal of this section is to develop time- and space-efficient algorithms for solving equations of the form

$$\tau(\gamma) = \delta; \tag{18.4}$$

that is, given τ and $\delta \in V$, find $\gamma \in V$ satisfying (18.4). The algorithms we develop will have the following properties: they will be probabilistic, and will use an expected number of $O(\ell^2)$ operations in F, an expected number of $O(\ell)$ evaluations of τ, and space for $O(\ell)$ elements of F. By an "evaluation of τ," we mean the computation of $\tau(\alpha)$ for a given $\alpha \in V$.

We shall assume that elements of V are represented as coordinate vectors with respect to some fixed basis for V. This means that a single element of V is represented as a vector of ℓ elements of F. Now, if the matrix of τ with respect to the given basis is sparse, having, say, $\ell^{1+o(1)}$ non-zero entries, then the space required to represent τ is $\ell^{1+o(1)}$ elements of F, and the time required to evaluate τ is $\ell^{1+o(1)}$ operations in F. Under these assumptions, our algorithms to solve (18.4) use an expected number of $\ell^{2+o(1)}$ operations in F, and space for $\ell^{1+o(1)}$ elements of F. This is to be compared with standard Gaussian elimination: even if the original matrix is sparse, during the execution of the algorithm, most of the entries in the matrix may eventually be "filled in" with non-zero field elements, leading to a running time of $\Omega(\ell^3)$ operations in F, and a space requirement of $\Omega(\ell^2)$ elements of F. Thus, the algorithms presented here will be much more efficient than Gaussian elimination when the matrix of τ is sparse.

We hasten to point out that the algorithms presented here may be more efficient than Gaussian elimination in other cases, as well. All that matters is that τ can be evaluated using $o(\ell^2)$ operations in F and/or represented using space for $o(\ell^2)$ elements of F—in either case, we obtain a time and/or space improvement over Gaussian elimination. Indeed, there are applications where the matrix of the linear map τ may not be sparse, but nevertheless has special structure that allows it to be represented and evaluated in subquadratic time and/or space.

We shall only present algorithms that work in two special, but important, cases:

- the first case is where τ is bijective,
- the second case is where τ is not bijective, $\delta = 0$, and a non-zero solution γ to (18.4) is required (i.e., we are looking for a non-zero element of Ker τ).

In both cases, the key will be to use Algorithm MP in §18.3 to find the minimal polynomial ϕ of the linearly generated sequence

$$\Psi := \{\alpha_i\}_{i=0}^{\infty} \quad (\alpha_i := \tau^i(\beta), \ i = 0, 1, \ldots), \tag{18.5}$$

where β is a suitably chosen element of V. From the discussion in Example 18.3, this sequence has full rank, and so we may use Algorithm MP. We may use $M := \ell$ as an upper bound on the degree of ϕ (assuming we know nothing more about τ and β that would allow us to use a smaller upper bound). In using Algorithm MP in this application, note that we do not want to store $\alpha_0, \ldots, \alpha_{2\ell-1}$ — if we did, we would not satisfy our stated space bound. Instead of storing the α_i's in a "warehouse," we use a "just in time" strategy for computing them, as follows:

- In the body of the main loop of Algorithm MP, where we calculate the projections $z_i := \pi(\alpha_i)$, for $i = 0 \ldots 2\ell - 1$, we perform the computation as follows:

$$\alpha \leftarrow \beta$$
$$\text{for } i \leftarrow 0 \text{ to } 2\ell - 1 \text{ do}$$
$$\quad z_i \leftarrow \pi(\alpha), \ \alpha \leftarrow \tau(\alpha)$$

- In the test at the bottom of the main loop of Algorithm MP, if $g = \sum_{j=0}^{k} a_j X^j$, we compute $v := g \star \Psi \in V$ using the following Horner-like scheme:

$$v \leftarrow 0$$
$$\text{for } j \leftarrow k \text{ down to } 0 \text{ do}$$
$$\quad v \leftarrow \tau(v) + a_j \cdot \beta$$

With this implementation, Algorithm MP uses an expected number of $O(\ell^2)$ operations in F, an expected number of $O(\ell)$ evaluations of τ, and space for $O(\ell)$ elements of F. Of course, the "warehouse" strategy is faster than the "just in time" strategy by a constant factor, but it uses about ℓ times as much space; thus, for large ℓ, using the "just in time" strategy is a very good time/space trade-off.

The bijective case. Now consider the case where τ is bijective, and we want to solve (18.4) for a given $\delta \in V$. We may as well assume that $\delta \neq 0$, since otherwise, $\gamma = 0$ is the unique solution to (18.4). We proceed as follows. First, using Algorithm MP as discussed above, compute the minimal polynomial ϕ of the sequence Ψ defined in (18.5), using $\beta := \delta$. Let $\phi = \sum_{j=0}^{m} c_j X^j$, where $c_m = 1$ and $m > 0$. Then we have

$$c_0 \delta + c_1 \tau(\delta) + \cdots + c_m \tau^m(\delta) = 0. \tag{18.6}$$

We claim that $c_0 \neq 0$. To prove the claim, suppose that $c_0 = 0$. Then applying τ^{-1} to (18.6), we would obtain

$$c_1 \delta + \cdots + c_m \tau^{m-1}(\delta) = 0,$$

which would imply that ϕ/X is a generating polynomial for Ψ, contradicting the minimality of ϕ. That proves the claim.

Since $c_0 \neq 0$, we can apply τ^{-1} to (18.6), and solve for $\gamma = \tau^{-1}(\delta)$ as follows:

$$\gamma = -c_0^{-1}(c_1 \delta + \cdots + c_m \tau^{m-1}(\delta)).$$

To actually compute γ, we use the same "just in time" strategy as was used in the implementation of the computation of $g \star \Psi$ in Algorithm MP, which costs $O(\ell^2)$ operations in F, $O(\ell)$ evaluations of τ, and space for $O(\ell)$ elements of F.

The non-bijective case. Now consider the case where τ is not bijective, and we want to find non-zero $\gamma \in V$ such that $\tau(\gamma) = 0$. The idea is this. Suppose we choose an arbitrary, non-zero element β of V, and use Algorithm MP to compute the minimal polynomial ϕ of the sequence Ψ defined in (18.5), using this value of β. Let $\phi = \sum_{j=0}^{m} c_j X^j$, where $m > 0$ and $c_m = 1$. Then we have

$$c_0 \beta + c_1 \tau(\beta) + \cdots + c_m \tau^m(\beta) = 0. \tag{18.7}$$

Let

$$\gamma := c_1 \beta + \cdots + c_m \tau^{m-1}(\beta).$$

We must have $\gamma \neq 0$, since $\gamma = 0$ would imply that $\lfloor \phi/X \rfloor$ is a non-zero generating polynomial for Ψ, contradicting the minimality of ϕ. If it happens that $c_0 = 0$, then equation (18.7) implies that $\tau(\gamma) = 0$, and we are done. As before, to actually compute γ, we use the same "just in time" strategy as was used in the implementation of the computation of $g \star \Psi$ in Algorithm MP, which costs $O(\ell^2)$ operations in F, $O(\ell)$ evaluations of τ, and space for $O(\ell)$ elements of F.

The above approach fails if $c_0 \neq 0$. However, in this "bad" case, equation (18.7) implies that $\beta = -c_0^{-1} \tau(\gamma)$; in particular, $\beta \in \text{Im } \tau$. One way to avoid such a "bad" β is to randomize: as τ is not surjective, the image of τ is a subspace of V of dimension strictly less than ℓ, and therefore, a *randomly* chosen β lies in the image of τ with probability at most $1/|F|$. So a simple technique is to choose repeatedly β at random until we get a "good" β. The overall complexity of the resulting algorithm will be as required: $O(\ell^2)$ expected operations in F, $O(\ell)$ expected evaluations of τ, and space for $O(\ell)$ elements of F.

As a special case of this situation, consider the problem that arose in Chapter 15 in connection with algorithms for computing discrete logarithms and factoring. We had to solve the following problem: given an $\ell \times (\ell - 1)$ matrix A with entries in a finite field F, containing $\ell^{1+o(1)}$ non-zero entries, find non-zero $v \in F^{1 \times \ell}$ such that $vA = 0$. To solve this problem, we can augment the matrix A, adding an extra column of zeros, to get an $\ell \times \ell$ matrix A'. Now, let $V = F^{1 \times \ell}$ and let τ be the F-linear map on V that sends $\gamma \in V$ to $\gamma A'$. A non-zero solution γ to the equation $\tau(\gamma) = 0$ will provide us with the solution to our original problem; thus, we can apply the above technique directly, solving this problem using $\ell^{2+o(1)}$ expected operations in F, and space for $\ell^{1+o(1)}$ elements of F. As a side remark, in this particular application, we can choose a "good" β in the above algorithm without randomization: just choose $\beta := (0, \ldots, 0, 1)$, which is clearly not in the image of τ.

18.5 Computing minimal polynomials in $F[X]/(f)$ (II)

Let us return to the problem discussed in §17.2: F is a field, $f \in F[X]$ is a monic polynomial of degree $\ell > 0$, and $E := F[X]/(f)$; we are given an element $\alpha \in E$, and want to compute the minimal polynomial $\phi \in F[X]$ of α over F. As discussed in Example 18.2, this problem is equivalent to the problem of computing the minimal polynomial of the sequence

$$\Psi := \{\alpha_i\}_{i=0}^{\infty} \quad (\alpha_i := \alpha^i, \ i = 0, 1, \ldots),$$

and the sequence has full rank; therefore, we can use Algorithm MP in §18.3 directly to solve this problem, assuming F is a finite field.

If we use the "just in time" strategy in the implementation of Algorithm MP, as was used in §18.4, we get an algorithm that computes the minimal polynomial of α using $O(\ell^3)$ expected operations in F, but space for just $O(\ell^2)$ elements of F. Thus, in terms of space, this approach is far superior to the algorithm in §17.2, based on Gaussian elimination. In terms of time complexity, the algorithm based on linearly generated sequences is a bit slower than the one based on Gaussian elimination (but only by a constant factor). However, if we use any subquadratic-time algorithm for polynomial arithmetic (see §17.6 and §17.7), we immediately get an algorithm that runs in subcubic time, while still using linear space. In the exercises below, you are asked to develop an algorithm that computes the minimal polynomial of α using just $O(\ell^{2.5})$ operations in F, at the expense of requiring space for $O(\ell^{1.5})$ elements of F—this algorithm does not rely on fast polynomial arithmetic, and can be made even faster if such arithmetic is used.

EXERCISE 18.8. Let $f \in F[X]$ be a monic polynomial of degree $\ell > 0$ over a field F, and let $E := F[X]/(f)$. Also, let $\xi := [X]_f \in E$. For computational purposes, we assume that elements of E and $\mathcal{D}_F(E)$ are represented as coordinate vectors with respect to the usual "polynomial" basis $\{\xi^{i-1}\}_{i=1}^{\ell}$. For $\beta \in E$, let M_β denote the β-multiplication map on E that sends $\alpha \in E$ to $\alpha\beta \in E$, which is an F-linear map from E into E.

   (a) Given as input the polynomial f defining E, along with a projection $\pi \in \mathcal{D}_F(E)$ and an element $\beta \in E$, show how to compute the projection $\pi \circ M_\beta \in \mathcal{D}_F(E)$, using $O(\ell^2)$ operations in F.

   (b) Given as input the polynomial f defining E, along with a projection $\pi \in \mathcal{D}_F(E)$, an element $\alpha \in E$, and a parameter $k > 0$, show how to compute $(\pi(1), \pi(\alpha), \ldots, \pi(\alpha^{k-1}))$ using just $O(k\ell + k^{1/2}\ell^2)$ operations in F, and space for $O(k^{1/2}\ell)$ elements of F. Hint: use the same hint as in Exercise 17.3.

EXERCISE 18.9. Let $f \in F[X]$ be a monic polynomial over a finite field F of degree $\ell > 0$, and let $E := F[X]/(f)$. Show how to use the result of the previous exercise, as well as Exercise 17.3, to get an algorithm that computes the minimal polynomial of $\alpha \in E$ over F using $O(\ell^{2.5})$ expected operations in F, and space for $O(\ell^{1.5})$ operations in F.

EXERCISE 18.10. Let $f \in F[X]$ be a monic polynomial of degree $\ell > 0$ over a field F (not necessarily finite), and let $E := F[X]/(f)$. Further, suppose that f is irreducible, so that E is itself a field. Show how to compute the minimal polynomial of $\alpha \in E$ over F *deterministically*, using algorithms that satisfy the following complexity bounds:

   (a) $O(\ell^3)$ operations in F and space for $O(\ell)$ elements of F;
   (b) $O(\ell^{2.5})$ operations in F and space for $O(\ell^{1.5})$ elements of F.

18.6 The algebra of linear transformations (∗)

Throughout this chapter, one could hear the whispers of the algebra of linear transformations. We develop some of the aspects of this theory here, leaving a number of details as exercises. It will not play a role in any material that follows, but it serves to provide the reader with a "bigger picture."

Let F be a field and V be an F-vector space. We denote by $\mathcal{L}_F(V)$ the set of all F-linear maps from V into V. Thus, $\mathcal{L}_F(V) = \mathrm{Hom}_F(V, V)$, and is a vector space over F, with addition and scalar multiplication defined point-wise (see Theorem 13.12). Elements of $\mathcal{L}_F(V)$ are called **linear transformations**.

   For $\tau, \tau' \in \mathcal{L}_F(V)$, the composed map, $\tau \circ \tau'$, which sends $\alpha \in V$ to $\tau(\tau'(\alpha))$

is also an element of $\mathcal{L}_F(V)$. As always, function composition is associative (i.e., for $\tau, \tau', \tau'' \in \mathcal{L}_F(V)$, we have $\tau \circ (\tau' \circ \tau'') = (\tau \circ \tau') \circ \tau''$); however, function composition is not in general commutative (i.e., we may have $\tau \circ \tau' \neq \tau' \circ \tau$ for some $\tau, \tau' \in \mathcal{L}_F(V)$). The following theorem considers the interaction between composition, addition, and scalar multiplication.

Theorem 18.6. *For all $\tau, \tau', \tau'' \in \mathcal{L}_F(V)$, and for all $c \in F$, we have:*

    (i) $\tau \circ (\tau' + \tau'') = \tau \circ \tau' + \tau \circ \tau''$;

    (ii) $(\tau' + \tau'') \circ \tau = \tau' \circ \tau + \tau'' \circ \tau$;

    (iii) $(c\tau) \circ \tau' = c(\tau \circ \tau') = \tau \circ (c\tau')$.

Proof. Exercise. $\square$

Under the addition operation and scalar multiplication of the vector space $\mathcal{L}_F(V)$, and defining multiplication on $\mathcal{L}_F(V)$ using the "$\circ$" operation, the previous theorem implies that $\mathcal{L}_F(V)$ satisfies all the properties of an F-algebra (see Definition 16.1), except for the fact that multiplication is not commutative (the identity map acts as the multiplicative identity). Thus, we can think of $\mathcal{L}_F(V)$ as a *non-commutative F-algebra*.

Let $\tau \in \mathcal{L}_F(V)$ be a linear transformation. For each integer $i \geq 0$, the map τ^i (i.e., the i-fold composition of τ) is also an element of $\mathcal{L}_F(V)$. Note that τ^0 is by definition just the identity map on V. For each polynomial $g \in F[X]$, with $g = \sum_i a_i X^i$, we denote by $g(\tau)$ the linear transformation

$$g(\tau) := \sum_i a_i \tau^i \in \mathcal{L}_F(V).$$

Thus, for $\alpha \in V$, the value of $g(\tau)$ at α is $\sum_i a_i \tau^i(\alpha)$.

Theorem 18.7. *For all $\tau \in \mathcal{L}_F(V)$, for all $c \in F$, and for all $g, h \in F[X]$, we have:*

    (i) $g(\tau) + h(\tau) = (g + h)(\tau)$;

    (ii) $c \cdot g(\tau) = (cg)(\tau)$;

    (iii) $g(\tau) \circ h(\tau) = (gh)(\tau) = h(\tau) \circ g(\tau)$.

Proof. Exercise. $\square$

Let $\tau \in \mathcal{L}_F(V)$ be a linear transformation. We define

$$F[\tau] := \{g(\tau) : g \in F[X]\},$$

which is a subset of $\mathcal{L}_F(V)$. By the previous theorem, it is clear that $F[\tau]$ is closed under addition, multiplication (i.e., composition), and scalar multiplication, and that $F[\tau]$ is in fact an F-algebra in the usual sense (i.e., multiplication is commutative). Moreover, the expressions $F[\tau]$ and $g(\tau)$ (for $g \in F[X]$) have the same meaning as in §16.1.

Let ϕ_τ be the minimal polynomial of τ over F, so that $F[\tau]$ is isomorphic as an F-algebra to $F[X]/(\phi_\tau)$. We can also characterize ϕ_τ as follows:

> if there exists a non-zero polynomial $g \in F[X]$ such that $g(\tau) = 0$, then ϕ_τ is the monic polynomial of least degree with this property; otherwise, $\phi_\tau = 0$.

Another way to characterize ϕ_τ is as follows:

> ϕ_τ is the minimal polynomial of the sequence $\{\tau^i\}_{i=0}^\infty$.

If V has finite dimension $\ell > 0$, then by Theorem 14.4, $\mathcal{L}_F(V)$ is isomorphic as an F-vector space to $F^{\ell \times \ell}$, and so in particular, has dimension ℓ^2. Therefore, there must be a linear dependence among $1, \tau, \ldots, \tau^{\ell^2}$, which implies that the minimal polynomial of τ is non-zero with degree at most ℓ^2 (and at least 1). We shall show below that in this case, the minimal polynomial of τ actually has degree at most ℓ.

For a fixed $\tau \in \mathcal{L}_F(V)$, we can define a "scalar multiplication" operation $\odot$, that maps $g \in F[X]$ and $\alpha \in V$ to

$$g \odot \alpha := g(\tau)(\alpha) \in V;$$

that is, if $g = \sum_i a_i X^i$, then

$$g \odot \alpha = \sum_i a_i \tau^i(\alpha).$$

Theorem 18.8. *The scalar multiplication $\odot$, together with the usual addition operation on V, makes V into an $F[X]$-module; that is, for all $g, h \in F[X]$ and $\alpha, \beta \in V$, we have*

$$g \odot (h \odot \alpha) = (gh) \odot \alpha, \ (g + h) \odot \alpha = g \odot \alpha + h \odot \alpha,$$
$$g \odot (\alpha + \beta) = g \odot \alpha + g \odot \beta, \ 1 \odot \alpha = \alpha.$$

Proof. Exercise. □

Note that each choice of τ gives rise to a different $F[X]$-module structure, but all of these structures are extensions of the usual vector space structure, in the sense that for all $c \in F$ and $\alpha \in V$, we have $c \odot \alpha = c\alpha$.

Now, for fixed $\tau \in \mathcal{L}_F(V)$ and $\alpha \in V$, consider the $F[X]$-linear map $\rho_{\tau,\alpha}$: $F[X] \to V$ that sends $g \in F[X]$ to $g \odot \alpha = g(\tau)(\alpha)$. The kernel of this map must be a submodule, and hence an ideal, of $F[X]$; since every ideal of $F[X]$ is principal, it follows that $\mathrm{Ker}\, \rho_{\tau,\alpha}$ is the ideal of $F[X]$ generated by some polynomial $\phi_{\tau,\alpha}$, which we can make unique by insisting that it is monic or zero. We call $\phi_{\tau,\alpha}$ the **minimal polynomial of α under τ**. We can also characterize $\phi_{\tau,\alpha}$ as follows:

> if there exists a non-zero polynomial $g \in F[X]$ such that $g(\tau)(\alpha) = 0$, then $\phi_{\tau,\alpha}$ the monic polynomial of least degree with this property; otherwise, $\phi_{\tau,\alpha} = 0$.

Another way to characterize $\phi_{\tau,\alpha}$ is as follows:

> $\phi_{\tau,\alpha}$ is the minimal polynomial of the sequence $\{\tau^i(\alpha)\}_{i=0}^{\infty}$.

Note that since $\phi_\tau(\tau)$ is the zero map, we have

$$\phi_\tau \odot \alpha = \phi_\tau(\tau)(\alpha) = 0,$$

and hence $\phi_\tau \in \text{Ker} \, \rho_{\tau,\alpha}$, which means that $\phi_{\tau,\alpha} \mid \phi_\tau$.

Now consider the image of $\rho_{\tau,\alpha}$, which we shall denote by $\langle \alpha \rangle_\tau$. As an $F[X]$-module, $\langle \alpha \rangle_\tau$ is isomorphic to $F[X]/(\phi_{\tau,\alpha})$. In particular, if $\phi_{\tau,\alpha}$ is non-zero and has degree m, then $\langle \alpha \rangle_\tau$ is a vector space of dimension m over F; indeed, the elements $\alpha, \tau(\alpha), \ldots, \tau^{m-1}(\alpha)$ form a basis for $\langle \alpha \rangle_\tau$ over F; moreover, m is the smallest non-negative integer such that $\{\tau^i(\alpha)\}_{i=0}^{m}$ is linearly dependent.

Observe that for every $\beta \in \langle \alpha \rangle_\tau$, we have $\phi_{\tau,\alpha} \odot \beta = 0$; indeed, if $\beta = g \odot \alpha$, then

$$\phi_{\tau,\alpha} \odot (g \odot \alpha) = (\phi_{\tau,\alpha} g) \odot \alpha = g \odot (\phi_{\tau,\alpha} \odot \alpha) = g \odot 0 = 0.$$

The following three theorems develop some simple facts; the proofs of these are straightforward, and left as exercises. In each theorem, τ is an element of $\mathcal{L}_F(V)$, and $\odot$ is the associated scalar multiplication that makes V into an $F[X]$-module.

Theorem 18.9. *Let $\alpha \in V$ have minimal polynomial $f \in F[X]$ under τ, and let $\beta \in V$ have minimal polynomial $g \in F[X]$ under τ. If $\gcd(f, g) = 1$, then $\langle \alpha \rangle_\tau \cap \langle \beta \rangle_\tau = \{0\}$, and $\alpha + \beta$ has minimal polynomial $f \cdot g$ under τ.*

Theorem 18.10. *Let $\alpha \in V$. Let $f \in F[X]$ be a monic irreducible polynomial such that $f^e \odot \alpha = 0$ but $f^{e-1} \odot \alpha \neq 0$ for some integer $e \geq 1$. Then f^e is the minimal polynomial of α under τ.*

Theorem 18.11. *Let $\alpha \in V$, and suppose that α has minimal polynomial $f \in F[X]$ under τ, with $f \neq 0$. Let $g \in F[X]$. Then $g \odot \alpha$ has minimal polynomial $f / \gcd(f, g)$ under τ.*

We are now ready to state the main result of this section, whose statement and proof are analogous to that of Theorem 6.41:

Theorem 18.12. *Let $\tau \in \mathcal{L}_F(V)$, and suppose that τ has non-zero minimal polynomial ϕ. Then there exists $\beta \in V$ such that the minimal polynomial of β under τ is ϕ.*

Proof. Let $\odot$ be the scalar multiplication associated with τ. Let $\phi = \phi_1^{e_1} \cdots \phi_r^{e_r}$ be the factorization of ϕ into monic irreducible polynomials in $F[X]$.

First, we claim that for each $i = 1, \ldots, r$, there exists $\alpha_i \in V$ such that $\phi/\phi_i \odot \alpha_i \neq 0$. Suppose the claim were false: then for some i, we would have $\phi/\phi_i \odot \alpha = 0$ for all $\alpha \in V$; however, this means that $(\phi/\phi_i)(\tau) = 0$, contradicting the minimality property in the definition of the minimal polynomial ϕ. That proves the claim.

Let $\alpha_1, \ldots, \alpha_r$ be as in the above claim. Then by Theorem 18.10, each $\phi/\phi_i^{e_i} \odot \alpha_i$ has minimal polynomial $\phi_i^{e_i}$ under τ. Finally, by Theorem 18.9,

$$\beta := \phi/\phi_1^{e_1} \odot \alpha_1 + \cdots + \phi/\phi_r^{e_r} \odot \alpha_r$$

has minimal polynomial ϕ under τ. $\square$

Theorem 18.12 says that if τ has minimal polynomial ϕ of degree $m \geq 0$, then there exists $\beta \in V$ such that $\{\tau^i(\beta)\}_{i=0}^{m-1}$ is linearly independent. From this, it immediately follows that:

Theorem 18.13. *If V has finite dimension $\ell > 0$, then for every $\tau \in \mathcal{L}_F(V)$, the minimal polynomial of τ is non-zero of degree at most ℓ.*

We close this section with a simple observation. Let V be an arbitrary $F[X]$-module with scalar multiplication $\odot$. Restricting the scalar multiplication from $F[X]$ to F, we can naturally view V as an F-vector space. Let $\tau : V \to V$ be the map that sends $\alpha \in V$ to $X \odot \alpha$. It is easy to see that $\tau \in \mathcal{L}_F(V)$, and that for all polynomials $g \in F[X]$, and all $\alpha \in V$, we have $g \odot \alpha = g(\tau)(\alpha)$. Thus, instead of starting with a vector space and defining an $F[X]$-module structure in terms of a given linear map, we can go the other direction, starting from an $F[X]$-module and obtaining a corresponding linear map. Furthermore, using the language introduced in Examples 13.19 and 13.20, we see that the $F[X]$-exponent of V is the ideal of $F[X]$ generated by the minimal polynomial of τ, and the $F[X]$-order of any element $\alpha \in V$ is the ideal of $F[X]$ generated by the minimal polynomial of α under τ. Theorem 18.12 says that there exists an element in V whose $F[X]$-order is equal to the $F[X]$-exponent of V, assuming the latter is non-zero.

So depending on one's mood, one can place emphasis either on the linear map τ, or just talk about $F[X]$-modules without mentioning any linear maps.

EXERCISE 18.11. Let $\tau \in \mathcal{L}_F(V)$ have non-zero minimal polynomial ϕ of degree m, and let $\phi = \phi_1^{e_1} \cdots \phi_r^{e_r}$ be the factorization of ϕ into monic irreducible polynomials in $F[X]$. Let $\odot$ be the scalar multiplication associated with τ. Show that $\beta \in V$ has minimal polynomial ϕ under τ if and only if $\phi/\phi_i \odot \beta \neq 0$ for $i = 1, \ldots, r$.

EXERCISE 18.12. Let $\tau \in \mathcal{L}_F(V)$ have non-zero minimal polynomial ϕ. Show that τ is bijective if and only if $X \nmid \phi$.

EXERCISE 18.13. Let F be a finite field, and let V have finite dimension $\ell > 0$ over F. Let $\tau \in \mathcal{L}_F(V)$ have minimal polynomial ϕ, with $\deg(\phi) = m$ (and of course, by Theorem 18.13, we have $m \leq \ell$). Suppose that $\alpha_1, \ldots, \alpha_s$ are randomly chosen elements of V. Let g_j be the minimal polynomial of α_j under τ, for $j = 1, \ldots, s$. Let Q be the probability that $\mathrm{lcm}(g_1, \ldots, g_s) = \phi$. The goal of this exercise is to show that $Q \geq \Lambda_F^\phi(s)$, where $\Lambda_F^\phi(s)$ is as defined in §18.3.

   (a) Using Theorem 18.12 and Theorem 18.11, show that if $m = \ell$, then $Q = \Lambda_F^\phi(s)$.

   (b) Without the assumption that $m = \ell$, things are a bit more challenging. Adopting the matrix-oriented point of view discussed at the end of §18.3, and transposing everything, show that

   – there exists $\pi \in \mathcal{D}_F(V)$ such that the sequence $\{\pi \circ \tau^i\}_{i=0}^\infty$ has minimal polynomial ϕ, and

   – if, for $j = 1, \ldots, s$, we define h_j to be the minimal polynomial of the sequence $\{\pi(\tau^i(\alpha_j))\}_{i=0}^\infty$, then the probability that $\mathrm{lcm}(h_1, \ldots, h_s) = \phi$ is equal to $\Lambda_F^\phi(s)$.

   (c) Show that $h_j \mid g_j$, for $j = 1, \ldots, s$, and conclude that $Q \geq \Lambda_F^\phi(s)$.

EXERCISE 18.14. Let $f, g \in F[X]$ with $f \neq 0$, and let $h := f/\gcd(f, g)$. Show that $g \cdot F[X]/(f)$ and $F[X]/(h)$ are isomorphic as $F[X]$-modules.

EXERCISE 18.15. In this exercise, you are to derive the **fundamental theorem of finite dimensional $F[X]$-modules**, which is completely analogous to the fundamental theorem of finite abelian groups. Both of these results are really special cases of a more general decomposition theorem for modules over a principal ideal domain. Let V be an $F[X]$-module. Assume that as an F-vector space, V has finite dimension $\ell > 0$, and that the $F[X]$-exponent of V is generated by the monic polynomial $\phi \in F[X]$ (note that $1 \leq \deg(\phi) \leq \ell$). Show that there exist monic, non-constant polynomials $\phi_1, \ldots, \phi_t \in F[X]$ such that

   • $\phi_i \mid \phi_{i+1}$ for $i = 1, \ldots, t-1$, and
   • V is isomorphic, as an $F[X]$-module, to the direct product of $F[X]$-modules

$$V' := F[X]/(\phi_1) \times \cdots \times F[X]/(\phi_t).$$

Moreover, show that the polynomials $\phi_1, \ldots, \phi_t$ satisfying these conditions are uniquely determined, and that $\phi_t = \phi$. Hint: one can just mimic the proof of Theorem 6.45, where the exponent of a group corresponds to the $F[X]$-exponent of an $F[X]$-module, and the order of a group element corresponds to the $F[X]$-order of an element of an $F[X]$-module—everything translates rather directly, with just a few minor, technical differences, and the previous exercise is useful in proving the uniqueness part of the theorem.

EXERCISE 18.16. Let us adopt the same assumptions and notation as in Exercise 18.15, and let $\tau \in \mathcal{L}_F(V)$ be the map that sends $\alpha \in V$ to $X \odot \alpha$. Further, let $\sigma : V \to V'$ be the isomorphism of that exercise, and let $\tau' \in \mathcal{L}_F(V')$ be the X-multiplication map on V'.

(a) Show that $\sigma \circ \tau = \tau' \circ \sigma$.

(b) From part (a), derive the following: there exists a basis for V over F, with respect to which the matrix of τ is the "block diagonal" matrix

$$T = \begin{pmatrix} C_1 & & & \\ & C_2 & & \\ & & \ddots & \\ & & & C_t \end{pmatrix},$$

where each C_i is the companion matrix of ϕ_i (see Example 14.1).

EXERCISE 18.17. Let us adopt the same assumptions and notation as in Exercise 18.15.

(a) Using the result of that exercise, show that V is isomorphic, as an $F[X]$-module, to a direct product of $F[X]$-modules

$$F[X]/(f_1^{e_1}) \times \cdots \times F[X]/(f_r^{e_r}),$$

where the f_i's are monic irreducible polynomials (not necessarily distinct) and the e_i's are positive integers, and this direct product is unique up to the order of the factors.

(b) Using part (a), show that there exists a basis for V over F, with respect to which the matrix of τ is the "block diagonal" matrix

$$T' = \begin{pmatrix} C'_1 & & & \\ & C'_2 & & \\ & & \ddots & \\ & & & C'_r \end{pmatrix},$$

where each C'_i is the companion matrix of $f_i^{e_i}$.

EXERCISE 18.18. Let us adopt the same assumptions and notation as in Exercise 18.15.

(a) Suppose $\alpha \in V$ corresponds to $([g_1]_{\phi_1}, \ldots, [g_t]_{\phi_t}) \in V'$ under the isomorphism of that exercise. Show that the $F[X]$-order of α is generated by the polynomial

$$\text{lcm}(\phi_1/\gcd(g_1, \phi_1), \ldots, \phi_t/\gcd(g_t, \phi_t)).$$

(b) Using part (a), give a short and simple proof of the result of Exercise 18.13.

18.7 Notes

Berlekamp [15] and Massey [64] discuss an algorithm for finding the minimal polynomial of a linearly generated sequence that is closely related to the one presented in §18.2, and which has a similar complexity. This connection between Euclid's algorithm and finding minimal polynomials of linearly generated sequences has been observed by many authors, including Mills [68], Welch and Scholtz [108], and Dornstetter [36].

The algorithm presented in §18.3 is due to Wiedemann [109], as are the algorithms for solving sparse linear systems in §18.4, as well as the statement and proof outline of the result in Exercise 18.13.

Our proof of Theorem 18.4 is based on an exposition by Morrison [69].

Using fast matrix and polynomial arithmetic, Shoup [96] shows how to implement the algorithms in §18.5 so as to use just $O(\ell^{(\omega+1)/2})$ operations in F, where ω is the exponent for matrix multiplication (see §14.6), and so $(\omega + 1)/2 < 1.7$.†

---

† The running times of these algorithms can be improved using faster algorithms for modular composition — see footnote on p. 485.

19

Finite fields

This chapter develops some of the basic theory of finite fields. As we already know (see Theorem 7.7), every finite field must be of cardinality p^w, for some prime p and positive integer w. The main results of this chapter are:

- for every prime p and positive integer w, there exists a finite field of cardinality p^w, and

- any two finite fields of the same cardinality are isomorphic.

19.1 Preliminaries

We begin by stating some simple but useful divisibility criteria for polynomials over an arbitrary field. These will play a crucial role in the development of the theory.

Let F be a field. A polynomial $f \in F[X]$ is called **square-free** if it is not divisible by the square of any polynomial of degree greater than zero. Using formal derivatives (see §16.7), we obtain the following useful criterion for establishing that a polynomial is square-free:

Theorem 19.1. *If F is a field, and $f \in F[X]$ with $\gcd(f, \mathbf{D}(f)) = 1$, then f is square-free.*

Proof. Suppose f is not square-free, and write $f = g^2 h$, for $g, h \in F[X]$ with $\deg(g) > 0$. Taking formal derivatives, we have

$$\mathbf{D}(f) = 2g\mathbf{D}(g)h + g^2\mathbf{D}(h),$$

and so clearly, g is a common divisor of f and $\mathbf{D}(f)$. $\square$

Theorem 19.2. *Let F be a field, and let k, ℓ be positive integers. Then $X^k - 1$ divides $X^\ell - 1$ in $F[X]$ if and only if k divides ℓ.*

Proof. Let $\ell = kq + r$, with $0 \le r < k$. We have

$$X^\ell \equiv X^{kq}X^r \equiv X^r \pmod{X^k - 1},$$

and $X^r \equiv 1 \pmod{X^k - 1}$ if and only if $r = 0$. $\square$

Theorem 19.3. *Let $a \ge 2$ be an integer and let k, ℓ be positive integers. Then $a^k - 1$ divides $a^\ell - 1$ if and only if k divides ℓ.*

Proof. The proof is analogous to that of Theorem 19.2. We leave the details to the reader. $\square$

One may combine these last two theorems, obtaining:

Theorem 19.4. *Let $a \ge 2$ be an integer, k, ℓ be positive integers, and F a field. Then $X^{a^k} - X$ divides $X^{a^\ell} - X$ in $F[X]$ if and only if k divides ℓ.*

Proof. Now, $X^{a^k} - X$ divides $X^{a^\ell} - X$ if and only if $X^{a^k-1} - 1$ divides $X^{a^\ell-1} - 1$. By Theorem 19.2, this happens if and only if $a^k - 1$ divides $a^\ell - 1$. By Theorem 19.3, this happens if and only if k divides ℓ. $\square$

We end this section by recalling some concepts discussed earlier, mainly in §16.1, §16.5, and §16.6, that will play an important role in this chapter.

Suppose F is a field, and E is an extension field of F; that is, F is a subfield of E (or, more generally, F is embedded in E via some canonical embedding, and we identify elements of F with their images in E under this embedding). We may view E as an F-algebra via inclusion, and in particular, as an F-vector space. If E' is also an extension field of F, and $\rho : E \to E'$ is a ring homomorphism, then ρ is an F-algebra homomorphism if and only if $\rho(a) = a$ for all $a \in F$.

Let us further assume that as an F-vector space, E has finite dimension ℓ. This dimension ℓ is called the degree of E over F, and is denoted $(E : F)$, and E is called a finite extension of F. Now consider an element $\alpha \in E$. Then α is algebraic over F, which means that there exists a non-zero polynomial $g \in F[X]$ such that $g(\alpha) = 0$. The monic polynomial $\phi \in F[X]$ of least degree such that $\phi(\alpha) = 0$ is called the minimal polynomial of α over F. The polynomial ϕ is irreducible over F, and its degree $m := \deg(\phi)$ is called the degree of α over F. The ring $F[\alpha] = \{g(\alpha) : g \in F[X]\}$, which is the smallest subring of E containing F and α, is actually a field, and is isomorphic, as an F-algebra, to $F[X]/(\phi)$, via the map that sends $g(\alpha) \in F[\alpha]$ to $[g]_\phi \in F[X]/(\phi)$. In particular, $(F[\alpha] : F) = m$, and the elements $1, \alpha, \ldots, \alpha^{m-1}$ form a basis for $F[\alpha]$ over F. Moreover, m divides ℓ.

19.2 The existence of finite fields

Let F be a finite field. As we saw in Theorem 7.7, F must have cardinality p^w, where p is prime and w is a positive integer, and p is the characteristic of F. However, we can say a bit more than this. As discussed in Example 7.53, the field $\mathbb{Z}_p$ is embedded in F, and so we may simply view $\mathbb{Z}_p$ as a subfield of F. Moreover, it must be the case that w is equal to $(F : \mathbb{Z}_p)$.

We want to show that there exist finite fields of every prime-power cardinality. Actually, we shall prove a more general result:

> *If F is a finite field, then for every integer $\ell \geq 1$, there exists an extension field E of degree ℓ over F.*

For the remainder of this section, F denotes a finite field of cardinality $q = p^w$, where p is prime and $w \geq 1$.

Suppose for the moment that E is an extension of degree ℓ over F. Let us derive some basic facts about E. First, observe that E has cardinality q^ℓ. By Theorem 7.29, E^* is cyclic, and the order of E^* is $q^\ell - 1$. If $\gamma \in E^*$ is a generator for E^*, then every non-zero element of E can be expressed as a power of γ; in particular, every element of E can be expressed as a polynomial in γ with coefficients in F; that is, $E = F[\gamma]$. Let $\phi \in F[X]$ be the minimal polynomial of γ over F, which is an irreducible polynomial of degree ℓ. It follows that E is isomorphic (as an F-algebra) to $F[X]/(\phi)$.

So we have shown that every extension of degree ℓ over F must be isomorphic, as an F-algebra, to $F[X]/(f)$ for some irreducible polynomial $f \in F[X]$ of degree ℓ. Conversely, given any irreducible polynomial f over F of degree ℓ, we can construct the finite field $F[X]/(f)$, which has degree ℓ over F. Thus, the question of the existence of a finite field of degree ℓ over F reduces to the question of the existence of an irreducible polynomial over F of degree ℓ.

We begin with a simple generalization of Fermat's little theorem:

Theorem 19.5. *For every $a \in F$, we have $a^q = a$.*

Proof. The multiplicative group of units F^* of F has order $q - 1$, and hence, every $a \in F^*$ satisfies the equation $a^{q-1} = 1$. Multiplying this equation by a yields $a^q = a$ for all $a \in F^*$, and this latter equation obviously holds for $a = 0$ as well. $\square$

This simple fact has a number of consequences.

Theorem 19.6. *We have*

$$X^q - X = \prod_{a \in F}(X - a).$$

Proof. Since each $a \in F$ is a root of $X^q - X$, by Theorem 7.13, the polynomial

$\prod_{a \in F}(X - a)$ divides the polynomial $X^q - X$. Since the degrees and leading coefficients of these two polynomials are the same, the two polynomials must be equal. $\square$

Theorem 19.7. *Let E be an F-algebra. Then the map $\sigma : E \to E$ that sends $\alpha \in E$ to α^q is an F-algebra homomorphism.*

Proof. By Theorem 16.3, either E is trivial or contains an isomorphic copy of F as a subring. In the former case, there is nothing to prove. So assume that E contains an isomorphic copy of F as a subring. It follows that E must have characteristic p.

Since $q = p^w$, we see that $\sigma = \tau^w$, where $\tau(\alpha) := \alpha^p$. By the discussion in Example 7.48, the map τ is a ring homomorphism, and hence so is σ. Moreover, by Theorem 19.5, we have

$$\sigma(c1_E) = (c1_E)^q = c^q 1_E^q = c1_E$$

for all $c \in F$. Thus (see Theorem 16.5), σ is an F-algebra homomorphism. $\square$

The map σ defined in Theorem 19.7 is called the **Frobenius map on E over F**. In the case where E is a finite field, we can say more about it:

Theorem 19.8. *Let E be a finite extension of F, and let σ be the Frobenius map on E over F. Then σ is an F-algebra automorphism on E. Moreover, for all $\alpha \in E$, we have $\sigma(\alpha) = \alpha$ if and only if $\alpha \in F$.*

Proof. The fact that σ is an F-algebra homomorphism follows from the previous theorem. Any ring homomorphism from a field into a field is injective (see Exercise 7.47). Surjectivity follows from injectivity and finiteness.

For the second statement, observe that $\sigma(\alpha) = \alpha$ if and only if α is a root of the polynomial $X^q - X$, and since all q elements of F are already roots, by Theorem 7.14, there can be no other roots. $\square$

As the Frobenius map on finite fields plays a fundamental role in the study of finite fields, let us develop a few simple properties right away. Suppose E is a finite extension of F, and let σ be the Frobenius map on E over F. Since the composition of two F-algebra automorphisms is also an F-algebra automorphism, for every $i \geq 0$, the i-fold composition σ^i, which sends $\alpha \in E$ to $\alpha^{q^i} \in E$, is also an F-algebra automorphism. Since σ is an F-algebra automorphism, the inverse function σ^{-1} is also an F-algebra automorphism. Hence, σ^i is an F-algebra automorphism for all $i \in \mathbb{Z}$. If E has degree ℓ over F, then applying Theorem 19.5 to the field E, we see that σ^ℓ is the identity map. More generally, we have:

Theorem 19.9. *Let E be a extension of degree ℓ over F, and let σ be the Frobenius map on E over F. Then for all integers i and j, we have $\sigma^i = \sigma^j$ if and only if $i \equiv j \pmod{\ell}$.*

Proof. We may assume $i \geq j$. We have

$$\sigma^i = \sigma^j \iff \sigma^{i-j} = \sigma^0 \iff \alpha^{q^{i-j}} - \alpha = 0 \text{ for all } \alpha \in E$$

$$\iff \left(\prod_{\alpha \in E}(X - \alpha)\right) \mid (X^{q^{i-j}} - X) \text{ (by Theorem 7.13)}$$

$$\iff (X^{q^\ell} - X) \mid (X^{q^{i-j}} - X) \text{ (by Theorem 19.6, applied to } E)$$

$$\iff \ell \mid (i - j) \text{ (by Theorem 19.4)}$$

$$\iff i \equiv j \pmod{\ell}. \quad \square$$

From the above theorem, it follows that every power of the Frobenius map σ can be written uniquely as σ^i for some $i = 0, \ldots, \ell - 1$.

The following theorem generalizes Theorem 19.6:

Theorem 19.10. *For $k \geq 1$, let P_k denote the product of all the monic irreducible polynomials in $F[X]$ of degree k. For all positive integers ℓ, we have*

$$X^{q^\ell} - X = \prod_{k \mid \ell} P_k,$$

where the product is over all positive divisors k of ℓ.

Proof. First, we claim that the polynomial $X^{q^\ell} - X$ is square-free. This follows immediately from Theorem 19.1, since $\mathbf{D}(X^{q^\ell} - X) = q^\ell X^{q^\ell - 1} - 1 = -1$.

Thus, we have reduced the proof to showing that if f is a monic irreducible polynomial of degree k, then f divides $X^{q^\ell} - X$ if and only if k divides ℓ.

So let f be a monic irreducible polynomial of degree k. Let $E := F[X]/(f) = F[\xi]$, where $\xi := [X]_f \in E$. Observe that E is an extension field of degree k over F. Let σ be the Frobenius map on E over F.

First, we claim that f divides $X^{q^\ell} - X$ if and only if $\sigma^\ell(\xi) = \xi$. Indeed, f is the minimal polynomial of ξ over F, and so f divides $X^{q^\ell} - X$ if and only if ξ is a root of $X^{q^\ell} - X$, which is the same as saying $\xi^{q^\ell} = \xi$, or equivalently, $\sigma^\ell(\xi) = \xi$.

Second, we claim that $\sigma^\ell(\xi) = \xi$ if and only if $\sigma^\ell(\alpha) = \alpha$ for all $\alpha \in E$. To see this, first suppose that $\sigma^\ell(\alpha) = \alpha$ for all $\alpha \in E$. Then in particular, this holds for $\alpha = \xi$. Conversely, suppose that $\sigma^\ell(\xi) = \xi$. Every $\alpha \in E$ can be written as $\alpha = g(\xi)$ for some $g \in F[X]$, and since σ^ℓ is an F-algebra homomorphism, by Theorem 16.7 we have

$$\sigma^\ell(\alpha) = \sigma^\ell(g(\xi)) = g(\sigma^\ell(\xi)) = g(\xi) = \alpha.$$

Finally, we see that $\sigma^\ell(\alpha) = \alpha$ for all $\alpha \in E$ if and only if $\sigma^\ell = \sigma^0$, which by Theorem 19.9 holds if and only if $k \mid \ell$. $\quad \square$

For $\ell \geq 1$, let $\Pi_F(\ell)$ denote the number of monic irreducible polynomials of degree ℓ in $F[X]$.

Theorem 19.11. *For all $\ell \geq 1$, we have*

$$q^\ell = \sum_{k \mid \ell} k \Pi_F(k). \tag{19.1}$$

Proof. Just equate the degrees of both sides of the identity in Theorem 19.10. $\square$

From Theorem 19.11 it is easy to deduce that $\Pi_F(\ell) > 0$ for all ℓ, and in fact, one can prove a density result—essentially a "prime number theorem" for polynomials over finite fields:

Theorem 19.12. *For all $\ell \geq 1$, we have*

$$\frac{q^\ell}{2\ell} \leq \Pi_F(\ell) \leq \frac{q^\ell}{\ell}, \tag{19.2}$$

and

$$\Pi_F(\ell) = \frac{q^\ell}{\ell} + O\left(\frac{q^{\ell/2}}{\ell}\right). \tag{19.3}$$

Proof. First, since all the terms in the sum on the right hand side of (19.1) are non-negative, and $\ell \Pi_F(\ell)$ is one of these terms, we may deduce that $\ell \Pi_F(\ell) \leq q^\ell$, which proves the second inequality in (19.2). Since this holds for all ℓ, we have

$$\ell \Pi_F(\ell) = q^\ell - \sum_{\substack{k \mid \ell \\ k < \ell}} k \Pi_F(k) \geq q^\ell - \sum_{\substack{k \mid \ell \\ k < \ell}} q^k \geq q^\ell - \sum_{k=1}^{\lfloor \ell/2 \rfloor} q^k.$$

Let us set

$$S(q, \ell) := \sum_{k=1}^{\lfloor \ell/2 \rfloor} q^k = \frac{q}{q-1}(q^{\lfloor \ell/2 \rfloor} - 1),$$

so that $\ell \Pi_F(\ell) \geq q^\ell - S(q, \ell)$. It is easy to see that $S(q, \ell) = O(q^{\ell/2})$, which proves (19.3). For the first inequality of (19.2), it suffices to show that $S(q, \ell) \leq q^\ell/2$. One can verify this directly for $\ell \in \{1, 2, 3\}$, and for $\ell \geq 4$, we have

$$S(q, \ell) \leq q^{\ell/2+1} \leq q^{\ell-1} \leq q^\ell/2. \quad \square$$

We note that the inequalities in (19.2) are tight, in the sense that $\Pi_F(\ell) = q^\ell/2\ell$ when $q = 2$ and $\ell = 2$, and $\Pi_F(\ell) = q^\ell$ when $\ell = 1$. The first inequality in (19.2) implies not only that $\Pi_F(\ell) > 0$, but that the fraction of all monic degree ℓ polynomials that are irreducible is at least $1/2\ell$, while (19.3) says that this fraction gets arbitrarily close to $1/\ell$ as either q or ℓ are sufficiently large.

EXERCISE 19.1. Starting from Theorem 19.11, show that

$$\Pi_F(\ell) = \ell^{-1} \sum_{k|\ell} \mu(k) q^{\ell/k},$$

where μ is the Möbius function (see §2.9).

EXERCISE 19.2. How many irreducible polynomials of degree 30 over $\mathbb{Z}_2$ are there?

19.3 The subfield structure and uniqueness of finite fields

Let E be an extension of degree ℓ over a field F. If K is an intermediate field, that is, a subfield of E containing F, then Theorem 16.23 says that $(E : F) = (E : K)(K : F)$, and so in particular, the degree of K over F divides ℓ.

In the case where F is a finite field, we can say much more about such intermediate fields. Recall that if $\rho : E \to E$ be an F-algebra homomorphism, then the subalgebra of E fixed by ρ is defined as $K := \{\alpha \in E : \rho(\alpha) = \alpha\}$ (see Theorem 16.6). Not only is K a subalgebra of E, but it is also a field, and so K is itself an intermediate field.

Theorem 19.13. *Let E be an extension of degree ℓ over a finite field F. Let σ be the Frobenius map on E over F. Then the intermediate fields K, with $F \subseteq K \subseteq E$, are in one-to-one correspondence with the divisors k of ℓ, where the divisor k corresponds to the subalgebra of E fixed by σ^k, which has degree k over F.*

Proof. Let q be the cardinality of F.

Suppose k is a divisor of ℓ. By Theorem 19.6 (applied to E), the polynomial $X^{q^\ell} - X$ splits into distinct monic linear factors over E. By Theorem 19.4, the polynomial $X^{q^k} - X$ divides $X^{q^\ell} - X$. Hence, $X^{q^k} - X$ also splits into distinct monic linear factors over E. This says that the subalgebra of E fixed by σ^k, which consists of the roots of $X^{q^k} - X$, has precisely q^k elements, and hence is an extension of degree k over F.

Now let K be an arbitrary intermediate field, and let k be the degree of K over F. As already mentioned, we must have $k \mid \ell$. Also, by Theorem 19.8 (applied with K in place of F), K is the subalgebra of E fixed by σ^k. $\square$

The next theorem shows that up to isomorphism, there is only one finite field of a given cardinality.

Theorem 19.14. *Let E and E' be finite extensions of the same degree over a finite field F. Then E and E' are isomorphic as F-algebras.*

Proof. Let q be the cardinality of F, and let ℓ be the degree of the extensions.

As we have argued before, we have $E' = F[\alpha']$ for some $\alpha' \in E'$, and so E' is isomorphic as an F-algebra to $F[X]/(\phi)$, where ϕ is the minimal polynomial of α' over F. As ϕ is an irreducible polynomial of degree ℓ, by Theorem 19.10, ϕ divides $X^{q^\ell} - X$, and by Theorem 19.6 (applied to E), $X^{q^\ell} - X = \prod_{\alpha \in E}(X - \alpha)$, from which it follows that ϕ has a root $\alpha \in E$. Since ϕ is irreducible, ϕ is the minimal polynomial of α over F, and hence $F[\alpha]$ is isomorphic as an F-algebra to $F[X]/(\phi)$. Since α has degree ℓ over F, we must have $E = F[\alpha]$. Thus, $E = F[\alpha] \cong F[X]/(\phi) \cong F[\alpha'] = E'$. $\square$

EXERCISE 19.3. This exercise develops an alternative proof for the existence of finite fields—however, it does not yield a density result for irreducible polynomials. Let F be a finite field of cardinality q, and let $\ell \geq 1$ be an integer. Let E be a splitting field for the polynomial $X^{q^\ell} - X \in F[X]$ (see Theorem 16.25), and let σ be the Frobenius map on E over F. Let K be the subalgebra of E fixed by σ^ℓ. Show that K is an extension of F of degree ℓ.

EXERCISE 19.4. Let E be an extension of degree ℓ over a finite field F of cardinality q. Show that at least half the elements of E have degree ℓ over F, and that the total number of elements of degree ℓ over F is $q^\ell + O(q^{\ell/2})$.

EXERCISE 19.5. Let E be a finite extension of a finite field F, and suppose $\alpha, \beta \in E$, where α has degree a over F, β has degree b over F, and $\gcd(a, b) = 1$. Show that β has degree b over $F[\alpha]$, that α has degree a over $F[\beta]$, and that $\alpha+\beta$ has degree ab over F. Hint: consider the subfields $F[\alpha]$, $F[\beta]$, $F[\alpha][\beta] = F[\alpha, \beta] = F[\beta][\alpha]$, and $F[\alpha + \beta]$, and their degrees over F.

19.4 Conjugates, norms and traces

Throughout this section, F denotes a finite field of cardinality q, E denotes an extension of degree ℓ over F, and σ denotes the Frobenius map on E over F.

Consider an element $\alpha \in E$. We say that $\beta \in E$ is **conjugate to** α (**over** F) if $\beta = \sigma^i(\alpha)$ for some $i \in \mathbb{Z}$. The reader may verify that the "conjugate to" relation is an equivalence relation. We call the equivalence classes of this relation **conjugacy classes**, and we call the elements of the conjugacy class containing α the **conjugates of** α.

Starting with α, we can start listing conjugates:

$$\alpha, \sigma(\alpha), \sigma^2(\alpha), \ldots.$$

As σ^ℓ is the identity map, this list will eventually start repeating. Let k be the smallest positive integer such that $\sigma^k(\alpha) = \sigma^i(\alpha)$ for some $i = 0, \ldots, k - 1$. It must be the case that $i = 0$ — otherwise, applying σ^{-1} to the equation $\sigma^k(\alpha) = \sigma^i(\alpha)$ would yield $\sigma^{k-1}(\alpha) = \sigma^{i-1}(\alpha)$, and since $0 \le i - 1 < k - 1$, this would contradict the minimality of k.

Thus, $\alpha, \sigma(\alpha), \ldots, \sigma^{k-1}(\alpha)$ are all distinct, and $\sigma^k(\alpha) = \alpha$. Moreover, for every integer i, we have $\sigma^i(\alpha) = \sigma^j(\alpha)$, where $j = i \bmod k$. Therefore, the k distinct elements $\alpha, \sigma(\alpha), \ldots, \sigma^{k-1}(\alpha)$ are all the conjugates of α. Also, $\sigma^i(\alpha) = \alpha$ if and only if k divides i, and since $\sigma^\ell(\alpha) = \alpha$, it must be the case that k divides ℓ. In addition, the conjugates of α are powers of α, and in particular, they all belong to $F[\alpha]$.

With α and k as above, consider the polynomial

$$\phi := \prod_{i=0}^{k-1} (X - \sigma^i(\alpha)).$$

The coefficients of ϕ obviously lie in E, but we claim that in fact, they lie in F. This is easily seen as follows. Extend the domain of definition of σ from E to $E[X]$ by applying σ coefficient-wise to polynomials; this yields a ring homomorphism from $E[X]$ into $E[X]$, which we also denote by σ (see Example 7.46). Applying σ to ϕ, we obtain

$$\sigma(\phi) = \prod_{i=0}^{k-1} \sigma(X - \sigma^i(\alpha)) = \prod_{i=0}^{k-1} (X - \sigma^{i+1}(\alpha)) = \prod_{i=0}^{k-1} (X - \sigma^i(\alpha)),$$

since $\sigma^k(\alpha) = \alpha$. Thus we see that $\sigma(\phi) = \phi$. Writing $\phi = \sum_i c_i X^i$, it follows that $\sigma(c_i) = c_i$ for all i, and hence by Theorem 19.8, $c_i \in F$ for all i. Hence $\phi \in F[X]$. We further claim that ϕ is the minimal polynomial of α. To see this, let $f \in F[X]$ be any polynomial over F for which α is a root. Then for every integer i, by Theorem 16.7, we have

$$0 = \sigma^i(0) = \sigma^i(f(\alpha)) = f(\sigma^i(\alpha)).$$

Thus, all the conjugates of α are also roots of f, and so ϕ divides f. That proves that ϕ is the minimal polynomial of α. Since ϕ is the minimal polynomial of α and $\deg(\phi) = k$, it follows that the number k is none other than the degree of α over F.

Let us summarize the above discussion as follows:

Theorem 19.15. *Let $\alpha \in E$ be of degree k over F, and let ϕ be the minimal polynomial of α over F. Then k is the smallest positive integer such that $\sigma^k(\alpha) = \alpha$, the distinct conjugates of α are $\alpha, \sigma(\alpha), \ldots, \sigma^{k-1}(\alpha)$, and ϕ factors over E (in fact, over $F[\alpha]$) as*

$$\phi = \prod_{i=0}^{k-1} (X - \sigma^i(\alpha)).$$

Another useful way of reasoning about conjugates is as follows. First, if $\alpha = 0$, then the degree of α over F is 1, and there is nothing more to say, so let us assume that $\alpha \in E^*$. If r is the multiplicative order of α, then note that every conjugate $\sigma^i(\alpha)$ also has multiplicative order r — this follows from the fact that for every positive integer s, $\alpha^s = 1$ if and only if $(\sigma^i(\alpha))^s = 1$. Also, note that we must have $r \mid |E^*| = q^\ell - 1$, or equivalently, $q^\ell \equiv 1 \pmod{r}$. Focusing now on the fact that σ is the q-power map, we see that the degree k of α is the smallest positive integer such that $\alpha^{q^k} = \alpha$, which holds if and only if $\alpha^{q^k-1} = 1$, which holds if and only if $q^k \equiv 1 \pmod{r}$. Thus, the degree of α over F is simply the multiplicative order of q modulo r. Again, we summarize these observations as a theorem:

Theorem 19.16. *If $\alpha \in E^*$ has multiplicative order r, then the degree of α over F is equal to the multiplicative order of q modulo r.*

For $\alpha \in E$, define the polynomial

$$\chi := \prod_{i=0}^{\ell-1} (X - \sigma^i(\alpha)).$$

It is easy to see, using the same type of argument as was used to prove Theorem 19.15, that $\chi \in F[X]$, and indeed, that

$$\chi = \phi^{\ell/k},$$

where k is the degree of α over F. The polynomial χ is called the **characteristic polynomial of α (from E to F)**.

Two functions that are often useful are the "norm" and "trace." The **norm of α (from E to F)** is defined as

$$\mathbf{N}_{E/F}(\alpha) := \prod_{i=0}^{\ell-1} \sigma^i(\alpha),$$

while the **trace of α (from E to F)** is defined as

$$\mathbf{Tr}_{E/F}(\alpha) := \sum_{i=0}^{\ell-1} \sigma^i(\alpha).$$

It is easy to see that both the norm and trace of α are elements of F, as they are fixed by σ; alternatively, one can see this by observing that they appear, possibly with a minus sign, as coefficients of the characteristic polynomial χ — indeed, the constant term of χ is equal to $(-1)^\ell \mathbf{N}_{E/F}(\alpha)$, and the coefficient of $X^{\ell-1}$ in χ is $-\mathbf{Tr}_{E/F}(\alpha)$.

The following two theorems summarize the most important facts about the norm and trace functions.

Theorem 19.17. *The function* $\mathbf{N}_{E/F}$, *restricted to* E^*, *is a group homomorphism from* E^* *onto* F^*.

Proof. We have

$$\mathbf{N}_{E/F}(\alpha) = \prod_{i=0}^{\ell-1} \alpha^{q^i} = \alpha^{\sum_{i=0}^{\ell-1} q^i} = \alpha^{(q^\ell-1)/(q-1)}.$$

Since E^* is a cyclic group of order $q^\ell - 1$, the image of the $(q^\ell - 1)/(q - 1)$-power map on E^* is the unique subgroup of E^* of order $q - 1$ (see Theorem 6.32). Since F^* is a subgroup of E^* of order $q - 1$, it follows that the image of this power map is F^*. $\square$

Theorem 19.18. *The function* $\mathbf{Tr}_{E/F}$ *is an* F-linear map from E onto F.

Proof. The fact that $\mathbf{Tr}_{E/F}$ is an F-linear map is a simple consequence of the fact that σ is an F-linear map. As discussed above, $\mathbf{Tr}_{E/F}$ maps into F. Since the image of $\mathbf{Tr}_{E/F}$ is a subspace of F, the image is either $\{0\}$ or F, and so it suffices to show that $\mathbf{Tr}_{E/F}$ does not map all of E to zero. But an element $\alpha \in E$ is in the kernel of $\mathbf{Tr}_{E/F}$ if and only if α is a root of the polynomial

$$X + X^q + \cdots + X^{q^{\ell-1}},$$

which has degree $q^{\ell-1}$. Since E contains q^ℓ elements, not all elements of E can lie in the kernel of $\mathbf{Tr}_{E/F}$. $\square$

Example 19.1. As an application of some of the above theory, let us investigate the factorization of the polynomial $X^r - 1$ over F, a finite field of cardinality q. Let us assume that $r > 0$ and is relatively prime to q. Let E be a splitting field of $X^r - 1$ (see Theorem 16.25), so that E is a finite extension of F in which $X^r - 1$ splits into linear factors:

$$X^r - 1 = \prod_{i=1}^{r} (X - \alpha_i).$$

We claim that the roots α_i of $X^r - 1$ are distinct—this follows from the Theorem 19.1 and the fact that $\gcd(X^r - 1, rX^{r-1}) = 1$.

Next, observe that the r roots of $X^r - 1$ in E actually form a subgroup of E^*, and since E^* is cyclic, this subgroup must be cyclic as well. So the roots of $X^r - 1$ form a cyclic subgroup of E^* of order r. Let ζ be a generator for this group. Then all the roots of $X^r - 1$ are contained in $F[\zeta]$, and so we may as well assume that $E = F[\zeta]$.

Let us compute the degree of ζ over F. By Theorem 19.16, the degree ℓ of ζ over F is the multiplicative order of q modulo r. Moreover, the $\varphi(r)$ roots of

$X^r - 1$ of multiplicative order r are partitioned into $\varphi(r)/\ell$ conjugacy classes, each of size ℓ (here, φ is Euler's phi function); indeed, as the reader is urged to verify, these conjugacy classes are in one-to-one correspondence with the cosets of the subgroup of $\mathbb{Z}_r^*$ generated by $[q]_r$, where each such coset $C \subseteq \mathbb{Z}_r^*$ corresponds to the conjugacy class $\{\zeta^a : [a]_r \in C\}$.

More generally, for every $s \mid r$, every root of $X^r - 1$ whose multiplicative order is s has degree k over F, where k is the multiplicative order of q modulo s. As above, the $\varphi(s)$ roots of multiplicative order s are partitioned into $\varphi(s)/k$ conjugacy classes, which are in one-to-one correspondence with the cosets of the subgroup of $\mathbb{Z}_s^*$ generated by $[q]_s$.

This tells us exactly how $X^r - 1$ splits into irreducible factors over F. Things are a bit simpler when r is prime, in which case, from the above discussion, we see that

$$X^r - 1 = (X - 1) \prod_{i=1}^{(r-1)/\ell} f_i,$$

where the f_i's are distinct monic irreducible polynomials, each of degree ℓ, and ℓ is the multiplicative order of q modulo r.

In the above analysis, instead of constructing the field E using Theorem 16.25, one could instead simply construct E as $F[X]/(f)$, where f is any irreducible polynomial of degree ℓ, and where ℓ is the multiplicative order of q modulo r. We know that such a polynomial f exists by Theorem 19.12, and since E has cardinality q^ℓ, and $r \mid (q^\ell - 1) = |E^*|$, and E^* is cyclic, we know that E^* contains an element ζ of multiplicative order r, and each of the r distinct powers $1, \zeta, \ldots, \zeta^{r-1}$ are roots of $X^r - 1$, and so this E is a splitting field of $X^r - 1$ over F. $\square$

EXERCISE 19.6. Let E be an extension of degree ℓ over a finite field F. Show that for $a \in F$, we have $\mathbf{N}_{E/F}(a) = a^\ell$ and $\mathbf{Tr}_{E/F}(a) = \ell a$.

EXERCISE 19.7. Let E be a finite extension of a finite field F. Let K be an intermediate field, $F \subseteq K \subseteq E$. Show that for all $\alpha \in E$

(a) $\mathbf{N}_{E/F}(\alpha) = \mathbf{N}_{K/F}(\mathbf{N}_{E/K}(\alpha))$, and

(b) $\mathbf{Tr}_{E/F}(\alpha) = \mathbf{Tr}_{K/F}(\mathbf{Tr}_{E/K}(\alpha))$.

EXERCISE 19.8. Let F be a finite field, and let $f \in F[X]$ be a monic irreducible polynomial of degree ℓ. Let $E = F[X]/(f) = F[\xi]$, where $\xi := [X]_f$.

(a) Show that

$$\frac{\mathbf{D}(f)}{f} = \sum_{j=1}^{\infty} \mathbf{Tr}_{E/F}(\xi^{j-1})X^{-j}.$$

(b) From part (a), deduce that the sequence of elements

$$\mathbf{Tr}_{E/F}(\xi^{j-1}) \quad (j = 1, 2, \ldots)$$

is linearly generated over F with minimal polynomial f.

(c) Show that one can always choose a polynomial f so that sequence in part (b) is purely periodic with period $q^\ell - 1$.

EXERCISE 19.9. Let F be a finite field, and $f \in F[X]$ a monic irreducible polynomial of degree k over F. Let E be an extension of degree ℓ over F. Show that over E, f factors as the product of d distinct monic irreducible polynomials, each of degree k/d, where $d := \gcd(k, \ell)$.

EXERCISE 19.10. Let E be a finite extension of a finite field F of characteristic p. Show that if $\alpha \in E$ and $0 \neq a \in F$, and if α and $\alpha + a$ are conjugate over F, then p divides the degree of α over F.

EXERCISE 19.11. Let F be a finite field of characteristic p. For $a \in F$, consider the polynomial $f := X^p - X - a \in F[X]$.

(a) Show that if $F = \mathbb{Z}_p$ and $a \neq 0$, then f is irreducible.

(b) More generally, show that if $\mathbf{Tr}_{F/\mathbb{Z}_p}(a) \neq 0$, then f is irreducible, and otherwise, f splits into distinct monic linear factors over F.

EXERCISE 19.12. Let E be a finite extension of a finite field F. Show that every F-algebra automorphism on E must be a power of the Frobenius map on E over F.

EXERCISE 19.13. Show that for all primes p, the polynomial $X^4 + 1$ is reducible in $\mathbb{Z}_p[X]$. (Contrast this to the fact that this polynomial is irreducible in $\mathbb{Q}[X]$, as discussed in Exercise 16.49.)

EXERCISE 19.14. This exercise depends on the concepts and results in §18.6. Let E be an extension of degree ℓ over a finite field F. Let σ be the Frobenius map on E over F.

(a) Show that the minimal polynomial of σ over F is $X^\ell - 1$.

(b) Show that there exists $\beta \in E$ such that the minimal polynomial of β under σ is $X^\ell - 1$.

(c) Conclude that $\beta, \sigma(\beta), \ldots, \sigma^{\ell-1}(\beta)$ form a basis for E over F. This type of basis is called a **normal basis**.

20

Algorithms for finite fields

This chapter discusses efficient algorithms for factoring polynomials over finite fields, and related problems, such as testing if a given polynomial is irreducible, and generating an irreducible polynomial of given degree.

> *Throughout this chapter, F denotes a finite field of characteristic p and cardinality $q = p^w$.*

In addition to performing the usual arithmetic and comparison operations in F, we assume that our algorithms have access to the numbers p, w, and q, and have the ability to generate random elements of F. Generating such a random field element will count as one "operation in F," along with the usual arithmetic operations. Of course, the "standard" ways of representing F as either $\mathbb{Z}_p$ (if $w = 1$), or as the ring of polynomials modulo an irreducible polynomial over $\mathbb{Z}_p$ of degree w (if $w > 1$), satisfy the above requirements, and also allow for the implementation of arithmetic operations in F that take time $O(\text{len}(q)^2)$ on a RAM (using simple, quadratic-time arithmetic for polynomials and integers).

20.1 Tests for and constructing irreducible polynomials

Let $f \in F[X]$ be a monic polynomial of degree $\ell > 0$. We develop here an efficient algorithm that determines if f is irreducible.

The idea is a simple application of Theorem 19.10. That theorem says that for every integer $k \geq 1$, the polynomial $X^{q^k} - X$ is the product of all monic irreducibles whose degree divides k. Thus, $\gcd(X^q - X, f)$ is the product of all the distinct linear factors of f. If f has no linear factors, then $\gcd(X^{q^2} - X, f)$ is the product of all the distinct quadratic irreducible factors of f. And so on. Now, if f is not irreducible, it must be divisible by some irreducible polynomial of degree at most $\ell/2$, and if g is an irreducible factor of f of minimal degree, say k, then we have $k \leq \ell/2$ and $\gcd(X^{q^k} - X, f) \neq 1$. Conversely, if f is irreducible, then $\gcd(X^{q^k} - X, f) = 1$ for all positive integers k up to $\ell/2$. So to test if f is irreducible, it suffices to check if $\gcd(X^{q^k} - X, f) = 1$ for all positive integers k up to $\ell/2$—if so, we may conclude that f is irreducible, and otherwise, we may conclude that f is not irreducible. To carry out the computation efficiently, we note that if $h \equiv X^{q^k} \pmod{f}$, then $\gcd(h - X, f) = \gcd(X^{q^k} - X, f)$.

The above observations suggest the following algorithm.

Algorithm IPT. On input f, where $f \in F[X]$ is a monic polynomial of degree $\ell > 0$, determine if f is irreducible as follows:

> $h \leftarrow X \bmod f$
> for $k \leftarrow 1$ to $\lfloor \ell/2 \rfloor$ do
> $\quad h \leftarrow h^q \bmod f$
> $\quad$ if $\gcd(h - X, f) \neq 1$ then return *false*
> return *true*

The correctness of Algorithm IPT follows immediately from the above discussion. As for the running time, we have:

Theorem 20.1. *Algorithm* IPT *uses* $O(\ell^3 \operatorname{len}(q))$ *operations in* F.

Proof. Consider an execution of a single iteration of the main loop. The cost of the qth-powering step (using a standard repeated-squaring algorithm) is $O(\operatorname{len}(q))$ multiplications modulo f, and so $O(\ell^2 \operatorname{len}(q))$ operations in F. The cost of the gcd computation is $O(\ell^2)$ operations in F. Thus, the cost of a single loop iteration is $O(\ell^2 \operatorname{len}(q))$ operations in F, from which it follows that the cost of the entire algorithm is $O(\ell^3 \operatorname{len}(q))$ operations in F. $\square$

Using a standard representation for F, each operation in F takes time $O(\operatorname{len}(q)^2)$ on a RAM, and so the running time of Algorithm IPT on a RAM is $O(\ell^3 \operatorname{len}(q)^3)$, which means that it is a polynomial-time algorithm.

Let us now consider the related problem of constructing an irreducible polynomial of specified degree $\ell > 0$. To do this, we can simply use the result of Theorem 19.12, which has the following probabilistic interpretation: if we choose a random, monic polynomial f of degree ℓ over F, then the probability that f is irreducible is at least $1/2\ell$. This suggests the following probabilistic algorithm:

Algorithm RIP. On input ℓ, where ℓ is a positive integer, generate a monic irreducible polynomial $f \in F[X]$ of degree ℓ as follows:

repeat
    choose $c_0, \ldots, c_{\ell-1} \in F$ at random
    set $f \leftarrow X^\ell + \sum_{i=0}^{\ell-1} c_i X^i$
    test if f is irreducible using Algorithm IPT
until f is irreducible
output f

Theorem 20.2. *Algorithm RIP uses an expected number of $O(\ell^4 \operatorname{len}(q))$ operations in F, and its output is uniformly distributed over all monic irreducibles of degree ℓ.*

Proof. This is a simple application of the generate-and-test paradigm (see Theorem 9.3, and Example 9.10 in particular). Because of Theorem 19.12, the expected number of loop iterations of the above algorithm is $O(\ell)$. Since Algorithm IPT uses $O(\ell^3 \operatorname{len}(q))$ operations in F, the statement about the running time of Algorithm RIP is immediate. The statement about its output distribution is clear. $\square$

The expected running-time bound in Theorem 20.2 is actually a bit of an overestimate. The reason is that if we generate a random polynomial of degree ℓ, it is likely to have a small irreducible factor, which will be discovered very quickly by Algorithm IPT. In fact, it is known (see §20.7) that the expected value of the degree of the least degree irreducible factor of a random monic polynomial of degree ℓ over F is $O(\operatorname{len}(\ell))$, from which it follows that the expected number of operations in F performed by Algorithm RIP is actually $O(\ell^3 \operatorname{len}(\ell) \operatorname{len}(q))$.

EXERCISE 20.1. Let $f \in F[X]$ be a monic polynomial of degree $\ell > 0$. Also, let $\xi := [X]_f \in E$, where E is the F-algebra $E := F[X]/(f)$.

(a) Given as input $\alpha \in E$ and $\xi^{q^m} \in E$ (for some integer $m > 0$), show how to compute the value $\alpha^{q^m} \in E$, using just $O(\ell^{2.5})$ operations in F, and space for $O(\ell^{1.5})$ elements of F. Hint: see Theorems 16.7 and 19.7, as well as Exercise 17.3.

(b) Given as input $\xi^{q^m} \in E$ and $\xi^{q^{m'}} \in E$, where m and m' are positive integers, show how to compute the value $\xi^{q^{m+m'}} \in E$, using $O(\ell^{2.5})$ operations in F, and space for $O(\ell^{1.5})$ elements of F.

(c) Given as input $\xi^q \in E$ and a positive integer m, show how to compute the value $\xi^{q^m} \in E$, using $O(\ell^{2.5} \operatorname{len}(m))$ operations in F, and space for $O(\ell^{1.5})$ elements of F. Hint: use a repeated-squaring-like algorithm.

EXERCISE 20.2. This exercise develops an alternative irreducibility test.

(a) Show that a monic polynomial $f \in F[X]$ of degree $\ell > 0$ is irreducible if and only if $X^{q^\ell} \equiv X \pmod{f}$ and $\gcd(X^{q^{\ell/s}} - X, f) = 1$ for all primes $s \mid \ell$.

(b) Using part (a) and the result of the previous exercise, show how to determine if f is irreducible using $O(\ell^{2.5} \operatorname{len}(\ell)\omega(\ell) + \ell^2 \operatorname{len}(q))$ operations in F, where $\omega(\ell)$ is the number of distinct prime factors of ℓ.

(c) Show that the operation count in part (b) can be reduced to

$$O(\ell^{2.5} \operatorname{len}(\ell) \operatorname{len}(\omega(\ell)) + \ell^2 \operatorname{len}(q)).$$

Hint: see Exercise 3.39.

EXERCISE 20.3. Design and analyze a *deterministic* algorithm that takes as input a list of irreducible polynomials $f_1, \ldots, f_r \in F[X]$, where $\ell_i := \deg(f_i)$ for $i = 1, \ldots, r$, and assume that $\{\ell_i\}_{i=1}^r$ is pairwise relatively prime. Your algorithm should output an irreducible polynomial $f \in F[X]$ of degree $\ell := \prod_{i=1}^r \ell_i$ using $O(\ell^3)$ operations in F. Hint: use Exercise 19.5.

EXERCISE 20.4. Design and analyze a probabilistic algorithm that, given a monic irreducible polynomial $f \in F[X]$ of degree ℓ as input, generates as output a random monic irreducible polynomial $g \in F[X]$ of degree ℓ (i.e., g should be uniformly distributed over all such polynomials), using an expected number of $O(\ell^{2.5})$ operations in F. Hint: use Exercise 18.9 (or alternatively, Exercise 18.10).

EXERCISE 20.5. Let $f \in F[X]$ be a monic irreducible polynomial of degree ℓ, let $E := F[X]/(f)$, and let $\xi := [X]_f \in E$. Design and analyze a deterministic algorithm that takes as input the polynomial f defining the extension E, and outputs the values

$$s_j := \mathbf{Tr}_{E/F}(\xi^j) \in F \quad (j = 0, \ldots, \ell - 1),$$

using $O(\ell^2)$ operations in F. Here, $\mathbf{Tr}_{E/F}$ is the trace from E to F (see §19.4). Show that given an arbitrary $\alpha \in E$, along with the values $s_0, \ldots, s_{\ell-1}$, one can compute $\mathbf{Tr}_{E/F}(\alpha)$ using just $O(\ell)$ operations in F.

20.2 Computing minimal polynomials in $F[X]/(f)$ (III)

We consider, for the third and final time, the problem considered in §17.2 and §18.5: $f \in F[X]$ is a monic polynomial of degree $\ell > 0$, and $E := F[X]/(f) = F[\xi]$, where $\xi := [X]_f$; we are given an element $\alpha \in E$, and want to compute the minimal polynomial $\phi \in F[X]$ of α over F. We develop an alternative algorithm, based on the theory of finite fields. Unlike the algorithms in §17.2 and §18.5, this algorithm only works when F is finite and the polynomial f is irreducible, so that E is also a finite field.

From Theorem 19.15, we know that the degree of α over F is the smallest positive integer k such that $\alpha^{q^k} = \alpha$. By successive qth powering, we can determine the degree k and compute the conjugates $\alpha, \alpha^q, \ldots, \alpha^{q^{k-1}}$ of α, using $O(k \operatorname{len}(q))$ operations in E, and hence $O(k\ell^2 \operatorname{len}(q))$ operations in F.

Now, we could simply compute the minimal polynomial ϕ by directly using the formula

$$\phi(Y) = \prod_{i=0}^{k-1} (Y - \alpha^{q^i}). \tag{20.1}$$

This would involve computations with polynomials in the variable Y whose coefficients lie in the extension field E, although at the end of the computation, we would end up with a polynomial all of whose coefficients lie in F. The cost of this approach would be $O(k^2)$ operations in E, and hence $O(k^2\ell^2)$ operations in F.

A more efficient approach is the following. Substituting ξ for Y in the identity (20.1), we have

$$\phi(\xi) = \prod_{i=0}^{k-1} (\xi - \alpha^{q^i}).$$

Using this formula, we can compute (given the conjugates of α) the value $\phi(\xi) \in E$ using $O(k)$ operations in E, and hence $O(k\ell^2)$ operations in F. Now, $\phi(\xi)$ is an element of E, and for computational purposes, it is represented as $[g]_f$ for some polynomial $g \in F[X]$ of degree less than ℓ. Moreover, $\phi(\xi) = [\phi]_f$, and hence $\phi \equiv g \pmod{f}$. In particular, if $k < \ell$, then $g = \phi$; otherwise, if $k = \ell$, then $g = \phi - f$. In either case, we can recover ϕ from g with an additional $O(\ell)$ operations in F.

Thus, given the conjugates of α, we can compute ϕ using $O(k\ell^2)$ operations in F. Adding in the cost of computing the conjugates, this gives rise to an algorithm that computes the minimal polynomial of α using $O(k\ell^2 \operatorname{len}(q))$ operations in F.

In the worst case, then, this algorithm uses $O(\ell^3 \operatorname{len}(q))$ operations in F. A reasonably careful implementation needs space for storing a constant number of elements of E, and hence $O(\ell)$ elements of F. For very small values of q, the efficiency of this algorithm will be comparable to that of the algorithm in §18.5, but for large q, it will be much less efficient. Thus, this approach does not really yield a better algorithm, but it does serve to illustrate some of the ideas of the theory of finite fields.

20.3 Factoring polynomials: square-free decomposition

In the remaining sections of this chapter, we develop efficient algorithms for factoring polynomials over the finite field F. We begin in this section with a simple and efficient preprocessing step. Recall that a polynomial is called square-free if it is not divisible by the square of any polynomial of degree greater than zero. This preprocessing algorithm takes the polynomial to be factored, and partially factors it into a product of square-free polynomials. Given this algorithm, we can focus our attention on the problem of factoring square-free polynomials.

Let $f \in F[X]$ be a monic polynomial of degree $\ell > 0$. Suppose that f is not square-free. According to Theorem 19.1, $d := \gcd(f, \mathbf{D}(f)) \neq 1$, where $\mathbf{D}(f)$ is the formal derivative of f; thus, we might hope to get a non-trivial factorization of f by computing d. However, we have to consider the possibility that $d = f$. Can this happen? The answer is "yes," but if it does happen that $d = f$, we can still get a non-trivial factorization of f by other means:

Theorem 20.3. *Suppose that $f \in F[X]$ is a monic polynomial of degree $\ell > 0$, and that $\gcd(f, \mathbf{D}(f)) = f$. Then $f = g(X^p)$ for some $g \in F[X]$. Moreover, if $g = \sum_i a_i X^i$, then $f = h^p$, where*

$$h = \sum_i a_i^{p^{(w-1)}} X^i. \tag{20.2}$$

Proof. Since $\deg(\mathbf{D}(f)) < \deg(f)$ and $\gcd(f, \mathbf{D}(f)) = f$, we must have $\mathbf{D}(f) = 0$. If $f = \sum_i c_i X^i$, then $\mathbf{D}(f) = \sum_i i c_i X^{i-1}$. Since this derivative must be zero, it follows that all the coefficients c_i with $i \not\equiv 0 \pmod{p}$ must be zero to begin with. That proves that $f = g(X^p)$ for some $g \in F[X]$. Furthermore, if h is defined as above, then

$$h^p = \left(\sum_i a_i^{p^{(w-1)}} X^i \right)^p = \sum_i a_i^{p^w} X^{ip} = \sum_i a_i (X^p)^i = g(X^p) = f. \quad \square$$

Our goal now is to design an efficient algorithm that takes as input a monic polynomial $f \in F[X]$ of degree $\ell > 0$, and outputs a list of pairs $((g_1, s_1), \ldots, (g_t, s_t))$, where

- each $g_i \in F[X]$ is monic, non-constant, and square-free,
- each s_i is a positive integer,
- the family of polynomials $\{g_i\}_{i=1}^t$ is pairwise relatively prime, and
- $f = \prod_{i=1}^t g_i^{s_i}$.

We call such a list a **square-free decomposition** of f. There are a number of ways to do this. The algorithm we present is based on the following theorem, which itself is a simple consequence of Theorem 20.3.

Theorem 20.4. *Let $f \in F[X]$ be a monic polynomial of degree $\ell > 0$. Suppose that the factorization of f into irreducibles is $f = f_1^{e_1} \cdots f_r^{e_r}$. Then*

$$\frac{f}{\gcd(f, \mathbf{D}(f))} = \prod_{\substack{1 \leq i \leq r \\ e_i \not\equiv 0 \pmod{p}}} f_i.$$

Proof. The theorem can be restated in terms of the following claim: for each $i = 1, \ldots, r$, we have

- $f_i^{e_i} \mid \mathbf{D}(f)$ if $e_i \equiv 0 \pmod{p}$, and
- $f_i^{e_i-1} \mid \mathbf{D}(f)$ but $f_i^{e_i} \nmid \mathbf{D}(f)$ if $e_i \not\equiv 0 \pmod{p}$.

To prove the claim, we take formal derivatives using the usual rule for products, obtaining

$$\mathbf{D}(f) = \sum_j e_j f_j^{e_j-1} \mathbf{D}(f_j) \prod_{k \neq j} f_k^{e_k}. \qquad (20.3)$$

Consider a fixed index i. Clearly, $f_i^{e_i}$ divides every term in the sum on the right-hand side of (20.3), with the possible exception of the term with $j = i$. In the case where $e_i \equiv 0 \pmod{p}$, the term with $j = i$ vanishes, and that proves the claim in this case. So assume that $e_i \not\equiv 0 \pmod{p}$. By the previous theorem, and the fact that f_i is irreducible, and in particular, not the pth power of any polynomial, we see that $\mathbf{D}(f_i)$ is non-zero, and (of course) has degree strictly less than that of f_i. From this, and (again) the fact that f_i is irreducible, it follows that the term with $j = i$ is divisible by $f_i^{e_i-1}$, but not by $f_i^{e_i}$, from which the claim follows. $\square$

This theorem provides the justification for the following square-free decomposition algorithm.

Algorithm SFD. On input f, where $f \in F[X]$ is a monic polynomial of degree $\ell > 0$, compute a square-free decomposition of f as follows:

> initialize an empty list L
> $s \leftarrow 1$
> repeat
> > $j \leftarrow 1$, $g \leftarrow f/\gcd(f, \mathbf{D}(f))$
> > while $g \neq 1$ do
> > > $f \leftarrow f/g$, $h \leftarrow \gcd(f, g)$, $m \leftarrow g/h$
> > > if $m \neq 1$ then append (m, js) to L
> > > $g \leftarrow h$, $j \leftarrow j+1$
> >
> > if $f \neq 1$ then // *f is a pth power*
> > > // *compute a pth root as in (20.2)*
> > > $f \leftarrow f^{1/p}$, $s \leftarrow ps$
> until $f = 1$
> output L

Theorem 20.5. *Algorithm* SFD *correctly computes a square-free decomposition of f using $O(\ell^2 + \ell(w-1)\operatorname{len}(p)/p)$ operations in F.*

Proof. Let $f = \prod_i f_i^{e_i}$ be the factorization of the input f into irreducibles. Let S

be the set of indices i such that $e_i \not\equiv 0 \pmod{p}$, and let S' be the set of indices i such that $e_i \equiv 0 \pmod{p}$. Also, for $j \geq 1$, let $S_{\geq j} := \{i \in S : e_i \geq j\}$ and $S_{=j} := \{i \in S : e_i = j\}$.

Consider the first iteration of the main loop. By Theorem 20.4, the value first assigned to g is $\prod_{i \in S} f_i$. It is straightforward to prove by induction on j that at the beginning of the jth iteration of the inner while loop, the value assigned to g is $\prod_{i \in S_{\geq j}} f_i$, and the value assigned to f is $\prod_{i \in S_{\geq j}} f_i^{e_i - j + 1} \cdot \prod_{i \in S'} f_i^{e_i}$. Moreover, in the jth loop iteration, the value assigned to m is $\prod_{i \in S_{=j}} f_i$. It follows that when the while loop terminates, the value assigned to f is $\prod_{i \in S'} f_i^{e_i}$, and the value assigned to L is a square-free decomposition of $\prod_{i \in S} f_i^{e_i}$; if f does not equal 1 at this point, then subsequent iterations of the main loop will append to L a square-free decomposition of $\prod_{i \in S'} f_i^{e_i}$.

That proves the correctness of the algorithm. Now consider its running time. Again, consider just the first iteration of the main loop. The cost of computing $f / \gcd(f, \mathbf{D}(f))$ is at most $C_1 \ell^2$ operations in F, for some constant C_1. Now consider the cost of the inner while loop. It is not hard to see that the cost of the jth iteration of the inner while loop is at most

$$C_2 \ell \sum_{i \in S_{\geq j}} \deg(f_i)$$

operations in F, for some constant C_2. This follows from the observation in the previous paragraph that the value assigned to g is $\prod_{i \in S_{\geq j}} f_i$, along with our usual cost estimates for division and Euclid's algorithm. Therefore, the total cost of all iterations of the inner while loop is at most

$$C_2 \ell \sum_{j \geq 1} \sum_{i \in S_{\geq j}} \deg(f_i)$$

operations in F. In this double summation, for each $i \in S$, the term $\deg(f_i)$ is counted exactly e_i times, and so we can write this cost estimate as

$$C_2 \ell \sum_{i \in S} e_i \deg(f_i) \leq C_2 \ell^2.$$

Finally, it is easy to see that in the if-then statement at the end of the main loop body, if the algorithm does in fact compute a pth root, then this takes at most

$$C_3 \ell (w - 1) \operatorname{len}(p) / p$$

operations in F, for some constant C_3. Thus, we have shown that the total cost of the first iteration of the main loop is at most

$$(C_1 + C_2) \ell^2 + C_3 \ell (w - 1) \operatorname{len}(p) / p$$

operations in F. If the main loop is executed a second time, the degree of f at the start of the second iteration is at most ℓ/p, and hence the cost of the second loop iteration is at most

$$(C_1 + C_2)(\ell/p)^2 + C_3(\ell/p)(w-1)\operatorname{len}(p)/p$$

operations in F. More generally, for $t = 1, 2, \ldots$, the cost of loop iteration t is at most

$$(C_1 + C_2)(\ell/p^{t-1})^2 + C_3(\ell/p^{t-1})(w-1)\operatorname{len}(p)/p,$$

operations in F, and summing over all $t \geq 1$ yields the stated bound. $\quad\square$

20.4 Factoring polynomials: the Cantor–Zassenhaus algorithm

In this section, we present an algorithm due to Cantor and Zassenhaus for factoring a given polynomial over the finite field F into irreducibles. We shall assume that the input polynomial is square-free, using Algorithm SFD in §20.3 as a preprocessing step, if necessary. The algorithm has two stages:

Distinct Degree Factorization: The input polynomial is decomposed into factors so that each factor is a product of distinct irreducibles of the same degree (and the degree of those irreducibles is also determined).

Equal Degree Factorization: Each of the factors produced in the distinct degree factorization stage are further factored into their irreducible factors.

The algorithm we present for distinct degree factorization is a deterministic, polynomial-time algorithm. The algorithm we present for equal degree factorization is a *probabilistic* algorithm that runs in expected polynomial time (and whose output is always correct).

20.4.1 Distinct degree factorization

The problem, more precisely stated, is this: given a monic, square-free polynomial $f \in F[X]$ of degree $\ell > 0$, produce a list of pairs $((g_1, k_1), \ldots, (g_t, k_t))$ where

- each g_i is the product of monic irreducible polynomials of degree k_i, and
- $f = \prod_{i=1}^{t} g_i$.

This problem can be easily solved using Theorem 19.10, using a simple variation of the algorithm we discussed in §20.1 for irreducibility testing. The basic idea is this. We can compute $g := \gcd(X^q - X, f)$, so that g is the product of all the linear factors of f. After removing all linear factors from f, we next compute $\gcd(X^{q^2} - X, f)$, which will be the product of all the quadratic irreducibles dividing f, and we can remove these from f—although $X^{q^2} - X$ is the product of all linear and quadratic irreducibles, since we have already removed the linear factors from f, the gcd will give us just the quadratic factors of f. In general, for $k = 1, \ldots, \ell$, having removed all the irreducible factors of degree less than k from f, we compute $\gcd(X^{q^k} - X, f)$ to obtain the product of all the irreducible factors of f of degree k, and then remove these from f.

The above discussion leads to the following algorithm for distinct degree factorization.

Algorithm DDF. On input f, where $f \in F[X]$ is a monic square-free polynomial of degree $\ell > 0$, do the following:

> initialize an empty list L
> $h \leftarrow X \bmod f$
> $k \leftarrow 0$
> while $f \neq 1$ do
> > $h \leftarrow h^q \bmod f$, $k \leftarrow k + 1$
> > $g \leftarrow \gcd(h - X, f)$
> > if $g \neq 1$ then
> > > append (g, k) to L
> > > $f \leftarrow f/g$
> > > $h \leftarrow h \bmod f$
>
> output L

The correctness of Algorithm DDF follows from the discussion above. As for the running time:

Theorem 20.6. *Algorithm DDF uses $O(\ell^3 \operatorname{len}(q))$ operations in F.*

Proof. Note that the body of the main loop is executed at most ℓ times, since after ℓ iterations, we will have removed all the factors of f. Thus, we perform at most ℓ qth-powering steps, each of which takes $O(\ell^2 \operatorname{len}(q))$ operations in F, and so the total contribution to the running time of these is $O(\ell^3 \operatorname{len}(q))$ operations in F. We also have to take into account the cost of the gcd and division computations. The cost per loop iteration of these is $O(\ell^2)$ operations in F, contributing a term of $O(\ell^3)$ to the total operation count. This term is dominated by the cost of the qth-powering steps, and so the total cost of Algorithm DDF is $O(\ell^3 \operatorname{len}(q))$ operations in F. $\square$

20.4.2 Equal degree factorization

The problem, more precisely stated, is this: given a monic polynomial $f \in F[X]$ of degree $\ell > 0$, and an integer $k > 0$, such that f is of the form

$$f = f_1 \cdots f_r$$

for distinct monic irreducible polynomials $f_1, \ldots, f_r$, each of degree k, compute these irreducible factors of f. Note that given f and k, the value of r is easily determined, since $r = \ell/k$.

We begin by discussing the basic mathematical ideas that will allow us to efficiently split f into two non-trivial factors, and then we present a somewhat more elaborate algorithm that completely factors f.

By the Chinese remainder theorem, we have an F-algebra isomorphism

$$\theta: \quad E \to E_1 \times \cdots \times E_r$$
$$[g]_f \mapsto ([g]_{f_1}, \ldots, [g]_{f_r}),$$

where E is the F-algebra $F[X]/(f)$, and for $i = 1, \ldots, r$, E_i is the extension field $F[X]/(f_i)$ of degree k over F.

Recall that $q = p^w$. We have to treat the cases $p = 2$ and $p > 2$ separately. We first treat the case $p = 2$. Let us define the polynomial

$$M_k := \sum_{j=0}^{wk-1} X^{2^j} \in F[X]. \tag{20.4}$$

(The algorithm in the case $p > 2$ will only differ in the definition of M_k.)

For $\alpha \in E$, if $\theta(\alpha) = (\alpha_1, \ldots, \alpha_r)$, then we have

$$\theta(M_k(\alpha)) = M_k(\theta(\alpha)) = (M_k(\alpha_1), \ldots, M_k(\alpha_r)).$$

Note that each E_i is an extension of $\mathbb{Z}_2$ of degree wk, and that

$$M_k(\alpha_i) = \sum_{j=0}^{wk-1} \alpha_i^{2^j} = \mathrm{Tr}_{E_i/\mathbb{Z}_2}(\alpha_i),$$

where $\mathrm{Tr}_{E_i/\mathbb{Z}_2} : E_i \to \mathbb{Z}_2$ is the trace from E_i to $\mathbb{Z}_2$, which is a surjective, $\mathbb{Z}_2$-linear map (see §19.4).

Now, suppose we choose $\alpha \in E$ at random. Then if $\theta(\alpha) = (\alpha_1, \ldots, \alpha_r)$, the family of random variables $\{\alpha_i\}_{i=1}^r$ is mutually independent, with each α_i uniformly distributed over E_i. It follows that the family of random variables $\{M_k(\alpha_i)\}_{i=1}^r$ is mutually independent, with each $M_k(\alpha_i)$ uniformly distributed over $\mathbb{Z}_2$. Thus, if $g := \mathrm{rep}(M_k(\alpha))$ (i.e., $g \in F[X]$ is the polynomial of degree less than ℓ such that $M_k(\alpha) = [g]_f$), then $\gcd(g, f)$ will be the product of those factors f_i of f such that $M_k(\alpha_i) = 0$. We will fail to get a non-trivial factorization only if the $M_k(\alpha_i)$

are either all 0 or all 1, which for $r \geq 2$ happens with probability at most $1/2$ (the worst case being when $r = 2$).

That is our basic splitting strategy. The algorithm for completely factoring f works as follows. The algorithm proceeds in stages. At any stage, we have a partial factorization $f = \prod_{h \in H} h$, where H is a set of non-constant, monic polynomials. Initially, $H = \{f\}$. With each stage, we attempt to get a finer factorization of f by trying to split each $h \in H$ using the above splitting strategy—if we succeed in splitting h into two non-trivial factors, then we replace h by these two factors. We continue in this way until $|H| = r$.

Here is the full equal degree factorization algorithm.

Algorithm EDF. On input f, k, where $f \in F[X]$ is a monic polynomial of degree $\ell > 0$, and k is a positive integer, such that f is the product of $r := \ell/k$ distinct monic irreducible polynomials, each of degree k, do the following, with M_k as defined in (20.4):

> $H \leftarrow \{f\}$
> while $|H| < r$ do
> > $H' \leftarrow \emptyset$
> > for each $h \in H$ do
> > > choose $\alpha \in F[X]/(h)$ at random
> > > $d \leftarrow \gcd(\mathrm{rep}(M_k(\alpha)), h)$
> > > if $d = 1$ or $d = h$
> > > > then $H' \leftarrow H' \cup \{h\}$
> > > > else $H' \leftarrow H' \cup \{d, h/d\}$
> > $H \leftarrow H'$
> output H

The correctness of the algorithm is clear from the above discussion. As for its expected running time, we can get a quick-and-dirty upper bound as follows:

- For a given h and $\alpha \in F[X]/(h)$, the value $M_k(\alpha)$ can be computed using $O(k \deg(h)^2 \, \mathrm{len}(q))$ operations in F, and so the number of operations in F performed in each iteration of the main loop is at most a constant times

$$k \, \mathrm{len}(q) \sum_{h \in H} \deg(h)^2 \leq k \, \mathrm{len}(q) \left(\sum_{h \in H} \deg(h) \right)^2 = k \ell^2 \, \mathrm{len}(q).$$

- The expected number of iterations of the main loop until we get some non-trivial split is $O(1)$.
- The algorithm finishes after getting $r - 1$ non-trivial splits.

- Therefore, the total expected cost is $O(rk\ell^2 \operatorname{len}(q))$, or $O(\ell^3 \operatorname{len}(q))$, operations in F.

This analysis gives a bit of an over-estimate — it does not take into account the fact that we expect to get fairly "balanced" splits. For the purposes of analyzing the overall running time of the Cantor–Zassenhaus algorithm, this bound suffices; however, the following analysis gives a tight bound on the complexity of Algorithm EDF.

Theorem 20.7. *In the case $p = 2$, Algorithm EDF uses an expected number of $O(k\ell^2 \operatorname{len}(q))$ operations in F.*

Proof. We may assume $r \geq 2$. Let L be the random variable that represents the number of iterations of the main loop of the algorithm. For $n \geq 1$, let H_n be the random variable that represents the value of H at the beginning of the nth loop iteration. For $i, j = 1, \ldots, r$, we define L_{ij} to be the largest value of n (with $1 \leq n \leq L$) such that $f_i \mid h$ and $f_j \mid h$ for some $h \in H_n$.

We first claim that $E[L] = O(\operatorname{len}(r))$. To prove this claim, we make use of the fact (see Theorem 8.17) that

$$E[L] = \sum_{n \geq 1} P[L \geq n].$$

Now, $L \geq n$ if and only if for some i, j with $1 \leq i < j \leq r$, we have $L_{ij} \geq n$. Moreover, if f_i and f_j have not been separated at the beginning of one loop iteration, then they will be separated at the beginning of the next with probability $1/2$. It follows that

$$P[L_{ij} \geq n] = 2^{-(n-1)}.$$

So we have

$$P[L \geq n] \leq \sum_{i<j} P[L_{ij} \geq n] \leq r^2 2^{-n}.$$

Therefore,

$$E[L] = \sum_{n \geq 1} P[L \geq n] = \sum_{n \leq 2 \log_2 r} P[L \geq n] + \sum_{n > 2 \log_2 r} P[L \geq n]$$

$$\leq 2 \log_2 r + \sum_{n > 2 \log_2 r} r^2 2^{-n} \leq 2 \log_2 r + \sum_{n \geq 0} 2^{-n} = 2 \log_2 r + 2,$$

which proves the claim.

As discussed in the paragraph above this theorem, the cost of each iteration of the main loop is $O(k\ell^2 \operatorname{len}(q))$ operations in F. Combining this with the fact that $E[L] = O(\operatorname{len}(r))$, it follows that the expected number of operations in F for the entire algorithm is $O(\text{len}(r)k\ell^2 \text{ len}(q))$. This is significantly better than the above quick-and-dirty estimate, but is not quite the result we are after. For this, we have to work a little harder.

For each polynomial h dividing f, define $\omega(h)$ to be the number of irreducible factors of h. Let us also define the random variable

$$S := \sum_{n=1}^{L} \sum_{h \in H_n} \omega(h)^2.$$

It is easy to see that the total number of operations performed by the algorithm is $O(Sk^3 \text{ len}(q))$, and so it will suffice to show that $E[S] = O(r^2)$.

We claim that

$$S = \sum_{i,j} L_{ij},$$

where the sum is over all $i, j = 1, \ldots, r$. To see this, define $\delta_{ij}(h)$ to be 1 if both f_i and f_j divide h, and 0 otherwise. Then we have

$$S = \sum_{n} \sum_{h \in H_n} \sum_{i,j} \delta_{ij}(h) = \sum_{i,j} \sum_{n} \sum_{h \in H_n} \delta_{ij}(h) = \sum_{i,j} L_{ij},$$

which proves the claim.

We can write

$$S = \sum_{i \neq j} L_{ij} + \sum_{i} L_{ii} = \sum_{i \neq j} L_{ij} + rL.$$

For $i \neq j$, we have

$$E[L_{ij}] = \sum_{n \geq 1} P[L_{ij} \geq n] = \sum_{i \geq 1} 2^{-(n-1)} = 2,$$

and so

$$E[S] = \sum_{i \neq j} E[L_{ij}] + r\, E[L] = 2r(r-1) + O(r \text{ len}(r)) = O(r^2).$$

That proves the theorem. $\square$

That completes the discussion of Algorithm EDF in the case $p = 2$. Now assume that $p > 2$, so that p, and hence also q, is odd. Algorithm EDF in this case is exactly the same as above, except that in this case, we define the polynomial M_k as

$$M_k := X^{(q^k-1)/2} - 1 \in F[X]. \tag{20.5}$$

Just as before, for $\alpha \in E$ with $\theta(\alpha) = (\alpha_1, \ldots, \alpha_r)$, we have

$$\theta(M_k(\alpha)) = M_k(\theta(\alpha)) = (M_k(\alpha_1), \ldots, M_k(\alpha_r)).$$

Note that each group E_i^* is a cyclic group of order $q^k - 1$, and therefore, the image of the $(q^k - 1)/2$-power map on E_i^* is $\{\pm 1\}$.

Now, suppose we choose $\alpha \in E$ at random. Then if $\theta(\alpha) = (\alpha_1, \dots, \alpha_r)$, the family of random variables $\{\alpha_i\}_{i=1}^r$ is mutually independent, with each α_i uniformly distributed over E_i. It follows that the family of random variables $\{M_k(\alpha_i)\}_{i=1}^r$ is mutually independent. If $\alpha_i = 0$, which happens with probability $1/q^k$, then $M_k(\alpha_i) = -1$; otherwise, $\alpha_i^{(q^k-1)/2}$ is uniformly distributed over $\{\pm 1\}$, and so $M_k(\alpha_i)$ is uniformly distributed over $\{0, -2\}$. That is to say,

$$M_k(\alpha_i) = \begin{cases} 0 & \text{with probability } (q^k - 1)/2q^k, \\ -1 & \text{with probability } 1/q^k, \\ -2 & \text{with probability } (q^k - 1)/2q^k. \end{cases}$$

Thus, if $g := \mathrm{rep}(M_k(\alpha))$, then $\gcd(g, f)$ will be the product of those factors f_i of f such that $M_k(\alpha_i) = 0$. We will fail to get a non-trivial factorization only if the $M_k(\alpha_i)$ are either all zero or all non-zero. Assume $r \geq 2$. Consider the worst case, namely, when $r = 2$. In this case, a simple calculation shows that the probability that we fail to split these two factors is

$$\left(\frac{q^k - 1}{2q^k}\right)^2 + \left(\frac{q^k + 1}{2q^k}\right)^2 = \frac{1}{2}(1 + 1/q^{2k}).$$

The (very) worst case is when $q^k = 3$, in which case the probability of failure is at most $5/9$.

The same quick-and-dirty analysis given just above Theorem 20.7 applies here as well, but just as before, we can do better:

Theorem 20.8. *In the case $p > 2$, Algorithm EDF uses an expected number of $O(k\ell^2 \operatorname{len}(q))$ operations in F.*

Proof. The analysis is essentially the same as in the case $p = 2$, except that now the probability that we fail to split a given pair of irreducible factors is at most $5/9$, rather than equal to $1/2$. The details are left as an exercise for the reader. $\square$

20.4.3 Analysis of the whole algorithm

Given an arbitrary monic square-free polynomial $f \in F[X]$ of degree $\ell > 0$, the distinct degree factorization step takes $O(\ell^3 \operatorname{len}(q))$ operations in F. This step produces a number of polynomials that must be further subjected to equal degree factorization. If there are t such polynomials, where the ith polynomial has degree ℓ_i, for $i = 1, \dots, t$, then $\sum_{i=1}^t \ell_i = \ell$. Now, the equal degree factorization step for the ith polynomial takes an expected number of $O(\ell_i^3 \operatorname{len}(q))$ operations in F (actually, our initial, "quick and dirty" estimate is good enough here), and so it follows that the total expected cost of all the equal degree factorization steps is $O(\sum_i \ell_i^3 \, \text{len}(q))$, which is $O(\ell^3 \, \text{len}(q))$, operations in F. Putting this all together, we conclude:

Theorem 20.9. *The Cantor–Zassenhaus factoring algorithm uses an expected number of $O(\ell^3 \, \text{len}(q))$ operations in F.*

This bound is tight, since in the worst case, when the input is irreducible, the algorithm really does do this much work. Also, we have assumed the input to the Cantor–Zassenhaus is a square-free polynomial. However, we may use Algorithm SFD as a preprocessing step to ensure that this is the case. Even if we include the cost of this preprocessing step, the running time estimate in Theorem 20.9 remains valid.

EXERCISE 20.6. Show how to modify Algorithm DDF so that the main loop halts as soon as $2k > \deg(f)$.

EXERCISE 20.7. Suppose that in Algorithm EDF, we replace the two lines

> for each $h \in H$ do
> > choose $\alpha \in F[X]/(h)$ at random

by the following:

> choose $a_0, \dots, a_{2k-1} \in F$ at random
> $g \leftarrow \sum_{j=0}^{2k-1} a_j X^j \in F[X]$
> for each $h \in H$ do
> > $\alpha \leftarrow [g]_h \in F[X]/(h)$

Show that the expected running time bound of Theorem 20.6 still holds (you may assume $p = 2$ for simplicity).

EXERCISE 20.8. This exercise extends the techniques developed in Exercise 20.1. Let $f \in F[X]$ be a monic polynomial of degree $\ell > 0$, and let $\xi := [X]_f \in E$, where $E := F[X]/(f)$. For each integer $m > 0$, define polynomials

$$T_m := X + X^q + \cdots + X^{q^{m-1}} \in F[X] \quad \text{and} \quad N_m := X \cdot X^q \cdots \cdots X^{q^{m-1}} \in F[X].$$

(a) Given as input $\xi^{q^m} \in E$ and $\xi^{q^{m'}} \in E$, where m and m' are positive integers, along with $T_m(\alpha)$ and $T_{m'}(\alpha)$, for some $\alpha \in E$, show how to compute the values $\xi^{q^{m+m'}}$ and $T_{m+m'}(\alpha)$, using $O(\ell^{2.5})$ operations in F, and space for $O(\ell^{1.5})$ elements of F.

(b) Given as input $\xi^q \in E$, $\alpha \in E$, and a positive integer m, show how to compute (using part (a)) the value $T_m(\alpha)$, using $O(\ell^{2.5} \operatorname{len}(m))$ operations in F, and space for $O(\ell^{1.5})$ elements of F.

(c) Repeat parts (a) and (b), except with "N" in place of "T."

EXERCISE 20.9. Using the result of the previous exercise, show how to implement Algorithm EDF so that it uses an expected number of

$$O(\operatorname{len}(k)\ell^{2.5} + \ell^2 \operatorname{len}(q))$$

operations in F, and space for $O(\ell^{1.5})$ elements of F.

EXERCISE 20.10. This exercise depends on the concepts and results in §18.6. Let E be an extension field of degree ℓ over F, specified by an irreducible polynomial of degree ℓ over F. Design and analyze an efficient probabilistic algorithm that finds a normal basis for E over F (see Exercise 19.14). Hint: there are a number of approaches to solving this problem; one way is to start by factoring $X^\ell - 1$ over F, and then turn the construction in Theorem 18.12 into an efficient probabilistic procedure; if you mimic Exercise 11.2, your entire algorithm should use $O(\ell^3 \operatorname{len}(\ell) \operatorname{len}(q))$ operations in F (or $O(\operatorname{len}(r)\ell^3 \operatorname{len}(q))$ operations, where r is the number of distinct irreducible factors of $X^\ell - 1$ over F).

20.5 Factoring polynomials: Berlekamp's algorithm

We now develop an alternative algorithm, due to Berlekamp, for factoring a polynomial over the finite field F into irreducibles. We shall assume that the input polynomial is square-free, using Algorithm SFD in §20.3 as a preprocessing step, if necessary.

Let us now assume we have a monic square-free polynomial $f \in F[X]$ of degree $\ell > 0$ that we want to factor into irreducibles. We first present the mathematical ideas underpinning the algorithm.

Let E be the F-algebra $F[X]/(f)$. Let σ be the Frobenius map on E over F, which maps $\alpha \in E$ to $\alpha^q \in E$. We know that σ is an F-algebra homomorphism (see Theorem 19.7). Consider the subalgebra B of E fixed by σ (see Theorem 16.6). Thus,

$$B = \{\alpha \in E : \alpha^q = \alpha\}.$$

The subalgebra B is called the **Berlekamp subalgebra of** E. Let us take a closer look at it. Suppose that f factors into irreducibles as

$$f = f_1 \cdots f_r,$$

and let

$$\theta: \quad E \to E_1 \times \cdots \times E_r$$

$$[g]_f \mapsto ([g]_{f_1}, \ldots, [g]_{f_r})$$

be the F-algebra isomorphism from the Chinese remainder theorem, where $E_i := F[X]/(f_i)$ is an extension field of F of finite degree for $i = 1, \ldots, r$. Now, for $\alpha \in E$, if $\theta(\alpha) = (\alpha_1, \ldots, \alpha_r)$, then we have $\alpha^q = \alpha$ if and only if $\alpha_i^q = \alpha_i$ for $i = 1, \ldots, r$; moreover, by Theorem 19.8, we know that for all $\alpha_i \in E_i$, we have $\alpha_i^q = \alpha_i$ if and only if $\alpha_i \in F$. Thus, we may characterize B as follows:

$$B = \{\theta^{-1}(c_1, \ldots, c_r) : c_1, \ldots, c_r \in F\}.$$

Since B is a subalgebra of E, then as F-vector spaces, B is a subspace of E. Of course, E has dimension ℓ over F, with the natural basis $\{\xi^{i-1}\}_{i=1}^{\ell}$, where $\xi := [X]_f$. As for the Berlekamp subalgebra, from the above characterization of B, it is evident that the elements

$$\theta^{-1}(1, 0, \ldots, 0), \; \theta^{-1}(0, 1, 0, \ldots, 0), \; \ldots, \; \theta^{-1}(0, \ldots, 0, 1)$$

form a basis for B over F, and hence, B has dimension r over F.

Now we come to the actual factoring algorithm.

Stage 1: Construct a basis for B

The first stage of Berlekamp's factoring algorithm constructs a basis for B over F. We can easily do this using Gaussian elimination, as follows. Let $\rho : E \to E$ be the map that sends $\alpha \in E$ to $\sigma(\alpha) - \alpha = \alpha^q - \alpha$. Since σ is an F-linear map, the map ρ is also F-linear. Moreover, the kernel of ρ is none other than the Berlekamp subalgebra B. So to find a basis for B, we simply need to find a basis for the kernel of ρ using Gaussian elimination over F, as in §14.4.

To perform the Gaussian elimination, we need to choose a basis S for E over F, and construct the matrix $Q := \text{Mat}_{S,S}(\rho) \in F^{\ell \times \ell}$, that is, the matrix of ρ with respect to this basis, as in §14.2, so that evaluation of ρ corresponds to multiplying a row vector on the right by Q. We are free to choose a basis in any convenient way, and the most convenient basis, of course, is $S := \{\xi^{i-1}\}_{i=1}^{\ell}$, since for computational purposes, we already represent an element $\alpha \in E$ by its coordinate vector $\text{Vec}_S(\alpha)$. The matrix Q, then, is the $\ell \times \ell$ matrix whose ith row, for $i = 1, \ldots, \ell$, is $\text{Vec}_S(\rho(\xi^{i-1}))$. Note that if $\alpha = \xi^q$, then $\rho(\xi^{i-1}) = (\xi^{i-1})^q - \xi^{i-1} = (\xi^q)^{i-1} - \xi^{i-1} = \alpha^{i-1} - \xi^{i-1}$. This observation allows us to construct the rows of Q by first computing ξ^q via repeated squaring, and then just computing successive powers of ξ^q.

After we construct the matrix Q, we apply Gaussian elimination to get row vectors $v_1, \ldots, v_r$ that form a basis for the row null space of Q. It is at this point that our algorithm actually discovers the number r of irreducible factors of f. Our basis for B is $\{\beta_i\}_{i=1}^r$, where $\mathrm{Vec}_S(\beta_i) = v_i$ for $i = 1, \ldots, r$.

Putting this all together, we have the following algorithm to compute a basis for the Berlekamp subalgebra.

Algorithm B1. On input f, where $f \in F[X]$ is a monic square-free polynomial of degree $\ell > 0$, do the following, where $E := F[X]/(f)$, $\xi := [X]_f \in E$, and $S := \{\xi^{i-1}\}_{i=1}^\ell$:

> let Q be an $\ell \times \ell$ matrix over F (initially with undefined entries)
> compute $\alpha \leftarrow \xi^q$ using repeated squaring
> $\beta \leftarrow 1_E$
> for $i \leftarrow 1$ to ℓ do   // *invariant:* $\beta = \alpha^{i-1} = (\xi^{i-1})^q$
> $\quad$ $\mathrm{Row}_i(Q) \leftarrow \mathrm{Vec}_S(\beta)$, $Q(i,i) \leftarrow Q(i,i) - 1$, $\beta \leftarrow \beta\alpha$
> compute a basis $\{v_i\}_{i=1}^r$ of the row null space of Q using
> $\quad$ Gaussian elimination
> for $i = 1, \ldots, r$ do $\beta_i \leftarrow \mathrm{Vec}_S^{-1}(v_i)$
> output $\{\beta_i\}_{i=1}^r$

The correctness of Algorithm B1 is clear from the above discussion. As for the running time:

Theorem 20.10. *Algorithm B1 uses $O(\ell^3 + \ell^2 \, \mathrm{len}(q))$ operations in F.*

Proof. This is just a matter of counting. The computation of α takes $O(\mathrm{len}(q))$ operations in E using repeated squaring, and hence $O(\ell^2 \, \mathrm{len}(q))$ operations in F. To build the matrix Q, we have to perform an additional $O(\ell)$ operations in E to compute the successive powers of α, which translates into $O(\ell^3)$ operations in F. Finally, the cost of Gaussian elimination is an additional $O(\ell^3)$ operations in F. $\square$

Stage 2: Splitting with a basis for B

The second stage of Berlekamp's factoring algorithm is a probabilistic procedure that factors f using a basis $\{\beta_i\}_{i=1}^r$ for B. As we did with Algorithm EDF in §20.4.2, we begin by discussing how to efficiently split f into two non-trivial factors, and then we present a somewhat more elaborate algorithm that completely factors f.

Let $M_1 \in F[X]$ be the polynomial defined by (20.4) and (20.5); that is,

$$M_1 := \begin{cases} \sum_{j=0}^{w-1} X^{2^j} & \text{if } p = 2, \\ X^{(q-1)/2} - 1 & \text{if } p > 2. \end{cases}$$

Using our basis for B, we can easily generate a random element β of B by simply choosing $c_1, \ldots, c_r$ at random, and computing $\beta := \sum_i c_i \beta_i$. If $\theta(\beta) = (b_1, \ldots, b_r)$, then the family of random variables $\{b_i\}_{i=1}^r$ is mutually independent, with each b_i uniformly distributed over F. Just as in Algorithm EDF, $\gcd(\mathrm{rep}(M_1(\beta)), f)$ will be a non-trivial factor of f with probability at least $1/2$, if $p = 2$, and probability at least $4/9$, if $p > 2$.

That is the basic splitting strategy. We turn this into an algorithm to completely factor f using the same technique of iterative refinement that was used in Algorithm EDF. That is, at any stage of the algorithm, we have a partial factorization $f = \prod_{h \in H} h$, which we try to refine by attempting to split each $h \in H$ using the strategy outlined above. One technical difficulty is that to split such a polynomial h, we need to efficiently generate a random element of the Berlekamp subalgebra of $F[X]/(h)$. A particularly efficient way to do this is to use our basis for the Berlekamp subalgebra of $F[X]/(f)$ to generate a random element of the Berlekamp subalgebra of $F[X]/(h)$ for all $h \in H$ simultaneously. Let $g_i := \mathrm{rep}(\beta_i)$ for $i = 1, \ldots, r$. If we choose $c_1, \ldots, c_r \in F$ at random, and set $g := c_1 g_1 + \cdots + c_r g_r$, then $[g]_f$ is a random element of the Berlekamp subalgebra of $F[X]/(f)$, and by the Chinese remainder theorem, it follows that the family of random variables $\{[g]_h\}_{h \in H}$ is mutually independent, with each $[g]_h$ uniformly distributed over the Berlekamp subalgebra of $F[X]/(h)$.

Here is the algorithm for completely factoring a polynomial, given a basis for the corresponding Berlekamp subalgebra.

Algorithm B2. On input $f, \{\beta_i\}_{i=1}^r$, where $f \in F[X]$ is a monic square-free polynomial of degree $\ell > 0$, and $\{\beta_i\}_{i=1}^r$ is a basis for the Berlekamp subalgebra of $F[X]/(f)$, do the following, where $g_i := \mathrm{rep}(\beta_i)$ for $i = 1, \ldots, r$:

$$H \leftarrow \{f\}$$
while $|H| < r$ do
        choose $c_1, \ldots, c_r \in F$ at random
        $g \leftarrow c_1 g_1 + \cdots + c_r g_r \in F[X]$
        $H' \leftarrow \emptyset$
        for each $h \in H$ do
                $\beta \leftarrow [g]_h \in F[X]/(h)$
                $d \leftarrow \gcd(\mathrm{rep}(M_1(\beta)), h)$
                if $d = 1$ or $d = h$
                        then $H' \leftarrow H' \cup \{h\}$
                        else $H' \leftarrow H' \cup \{d, h/d\}$
        $H \leftarrow H'$
output H

The correctness of the algorithm is clear. As for its expected running time, we can get a quick-and-dirty upper bound as follows:

- The cost of generating g in each loop iteration is $O(r\ell)$ operations in F. For a given h, the cost of computing $\beta := [g]_h \in F[X]/(h)$ is $O(\ell \deg(h))$ operations in F, and the cost of computing $M_1(\beta)$ is $O(\deg(h)^2 \operatorname{len}(q))$ operations in F. Therefore, the number of operations in F performed in each iteration of the main loop is at most a constant times

$$r\ell + \ell \sum_{h \in H} \deg(h) + \operatorname{len}(q) \sum_{h \in H} \deg(h)^2$$

$$\leq 2\ell^2 + \operatorname{len}(q) \left(\sum_{h \in H} \deg(h) \right)^2 = O(\ell^2 \operatorname{len}(q)).$$

- The expected number of iterations of the main loop until we get some nontrivial split is $O(1)$.

- The algorithm finishes after getting $r - 1$ non-trivial splits.

- Therefore, the total expected cost is $O(r\ell^2 \operatorname{len}(q))$ operations in F.

A more careful analysis reveals:

Theorem 20.11. *Algorithm B2 uses an expected number of*

$$O(\operatorname{len}(r)\ell^2 \operatorname{len}(q))$$

operations in F.

Proof. The proof follows the same line of reasoning as the analysis of Algorithm EDF. Indeed, using the same argument as was used there, the expected number of iterations of the main loop is $O(\operatorname{len}(r))$. As discussed in the paragraph above this theorem, the cost per loop iteration is $O(\ell^2 \operatorname{len}(q))$ operations in F. The theorem follows. $\square$

The bound in the above theorem is tight (see Exercise 20.11 below): unlike Algorithm EDF, we cannot make the multiplicative factor of $\operatorname{len}(r)$ go away.

Putting together Algorithms B1 and B2, we get Berlekamp's complete factoring algorithm. The running time bound is easily estimated from the results already proved:

Theorem 20.12. *Berlekamp's factoring algorithm uses an expected number of* $O(\ell^3 + \ell^2 \operatorname{len}(\ell) \operatorname{len}(q))$ *operations in F.*

We have assumed the input to Berlekamp's algorithm is a square-free polynomial. However, we may use Algorithm SFD as a preprocessing step to ensure that this is the case. Even if we include the cost of this preprocessing step, the running time estimate in Theorem 20.12 remains valid.

So we see that Berlekamp's algorithm is faster than the Cantor–Zassenhaus algorithm, whose expected operation count is $O(\ell^3 \operatorname{len}(q))$. The speed advantage of Berlekamp's algorithm grows as q gets large. The one disadvantage of Berlekamp's algorithm is space: it requires space for $\Theta(\ell^2)$ elements of F, while the Cantor–Zassenhaus algorithm requires space for only $O(\ell)$ elements of F. One can in fact implement the Cantor–Zassenhaus algorithm so that it uses $O(\ell^3 + \ell^2 \operatorname{len}(q))$ operations in F, while using space for only $O(\ell^{1.5})$ elements of F—see Exercise 20.13 below.

EXERCISE 20.11. Give an example of a family of input polynomials that cause Algorithm B2 to use an expected number of at least $\Omega(\ell^2 \operatorname{len}(\ell) \operatorname{len}(q))$ operations in F. Assume that computing $M_1(\beta)$ for $\beta \in F[X]/(h)$ takes $\Omega(\deg(h)^2 \operatorname{len}(q))$ operations in F.

EXERCISE 20.12. Using the ideas behind Berlekamp's factoring algorithm, devise a deterministic irreducibility test that, given a monic polynomial of degree ℓ over F, uses $O(\ell^3 + \ell^2 \operatorname{len}(q))$ operations in F.

EXERCISE 20.13. This exercise develops a variant of the Cantor–Zassenhaus algorithm that uses $O(\ell^3 + \ell^2 \operatorname{len}(q))$ operations in F, while using space for only $O(\ell^{1.5})$ elements of F. By making use the variant of Algorithm EDF discussed in Exercise 20.9, our problem is reduced to that of implementing Algorithm DDF within the stated time and space bounds, assuming that the input polynomial is square-free.

(a) Show that for all non-negative integers i, j, with $i \neq j$, the irreducible polynomials in $F[X]$ that divide $X^{q^i} - X^{q^j}$ are precisely those whose degree divides $i - j$.

(b) Let $f \in F[X]$ be a monic polynomial of degree $\ell > 0$, and let $m = O(\ell^{1/2})$. Let $\xi := [X]_f \in E$, where $E := F[X]/(f)$. Show how to compute

$$\xi^q, \xi^{q^2}, \ldots, \xi^{q^{m-1}} \in E \text{ and } \xi^{q^m}, \xi^{q^{2m}}, \ldots, \xi^{q^{(m-1)m}} \in E$$

using $O(\ell^3 + \ell^2 \operatorname{len}(q))$ operations in F, and space for $O(\ell^{1.5})$ elements of F.

(c) Combine the results of parts (a) and (b) to implement Algorithm DDF on square-free inputs of degree ℓ, so that it uses $O(\ell^3 + \ell^2 \operatorname{len}(q))$ operations in F, and space for $O(\ell^{1.5})$ elements of F.

20.6 Deterministic factorization algorithms (∗)

The algorithms of Cantor and Zassenhaus and of Berlekamp are probabilistic. The exercises below develop a deterministic variant of the Cantor–Zassenhaus algorithm. (One can also develop deterministic variants of Berlekamp's algorithm, with similar complexity.)

This algorithm is only practical for finite fields of small characteristic, and is anyway mainly of theoretical interest, since from a practical perspective, there is nothing wrong with the above probabilistic method. In all of these exercises, we assume that we have access to a basis $\{\varepsilon_i\}_{i=1}^{w}$ for F as a vector space over $\mathbb{Z}_p$.

To make the Cantor–Zassenhaus algorithm deterministic, we only need to develop a deterministic variant of Algorithm EDF, as Algorithm DDF is already deterministic.

EXERCISE 20.14. Let $f = f_1 \cdots f_r$, where the f_i's are distinct monic irreducible polynomials in $F[X]$. Assume that $r > 1$, and let $\ell := \deg(f)$. For this exercise, the degrees of the f_i's need not be the same. For an intermediate field F', with $\mathbb{Z}_p \subseteq F' \subseteq F$, let us call a set $S = \{\lambda_1, \ldots, \lambda_s\}$, where each $\lambda_u \in F[X]$ with $\deg(\lambda_u) < \ell$, a **separating set for** f **over** F' if the following conditions hold:

- for $i = 1, \ldots, r$ and $u = 1, \ldots, s$, there exists $c_{ui} \in F'$ such that $\lambda_u \equiv c_{ui} \pmod{f_i}$, and

- for every pair of distinct indices i, j, with $1 \leq i < j \leq r$, there exists $u = 1, \ldots, s$ such that $c_{ui} \neq c_{uj}$.

Show that if S is a separating set for f over $\mathbb{Z}_p$, then the following algorithm completely factors f using $O(p|S|\ell^2)$ operations in F.

$$H \leftarrow \{f\}$$
$$\text{for each } \lambda \in S \text{ do}$$
$$\quad \text{for each } a \in \mathbb{Z}_p \text{ do}$$
$$\quad\quad H' \leftarrow \emptyset$$
$$\quad\quad \text{for each } h \in H \text{ do}$$
$$\quad\quad\quad d \leftarrow \gcd(\lambda - a, h)$$
$$\quad\quad\quad \text{if } d = 1 \text{ or } d = h$$
$$\quad\quad\quad\quad \text{then } H' \leftarrow H' \cup \{h\}$$
$$\quad\quad\quad\quad \text{else } H' \leftarrow H' \cup \{d, h/d\}$$
$$\quad\quad H \leftarrow H'$$
$$\text{output } H$$

EXERCISE 20.15. Let f be as in the previous exercise. Show that if S is a separating set for f over F, then the set

$$S' := \left\{ \sum_{i=0}^{w-1} (\varepsilon_j \lambda)^{p^i} \bmod f : 1 \le j \le w, \ \lambda \in S \right\}$$

is a separating set for f over $\mathbb{Z}_p$. Show how to compute this set using $O(|S|\ell^2 \operatorname{len}(p) w(w-1))$ operations in F.

EXERCISE 20.16. Let f be as in the previous two exercises, but further suppose that each irreducible factor of f is of the same degree, say k. Let $E := F[X]/(f)$ and $\xi := [X]_f \in E$. Define the polynomial $\phi \in E[Y]$ as follows:

$$\phi := \prod_{i=0}^{k-1} (Y - \xi^{q^i}).$$

If

$$\phi = Y^k + \alpha_{k-1} Y^{k-1} + \cdots + \alpha_0,$$

with $\alpha_0, \ldots, \alpha_{k-1} \in E$, show that the set

$$S := \{\operatorname{rep}(\alpha_i) : 0 \le i \le k-1\}$$

is a separating set for f over F, and can be computed deterministically using $O(k^2 + k \operatorname{len}(q))$ operations in E, and hence $O(k^2\ell^2 + k\ell^2 \operatorname{len}(q))$ operations in F.

EXERCISE 20.17. Put together all of the above pieces, together with Algorithms SFD and DDF, so as to obtain a deterministic algorithm for factoring polynomials over F that runs in time at most p times a polynomial in the size of the input, and make a careful estimate of the running time of your algorithm.

EXERCISE 20.18. It is a fact that when our prime p is odd, then for all integers a, b, with $a \not\equiv b \pmod{p}$, there exists a non-negative integer $i \le p^{1/2} \log_2 p$ such that $(a + i \mid p) \ne (b + i \mid p)$ (here, "$(\cdot \mid \cdot)$" is the Legendre symbol). Using this fact, design and analyze a deterministic algorithm for factoring polynomials over F that runs in time at most $p^{1/2}$ times a polynomial in the size of the input.

The following two exercises show that the problem of factoring polynomials over F reduces in deterministic polynomial time to the problem of finding roots of polynomials over $\mathbb{Z}_p$.

EXERCISE 20.19. Let f be as in Exercise 20.14. Suppose that $S = \{\lambda_1, \ldots, \lambda_s\}$ is a separating set for f over $\mathbb{Z}_p$, and $\phi_u \in F[X]$ is the minimal polynomial over F of $[\lambda_u]_f \in F[X]/(f)$ for $u = 1, \ldots, s$. Show that each ϕ_u is the product of linear factors over $\mathbb{Z}_p$, and that given S, along with the roots of all the ϕ_u's, we can deterministically factor f using $(|S| + \ell)^{O(1)}$ operations in F. Hint: see Exercise 16.9.

EXERCISE 20.20. Using the previous exercise, show that the problem of factoring a polynomial over F reduces in deterministic polynomial time to the problem of finding roots of polynomials over $\mathbb{Z}_p$.

20.7 Notes

The average-case analysis of Algorithm IPT, assuming its input is random, and the application to the analysis of Algorithm RIP, is essentially due to Ben-Or [14]. If one implements Algorithm RIP using fast polynomial arithmetic, one gets an expected cost of $O(\ell^{2+o(1)} \operatorname{len}(q))$ operations in F. Note that Ben-Or's analysis is a bit incomplete — see Exercise 32 in Chapter 7 of Bach and Shallit [11] for a complete analysis of Ben-Or's claims.

The asymptotically fastest probabilistic algorithm for constructing an irreducible polynomial over F of given degree ℓ is due to Shoup [96]. That algorithm uses an expected number of $O(\ell^{2+o(1)} + \ell^{1+o(1)} \operatorname{len}(q))$ operations in F, and in fact does not follow the "generate and test" paradigm of Algorithm RIP, but uses a completely different approach.

As far as *deterministic* algorithms for constructing irreducible polynomials of given degree over F, the only known methods are efficient when the characteristic p of F is small (see Chistov [26], Semaev [88], and Shoup [94]), or under a generalization of the Riemann hypothesis (see Adleman and Lenstra [4]). Shoup [94] in fact shows that the problem of constructing an irreducible polynomial of given degree over F is deterministic, polynomial-time reducible to the problem of factoring polynomials over F.

The algorithm in §20.2 for computing minimal polynomials over finite fields is due to Gordon [43].

The square-free decomposition of a polynomial over a field of characteristic zero can be computed using an algorithm of Yun [111] using $O(\ell^{1+o(1)})$ field operations. Yun's algorithm can be adapted to work over finite fields as well (see Exercise 14.30 in von zur Gathen and Gerhard [39]).

The Cantor–Zassenhaus algorithm was initially developed by Cantor and Zassenhaus [24], although many of the basic ideas can be traced back quite a ways. A straightforward implementation of this algorithm using fast polynomial arithmetic uses an expected number of $O(\ell^{2+o(1)} \operatorname{len}(q))$ operations in F.

Berlekamp's algorithm was initially developed by Berlekamp [15, 16], but again, the basic ideas go back a long way. A straightforward implementation using fast polynomial arithmetic uses an expected number of $O(\ell^3 + \ell^{1+o(1)} \operatorname{len}(q))$ operations in F; the term ℓ^3 may be replaced by ℓ^ω, where ω is the exponent of matrix multiplication (see §14.6).

There are no known efficient, deterministic algorithms for factoring polynomials over F when the characteristic p of F is large (even under a generalization of the Riemann hypothesis, except in certain special cases).

The asymptotically fastest algorithms for factoring polynomials over F are due to von zur Gathen, Kaltofen, and Shoup:† the algorithm of von zur Gathen and Shoup [40] uses an expected number of $O(\ell^{2+o(1)} + \ell^{1+o(1)} \operatorname{len}(q))$ operations in F; the algorithm of Kaltofen and Shoup [53] has a cost that is subquadratic in the degree—it uses an expected number of $O(\ell^{1.815} \operatorname{len}(q)^{0.407})$ operations in F when $\operatorname{len}(q) = O(\ell^{1.375})$. Exercises 20.1, 20.8, and 20.9 are based on [40]. Although the "fast" algorithms in [40] and [53] are mainly of theoretical interest, a variant in [53], which uses $O(\ell^{2.5} + \ell^{1+o(1)} \operatorname{len}(q))$ operations in F, and space for $O(\ell^{1.5})$ elements of F, has proven to be quite practical (Exercise 20.13 develops some of these ideas; see also Shoup [97]).

---

† The running times of these algorithms can be improved using faster algorithms for modular composition—see footnote on p. 485.

21

Deterministic primality testing

For many years, despite much research in the area, there was no known deterministic, polynomial-time algorithm for testing whether a given integer $n > 1$ is a prime. However, that is no longer the case — the breakthrough algorithm of Agrawal, Kayal, and Saxena, or Algorithm AKS for short, is just such an algorithm. Not only is the result itself remarkable, but the algorithm is striking both in its simplicity, and in the fact that the proof of its running time and correctness are completely elementary (though ingenious).

We should stress at the outset that although this result is an important theoretical result, as of yet, it has no real practical significance: probabilistic tests, such as the Miller–Rabin test discussed in Chapter 10, are *much* more efficient, and a practically minded person should not at all be bothered by the fact that such algorithms may in theory make a mistake with an incredibly small probability.

21.1 The basic idea

The algorithm is based on the following fact:

Theorem 21.1. *Let $n > 1$ be an integer. If n is prime, then for all $a \in \mathbb{Z}_n$, we have the following identity in the ring $\mathbb{Z}_n[X]$:*

$$(X + a)^n = X^n + a. \tag{21.1}$$

Conversely, if n is composite, then for all $a \in \mathbb{Z}_n^*$, the identity (21.1) does not hold.

Proof. Note that

$$(X + a)^n = X^n + a^n + \sum_{i=1}^{n-1} \binom{n}{i} a^i X^{n-i}.$$

If n is prime, then by Fermat's little theorem (Theorem 2.14), we have $a^n = a$, and by Exercise 1.14, all of the binomial coefficients $\binom{n}{i}$, for $i = 1, \ldots, n - 1$, are

548

divisible by n, and hence their images in the ring $\mathbb{Z}_n$ vanish. That proves that the identity (21.1) holds when n is prime.

Conversely, suppose that n is composite and that $a \in \mathbb{Z}_n^*$. Consider any prime factor p of n, and suppose $n = p^k m$, where $p \nmid m$.

We claim that $p^k \nmid \binom{n}{p}$. To prove the claim, one simply observes that

$$\binom{n}{p} = \frac{n(n-1)\cdots(n-p+1)}{p!},$$

and the numerator of this fraction is an integer divisible by p^k, but no higher power of p, and the denominator is divisible by p, but no higher power of p. That proves the claim.

From the claim, and the fact that $a \in \mathbb{Z}_n^*$, it follows that the coefficient of X^{n-p} in $(X + a)^n$ is not zero, and hence the identity (21.1) does not hold. $\square$

Of course, Theorem 21.1 does not immediately give rise to an efficient primality test, since just evaluating the left-hand side of the identity (21.1) takes time $\Omega(n)$ in the worst case. The key observation of Agrawal, Kayal, and Saxena is that if (21.1) holds modulo $X^r - 1$ for a suitably chosen value of r, and for sufficiently many a, then n must be prime. To make this idea work, one must show that a suitable r exists that is bounded by a polynomial in $\text{len}(n)$, and that the number of different values of a that must be tested is also bounded by a polynomial in $\text{len}(n)$.

21.2 The algorithm and its analysis

The algorithm is shown in Fig. 21.1. A few remarks on implementation are in order:

- In step 1, we can use the algorithm for perfect-power testing discussed in Exercise 3.31.

- The search for r in step 2 can just be done by brute-force search; likewise, the determination of the multiplicative order of $[n]_r \in \mathbb{Z}_r^*$ can be done by brute force: after verifying that $\gcd(n, r) = 1$, compute successive powers of n modulo r until we get 1.

We want to prove that Algorithm AKS runs in polynomial time and is correct. To prove that it runs in polynomial time, it clearly suffices to prove that there exists an integer r satisfying the condition in step 2 that is bounded by a polynomial in $\text{len}(n)$, since all other computations can be carried out in time $(r + \text{len}(n))^{O(1)}$. Correctness means that it outputs *true* if and only if n is prime.

On input n, where n is an integer and $n > 1$, do the following:

1. if n is of the form a^b for integers $a > 1$ and $b > 1$ then
        return *false*
2. find the smallest integer $r > 1$ such that either
        $\gcd(n, r) > 1$

    or

        $\gcd(n, r) = 1$ and
        $[n]_r \in \mathbb{Z}_r^*$ has multiplicative order $> 4 \operatorname{len}(n)^2$
3. if $r = n$ then return *true*
4. if $\gcd(n, r) > 1$ then return *false*
5. for $j \leftarrow 1$ to $2 \operatorname{len}(n) \lfloor r^{1/2} \rfloor + 1$ do
        if $(X + j)^n \not\equiv X^n + j \pmod{X^r - 1}$ in the ring $\mathbb{Z}_n[X]$ then
            return *false*
6. return *true*

Fig. 21.1. Algorithm AKS

21.2.1 Running time analysis

The question of the running time of Algorithm AKS is settled by the following fact:

Theorem 21.2. *For integers $n > 1$ and $m \geq 1$, the least prime r such that $r \nmid n$ and the multiplicative order of $[n]_r \in \mathbb{Z}_r^*$ is greater than m is $O(m^2 \operatorname{len}(n))$.*

Proof. Call a prime r "good" if $r \nmid n$ and the multiplicative order of $[n]_r \in \mathbb{Z}_r^*$ is greater than m, and otherwise call r "bad." If r is bad, then either $r \mid n$ or $r \mid (n^d - 1)$ for some $d = 1, \ldots, m$. Thus, any bad prime r satisfies

$$r \mid n \prod_{d=1}^{m} (n^d - 1).$$

If all primes r up to some given bound $x \geq 2$ are bad, then the product of all primes up to x divides $n \prod_{d=1}^{m} (n^d - 1)$, and so in particular,

$$\prod_{r \leq x} r \leq n \prod_{d=1}^{m} (n^d - 1),$$

where the first product is over all primes r up to x. Taking logarithms, we obtain

$$\sum_{r \leq x} \log r \leq \log\left(n \prod_{d=1}^{m}(n^d - 1)\right) \leq (\log n)\left(1 + \sum_{d=1}^{m} d\right)$$

$$= (\log n)(1 + m(m+1)/2).$$

But by Theorem 5.7, we have

$$\sum_{r \leq x} \log r \geq cx$$

for some constant $c > 0$, from which it follows that

$$x \leq c^{-1}(\log n)(1 + m(m+1)/2),$$

and the theorem follows. $\square$

From this theorem, it follows that the value of r found in step 2—which need not be prime—will be $O(\mathrm{len}(n)^5)$. From this, we obtain:

Theorem 21.3. *Algorithm AKS can be implemented so that its running time is* $O(\mathrm{len}(n)^{16.5})$.

Proof. As discussed above, the value of r determined in step 2 will be $O(\mathrm{len}(n)^5)$. It is fairly straightforward to see that the running time of the algorithm is dominated by the running time of step 5. Here, we have to perform $O(r^{1/2}\,\mathrm{len}(n))$ exponentiations to the power n in the ring $\mathbb{Z}_n[X]/(X^r - 1)$. Each of these exponentiations takes $O(\mathrm{len}(n))$ operations in $\mathbb{Z}_n[X]/(X^r - 1)$, each of which takes $O(r^2)$ operations in $\mathbb{Z}_n$, each of which takes time $O(\mathrm{len}(n)^2)$. This yields a running time bounded by a constant times

$$r^{1/2}\,\mathrm{len}(n) \times \mathrm{len}(n) \times r^2 \times \mathrm{len}(n)^2 = r^{2.5}\,\mathrm{len}(n)^4.$$

Substituting the bound $O(\mathrm{len}(n)^5)$ for r, we obtain the desired bound. $\square$

21.2.2 Correctness

As for the correctness of Algorithm AKS, we first show:

Theorem 21.4. *If the input to Algorithm AKS is prime, then the output is true.*

Proof. Assume that the input n is prime. The test in step 1 will certainly fail. If the algorithm does not return *true* in step 3, then certainly the test in step 4 will fail as well. If the algorithm reaches step 5, then all of the tests in the loop in step 5 will fail—this follows from Theorem 21.1. $\square$

The interesting case is the following:

Theorem 21.5. *If the input to Algorithm AKS is composite, then the output is false.*

The proof of this theorem is rather long, and is the subject of the remainder of this section.

Suppose the input n is composite. If n is a prime power, then this will be detected in step 1, so we may assume that n is not a prime power. Assume that the algorithm has found a suitable value of r in step 2. Clearly, the test in 3 will fail. If the test in step 4 passes, we are done, so we may assume that this test fails; that is, we may assume that all prime factors of n are greater than r. Our goal now is to show that one of the tests in the loop in step 5 must pass. The proof will be by contradiction: we shall assume that none of the tests pass, and derive a contradiction.

The assumption that none of the tests in step 5 fail means that in the ring $\mathbb{Z}_n[X]$, the following congruences hold:

$$(X + j)^n \equiv X^n + j \pmod{X^r - 1} \quad (j = 1, \dots, 2\operatorname{len}(n)\lfloor r^{1/2}\rfloor + 1). \qquad (21.2)$$

For the rest of the proof, we fix a particular prime divisor p of n — the choice of p does not matter. Since $p \mid n$, we have a natural ring homomorphism from $\mathbb{Z}_n[X]$ to $\mathbb{Z}_p[X]$ (see Examples 7.52 and 7.46), which implies that the congruences (21.2) hold in the ring of polynomials over $\mathbb{Z}_p$ as well. *From now on, we shall work exclusively with polynomials over $\mathbb{Z}_p$.*

Let us state in somewhat more abstract terms the precise assumptions we are making in order to derive our contradiction:

(A0) $n > 1$, $r > 1$, and $\ell \geq 1$ are integers, p is a prime dividing n, and $\gcd(n, r) = 1$;

(A1) n is not a prime power;

(A2) $p > r$;

(A3) the congruences

$$(X + j)^n \equiv X^n + j \pmod{X^r - 1} \quad (j = 1, \dots, \ell)$$

hold in the ring $\mathbb{Z}_p[X]$;

(A4) the multiplicative order of $[n]_r \in \mathbb{Z}_r^*$ is greater than $4\operatorname{len}(n)^2$;

(A5) $\ell > 2\operatorname{len}(n)\lfloor r^{1/2}\rfloor$.

The rest of the proof will rely only on these assumptions, and not on any other details of Algorithm AKS. From now on, only assumption (A0) will be implicitly in force. The other assumptions will be explicitly invoked as necessary. Our goal is to show that assumptions (A1), (A2), (A3), (A4), and (A5) cannot all be true simultaneously.

Define the $\mathbb{Z}_p$-algebra $E := \mathbb{Z}_p[X]/(X^r - 1)$, and let $\xi := [X]_{X^r-1} \in E$, so that $E = \mathbb{Z}_p[\xi]$. Every element of E can be expressed uniquely as $g(\xi) = [g]_{X^r-1}$, for $g \in \mathbb{Z}_p[X]$ of degree less than r, and for an arbitrary polynomial $g \in \mathbb{Z}_p[X]$, we have $g(\xi) = 0$ if and only if $(X^r - 1) \mid g$. Note that $\xi \in E^*$ and has multiplicative order r: indeed, $\xi^r = 1$, and $\xi^s - 1$ cannot be zero for $s < r$, since $X^s - 1$ has degree less than r.

Assumption (A3) implies that we have a number of interesting identities in the $\mathbb{Z}_p$-algebra E:

$$(\xi + j)^n = \xi^n + j \quad (j = 1, \ldots, \ell).$$

For the polynomials $g_j := X + j \in \mathbb{Z}_p[X]$, with j in the given range, these identities say that $g_j(\xi)^n = g_j(\xi^n)$.

In order to exploit these identities, we study more generally functions σ_k, for various integer values k, that send $g(\xi) \in E$ to $g(\xi^k)$, for arbitrary $g \in \mathbb{Z}_p[X]$, and we investigate the implications of the assumption that such functions behave like the k-power map on certain inputs. To this end, let $\mathbb{Z}^{(r)}$ denote the set of all positive integers k such that $\gcd(r, k) = 1$. Note that the set $\mathbb{Z}^{(r)}$ is **multiplicative**, by which we mean $1 \in \mathbb{Z}^{(r)}$, and $kk' \in \mathbb{Z}^{(r)}$ for all $k, k' \in \mathbb{Z}^{(r)}$. Also note that because of our assumption (A0), both n and p are in $\mathbb{Z}^{(r)}$. For $k \in \mathbb{Z}^{(r)}$, let $\hat{\sigma}_k : \mathbb{Z}_p[X] \to E$ be the polynomial evaluation map that sends $g \in \mathbb{Z}_p[X]$ to $g(\xi^k)$. This is of course a $\mathbb{Z}_p$-algebra homomorphism, and we have:

Lemma 21.6. *For all $k \in \mathbb{Z}^{(r)}$, the kernel of $\hat{\sigma}_k$ is $(X^r - 1)$, and the image of $\hat{\sigma}_k$ is E.*

Proof. Let $J := \text{Ker} \, \hat{\sigma}_k$, which is an ideal of $\mathbb{Z}_p[X]$. Let k' be a positive integer such that $kk' \equiv 1 \pmod{r}$, which exists because $\gcd(r, k) = 1$.

To show that $J = (X^r - 1)$, we first observe that

$$\hat{\sigma}_k(X^r - 1) = (\xi^k)^r - 1 = (\xi^r)^k - 1 = 1^k - 1 = 0,$$

and hence $(X^r - 1) \subseteq J$.

Next, we show that $J \subseteq (X^r - 1)$. Let $g \in J$. We want to show that $(X^r - 1) \mid g$. Now, $g \in J$ means that $g(\xi^k) = 0$. If we set $h := g(X^k)$, this implies that $h(\xi) = 0$, which means that $(X^r - 1) \mid h$. So let us write $h = (X^r - 1)f$, for some $f \in \mathbb{Z}_p[X]$. Then

$$g(\xi) = g(\xi^{kk'}) = h(\xi^{k'}) = (\xi^{k'r} - 1)f(\xi^{k'}) = 0,$$

which implies that $(X^r - 1) \mid g$.

That finishes the proof that $J = (X^r - 1)$.

Finally, to show that $\hat{\sigma}_k$ is surjective, suppose we are given an arbitrary element of E, which we can express as $g(\xi)$ for some $g \in \mathbb{Z}_p[X]$. Now set $h := g(X^{k'})$, and observe that

$$\hat{\sigma}_k(h) = h(\xi^k) = g(\xi^{kk'}) = g(\xi). \quad \square$$

Because of Lemma 21.6, then by Theorem 7.26, the map $\sigma_k : E \to E$ that sends $g(\xi) \in E$ to $g(\xi^k)$, for $g \in \mathbb{Z}_p[X]$, is well defined, and is a ring automorphism — indeed, a $\mathbb{Z}_p$-*algebra* automorphism — on E. Note that for all $k, k' \in \mathbb{Z}^{(r)}$, we have

- $\sigma_k = \sigma_{k'}$ if and only if $\xi^k = \xi^{k'}$ if and only if $k \equiv k' \pmod{r}$, and

- $\sigma_k \circ \sigma_{k'} = \sigma_{k'} \circ \sigma_k = \sigma_{kk'}$.

So in fact, the set $\{\sigma_k : k \in \mathbb{Z}^{(r)}\}$ under composition forms an abelian group that is isomorphic to $\mathbb{Z}_r^*$.

> **Remark.** It is perhaps helpful (but not necessary for the proof) to examine the behavior of the map σ_k in a bit more detail. Let $\alpha \in E$, and let
>
> $$\alpha = \sum_{i=0}^{r-1} a_i \xi^i$$
>
> be the canonical representation of α. Since $\gcd(r, k) = 1$, the map $\pi : \{0, \ldots, r-1\} \to \{0, \ldots, r-1\}$ that sends i to $ki \bmod r$ is a permutation whose inverse is the permutation π' that sends i to $k'i \bmod r$, where k' is a multiplicative inverse of k modulo r. Then we have
>
> $$\sigma_k(\alpha) = \sum_{i=0}^{r-1} a_i \xi^{ki} = \sum_{i=0}^{r-1} a_i \xi^{\pi(i)} = \sum_{i=0}^{r-1} a_{\pi'(i)} \xi^i.$$
>
> Thus, the action of σ_k is to permute the coordinate vector $(a_0, \ldots, a_{r-1})$ of α, sending α to the element in E whose coordinate vector is $(a_{\pi'(0)}, \ldots, a_{\pi'(r-1)})$. So we see that although we defined the maps σ_k in a rather "highbrow" algebraic fashion, their behavior in concrete terms is actually quite simple.

Recall that the p-power map on E is a $\mathbb{Z}_p$-algebra homomorphism (see Theorem 19.7), and so for all $\alpha \in E$, if $\alpha = g(\xi)$ for $g \in \mathbb{Z}_p[X]$, then (by Theorem 16.7) we have

$$\alpha^p = g(\xi)^p = g(\xi^p) = \sigma_p(\alpha).$$

Thus, σ_p acts just like the p-power map on all elements of E.

We can restate assumption (A3) as follows:

$$\sigma_n(\xi + j) = (\xi + j)^n \quad (j = 1, \ldots, \ell).$$

That is to say, the map σ_n acts just like the n-power map on the elements $\xi + j$ for $j = 1, \ldots, \ell$.

Now, although the σ_p map must act like the p-power map on all of E, there is no good reason why the σ_n map should act like the n-power map on any particular element of E, and so the fact that it does so on all the elements $\xi + j$ for $j = 1, \ldots, \ell$ looks decidedly suspicious. To turn our suspicions into a contradiction, let us start by defining some notation. For $\alpha \in E$, let us define

$$C(\alpha) := \{k \in \mathbb{Z}^{(r)} : \sigma_k(\alpha) = \alpha^k\},$$

and for $k \in \mathbb{Z}^{(r)}$, let us define

$$D(k) := \{\alpha \in E : \sigma_k(\alpha) = \alpha^k\}.$$

In words: $C(\alpha)$ is the set of all k for which σ_k acts like the k-power map on α, and $D(k)$ is the set of all α for which σ_k acts like the k-power map on α. From the discussion above, we have $p \in C(\alpha)$ for all $\alpha \in E$, and it is also clear that $1 \in C(\alpha)$ for all $\alpha \in E$. Also, it is clear that $\alpha \in D(p)$ for all $\alpha \in E$, and $1_E \in D(k)$ for all $k \in \mathbb{Z}^{(r)}$.

The following two simple lemmas say that the sets $C(\alpha)$ and $D(k)$ are multiplicative.

Lemma 21.7. *For every $\alpha \in E$, if $k \in C(\alpha)$ and $k' \in C(\alpha)$, then $kk' \in C(\alpha)$.*

Proof. If $\sigma_k(\alpha) = \alpha^k$ and $\sigma_{k'}(\alpha) = \alpha^{k'}$, then

$$\sigma_{kk'}(\alpha) = \sigma_k(\sigma_{k'}(\alpha)) = \sigma_k(\alpha^{k'}) = (\sigma_k(\alpha))^{k'} = (\alpha^k)^{k'} = \alpha^{kk'},$$

where we have made use of the homomorphic property of σ_k. $\square$

Lemma 21.8. *For every $k \in \mathbb{Z}^{(r)}$, if $\alpha \in D(k)$ and $\beta \in D(k)$, then $\alpha\beta \in D(k)$.*

Proof. If $\sigma_k(\alpha) = \alpha^k$ and $\sigma_k(\beta) = \beta^k$, then

$$\sigma_k(\alpha\beta) = \sigma_k(\alpha)\sigma_k(\beta) = \alpha^k\beta^k = (\alpha\beta)^k,$$

where again, we have made use of the homomorphic property of σ_k. $\square$

Let us define

- s to be the multiplicative order of $[p]_r \in \mathbb{Z}_r^*$, and
- t to be the order of the subgroup of $\mathbb{Z}_r^*$ generated by $[p]_r$ and $[n]_r$.

Since $r \mid (p^s - 1)$, if we take any extension field F of degree s over $\mathbb{Z}_p$ (which we know exists by Theorem 19.12), then since F^* is cyclic (Theorem 7.29) and has order $p^s - 1$, we know that there exists an element $\zeta \in F^*$ of multiplicative order r (Theorem 6.32). Let us define the polynomial evaluation map $\hat{\tau} : \mathbb{Z}_p[X] \to F$ that sends $g \in \mathbb{Z}_p[X]$ to $g(\zeta) \in F$. Since $X^r - 1$ is clearly in the kernel of $\hat{\tau}$, then by Theorem 7.27, the map $\tau : E \to F$ that sends $g(\xi)$ to $g(\zeta)$, for $g \in \mathbb{Z}_p[X]$, is a well-defined ring homomorphism, and actually, it is a $\mathbb{Z}_p$-algebra homomorphism.

For concreteness, one could think of F as $\mathbb{Z}_p[X]/(f)$, where f is an irreducible factor of $X^r - 1$ of degree s. In this case, we could simply take ζ to be $[X]_f$ (see

Example 19.1), and the map $\hat{\tau}$ above would be just the natural map from $\mathbb{Z}_p[X]$ to $\mathbb{Z}_p[X]/(f)$.

The key to deriving our contradiction is to examine the set $S := \tau(D(n))$, that is, the image under τ of the set $D(n)$ of all elements $\alpha \in E$ for which σ_n acts like the n-power map.

Lemma 21.9. *Under assumption (A1), we have*

$$|S| \leq n^{2\lfloor t^{1/2} \rfloor}.$$

Proof. Consider the set of integers

$$I := \{n^u p^v : u, v = 0, \ldots, \lfloor t^{1/2} \rfloor\}.$$

We first claim that $|I| > t$. To prove this, we first show that each distinct pair (u, v) gives rise to a distinct value $n^u p^v$. To this end, we make use of our assumption (A1) that n is not a prime power, and so is divisible by some prime q other than p. Thus, if $(u', v') \neq (u, v)$, then either

- $u \neq u'$, in which case the power of q in the prime factorization of $n^u p^v$ is different from that in $n^{u'} p^{v'}$, or

- $u = u'$ and $v \neq v'$, in which case the power of p in the prime factorization of $n^u p^v$ is different from that in $n^{u'} p^{v'}$.

The claim now follows from the fact that both u and v range over a set of size $\lfloor t^{1/2} \rfloor + 1 > t^{1/2}$, and so there are strictly more than t such pairs (u, v).

Next, recall that t was defined to be the order of the subgroup of $\mathbb{Z}_r^*$ generated by $[n]_r$ and $[p]_r$; equivalently, t is the number of distinct residue classes of the form $[n^u p^v]_r$, where u and v range over all non-negative integers. Since each element of I is of the form $n^u p^v$, and $|I| > t$, we may conclude that there must be two distinct elements of I, call them k and k', that are congruent modulo r. Furthermore, any element of I is a product of two positive integers each of which is at most $n^{\lfloor t^{1/2} \rfloor}$, and so both k and k' lie in the range $1, \ldots, n^{2\lfloor t^{1/2} \rfloor}$.

Now, let $\alpha \in D(n)$. This is equivalent to saying $n \in C(\alpha)$. We always have $1 \in C(\alpha)$ and $p \in C(\alpha)$, and so by Lemma 21.7, we have $n^u p^v \in C(\alpha)$ for all non-negative integers u, v, and so in particular, $k, k' \in C(\alpha)$.

Since both k and k' are in $C(\alpha)$, we have

$$\sigma_k(\alpha) = \alpha^k \quad \text{and} \quad \sigma_{k'}(\alpha) = \alpha^{k'}.$$

Since $k \equiv k' \pmod{r}$, we have $\sigma_k = \sigma_{k'}$, and hence

$$\alpha^k = \alpha^{k'}.$$

Now apply the homomorphism τ, obtaining

$$\tau(\alpha)^k = \tau(\alpha)^{k'}.$$

Since this holds for all $\alpha \in D(n)$, we conclude that all elements of S are roots of the polynomial $X^k - X^{k'}$. Since $k \neq k'$, we see that $X^k - X^{k'}$ is a non-zero polynomial of degree at most $\max\{k, k'\} \leq n^{2\lfloor t^{1/2} \rfloor}$, and hence can have at most $n^{2\lfloor t^{1/2} \rfloor}$ roots in the field F (Theorem 7.14). $\square$

Lemma 21.10. *Under assumptions (A2) and (A3), we have*

$$|S| \geq 2^{\min(t,\ell)} - 1.$$

Proof. Let $m := \min(t, \ell)$. Under assumption (A3), we have $\xi + j \in D(n)$ for $j = 1, \ldots, m$. Under assumption (A2), we have $p > r > t \geq m$, and hence the integers $j = 1, \ldots, m$ are distinct modulo p. Define

$$P := \Big\{ \prod_{j=1}^{m} (X + j)^{e_j} \in \mathbb{Z}_p[X] : e_j \in \{0, 1\} \text{ for } j = 1, \ldots, m, \text{ and } \sum_{j=1}^{m} e_j < m \Big\}.$$

That is, we form P by taking products over all subsets $S \subsetneq \{X + j : j = 1, \ldots, m\}$. Clearly, $|P| = 2^m - 1$.

Define $P(\xi) := \{f(\xi) \in E : f \in P\}$ and $P(\zeta) := \{f(\zeta) \in F : f \in P\}$. Note that $\tau(P(\xi)) = P(\zeta)$, and that by Lemma 21.8, $P(\xi) \subseteq D(n)$.

Therefore, to prove the lemma, it suffices to show that $|P(\zeta)| = 2^m - 1$. Suppose that this is not the case. This would give rise to distinct polynomials $g, h \in \mathbb{Z}_p[X]$, both of degree at most $t - 1$, such that

$$g(\xi) \in D(n), \ h(\xi) \in D(n), \text{ and } \tau(g(\xi)) = \tau(h(\xi)).$$

So we have $n \in C(g(\xi))$ and (as always) $1, p \in C(g(\xi))$. Likewise, we have $1, n, p \in C(h(\xi))$. By Lemma 21.7, for all integers k of the form $n^u p^v$, where u and v range over all non-negative integers, we have

$$k \in C(g(\xi)) \text{ and } k \in C(h(\xi)).$$

For each such k, since $\tau(g(\xi)) = \tau(h(\xi))$, we have $\tau(g(\xi))^k = \tau(h(\xi))^k$, and hence

$$\begin{aligned}
0 &= \tau(g(\xi))^k - \tau(h(\xi))^k \\
&= \tau(g(\xi)^k) - \tau(h(\xi)^k) \quad (\tau \text{ is a homomorphism}) \\
&= \tau(g(\xi^k)) - \tau(h(\xi^k)) \quad (k \in C(g(\xi)) \text{ and } k \in C(h(\xi))) \\
&= g(\zeta^k) - h(\zeta^k) \quad (\text{definition of } \tau).
\end{aligned}$$

Thus, the polynomial $f := g - h \in \mathbb{Z}_p[X]$ is a non-zero polynomial of degree at most $t - 1$, having roots ζ^k in the field F for all k of the form $n^u p^v$. Now, t is by definition the number of distinct residue classes of the form $[n^u p^v]_r \in \mathbb{Z}_r^*$. Also, since ζ has multiplicative order r, for all integers k, k', we have $\zeta^k = \zeta^{k'}$ if and only if $k \equiv k' \pmod{r}$. Therefore, as k ranges over all integers of the form $n^u p^v$,

ζ^k ranges over precisely t distinct values in F. But since all of these values are roots of the polynomial f, which is non-zero and of degree at most $t - 1$, this is impossible (Theorem 7.14). □

We are now (finally!) in a position to complete the proof of Theorem 21.5. Under assumptions (A1), (A2), and (A3), Lemmas 21.9 and 21.10 imply that

$$2^{\min(t,\ell)} - 1 \le |S| \le n^{2\lfloor t^{1/2} \rfloor}. \tag{21.3}$$

The contradiction is provided by the following:

Lemma 21.11. *Under assumptions (A4) and (A5), we have*

$$2^{\min(t,\ell)} - 1 > n^{2\lfloor t^{1/2} \rfloor}.$$

Proof. Observe that $\log_2 n \le \text{len}(n)$, and so it suffices to show that

$$2^{\min(t,\ell)} - 1 > 2^{2\,\text{len}(n)\lfloor t^{1/2} \rfloor},$$

and for this, it suffices to show that

$$\min(t, \ell) > 2\,\text{len}(n)\lfloor t^{1/2} \rfloor,$$

since for all integers a, b with $a > b \ge 1$, we have $2^a > 2^b + 1$.

To show that $t > 2\,\text{len}(n)\lfloor t^{1/2} \rfloor$, it suffices to show that $t > 2\,\text{len}(n)t^{1/2}$, or equivalently, that $t > 4\,\text{len}(n)^2$. But observe that by definition, t is the order of the subgroup of $\mathbb{Z}_r^*$ generated by $[n]_r$ and $[p]_r$, which is at least as large as the multiplicative order of $[n]_r$ in $\mathbb{Z}_r^*$, and by assumption (A4), this is larger than $4\,\text{len}(n)^2$.

Finally, directly by assumption (A5), we have $\ell > 2\,\text{len}(n)\lfloor t^{1/2} \rfloor$. □

That concludes the proof of Theorem 21.5.

EXERCISE 21.1. Show that if Conjecture 5.24 is true, then the value of r discovered in step 2 of Algorithm AKS satisfies $r = O(\text{len}(n)^2)$.

21.3 Notes

The algorithm presented here is due to Agrawal, Kayal, and Saxena [6].

If fast algorithms for integer and polynomial arithmetic are used, then using the analysis presented here, it is easy to see that the algorithm runs in time $O(\text{len}(n)^{10.5+o(1)})$. More generally, it is easy to see that the algorithm runs in time $O(r^{1.5+o(1)}\,\text{len}(n)^{3+o(1)})$, where r is the value determined in step 2 of the algorithm. In our analysis of the algorithm, we were able to obtain the bound $r = O(\text{len}(n)^5)$, leading to the running-time bound $O(\text{len}(n)^{10.5+o(1)})$. Using a result of Fouvry [37], one can show that $r = O(\text{len}(n)^3)$, leading to a running-time bound of $O(\text{len}(n)^{7.5+o(1)})$. Moreover, if Conjecture 5.24 on the density of Sophie Germain primes were true, then one could show that $r = O(\text{len}(n)^2)$ (see Exercise 21.1), which would lead to a running-time bound of $O(\text{len}(n)^{6+o(1)})$. This running-time bound can be achieved rigorously by a different algorithm, due to Lenstra and Pomerance [62].

Prior to this algorithm, the fastest deterministic, rigorously proved primality test was one introduced by Adleman, Pomerance, and Rumely [5], called the **Jacobi sum test**, which runs in time

$$O(\text{len}(n)^{c\,\text{len}(\text{len}(\text{len}(n)))})$$

for some constant c. Note that for numbers n with less than 2^{256} bits, the value of $\text{len}(\text{len}(\text{len}(n)))$ is at most 8, and so this algorithm runs in time $O(\text{len}(n)^{8c})$ for any n that one could ever actually write down.

We also mention the earlier work of Adleman and Huang [3], who gave a probabilistic algorithm whose output is always correct, and which runs in expected polynomial time (i.e., a *Las Vegas* algorithm, in the parlance of §9.7).

Appendix: Some useful facts

A1. Some handy inequalities. The following inequalities involving exponentials and logarithms are very handy.

(i) For all real numbers x, we have

$$1 + x \leq e^x,$$

or, taking logarithms, for $x > -1$, we have

$$\log(1 + x) \leq x.$$

(ii) For all real numbers $x \geq 0$, we have

$$e^{-x} \leq 1 - x + x^2/2,$$

or, taking logarithms,

$$-x \leq \log(1 - x + x^2/2).$$

(iii) For all real numbers x with $0 \leq x \leq 1/2$, we have

$$1 - x \geq e^{-x-x^2} \geq e^{-2x},$$

or, taking logarithms,

$$\log(1 - x) \geq -x - x^2 \geq -2x.$$

(i) and (ii) follow easily from Taylor's formula with remainder, applied to the function e^x, while (iii) may be proved by expanding $\log(1 - x)$ as a Taylor series, and making a simple calculation.

A2. Binomial coefficients. For integers n and k, with $0 \leq k \leq n$, one defines the **binomial coefficient**

$$\binom{n}{k} := \frac{n!}{k!(n-k)!} = \frac{n(n-1)\cdots(n-k+1)}{k!}.$$

We have the identities

$$\binom{n}{n} = \binom{n}{0} = 1,$$

and for $0 < k < n$, we have **Pascal's identity**

$$\binom{n}{k} = \binom{n-1}{k-1} + \binom{n-1}{k},$$

which may be verified by direct calculation. From these identities, it follows that $\binom{n}{k}$ is an integer, and indeed, is equal to the number of subsets of $\{1, \ldots, n\}$ of cardinality k. The usual **binomial theorem** also follows as an immediate consequence: for all numbers a, b, and for all positive integers n, we have the **binomial expansion**

$$(a + b)^n = \sum_{k=0}^{n} \binom{n}{k} a^{n-k} b^k.$$

It is also easily verified, directly from the definition, that

$$\binom{n}{k} < \binom{n}{k+1} \quad \text{for } 0 \le k < (n-1)/2,$$

$$\binom{n}{k} > \binom{n}{k+1} \quad \text{for } (n-1)/2 < k < n, \text{ and}$$

$$\binom{n}{k} = \binom{n}{n-k} \quad \text{for } 0 \le k \le n.$$

In other words, if we fix n, and view $\binom{n}{k}$ as a function of k, then this function is increasing on the interval $[0, n/2]$, decreasing on the interval $[n/2, n]$, and its graph is symmetric with respect to the line $k = n/2$.

A3. Countably infinite sets. Let $\mathbb{Z}^+ := \{1, 2, 3, \ldots\}$, the set of positive integers. A set S is called **countably infinite** if there is a bijection $f : \mathbb{Z}^+ \to S$; in this case, we can enumerate the elements of S as $x_1, x_2, x_3, \ldots$, where $x_i := f(i)$.

A set S is called **countable** if it is either finite or countably infinite.

For a set S, the following conditions are equivalent:

- S is countable;
- there is a surjective function $g : \mathbb{Z}^+ \to S$;
- there is an injective function $h : S \to \mathbb{Z}^+$.

The following facts can be easily established:

(i) if $S_1, \ldots, S_n$ are countable sets, then so are $S_1 \cup \cdots \cup S_n$ and $S_1 \times \cdots \times S_n$;

(ii) if $S_1, S_2, S_3, \ldots$ are countable sets, then so is $\bigcup_{i=1}^{\infty} S_i$;

(iii) if S is a countable set, then so is the set $\bigcup_{i=0}^{\infty} S^{\times i}$ of all finite sequences of elements in S.

Some examples of countably infinite sets: $\mathbb{Z}$, $\mathbb{Q}$, the set of all finite bit strings. Some examples of uncountable sets: $\mathbb{R}$, the set of all infinite bit strings.

A4. Integrating piece-wise continuous functions. In discussing the Riemann integral $\int_a^b f(t)\, dt$, many introductory calculus texts only discuss in any detail the case where the integrand f is continuous on the closed interval $[a, b]$, in which case the integral is always well defined. However, the Riemann integral is well defined for much broader classes of functions. For our purposes in this text, it is convenient and sufficient to work with integrands that are **piece-wise continuous** on $[a, b]$, which means that there exist real numbers $x_0, x_1, \ldots, x_k$ and functions $f_1, \ldots, f_k$, such that $a = x_0 \le x_1 \le \cdots \le x_k = b$, and for each $i = 1, \ldots, k$, the function f_i is continuous on the *closed* interval $[x_{i-1}, x_i]$, and agrees with f on the *open* interval (x_{i-1}, x_i). In this case, f is integrable on $[a, b]$, and indeed

$$\int_a^b f(t)\, dt = \sum_{i=1}^{k} \int_{x_{i-1}}^{x_i} f_i(t)\, dt.$$

It is not hard to prove this equality, using the basic definition of the Riemann integral; however, for our purposes, we can also just take the value of the expression on the right-hand side as the definition of the integral on the left-hand side.

If f is piece-wise continuous on $[a, b]$, then it is also bounded on $[a, b]$, meaning that there exists a positive number M such that $|f(t)| \le M$ for all $t \in [a, b]$, from which it follows that $|\int_a^b f(t)\, dt| \le M(b - a)$.

We also say that f is piece-wise continuous on $[a, \infty)$ if for all $b \ge a$, f is piece-wise continuous on $[a, b]$. In this case, we may define the improper integral $\int_a^{\infty} f(t)\, dt$ as the limit, as $b \to \infty$, of $\int_a^b f(t)\, dt$, provided the limit exists.

A5. Estimating sums by integrals. Using elementary calculus, it is easy to estimate a sum over a monotone sequence in terms of a definite integral, by interpreting the integral as the area under a curve. Let f be a real-valued function that is (at least piece-wise) continuous and monotone on the closed interval $[a, b]$, where a and b are integers. Then we have

$$\min(f(a), f(b)) \le \sum_{i=a}^{b} f(i) - \int_{a}^{b} f(t)\, dt \le \max(f(a), f(b)).$$

A6. Infinite series. Consider an infinite series $\sum_{i=1}^{\infty} x_i$. It is a basic fact from calculus that if the x_i's are non-negative and $\sum_{i=1}^{\infty} x_i$ converges to a value y, then any infinite series whose terms are a rearrangement of the x_i's converges to the same value y.

If we drop the requirement that the x_i's are non-negative, but insist that the series $\sum_{i=1}^{\infty} |x_i|$ converges, then the series $\sum_{i=1}^{\infty} x_i$ is called **absolutely convergent**. In this case, then not only does the series $\sum_{i=1}^{\infty} x_i$ converge to some value y, but any infinite series whose terms are a rearrangement of the x_i's also converges to the same value y.

A7. Double infinite series. The topic of **double infinite series** may not be discussed in a typical introductory calculus course; we summarize here the basic facts that we need.

Suppose that $\{x_{ij}\}_{i,j=1}^{\infty}$ is a family non-negative real numbers such that for each i, the series $\sum_j x_{ij}$ converges to a value r_i, and for each j the series $\sum_i x_{ij}$ converges to a value c_j. Then we can form the double infinite series $\sum_i \sum_j x_{ij} = \sum_i r_i$ and the double infinite series $\sum_j \sum_i x_{ij} = \sum_j c_j$. If $(i_1, j_1), (i_2, j_2), \ldots$ is an enumeration of all pairs of indices (i, j), we can also form the single infinite series $\sum_k x_{i_k j_k}$. We then have $\sum_i \sum_j x_{ij} = \sum_j \sum_i x_{ij} = \sum_k x_{i_k j_k}$, where the three series either all converge to the same value, or all diverge. Thus, we can reverse the order of summation in a double infinite series of non-negative terms. If we drop the non-negativity requirement, the same result holds provided $\sum_k |x_{i_k j_k}| < \infty$.

Now suppose $\sum_i a_i$ is an infinite series of non-negative terms that converges to A, and that $\sum_j b_j$ is an infinite series of non-negative terms that converges to B. If $(i_1, j_1), (i_2, j_2), \ldots$ is an enumeration of all pairs of indices (i, j), then $\sum_k a_{i_k} b_{j_k}$ converges to AB. Thus, we can multiply term-wise infinite series with non-negative terms. If we drop the non-negativity requirement, the same result holds provided $\sum_i a_i$ and $\sum_j b_j$ converge absolutely.

A8. Convex functions. Let I be an interval of the real line (either open, closed, or half open, and either bounded or unbounded), and let f be a real-valued function defined on I. The function f is called **convex on I** if for all $x_0, x_2 \in I$, and for all $t \in [0, 1]$, we have

$$f(tx_0 + (1 - t)x_2) \le tf(x_0) + (1 - t)f(x_2).$$

Geometrically, convexity means that for every three points $P_i = (x_i, f(x_i))$, $i = 0, 1, 2$, where each $x_i \in I$ and $x_0 < x_1 < x_2$, the point P_1 lies on or below the line through P_0 and P_2.

We state here the basic analytical facts concerning convex functions:

(i) if f is convex on I, then f is continuous on the interior of I (but not necessarily at the endpoints of I, if any);

(ii) if f is continuous on I and differentiable on the interior of I, then f is convex on I if and only if its derivative is non-decreasing on the interior of I.

Bibliography

[1] L. M. Adleman. A subexponential algorithm for the discrete logarithm problem with applications to cryptography. In *20th Annual Symposium on Foundations of Computer Science*, pages 55–60, 1979.

[2] L. M. Adleman. The function field sieve. In *Proc. 1st International Symposium on Algorithmic Number Theory (ANTS-I)*, pages 108–121, 1994.

[3] L. M. Adleman and M.-D. Huang. *Primality Testing and Two Dimensional Abelian Varieties over Finite Fields (Lecture Notes in Mathematics No. 1512).* Springer-Verlag, 1992.

[4] L. M. Adleman and H. W. Lenstra, Jr. Finding irreducible polynomials over finite fields. In *18th Annual ACM Symposium on Theory of Computing*, pages 350–355, 1986.

[5] L. M. Adleman, C. Pomerance, and R. S. Rumely. On distinguishing prime numbers from composite numbers. *Annals of Mathematics*, 117:173–206, 1983.

[6] M. Agrawal, N. Kayal, and N. Saxena. PRIMES is in P. *Annals of Mathematics*, 160(2):781–793, 2004.

[7] W. Alford, A. Granville, and C. Pomerance. There are infinitely many Carmichael numbers. *Annals of Mathematics*, 140:703–722, 1994.

[8] T. M. Apostol. *Introduction to Analytic Number Theory.* Springer-Verlag, 1973.

[9] E. Bach. How to generate factored random numbers. *SIAM Journal on Computing*, 17:179–193, 1988.

[10] E. Bach. Explicit bounds for primality testing and related problems. *Mathematics of Computation*, 55:355–380, 1990.

[11] E. Bach and J. Shallit. *Algorithmic Number Theory*, volume 1. MIT Press, 1996.

[12] P. Bateman and R. Horn. A heuristic asymptotic formula concerning the distribution of prime numbers. *Mathematics of Computation*, 16:363–367, 1962.

[13] M. Bellare and P. Rogaway. Random oracles are practical: a paradigm for designing efficient protocols. In *First ACM Conference on Computer and Communications Security*, pages 62–73, 1993.

[14] M. Ben-Or. Probabilistic algorithms in finite fields. In *22nd Annual Symposium on Foundations of Computer Science*, pages 394–398, 1981.

[15] E. R. Berlekamp. *Algebraic Coding Theory.* McGraw-Hill, 1968.

[16] E. R. Berlekamp. Factoring polynomials over large finite fields. *Mathematics of Computation*, 24(111):713–735, 1970.

[17] L. Blum, M. Blum, and M. Shub. A simple unpredictable pseudo-random number generator. *SIAM Journal on Computing*, 15:364–383, 1986.

[18] D. Boneh. The Decision Diffie-Hellman Problem. In *Proc. 3rd International Symposium on Algorithmic Number Theory (ANTS-III)*, pages 48–63, 1998. Springer LNCS 1423.

[19] D. Boneh and G. Durfee. Cryptanalysis of RSA with private key d less than $N^{0.292}$. *IEEE Transactions on Information Theory*, IT-46:1339–1349, 2000.

[20] R. P. Brent and H. T. Kung. Fast algorithms for manipulating formal power series. *Journal of the ACM*, 25:581–595, 1978.

[21] J. P. Buhler, H. W. Lenstra, Jr., and C. Pomerance. Factoring integers with the number field sieve. In A. K. Lenstra and H. W. Lenstra, Jr., editors, *The Development of the Number Field Sieve*, pages 50–94. Springer-Verlag, 1993.

[22] D. A. Burgess. The distribution of quadratic residues and non-residues. *Mathematika*, 4:106–112, 1957.

[23] E. Canfield, P. Erdős, and C. Pomerance. On a problem of Oppenheim concerning 'Factorisatio Numerorum'. *Journal of Number Theory*, 17:1–28, 1983.

[24] D. G. Cantor and E. Kaltofen. On fast multiplication of polynomials over arbitrary rings. *Acta Informatica*, 28:693–701, 1991.

[25] J. L. Carter and M. N. Wegman. Universal classes of hash functions. *Journal of Computer and System Sciences*, 18:143–154, 1979.

[26] A. L. Chistov. Polynomial time construction of a finite field. In *Abstracts of Lectures at 7th All-Union Conference in Mathematical Logic, Novosibirsk*, page 196, 1984. In Russian.

[27] D. Coppersmith. Modifications to the number field sieve. *Journal of Cryptology*, 6:169–180, 1993.

[28] D. Coppersmith and S. Winograd. Matrix multiplication via arithmetic progressions. *Journal of Symbolic Computation*, 9(3):23–52, 1990.

[29] T. H. Cormen, C. E. Leiserson, R. L. Rivest, and C. Stein. *Introduction to Algorithms*. MIT Press, second edition, 2001.

[30] R. Crandall and C. Pomerance. *Prime Numbers: A Computational Perspective*. Springer, 2001.

[31] I. Damgård and G. Frandsen. Efficient algorithms for gcd and cubic residuosity in the ring of Eisenstein integers. In *14th International Symposium on Fundamentals of Computation Theory, Springer LNCS 2751*, pages 109–117, 2003.

[32] I. Damgård, P. Landrock, and C. Pomerance. Average case error estimates for the strong probable prime test. *Mathematics of Computation*, 61:177–194, 1993.

[33] L. E. Dickson. A new extension of Dirichlet's theorem on prime numbers. *Messenger of Mathematics*, 33:151–161, 1904.

[34] W. Diffie and M. E. Hellman. New directions in cryptography. *IEEE Transactions on Information Theory*, IT-22:644–654, 1976.

[35] J. Dixon. Asymptotically fast factorization of integers. *Mathematics of Computation*, 36:255–260, 1981.

[36] J. L. Dornstetter. On the equivalence between Berlekamp's and Euclid's algorithms. *IEEE Transactions on Information Theory*, IT-33:428–431, 1987.

[37] E. Fouvry. Théorème de Brun-Titchmarsh; application au théorème de Fermat. *Inventiones Mathematicae*, 79:383–407, 1985.

[38] M. Fürer. Faster integer multiplication. In *39th Annual ACM Symposium on Theory of Computing*, pages 57–66, 2007.

[39] J. von zur Gathen and J. Gerhard. *Modern Computer Algebra*. Cambridge University Press, second edition, 2003.

[40] J. von zur Gathen and V. Shoup. Computing Frobenius maps and factoring polynomials. *Computational Complexity*, 2:187–224, 1992.

[41] S. Goldwasser and S. Micali. Probabilistic encryption. *Journal of Computer and System Sciences*, 28:270–299, 1984.

[42] D. M. Gordon. Discrete logarithms in GF(p) using the number field sieve. *SIAM Journal on Discrete Mathematics*, 6:124–138, 1993.

[43] J. Gordon. Very simple method to find the minimal polynomial of an arbitrary nonzero element of a finite field. *Electronic Letters*, 12:663–664, 1976.

[44] H. Halberstam and H. Richert. *Sieve Methods*. Academic Press, 1974.

[45] G. H. Hardy and J. E. Littlewood. Some problems of partito numerorum. III. On the expression of a number as a sum of primes. *Acta Mathematica*, 44:1–70, 1923.

[46] G. H. Hardy and E. M. Wright. *An Introduction to the Theory of Numbers*. Oxford University Press, fifth edition, 1984.

[47] D. Heath-Brown. Zero-free regions for Dirichlet L-functions and the least prime in an arithmetic progression. *Proceedings of the London Mathematical Society*, 64:265–338, 1992.

[48] R. Impagliazzo, L. Levin, and M. Luby. Pseudo-random number generation from any one-way function. In *21st Annual ACM Symposium on Theory of Computing*, pages 12–24, 1989.

[49] R. Impagliazzo and D. Zuckermann. How to recycle random bits. In *30th Annual Symposium on Foundations of Computer Science*, pages 248–253, 1989.

[50] H. Iwaniec. On the error term in the linear sieve. *Acta Arithmetica*, 19:1–30, 1971.

[51] H. Iwaniec. On the problem of Jacobsthal. *Demonstratio Mathematica*, 11:225–231, 1978.

[52] A. Kalai. Generating random factored numbers, easily. In *Proc. 13th ACM-SIAM Symposium on Discrete Algorithms*, page 412, 2002.

[53] E. Kaltofen and V. Shoup. Subquadratic-time factoring of polynomials over finite fields. In *27th Annual ACM Symposium on Theory of Computing*, pages 398–406, 1995.

[54] A. A. Karatsuba and Y. Ofman. Multiplication of multidigit numbers on automata. *Soviet Physics Doklady*, 7:595–596, 1963.

[55] S. H. Kim and C. Pomerance. The probability that a random probable prime is composite. *Mathematics of Computation*, 53(188):721–741, 1989.

[56] D. E. Knuth. *The Art of Computer Programming*, volume 2. Addison-Wesley, second edition, 1981.

[57] T. Krovetz and P. Rogaway. Variationally universal hashing. *Information Processing Letters*, 100:36–39, 1996.

[58] D. Lehmann. On primality tests. *SIAM Journal on Computing*, 11:374–375, 1982.

[59] D. Lehmer and R. Powers. On factoring large numbers. *Bulletin of the AMS*, 37:770–776, 1931.

[60] H. W. Lenstra, Jr. Factoring integers with elliptic curves. *Annals of Mathematics*, 126:649–673, 1987.

[61] H. W. Lenstra, Jr. and C. Pomerance. A rigorous time bound for factoring integers. *Journal of the AMS*, 4:483–516, 1992.

[62] H. W. Lenstra, Jr. and C. Pomerance. Primality testing with Gaussian periods. Manuscript, `math.dartmouth.edu/~carlp`, 2005.

[63] M. Luby. *Pseudorandomness and Cryptographic Applications*. Princeton University Press, 1996.

[64] J. Massey. Shift-register synthesis and BCH coding. *IEEE Transactions on Information Theory*, IT-15:122–127, 1969.

[65] U. Maurer. Fast generation of prime numbers and secure public-key cryptographic parameters. *Journal of Cryptology*, 8:123–155, 1995.

[66] A. Menezes, P. van Oorschot, and S. Vanstone. *Handbook of Applied Cryptography*. CRC Press, 1997.

[67] G. L. Miller. Riemann's hypothesis and tests for primality. *Journal of Computer and System Sciences*, 13:300–317, 1976.

[68] W. Mills. Continued fractions and linear recurrences. *Mathematics of Computation*, 29:173–180, 1975.

[69] K. Morrison. Random polynomials over finite fields. Manuscript, `www.calpoly.edu/~kmorriso/Research/RPFF.pdf`, 1999.

[70] M. Morrison and J. Brillhart. A method of factoring and the factorization of F_7. *Mathematics of Computation*, 29:183–205, 1975.

[71] V. I. Nechaev. Complexity of a determinate algorithm for the discrete logarithm. *Mathematical Notes*, 55(2):165–172, 1994. Translated from *Matematicheskie Zametki*, 55(2):91–101, 1994.

[72] I. Niven and H. Zuckerman. *An Introduction to the Theory of Numbers*. John Wiley and Sons, Inc., second edition, 1966.

[73] J. Oesterlé. Versions effectives du théorème de Chebotarev sous l'hypothèse de Riemann généralisée. *Astérisque*, 61:165–167, 1979.

[74] P. van Oorschot and M. Wiener. On Diffie-Hellman key agreement with short exponents. In *Advances in Cryptology–Eurocrypt '96, Springer LNCS 1070*, pages 332–343, 1996.

[75] S. Pohlig and M. Hellman. An improved algorithm for computing logarithms over GF(p) and its cryptographic significance. *IEEE Transactions on Information Theory*, IT-24:106–110, 1978.

[76] J. M. Pollard. Monte Carlo methods for index computation mod p. *Mathematics of Computation*, 32:918–924, 1978.

[77] J. M. Pollard. Factoring with cubic integers. In A. K. Lenstra and H. W. Lenstra, Jr., editors, *The Development of the Number Field Sieve*, pages 4–10. Springer-Verlag, 1993.

[78] C. Pomerance. Analysis and comparison of some integer factoring algorithms. In H. W. Lenstra, Jr. and R. Tijdeman, editors, *Computational Methods in Number Theory, Part I*, pages 89–139. Mathematisch Centrum, 1982.

[79] M. O. Rabin. Probabilistic algorithms. In *Algorithms and Complexity, Recent Results and New Directions*, pages 21–39. Academic Press, 1976.

[80] D. Redmond. *Number Theory — An Introduction*. Marcel Dekker, 1996.

[81] I. Reed and G. Solomon. Polynomial codes over certain finite fields. *SIAM Journal on Applied Mathematics*, pages 300–304, 1960.

[82] R. L. Rivest, A. Shamir, and L. M. Adleman. A method for obtaining digital signatures and public-key cryptosystems. *Communications of the ACM*, 21(2):120–126, 1978.

[83] J. Rosser and L. Schoenfeld. Approximate formulas for some functions of prime numbers. *Illinois Journal of Mathematics*, 6:64–94, 1962.

[84] O. Schirokauer, D. Weber, and T. Denny. Discrete logarithms: the effectiveness of the index calculus method. In *Proc. 2nd International Symposium on Algorithmic Number Theory (ANTS-II)*, pages 337–361, 1996.

[85] A. Schönhage. Schnelle Berechnung von Kettenbruchentwicklungen. *Acta Informatica*, 1:139–144, 1971.

[86] A. Schönhage and V. Strassen. Schnelle Multiplikation grosser Zahlen. *Computing*, 7:281–282, 1971.

[87] R. Schoof. Elliptic curves over finite fields and the computation of square roots mod *p*. *Mathematics of Computation*, 44:483–494, 1985.

[88] I. A. Semaev. Construction of irreducible polynomials over finite fields with linearly independent roots. *Mat. Sbornik*, 135:520–532, 1988. In Russian; English translation in *Math. USSR–Sbornik*, 63(2):507–519, 1989.

[89] A. Shamir. Factoring numbers in $O(\log n)$ arithmetic steps. *Information Processing Letters*, 8:28–31, 1979.

[90] A. Shamir. How to share a secret. *Communications of the ACM*, 22:612–613, 1979.

[91] D. Shanks. Class number, a theory of factorization, and genera. In *Proceedings of Symposia in Pure Mathematics*, volume 20, pages 415–440, 1969.

[92] P. Shor. Algorithms for quantum computation: discrete logarithms and factoring. In *35th Annual Symposium on Foundations of Computer Science*, pages 124–134, 1994.

[93] P. Shor. Polynomial-time algorithms for prime factorization and discrete logarithms on a quantum computer. *SIAM Review*, 41:303–332, 1999.

[94] V. Shoup. New algorithms for finding irreducible polynomials over finite fields. *Mathematics of Computation*, 54(189):435–447, 1990.

[95] V. Shoup. Searching for primitive roots in finite fields. *Mathematics of Computation*, 58:369–380, 1992.

[96] V. Shoup. Fast construction of irreducible polynomials over finite fields. *Journal of Symbolic Computation*, 17(5):371–391, 1994.

[97] V. Shoup. A new polynomial factorization algorithm and its implementation. *Journal of Symbolic Computation*, 20(4):363–397, 1995.

[98] V. Shoup. Lower bounds for discrete logarithms and related problems. In *Advances in Cryptology–Eurocrypt '97*, pages 256–266, 1997.

[99] R. Solovay and V. Strassen. A fast Monte-Carlo test for primality. *SIAM Journal on Computing*, 6:84–85, 1977.

[100] J. Stein. Computational problems associated with Racah algebra. *Journal of Computational Physics*, 1:397–405, 1967.

[101] D. R. Stinson. Universal hashing and authentication codes. *Designs, Codes, and Cryptography*, 4:369–380, 1994.

[102] A. Walfisz. *Weylsche Exponentialsummen in der neueren Zahlentheorie.* VEB Deutscher Verlag der Wissenschaften, 1963.

[103] P. Wang, M. Guy, and J. Davenport. *p*-adic reconstruction of rational numbers. *SIGSAM Bulletin*, 16:2–3, 1982.

[104] Y. Wang. On the least primitive root of a prime. *Scientia Sinica*, 10(1):1–14, 1961.

[105] M. N. Wegman and J. L. Carter. New hash functions and their use in authentication and set equality. *Journal of Computer and System Sciences*, 22:265–279, 1981.

[106] A. Weilert. $(1 + i)$-ary GCD computation in $\mathbf{Z}[i]$ as an analogue to the binary GCD algorithm. *Journal of Symbolic Computation*, 30:605–617, 2000.

[107] A. Weilert. Asymptotically fast GCD computation in $\mathbf{Z}[i]$. In *Proc. 4th International Symposium on Algorithmic Number Theory (ANTS-IV)*, pages 595–613, 2000.

[108] L. Welch and R. Scholtz. Continued fractions and Berlekamp's algorithm. *IEEE Transactions on Information Theory*, IT-25:19–27, 1979.

[109] D. Wiedemann. Solving sparse linear systems over finite fields. *IEEE Transactions on Information Theory*, IT-32:54–62, 1986.

[110] M. Wiener. Cryptanalysis of short RSA secret exponents. *IEEE Transactions on Information Theory*, IT-44:553–558, 1990.

[111] D. Y. Y. Yun. On square-free decomposition algorithms. In *Proc. ACM Symposium on Symbolic and Algebraic Computation*, pages 26–35, 1976.

Index of notation

Entries are listed in order of appearance.

log: natural logarithm, xiv

exp: exponential function, xiv

$\emptyset, \in, \subseteq, \subsetneq, \cup, \cap, \setminus, |\cdot|$: set notation, xiv

$S_1 \times \cdots \times S_n$, $S^{\times n}$: Cartesian product, xiv

$\{x_i\}_{i \in I}$: family, xv

$\{x_i\}_{i=m}^n$, $\{x_i\}_{i=m}^\infty$: sequence, xv

$\mathbb{Z}$: the integers, xv

$\mathbb{Q}$: the rationals, xv

$\mathbb{R}$: the reals, xv

$\mathbb{C}$: the complex numbers, xv

∞: arithmetic with infinity, xvi

$[a, b], (a, b)$, etc.: interval notation, xvi

$f(S)$: image of a set, xvi

f^{-1}: pre-image of a set/inverse function, xvi

$f \circ g$: function composition, xvi

$a \mid b$: a divides b, 1

$\lfloor x \rfloor$: floor of x, 4

$\lceil x \rceil$: ceiling of x, 4

$a \bmod b$: integer remainder, 4

$a\mathbb{Z}$: ideal generated by a, 5

$I_1 + I_2$: sum of ideals, 6

gcd: greatest common divisor, 7

$v_p(n)$: largest power to which p divides n, 10

lcm: least common multiple, 11

$a \equiv b \pmod{n}$: a congruent to b modulo n, 16

$b/a \bmod n$: integer remainder, 22

$a^{-1} \bmod n$: integer modular inverse, 22

$[a]_n, [a]$: residue class of a modulo n, 25

$\mathbb{Z}_n$: residue classes modulo n, 25

$\mathbb{Z}_n^*$: invertible residue classes, 28

$\varphi(n)$: Euler's phi function, 31

$(\mathbb{Z}_n^*)^m$: mth powers in $\mathbb{Z}_n^*$, 36

$\mu(n)$: Möbius function, 46

$O, \Omega, \Theta, o, \sim$: asymptotic notation, 50

len(a): length (in bits) of an integer, 62

rep(α): canonical representative of $\alpha \in \mathbb{Z}_n$, 65

$\pi(x)$: number of primes up to x, 104

ϑ: Chebyshev's theta function, 107

li: logarithmic integral, 117

$\zeta(s)$: Riemann's zeta function, 118

Map(I, G): group of functions $f : I \to G$, 131

mG: the subgroup $\{ma : a \in G\}$, 133

$G\{m\}$: the subgroup $\{a \in G : ma = 0_G\}$, 133

G^m: multiplicative subgroup $\{a^m : a \in G\}$, 133

$H_1 + H_2$: sum of subgroups, 136

$H_1 H_2$: product of subgroups, 136

$a \equiv b \pmod{H}$: $a - b \in H$, 137

$[a]_H$: coset of H containing a, 138

G/H: quotient group, 140

$[G : H]$: index, 140

Ker ρ: kernel, 143

Im ρ: image, 143

$G \cong G'$: isomorphic groups, 146

Hom(G, G'): group homomorphisms $G \to G'$, 151

$\langle a \rangle$: subgroup generated by a, 153

$\langle a_1, \ldots, a_k \rangle$: subgroup generated by $a_1, \ldots, a_k$, 153

$\bar{\alpha}$: complex conjugate of α, 167

$N(\alpha)$: norm of $\alpha \in \mathbb{C}$, 167

Map(I, R): ring of functions $f : I \to R$, 168

AB: ring-theoretic product, 169

$a \mid b$: a divides b, 170

R^*: multiplicative group of units of R, 170

$\mathbb{Z}[i]$: Gaussian integers, 174

$\mathbb{Q}^{(m)}$: $\{a/b : \gcd(b, m) = 1\}$, 174

$R[X]$: ring of polynomials, 176

deg(g): degree of a polynomial, 177

lc(g): leading coefficient of a polynomial, 177

$g \bmod h$: polynomial remainder, 178

aR: ideal generated by a, 186

$(a_1, \ldots, a_k)$: ideal generated by $a_1, \ldots, a_k$, 186

R/I: quotient ring, 187

$a \equiv b \pmod{d}$: $a - b \in dR$, 187

$[a]_d$: the residue class $[a]_{dR}$, 187

$R[\alpha]$: smallest subring containing R and α, 192

$R[\alpha_1, \ldots, \alpha_n]$: smallest subring containing R and $\alpha_1, \ldots, \alpha_n$, 193

$R \cong R'$: isomorphic rings, 195

P: probability distribution, 207

$P_1 P_2, P_1^n$: product distribution, 211

$P[\mathcal{A} \mid \mathcal{B}]$: conditional probability of $\mathcal{A}$ given $\mathcal{B}$, 214

$E[X]$: expected value of X, 233

$Var[X]$: variance of X, 235

$E[X \mid \mathcal{B}]$: conditional expectation of X given $\mathcal{B}$, 237

$\Delta[X; Y]$: statistical distance, 260

$y \xleftarrow{\mathcal{E}} \{0, 1\}$, $y \xleftarrow{\mathcal{E}} \{0, 1\}^{\times \ell}$: assign random bit(s), 278

$y \xleftarrow{\mathcal{E}} T$: assign random element of T, 287

$\log_\gamma \alpha$: discrete logarithm, 327

$(a \mid p)$: Legendre symbol, 342

$(a \mid n)$: Jacobi symbol, 346

J_n: Jacobi map, 347

$Map(I, M)$: R-module of functions $f : I \to M$, 360

cM: submodule $\{c\alpha : \alpha \in M\}$, 361

$M\{c\}$: submodule $\{\alpha \in M : c\alpha = 0_M\}$, 361

$R\alpha$: submodule $\{c\alpha : c \in R\}$, 361

$\langle \alpha_1, \ldots, \alpha_k \rangle_R$: submodule generated by $\alpha_1, \ldots, \alpha_k$, 361

$R[X]_{<\ell}$: polynomials of degree less than ℓ, 361

M/N: quotient module, 362

$M \cong M'$: isomorphic modules, 365

$Hom_R(M, M')$: R-linear maps $M \to M'$, 366

$\dim_F(V)$: dimension, 372

$A(i, j)$: (i, j) entry of A, 378

$Row_i(A)$: ith row of A, 378

$Col_j(A)$: jth column of A, 378

$R^{m \times n}$: $m \times n$ matrices over R, 378

$0_R^{m \times n}$: $m \times n$ zero matrix, 378

A^T: transpose of A, 380

$Vec_S(\alpha)$: coordinate vector, 382

$Mat_{S,T}(\rho)$: matrix of linear map, 383

$\Psi(y, x)$: number of y-smooth integers up to x, 399

$Map(I, E)$: R-algebra of functions $f : I \to E$, 423

$R[\alpha]$: subalgebra generated by α, 426

gcd: greatest common divisor (polynomial), 432

lcm: least common multiple (polynomial), 433

$h/g \bmod f$: polynomial remainder, 435

$g^{-1} \bmod f$: polynomial modular inverse, 435

$(E : F)$: degree of an extension, 440

$F(\alpha)$: smallest subfield containing F and α, 441

$R[\![X]\!]$: formal power series, 446

$R(\!(X)\!)$: formal Laurent series, 447

$R(\!(X^{-1})\!)$: reversed Laurent series, 449

$\deg(g)$: degree of $g \in R(\!(X^{-1})\!)$, 449

$lc(g)$: leading coefficient of $g \in R(\!(X^{-1})\!)$, 449

$\lfloor g \rfloor$: floor of $g \in R(\!(X^{-1})\!)$, 449

$len(g)$: length of a polynomial, 466

$rep(\alpha)$: canonical representative of $\alpha \in R[X]/(f)$, 466

$\mathcal{D}_F(V)$: dual space, 492

$\mathcal{L}_F(V)$: space of linear transformations, 501

$\mathbf{N}_{E/F}(\alpha)$: norm, 518

$\mathbf{Tr}_{E/F}(\alpha)$: trace, 518

Index

Abel's identity, 112
abelian group, 126
additive identity, 27
additive inverse, 27
additive subgroup, 169
Adleman, L. M., 99, 103, 419, 420, 546, 559
Agrawal, M., 548, 558
Alford, W., 325
algebra, 421
algebraic
    element, 441
    extension, 441
almost universal hash functions, 258
Apostol, T. M., 125
approximately computes, 302
arithmetic function, 45
arithmetic/geometric mean, 240
Artin's conjecture, 99
associate
    elements of an integral domain, 451
    polynomials, 430
associative binary operation, xvii
asymptotic notation, 50
Atlantic City algorithm, 303
automorphism
    algebra, 424
    group, 146
    module, 365
    ring, 195
    vector space, 370

baby step/giant step method, 330
Bach, E., 125, 305, 325, 340, 357, 546
basis, 367
Bateman, P., 125
Bayes' theorem, 215
Bellare, M., 420
Ben-Or, M., 546
Berlekamp subalgebra, 538
Berlekamp's algorithm, 538
Berlekamp, E. R., 508, 546
Bernoulli trial, 208
Bernstein approximation, 240

Bertrand's postulate, 109
big-O, -Omega, -Theta, 50
bijection, xvi
bijective, xvi
binary gcd algorithm, 77
binary operation, xvii
binomial coefficient, 561
binomial distribution, 223
binomial expansion, 562
binomial theorem, 169, 562
birthday paradox, 248
bivariate polynomial, 183
Blum, L., 103
Blum, M., 103
Boneh, D., 103, 341
Bonferroni's inequalities, 213
Boole's equality, 210
Boole's inequality, 210
Boolean circuits, 72
Brent, R. P., 485
Brillhart, J., 418
Buhler, J. P., 419
Burgess, D. A., 357

$\mathbb{C}$, xv
cancellation law, 2, 21, 28, 129, 171, 435
Canfield, E., 418
canonical representative
    integer, 65
    polynomial, 466
Cantor, D. G., 546
Cantor–Zassenhaus algorithm, 530
cardinality, xiv
Carmichael number, 308
Carmichael, R. D., 325
Carter, J. L., 275
Cartesian product, xiv
ceiling, 4
characteristic of a ring, 169
characteristic polynomial, 518
Chebyshev's inequality, 241
Chebyshev's theorem, 105
Chebyshev's theta function, 107

Chernoff bound, 242
Chinese remainder theorem
  general, 202
  integer, 23, 82
  polynomial, 435, 473
Chistov, A. L., 546
classification of cyclic groups, 156
closed under, xvii
column null space, 395
column rank, 394
column space, 394
column vector, 378
common divisor
  in an integral domain, 452
  integer, 6
  polynomial, 431
common multiple
  in an integral domain, 452
  integer, 11
  polynomial, 433
commutative binary operation, xvii
commutative ring with unity, 166
companion matrix, 385
complex conjugation, 167, 198
composite, 2
composition, xvi
conditional distribution, 213, 224
conditional expectation, 237
conditional probability, 214
congruence, 16, 137
conjugacy class, 516
conjugate, 516
constant polynomial, 176
constant term, 177
continued fraction method, 418
continued fractions, 103
convex function, 564
coordinate vector, 382
  of a projection, 492
Coppersmith, D., 419
Cormen, T. H., 340
coset, 138
countable, 562
countably infinite, 562
covariance, 240
Crandall, R., 73, 125, 420
cyclic, 153

Damgård, I., 325, 464
Davenport, J., 103
De Morgan's law, 208
decisional Diffie–Hellman problem, 338
degree
  of a polynomial, 177
  of a reversed Laurent series, 449
  of an element in an extension field, 441
  of an extension, 440
Denny, T., 420
derivative, 444
deterministic algorithm, 278
deterministic poly-time equivalent, 336

deterministic poly-time reducible, 336
Dickson, L. E., 125
Diffie, W., 341
Diffie–Hellman key establishment protocol, 334
Diffie–Hellman problem, 335
dimension, 372
direct product
  of algebras, 422
  of groups, 130
  of modules, 360
  of rings, 168
Dirichlet inverse, 49
Dirichlet product, 45
Dirichlet series, 120
Dirichlet's theorem, 121
Dirichlet, G., 125
discrete logarithm, 327
  algorithm for computing, 329, 400
discrete probability distribution, 270
discriminant, 182
disjoint, xv
distinct degree factorization, 530, 543
distributive law
  Boolean, 208
divides, 1, 170
divisible by, 1, 170
division with remainder property
  integer, 3
  polynomial, 178
divisor, 1, 170
Dixon, J., 418
Dornstetter, J. L., 508
dual space, 492
Durfee, G., 103

Eisenstein integers, 456
Eisenstein's criterion, 462
elementary row operation, 389
elliptic curve method, 419
equal degree factorization, 532, 537
equivalence class, 15
equivalence relation, 15
Eratosthenes
  sieve of, 115
Erdős, P., 418
error correcting code, 97, 477
error probability, 302
essentially equal
  probability distributions, 212
Euclidean algorithm
  extended
    integer, 78
    polynomial, 470
  integer, 74
  polynomial, 469
Euclidean domain, 454
Euler's criterion, 38, 165, 205
Euler's identity, 118
Euler's phi function, 31
  and factoring, 320
Euler's summation formula, 114

Euler's theorem, 34, 157
Euler's totient function, 31
Euler, L., 125
event, 208
eventually positive, 50
exp, xiv
expectation, 233
expected polynomial time, 283
expected running time, 283
expected value, 233
exponent, 160
    module, 363
extended Euclidean algorithm
    integer, 78
    polynomial, 470
extended Gaussian elimination, 391
extension, xvi
extension field, 175, 440
extension ring, 174

factoring
    and Euler's phi function, 320
factoring algorithm
    integer, 407, 414
        deterministic, 484
    polynomial, 530, 538
        deterministic, 544
family, xv
fast Fourier transform, 480
Fermat's little theorem, 34, 35
FFT, 480
field, 170
field of fractions, 427
finite dimensional, 372
finite expectation, 272
finite extension, 440
finite fields
    existence, 511
    subfield structure, 515
    uniqueness, 515
finitely generated
    abelian group, 153
    module, 361
floor, 4
    of reversed Laurent series, 449
formal derivative, 444
formal Laurent series, 447
formal power series, 446
Fouvry, E., 559
Frandsen, G., 464
Frobenius map, 512
fundamental theorem
    of arithmetic, 2
    of finite abelian groups, 163
    of finite dimensional $F[X]$-modules, 506
Fürer, M., 72

von zur Gathen, J., 485, 546, 547
Gauss' lemma, 344
Gaussian elimination, 389
Gaussian integers, 174, 199, 454, 457

gcd
    integer, 7
    polynomial, 432
generating polynomial, 487
generator, 153
    algorithm for finding, 327
geometric distribution, 271, 274
Gerhard, J., 485, 546
Goldwasser, S., 357
Gordon, D. M., 420
Gordon, J., 546
Granville, A., 325
greatest common divisor
    in an integral domain, 452
    integer, 6
    polynomial, 431
group, 126
Guy, M., 103

Hadamard, J., 124
Halberstam, H., 325
Hardy, G. H., 103, 124, 125
hash function, 252
Heath-Brown, D., 125
Hellman, M., 340, 341
Hensel lifting, 351
homomorphism
    algebra, 424
    group, 142
    module, 363
    ring, 192
    vector space, 370
Horn, R., 125
Horner's rule, 467
Huang, M.-D., 559
hybrid argument, 264

ideal, 5, 185
    generated by, 5, 186
    maximal, 190
    prime, 189
    principal, 5, 186
identity element, 126
identity map, xvi
identity matrix, 379
image, xvi
image of a random variable, 221
Impagliazzo, R., 276
inclusion map, xvi
inclusion/exclusion principle, 210
independent, 214, 224
    k-wise, 218, 225
    mutually, 218, 225
indeterminate, 176
index, 140
index calculus method, 420
index set, xv
indicator variable, 222
infinite extension, 440
infinite order, 130
injective, xvi

integral domain, 171
internal direct product, 148
inverse
    multiplicative, 170
    of a group element, 126
    of a matrix, 386
inverse function, xvi
invertible matrix, 386
irreducible element, 451
irreducible polynomial, 430
    algorithm for generating, 523
    algorithm for testing, 522
    number of, 514
isomorphism
    algebra, 424
    group, 146
    module, 365
    ring, 195
    vector space, 370
Iwaniec, H., 340

Jacobi map, 347
Jacobi sum test, 559
Jacobi symbol, 346
    algorithm for computing, 348
Jensen's inequality, 239, 275

Kalai, A., 305
Kaltofen, E., 547
Karatsuba, A. A., 71
Kayal, N., 548, 558
Kedlaya, K., 485
kernel, 143
kills, 160
Kim, S. H., 325
Knuth, D. E., 72, 73, 103
von Koch, H., 125
Kronecker substitution, 479
Krovetz, T., 276
Kung, H. T., 485

Lagrange interpolation formula, 436
Las Vegas algorithm, 303
Latin square, 131
law of large numbers, 242
law of quadratic reciprocity, 343
law of total expectation, 237
law of total probability, 215
lcm
    integer, 11
    polynomial, 433
leading coefficient, 177
    of a reversed Laurent series, 449
least common multiple
    in an integral domain, 452
    integer, 11
    polynomial, 433
leftover hash lemma, 267
Legendre symbol, 342
Lehmann, D., 325
Lehmer, D., 418

Leiserson, C. E., 340
len, 62, 466
length
    of a polynomial, 466
    of an integer, 62
Lenstra, Jr., H. W., 419, 546, 559
Levin, L., 276
li, 117
linear combination, 361
linear map, 363
linear transformation, 501
linearly dependent, 367
linearly generated sequence, 486
    minimal polynomial of, 487
    of full rank, 492
linearly independent, 367
little-o, 50
Littlewood, J. E., 125
log, xiv
logarithmic integral, 117
lowest terms, 12
Luby, M., 276, 305

map, xvi
Markov's inequality, 241
Massey, J., 508
matrix, 377
matrix of a linear map, 383
Maurer, U., 325
maximal ideal, 190
memory cells, 53
Menezes, A., 103
Mertens' theorem, 113
Micali, S., 357
Miller, G. L., 324, 325
Miller–Rabin test, 307
Mills, W., 484, 508
min entropy, 266
minimal polynomial, 438
    algorithm for computing, 468, 500, 525
    of a linear transformation, 503
    of a linearly generated sequence, 487
    of an element under a linear transformation, 504
Möbius function (μ), 46
Möbius inversion formula, 47
mod, 4, 16, 22, 178, 435
modular composition, 467, 485
modular square root
    algorithm for computing, 350
module, 358
modulus, 16
monic associate, 430
monic polynomial, 177
monomial, 183
Monte Carlo algorithm, 303
Morrison, K., 508
Morrison, M., 418
multi-variate polynomial, 184
multiple, 1, 170
multiple root, 182
multiplication map, 143, 166, 363

multiplicative function, 46
multiplicative group of units, 170
multiplicative identity, 27
multiplicative inverse
  in a ring, 170
  modulo integers, 21
  modulo polynomials, 435
multiplicative order, 33, 153
multiplicative order modulo n, 33
multiplicity, 182
mutually independent, 218, 225

natural map, 143, 192
Newton interpolation, 474
Newton's identities, 450
Niven, I., 357
norm, 167, 518
normal basis, 521
number field sieve, 419

Oesterlé, J., 125
one-sided error, 304
one-time pad, 229
one-to-one correspondence, xvi
van Oorschot, P., 103, 341
order
  in a module, 364
  of a group element, 153
  of an abelian group, 130

pairwise disjoint, xv
pairwise independent
  events, 218
  hash functions, 252
  random variables, 225
pairwise relatively prime
  integers, 11
  polynomials, 434
parity check matrix, 385
partition, xv
Pascal's identity, 562
Penk, M., 103
perfect power, 64
period, 98
periodic sequence, 98
PID, 455
pivot element, 389
pivot sequence, 388
Pohlig, S., 340
Pollard, J. M., 340, 419
polynomial
  associate, 430
  irreducible, 430
  monic, 177
  primitive, 459
  reducible, 430
polynomial evaluation map, 192, 426
polynomial time, 55
  expected, 283
  strict, 283
Pomerance, C., 73, 125, 325, 418–420, 559

de la Vallée Poussin, C.-J., 124, 125
power map, 143
pre-image, xvi
pre-period, 98
primality test
  deterministic, 548
  probabilistic, 306
prime
  ideal, 189
  number, 2
prime number theorem, 116
  irreducible polynomials over a finite field, 514
primitive polynomial, 459
principal ideal, 5, 186
principal ideal domain, 455
probabilistic algorithm, 278
probability distribution
  conditional, 213
  discrete, 270
  finite, 207
product distribution, 211
program, 53
projection, 492
public key cryptography, 341
public key cryptosystem, 99
purely periodic, 98

$\mathbb{Q}$, xv
quadratic formula, 182
quadratic reciprocity, 343
quadratic residue, 36
quadratic residuosity
  algorithm for testing, 349
  assumption, 355
quadratic sieve, 415
quantum computer, 420
quotient algebra, 423
quotient group, 140
quotient module, 362
quotient ring, 187
quotient space, 370

$\mathbb{R}$, xv
Rabin, M. O., 324
Rackoff, C., 418
RAM, 53
random access machine, 53
random self-reduction, 337
random variable, 221
  conditional distribution of, 224
  conditional expectation, 237
  distribution of, 222
  expected value, 233
  image, 221
  independent, 224
  real valued, 221
  variance, 235
random walk, 244
randomized algorithm, 278
rank, 395
rational function field, 429

rational function reconstruction, 474
rational reconstruction problem, 90
recursion tree, 332, 340
Redmond, D., 125
reduced row echelon form, 388
reducible polynomial, 430
Reed, I., 484
Reed–Solomon code, 97, 477
regular function, 223
relatively prime
    in an integral domain, 452
    integers, 7
    polynomials, 432
Renyi entropy, 266
rep, 65, 466
repeated-squaring algorithm, 65
representation, 337
representative
    of a coset, 138
    of a residue class, 25
    of an equivalence class, 16
residue class, 25, 187
residue class ring, 187
restriction, xvi
reversed Laurent series, 448
Richert, H., 325
Riemann hypothesis, 118, 120, 122, 125, 324, 326,
        340, 357, 482, 546, 547
Riemann's zeta function, 118
Riemann, B., 125
ring, 166
ring of polynomials, 176
ring-theoretic product, 169
Rivest, R. L., 99, 103, 340
Rogaway, P., 276, 420
root of a polynomial, 179, 425
Rosser, J., 124
row echelon form, 397
row null space, 393
row rank, 394
row space, 392
row vector, 378
RSA cryptosystem, 99
Rumely, R. S., 559
running time
    expected, 283

sample mean, 241
sample space, 207
Saxena, N., 548, 558
scalar, 358
scalar multiplication map, 358
Schirokauer, O., 420
Schoenfeld, L., 124
Schönhage, A., 72, 102
Scholtz, R., 508
Schoof, R., 103
secret sharing, 230
Semaev, I. A., 546
separating set, 544
sequence, xv

Shallit, J., 125, 357, 546
Shamir, A., 71, 99, 103, 275
Shanks, D., 340
shift register sequence, 488
Shor, P., 420
Shoup, V., 340, 508, 546, 547
Shub, M., 103
sieve of Eratosthenes, 115
simple root, 182
smooth number, 399, 414
Solomon, G., 484
Solovay, R., 325, 357
Sophie Germain prime, 123
spans, 367
splitting field, 443
square root (modular)
    algorithm for computing, 350
square-free
    integer, 9
    polynomial, 509
square-free decomposition, 527
square-free decomposition algorithm, 527
standard basis, 368
statistical distance, 260
Stein, C., 340
Stein, J., 103
Stinson, D. R., 276
Stirling's approximation, 114
Strassen, V., 72, 325, 357
strict polynomial time, 283
subalgebra, 423
    fixed by, 425
    generated by, 426
subfield, 175
subgroup, 132
    generated by, 153
submodule, 360
subring, 173
subspace, 370
sufficiently large, 50
surjective, xvi

theta function of Chebyshev, 107
total degree, 183
total expectation
    law of, 237
total probability
    law of, 215
trace, 518
transcendental element, 441
transpose, 380
trial division, 306
trivial
    group, 130
    module, 359
    ring, 168
twin primes conjecture, 124
two-sided error, 304

UFD, 451
ultimately periodic sequence, 98

Umans, C., 485
union bound, 210
unique factorization
    in a Euclidean domain, 454
    in a PID, 455
    in $D[X]$, 459
    in $F[X]$, 430
    in $\mathbb{Z}$, 2
unique factorization domain, 451
unit, 170
universal hash functions, 252

Vandermonde matrix, 385
Vanstone, S., 103
variance, 235
vector space, 370

Walfisz, A., 124
Wang, P., 103
Wang, Y., 340
Weber, D., 420
Wegman, N. M., 275
Weilert, A., 464
Welch, L., 508
well-behaved complexity function, 69
Wiedemann, D., 508
Wiener, M., 103, 341
Wright, E. M., 103, 124, 125

Yun, D. Y. Y., 546

$\mathbb{Z}$, xv
Zassenhaus, H., 546
zero divisor, 171
zero element, 129
zero matrix, 378
zero-sided error, 304
zeta function of Riemann, 118
Zuckerman, H., 357
Zuckermann, D., 276